The Enterobacteria

The Enterobacteria

J. Michael Janda, PhD (ABMM)
Chief, Enterics and Special Pathogens Section
Microbial Diseases Laboratory
California Department of Health Services
Berkeley, California

Sharon L. Abbott, BA
Supervisor, Enteric Bacteriology
Microbial Diseases Laboratory
California Department of Health Services
Berkeley, California

Lippincott - Raven
PUBLISHERS
Philadelphia • New York

Acquisitions Editor: Ruth W. Weinberg
Manufacturing Manager: Dennis Teston
Supervising Editor: Kimberly Swan
Production Service: P. M. Gordon Associates
Cover Designer: David Levy
Indexer: Grant Hackett
Compositor: Lippincott–Raven Electronic Production
Printer: Maple Press

Printed in the United States of America

9 8 7 6 5 4 3 2 1

Library of Congress Cataloging-in-Publication Data

Janda, J. Michael.
The enterobacteria / J. Michael Janda and Sharon L. Abbott.
p. cm.
Includes bibliographical references and index.
ISBN 0–397–51723–8
1. Enterobacteriaceae. I. Abbott, Sharon L. II. Title.
[DNLM: 1. Enterobacteriaceae. QW 138 J33e 1998]
QR82.E6J36 1998
616′.0144—dc21
DNLM/DLC
For Library of Congress 97-42726
CIP

Care has been taken to confirm the accuracy of the information presented and to describe generally accepted practices. However, the authors, editors, and publisher are not responsible for errors or omissions or for any consequences from application of the information in this book and make no warranty, express or implied, with respect to the contents of the publication.

The authors, editors, and publisher have exerted every effort to ensure that drug selection and dosage set forth in this text are in accordance with current recommendations and practice at the time of publication. However, in view of ongoing research, changes in government regulations, and the constant flow of information relating to drug therapy and drug reactions, the reader is urged to check the package insert for each drug for any change in indications and dosage and for added warnings and precautions. This is particularly important when the recommended agent is a new or infrequently employed drug.

Some drugs and medical devices presented in this publication have Food and Drug Administration (FDA) clearance for limited use in restricted research settings. It is the responsibility of the health care provider to ascertain the FDA status of each drug or device planned for use in their clinical practice.

To my parents Bernie and Ellis, my wife Claudia, and my children Michael, Matthew, and Jennifer, without whose support, encouragement, and ultimate patience this project would not have been possible.

J. Michael Janda

Thanks to my loving family and caring friends for their infinite patience and understanding.

Sharon L. Abbott

Contents

Preface ... ix
Acknowledgment ... xi
1. Historical Perspectives on the Family *Enterobacteriaceae* ... 1
2. Medical Significance of the Family *Enterobacteriaceae* ... 8
3. *Escherichia coli* ... 13
4. Other *Escherichia* Species ... 61
5. The Genus *Shigella* ... 66
6. The Salmonellae ... 80
7. *Salmonella* Typhi and *Salmonella* Paratyphi A, B, and C ... 101
8. The Genus *Klebsiella* ... 110
9. The Genus *Serratia* ... 131
10. The Genus *Enterobacter* ... 148
11. The Genus *Pantoea* ... 178
12. The Genus *Citrobacter* ... 183
13. The Genus *Yersinia* ... 202
14. The Genus *Proteus* ... 245
15. The Genus *Morganella* ... 270
16. The Genus *Providencia* ... 284
17. The Genus *Edwardsiella* ... 304
18. The Genus *Hafnia* ... 322
19. The Genera *Cedecea* and *Kluyvera* ... 336
20. Miscellaneous Genera Associated with Clinical Specimens ... 344
21. *Enterobacteriaceae* Not Associated with Human Disease ... 361
Subject Index ... 367

Preface

No taxonomically defined group of bacteria has had a greater impact on infectious diseases, medical and clinical microbiology, and public health than the enterobacteria. One has only to look as far as newspaper headlines to chronicle the recent history of some of the more famous members of this family—a beef recall caused by contamination of a processing plant with *Escherichia coli* O157:H7 and a nationwide outbreak of diarrheal illnesses associated with contaminated eggs containing *Salmonella* Enteritidis, just to name two. Such outbreaks have had major political, economic, legal, social, and medical ramifications that extend far beyond those individuals directly affected by ingestion of adulterated foods. However, these examples pale in comparison to the role of enterobacteria as nosocomial pathogens. Nosocomial data indicate that many genera of *Enterobacteriaceae,* such as *Escherichia, Enterobacter, Klebsiella,* and *Serratia,* cause a significant proportion of hospital-acquired bacteremias, urinary tract illnesses, respiratory tract diseases, and wound infections. Even more frightening has been the rapid rate at which they develop antimicrobial resistance in this setting, particularly to extended-spectrum β-lactam compounds. Such collective developments require accurate and rapid identification of the enterobacteria, often to the subspecies level, to detect specific pathogenic biotypes or groups.

As technology advances, analysis of enterobacteria has become more complex. In no area of infectious diseases is this more apparent than gastroenteritis. What was once a limited field of etiologic agents now contains more than 40 different species or groups associated with diarrheal disease. Many of these agents are members of the family *Enterobacteriaceae.* Likewise, it has been difficult for the medical and scientific community to keep up with the taxonomic revolution which includes an ever-expanding list of genera and species within the enterobacteria. Laboratories geared to rapidly isolate and identify gram-negative pathogens with the use of automated commercial systems have a difficult task: they often must confirm their primary identification by another method, they must keep abreast of new agents and new disease associations, and they must be aware of the risk factors related to nosocomial or community-acquired outbreaks of disease.

The clinical diagnosis of enterobacterial infections necessitates close working relationships between microbiologists, physicians, infectious disease specialists, pathologists, epidemiologists, and infection control personnel. It is hoped this book will serve as a reference source to achieve this goal. The book is divided primarily into chapters on individual genera within the *Enterobacteriaecae,* with specific subheadings on topics of interest to medical staff. We have attempted to make this book as comprehensive and up-to-date as possible with the recognition that certain groups, such as *Escherichia coli,* could easily fill two books of this size. Most references date from 1980; although many important contributions to the field occurred before that date, we believe that most of these studies can be traced back from more recently cited literature.

Above all, we have attempted to provide a useful and comprehensive book that will be an aid in the isolation and identification of this important group of microorganisms.

J. Michael Janda
Sharon L. Abbott

Acknowledgment

We would like to acknowledge the excellent photographic expertise of David Lindquist and graphic talents of Linda Johnson, who made parts of this book possible.

1

Historical Perspectives on the Family *Enterobacteriaceae*

The origins of the family *Enterobacteriaceae* predate the "golden era" of bacteriology, when such luminaries as Koch, Pasteur, Roux, Metchnikoff, and Kitasato made their epoch-making discoveries.[1] The first bona fide member of the family was described in 1823 by Bizio,[2] who found *Serratia marcescens* growing on an Italian barley dish (polenta). *Klebsiella* and *Proteus* were described more than half a century later, in the mid-1880s.[3,4] The discovery of these organisms as distinct entities was no doubt partly attributable to their unusual morphologic and cultural features (eg, capsulation, swarming motility). By the turn of the century, many additional groups that would later be included within the *Enterobacteriaceae* had been discovered. Many of these originally were identified because of their association with outbreaks of diarrheal disease in children. Rules of taxonomic nomenclature did not exist, and methods of ascertaining the uniqueness of each isolate were frequently hindered by a limited knowledge of cultural, morphologic, and biochemical properties, so new isolates (groups) were often referred to by the surname of the discoverer. For example, Morgan No. 1 Bacillus in 1906 (London) and Flexner's Bacillus in 1902–1903 (Rockefeller Institute, NY) were originally recovered from fecal specimens linked to episodes of summer diarrhea in pediatric patients.[5] The uniqueness of each of these groups was definitively established years later, resulting in their ascension to the levels of genus *(Morganella)* and species *(Shigella flexneri),* respectively.

During this same era, it became apparent that the gastrointestinal tract of humans and other vertebrates serves as a major reservoir for gram-negative, facultatively anaerobic bacilli. However, because of the lack of a defined bacterial classification system and the inability to conclusively establish the legitimacy of any described taxon with the tests then available, taxonomic uncertainty led to the creation of many ambiguous classification proposals in the period 1900–1945. With the increasing realization that more biochemical, antigenic, cultural, and morphologic diversity existed within the microbial world than was previously recognized, many species were renamed or redescribed as new taxa, based on a limited number of phenotypic differences. A common solution to these taxonomic problems was to place such groups into one of two nondescript genera, *Bacterium* or *Bacillus*.[6] Examples of such designations include *Escherichia coli ("Bacterium coli"), Enterobacter cloacae ("Bacillus cloacae"),* and *Klebsiella pneumoniae ("Bacillus mucosum capsulatus").* Inclusion in either genus simply reflected the imprecise nature of classification during this period. Other groups received their species names based on the disease they caused or an unusual feature of the microorganism (Table 1). As late as the mid-1950s controversy still existed regarding the genus *Bacterium,* with British bacteriologists favoring retention of most groups under a single genus and their American counterparts supporting creation of multiple genera.[6]

TABLE 1. *Previous designations for Klebsiella pneumoniae*

Designation	Year
Friedländer's pneumobacillus	1883
Bacterium pneumoniae crouposae	1885
Klebsiella couposa	1885
Hyalococcus pneumoniae	1886
Bacillus pneumoniae	1886
Klebsiella friedlanderi	1889
Bacillus mucosus capsulatus	1893
Bacterium pneumonicum	1900
Encapsulatus pneumoniae	1919
Proteus pneumoniae	1926
Bacterium friedlander	1928

Data from Chapman.[7]

By far the phenotypic feature relied on most in creating the various classification schemes was lactose fermentation. This single marker resulted in the introduction of a number of perplexing and equivocal terms into the scientific and medical literature. In 1893, Theobald Smith noted that most gram-negative bacteria could be separated into two major groups based on lactose fermentation.[1] Lactose-negative organisms were often found associated with pathogenic processes (dysentery), whereas lactose-positive bacteria appeared to be more benign in nature. Such observations gave rise to the description of the latter group as "coliform" bacilli. Four years later, Widal and Nobecourt used the term "paracolon" to describe a bacterium that was associated with a case of gastroenteritis but had delayed lactose fermentation.[1] The word "paracolon" later became a synonym for intermediate bacteria belonging neither to the rapid–lactose-fermenting coliform group nor to the non–lactose-fermenting species exemplified by salmonellae and shigellae. Still other terms emerged, such as "aberrant coliforms," which referred to isolates unable to produce acid and/or gas (>20%) from lactose within 48 hours of incubation.[8] After several decades, the term "paracolon" was being used increasingly in both European and American circles. Borman and colleagues subdivided "coliforms" into late–lactose-fermenting "paracolons" and true non–lactose-fermenting bacilli, thus elevating the paracolon group to generic rank as the genus *Paracolobactrum*.[9] However, this proposal did not meet with widespread acceptance, and most investigators continued to refer to these isolates as part of the paracolon group. As new biochemical and antigenic types were discovered, the paracolon group was often subdivided on the basis of reactions to the indole, methyl red, Voges-Proskauer, and citrate tests (IMViC). Names for these subgroups commonly originated from the location at which they were first isolated, such as the Bethesda-Ballerup and Providence subgroups. Many of these subdivisions eventually evolved into the current genera and species when new and more sophisticated systematic methods became available. An example of this evolutionary process can be seen in Fig. 1. Despite this eventual outcome, the term "paracolon" persisted in the literature for a prolonged period. It was still in use in the late 1960s to refer collectively to multiple taxa that had once been assembled under this single epithet.[10]

The birth of the family *Enterobacteriaceae* can be traced to the year 1937, when Otto Rahn proposed this name for a group of biochemically and morphologically similar organisms.[11] In his article on new principles for the classification of bacteria Rahn stated,

> The need for a genus name for the entire colon-typhoid group has been felt by those bacteriologists who do not agree with Castellani and Chalmers' many genera. The name *Enterobacteriaceae* as a family name with the single genus *Enterobacter* is here proposed.

Although Rahn created only a single genus to house 112 species, many present-day genera (ie, *Escherichia, Salmonella, Klebsiella, Proteus, Erwinia, Serratia,* and *Shigella*) were represented under *Enterobacter*. Thus, the family name was firmly established and defined by Rahn by the inclusion of so many currently accepted *Enterobacteriaceae* genera within this original description. Nevertheless, sporadic episodes of taxonomic controversy muddled the legitimacy of this family name for more than 40 years after Rahn's proposal. In a 1980 opposition statement to the rejection of the family name *Enterobacteriaceae,*

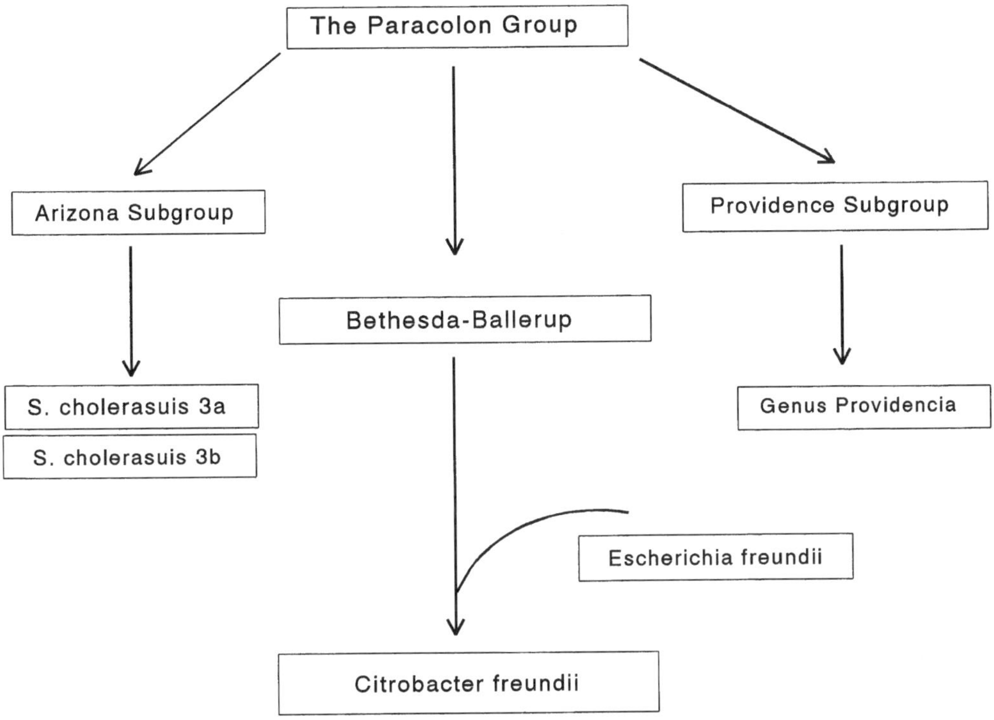

FIG. 1. Taxonomic evolution of some subgroups within the paracolon group.

Farmer and associates[12] chronicled the taxonomic history of this family from its inception in 1937. The salient points of this historical review are summarized as follows. Both the family name *Enterobacteriaceae* Rahn 1937 and the genus name *Enterobacter* Rahn 1937 became illegitimate with the publication of the first *International Code of Nomenclature of Bacteria* ("Bacteriological Code") in 1948.[13] In 1957, a formal request was made to conserve this family name based on its widespread acceptance by the scientific community and to establish *E. coli* as the type genus and species.[14] In Opinion No. 15 rendered by the Judicial Commission of the International Committee on Bacteriological Nomenclature in 1958, four major points were addressed:[15] (1) that the family named *Enterobacteriaceae* Rahn 1937 is conserved, (2) that *Escherichia* is designated the type genus of the family, (3) that the genus name *Escherichia* is conserved, and (4) that the type species for the genus *Escherichia* is designated *Escherichia coli.* For the next two decades these names remained a stable fixture in the medical and scientific literature, and the 1976 revision of the Bacteriological Code again reaffirmed the 1958 decision.[16] However, in 1979, Lapage[17] challenged the validity of the family name by requesting an opinion on his proposal to replace *Enterobacteriaceae* with *Enterobacteraceae* (no "i"). The major justification for this proposed change came from Rules 9 and 21a of the Bacteriological Code, which state that the name must be formed by adding the suffix "*-aceae*" to the root word[18]; other arguments were also brought forward.[12,17] To further complicate matters, both family names were inadvertently left off the 1980 *Approved List of Bacterial Names.* However, Farmer and associates[12] made an eloquent case for retention of both the family and type genus and species names based on their universal acceptance in the literature, the desirability of taxonomic stability, and the previously published Opinion No. 15. They also pointed out that *E. coli,*

TABLE 2. *Major phenotypic features of the* Enterobacteriaceae

Trait or Characteristic	Phenotype or Value[a]
Gram-negative rods, small	+
Aerobic/facultatively anaerobic metabolism	+
Oxidase	–
Catalase	+
Nitrates→nitrites	+
Requires Na^+ for growth	–
Fermentation of D-glucose	+
Contains enterobacterial common antigen	+
DNA relatedness (between genera)	>20%
G+C content (moles %)	38–60

[a]Minor exceptions occur occasionally within species or groups.

and not *Enterobacter*, was the best-described bacterial species at the molecular level, a fact even more true today. They argued that because *Escherichia* is one of the "core genera" within the *Enterobacteriaceae* and *E. coli* is the prototype organism in the family to which all other members are compared, it should remain the type genus and species.[12] In a companion article to this opposition statement, Ewing and colleagues[18] proposed the name *Enterobacteriaceae* fam. nov., nov. rev. to replace *Enterobacteriaceae* Rahn 1937, which lost standing on January 1, 1980, based on Rule 28a of the Bacteriological Code. In 1981, the Judicial Commission of the International Committee on Systematic Bacteriology stated that the family name *Enterobacteriaceae* Rahn 1937 (type genus *Escherichia* Castellani and Chalmers 1919) did have standing and belonged on the *Approved List.*[19] One year later, the same Judicial Commission formally rejected the proposal of Lapage, bringing this tumultuous taxonomic controversy to an end.[20]

The principal features identifying isolates as members of the family *Enterobacteriaceae* are listed in Table 2. Although most of these unifying characteristics have been known for decades, the assignment of individual groups (or species) has vacillated between genera for prolonged periods. This can best be illustrated by following the evolution of naming of the genera and species within the *Enterobacteriaceae* in various editions of *Bergey's Manual of Determinative Bacteriology* (Table 3). First, there was a general decrease in the number of recognized species between 1939 and 1974 (from 67 to 37 species, not including salmonellae). This trend reflects several factors. In its infancy, the *Enterobacteriaceae* were overclassified because both antigenic and biochemical variants of previously de-

TABLE 3. *Taxonomic evolution of groups within the* Enterobacteriaceae *according to* Bergey's Manual of Determinative Bacteriology

Trait	Fifth Edition (1939)	Sixth Edition (1948)	Eighth Edition (1974)
Tribes (no.)	5	5	5
Genera[a]	*Escherichia* (2)	*Escherichia* (3)	*Escherichia* (1)
	Aerobacter (2)	*Aerobacter* (2)	*Enterobacter* (2)
	Klebsiella (6)	*Klebsiella* (1)	*Klebsiella* (3)
	Erwinia (13)	*Erwinia* (22)	*Erwinia* (13)
	Serratia (6)	*Serratia* (5)	*Serratia* (1)
	Proteus (8)	*Proteus* (4)	*Proteus* (5)
	Salmonella	*Salmonella*	*Salmonella*
	Eberthella (14)	*Shigella* (11)	*Shigella* (4)
	Shigella (15)		*Edwardsiella* (1)
			Citrobacter (2)
			Hafnia (1)
			Yersinia (3)
Species[b]	67	49	37

[a]Number in parentheses indicates number of species in each respective genus; numbers for *Salmonella* are not included because within this genus the concept of a species varied dramatically from edition to edition.
[b]Total number of species in family (excluding *Salmonella*).

scribed groups were often thought to represent new taxa. Less than 2 years after Rahn's original description of 112 species within this family, the fifth edition of *Bergey's Manual* listed only 67, a 40% reduction. By 1974 (eighth edition), only 37 species remained, fewer than one third of the original total. *Eberthella,* a genus appearing in the fifth but not the sixth edition, originally was constructed to house the typhoid bacillus; later it became clear that these organisms belonged in the genus *Salmonella,* and *Eberthella* lost its standing.[6] The genus *Aerobacter* was replaced because the name previously had been used to refer to organisms contained in multiple genera such as *Klebsiella.*[21] Still other groups at the species or subspecies level were merged as a result of close similarity in biochemical properties (as with *Eberthella*). For instance, the Bethesda-Ballerup group was combined with the species *Escherichia freundii* (fifth edition) because they differed from each other only in the rapidity of lactose fermentation.[22] Later, *E. freundii* became *Citrobacter freundii* in the eighth edition of *Bergey's Manual* (see Fig. 1).

A second major feature of these taxonomic changes concerns the increase in the number of genera within the family to 12 by 1974 (see Table 3). This expansion was largely the result of two independent systematic methods, namely numerical taxonomy and DNA–DNA reassociation kinetics. Use of such taxonomic tools led to the creation of new taxa from unnamed groups (eg, *Edwardsiella, Yersinia*) or by removal of species from previously recognized genera (eg, *Hafnia "Enterobacter" alvei*). By the late 1960s and early 1970s Don Brenner and colleagues, first at the Walter Reed Army Institute of Research and later at the Centers for Disease Control and Prevention, pioneered the use of DNA–DNA hybridization as the "gold standard" for defining relatedness at the species level.[23] By the mid-1970s investigations relying on numerical taxonomy methods or DNA hybridization were helping to define relatedness at both the genus and species levels within the family.[24,25] The result was the definition or redefinition of many genera, species, and groups within the *Enterobacteriaceae,* not on the basis of phenotypic similarity but on a solid genetic (molecular) foundation.[26]

This taxonomic revolution continued throughout the 1980s and 1990s. Today, there are more than 30 distinct genera and more than 100 species within the family, ranging from *Serratia,* recognized in 1823, to the *Photorhabdus,* described in 1993, a period spanning 170 years.[27] *Enterobacteriaceae,* however, is clearly an "extended" family in the genetic sense. Some genera *(Proteus, Providencia, Yersinia)* exhibit only a minimal amount of relatedness (10% to 15%) to the core genera (*Salmonella, Citrobacter,* and *Shigella*), particularly *E. coli.*[28] This has led some to state that overclassification within this group is equivalent in taxonomic spread to a single genus *(Bacillus)* elsewhere.[27] Nevertheless, the medical and ecologic importance of this group of organisms cannot be denied (see Chapter 2). Furthermore, because criteria for genetic relatedness are somewhat subjective for species assignment (~70% relatedness), and even more so at the genus and family levels, the essential fact of this classification system is that it works functionally. To demonstrate this point is the fact that *E. coli* and *Shigella* are genetically indistinguishable at the species level; strains from each group are 70% to 100% related to one another. However, the genus *Shigella* remains separate from *Escherichia* because of its unique clinical and public health importance as a cause of bacillary dysentery and shigellosis; in contrast, many enteric *E. coli* strains produce no disease whatsoever. Therefore, genera and species within this family have been retained not only on a genetic basis but also on a functional one, with taxonomic rules superseded in favor of practical importance.

Because taxonomy is a dynamic process, the concept of inclusionary groups and new taxa within the *Enterobacteriaceae* will continue to evolve as further taxonomic studies redefine members of this family. Two examples highlight this statement. A large number of named groups within the *Enterobacteri-*

TABLE 4. *Selected genetically heterogeneous groups within the family* Enterobacteriaceae

Group	Proposed Species	Unnamed Genomospecies[a]
Citrobacter freundii	*C. youngae, C. braakii, C. werkmanii, C. sedlakii, C. rodentium*	2
Serratia liquefaciens	*S. grimesii*[b], *S. proteamaculans* ssp. *proteamaculans*, ssp. *quinavora, S. liquefaciens*	0
Proteus vulgaris	—	4
Enterobacter cloacae	—	3
Enterobacter agglomerans	*Pantoea agglomerans, P. dispersa, P. ananas, Leclericia adecarboxylata*	8
Hafnia alvei	—	2

[a]Minimum number of distinct DNA groups within presently designated species.
[b]Contains small group of strains that approach species definition, but numbers are at present too small.

aceae are known to contain multiple species although at present they cannot be unambiguously separated from one another by simple phenotypic tests. Examples of such phenospecies are listed in Table 4. In some instances these have been partially resolved into new taxa that can be separated by readily available biochemical methods. The phenospecies *C. freundii* has been subdivided into a number of new, clinically relevant genomospecies[29] plus one species exclusively associated with rodent populations.[30] Other groups, such as *E. cloacae,* presently contain an unknown number of new groups that cannot be distinguished from one another except by molecular techniques. However, it is apparent from Table 4 that many commonly identified groups and species in the clinical laboratory are actually heterogenous and will soon yield new genomospecies in the future.

The second issue related to this unfolding taxonomy concerns the bacterium *Plesiomonas shigelloides,* presently a member of the family *Vibrionaceae.* An increasing body of information indicates that *P. shigelloides* is genetically more closely related to the enterobacteria (eg, *Proteus*) than to the *Vibrionaceae.* These data include results of phylogenetic analysis[31,32] and chemotaxonomy[33] and the fact that the species possesses the common enterobacterial antigen.[34] Whether a new designation will be made is speculative. However, if *P. shigelloides* does become a member of the *Enterobacteriaceae,* the change will require redefinition of the cardinal features of this family by virtue of that organism's positive oxidase reaction. Such unanswered problems and questions will continue to intrigue taxonomists and microbiologists as the *Enterobacteriaceae* undergo a continued morphogenesis.

REFERENCES

1. Wilson GS, Miles A. The enterobacteria. In: Wilson GS, Miles A, eds. *Topley and Wilson's principles of bacteriology and immunology.* 6th ed. Baltimore: Williams & Wilkins, 1975;839–857.
2. Bizio B. Lettera di Bartolomeo Bizio al chiarissimo canonico Angelo Bellani sopra il fenomeno della polenta porporina. *Biblioteca Italiana o sia Giornale di Letteratura Scienze a Arti* 1823;30:275–295.
3. Ørskov I. Genus V: The genus *Klebsiella* Trevisan 1885, 105[AL]. In: Krieg NR, Holt JG, eds. *Bergey's manual of systematic bacteriology.* Vol 1. Baltimore: Williams & Wilkins, 1984;461–465.
4. Penner JL. Genus XI: *Proteus* Hauser 1885, 12[AL]. In: Krieg NR, Holt JG, eds. *Bergey's manual of systematic bacteriology.* Baltimore: Williams & Wilkins, 1984; 491–494.
5. Walker-Smith JA. Evolution of concept of infective origin of infantile diarrhoea from Ballard to Bray (1887–1945). *J Infect* 1996;33:213–218.
6. Wilson GS, Miles AA. *Bacterium.* In: Wilson GS, Miles AA, eds. *Topley and Wilson's principles of bacteriology and immunity.* 4th ed. Baltimore: Williams & Wilkins, 1955;751–783.
7. Chapman OD. Genus III: *Klebsiella* Trevisan. In: Breed RS, Murray EGD, Hitchens AP, eds. *Bergey's manual of determinative bacteriology.* 6th ed. Baltimore: Williams & Wilkins, 1948;457–458.
8. Stuart CA, Mickle FL, Borman EK. Suggested grouping of slow lactose fermenting organisms. *Am J Public Health* 1940;30:499–508.
9. Borman EK, Stuart CA, Wheeler KM. Taxonomy of the family *Enterobacteriaceae. J Bacteriol* 1944;48:351–367.

10. Fields BN, Uwaydah M, Kunz LJ, Swartz MN. The so-called "paracolon" bacteria: A bacteriologic and clinical reappraisal. *Am J Med* 1967;42:89–106.
11. Rahn O. New principles for the classification of bacteria. *Zentrabl Bakteriol Parasitenkd Infektionskr Hyg Abt* 1937;96:273–286.
12. Farmer JJ III, Brenner DJ, Ewing WH. Opposition to recent proposals which would reject the family name *Enterobacteriaceae* and *Escherichia* as its type genus. *Int J Syst Bacteriol* 1980;30:660–673.
13. Buchanan RE, St. John-Brooks R, Breed RS. International bacteriological code of nomenclature. *J Bacteriol* 1948;55:287–306.
14. Editorial Board. Proposal to conserve the family name *Enterobacteriaceae* and to designate the type genus. *Int Bull Bacteriol Nomencl Taxon* 1957;7:165–166.
15. Judicial Commission of the International Committee on Bacteriological Nomenclature. Conservation of the family name *Enterobacteriaceae*, of the name of the type genus, and designation of the type species. *Int Bull Bacteriol Nomencl Taxon* 1958;8:73–74.
16. Lapage SP, Sneath PHA, Lessel EF, Skerman VBD, Seeliger HPR, Clark WA, eds. *International code of nomenclature of bacteria: 1976 revision.* Washington DC: American Society for Microbiology, 1975.
17. Lapage SP. Proposal of *Enterobacteraceae* nom. nov. as a substitute for the illegitimate but conserved name *Enterobacteriaceae* Rahn 1937: Request for an opinion. *Int J Syst Bacteriol* 1979;29:265–266.
18. Ewing WH, Farmer JJ III, Brenner DJ. Proposal of *Enterobacteriaceae* fam. nov., nom. rev. to replace *Enterobacteriaceae* Rahn 1937, nom. fam. cons. (Opin. 15, Jud. Comm. 1958), which lost standing in nomenclature on 1 January 1980. *Int J Syst Bacteriol* 1980;30: 674–675.
19. Judicial Commission of the International Committee on Systematic Bacteriology. Present standing of the family name *Enterobacteriaceae* Rahn 1937. *Int J Syst Bacteriol* 1981;31:104.
20. Wayne LG. Actions of the Judicial Commission of the International Committee on Systematic Bacteriology on requests for opinions published between July 1979 and April 1981. *Int J Syst Bacteriol* 1982;32:464–465.
21. Hormaeche E, Edwards PR. A proposed genus *Enterobacter. Int Bull Bacteriol Nomencl Taxon* 1960;10: 71–74.
22. Carlquist PR. A biochemical test for separating paracolon groups. *J Bacteriol* 1956;71:339–341.
23. Brenner DJ. Characterization and clinical identification of *Enterobacteriaceae* by DNA hybridization. *Prog Clin Pathol* 1977;7:71–117.
24. Johnson R, Colwell RR, Sakazaki R, Tamura K. Numerical taxonomy study of the *Enterobacteriaceae. Int J Syst Bacteriol* 1975;25:12–37.
25. Steigerwalt AG, Fanning GR, Fife-Asbury MA, Brenner DJ. DNA relatedness among species of *Enterobacter* and *Serratia. Can J Microbiol* 1976;22:121–137.
26. Ewing EH. *Edwards and Ewing's identification of* Enterobacteriaceae. 4th ed. New York: Elsevier, 1986.
27. Logan NA. *Bacterial systematics.* Oxford: Blackwell Scientific Publications, 1994.
28. Brenner DJ. Family 1: *Enterobacteriaceae* Rahn 1937. In: Krieg NR, Holt JG, eds. *Bergey's manual of systematic bacteriology.* Baltimore: Williams & Wilkins, 1984;408–411.
29. Brenner DJ, Grimont PAD, Steigerwalt AG, Fanning GR, Ageron E, Riddle CF. Classification of citrobacteria by DNA hybridization: Designation of *Citrobacter farmeri* sp. nov., *Citrobacter youngae* sp. nov., *Citrobacter braakii* sp. nov., *Citrobacter werkmanii* sp. nov., *Citrobacter sedlakii* sp. nov., and three unnamed *Citrobacter* genomospecies. *Int J Syst Bacteriol* 1993;645–658.
30. Schauer DB, Zabel BA, Pedraza IF, O'Hara CM, Steigerwalt AG, Brenner DJ. Genetic and biochemical characterization of *Citrobacter rodentium* sp. nov. *J Clin Microbiol* 1995;33:2064–2068.
31. MacDonell MT, Swartz DG, Ortiz-Conde BA, Last GA, Colwell RR. Ribosomal RNA phylogenies for the vibrio-enteric group of eubacteria. *Microbiol Sci* 1986;3: 172–178.
32. Ruimy R, Breitmayer V, Elbaze P, et al. Phylogenetic analysis and assessment of the genera *Vibrio, Photobacterium, Aeromonas,* and *Plesiomonas* deduced from small-subunit rRNA sequences. *Int J Syst Bacteriol* 1994;44:416–426.
33. Basu S, Tharanathan RN, Kontrohr T, Mayer H. Chemical structure of the lipid A component of *Plesiomonas shigelloides* and its taxonomical significance. *FEMS Microbiol Lett* 1985;28:7–10.
34. Whang HY, Heller ME, Neter E. Production by *Aeromonas* of common enterobacterial antigen and its possible taxonomic significance. *J Bacteriol* 1972;110:161–164.

2

Medical Significance of the Family *Enterobacteriaceae*

No single group of taxonomically defined microorganisms has received greater attention from the medical and scientific communities than the *Enterobacteriaceae.* The family from its inception has been linked to colonization of the gastrointestinal tract of humans and other vertebrates and to the pathologic processes that result from subsequent infection of the gut. Initially many of these groups were recognized and identified based on their recovery from the feces of symptomatic patients who presented with a variety of gastrointestinal syndromes, including bacillary dysentery *(Shigella),* typhoid fever *(Salmonella typhi),* and summer episodes of infantile diarrhea (various agents). Later it became increasingly apparent that the enterobacteria also play a significant role in extraintestinal disease. In the past two decades, *Enterobacteriaceae* have come to be recognized as paramount in the causation of systemic illnesses. The emergence of this family as a major agent in the development of serious and life-threatening complications can be attributed to several interrelated factors.

First, vital statistics indicate that life expectancy continues to rise. Crossley and Peterson[1] estimated that only 15 million people worldwide were 65 years of age or older in 1900. By 1992, the figure had risen to 342 million, and it is projected to reach 2.5 billion by the year 2050. The percentage of the world's population represented by this group is increasing similarly (from <1% in 1900 to 6.2% in 1992 and an estimated 20% in 2050). Such outcomes are the direct result of better nutrition, exercise, an increasing standard of living (hygiene), and general improvements in the quality of health care delivery systems. However, along with this trend toward an improved life expectancy come a number of indirect consequences. For example, a greater percentage of immunocompromised people survive life-threatening conditions (eg, cancer, liver disease). Such persons are at increased risk for development of serious endogenous enterobacterial infections after surgery, immunosuppressive treatments (eg, chemotherapy, corticosteroid therapy), or other medical procedures. A second indirect consequence of the increasing geriatric population is the fact that many such persons will eventually require extensive long-term care because of circulatory complications, chronic diseases (eg, diabetes, pulmonary diseases), and urinary problems that arise with advancing age. Catheterization, a repercussion of short- or long-term urinary tract incontinence, predisposes patients to enteric infections, particularly those associated with the tribe *Proteeae* (*Proteus, Morganella,* and *Providencia*). It is not then surprising that the scope and the number of medical reports related to the family *Enterobacteriaceae* in the literature continue to rise. During the 5 years ending in March of 1997, more than 1300 separate entries were recorded in *Index Medicus* on various aspects of the *Enterobacteriaceae,* including infections, epidemiology, taxonomy, and microbiology. Such numbers are a testament to the overall medical importance of this single group.

Furthermore, the *Enterobacteriaceae* form the backbone of the database matrices to most commonly used manual, semiautomated, and automated microbial identification systems. These panels include such popular commercial products as the MicroScan WalkAway System, Vitek, API 20E (E is for enteric) strips, and the BBL Crystal Rapid Stool/Enteric ID kit. In addition, no group has had more selective and differential media devised for the specific isolation of various members from both clinical and environmental samples than the *Enterobacteriaceae.* It is no wonder then that this taxa occupies such a preeminent position in the fields of infectious disease and medical microbiology.

The hospital environment is one of the major settings in which *Enterobacteriaceae* infections occur. Schaberg and Culver[2] reviewed in 1991 the latest cumulative surveillance data on nosocomial infections reported to the National Nosocomial Infections Surveillance System (NNIS) for the years 1986 through 1989. The results, which are depicted in Table 1, dramatically illustrate how common certain enterobacteria are in hospital-acquired infections. Four of the top 10 nosocomial pathogens are members of the *Enterobacteriaceae,* and *Escherichia coli* continues to be the dominant nosocomial pathogen, although the percentage of infections caused by this organism declined from 23% to 16% during the 1980s.[2] Their leading role as nosocomial agents is a reflection of their natural habitat (the gastrointestinal tract) and their close proximity to the urogenital system. The urinary tract remains the single most common site involved in enterobacteria-associated disease; 46% of all reported cases of urinary tract infection in the latest NNIS survey involve members of this family. Enterobacteria are commonly involved in hospital-acquired pneumonias and wound infections (32% and 28% of cases, respectively). Although some species (eg, klebsiellae) have declined in overall incidence, others have taken their place. The role of *Enterobacter* spp. in causing nosocomial infections rose markedly between 1980 (cumulative data) and 1986–1989. This increase was observed primarily in hospital-acquired pneumonias (which rose from 9% to 11% of cases) and in wound infections (from 5% to 8%).

The association of enterobacteria with extraintestinal disease affords the best opportunity to understand the clinical impact of this family on medicine. Septicemia is the hallmark of invasive disease, and the *Enterobacteriaceae* are common etiologic agents of such systemic infections. Although the latest NNIS survey estimates the percentage of nosocomial bloodborne infections caused by this group to be about 20%, the figure fluctuates from study to study. Enterobacteria (particularly *E. coli*) also can cause septicemia in community settings. Ehni and colleagues[3] reviewed data on the incidence of bacteremia/fungemia in granulocytopenic patients over an 11-year period at a tertiary-care hospital in Denver, Colorado. The incidence of bacter-

TABLE 1. *Frequency of some members of the* Enterobacteriaceae *as nosocomial pathogens*

Organism	Rank[a]				
	Urinary Tract	Wound	Pneumonia	Bloodstream	Overall
E. coli	1	4	5	4	1
Enterobacter sp.	5	5	3	5	6
K. pneumoniae	5	8	4	7	7
Proteus mirabilis	7	7	8	11	9
Citrobacter sp.	9	10	10	11	11
S. marcescens	12	11	6	11	13

Data from Schaberg and Culver.[2]
[a]Ranking among all major nosocomial pathogens by site.

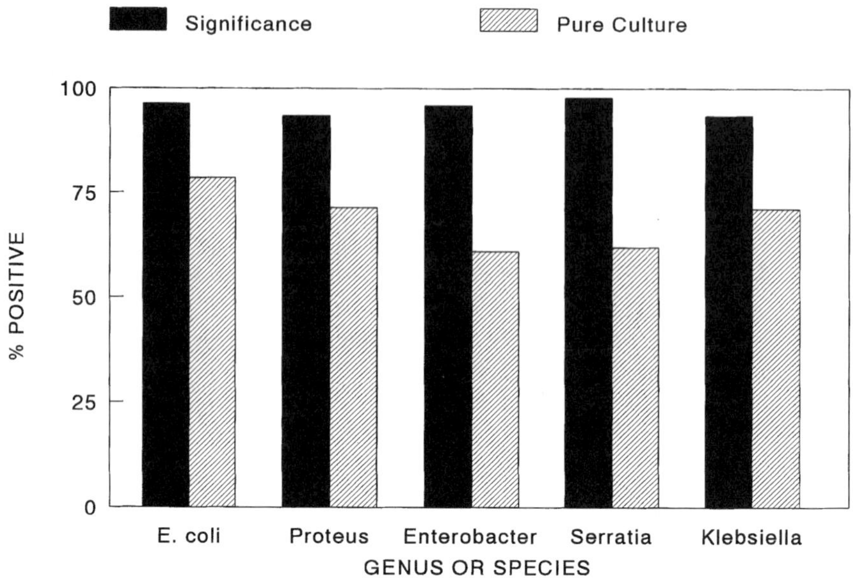

FIG. 1. Clinical significance *(black)* and frequency of isolation in pure culture *(hatched)* of common *Enterobacteriaceae* groups involved in bacteremia. Data from Roberts et al.[4]

emia caused by *Enterobacteriaceae* rose from 34% (1975–1977) to 44% (1985–1986). Three family members—*E. coli, Klebsiella,* and *Enterobacter*—were responsible for all bloodborne infections. Two of the three highest fatality rates associated with bacterial sepsis in this study involved members of the *Enterobacteriaceae* (*E. coli,* 33%; *Klebsiella,* 25%).

In a 1991 study, Roberts and colleagues[4] reviewed positive blood culture records in adult patients at Vancouver General Hospital (Canada) between 1984 and 1987. The *Enterobacteriaceae* represented 67.4% of all gram-negative bacillary isolates and 22.5% of all positive culture isolates. Even more significant was the relative importance of finding enterobacteria associated with bacteremic episodes (Fig. 1). Virtually all strains isolated from blood were judged to be significant by clinical criteria regardless of their genus or species designation; significance rates ranged from a low of 93.3% for *Enterobacter* and *Proteus* to a high of 100% for *Salmonella, Citrobacter,* and *Providencia* blood isolates. The inherent pathogenicity of this group was further substantiated by the high percentage of enterobacterial isolates recovered from monomicrobic infections. Of the eight genera isolated from blood, only *Citrobacter* was recovered less than 50% of the time in pure culture. Fatality rates associated with enterobacterial sepsis ranged from 20% to 54%. The latter value, associated with *Serratia* bacteremia, was the highest fatality rate observed for any cause of bacterial sepsis in the investigation.[4] Similar findings were reported from Israel in a 6-year prospective survey of gram-negative bacteremia in pediatric patients.[5] Overall, *Enterobacteriaceae* were responsible for 68.7% of the 374 episodes of gram-negative sepsis observed in 304 children and infants (Fig. 2). *Klebsiella pneumoniae* was the predominant pathogen and accounted for 25.8% of all cases of gram-negative sepsis in children.

Even for invasive infections not traditionally linked to enterobacteria, this family cannot be ignored. A Brazilian study[6] of bacterial meningitis in children and adults found that, although *Enterobacteriaceae* accounted for only 3.6% of total cases, the fatality rate for enterobacteria-associated meningitis was 86%, the

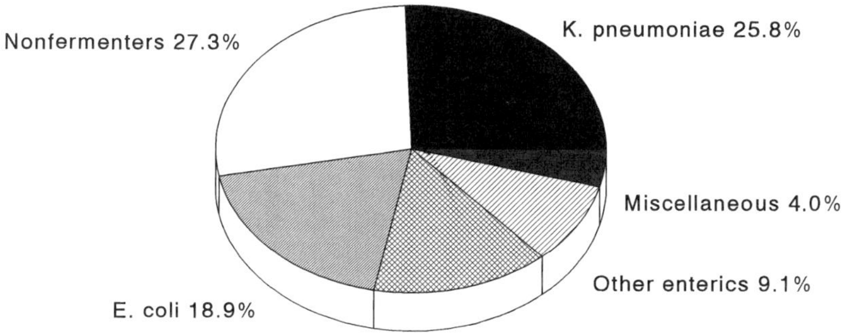

FIG. 2. Frequency of various bacterial groups as a cause of bacteremia in children and infants. Data from Levy et al.[5]

highest recorded for any group including *Pseudomonas*. The composite data from these clinical and epidemiologic investigations unequivocally establish that the enterobacteria play an important global role in the causation of serious and life-threatening invasive infections in both adults and children.

One has only to look as far as gastrointestinal disease to discover how important the family *Enterobacteriaceae* is in community-acquired infections. The genus *Salmonella* is one of the two premier pathogens (along with *Campylobacter*) responsible for most cases of bacterial gastroenteritis in the United States, and it is a leading cause of diarrheal disease throughout the world. In 1995, more than 41,000 cases of salmonellosis were reported to the Centers for Disease Control and Prevention.[7] Projected estimates suggest that between 1 and 4 million cases of *Salmonella* enteritis occur each year in the United States.[8] Virtually all of these illnesses occur after the consumption of contaminated poultry, dairy, or other food products, producing significant morbidity and mortality as well as an enormous economic effect.[9,10] The financial impact of salmonellosis in the United States alone is staggering, with yearly estimates ranging from $983 million to $1.4 billion, including physician visits, hospitalization, and death.[10] Such figures, moreover, do not encompass indirect costs such as lost income, inconvenience, regulatory programs, and agricultural issues.[10]

The role of enterobacteria in diarrheal disease does not stop with salmonellosis. Global travelers are prone to develop diarrheal disease after excursions to exotic locations. The leading cause of traveler's diarrhea is enterotoxigenic *E. coli* (ETEC), which has been estimated to cause 36% of all episodes worldwide.[11] ETEC-associated diarrhea has been reported from virtually every corner of the globe, including Africa, Asia, Latin America, and the Middle East. Estimates of the percentage of total cases of traveler's diarrhea caused by ETEC in these locales range from a low of 20% to a high of 75%.[11,12] The shiga toxin-producing *E. coli* (STEC), typified by serotype O157:H7, represent an emerging cause of bloody diarrhea both domestically and in foreign countries such as South Africa and Swaziland.[13] Today in the United States, STEC is a major cause of multistate outbreaks of hemorrhagic colitis and hemolytic-uremic syndrome in adults and children who consume adulterated food products, especially contaminated beef.[14]

Even more fascinating is the potential role played by many other *Enterobacteriaceae* groups in bacterial gastroenteritis. It is well recognized that even the best-controlled epidemiologic investigations fail to detect a potential enteropathogen in 30% to 60% of all cases, indicating that many unrecognized causes of bacterial diarrhea exist. Experimental data have identified new enterobacterial groups that may play a role in diarrheal dis-

TABLE 2. Enterobacteriaceae *groups implicated as causative agents of diarrheal disease*

Recognized or Classic Pathogens	Reputed or Potential Enteropathogens
Salmonella	*Edwardsiella tarda*
Shigella	Enteroadherent *E. coli* (EAEC)
Enteropathogenic *E. coli* (EPEC)	*Citrobacter freundii*
Enterotoxigenic *E. coli* (ETEC)	*Enterobacter cloacae*
Enteroinvasive *E. coli* (EIEC)	*Klebsiella pneumoniae*
Shiga toxin-producing *E. coli* (STEC)	*Klebsiella oxytoca*
Yersinia enterocolitica	*Hafnia alvei*
	Providencia alcalifaciens

ease.[15] Most of these taxa were assumed previously to be simple colonizers (commensals) of the gastrointestinal tract of humans (Table 2). Further clinical, epidemiologic, and molecular studies may provide direct evidence that some or all of these reputed agents cause gastroenteritis.

These few examples illustrate the clinical significance and medical importance of the family *Enterobacteriaceae.* Its overall importance will probably continue to increase for decades. Surprisingly, outside of four genera (*Escherichia, Salmonella, Shigella,* and *Yersinia*), we know very little about the pathogenicity of the taxa in this family. Many such groups have been discovered only in the past decade. As our knowledge increases regarding the epidemiology and disease spectrum of these groups, their relative importance will increase also. This has already been observed for such bacteria as *Citrobacter koseri* in neonatal meningitis and *Klebsiella rhinoscleromatis* in rhinoscleroderma. Other major species (eg, *Enterobacter cloacae, Proteus vulgaris*) are known to be genetically heterogeneous (see Chapter 1) and will undoubtedly yield additional new taxa of medical significance. Finally, because people now tend to live longer and to survive devastating diseases, the overall population will become increasingly susceptible to enterobacterial infections with advancing age. Thus, the family *Enterobacteriaceae* should continue to hold its paramount position among medically significant procaryotic organisms for the foreseeable future.

REFERENCES

1. Crossley KB, Peterson PK. Infections in the elderly. *Clin Infect Dis* 1996;22:209–215.
2. Schaberg DR, Culver DH. Major trends in the microbial etiology of nosocomial infection. *Am J Med* 1991;91 (Suppl 3B):72S–75S.
3. Ehni WF, Reller LB, Ellison RT III. Bacteremia in granulocytopenic patients in a tertiary-care general hospital. *Rev Infect Dis* 1991;13:613–619.
4. Roberts FJ, Geere IW, Coldman A. A three-year study of positive blood cultures, with emphasis on prognosis. *Rev Infect Dis* 1991;13:34–46.
5. Levy I, Leibovici L, Drucker M, Samra Z, Konisberger H, Ashkenazi S. A prospective study of gram-negative bacteremia in children. *Pediatr Infect Dis J* 1996;15: 117–122.
6. Bryan JP, de Silva HR, Rocha H, Scheld WM. Etiology and mortality of bacterial meningitis in northeastern Brazil. *Rev Infect Dis* 1990;12:128–135.
7. Anonymous. Salmonella *surveillance: Annual tabulation summary 1993–1995.* Atlanta: US Department of Health and Human Services, 1996.
8. L'Ecuyer PB, Diego J, Murphy D, et al. Nosocomial outbreak of gastroenteritis due to *Salmonella senftenberg. Clin Infect Dis* 1996;23:734–742.
9. Todd ECD. Costs of acute bacterial foodborne disease in Canada and the United States. *Int J Food Microbiol* 1989;9:313–326.
10. Rubino JR. The economic impact of human *Salmonella* infection. *Clin Microbiol Newslett* 1997;19:17–22.
11. Okhuysen PC, Ericsson CD. Travelers' diarrhea: Prevention and treatment. *Med Clin North Am* 1992;76: 1357–1373.
12. Barry M. Traveler's diarrhea: New perspectives. *Infect Agents Dis* 1992;1:114–118.
13. Hughes JM, La Montagne JR. The challenges posed by emerging infectious diseases. *ASM News* 1994;60: 248–250.
14. Armstrong GL, Hollingsworth J, Morris JG Jr. Emerging foodborne pathogens: *Escherichia coli* O157:H7 as a model of entry of a new pathogen into the food supply of the developed world. *Epidemiol Rev* 1996;18:29–51.
15. Abbott SL, Janda JM. Bacterial gastroenteritis I: Incidence and etiologic agents. *Clin Microbiol Newslett* 1992;14:17–21.

3
Escherichia coli

TAXONOMY

In 1885, a young German physician named Theodor Escherich began a series of microbiologic investigations aimed at the analysis of the gastrointestinal contents of infants.[1] Using aseptic techniques learned from a disciple of Robert Koch, Escherich was able to isolate several microorganisms from feces, one of which he named *Bacterium coli commune.*[1] *Bacterium coli* was found early on to be a very commonly isolated microorganism, particularly from intestinal samples.[2] During a 15-year period (1887–1902), considerable controversy existed as to whether *Bacterium coli* was identical to the causative agent of typhoid fever, *Bacillus typhi (Salmonella typhi).* This led to the discovery of a number of key tests, including lactose fermentation, that clearly separated these two organisms.[3]

By 1919, Castellani and Chalmers recognized that *Bacterium coli* did not meet the biochemical and morphologic criteria for inclusion within this genus.[2] As a consequence of these observations, the genus *Escherichia* was proposed by the same authors to honor Theodor Escherich, the discoverer of *Bacterium coli.* Since 1937, when Rahn proposed the family *Enterobacteriaceae,* the genus *Escherichia* has resided within that group. In 1958, the Judicial Commission of the International Committee on Bacteriological Nomenclature rendered Opinion No. 15, which stated, among other things, that the genus *Escherichia* was the type genus for the family and *Escherichia coli* was the type species.[4] Although some controversy regarding this decision occurred over the next several decades,[5] it remains valid today (see Chapter 1). The seminal discovery of Escherich more than a century ago thus resulted in the biochemical and molecular definition of a species that became the cornerstone on which the family *Enterobacteriaceae* was built.

EPIDEMIOLOGY

The intestinal tract of homeothermic animals (including humans) serves as the natural reservoir for *E. coli.*[2] Birds may also harbor *E. coli* as part of their normal bowel flora.[6] Estimates suggest that *E. coli* constitutes approximately 1% of the total microbial biomass of feces, with quantitative plate counts yielding coliforms in numbers exceeding 10^9 cfu per gram of human feces.[2] Virtually all healthy domesticated animals and humans contain *E. coli* as part of their normal gut flora, usually as a commensalistic association. Despite this benign relationship, the resident *E. coli* flora is in a constant state of flux, with new serotypes or biotypes being continually introduced, presumably through oral ingestion of contaminated food or water.[2]

Under certain circumstances (eg, age-related susceptibility), *E. coli* can produce disease in animals. The general term used to describe such conditions is colibacillosis. Colibacillosis may involve infections that originate primarily from the gut (enteric colibacillosis) or those that commence as extraintestinal manifestations of disease without preceding intestinal symptomatology (systemic colibacillosis). This distinction is somewhat important because certain groups (eg, poultry) are prone to develop only the systemic form of disease whereas others (eg, lambs) exhibit both types in approximately equal numbers.[7]

Enteric colibacillosis can be subdivided into an enteric form and a more aggressive disease called colibacillary toxemia.[7,8] The enteric form commonly is manifested at presentation as an acute diarrhea, and it is characterized by the presence of enterotoxigenic *E. coli* (ETEC) in the feces of infected animals. In colibacillary toxemia, a disease associated with the neonatal state, infection results primarily from the massive release of endotoxin (and possibly other products) rather than enterotoxin. Collapse and sudden death (without other significant morbidity) are often the prominent findings in animals with colibacillary toxemia.[7] In piglets, three distinct entities have been described: shock in weaner syndrome, hemorrhagic enteritis, and edema disease.[7] Edema disease is by far the most serious complication; it is characterized by edema of the subcutaneous tissues of the head and orbit and neurologic abnormalities such as flaccid paralysis.[7] Again, low morbidity with high mortality is the keynote of this infection.

Systemic colibacillosis (colisepticemia) is an invasive disease found principally in calves, lambs, and poultry.[7,8] The disease may be manifested at presentation as frank septicemia with shock-like syndromes (in calves) or as focal extraintestinal infections of the central nervous system (meningitis), bones (arthritis), or other organ tissues. In poultry (eg, chickens, turkeys), the initial findings of systemic colibacillosis are respiratory distress accompanied by severe green diarrhea and lameness.[7,9] Infection is thought to occur via the inhalation of feces-contaminated dust.[10] Concurrent *Mycoplasma* infection is thought to be a predisposing factor to infection.[7] The initial presentation progresses into later conditions such as air sacculitis, pneumonitis, and pericarditis.[7,10]

Because animals are naturally colonized with *E. coli,* it is not surprising that these organisms can find their way into the food chain and therefore into consumable products. This point is dramatically illustrated by the recent emergence of *E. coli* O157:H7 as a major foodborne pathogen causing outbreaks of hemorrhagic colitis linked to the consumption of fast food hamburgers, upscale apple juice blends, and other products. There are at least two mechanisms by which adulteration can take place. During the processing of slaughtered animals at abattoirs, meats destined for retail consumption can become fecally contaminated with *E. coli.* If processed meats are improperly stored or cooked, the degree of initial contamination can become magnified. A second method leading to adulteration involves the excretion of biologic wastes by both domesticated and wild animals onto lands used for agricultural purposes. Such excreta, whether present accidentally or intentionally (ie, manure), can serve as a potential source for the defilement of fruits and vegetables. Furthermore, grazing of herds on lands adjacent to sources of fresh water can lead to the accidental contamination of rivers, lakes, and estuaries due to fecal runoff precipitated by heavy periods of rain. These facts are highlighted by a 1996 article on the frequency of *E. coli* contamination in commercial foods.[11] Of 50 food samples collected from retail establishments, 80% contained coliforms and 46% specifically contained *E. coli.* Foods most often found to harbor *E. coli* were meat and meat products (72%) and vegetables (55%); even 2 of 13 finfish specimens and 1 of 4 shellfish samples contained *E. coli.* Because *E. coli* is not a free-living microbe, its presence in potable water is an indication of fecal contamination. As an indicator organism, it forms the basis for drinking water standards in the United States.[12]

Fecal contamination of food and potable water sources with pathogenic *E. coli* can lead to devastating outbreaks of diarrheal disease. One study from Taiwan of foodborne disease of bacterial origin (1986–1995) found that *E. coli* was the cause of 6.5% of such outbreaks and 6% of all cases.[13] Todd[14] estimated that foodborne disease caused by *E. coli* O157:H7 in Canada and the United States alone costs between $47 million and $84 million annually.

E. coli is also a major cause of nosocomial infections. A majority of these hospital-acquired illnesses arise from the spread of endogenous strains of *E. coli* from the gastrointestinal tract to systemic sites after surgery, invasive diagnostic procedures, catheterization, or chemotherapy. The most recent report of the National Nosocomial Infections Surveillance (NNIS) system[15,16] still found *E. coli* to be the microbe causing the most illnesses (16%) at major sites of infection (eg, urine, blood, wounds, respiratory sites). Most of this nosocomial disease involved the urinary tract, where *E. coli* accounted for approximately 26% of all reported infections.[15] In long-term care facilities, *E. coli* is a particularly troublesome pathogen. Studies suggest that oropharyngeal colonization by *E. coli* occurs in 10% to 24% of nursing home residents, leading to a rate of nosocomial pneumonia in this population that is estimated to range from 6.4% to 17%.[17] *E. coli* can cause bacteriuria in up to 54% of all persons in long-term care facilities,[17] and such asymptomatic conditions often lead to *E. coli* bacteremia (13%–37%). Therefore, unlike many other *Enterobacteriaceae, E. coli* is a major community-acquired and nosocomial pathogen causing significant morbidity and mortality in both healthy and immunocompromised persons.

CLINICAL SIGNIFICANCE

E. coli causes a panoply of human infections ranging from gastrointestinal disturbances to extraintestinal illnesses. These diseases include both well-recognized diarrheal syndromes and less frequently encountered entities such as wound infections, appendicitis, and endophthalmitis. Most *E. coli* infections fall into one of the four major categories described in the following sections.[18] These organotropic diseases are not the result of random infection, but rather are caused by strains possessing specific somatic (O-antigen) determinants and virulence-associated characteristics. *E. coli* is responsible for about 85% of community-acquired urinary tract infections (UTIs) and up to 50% of those that are nosocomially transmitted.[19] In children *E. coli* accounts for 80% to 90% of UTIs.[20,21] *E. coli* is also a major cause of nosocomial bacteremia, with studies reporting incidence rates ranging from 10% to 20%. Although the frequency of *E. coli* septicemia is on the decline in the latest national survey,[15] it is still one of the five leading causes of nosocomial bloodstream infections. One of the secondary complications of *E. coli* septicemia is meningitis in young infants; *E. coli* is the leading cause of gram-negative meningitis during the neonatal period.[22] The most alarming aspect of *E. coli* neonatal meningitis is the high rates of neurologic sequelae and death (~30% each) associated with infection.[23] By far the *E. coli* group that receives the greatest amount of attention comprises those strains associated with bacterial gastroenteritis.[24] *E. coli* is a major cause of gastroenteritis on a global basis, particularly in persons with recent histories of foreign travel. However, diarrheagenic strains of *E. coli* are difficult to diagnose in the clinical laboratory and present a formidable challenge regarding their rapid and accurate identification.[25] Clearly, the medical importance of this bacterium will continue to increase as new clinical manifestations of disease are linked to specific groups of pathogenic *E. coli.*

Gastroenteritis

Since the infancy of modern bacteriology (1890–1910), *E. coli* has been suspected to cause diarrheal syndromes in humans. Early studies, often involving outbreaks of summer diarrhea in Europe, linked *E. coli* as well as a number of other agents (eg, proteeae) to episodes of gastrointestinal disease. The multiplicity of species detected in such outbreaks left many doubting the potential role that many of these agents would later be shown conclusively to play in the causation of diarrheal disease. It took the antigenic typing scheme of Kauffmann coupled with epidemiologic surveys of *E. coli*–associated infantile diarrhea in the 1940s and 1950s to establish unequivocally this species' enteropathogenic capability.[26,27] Excellent articles detailing historical aspects of early studies on diarrheal disease and the association of coliforms with pediatric gastroenteritis can be found in the reviews of Robins-Browne[26] and Levine.[27] Today, the number of types of *E. coli* associated with gastroenteritis has greatly expanded to include both well-recognized categories (eg, enteropathogenic *E. coli* [EPEC]) and newly described groups implicated as possible causes of gastroenteritis (eg, cytotoxic necrotizing factor [CNF]–producing *E. coli*).[28] Current evidence suggests that there is no foreseeable end to the discovery of new *E. coli* groups associated with enteritis.

No area of clinical microbiology is more problematic than that involving the association of certain pathogenic groups of *E. coli* with gastroenteritis. This perplexing situation has often led to confusion among physicians and microbiologists alike regarding conflicting terminology, pathogen group–disease associations, and laboratory test selection. Table 1 lists the current and common former designations for the major groups of *E. coli* proven or thought to cause diarrhea. Presently, four major pathogenic classes—enteropathogenic (EPEC), enterotoxigenic (ETEC), enteroinvasive (EIEC), and Shiga toxin–producing (STEC) *E. coli*—are bona fide causes of gastrointestinal disturbances. One provisional group, the enteroadherent *E. coli* (EAEC), is implicated in eliciting enteritis, and a sixth, less well-defined cluster of newly described agents is potentially associated with diarrheal disease.

The terminology for three of these enterovirulent groups (EPEC, ETEC, and EIEC) has remained essentially unchanged for many years, reflecting the facts that they were discovered decades ago and their microbiology, epidemiology, and pathogenicity have been well characterized. STEC, a more recently discovered group, has undergone an evolution in terminology (see later discussion) that resulted from the recognition that many different serotypes of *E. coli* other than O157:H7 elaborate a toxin identical or almost identical to that produced by Shiga's bacillus *(Shigella dysenteriae)*. Eventually these organisms were reclassified according to toxin type rather than disease manifestation (hemorrhagic colitis). A similar situation exists for EAEC. In addition to the terms listed in Table 1, a number of other acronyms have been invoked on a less frequent basis for certain *E. coli* associated with gastroenteritis.[25,27]

Distinguishing features of the currently recognized groups are listed in Table 2. Although all five groups initially cause a secretory type of gastroenteritis, only EIEC and STEC routinely go on to produce bloody diarrhea in the later stages of illness, a major feature that distinguishes them from other enterovirulent groups of *E. coli*. These groups can also be separated from one another by pathogenic mechanisms, which provide a basis for differential laboratory tests. Other notable differences among these classes of pathogenic *E. coli* are listed in subsequent tables.

Enteropathogenic E. coli *(EPEC)*

The EPEC are the oldest known group of coliforms recognized to cause diarrheal disease in humans. Beginning in the early 1940s in

TABLE 1. *Pathogenic groups of* E. coli *associated with gastroenteritis*

Current Terminology	Acronym	Prior Designations/Inclusive Groups
Enteropathogenic *E. coli*	EPEC	—
Enterotoxigenic *E. coli*	ETEC	—
Enteroinvasive *E. coli*	EIEC	—
Shiga toxin–producing *E. coli*	STEC	Enterohemorrhagic *E. coli* (EHEC), verocytotoxin-producing *E. coli* (VTEC), Shiga-like toxin–producing *E. coli* (SLTEC)
Enteroadherent *E. coli*	EAEC	Enteroaggregative *E. coli* (EAggEC), diffuse adhering *E. coli* (DAEC)
"Reputed agents"	—	Cytotoxic necrotizing factor (CNF)–producing *E. coli*, cytolethal distending toxin (CLDT)–producing *E. coli*

TABLE 2. *Major characteristics of diarrheagenic* E. coli *groups*

Group	Disease Produced	Nature of Diarrhea: Early	Nature of Diarrhea: Late	Frequency in U.S.	Pathogenic Mechanism[a]
EPEC	Enteritis	Watery	Watery	+	Adherence
ETEC	Enteritis	Watery	Watery	++	Toxin(s)
EIEC	Dysentery	Watery	Bloody	+	Invasion
STEC	Colitis	Watery	Bloody	+++	Toxin(s)
EAEC	Enteritis	Watery	Watery	Unknown	Adherence

[a]Virulence mechanism or diagnostic feature of each pathogenic group.
+++, common; ++, occasionally isolated; +, rare.

Great Britain and subsequently throughout Europe and the United States in the 1950s, outbreaks of infantile gastroenteritis began to surface that were associated with the recovery of certain serogroups of *E. coli* (O111, O55) from diarrheal specimens.[24,26] This circumstantial association became stronger as time passed. Convincing evidence that these serogroups were indeed the cause of infantile gastroenteritis came from the observations that (1) strains of classic serotypes associated with diarrheal disease could be isolated from the stools of sick children but not routinely from healthy controls, and (2) strains associated with infantile gastroenteritis were able to induce diarrheal disease on challenge in neonates and adult volunteers.[26,28,29] In 1955, Neter and colleagues[30] coined the phrase "enteropathogenic *E. coli*" to refer collectively to this specific group of serotypes linked with diarrhea in young children. The acronym EPEC has persisted for more than 40 years. Although controversy continued to occur regarding the enteropathogenicity and uniqueness of this group for several decades, the debate was finally put to rest in the late 1970s when Levine and others fed previously isolated EPEC strains (O127:H6, O142:H6) from outbreaks of infantile gastroenteritis to adult volunteers,[27] who subsequently developed diarrhea. In 1979, Cravioto and coinvestigators[31] at the Central Public Health Laboratory in London identified the first virulence-associated marker (adhesive factor) for EPEC strains, thus providing the missing piece to the puzzle.

EPEC-associated gastroenteritis is a disease that manifests primarily in young children (<2 years), particularly those younger than 6 months of age. These illnesses are commonly associated with nursery outbreaks of enteritis or nosocomial infections in which patient-to-patient spread occurs, most often via the contaminated hands of medical personnel or relatives. Reservoirs of EPEC infection are thought to include both symptomatic and asymptomatic infants and children, asymptomatic adults, and possibly cattle.[29] Although there is an age predilection for this disease, this representation may not be completely factual, because EPEC outbreaks of gastrointestinal disease have been documented in older children and adults.[32] The most common symptoms associated with EPEC infections are diarrhea, abdominal pain, low-grade fever, and malaise.[29,32,33] Mucus in stools is a frequent but not universal finding.[28,29,33] One of the interesting aspects of EPEC diarrhea is its relative severity. In comparison with shigellosis, EPEC diarrhea can involve a more protracted course.[33–35] Secondary complications arising from primary infection are known to include dehydration, zinc deficiency, and the inability to tolerate oral feedings.[33] Rarely, EPEC disease progresses to more fulminant conditions such as septicemia and necrotizing enterocolitis.[34,36] Additional features of EPEC infections, including infective doses and duration of symptoms, are listed in Table 3.

During the period 1950–1970, many devastating outbreaks of EPEC disease occurred worldwide. The Centers for Disease Control and Prevention (CDC) reported that most *E. coli* outbreaks in children in the United States between 1960 and 1972 were caused by

TABLE 3. *Properties of* E. coli *groups associated with gastroenteritis*

Group	Infectious Dose (cfu)	Incubation Period (h)	Duration of Illness (d)		Attack Rate (%)	Presence in Stool	
			Mean	Range		Mucus	Blood
EPEC	10^6–10^{10}	9–19	5	3–14	33–72	+	–
EIEC	≥10^8	<24	2–4	1–12	33–56	+	+
ETEC	10^8–10^{10}	3–166	4–7	1–53	16–77	–	–
STEC	50–700	24–336	4–5	6–9	28–75	–	+
EAEC	~10^{10}	14–46	5	3–>14	?	–	–

Data from references 28, 29, 32–34, 43, 44, 52–54, 76, 77, 79, 129–131, and 134.

EPEC serotypes.[34] However, since that time the incidence of EPEC infections has changed dramatically in industrialized countries, so that it is no longer considered to be a major threat. Reasons for this declining trend are unclear but may include sterilization of infant bottles, pasteurization of milk or formula, better infection control procedures, and changes in the virulence characteristics of EPEC strains.[24] However, outbreaks of disease still do occur (Table 4).[37] Probably the most interesting of these was a large outbreak of gastroenteritis caused by *E. coli* O111:B4 in 611 students and 39 adults at a school complex in Halikko, Finland.[32] In addition to a high attack rate (72%), virtually all infections occurred in children older than 2 years of age (range 7–19 years). Prominent symptoms included abdominal pain (81%–95%), headache (47%), vertigo (23%), sore throat (18%), myalgia (16%), and vomiting (9%); 1% of patients had recurring episodes of diarrhea for up to 4 months after the initial onset of disease. Although food was the suspected vehicle of infection, the source of these illnesses was never proven.

Despite the decline in the overall prevalence of EPEC disease in industrialized countries, EPEC is still a global concern. EPEC has been demonstrated to cause sporadic disease on a routine basis in many developing countries.[34] Epidemiologic investigations within the last decade have found high prevalence rates of EPEC infection in Iran (31%), Thailand (13%), Venezuela (9%), New Caledonia (8%), and Bangkok (6%) in various populations.[38–42]

Enteroinvasive E. coli *(EIEC)*

Dysentery is the classic clinical presentation associated with EIEC infection. Like *Shigella,* with which it shares biochemical properties (eg, anaerogenic, nonmotile) and genetic relatedness, EIEC produce disease by invading and replicating in the colonic mucosa, the preferred site of infection. Symptoms commonly associated with EIEC infection include fever, severe abdominal cramps, malaise, toxemia, and watery diarrhea.[27,28] Other signs of EIEC illness include chills, tenesmus, and myalgia.[33] Although most reviews indicate that EIEC infections progress from a secretory diarrhea to gross dysentery characterized by scanty stools with blood and

TABLE 4. *Recent outbreaks of EPEC-associated diarrhea*

Date	Location	Age	Number		Serotype	Vehicle of Infection	Ref.
			Infected	Deaths			
1980	Belfast	Infants	30	2	O111ab:H2	Cross-infection	34,37
1986	Chester	Infants	?	?	O55:H⁻	?	37
1987	Chongqing	Neonates	69	2	O127:H6	Medical staff hands	36
1987	Halikko	School-age children; adults	650	0	O111:B4	Food (?)	32

mucus, Cohen and Giannella[33] found this to be a rare rather than a common scenario. Their viewpoint was supported by evidence from several outbreaks of EIEC-associated gastroenteritis in which bloody diarrhea was not a prominent finding.[43,44] EIEC infection has also been associated with the carrier state. One German and Danish study found 9 of 11 persons from whom EIEC organisms were isolated to be asymptomatic.[45] Most of these strains were detected by routine fecal examination of refugees from Romania and Yugoslavia between 1989 and 1995. Laboratory-acquired infections caused by EIEC have also been documented with the use of molecular fingerprinting techniques.[45]

EIEC is considered to be a rare enteric pathogen in the United States. Low prevalence rates may be a true reflection of the overall incidence, or they may be partly attributable to the difficulty in isolating and identifying this organism and its frequent confusion with *Shigella* species. Wanger and associates[46] studied returning U.S. students who had attended language classes in Guadalajara, Mexico, during the summers of 1985 and 1986 and who had become ill with diarrhea (four or more unformed stools per day). Of cultures from 273 students, 15 (5.5%) were positive for EIEC. Nine (60%) of these were positive for fecal leukocytes, and six students (40%) had occult blood in their stools (see Table 3). Almost 50% of these students were coinfected with another recognized enteric pathogen such as *Salmonella, Shigella,* or EPEC. A similar epidemiologic survey of 200 Thai children ranging in age from 1 to 10 years with a history of bloody diarrhea or dysentery found an EIEC prevalence rate of 5%.[47] Again, in only 30% of these cases was EIEC found to be the only pathogen.

Outbreaks of diarrheal disease caused by EIEC, although rare, have been reported. The CDC reported an outbreak of EIEC NT:H⁻ on board a cruise ship in 1981, which was linked to the consumption of potato salad and cold buffet items.[43] Common symptoms in 47 suspected cases of EIEC infection included nausea (72%), abdominal cramps (68%), headaches (68%), chills (60%), and dizziness (53%). A second 1981 outbreak of EIEC infection was traced to adults and children attending a school for the mentally handicapped.[44] An estimated 86 persons were infected, and *E. coli* O124:H30 was isolated from 6 student-care staff members and 13 students residing in two dormitories. The outbreak appeared to spread by person-to-person transmission. A 1983–1984 outbreak of dysentery in India caused by *E. coli* O152 was described by Chatterjee and Sanyal.[48] *E. coli* O152 was the single most common cause of bacillary dysentery during this period and accounted for 11.4% of all cases of diarrhea.[48] Finally, Wanger and colleagues[46] anecdotally reported a foodborne outbreak of EIEC (*E. coli* O143:H⁻) in Houston, Texas, in 1985. More than 370 cases were detected. The outbreak was associated with food from a Mexican restaurant in the Houston area; the strain may have been transported to Texas from Mexico by one of the restaurant's Mexican employees.

As with EPEC, extraintestinal disease caused by EIEC is an extremely rare event.[33] A fatal case of encephalopathy and septicemia caused by EIEC O152 in a 2-year-old child was reported in 1984.[48] In 1996, Ephros and companions[49] described two cases of encephalopathy in Israeli children subsequent to diarrhea associated with EIEC O144:H⁻ infection. Both children exhibited an altered mental status (stupor in one, obtundation and meningismus in the other), and no other explanation for their neurologic problems besides the *E. coli* O144 illness could be found.

Enterotoxigenic E. coli *(ETEC)*

ETEC is generally regarded as the second-oldest group of *E. coli* known to cause gastroenteritis. Epidemiologic surveys conducted in the late 1960s and early 1970s by several different investigators led to the discovery of a new group of *E. coli* strains responsible for diarrheal disease in young children. Unlike either EPEC (adhesive) or EIEC (invasive),

ETEC incites disease by the release of enterotoxins in the proximal portion of the small intestine where colonization takes place. At least two major enterotoxins are produced by ETEC. The first is a heat-labile enterotoxin (LT) that shows approximately 80% protein sequence identity with the heat-labile cytotonic cholera toxin of *Vibrio cholerae.*[50] A second toxin produced by ETEC is a heat-stable (ST) enterotoxin with a molecular mass of about 2 kDa.[50] Previous studies suggested that ETEC strains carrying both toxins (LT^+ST^+) predominate clinically, followed in decreasing frequency by strains carrying ST^+ only and those carrying LT^+ only. However, a 1996 review of 700 ETEC strains collected over a period of more than 30 years found that the relative prevalence of these toxin types is not dramatically different and that individual toxin-producing patterns may be related to serogroup specificity.[51]

The most common symptoms associated with ETEC infection are diarrhea (91%–97%) and abdominal cramps (80%–94%). Stools are typically watery in consistency, often yellow-tinged, and without the presence of mucus, pus, or fecal leukocytes.[33] The gastroenteritis induced by ETEC infection is typically indistinguishable from secretory diarrheas elicited by other gram-negative enteropathogens. Despite the fact that ETEC strains often produce an enterotoxin (LT) related to *V cholerae,* classic cholera ("rice-water stools") is almost never observed with ETEC infection. ETEC infections with bloody diarrhea are also uncommon, although this unusual manifestation of enteric infection has been reported.[52] Less often, nausea (39%–70%), headaches (35%–57%), myalgia (50%), weakness (35%), chills (31%), and low-grade fevers (13%–22%) are observed.[52–55] Vomiting (2%–13%) is not commonly associated with ETEC infection, a fact that helps distinguish this illness from gastrointestinal disturbances caused by Norwalk virus. Diarrheal disease induced by ETEC infection is usually self-limited (see Table 3).

ETEC infections commonly occur in four settings. In less developed countries, ETEC infections predominate in children younger than 3 years of age. These illnesses most often result from the consumption of contaminated drinking water, powdered milk, or other foods in areas where sanitation, hygiene, and public health standards are less than satisfactory. Some studies, particularly those originating from Bangladesh, suggest that children in this age group may experience up to three separate episodes of ETEC infection per child per year.[33] Unlike the relatively mild infections observed in industrialized countries, ETEC gastroenteritis in young children in underdeveloped nations can be severe at times, with dehydration and adverse nutritional consequences leading to retardation of normal growth development.[33] Dehydration has also been noted as a consequence of ETEC infection in industrialized countries.[55] As children mature, the incidence of ETEC disease apparently declines, suggesting that immunity may develop in local inhabitants with advancing age.[56] Nevertheless, ETEC is a major concern in these environments, with epidemiologic studies noting high prevalence rates in Chile (31.2%), Thailand (14%), and Spain (3.2%) in selected pediatric populations.[57–59] Household contacts of children with diarrhea, neighbors, and even members of households not associated with gastrointestinal symptomatology are often positive for ETEC on routine fecal examination, suggesting that these organisms are widely disseminated in the environment.[58]

ETEC infections in industrialized nations are most commonly associated with traveler's diarrhea. Such illnesses are often referred to by colloquial phrases such as "Montezuma's revenge" and "turista."[60] Various reviews have estimated that ETEC infections are responsible for 20% to 75% of all episodes of traveler's diarrhea worldwide,[61–63] with higher rates associated with travel to Africa and the Middle East.[63] Such extremely high rates typically occur in naive populations introduced to ETEC during their first few weeks of travel to endemic areas. Brown and colleagues[64] found that virtually all cases of ETEC infection in American citizens enrolled in academic programs in Mexico occurred within their first few weeks of arrival. Similar observations were made by Echeverria and associates[65] in American Peace Corps volunteers, 26% of whom became infected with ETEC during their first few weeks in Thailand.

A third specialized form of ETEC infection concerns traveler's diarrhea associated with military operations in foreign countries. Studies of American and British troops stationed in Saudi Arabia found rates of diarrheal disease due to ETEC to range from 21% to 26%.[66,67] Such high rates of ETEC disease often result in the inability of military forces to perform necessary logistic and tactical duties.

Despite the high standards of sanitation and hygiene in the United States, there have been at least 14 outbreaks of ETEC gastroenteritis reported in this country since 1975,[54] many since 1980 (Table 5). With the exception of one multistate outbreak of diarrheal disease caused by imported cheese,[52] all of these episodes were linked to domestic food products. Most were caused by ETEC strains of different serotypes and toxin patterns. However, two outbreaks in New England in 1993 were found to be caused by the same ETEC strain (O6:H$^-$). The only common food item used in both salad preparations was carrots, which originated from a single state but from multiple suppliers. Secondary transmission rates from outbreaks of diarrhea are low, owing to the high doses of ETEC required for infectivity (Table 3) and the relatively short excretion period for the bacterium in feces.[53]

Shiga Toxin–Producing **E. coli** *(STEC)*

Serotype O157:H7

No recently described enteric pathogen has received such intense medical and scientific scrutiny nor risen to such prominence through media exposure as *E. coli* O157:H7. *E. coli* O157:H7 is the archetypal serotype for a series of *E. coli* strains (STEC) found to elaborate potent extracellular cytolytic toxins. These cytotoxins, distinct from those produced by ETEC, were originally detected on Vero cells (African green monkey kidney) by Konowalchuk and colleagues.[69] At least two distinct verocytotoxins (VT1 and VT2) can be expressed independently or together by these organisms.

Since its discovery, this group has undergone numerous changes in nomenclature. The older literature refers to these organisms as verocytotoxin-producing *E. coli* (VTEC), but subsequent studies often refer to them as Shiga-like toxin–producing *E. coli* (SLTEC), a term that originated from the fact that VT1 is almost identical in structure to the Shiga toxin of *Shigella dysenteriae* and VT2 is 58% homologous.[70] Still others have referred to this group collectively as enterohemorrhagic *E. coli* (EHEC), based on the bloody colitis produced by these cytotoxigenic strains. In 1996, an international group of investigators[71] recommended streamlining this nomenclature by using the acronyms Stx1 and Stx2 to refer to the toxins previously called SLT-I (VT1) and SLT-II (VT2), respectively.

TABLE 5. *Foodborne outbreaks of ETEC-associated diarrheal disease, United States*

Date	Location	Occasion	No. Cases	Implicated Vehicles	Serotype	Toxin Type	Ref.
1980	Wisconsin	Mexican restaurant	415	Chips, sauce for chips, garnish, flour tortillas, guacamole, special sauce	O6:H16	LT+ST+	53
1981	Texas	Metropolitan hospital	282	Cafeteria food	O25:H$^-$	LT+	68
1983	Washington, DC	Office parties	45	French brie cheese	O27:H2	ST+	52
1993	Rhode Island	Airline flight	47	Garden salad	O6:H$^-$	LT+ST+	54
1993	New Hampshire	Buffet dinner	121	Tabouleh salad	O6:H$^-$	LT+ST+	54
1994	Wisconsin	Union banquet	220	Pan-fried potatoes	O153:H45	ST+	55

Since their discovery as etiologic agents of hemorrhagic colitis in 1982, the clinical importance of *E. coli* O157:H7 and STEC has escalated rapidly.[72] To illustrate this point, the CDC in 1994 constructed a logarithmic time line describing significant discoveries regarding *E. coli* O157:H7.[73] In little more than a decade, recognition of *E. coli* O157:H7 has undergone a major transformation from a newly described agent of diarrheal disease (1982) to the leading bacterial cause of bloody diarrhea in the United States (Fig. 1). Such monumental breakthroughs have placed *E. coli* O157:H7 on the list of emerging infectious diseases in the United States and in foreign lands such as South Africa and Swaziland.[74,75] A number of comprehensive reviews on various aspects of the disease spectrum, epidemiology, and microbiology of *E. coli* O157:H7 and STEC infections have been published.[70,76–79]

The colon is the primary site of *E. coli* O157:H7 infection. Early manifestations of disease include abdominal pain and watery diarrhea.[70,76,77] These symptoms typically last for 24 to 48 hours. Unlike other secretory diarrheas, however, the infection does not then begin to subside but rather progresses to a more severe illness accompanied by increasing cramping abdominal pain and bloody diarrhea.[70,76,77] Barium enemas of infected patients often produce radiographs with a characteristic "thumbprinting" pattern of the transverse colon suggestive of submucosal edema and infection with O157:H7.[72,76] Sigmoidoscopy usually reveals a friable, edematous mucosa with fresh blood.[33] Results of histologic examinations of colonic mucosa vary from minor pathologic changes to pseudomembrane formation.[33] At this stage of the disease it may be difficult to distinguish *E. coli* O157:H7 infection from other infectious sequelae or idiopathic inflammatory disorders involving the colon. Such a differential diagnosis can be critical, because establishment of STEC as the cause of colitis may

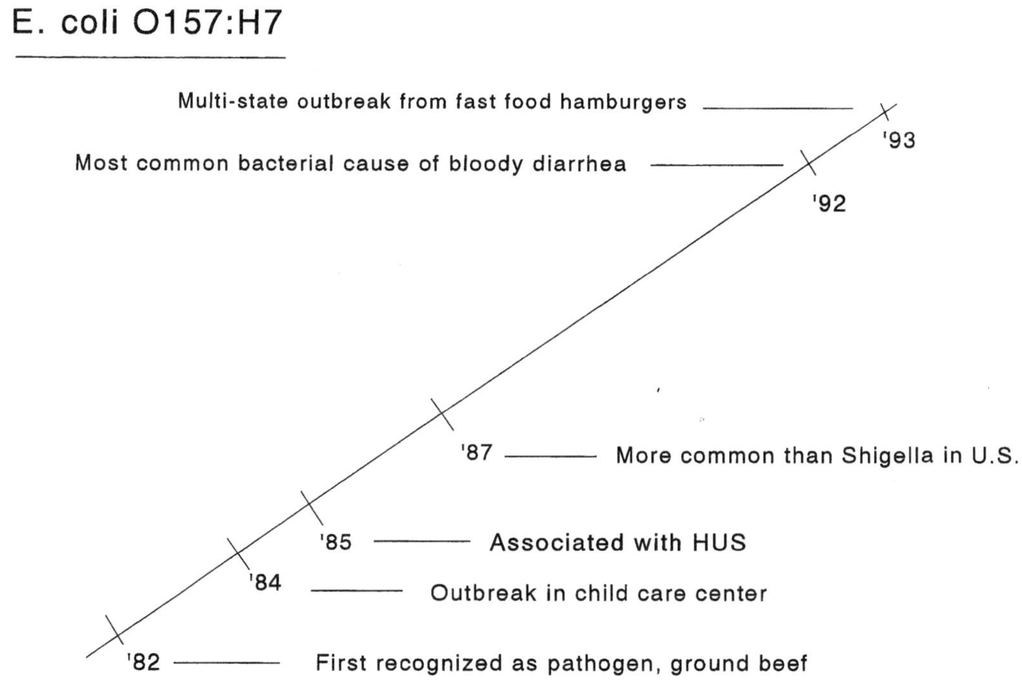

FIG. 1. Time line of major clinical and epidemiologic events associated with *E. coli* O157:H7 infections. (From *Addressing emerging infectious disease threats.*[73])

negate the need to perform additional invasive procedures such as a laparotomy or colonoscopy.[70] Characteristics that may be useful in making these distinctions for selected conditions are listed in Table 6. Unlike most of these conditions or disorders, *E. coli* O157:H7–associated hemorrhagic colitis is not often accompanied by prominent fever. Fewer than one third of patients with bloody diarrhea due to STEC have a low-grade fever.[77] In the original report linking *E. coli* O157:H7 to two outbreaks of hemorrhagic colitis in Oregon and Washington, only 7% of the collective total of 43 patients had temperatures higher than 38°C.[72] Another prominent difference between *E. coli* O157:H7 infection and *Shigella*-induced dysentery concerns the lack of fecal leukocytes or mucus in the stools of persons infected with the former agent.[27,28] However, a 1997 survey by Slutsker and associates[81] of 104 sporadic cases of *E. coli* O157:H7 diarrhea found 70.5% of stool specimens to contain fecal leukocytes (24% with ≥10 cells per high-power field). Unlike shigellae (which produces scant stools with blood), classic disease associated with O157:H7 infection involves copious bloody diarrhea; however, the entire gamut of presentations may be observed, from occult blood to blood-streaked (blood-tinged) stools to macroscopically bloody specimens. The number of bowel movements per person per day can range from 3 to 30.[33] Other symptoms useful in the differential diagnosis include vomiting and abdominal pain. In addition to the disorders listed in Table 6, other conditions that can be confused with O157:H7 infection include intussusception, appendicitis, ischemic colitis, and lower gastrointestinal infections caused by *Clostridium difficile, Salmonella, Campylobacter,* or *Yersinia*.[70]

Several other notable properties of O157:H7 infection make this agent unique among other groups of enterovirulent *E. coli* causing gastrointestinal disturbances. Unlike EPEC, EIEC, or ETEC, the estimated infectious dose of *E. coli* O157:H7 needed to cause diarrheal disease is extremely low and has been postulated to be fewer than 1000 organisms (see Table 3). These values are based on approximations resulting from several well-described outbreaks in which the O157:H7 microbial burden in implicated foods (eg, hamburgers) and individual records of consumption of suspect foods by symptomatic persons were determined.[79] Such low infectious doses, if true, rival those of *Shigella* species in respect to overt infectivity. A second cardinal feature of O157:H7 and STEC infections is a prolonged incubation period. Although early studies estimated this value to be about 4 days, Karmali[76] found a longer period to be required (mean 5.6 days) based on an analysis of five outbreaks. Such numbers complicate trace-back investigations and epidemiologic surveys. Frank bloody stools are another prominent characteristic of O157:H7 infection. Such devastating disease can lead to a variety of complications (eg, dehydration) necessitating hospitalization.[70] This hemor-

TABLE 6. *Clinical features associated with the differential diagnosis of* E. coli *O157:H7 infection*

	Clinical Features				
Disease/Disorder	Vomiting	Bloody Diarrhea	Mucus/Fecal Leukocytes	Abdominal Pain	Significant Fever
E. coli O157:H7 colitis	(+)	+	–	+	–
Shigella dysentery	–	+[a]	+	+	+
Inflammatory bowel disease					
Ulcerative colitis	–	+	+	(+)	+
Crohn's disease	+	–	+	+	+

Data from references 27, 28, 33, and 80.
[a] Scant blood stools
+, Typically positive; (+), occasionally positive; –, usually negative.

rhagic phase of the disease usually lasts 2 to 4 days, and the full duration of gastrointestinal symptomatology rarely exceeds 10 days.[77] Other symptoms sometimes noted in symptomatic persons include nausea (60%), vomiting (50%), and chills (30%). Fecal excretion of the organisms can continue for 4 to 9 days after infection,[33] but prolonged shedding (>21 days) is more common in children than in adults.[79,82,83]

On an annual basis in the United States alone, there are an estimated 20,000 infections and 250 deaths caused by *E. coli* O157:H7.[79] Tarr[70] reviewed and summarized the clinical situations from which O157:H7 can be recovered; these range from inapparent illnesses to severe secondary sequelae and associated intestinal and extraintestinal complications subsequent to primary infection. By far the least understood of these conditions is the carrier state. Screening tests have identified a number of asymptomatic contacts of infected persons who are culture-positive for O157:H7.[70,77] The incidence of this condition is unknown but is thought to be low; a long-term carrier state has not been reported. A second uncommon clinical presentation of O157:H7 infection is nonbloody diarrhea. Tarr[70] estimated that about 10% of clinically diagnosed cases of O157:H7 infection involve nonbloody diarrhea. Whether this condition is an accurate reflection of a milder illness that goes undetected in most instances because of failure to seek medical attention remains to be determined.[77] Alternatively, such illnesses may result from a rare combination of host and bacterial factors that produces a more self-limiting disease.

Approximately 90% to 95% of episodes of *E. coli* O157:H7 infection culminate in bloody diarrhea. Infections predominate during the summer months, and people at the extremes of age (ie, young children and the elderly) run the greatest risk of developing hemorrhagic colitis.[76,77,81] Other risk factors linked to O157 infection include recent use of antibiotics and gastrectomy.[77] Although cases of hemorrhagic colitis due to O157:H7 were first detected as part of a multistate outbreak of illness linked to the consumption of fast food hamburgers,[72] it became apparent early on that sporadic cases of infection contribute significantly to the overall incidence of the disease.[84] Today it is well recognized that this is the predominant form of disease seen in both the United States and Canada.[76,77] Heightened awareness of the fact that sporadic infections predominate has resulted from several intertwined factors, including improved isolation and screening methods to detect O157:H7 infection and the low-level, widespread dissemination of this clone in many consumable products.[85] Even some purported statewide outbreaks of O157:H7 disease have been determined to represent sporadic infections, based on use of fingerprinting techniques to establish strain relatedness and trace-back investigations.[86] Despite these observations, isolated cases are still a major concern. A 1997 epidemiologic survey of more than 100 sporadic cases of O157:H7 infection found a hospitalization rate of 47.1%.[81]

Probably the most frightening aspect of the low infective dose for *E. coli* O157:H7 is the massive number of outbreaks of hemorrhagic colitis reported in the literature. Many of these have reached national and international prominence through the activities of various news agencies. In 1989, Karmali[76] reviewed 20 such outbreaks that took place between 1982 and 1987 in Canada and the United States. These outbreaks occurred in many different settings (community, nursing homes, day care centers, restaurants) and involved between 2 and 73 persons each. However, these clusters of disease pale in comparison with the widespread outbreaks of O157:H7 infection described in the 1990s. Two examples will suffice. In 1993, a multistate outbreak of hemorrhagic colitis was detected in the western United States.[78,79] This outbreak involved more than 700 illnesses (>500 in the state of Washington alone[87]), and four persons subsequently died from hemolytic-uremic syndrome (HUS; see later discussion). Common symptoms included cramping abdominal pain (97%), bloody diarrhea (90%), subjective

TABLE 7. *Modes of transmission of* E. coli *O157:H7*

Foodborne illness
Water-related acquisition
Person-to-person transmission
Direct animal-to-human contact
Laboratory-acquired illness

fever (63%), and vomiting (54%); 31% of patients were hospitalized for a median of 4 days.[87] The illnesses were traced to the consumption of undercooked hamburgers from a fast food restaurant chain. The total impact from hospitalizations, disease investigation and surveillance, meat recall, and litigation was enormous. Between May and August of 1996, 19 outbreaks and 608 sporadic cases of disease caused by *E. coli* O157:H7 were detected in Japan.[88] The largest outbreak occurred in Sakai City, Osaka prefecture, where more than 9500 cases affecting more than 6000 children were reported.[89,90] Radish sprouts served in school lunches were thought but not proven to be the source of infection.[89] Three children and the executive director of the school lunch program died.[88,89]

A number of different modes of transmission of *E. coli* O157:H7 infection have been documented (Table 7). The foodborne route is by far the most common (Table 8). Although *E. coli* can be recovered from many animals, including dogs and cats,[96] sheep,[96,97] goats,[96] pigs,[96] and birds,[98] the strongest association between foodborne disease and O157:H7 infection involves cattle. *E. coli* O157:H7 can be isolated from both healthy and diseased cattle even though it probably does not play a pathogenic role in these animals.[79] The organisms are transiently shed for short periods (<1 month) by many cattle, particularly young calves.[99] Ground beef, in particular hamburger patties, has been found to be responsible for 58% of foodborne *E. coli* O157:H7 infections.[79,100] These observations are supported by other evidence linking cattle to a majority of O157:H7 infections in the United States and Canada, including direct isolation of the outbreak strain from implicated meat in at least six instances.[79] The incidence of cattle-related infections may be even greater than imagined because of contamination of vegetables by cattle manure used as fertilizer[101] and irrigation of produce with polluted water from streams and rivers near farms or grazing areas subject to fecal runoff during the rainy season. Infections can also result from the consumption of drinking water laced with O157:H7 bacilli or through exposure during the recreational use of aquatic facilities. A large waterborne outbreak (n=243) of *E. coli* O157:H7 in Cabool, Missouri, in 1989 apparently resulted from sewage contamination of the water distribution system at points where main breaks occurred during winter months.[91] Only 13% of those affected were hospitalized but four deaths resulted, all in persons 79 years of age or older. The estimated overall attack rate was 6.7%. *E. coli* O157:H7 outbreaks associated with recreational swimming have been reported by at least four states.[79] One of the best described of these outbreaks involved 21 persons at a lake near Portland, Oregon.[102] All of those affected reported swimming, and all were more likely than controls to have swallowed water. Microbiologic analysis of lake water indicated significant fecal contamination.[102] Small outbreaks attrib-

TABLE 8. *Foods implicated in the transmission of* E. coli *O157:H7 infections*

Meats	Vegetables	Fruits and Beverages	Other Foods
Hamburgers	Potatoes	Cantaloupes	Sandwiches
Ground beef	Lettuce	Apple cider	Mayonnaise
Dry-cured salami	Radish sprouts	Apple juice	Yogurt
Deer jerky	Parsley	Pasteurized milk	
Venison	Alfalfa sprouts	Water	
Mettwurst sausage	Salads		

Data from references 79, 88, and 91–95.

uted to inflatable wading pools apparently contaminated with O157:H7 bacteria from an infected child have also been described.[103]

Person-to-person transmission is a fairly common mechanism that promotes the unintentional spread of O157:H7 disease. Nursing homes, day care centers, facilities for the institutionalized mentally handicapped, and household contacts of infected persons are prime targets for secondary spread of O157:H7.[77,87,104] Attack rates as high as 22% have been reported.[83] Disease transmission in these settings almost invariably proceeds by the fecal-to-oral route through contamination of hands. Direct transmission of *E. coli* O157:H7 from cows to humans has also taken place on at least two occasions. One involved a 13-month-old boy who developed hemorrhagic colitis after repeatedly touching calves and placing his fingers in their mouths and his.[105] *E. coli* O157:H7 with the same phage type was isolated from both the child's feces and one of the calves. A second possible case of animal-to-human transmission occurred when a 15-month-old Scottish child reportedly became infected from cattle feces transmitted on the coat of the family dog.[79]

Because of the low infective dose of the O157:H7 strain, it is not surprising that laboratory-acquired infections can result. Booth and Rowe[106] published a report of *E. coli* O157:H7 infection in a 45-year-old laboratory technician whose job included subculturing strains, performing autoagglutination reactions, and opening broth cultures after steaming. Her fecal isolate had a toxin type and plasmid profile identical to one of 12 strains she had recently handled. A second laboratory-acquired infection was reported by Burnens and coworkers[107] to have occurred in a 46-year-old woman who performed cytotoxicity assays. Finally, Rao and colleagues[108] described a laboratory-associated illness in a 40-year-old male microbiology worker who developed bloody diarrhea after handling *E. coli* O157:H7 strains from Scotland 3 days before the onset of his illness. Again, phage and molecular typing linked the patient and laboratory isolates to each other.

The incidence of *E. coli* O157:H7 infection varies dramatically around the world. Areas where infection rates tend to be high include the United States, Canada, Great Britain, and Japan.[89,90,109–111] Su and Brandt[109] estimated that this organism is responsible for 0.6% to 2.4% of all cases of diarrhea in the United States; if one considers only instances of bloody diarrhea or hemorrhagic colitis in the diagnosis, the percentage of cases is much higher (15%–36%). Slutsker and associates[81] found the overall frequency of isolation of *E. coli* O157:H7 at 10 different laboratories throughout the United States to be 0.39%. The highest isolation rates occurred in Maine (2.13%), Wisconsin (1.54%), and Utah (1.03%); extremely low recovery rates were noted in Virginia (0.04%), Florida (0.07%), California (0.2%), Michigan (0.24%), New York (0.24%), and Arizona (0.29%). Although much of this geographic variation can be ascribed to temperature differences (ie, higher rates in colder climates), local differences in isolation rates observed in other foreign countries where *E. coli* O157:H7 infection is endemic cannot be explained satisfactorily by thermal gradients[110,111] and other reasons are not readily apparent. It is also clear that in many areas of the world (eg, central Europe, Australia) STEC disease due to *E. coli* O157:H7 infection is not a major public health problem and few outbreaks have been reported.[104,112]

The most serious potential consequence of hemorrhagic colitis is the development of one of two life-threatening complications, HUS or thrombotic thrombocytopenic purpura (TTP), subsequent to primary infection. The first direct evidence linking STEC to HUS came from a 1983 study by Karmali and coinvestigators,[113] in which 11 of 15 children with HUS (almost all ≤45 months of age) were found to have STEC infection. The data included isolation of verocytotoxigenic *E. coli* from stools (n=8), Shiga toxin–like activity only in fecal extracts (n=2), and serologic evidence of infection in one instance. Among the toxigenic strains of *E. coli* recovered were two isolates of O157:H7. Subsequent studies by numerous investigators[70,76,77,109] confirmed these seminal observations.

Classic HUS (one of a number of different forms) is by far the most common cause of acute renal failure in children.[76,109] The disease is characterized by a triad of symptoms: microangiopathic hemolytic anemia, thrombocytopenia, and acute renal failure.[70,76,109,114] Epidemiologic investigations indicate that 46% to 58% of all cases of HUS are caused by *E. coli* O157:H7 infection,[109] and in some studies values as high as 88% have been obtained.[76] Between 2% and 7% of children younger than 10 years of age with primary *E. coli* O157:H7 infection develop HUS as a complication. HUS appears 2 to 14 days (median 6–7 days) after the onset of diarrhea.[114] In addition to the classic signs of HUS, other symptoms may be present, such as pallor, fever, abdominal cramps, vomiting, oligoanuria, and edema.[70,109] Risk factors associated with HUS infection include age (young or elderly), female sex, bloody diarrhea, fever, leukocytosis, use of antimotility agents, and P antigen expression on erythrocytes.[77,114] Approximately 50% of HUS patients require dialysis, and 75% need transfusions of either platelets or blood.[70,114] Unusual manifestations of HUS disease occasionally occur, including hemorrhagic cystitis, balanitis, and anal dilation.[76] In one study cited by Karmali,[76] HUS developed in 12 (22%) of 55 persons with O157:H7 infection at a nursing home, and 11 died. Before the use of dialysis, the fatality rate associated with HUS in children was very high; now it is less than 10%.[76,114,115] Extrarenal involvement, now the main concern in cases of HUS, can include pancreatic thrombotic microangiopathy, hepatomegaly, and cholelithiasis.[116] Neurologic sequelae (eg, seizures, coma, hemiparesis, stroke) develop in about one fourth of children with HUS.[70,114] Currently most deaths result from encephalopathy.[115] High plasma elastase concentrations relative to the number of circulating neutrophils and neutrophilia are indicators of a poor prognosis.[117]

TTP is a disease most often seen in adults. A pentad of symptoms define TTP; these include fever and prominent neurologic abnormalities in addition to those features associated with HUS.[115] A prodromal diarrheal syndrome is usually absent.[114,115] Renal damage is typically less severe. In one outbreak described by Su and Brandt,[109] 8% of persons with hemorrhagic colitis caused by *E. coli* O157:H7 went on to develop TTP. However, accurate figures regarding the overall incidence of TTP relative to O157:H7 are not available, although it is thought to be low.

Immunity to O157:H7 infection apparently is not long lasting. Robson and colleagues[118] described bouts of *E. coli* O157:H7 disease that occurred in a young boy at 5 years of age and again 4 years later. In the second episode, neutralizing antibodies to Stx1 and Stx2 were detected in the boy's serum although the duration of their persistence was not determined. In another case, O157:H7 was recovered from the blood of a person 2 days before death[119]; further information was not provided. Other extraintestinal sites from which *E. coli* O157:H7 has been recovered include urine (two reports) and the penis.[77]

Non-O157:H7 Serotypes

An alarming concern regarding STEC is disease caused by serotypes other than O157:H7. More than 100 different serotypes of *E. coli* are known to exist that express either Stx1 or Stx2, or both.[114] Unlike *E. coli* O157:H7, which has phenotypic markers (eg, delayed sorbitol fermentation), non-O157:H7 strains generally lack unique biochemical characteristics and therefore are difficult to detect in feces. Lack of identifiers for non-O157 serotypes may mean that their overall incidence is underreported. Early studies suggested that non-O157:H7 serotypes were not a common cause of bloody diarrhea and that no outbreaks of STEC disease had ever been reported from the United States, Canada, or Great Britain.[77] This concept is now changing rapidly. In 1994, Caprioli and associates[120] described the first known outbreak of STEC disease resulting from an *E. coli* serotype other than O157:H7. This outbreak in Italy involved nine cases of

HUS caused by a nonmotile strain of *E. coli* O111 producing both Stx1 and Stx2; no vehicle for these infections was discovered. Two more outbreaks of STEC disease caused by non-O157 serotypes were reported in 1995. The first involved 18 confirmed or suspected cases of gastroenteritis in Helena, Montana, associated with the consumption of pasteurized milk; most patients (89%) had bloody diarrhea.[121] The outbreak was caused by a sorbitol-negative strain of *E. coli* O104:H21, a fact that aided its recognition. A second series of 23 cases of HUS in South Australia was tied to the ingestion of uncooked, semidry fermented sausage.[122] Again, most children (70%) had bloody diarrhea. *E. coli* O111:NM, producing both Shiga toxins, was identified as the causative agent. In 1996, a small familial outbreak of diarrhea and HUS in Ohio was again caused by *E. coli* O111:NM.[123] No source for these infections was detected.

These results suggest that disease caused by STEC serotypes other than O157:H7 is not an uncommon event but rather goes unrecognized in many instances. At least one additional serotype (O145:NM) has been implicated in outbreak disease.[123] In other cases, serotypes O26:H11 and O103:H2 have caused serious and fulminant cases of atypical HUS and hemorrhagic colitis, respectively.[124,125] In Argentina, the country with the highest incidence of HUS in the world, most cases of STEC-related HUS are caused by serotypes other than O157.[126] From these investigations it seems likely that STEC serotypes other than *E. coli* O157:H7 play important roles in causing hemorrhagic colitis, HUS, and outbreaks of disease, even though they are only beginning to be reported.

Enteroadherent E. coli *(EAEC)*

In 1985, Mathewson and coworkers[127] described a fifth reputed group of enterovirulent *E. coli* associated with traveler's diarrhea in Mexico. This new group was named enteroadherent *E. coli* (EAEC) because these strains attached to more than 40% of HEp-2 cells in tissue culture assays. Unlike adherent EPEC strains, EAEC members did not commonly belong to classic serotypes associated with infantile gastroenteritis, nor did they possess any of the known virulence mechanisms associated with EIEC, ETEC, or STEC. Of the more than 200 adults studied in this epidemiologic survey,[127] almost twice as many persons with diarrhea (14.9%) as asymptomatic controls (7.6%) were colonized with EAEC. In 1987, Nataro and colleagues[128] recognized at least three distinct adhesive patterns within the EAEC group. One, termed localized adherence, had previously been associated with EPEC serotypes. A second pattern caused uniform attachment over the surface of HEp-2 cells with bacilli one to two layers thick; organisms with this pattern were called diffuse adhering *E. coli* (DAEC). In the final type, termed aggregative adherence, bacilli subsequently named enteroaggregative *E. coli* (EAggEC) exhibited a "stacked brick" arrangement on both glass cover slides and HEp-2 cells.[128] Using these three adhesive patterns, Cravioto and companions[129] prospectively studied 72 Mexican children for the first 2 years of their lives for episodes of diarrheal disease. Over this interval more than 600 episodes of gastroenteritis were recorded. EAggEC strains were recovered significantly more often from children with persistent (>14 days) diarrhea than from asymptomatic controls (51% versus 8%, respectively). One third of all children colonized with EAggEC had bloody diarrhea[129]; mucus was occasionally present. Major differences in the frequency of DAEC among symptomatic and control groups were not detected during the course of this investigation.

Further evidence implicating EAEC and its subtypes as a cause of gastroenteritis came from several different avenues. Early studies found that the sera of some patients produced a fourfold rise in titer of antibodies directed against the O antigen of the infecting EAEC strain.[127] Challenge studies in volunteers with one of two EAEC strains at varying concentrations produced equivocal results. Diarrhea developed in only one of eight persons fed EAEC strain 199 (serotype O78:H33/35) displaying a DAEC phenotype.[130] A second EAEC isolate, strain 221 (serotype O78: H33/35), originally reported as exhibiting localized adherence but later confirmed to be an EAggEC,[128] elicited diarrhea in 5 of 16 test subjects. Overall, the disease noted in each symptomatic subject was mild and secretory in nature. Other volunteers who did not experience diarrhea did have some enteric symptoms, including fever, malaise, vomiting, and abdominal pain.[130] A volunteer study conducted almost 10 years later confirmed the heterogeneity of the in vivo pathogenicity of the EAggEC group.[131] Only one of four EAggEC strains fed to volunteers in this clinical trial caused diarrhea in three out of five patients. The causative strain, designated 042, belonged to serotype O44:H18. Loose stools were the predominant complaint. Fever, headaches, and other maladies associated with enteric disease were not present. Mean peak concentrations of EAggEC O42 in patient stools were 6.7×10^7 cfu per gram of feces. Three other EAggEC serotypes tested failed to evoke a diarrheal response.

In addition to volunteer challenge studies, reports from a number of outbreak investigations further support a probable role for some EAEC in gastroenteritis. Scotland and collaborators[132] found three small outbreaks of gastroenteritis associated with EAggEC O44:H18 infection. One of these pockets of infection involved four infants in a hospital, only two of whom had diarrhea (one bloody). A second cluster involved nine children and four staff members at a day care center; only three of the children had enteritis. A final outbreak described in this study concerned four older hospitalized adults (>68 years) with diarrhea apparently caused by EAggEC O44:H18.[132] Another publication by many of the same authors described nine outbreaks of diarrheal disease linked to O44:H18 infection in the United Kingdom.[133] Finally, Čobeljić and others[134] described an outbreak of gastroenteritis caused by EAggEC O44:H18 in a neonatal nursery ward; 19 neonates and infants were affected. Chief symptoms included watery diarrhea, fever, and weight loss.

Despite these results, there are a large number of epidemiologic investigations, including case-control studies, either supporting[135–142] or refuting[140,142–145] the association of EAggEC and DAEC with diarrhea. Interpretation of these studies is complicated by the types of diarrhea assessed (acute, chronic, persistent), patient populations analyzed (persons infected with the human immunodeficiency virus, aboriginal children), clinical settings used (urban, rural, slum), and geographic locales investigated (Brazil, Mexico, Australia, Bangladesh). Much of the controversy probably results from the fact that EAEC is a catchall acronym for a heterogeneous group of bacteria represented by diverse serotypes, adhesive characteristics, and other phenotypic markers. At least some of these EAEC strains appear to cause diarrhea, but further studies are needed to clarify their overall significance.

Reputed Agents of Gastroenteritis

In addition to the five major classes of enterovirulent *E. coli*, at least two other groups have been associated with gastroenteritis. Johnson and Lior[146] described a new toxin in the culture filtrates of 43 diarrheagenic strains of *E. coli*. This toxin, when exposed to cell lines such as Chinese hamster ovary (CHO), caused progressive cell distention and eventual cytotoxicity to monolayers over a 48- to 96-hour period. They coined the term cytolethal distending toxin (CLDT) to refer to this new group of toxigenic bacteria. CLDT+ strains were nonhemolytic and noninvasive and failed to produce ST; most were additionally negative for production of LT, Stx1, and Stx2. Serotyping of these CLDT+ strains indicated that most belonged to traditional EPEC serogroups. Most cases of CLDT-associated gastroenteritis were detected in children younger than 2 years of age. One outbreak associated with consumption of raw milk was linked to a CLDT+ strain of *E. coli* O53:H10.[146] A severe case of gastroenteritis in a 3-year-old child complicated by encephalopathy due to a CLDT+ strain of *E. coli*

O55:K59:H4 was reported in 1987[147]; the child suffered permanent residual side effects, including coordination problems. The association of CLDT with EPEC strains was confirmed by Bouzari and Varghese,[148] although only 16 (6.5%) of the 246 strains tested from 10 representative serogroups elaborated the toxin. In 1996, a case-control study[149] of gastroenteritis in Bangladeshi children failed to find a significant association between CLDT+ strains and enteritis, even though these isolates were recovered more frequently from symptomatic children than from controls (3.1% versus 0.93%, respectively). Symptomatic children with only CLDT+ *E. coli* in their stools typically presented with watery diarrhea (100%) and occasionally with vomiting (50%) and abdominal pain (38%). Serotyping of CLDT+ strains indicated that all belonged to either EPEC or EAggEC serogroups. At present it is unclear whether these strains truly represent a new group or are merely a specific subset of EPEC.

In 1983, Caprioli and coworkers[150] identified a new toxin produced by several strains of *E. coli* associated with enteritis in children. This toxin, liberated by sonication of whole cells, caused multinucleation of monolayers (CHO, Vero, HeLa) on prolonged (~72 hours) incubation; it was also determined to be cytotoxic for HeLa cells. The authors proposed the name cytotoxic necrotizing factor (CNF) for this activity based on these morphologic alterations. Since this study, CNF-producing strains have been found to be associated principally with extraintestinal disease.[151,152] However, Blanco and associates[153] reported two outbreaks of diarrheal disease and several sporadic infections caused by CNF-producing strains in Spain. One of these outbreaks involved newborn children, and the other involved both children and adults. Further studies are required to clarify the position of these organisms as agents of bacterial gastroenteritis.

Urinary Tract Infections

E. coli is the most common cause of both community-acquired and nosocomially transmitted UTIs.[15,154] Recent NNIS data indicate that 26% of all hospital-associated UTIs are caused by *E. coli*.[15] In community-acquired illnesses, between 80% and 90% of all UTIs are caused by *E. coli*.[20,155,156] The high incidence of *E. coli* bacteriuria and UTI is linked to the preeminent position of this organism among the facultatively anaerobic flora of the gastrointestinal tract. From there, seeding occurs to urogenital sites such as the vaginal introitus, the periurethral area, and the male prepuce, providing a local site for introduction and subsequent development of ascending UTI. Children, young adult women, and elderly adults are most prone to UTI with *E. coli*.

Approximately 1% of male children and 5% of pediatric females develop UTIs.[21] *E. coli* causes about 90% of first episodes of UTI in children.[20] Up to 80% of these children have recurrent bouts of UTI caused by new bacterial strains (including *E. coli*), which are acquired from exogenous sources.[21] Bacteriuria is also a frequent finding, with as many as 2.5% of young boys (<6 months of age) and 3% of young girls having *E. coli*–positive urine cultures and no overt signs of infection. The incidence of bacteriuria declines rapidly with age: reported prevalence rates among school-age children are 0.004% for boys and 1.2% for girls.[156]

In young women, *E. coli* UTIs are often associated with sexual activity, from whence the phrase "honeymoon cystitis" originated.[157] Risk factors associated with UTIs in this age group include intercourse within 24 hours of infection, use of a diaphragm and spermicide, postcoital micturition, and a recent history of UTI.[155,157] Transmission of *E. coli* UTIs between sex partners has also been documented.[158] Clinical conditions related to *E. coli* UTIs range from cystitis to pyelonephritis and complicated UTIs.[155] *E. coli* cystitis is a benign infection of the lower urinary tract characterized by acute dysuria and suprapubic or lower back pain; 20% of affected women have recurrent episodes of UTI.[155] Pyelonephritis is an infection of the upper urinary tract involving the kidneys or pelvis, or both. Symptoms can vary from a mild illness (no vomiting or nausea) to more severe disease (including urosepsis) requiring hospitalization.[155] Most isolates causing such disseminated disease belong to a particular subset of virulent *E. coli* organisms called uropathogenic strains. These strains possess a unique set of virulence determinants that include specialized fimbriae. The serogroups most commonly associated with uropathogenicity are O1, O2, O4, O6, O7, O18, and O75.[154] *E. coli* can also cause complicated UTIs in women. Most of these illnesses occur in women with anatomic defects in their urinary tract or other associated conditions such as stones, vesicoureteric reflux, and incomplete voiding.[154,155] Such infections often involve bacteria resistant to multiple antimicrobial agents and, if improperly treated, can lead to sepsis requiring hospitalization. Asymptomatic bacteriuria develops in approximately 3% to 8% of pregnant women.[159] Consequences of inadequate treatment of such asymptomatic conditions include pyelonephritis and increased risks of fetal morbidity and mortality.[159]

In adult men younger than 50 years of age, *E. coli* UTI is a relatively infrequent finding.[155] Ulleryd and associates[160] studied data from 88 cases of UTI in Swedish men collected over an 11-year period. Overall, pyelonephritis developed in 41 patients (47%), febrile UTIs in 33 (38%), and acute cystitis in another 14 (16%). The median ages of men in these groups were 57, 67, and 72 years, respectively. Patients were defined as having pyelonephritis if they presented with fever of at least 38°C plus flank pain and/or tenderness of the costovertebral angle; men with febrile illness had similar fever and/or dysuria (and urgency) without flank pain, and those with acute cystitis had no fever or low-grade fever and problems with dysuria and/or frequency. Twenty-eight patients (32%) had a history of UTI, and 28% had underlying complications, of which diabetes mellitus was the most common.[160] Some cases of uncomplicated *E. coli* cystitis may occur in younger men. Risk factors associated with such illnesses include homosexuality, lack of circumcision, and having sex partners with vaginal *E. coli* colonization.[155]

Bacteremia

E. coli is the most common cause of clinically significant gram-negative bacteremia. Early studies on the microbial etiology of bacteremia found *E. coli* to be the predominant pathogen, accounting for 17% to 33% of all septic episodes.[161] Today the significance of *E. coli* as a bloodborne pathogen has declined slightly because of the emergence of gram-positive organisms as agents of sepsis in the 1980s. The 1991 NNIS report found *E. coli* to be the fourth most common nosocomial bloodstream pathogen, trailing coagulase-negative staphylococci (CNS), *Staphylococcus aureus,* and enterococci in overall prevalence.[15] Table 9 lists the results of other

TABLE 9. *Relative importance of* E. coli *as a bloodborne pathogen in bacteremia*

Characteristic	Roberts et al[162]	Cockerill et al[163]	Weinstein et al[164]
Study site	Vancouver General Hospital	Mayo Clinic	Multicenter (NC, NJ, UT)
Study period	1984–1987	1984–1992	1992–1993
No. significant isolates[a]	1504	20,456	944
E. coli			
No. of isolates	261	2522	142
Percentage[b]	17.3	12.3	15.0
Overall rank[c]	2	2	3

[a]Includes all bacterial isolates judged to be clinically significant by various criteria.
[b]Percentage of significant *E. coli* isolates relative to total significant isolates of all microorganisms.
[c]Relative ranking of *E. coli* as a blood culture isolate (includes significant, contaminant, and transient isolates and those of undetermined significance).

general surveys on the relative frequency of *E. coli* in community-acquired and nosocomially transmitted episodes of septicemia.[162–164] Cockerill and associates[163] noted that the relative frequency of *E. coli* bacteremia declined in their institution between 1984–1988 and 1989–1992. The isolation from blood of both *E. coli* and many other members of the *Enterobacteriaceae* declined significantly. In most bacteremia studies, *E. coli* ranked behind either CNS or *S. aureus* (or both) in frequency, although most CNS blood isolates represented contamination rather than true infection. As with UTIs, a specific subset of *E. coli* strains expressing α-hemolysin and other virulence determinants cause a majority of septicemias. Serogroups linked to these septicemic strains include O6, O4, O2, O16, O18, O7, O1, and O75.[154]

E. coli bacteremia, like that caused by other gram-negative bacilli, is most often seen in older persons (median age 62–70 years) who either are immunocompromised or have one or more comorbid conditions.[161,165–167] Sepsis is most often observed in women, particularly in community settings where the initial focus of infection is the urinary tract.[165,167] Most studies indicate that 85% to 95% of persons with *E. coli* sepsis have underlying risk factors.[161,165,166] These factors commonly include liver disease (16%–35%), ethanol abuse (12%–39%), hematologic malignancies (15%–34%), solid tumors or other cancers (20%–32%), respiratory tract disease (8%–24%), diabetes (21%–26%), cardiovascular disease (29%–50%), renal disease (8%–25%), and use of corticosteroids and/or immunosuppressive therapy (14%–23%). The urinary tract is the single most common anatomic site from which *E. coli* septicemia originates (48%–58% of cases)[161,164–166]; this focus includes both uncomplicated UTIs and those involving medical procedures such as cystoscopy, transurethral resection, and bladder washout.[161,165] Other sites frequently reported as seeding *E. coli* septicemia include the gastrointestinal tract (9%–25%), the biliary tract (11%), peritoneal fluid (9%), the respiratory tract (6%), and wounds (4%); in about 12% to 27% of cases, no potential source for *E. coli* bacteremia can be found.[161,164–167]

In the largest series of cases reported to date (n=843), Grandsen and coinvestigators[165] prospectively analyzed data on cases of *E. coli* sepsis occurring over an 18-year period at one London hospital. Most patients had fever higher than 38ºC (81%), and many (60%) had an elevated leukocyte count (>10,000 cells/mm^3); 14.4% had signs of shock as evidenced by a systolic blood pressure lower than 90 mm Hg. In two large European investigations, the average incidence of *E. coli* bacteremia was respectively estimated to be 14 to 15 episodes and 24.4 episodes per 100,000 admissions per year.[165,167] A 1997 survey by Kuikka and others[161] found that complications developed in 43% of Finnish patients with *E. coli* bacteremia, including shock, respiratory insufficiency, acidosis, disseminated intravascular coagulation, and central nervous system symptoms. Reported mortality rates for *E. coli* bacteremia range from 12% to 33%.[161,162,164,166] Factors associated with a fatal outcome include pneumonia, no known focus of infection, shock, central nervous system disorder, thromboembolism, and a rapidly fatal underlying disease.[161] Other factors linked to a poor outcome include inappropriate initial therapy and an initial focus of infection other than the urinary tract.[165]

Virtually all isolates of *E. coli* recovered from blood are clinically significant. The results of two clinical studies involving a total of more than 400 *E. coli* blood isolates found 96.3% and 99.3% of these to represent true infection.[162,164] Most *E. coli* bacteremias (78%) involve monomicrobic infections.[162] One study of more than 850 cases of septicemia found 2.4% of all episodes of sepsis to involve two distinct *E. coli* strains.[165] A Danish study found *Klebsiella* spp., *S. aureus, Enterococcus faecalis,* CNS, and other *E. coli* strains to be the five microorganisms most frequently involved in cases of polymicrobic bacteremia.[167] Cockerill and colleagues[163] found that *E. coli* was isolated from blood significantly more often when the Septi-Chek system (Becton Dickinson, Sparks, MD) was used rather than nonvented tryptic soy broth bottles (Difco Laboratories, Detroit, MI). This same study found better recovery of *E. coli* from blood with the use of of aerobic bottles rather than nonvented ones (see Isolation).

Recurrent *E. coli* bacteremia is observed in 3% to 5% of patients.[161,168] These infections most often involve the same strain[168,169] and probably reflect inadequate therapy to eradicate an internal focus of infection. Rarely, enterovirulent groups of *E. coli* cause sepsis. Reported cases include EIEC septicemia in a 32-year-old man with the acquired immunodeficiency syndrome,[170] an ETEC bacteremia in a Turkish woman after an episode of watery diarrhea,[171] and a bloodborne infection caused by an atypical EPEC serotype in a patient who had undergone a cardiac bypass operation 2 days previously.[171] *E. coli* is less common but still an important cause of sepsis in children. One Israeli study of more than 400 gram-negative blood isolates from children found *E. coli* to be the third most common cause of bacteremia.[172] Neonates with underlying disease and neutropenia were especially prone to development of *E. coli* sepsis.[172] Most *E. coli* infections in this study were community acquired.

Meningitis

E. coli–associated meningitis is a life-threatening complication that can occur in any age group and is an especially serious condition in neonates. Estimates suggest that neonatal sepsis and meningitis develop in 1 to 5 infants per 1000 live births.[173] In this setting *E. coli* is the most common cause of neonatal meningitis. Infection results primarily from vertical transmission from mother to offspring during delivery or directly after birth.[173] Occasionally, *E. coli* meningitis can be acquired by horizontal transmission from staff members to infants in nurseries. By far the majority of these infections (~80%) involve strains carrying the K1 capsule.[23,165,173,174] This structure confers on such strains a number of properties, including resistance to serum killing and phagocytosis, which are thought to play important roles in the pathogenesis of the disease.[175] Serogroups most commonly associated with K1 expression include O7, O18, O1, and O6.[154] K1-positive organisms can be isolated from 20% to 40% of the feces of healthy persons of all age groups, including newborns, and from the cervix of 5% to 7% of pregnant women.[173,174] Surveys indicate that pregnant women colonized with *E. coli* K1–bearing strains transmit these organisms to about 70% of their offspring at birth or during the neonatal period[173]; disease subsequently develops in approximately 1% of these K1-colonized neonates. Mortality rates associated with meningitis caused by *E. coli* K1–bearing strains have been reported to be as high as 31%,[23] and long-term or permanent neurologic sequelae, including developmental disorders, occur in 29% to 50% of survivors.[23,173]

In adults, most cases of *E. coli* meningitis involve persons who have undergone neurologic procedures or sustained penetrating trauma to the head.[176] Less often (~20%), another illness predisposes the patient to meningitis (see Bacteremia). Christopher[176] described a case of *E. coli* meningitis in a 67-year-old man that was precipitated by a fall at home. The patient's underlying medical conditions included liver disease (hemochromatosis). An unusual case of *E. coli* meningitis in a 73-year-old woman has also been reported.[177] The woman was initially admitted to the hospital because of left thigh and leg pain, which became so severe in the lumbar area that an epidural catheter was inserted. Within 24 hours of insertion, she developed symptoms (nausea, numbness in legs) that progressed to pain in her neck and redness at the catheter site within an additional 33-hour period. Although the catheter was removed, her condition worsened 44 hours later with chief complaints of headache, neck pain, photophobia, nausea, fever, and chills. Culture of the cerebrospinal fluid produced *E. coli,* and the patient recovered fully after a 2-week course of antimicrobial therapy.

Other Infections

Because *E. coli* represents the predominant aerobic gram-negative flora of the gut and the most common cause of nosocomial and community-acquired UTIs, it is not surprising to find these organisms involved in a number of other infectious sequelae resulting from direct extension or hematogenous dissemination. Virtually no anatomic site is immune from *E. coli* colonization and/or infection. *E. coli* is the most frequently isolated pathogen from patients with primary peritonitis.[178] Infection is thought to result when the enteric bacilli gain access to the peritoneal cavity after translocation from the gut. In a study of pancreatic abscess, a secondary complication of pancreatic infection, Brook and Frazier[179] again found *E. coli* to be the most common pathogen isolated. Of 44 specimens analyzed over a 15-year period, *E. coli* was recovered on 18 occasions (41%). In most situations, *E. coli* was isolated as part of a mixed infection, but in four instances (22%) it was recovered in pure culture. Even in inflammatory conditions, where its role in pathogenesis is unclear, *E. coli* is still a frequently recovered microbe. Finnish studies on inflamed (n=119) and noninflamed appendices recovered *E. coli* from 88% of such anatomic specimens.[180] The *E. coli* strains isolated represented a number of diverse serotypes with distinctive adhesive characteristics.

Respiratory tract infection is another important complication of *E. coli* infections. The disease is characteristically found in older persons with underlying pulmonary conditions or other illnesses such as diabetes or cerebrovascular disease.[181,182] Factors potentially predisposing these patients to infection include therapy with broad-spectrum antibiotics, corticosteroids, or cytotoxic agents and assisted ventilation.[182] Berk and colleagues[181] studied 17 cases of *E. coli* pneumonia documented by transtracheal aspiration that occurred over a 2-year period at a Veterans Administration Medical Center in Tennessee. Recovered isolates were subdivided on the basis of presence or absence of the K1 capsule. Of six K1-associated pneumonias, most were community-acquired illnesses with the urinary tract as the probable source of infection. In contrast, non-K1 pneumonias (n=11) were more likely to be nosocomial in origin and to result from intraabdominal infection. Of the 17 cases, 6 (35%) appeared to result from aspiration. No difference in mortality (29%) or clinical disease (eg, bacteremia, shock) was noted between the two groups. A second study described a series of nine patients with bacteremic *E. coli* pneumonia.[182] In this investigation, a much higher fatality rate was observed (89%). However, this difference is probably attributable to patient selection (positive blood and sputum cultures), since four of the five bacteremic patients (80%) from Berk's study[181] also died.

Hematogenous dissemination of *E. coli* has been reported to cause a number of other clinical conditions, including infected aortic and carotid aneurysms,[183,184] subdural hematomas,[185] and endogenous endophthalmitis.[186,187] Almost all cases of endogenous endophthalmitis have occurred in persons with renal disease, including diabetes mellitus and polycystic kidney disease.[186] Ocular infection in most instances has resulted as a secondary consequence of UTI (pyelonephritis). Although this is a very rare complication, the overall prognosis for patients with such ocular disease is extremely poor.

LABORATORY IDENTIFICATION

Isolation

Isolation of *E. coli* from extraintestinal sources usually presents few difficulties. Cockerill and coworkers,[163] in a study of 281,797 consecutive blood cultures, found that 61% of *E. coli* (1053 of 1704 cases) were isolated by both aerobic (Septi-Chek) and anaerobic (nonvented tryptic soy broth) culture techniques. However, when *E. coli* was isolated by a single method only, aerobic culture was positive significantly more often than anaerobic culture (389 versus 262 cases, respectively, P=0.001). Isolation rates for the two aerobic methods used, Septi-Chek and the Isolator tube (Wampole Laboratories, Cranbury, NY), were statistically equivalent, with 59% of *E. coli* (1075 of 1836) cultured by both systems. The remaining 761 strains detected by only one system were almost evenly split (365 versus 396 respectively, statistically not significant).

Two exceptions that may present some difficulty in isolation (and identification) are cysteine- and thymidine-requiring *E. coli* strains, which make up 1.5% to 2.0% of all strains isolated from the urinary tract.[188] Although they are primarily involved in UTI, they also may seed the bloodstream, causing sepsis and abscesses. Cysteine-dependent strains are the most common auxotrophic group of *E. coli;* they grow as tiny, translucent colonies on routine media. Enhanced growth on media supplemented with L-cysteine (0.63 mmol/L) or around paper discs impregnated with the same reagent[188] identifies cysteine-deficient strains. Thymidine-dependent strains occur primarily in patients subjected to prolonged treatment with cotrimoxazole. These strains are easily recognized by their poor growth on Mueller-Hinton agar, which is thymidine-deficient to aid in testing of sulfa drugs. Both of these auxotrophic groups may require special handling in order to identify them biochemically (see Commercial Systems).

Isolation of *E. coli* from stools of patients with gastrointestinal infections is problematic. Essentially only O157 STEC strains possess markers that are readily usable in isolation media to separate them from normal fecal isolates of *E. coli*. For other pathogenic groups—EPEC, EIEC, ETEC, EAEC, and non-O157 STEC—five lactose-positive *E. coli* colonies should be picked from MacConkey (MAC) or other primary enteric plating agars and sent to a reference center for further workup. Isolates should be tested with any of the screening tests described later (see Biotypes of Enterovirulent *E. coli*) if they are available before strains are sent.

Although isolation of STEC strains can be a problem, one of the most common serogroups, O157, is a delayed sorbitol fermenter. To isolate this group, sorbitol is exchanged for lactose as the fermentable substrate in MAC agar, and the resulting medium is referred to as sorbitol-MacConkey (SMAC) agar.[189] On this medium, O157 colonies are colorless (Fig. 2A; Color Plate 1A). Because SMAC is not inhibitory, Zadik and associates[190] devised a medium incorporating tellurite and cefixime into SMAC (CT-SMAC) to increase selectivity. This medium increases O157 recovery significantly; more strains (26 versus 9, P=0.01) were recovered with the use of CT-SMAC than with another formulation of SMAC incorporating cefixime and rhamnose.[190] We use CT-SMAC in our laboratory, and with it we have isolated O157 STEC from patients from whom isolates could not be recovered with the use of SMAC (Fig. 2B; Color Plate 1B). The addition of rhamnose to CT-SMAC (CTR-SMAC) provides an additional differential characteristic. O157 STEC organisms are delayed or weak fermenters of rhamnose, and colonies on this medium are clear, whereas SOR− *E. coli,* which usually

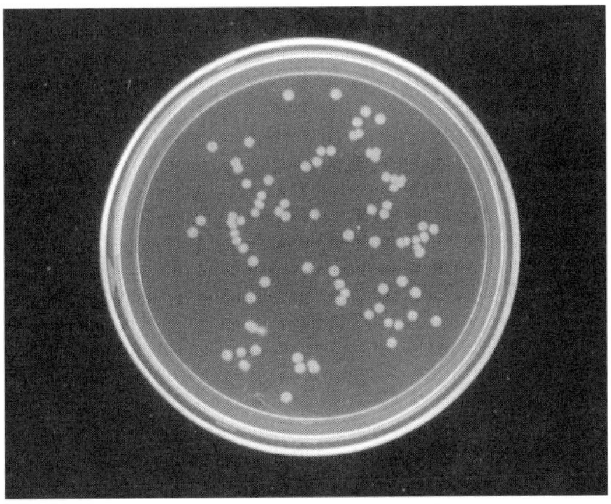

A

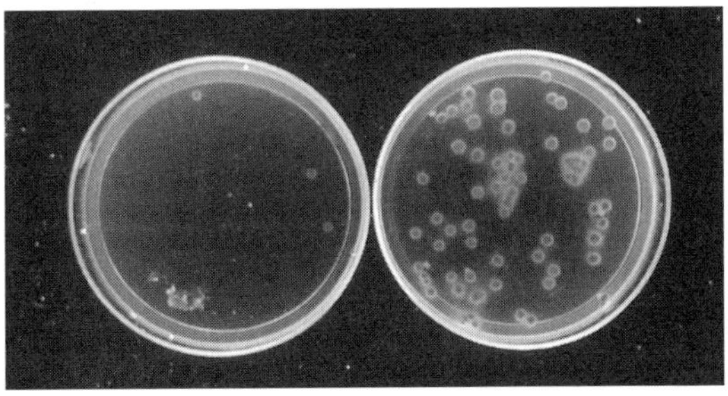

B

FIG. 2. A, *E. coli* O157:H7 on SMAC displaying sorbitol-negative colonies. **B,** CT-SMAC showing inhibition of normal *E. coli* flora with several sorbitol-negative O157:H7 colonies *(left);* SMAC alone *(right).* See also Color Plate 1.

rapidly ferment this substrate, form pink colonies. We have found CT-SMAC to be so inhibitory that the addition of rhamnose has not proved necessary; however, because approximately 20% of *E. coli* strains are SOR–, media containing rhamnose may be helpful. Although some O groups of STEC are also sorbitol delayed, most are not, and these strains are indistinguishable from nonpathogenic fecal *E. coli* on SMAC.

SMAC media incorporating 4-methylumbelliferyl-β-D-glucuronide (MUG) have not been successful for the isolation of O157 colonies in the presence of other *E. coli* colonies. *E. coli* that produce β-D-glucuronidase (GUR) cause so much fluorescence when they cleave this compound that nonfluorescing O157 colonies cannot be seen (Fig. 3; Color Plate 2).

In 1989, Beutin and colleagues[191] discovered a novel hemolysin they termed enterohemolysin, which correlates with the ability of a strain to produce Shiga toxin (Stx). This enterohemolysin is active only on washed red blood cells (RBCs) in the presence of calcium (10 mmol/L $CaCl_2$ in 5% defibrinated sheep blood washed three times). It requires at least 24 hours to produce a very narrow zone of hemolysis. The α-hemolysin of *E. coli,* on the other hand, is active on either washed or unwashed RBCs and forms easily distinguishable zones of hemolysis within 4 hours. Fig. 4 (Color Plate 3) shows the hemolytic activity of STEC, α-hemolytic, and nonhemolytic strains of *E. coli*. Washed RBC (enterohemolysin) plates can be used to screen for any STEC, not just O157 strains, in stools. This medium, however, has no inhibitory ingredients, and STEC colonies may be overgrown by other fecal flora.

A unique and sensitive isolation technique, immunomagnetic separation,[192] has also been applied to isolate O157 strains. Tiny magnetic beads coated with O157 antibody are added to a portion of stool and incubated with mixing for a short time. A powerful magnet then removes the beads, which are washed and plated on appropriate media (SMAC, CT-SMAC). This technique can also be used in conjunction with probes and polymerase chain reaction (PCR) to increase sensitivity; none of these techniques are approved by the FDA for use in clinical laboratories at this time.

Gicquelas and associates[193] have developed a rapid, practical, and economically feasible method for identifying diarrheagenic *E. coli* using a variety of probes. Probes for STEC, ETEC, EIEC, LT+ ETEC, and DAEC were biotinylated to eliminate the problems inherent in using radioactive probes. Biotinylated probes are stable for long periods and can be reused. However, they can be difficult to read, and positive controls should be included. If

A

B

FIG. 3. A, SMAC agar with β-D-glucuronide without fluorescence; circle marks *E. coli* O157:H7 colony. **B,** SMAC agar with β-D-glucuronide; O157:H7 colony masked by fluorescence of fecal *E. coli.* See also Color Plate 2.

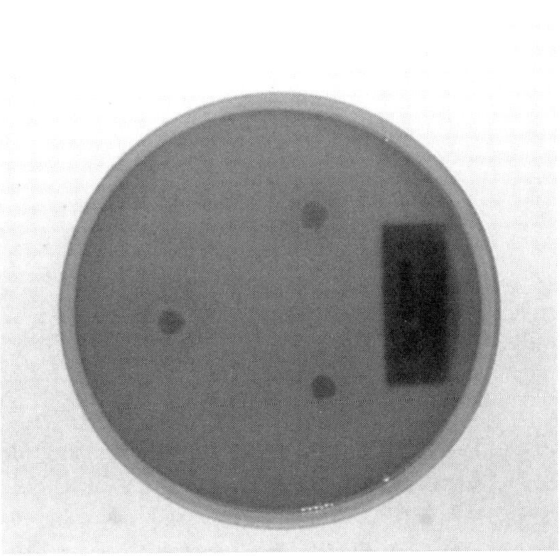

FIG. 4. Blood agar plate with washed sheep RBCs and $CaCl_2$; enterohemolysin-positive, Stx+ *E. coli (top);* α-hemolytic Stx– *E. coli (left);* nonhemolytic Stx– *E. coli (bottom).* See also Color Plate 3.

possible, the biotinylated probes should be run against radioactive probes once as a reading control for comparison.

Detectable levels of O157 or other STEC organisms in stool may disappear rapidly (within 7 days), and determination of disease resulting from Stx may require nonculture methods. Suitable methods include tissue cell assay, enzyme immunoassay, latex agglutination, DNA probes, and PCR.[194]

Screening or identification of *E. coli* from urine is frequently done by spot tests, one of the most useful of which has been the detection of GUR. However, some strains of *Salmonella, Shigella,* and *Citrobacter* can be positive with this test. Domínguez and coworkers[195] used a commercially available double-test tablet to test for indole production and GUR simultaneously. PGUA/Indole (Rosco, Taastrup, Denmark; Carr-Scarborough Microbiologicals, Decatur, GA) had a sensitivity of 94.1% and a specificity of 100% on 261 strains. The authors noted that a heavy inoculum is required and may not always be available from primary isolation plates. Other systems that simultaneously detect GUR, *o*-nitrophenyl-β-D-pyranogalactosidase (ONPG), and indole are also available.[196]

E. coli–like colonies picked from routine MAC cultures to detect diarrheagenic strains of EPEC, ETEC, EAEC, and non–O157 STEC appear as typical *E. coli* in triple sugar iron (TSI) agar, A/AG$^-$ or K/AG$^-$; lysine-iron (LIA) agar, K/K$^-$; and motility-indol-ornithine (MIO) medium, +/V/V. Colonies suspected to be EIEC appear *Shigella*-like in these media: K/A$^-$, K/A$^-$, and –/V/V in TSI, LIA, and MIO, respectively. Usually, any of these isolates must be referred to a reference center for further workup.

STEC belonging to serogroup O157 also appear as typical *E. coli* in TSI, LIA, and MIO media but may be screened with both specific antisera and the MUG test,[197] on which they are negative (no fluorescence). A number of commercial antisera, both polyclonal (Difco, Detroit, MI; Accurate Chemical and Scientific Corp., Westbury, NY) and latex reagents, are available. Strains may also be screened for toxin production in a number of commercial products. The list of available antisera (Pro-Lab Diagnostics, Austin, TX; Remel, Lenexa, KS; Oxoid Ltd., Hampshire, England) and toxin assays (Meridian Diagnostics, Cincinnati, OH; Denka Seiken, Tokyo, Japan) is expanding. Several companies are preparing to introduce new products, and changes are occurring at a rate too rapid to enumerate them here.

Identification

Commercial Systems

Published data suggest that *E. coli* isolates are identified with a fairly high degree of accuracy by most commercial systems (Table 10). However, this statement should be viewed with caution. With the exception of the first four kits listed in Table 10, most systems have not been rigorously challenged with a large number (>200) of *E. coli* isolates. Another problem concerns the amount of phenotypic diversity that exists within the *E. coli.* No reports have described in any detail the biochemical characteristics of *E. coli* strains used in challenge

TABLE 10. *Relative accuracy of commercial systems in identifying* E. coli

Commercial Systems	Incubation Period (h)	No. Tested	No. Correctly Identified (%)
Vitek	4–18	640	633 (98.9)
MicroScan W/A	2	626	615 (98.2)
API 20E	21–24	261	247 (94.6)
Crystal E/NF	18	208	205 (98.6)
RapID onE	4	44	44 (100)
Biolog GN	24	22	18 (81.8)
Cobas Micro ID-E/NF	21	19	15 (78.9)
Crystal RS/E	3	5	5 (100)

Data from references 198–210.

studies. Most evidence indicates that stock strains used for such investigations are often identified to species by other commercial products rather than by conventional (reference) methods. This severely limits the range of phenotypes tested, since *E. coli* strains identified with high confidence values by one system usually generate similar identification profiles on another. Because *E. coli* comprises many different groups of biochemically distinct strains (eg, nonreactive, Alkalescens-Dispar), it is highly questionable whether most systems can satisfactorily recognize these phenotypes. Furthermore, atypical *E. coli* strains deviating from the ideal phenotype in one or more biochemical characteristics apparently are not handled well by most kits. O'Hara and colleagues[200,202] at the CDC found that all four systems they evaluated had problems with one or more *E. coli* isolates with aberrant properties. *E. coli* strains that were indole-negative or urease-positive or lysine- and ornithine decarboxylase–negative and arginine dihydrolase–negative gave identifications ranging from *Salmonella/Arizona* to *Yersinia kristensenii* to *Klebsiella ozaenae* on such panels. These results underscore the need for further studies that take into account the full range of phenotypes that exist within *E. coli*. An additional problem arises with cysteine-dependent strains of *E. coli*. These auxotrophic strains may require 48 hours to yield an identification in some systems (API 20E API bioMérieux, Hazelwood, MO), and other systems may simply fail to correctly identify these strains.[211] McIver and Tapsall[211] attributed this result to poor growth in growth-dependent tests and recommended supplementing the organism suspension with 0.63 mmol/L of L-cysteine. There were no adverse effects to the test systems (API 20E; Vitek, Vitek Systems, Hazelwood, MO) with the addition of L-cysteine. Thymidine-deficient strains have not been studied in commercial identification systems. In cases in which such a strain was suspected and its identification was unclear, 95% to 100% have been found to be GUR+.

Ideally, it would be very useful if commercial systems could recognize specific pathogenic groups within *E. coli*. This ability would be predicated on the fact that unique biochemical (enzymatic) features exist within each pathogenic group, which enable them to be distinguished from nonpathogenic (saprophytic) isolates. A very limited amount of work has been performed in this area. Abbott and colleagues[212] found that more than 90% of *E. coli* O157:H7 strains generated two unique biochemical profile numbers when tested on MicroScan conventional (overnight) panels. One study found that most *E. coli* O157:H7 strains (91.9%) produced a single biotype number (5144172) on API 20E strips.[213] O157 is also listed in the database of the Crystal E/NF system, suggesting that this panel can accurately identify this serogroup. Therefore, if further studies support these initial observations, it appears that commercial screening systems based on unique biotypes are feasible, at least for *E. coli* O157.

Conventional Biochemicals

E. coli isolated from extraintestinal sites, if biochemically typical (Table 11), can be

TABLE 11. *Biochemical separation of* E. coli *and Shigella*

Organism	Motility	Gas from Glucose	Lysine Decarboxylase	Sodium Acetate	Mucate
E. coli	+	+	+	+	+
Inactive *E. coli*	–	–	V	V	V
Shigella	–	–[a]	–	–[b]	–[c]

+, >90% positive; –, <10% positive; V, 10–90% positive.
[a]Some *S. flexneri* 6 and *S. boydii* 13 and 14 strains.
[b]Rare strains of *S. flexneri* 4a.
[c]Rare strains of *S. sonnei.*

identified using only a small number of conventional media. However, inactive strains present a challenge even with conventional media when they are negative for any or all of the following tests: motility (MOT), production of gas from glucose, ONPG, lysine decarboxylase (LDC), and indole. The media most helpful in separating these inactive strains from *Shigella* are sodium acetate and mucate, either or both of which are positive in *E. coli* but only very rarely positive in *Shigella* (see Table 11; for separation of other species of *E. coli,* see Chapter 4). We occasionally encounter isolates that cannot be separated from *Shigella* even using as many as 60 different conventional media, necessitating a report as *Escherichia/Shigella* group. Serotyping should not be used as an adjunct test for the separation of such strains because cross-reactions frequently occur between antigens of these two groups. Additionally, inactive strains may fail to ferment any number of carbohydrates normally used by typical *E. coli,* including arabinose, rhamnose, xylose, maltose, and trehalose, either singly or in combination.

Although inactive strains are probably more common, strains of *E. coli* that utilize unusual substrates (for this genus) may also be encountered. Generally they are limited to one unusual trait but are otherwise fairly typical biochemically. The ability to produce hydrogen sulfide (in TSI and plating media) or urease and the utilization of citrate are the most common aberrant characteristics. Although few laboratories use potassium cyanide, it is a valuable tool in the separation of *E. coli* (which is negative) from many other *Enterobacteriaceae.* Occasionally strains that can grow in the presence of potassium cyanide may also be isolated. Strains that ferment cellobiose, adonitiol, or inositol may be encountered. In 1982, Brenner and coworkers[214] published biochemical profiles for atypical biogroups of *E. coli.* True-positive reactions by *E. coli* in Voges-Proskauer (VP), phenylpyruvic acid (PPA), or oxidase tests are rare. However, strains that produce a pinkish-orange reaction in VP, a slight green reaction in PPA especially at the top of the slant, or a light pink color more than 30 seconds after the addition of oxidase reagent may mistakenly be deemed to have given positive reactions. All of these reactions should be considered negative.

EIEC strains may be difficult to separate from *Shigella* because they generally do not produce gas in glucose, are LDC–, and are nonmotile. However, tests for sodium acetate and mucate are usually positive, differentiating these strains from *Shigella.* Strains of ETEC, EPEC, EAEC, and most serogroups of STEC are biochemically indistinguishable from routine *E. coli* except for those tests described in the next section. Strains of STEC that agglutinate in O157 antisera should be delayed in their fermentation of sorbitol and rhamnose and MUG– but otherwise biochemically typical. However, strains have been isolated in Europe that are both SOR+ and MUG+.[215] These strains must be identified serologically and/or by their ability to produce Stxs. A PCR–restriction fragment length polymorphism test has been developed that aids in the identification of these strains but is not commercially available.[216]

Biotypes of Enterovirulent E. coli

Probably the most difficult task of identification involves *E. coli* strains associated with diarrheal disease. Even during outbreak situations, nonpathogenic strains of *E. coli* may predominate over enterovirulent colonies on selective media designed for the isolation of enteric pathogens. To test large numbers of *E. coli*–like colonies for various virulence determinants (eg, toxins, adhesins) would be expensive, technically demanding, and time-consuming. Furthermore, if *E. coli* isolates are forwarded to reference laboratories for definitive characterization, failure to screen colonies at the primary laboratory for relevant characteristics may present further obstacles to the rapid and accurate identification of enterovirulent *E. coli* and conceivably could compromise the accuracy of the final diagnosis.

A number of biochemical markers or biotypes exist for several of these enterovirulent groups that aid in their initial isolation from stool. The major biotypes (markers) associated with each group are listed in Table 12. Probably the most important of these are the phenotypes associated with STEC infections caused by *E. coli* O157:H7. In pioneering studies conducted by Farmer and other public health microbiologists at the CDC[217] it was recognized that O157:H7 strains differed from routine *E. coli* isolates in their delayed fermentation of D-sorbitol. Of 64 O157:H7 isolates tested, 63 (98%) failed to ferment D-sorbitol after overnight incubation; prolonged incubation (≥72 hours) was required before fermentation of the alcoholic sugar occurred. This observation served as the basis for the development of selective and differential media such as SMAC, which aid in the initial isolation of O157:H7 from diarrheal stools. Although this phenotypic marker is extremely useful, a small percentage of nonpathogenic *E. coli* strains that reside in the colon exhibit the same marker (SOR–). Since this latter group may predominate over O157:H7 in feces (by as much as 4:1 based on incidence), screening of colonies for O157 antigen would be not only expensive but relatively unfruitful under routine conditions. However, the selection of colonies to test can be enhanced by use of a second biochemical marker. In 1987, Krishnan and associates[119] reported that all strains of serotype O157:H7 were negative for GUR activity even though most routine isolates[98] of *E. coli* were positive. Results from this Canadian study were independently confirmed by others. The fluorogenic substrate, MUG, could be incorporated into a number of media (eg, SMAC, MAC, nutrient agar), and a rapid 20-minute broth test to detect GUR activity was developed. Together, these two markers (SOR–, GUR–) eliminate most random (non-O157) *E. coli* isolates, which may express one but not both markers. Testing of isolates for GUR should be performed using plates without lactose because MAC agar and

TABLE 12. *Biotypes and markers of enterovirulent* E. coli

Group	Marker	Phenotype	Frequency of Phenotype (%): Routine Isolates	Frequency of Phenotype (%): Pathogenic Group
STEC (O157:H7)	SOR	–	2–5	~99
	GUR	–	4–8	~99
EIEC	LDC	–	10	~99
	MOT	–	5	~93
EAggEC	PEL	+	<2	~99
ETEC	SRB	–	39	~95
EPEC	None			

Data from references 197 and 217–221.

SOR, sorbitol fermentation; GUR, β-D-glucuronidase activity; LDC, lysine decarboxylase; MOT, motility; PEL, pellicle (scum) formation; SRB, sorbose fermentation.

TSI slant growth has been shown to inhibit MUG reactions in some instances.

One of the troubling aspects regarding screening methods for O157:H7 strains using the markers described is the emergence of phenotypic variants.[78] A 1992 German study by Gunzer and others[215] reported that 17 (39%) of 44 O157 strains isolated from patients with HUS or diarrhea were SOR+ (24 hours) and GUR+. This study was quickly followed by two European reports confirming the frequent isolation of SOR+ *E. coli* O157.[222,223] One of these surveys found 10 (26%) of 39 Stx-producing *E. coli* O157 to be SOR+ and GUR+; 9 were serotype O157:H⁻, and the 10th was found to be O157:H7.[222] Four O157:H⁻ strains expressing Stx2 only were found to be SOR+ also. The most recent addition to this list of phenotypic variants involves the first isolation of a GUR+ *E. coli* O157 strain within the United States.[224] This strain was recovered from the feces of a 20-year-old college student with bloody diarrhea. These isolations may portend the eventual widespread dissemination of such phenotypic variants, which would render current screening practices unsatisfactory.[78] O157 strains can also be biotyped using one of two published schema (Table 13). The reproducibility of biotypes in one of these studies has been questioned[119]; the overall usefulness of biotypes as an epidemiologic tool is presently unknown.

Markers for the presence of STEC serotypes other than O157:H7 are presently unavailable. Krishnan and associates[119] noted that all 12 non-O157 STEC isolates in their study were SOR+; 8 of these (including two O26 strains) were additionally GUR+. Ojeda and others,[225] studying STEC disease associated with HUS in Chile, found 83% of O26, 67% of O111, and the single O55 strain to be SOR–. This variation in phenotype expression within a serogroup is common. Most reported instances of STEC disease caused by non-O157:H7 strains involve SOR– isolates, a fact that aided in their initial recognition.

Markers (biotypes) have also been developed for other enterovirulent *E. coli* groups. EIEC strains have markers commonly associated with *Shigella* species, to which they are closely related on a genetic and pathogenic basis. In a study of more than 300 EIEC strains, failure to decarboxylate lysine (LDC–) and lack of motility (MOT–) independently correlated with a positive Serény (keratoconjunctivitis) test.[219] Together, these two markers (LDC–, MOT–) virtually eliminated the need to test further any isolates not bearing these phenotypes. A test to detect a subgroup of EAEC, proposed by Albert and collaborators,[221] is based on the observation that some EAggEC isolates form pellicles ("scum") on the surface of liquid media such as Mueller-Hinton or Luria broth.[221] Of 41 EAggEC strains evaluated, all were positive for pellicle formation (Fig. 5; Color Plate 4). None of the 51 non-EAggEC isolates screened, including DAEC strains, produced a similar reaction. Burke and colleagues[220] suggested that the inability to ferment sorbose may be a useful test to screen for ETEC. In their study of 111 ETEC isolates, 95.5% were found to be sorbose-negative. However, almost all of these isolates originated from Australia, and it is unclear whether this is a

TABLE 13. *Proposed biotyping schemes for* E. coli *O157*

Characteristic	Krishnan et al[119]	Aleksić et al[222]
Serotypes tested	O157:H7	O157:H7, O157:H⁻
Markers used		
Primary	DUL, RAF	SOR, GUR
Secondary	None	DUL, RHA, RAF, SUC
No. biotypes/biovars	4	11
Biotype designations	A–D	1–11

DUL, dulcitol fermentation; RAF, raffinose fermentation; SOR, sorbitol fermentation; GUR, β-D-glucuronidase activity; RHA, rhamnose fermentation; SUC, sucrose fermentation.

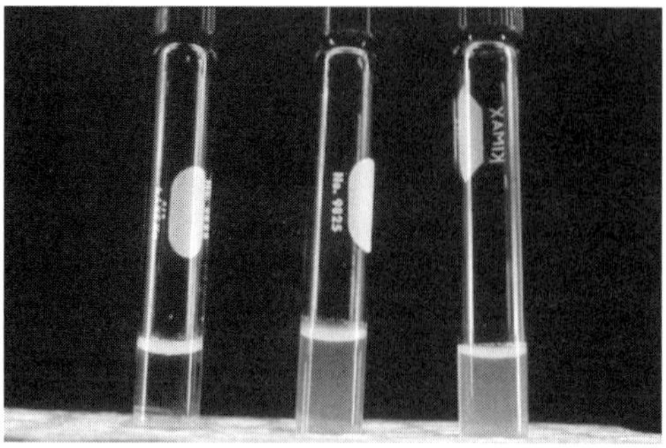

FIG. 5. Pellicle formation by three EAggEC strains in Mueller-Hinton broth; strongest reaction is indicated at far left (with autoagglutination), weakest reaction at right. See also Color Plate 4.

generalized trait for all ETEC from diverse geographic locales. Although the percentage of routine *E. coli* isolates that are sorbose-negative is high, the marker may still have some utility, if true, because it would eliminate approximately 60% of background strains.

Biotyping schemes have also been published for *E. coli* strains in general. A two-tiered system for biotyping of *E. coli* was proposed by Crichton and Old.[226] Using four tests (raffinose, sorbose, ornithine, and dulcitol), strains could be assigned to 1 of 16 primary biotypes. Six additional tests allowed for further discrimination. Of 599 strains tested in this study, 16 primary biotypes and 213 full biotypes were detected.

Immunodiagnostics

E. coli are, by far, the most difficult members of the *Enterobacteriaceae* to serotype. Ørskov and Ørskov[227] and Jann and Jann[228] have published excellent reviews on this subject. There are 173 known somatic (O), 80 capsular (K), and 56 flagellar (H) antigens, many of which can be subdivided into partial antigens. Besides the large number of antigenic types, serotyping is further complicated by the loss of surface (O, K) and flagellar factors. Even more problematic are cross-reactions, understanding of which requires an in-depth knowledge of the antigenic factors, which is available in only a few international reference centers. The International *Escherichia* and *Klebsiella* Centre at the Statens Seruminstitut in Copenhagen, Denmark, uses more than 300 cross-absorbed O sera to identify common O-group problems.[227] Also, all strains should be biochemically characterized as *E. coli* before serotyping is attempted because many other *Enterobacteriaceae* (eg, *Salmonella, Shigella, Klebsiella*) share identical antigens.

It is estimated that there are 50,000 to 100,000 serotypes of *E. coli*,[227] possibly more, but the number involved in human disease is limited. Certain serotypes are commonly associated with extraintestinal disease (Table 14). The notable O-type similarities among these diseases arise because sepsis and meningitis are usually seeded from the urinary tract. Still other serotypes are associated with each of the diarrheagenic groups. Although the serotypes involved in diarrheal disease are numerous, certain ones predominate; these are listed in Table 14.

Somatic typing is performed by slide and tube agglutination assays, the antigen being heated before testing to remove any capsular antigen that may be present. Testing for the K antigen is done by gel immunoprecipitation and phage typing. A latex reagent has been developed (Murex Diagnostics, Norcross, GA) for detection of the K1 antigen; this is the capsule that is most prevalent in extraintestinal infections, and it is also identical to the capsular polysaccharide B of *Neisseria meningitidis*. Demonstration of flagellar factors is complicated by sluggish motility and hence poor antigen formation. This can be improved by daily passages of the organism in a semisolid motility medium; usually several days are required (we have had O157:H7 strains that required up to 18 passages before the H antigen could be demonstrated). A tube agglutination assay is used for H typing.

Of the diarrheagenic groups of *E. coli*, those belonging to the STEC are most commonly encountered, and more than 100 serotypes have been found to produce Stxs. Several of the more common non-O157

TABLE 14. *Common antigenic factors of* E. coli *associated with extraintestinal and diarrheal disease*

Disease Group	Associated O Antigens	Associated K Antigens	Associated H Antigens
Neonatal meningitis	O2, O6, O7, O16, O18ac, O83	K1	–
Bacteremia	O1, O2, O6, O22, O75, O78	K1, K2, K5, K12, K15, K23	–
Urinary tract infections	O1, O4, O6, O16, O18	K1, K2, K5, K12, K13	–
Diarrheal disease			
ETEC	O6, O8, O15, O20, O25, O27, O63, O78, O115, O128, O148, O149, O159	–	H4, H7, H9, H11, H12, H19, H20, H21, H28, H40
EPEC	O18, O20, O26, O28, O44, O55, O86, O111, O119, O124, O125, O126, O128, O142, O158, O159	–	H2, H6, H7, H11, H12, H14, H18, H21, H27, H34
EIEC	O28, O112, O124, O136, O143, O144, O152, O164, O167	–	–
STEC	O26, O55, O111, O125, O128, O157	–	H6, H7, H8, H11

serotypes (eg, O26, O55, O111, O119, O126, O128) are also found in EPEC. Strains of O157:H7 or O157:HNM predominate in North America and are frequently responsible for outbreaks. This serotype, like many others, shares serologic cross-reactions not only with other *E. coli* isolates but with other genera[222,229,230] (Table 15).

Typing

As discussed by Ørskov and Ørskov,[227] other methods besides serotyping are required for epidemiologic investigations of *E. coli* outbreaks because unrelated clones may reside within the same serotype. Many methods have been employed successfully to this end, including bacteriophage typing,[231,232] multilocus enzyme electrophoresis,[233] pulsed-field gel electrophoresis,[88,234] and random amplification of polymorphic DNA.[235] We have successfully used plasmid analysis to link urinary tract infection and blood isolates in compromised patients (eg, human immunodeficiency virus infection, child with permanent ventricular shunt). Of the listed techniques, only bacteriophage typing of O157:H7 strains was not sufficiently discriminatory to be of use.[232]

TABLE 15. *Cross-reactions between* E. coli *serogroup O157 and other organisms*

12 *E. coli* O groups (O7, O8, O36, O59, O71, O79, O97, O103, O116, O147, O165, O168)
Escherichia hermanii
Yersinia enterocolitica O9
Salmonella group N (O30)
Citrobacter freundii
Brucella abortus
Vibrio cholerae O1

Data from references 222, 229, and 230.

Specialized Assays

Toxin Assays

E. coli strains produce a variety of toxins that require a number of different assays to detect[149,236–241] (Table 16). The toxins of *E. coli* cause a great deal of confusion for physicians, pathologists, and laboratory staff unused to working with them. The ETEC group produces two types of enterotoxins, LT and ST (Fig. 6; Color Plate 5), each with subtypes. The STEC produce Stx1 and Stx2 (Fig. 7; Color Plate 6). The symptoms produced by these two toxigenic groups are distinct (see Clinical Significance), and different methods (Table 16) must be used to determine their presence. In cell culture assays, specific neutralizing antitoxins must always be used to determine the specificity of the cytopathic effect observed, because nonspecific reactions occur and undoubtedly there are yet undiscovered toxins that have a similar cytopathic effect.

TABLE 16. *Toxins of* E. coli *and corresponding toxin assays*

Toxin	Abbreviation	Type	Method (Nonanimal)
α-Hemolysin	Hly	Cytotoxin	Hemolytic zone on BAP
Enterohemolysin	E-Hly	Cytotoxin	Hemolytic zone on BAP (washed RBCs with $CaCl_2$)
Heat-labile toxin (ETEC)	LTI: LTIh (human), LTIp (porcine); LTII	Enterotoxin	Y1 cell assay (cell rounding); CHO cell assay (elongation); latex agglutination; EIA; probes
Heat-stable toxin (ETEC)	STI (or Sta): STIh (human), STIp (porcine); STII	Enterotoxin	Latex agglutination; probes (not active on any known cell line)
Shiga toxin (STEC)	Stx1 (SLTI, VTI); Stx2 (SLTII, VTII); Stx2v	Cytotoxin	Vero cell assay; latex; EIA; probes
Cytotoxic necrotizing factor	CNFI; CNFII	Cytotoxin	Vero, HeLa cell assay (multinucleation)
Cytolethal distending toxin	CDLTI	Cytotoxin	Vero, HEp-2, HeLa, CHO cell assays (cell distention); EIA; probes

BAP, blood agar plate; CHO, Chinese hamster ovary; EIA, enzyme immunoassay.

There are commercially available toxin assay systems for the detection of ETEC (Oxoid; Pharmacia Fine Chemicals, Uppsala, Sweden) and STEC (Oxoid; Meridian Diagnostics). Because this is a rapidly expanding and changing area of microbiology, tests available at any given time may vary.

Invasion Assays

The invasion assay is a means of determining in vitro the ability of a bacterial strain to invade mammalian cells.[242] Generally, epithelial cell lines are used (HEp-2, HeLa), but other lines, including Caco-2, Henle 407, and Vero cells, are frequently employed. Bacteria are incubated with tissue cells (usually for 2–3 hours, but the time may vary with the genus tested) to allow attachment and penetration of the epithelial cell. Gentamicin, which is unable to penetrate mammalian cells,

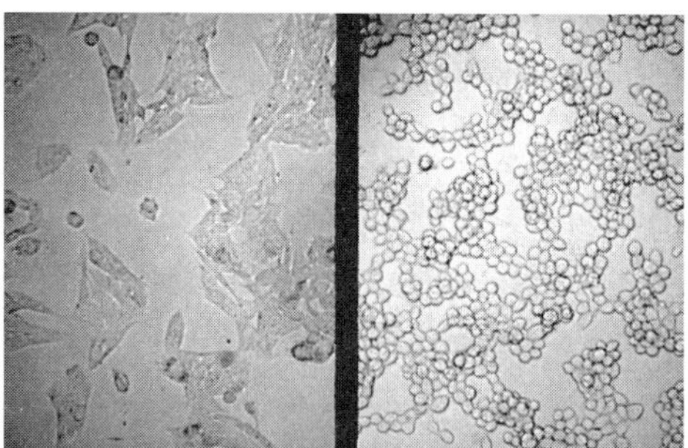

FIG. 6. Normal Y1 cell sheet *(left);* Y1 cells exposed to heat-labile (LT) enterotoxin for 24 hours *(right).* See also Color Plate 5.

A

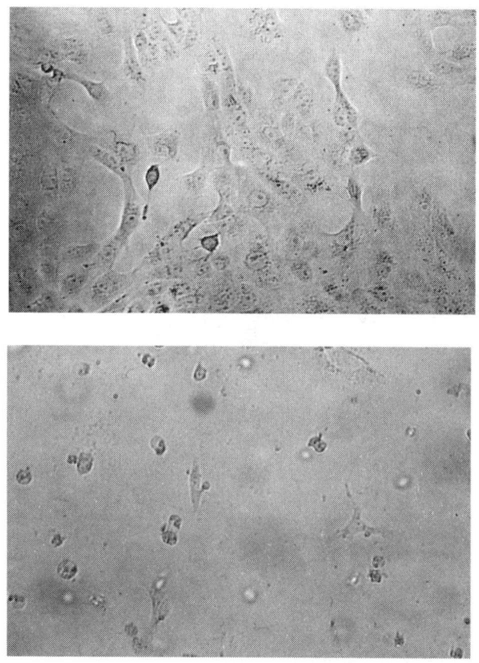

B

FIG. 7. A, Normal Vero cell sheet. **B,** Vero cells exposed to Shiga toxin (Stx) for 48 hours. See also Color Plate 6.

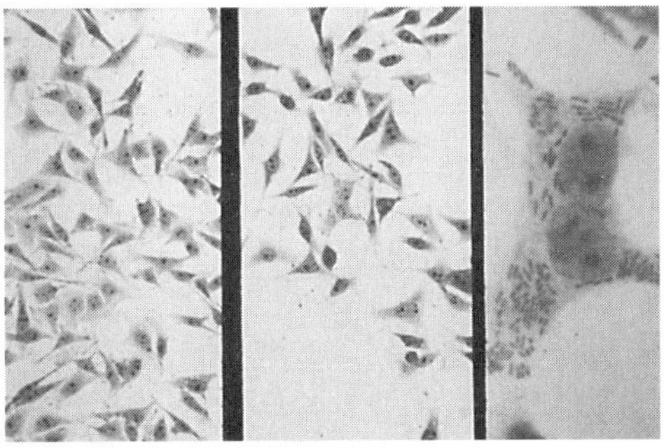

FIG. 8. Uninoculated HEp-2 cell sheet *(left)*; HEp-2 invaded by EIEC, ×20 *(center)*; enlarged view of invaded HEp-2 cells showing multiplication of cells intracellularly *(right)*. See also Color Plate 7.

is added to eliminate extracellular bacteria, and the tissue cell sheet is again incubated to allow multiplication of the bacteria that have invaded. The cell sheet is then washed and either fixed and Giemsa stained for visual examination of internalized bacteria or lysed with a detergent and subjected to plate counts for a quantitative assessment of invasion. The bacteria (Fig. 8; Color Plate 7) can be seen in Giemsa-stained preparations as dark blue forms within, and usually filling, the cytoplasm or within cytoplasmic vacuoles.

Adherence Assays

Adherence assays are necessary to detect or confirm the presence of EPEC and EAEC (EAggEC, DAEC). Common epithelial cell lines such as HEp-2 or HeLa are suitable for such purposes. Assays can be performed with either glass cover slips or tissue culture chamber slides, although the former method tends to give better results.[243] In early studies, discrepancies occurred when the adherence characteristics of certain *E. coli* strains were determined by two slightly different methods.[244] This difference was caused by the failure of one procedure, which employed organisms grown in tryptic soy broth with 1% D-mannose, to differentiate the localized adherence (LA) pattern from aggregative adherence (AA) in a 3-hour incubation period (both were reported as LA). This discrepancy was resolved by increasing the total length of incubation to 5 hours.[245] We typically use tryptone broth with 1% D-mannose and a 5-hour period (infection + incubation) to detect all three adherence patterns. Shorter incubation periods make it difficult to clearly distinguish the AA and DAEC patterns.

Three major patterns of adherent *E. coli* can be recognized on epithelial cell lines with the procedures described (Fig. 9). The LA pattern consists of packets of tightly adherent bacteria ("microcolonies") at a limited number of sites on the bacterial cell surface.[128,246] This pattern is historically associated with *E. coli* strains belonging to classic EPEC serotypes. The diffuse (DAEC) pattern was first recognized in the early 1980s and is characterized by a rather uni-

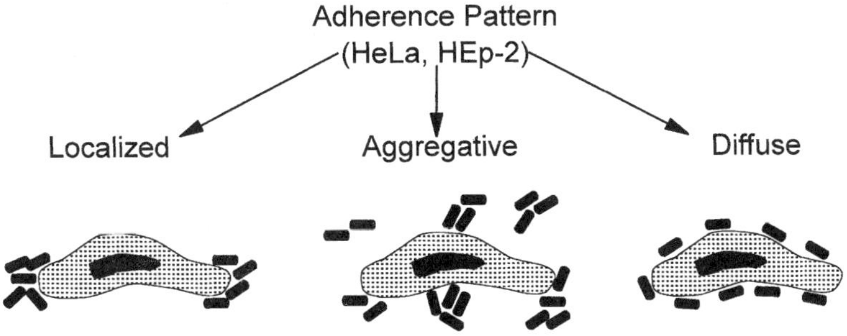

FIG. 9. Schematic representation of adherence patterns representing EPEC (localized) and EAEC (aggregative, diffuse) types.

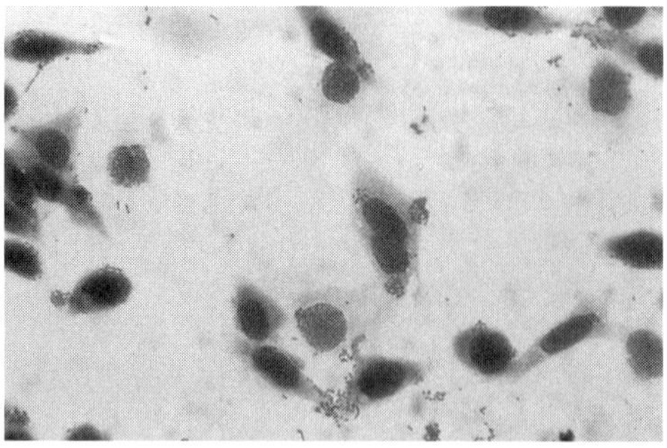

FIG. 10. EPEC showing localized adherence to HEp-2 cells. See also Color Plate 8.

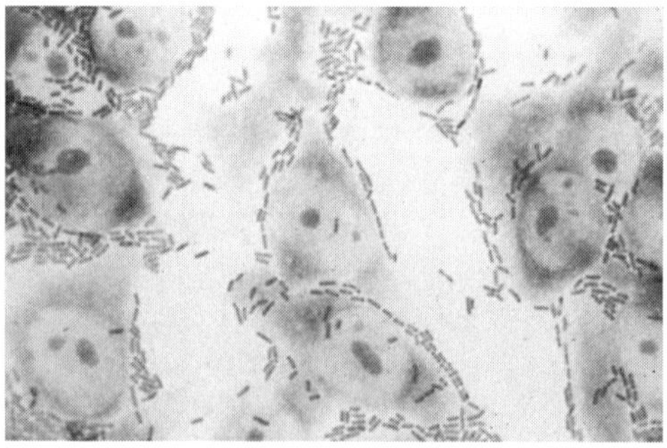

FIG. 12. DAEC showing diffuse adherence to HEp-2 cells. See also Color Plate 10.

form coating of the epithelial cell surface with bacteria one to two rows thick.[246] DAEC strains fall into both EPEC and non-EPEC serotypes. They are an extremely heterogeneous group at present, based on serologic properties and genetic characterization.[247] The final group (AA) causes large packets of bacteria to adhere in a "stacked brick" arrangement to HEp-2 or HeLa cells and the glass surface. This pattern is also extremely heterogeneous because a number of different subtypes exist, including AA bacteria that preferentially adhere to cells, to the glass slide, or to both. Some studies have recognized even more diversity within this adherence group with the reporting of a "honeycomb" pattern.[248] The relevance of these subtypes is not known. Most AA strains do not fall into serotypes traditionally associated with EPEC. Typical examples of these adherence phenotypes are shown in Figs. 10 through 12 (Color Plates 8 through 10).

Fluorescent Actin Stain Test

Both EPEC and STEC strains possess "attaching and effacing" (AE) adherence, which is characterized by destruction of the microvilli and formation of a pedestal by the enterocyte membrane that engulfs the bacterial cell. To do this, actin in the cell membrane is rearranged and accumulates beneath the attached bacteria. Knutton and colleagues[249] developed a *f*luorescent *a*ctin *s*tain (FAS) using a phallotoxin that binds specifically to actin in tissue cells (adherence assay) and has been polymerized (Fig. 13; Color Plate 11). Although this assay is excellent for identification of EPEC and STEC strains, there are disadvantages: tissue cell assay techniques and a

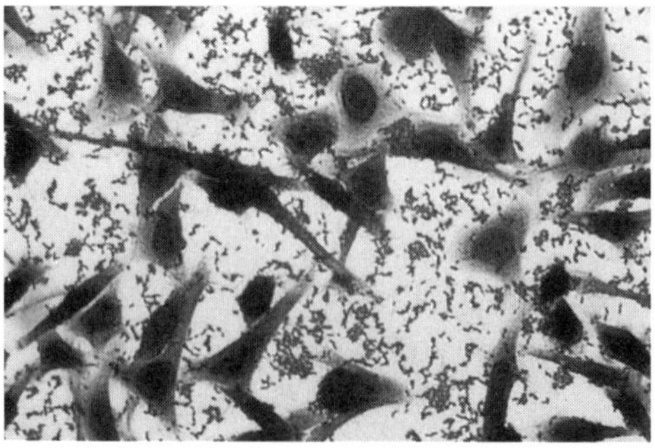

FIG. 11. EAggEC showing aggregative adherence to HEp-2 cells. See also Color Plate 9.

FIG. 13. Fluorescent actin stain (FAS) of EPEC-infected HEp-2 cells showing bundles of actin filaments. See also Color Plate 11.

fluorescence microscope are required, and the phallotoxin is extremely toxic.

PATHOGENESIS

An exhaustive amount of information exists on various aspects of *E. coli* pathogenicity in relation to human disease. These data include identification of virulence determinants, characterization, sequencing, and in vitro regulation and expression of bacterial genes and their products, as well as assessment of the proposed roles such factors play in pathogenicity using animal or cell culture models. An in-depth analysis of such topics is beyond the scope or purpose of this book. Many excellent reviews on selected aspects of *E. coli* pathogenicity are available.[50,250–258]

In its most simple terms, *E. coli* pathogenicity can be divided into intestinal and extraintestinal infections. Gastrointestinal disease involves the oral ingestion of pathogenic *E. coli* strains and their subsequent colonization and infection of the small intestine or colon, or both, with disease manifestations caused by expression of one or more virulence factors. In almost every instance, it is the coordinate expression of multiple virulence determinants by pathogenic bacteria that culminates in overt disease (diarrhea). Surprisingly little information exists on various aspects of infection (eg, colonization) for certain well-recognized groups of *E. coli* that are involved in gastroenteritis (eg, EIEC, STEC). In any case, at least two factors are required to produce enteric disease: mechanisms for attachment and colonization of the gastrointestinal mucosa and some mechanism for induction of diarrheal disease.

Table 17 lists some of the virulence factors that are thought to play important roles in gastrointestinal disease caused by various diarrheagenic groups. Many of these factors are plasmid encoded. Gastrointestinal infection by EPEC is thought to be a multistep process involving an initial nonspecific, localized adherence to the epithelial cell surface that is followed by a second, signal transduction phase causing effacement of microvilli.[50,253] A third period, characterized by an intimate association between bacteria and epithelial cells, results in cellular invasion (without multiplication) and a secretory response.[50] EPEC disease is classically associated with a histopathologic AE lesion of the microvilli, which can be detected by electron microscopic analysis of thin sections of infected tissues or monolayers. The AE lesion is associated with degenerated microvilli and cupping of attached bacteria ("pedestal formation"). This rearrangement of the microvillar cytoskeleton causes proliferation of actin filaments, which serve as the basis for a positive FAS test (see Specialized Assays). EPEC-induced disease involves at least two groups of virulence factors. An EAF plasmid encodes at least two virulence-associated determinants, one of which involves the expression of bundle-forming pili.[50] A second locus, termed LEE (locus of enterocyte effacement), is a 35-kb pathogenicity island required for expression of intimin and EPEC secreted protein (EspB).[50,259] The exact role each of these factors plays in disease is not completely understood.

ETEC cause secretory diarrhea by elaboration of an LT or ST enterotoxin (or both); the LT enterotoxin is similar but not identical to the cholera toxin of *V. cholerae,* although severe cholera-like diarrhea is not a classic presentation of ETEC infection.[50] Colonization of the gut presumably is associated with expression of colonization factor antigens or fimbrial coli surface–associated antigens. Little is known regarding EIEC pathogenesis. Invasion of intestinal cells, the hallmark of EIEC infection, is associated with a 200- to 250-kb (120–140 MDa) plasmid analogous to that described for *Shigella* species.[260] Specific colonization factors have not been identified, although an enterotoxin (ShET2) reputedly associated with watery diarrhea has been described.[50] As with EIEC, little has been definitively established regarding STEC pathogenicity despite its prominent role in hemorrhagic colitis. Although Stx1 and Stx2 are thought to play major contributory roles in disease pathogenesis, exactly how these cytotoxins function in vivo and elicit secretory

TABLE 17. *Pathogenic factors associated with diarrheagenic groups of* E. coli

Group	Factors	Genetic Location	Gene Products	Proposed Functions	Ref.
EPEC	LEE	Chromosome			259
	eaeA		Intimin	Attachment to intestinal epithelium	50
	espB (eaeB)		Secreted protein	Eukaryotic cell signal transduction; attachment to intestinal epithelium	50
	EAF	Plasmid	Bundle-forming pilus	Localized adherence to epithelial cells	50, 253
ETEC	CFA I	Plasmid	Fimbriae	GI tract colonization	254, 257
	CS1, CS3–CS6 (CFA II–CFA IV)	Plasmid	Fimbriae	GI tract colonization	254, 257
	CS2 (CFA II)	Chromosome	Fimbriae	GI tract colonization	254, 257
	LT	Plasmid	Cytotonic enterotoxin (heat-labile)	Secretory diarrhea	50
	ST	Plasmid	Enterotoxin (heat-stable)	Secretory diarrhea	50
EIEC	*pInv*	Plasmid	Factors promote invasion and multiplication within epithelial cells	Bloody diarrhea	260
	ShET2	Plasmid	Enterotoxin	Secretory diarrhea	50
STEC	Stx1, Stx2	Phage	Cytolytic toxins	Bloody and secretory diarrhea	50, 256
	rpoS	Chromosome	Regulator gene	Resistance to gastric acidity	261
EAEC					
EAggEC	EAST-1	Plasmid	Enterotoxin	Secretory diarrhea	241
	AAF/I	Plasmid	Fimbriae	GI tract colonization	254, 257
DAEC	F1845	Chromosome	Fimbriae	GI tract colonization	254, 257
	AIDA-I	Plasmid	Fimbriae	GI tract colonization	254, 257
	AFA I/III	Plasmid/chromosome	Afimbrial adhesins	GI tract colonization	254, 257

GI, gastrointestinal; LEE, locus of enterocyte effacement; *eae*, EPEC attaching-and-effacing A and B loci; EAF, EPEC adherence factor; CFA, colonization factor antigen; CS, coli surface antigen; *pInv*, invasion plasmid; ShET2, *Shigella* enterotoxin 2; EAST-1, EAggEC heat-stable enterotoxin; AAF, aggregative adherence factor; AIDA-1, adhesion involved in diffuse adherence; AFA, afimbrial adhesins.

and bloody diarrhea is not well appreciated. Finally, characterization of EAEC for reputed virulence-associated factors is only in its infancy. To date, several adhesins have been described that may be responsible for intestinal colonization.[254,257] At least three potential enterotoxic factors associated with EAEC have been reported, the best characterized of which is EAST-1.[50,241]

Selected pathogenic factors associated with systemic or extraintestinal disease are listed in Table 18. In contrast to virulence determinants associated with intestinal *E. coli* infections, many of these factors are chromosomally encoded. In 1991, Johnson[252] published a large and comprehensive review of virtually all aspects of *E. coli* pathogenicity associated with UTIs. *E. coli* produces a large number of fimbriae, which may be involved in bladder colonization and infection. Two of the better described of these adhesins are P and S fimbriae, whose specific biochemical receptors have been determined. Both fimbriae bind intensely to the kidney (Bowman's capsule, glomerulus, vessel walls) and bladder (vessel walls); in addition there is strong clinical, experimental, and epidemiologic information linking P fimbrial expression with *E. coli*–associated UTIs.[252] Among other potential virulence determinants, association with pathogenicity is based largely on a higher frequency of the phenotype (marker) in strains associated with UTIs than in those isolated from patients with asymptomatic bacteriuria or from other anatomic locations (feces). For both α-hemolysin and aerobactin production, possession of these virulence factors has been more commonly associated with upper UTIs and more severe disease.

Bacteremic strains of *E. coli* often contain virulence determinants not commonly found in routine fecal isolates. Strains that disseminate to the blood from local sites of infection (eg, urinary tract, gastrointestinal tract) must circumvent the normal bactericidal activity of serum (complement) and host phagocytic mechanisms. *E. coli* possesses several surface structures, including plasmid-encoded outer membrane proteins (OMPs), capsules, and somatic antigens that are antiphagocytic and resistant to the effects of normal human serum.[263] Several of the OMPs linked to serum resistance are carried by ColV, a virulence-associated plasmid, which is linked to strains causing extraintestinal disease.[265,266] ColV plasmids also carry the siderophore aerobactin, whose presence is most often found in blood isolates. In mice, aerobactin-positive strains are more virulent, presumably because in vivo bacterial growth is enhanced by iron acquisition.[267] As in the case of UTIs, secretion of α-hemolysin is significantly associated with bacteremic isolates. Expression of α-hemolysin in mice is linked to serum re-

TABLE 18. E. coli *virulence factors associated with extraintestinal infections*

Group	Factor	Genetic Location	Gene Product	Proposed Functions	Ref.
UTI	P	Chromosome	Fimbriae	Uroepithelial cell adherence	252, 257
	S	Chromosome	Fimbriae	Uroepithelial cell adherence	252, 257
	α-Hemolysin	Plasmid/ chromosome	Cytotoxin	Kidney damage, iron release from RBCs	251, 252, 257
	Aerobactin	Plasmid	Hydroxymate siderophore	Iron acquisition	252, 257
Bacteremia	TraT	Plasmid	OMP	Resistance to phagocytosis, serum resistance	252, 262
	iss	Plasmid	OMP	Serum resistance	252, 263
	K1, K5	Chromosome	Capsules	Serum resistance	264
	O antigens	Chromosome	LPS	Serum resistance	264
	Aerobactin	Plasmid	Hydroxamate siderophore	Iron acquisition, bloodstream invasion	267
	α–Hemolysin	Plasmid/ chromosome	Cytotoxin	Iron acquisition from RBCs	251, 268
Meningitis	K1	Chromosome	Capsule	Antiphagocytic, serum resistance	174, 175

OMP, outer membrane proteins; LPS, lipopolysaccharide.

sistance, certain capsular antigens, and virulence but is independent of aerobactin expression.[251] Molecular studies in which DNA containing the α-hemolysin gene of one isolate was recombined into a nonhemolytic fecal isolate resulted in development of enhanced virulence in the recipient in the rat peritonitis model.[268] Production of the virulence-related factor CNF may also be associated with bacteremic *E. coli* strains.[255] Finally, the K1 capsule is the major virulence determinant in *E. coli* strains causing neonatal meningitis.[174,175] It appears to produce its major effects by circumventing normal host cellular and immune responses.

ANTIMICROBIAL SUSCEPTIBILITY

Much of the resistance seen in *E. coli* isolates arises from the presence of β-lactamases (predominantly TEM-1 and/or TEM-2, occasionally SHV-1 or OXA-1) and extended-spectrum β-lactamases (ESBLs), which are plasmid mediated.[269,270] However, Sanders and Sanders[270] indicated that 1% to 24% of strains of ampicillin-resistant *E. coli* were resistant because of overproduction of a chromosomally encoded cephalosporinase. Although this is a constitutive enzyme in *E. coli* that usually is not expressed sufficiently to cause resistance, overproducing strains are frequently resistant to β-lactamase/β-lactam drug combinations, cephalosporins, cephamycin, and monobactams.

Ampicillin-resistant strains of *E. coli* are often isolated from younger patients with respiratory infections and from patients in tertiary-care hospitals.[270] Some of these ampicillin-resistant isolates are high-level TEM-1 β-lactamase producers, and these strains are resistant to amoxicillin-clavulanate (AUG), ampicillin-sulbactam (A/S), ticarcillin–clavulanic acid, and older-generation cephalosporins.[270] Although resistance to β-lactamase/β-lactam drugs existed before these pharmaceuticals were introduced into the medical community, it is not clear how much influence medical use has had on increasing the numbers of resistant strains. Sanders and Sanders[270] indicated in one survey that resistance in strains from community-acquired *E. coli* sepsis increased from 2% to 7% between 1970–1975 and 1985–1990, whereas resistance remained constant in hospital-acquired bacteremic strains for the same time intervals. They suggested that different forces may be at work, with high drug use in the community being responsible for the increased resistance in community-acquired strains and use of cephalosporins in the hospital setting perhaps increasing resistance in hospital strains.

Several large studies[271–275] in the United States indicate that *E. coli* remain susceptible to most of the commonly used antimicrobials. Of more than 20,000 *E. coli* tested at 15 sites across the country,[274] 93% to 100% of strains were susceptible to ceftazidime, amikacin, gentamicin, imipenem, and ciprofloxacin, and 87% to 96% were susceptible to cefazolin. For trimethoprim-sulfamethoxazole susceptibility, the rates fell somewhat to 75% to 90% overall, with one report of 100%. Nine of 15 laboratories reported susceptibilities to this drug in the 80% range and 3 reported a range in the high 70%s. Susceptibility to piperacillin was poor (58%–72%). Jones and coworkers,[272] using data from 43 medical centers, reported susceptibility of strains by site of isolation (ie, bloodstream, respiratory tract, urinary tract, and skin/soft tissue infections). Regardless of site, strains were 97% to 100% susceptible to ofloxacin, ciprofloxacin, ceftriaxone, ceftazidime, and imipenem. Good susceptibility rates were also found for cefuroxime (92%–97%) and gentamicin (94%–99%). Rates were moderate for cefazolin and ticarcillin–clavulanic acid at 83% to 91% and 84% to 90%, respectively but were dismal for ampicillin (54%–58%). In susceptibility studies on third-generation cephalosporins, both Jones and colleagues[271] and Fuchs and associates[275] found little resistance to ceftazidime (about 1%), ceftriaxone (0.3%–2.0%), or cefaperazone (3.5%) and essentially none for cefotaxime or ceftriaxone. In surveys by Jones and colleagues[272] and by Thornsberry,[273] resistance to fluoroquinolones was reported to average 2.6% (range

0.2%–27% in nine hospitals) and 1.6% (in 28,805 strains), respectively.

Although these data are encouraging, resistance to ceftazidime[274] indicates the possibility of ESBLs in U.S. strains of *E. coli;* one of three strains tested at the reference center for this study was an ESBL strain. High-level fluoroquinolone resistance has also been reported in Europe[276,277] and portends a possibly ominous future. Threlfall and associates[276] found 40 strains between 1993 and 1996 that were resistant to ciprofloxacin; 39 had minimal inhibitory concentrations (MICs) ranging from 4 to 256 µg/L. More than one half of the strains were obtained from cases of sepsis, approximately one fourth from the urinary tract, and the remainder from other sites of invasive disease. All strains were multiresistant, and 36 strains were resistant to four or more drugs, most commonly ampicillin, chloramphenicol, gentamicin, and trimethoprim. Lehn and coworkers[277] observed 2.4% high-level ciprofloxacin resistance in their strains except from the urology ward, where more than 10% of the strains fell into this category. High-level resistance was found to be caused by mutations in the *gyrA* gene, and the majority of patients had been treated previously with a fluoroquinolone. Pulsed-field gel electrophoresis of their isolates showed considerable diversity, indicating that resistance arose independently and not by interhospital spread.

Several articles regarding susceptibility testing methods merit attention. Awareness of the potential for an increase in ESBL-bearing strains has led to the necessity of devising satisfactory methods for their detection. Although studies of a variety of methods are currently being undertaken (eg, Etest, use of reduced susceptibility to cefpodoxime), Sanders and colleagues[278] found that both the two-disk test (AUG tested with any of the following four disks: cefotaxime, aztreonam, ceftriaxone, or ceftazidime) and the Vitek ESBL were reliable. However, the two-disk test was less easy to interpret and required repeat testing, whereas the Vitek did not require subjective interpretation.

Another problem arising in susceptibility testing methods centers on AUG and A/S, which are considered to be equivalent agents with equal clinical efficacy, thus eliminating the need to test both drugs. O'Shaughnessy and coworkers[279] tested 61 *E. coli* from a variety of sites for susceptibility to AUG and A/S by three different methods: the Kirby-Bauer disk test, MicroScan (Dade MicroScan, West Sacramento, CA), and Etest strips (AB Biodisk, Culver City, CA). They found poor correlation for the two drugs. They found 3%, 8%, and 10% resistance to AUG and 56%, 59%, and 44% resistance to A/S by the MicroScan, Etest, and Kirby-Bauer methods, respectively. In addition to the poor correlation between methods, these agents were poorly predictive for each other as well. The authors suggested that, if these drugs do possess equivalent clinical efficacy, susceptibility testing breakpoints need reevaluation.

The final point involves susceptibility testing of cysteine-requiring *E. coli* auxotrophs isolated from the urinary tract or from other extraintestinal sites. Although these strains require cysteine supplementation of commercial systems to achieve satisfactory biochemical identification, MIC testing with Mueller-Hinton or Iso-Sensitest agar does not require supplementation for accurate results. McIver and Tapsall[280] found consistent results in these two media and in Sensitest agar supplemented with 0.63 mmol/L cysteine hydrochloride, although growth on the Sensitest medium was superior. Strains grew poorly or not at all on unsupplemented Sensitest agar.

CONCLUSIONS

In the world of microbiology no other organism fills the role of Dr. Jekyll and Mr. Hyde quite as well as *E. coli.* Its crucial role as the predominant aerobic flora in the gastrointestinal tract of humans and animals and the many beneficial genetic techniques developed with the use of this organism attest to its Dr. Jekyll persona. However, it does have a dark, Mr. Hyde personality as well. It is the most common agent isolated from community- and hospital-acquired urinary tract infections. It ranks 6th as a cause of both nosocomial sepsis and pneumonia and 10th as a cause of wound infections.[15] It is the leading cause of diarrhea in developing nations and is associated with a high mortality rate in children younger than 5 years of age in these countries. The high morbidity and mortality rates in these countries, particularly in children, are undeniably linked to poor sanitation and fecally contaminated drinking water supplies. Although rectification of these problems would reduce the incidence of many microbiologic diseases simultaneously, major improvements are so expensive that they are slow to develop.

Furthermore, researchers are being confronted with additional potentially infectious diarrheagenic groups (CNF, CLDT, and EAEC; see Pathogenesis) while much work is yet to be done on already known pathogens (ETEC, EIEC, STEC, and EPEC). *E. coli* possess so many variable virulence characteristics that it is difficult to target those that would make good candidates for vaccine development.

Another area of concern is the development of high-level fluoroquinolone resistance and ESBLs in strains of *E. coli* due to increased use of associated drugs. The report of Johnson and coworkers[281] on *E. coli* isolates resistant to both gentamicin and apramycin, a veterinary aminoglycoside, is also disturbing. Their findings indicated spread of gentamicin-resistant organisms from animals to humans (interstrain and interspecies spread of resistant genes via transferable plasmids).

Clearly much research must be done before we can truly appreciate and fully understand this organism that ranks among the oldest known members of the *Enterobacteriaceae.*

REFERENCES

1. Sussman M. Theodor Escherich (1857–1911): A biographical note. In: Sussman M, ed. *The virulence of* Escherichia coli: *Reviews and methods.* London: Academic Press, 1985;1–4.
2. Bettelheim KA. The genus *Escherichia.* In: Balows A, Trüper HG, Dworkin M, Harder W, Schleifer K-H, eds. *The prokaryotes: A handbook on the biolology of bacteria. Ecophysiology, isolation, identification, applications.* New York: Springer-Verlag, 1991;2696–2736.
3. Ewing WH, Davis BR, Martin WJ. Biochemical characterization *Escherichia coli.* HHS Publication No. 1981 (CDC) 81-8109. Atlanta, GA: U.S. Dept. of Health and Human Services, 1981;1–15.
4. Judicial Commission of the International Committee on Bacteriological Nomenclature. Conservation of the family name *Enterobacteriaceae,* of the name of the type genus and designation of the type species. *Int Bull Bacteriol Nomencl Taxon* 1958;8:73–74.
5. Farmer JJ III, Brenner DJ, Ewing WH. Opposition to recent proposals which would reject the family name *Enterobacteriaceae* and *Escherichia* as its type genus. *Int J Syst Bacteriol* 1980;30:660–673.
6. Vidotto MC, Müller EE, de Freitas JC, Alfieri AA, Guimarães IG, Santos DS. Virulence factors of avian *Escherichia coli. Avian Dis* 1990;34:531–538.
7. Morris JA, Sojka WJ. *Escherichia coli* as a pathogen in animals. In: Sussman M, ed. *The virulence of* Escherichia coli: *Reviews and methods.* London: Academic Press, 1985;47–77.
8. Linton AH, Hinton MH. *Enterobacteriaceae* associated with animals in health and disease. *Soc Symposium Appl Bacteriol* 1988;17(Suppl):71S–85S.
9. Droual R, Chin RP, Rezvani M. Synovitis, osteomyelitis, and green liver in turkeys associated with *Escherichia coli. Avian Dis* 1996;40:417–424.
10. Provence DL, Curtiss R III. Role of *crl* in avian pathogenic *Escherichia coli:* A knockout mutation of *crl* does not affect hemagglutination activity, fibronectin binding, or curli production. *Infect Immun* 1992;60: 4460–4467.
11. Venkateswaran K, Murakoshi A, Satake M. Comparison of commercially available kits with standard methods for the detection of coliforms and *Escherichia coli* in foods. *Appl Environ Microbiol* 1996;62:2236–2243.
12. Brenner KP, Ranknin CC, Sivaganesan M, Scarpino PV. Comparison of the recoveries of *Escherichia coli* and total coliforms from drinking water by the MI agar method and the U.S. Environmental Protection Agency–approved membrane filter method. *Appl Environ Microbiol* 1996;62:203–208.
13. Pan TM, Wang TK, Lee CL, Chien SW, Horng CB. Food-borne disease outbreaks due to bacteria in Taiwan, 1986 to 1995. *J Clin Microbiol* 1997;35:1260–1262.
14. Todd ECD. Costs of acute bacterial foodborne disease in Canada and the United States. *Int J Food Microbiol* 1989;9:313–326.
15. Schaberg DR, Culver DH, Gaynes RP. Major trends in the microbial etiology of nosocomial infection. *Am J Med* 1991;91(Suppl 3B):72S–75S.
16. Jones RN. Impact of changing pathogens and antimicrobial susceptibility patterns in the treatment of serious infections in hospitalized patients. *Am J Med* 1996;100(Suppl 6A):3S–12S.
17. Nicolle LE, Strausbaugh LJ, Garibaldi RA. Infections and antibiotic resistance in nursing homes. *Clin Microbiol Rev* 1996;9:1–17.
18. Sussman M. *Escherichia coli* in human and animal disease. In: Sussman M, ed. *The virulence of* Escherichia coli: *Reviews and methods.* London: Academic Press, 1985;7–45.
19. Sobel JD. Bacterial etiologic agents in the pathogene-

sis of urinary tract infection. *Med Clin North Am* 1991;75:253–273.
20. Shapiro ED. Infections of the urinary tract. *Pediatr Infect Dis J* 1992;11:165–168.
21. Zelikovic I, Adelman RD, Nancarrow PA. Urinary tract infections in children: An update. *West J Med* 1992;157:554–561.
22. Kim KS, Itabashi H, Gemski P, Sadoff J, Warren RL, Cross AS. The K1 capsule is the critical determinant in the development of *Escherichia coli* meningitis in the rat. *J Clin Invest* 1992;90:897–905.
23. Guerina NG, Kessler TW, Guerina VJ, et al. The role of pili and capsule in the pathogenesis of neonatal infection with *Escherichia coli* K1. *J Infect Dis* 1983;148: 395–405.
24. Gross RJ, Rowe B. *Escherichia coli* diarrhoea. *J Hyg Camb* 1985;95:531–550.
25. Raj P. Pathogenesis and laboratory diagnosis of *Escherichia coli*–associated enteritis. *Clin Microbiol Newslett* 1993;15:89–93.
26. Robins-Browne RM. Traditional enteropathogenic *Escherichia coli* of infantile diarrhea. *Rev Infect Dis* 1987;9:28–53.
27. Levine MM. *Escherichia coli* that cause diarrhea: Enterotoxigenic, enteropathogenic, enteroinvasive, enterohemorrhagic, and enteroadherent. *J Infect Dis* 1987;155:377–389.
28. Schlager TA, Guerrant RL. Seven possible mechanisms for *Escherichia coli* diarrhea. *Infect Dis Clin North Am* 1988;2:607–624.
29. Levine MM, Edelman R. Enteropathogenic *Escherichia coli* of classic serotypes associated with infant diarrhea: Epidemiology and pathogenesis. *Epidemiol Rev* 1984;6:31–51.
30. Neter E, Westphal O, Lüderitz O, Gino RM, Gorzynski EA. Demonstration of antibodies against enteropathogenic *Escherichia coli* in sera of children of various ages. *Pediatrics* 1955;16:801–808.
31. Cravioto A, Gross RJ, Scotland SM, Rowe B. An adhesive factor found in strains of *Escherichia coli* belonging to the traditional infantile enteropathogenic serotypes. *Curr Microbiol* 1979;3:95–99.
32. Viljanen MK, Peltola T, Junnila SYT, et al. Outbreak of diarrhoea due to *Escherichia coli* O111:B4 in schoolchildren and adults: Association of Vi antigen–like reactivity. *Lancet* 1990;336:831–834.
33. Cohen MB, Giannella RA. Bacterial infections: Pathophysiology, clinical features, and treatment. In: Phillips SF, Pemberton JH, Shorter RG, eds. *The large intestine: Physiology, pathophysiology, and disease.* Mayo Foundation. New York: Raven Press, 1991; 395–428.
34. Edelman R, Levine MM. Summary of a workshop on enteropathogenic *Escherichia coli*. *J Infect Dis* 1983; 147:1108–1118.
35. Rothbaum R, McAdams AJ, Giannella R, Partin JC. A clinicopathologic study of enterocyte-adherent *Escherichia coli*: A cause of protracted diarrhea in infants. *Gastroenterology* 1982;83:441–454.
36. Wu S-X, Peng R-Q. Studies on an outbreak of neonatal diarrhea caused by EPEC O127:H6 with plasmid analysis restriction analysis and outer membrane protein determination. *Acta Paediatr* 1992;81:217–221.
37. Scotland SM, Willshaw GA, Smith HR, Gross RJ, Rowe B. Adhesion to cells in culture and plasmid profiles of enteropathogenic *Escherichia coli* isolated from outbreaks and sporadic cases of infant diarrhoea. *J Infect* 1989;19:237–249.
38. Katouli M, Jaafari A, Ketabi GR. The role of diarrhoeagenic *Escherichia coli* in acute diarrhoeal diseases in Bandar-Abbas, Iran. *J Med Microbiol* 1988; 27:71–74.
39. Echeverria P, Hoge CW, Bodhidatta L, et al. Etiology of diarrhea in a rural community in Western Thailand: Importance of enteric viruses and enterovirulent *Escherichia coli*. *J Infect Dis* 1994;169:916–919.
40. González R, Díaz C, Mariño M, Cloralt R, Pequeneze M, Pérez-Schael I. Age-specific prevalence of *Escherichia coli* with localized and aggregative adherence in Venezuelan infants with acute diarrhea. *J Clin Microbiol* 1997;35:1103–1107.
41. Germani Y, Bégaud E, Duval P, Le Bouguénec C. Prevalence of enteropathogenic, enteroaggregative, and diffusely adherent *Escherichia coli* among isolates from children with diarrhea in New Caledonia. *J Infect Dis* 1996;174:1124–1126.
42. Echeverria P, Ørskov F, Ørskov I, et al. Attaching and effacing enteropathogenic *Escherichia coli* as a cause of infantile diarrhea in Bangkok. *J Infect Dis* 1991; 164:550–554.
43. Snyder JD, Wells JG, Yashuk J, Puhr N, Blake PA. Outbreak of invasive *Escherichia coli* gastroenteritis on a cruise ship. *Am J Trop Med Hyg* 1984;32:281–284.
44. Harris JR, Mariano J, Wells JG, Payne BJ, Donnell HD, Cohen ML. Person-to-person transmission in an outbreak of enteroinvasive *Escherichia coli*. *Am J Epidemiol* 1985;122:245–252.
45. Beutin L, Gleier K, Kontny I, Echeverria P, Scheutz F. Origin and characteristics of enteroinvasive strains of *Escherichia coli* (EIEC) isolated from Germany. *Epidemiol Infect* 1997;118:199–205.
46. Wanger AR, Murray BE, Echeverria P, Mathewson JJ, DuPont HL. Enteroinvasive *Escherichia coli* in travelers with diarrhea. *J Infect Dis* 1988;158:640–642.
47. Taylor DN, Echeverria P, Pál T, et al. The role of *Shigella* spp., enteroinvasive *Escherichia coli,* and other enteropathogens as causes of childhood dysentery in Thailand. *J Infect Dis* 1986;153:1132–1138.
48. Chatterjee BD, Sanyal SN. Is it all shigellosis? *Lancet* 1984;2:574.
49. Ephros M, Cohen D, Yavzori M, Rotman N, Novic B, Ashkenazi S. Encephalopathy associated with enteroinvasive *Escherichia coli* O144:NM infection. *J Clin Microbiol* 1996;34:2432–2434.
50. Sears CL, Kaper JB. Enteric bacterial toxins: Mechanisms of action and linkage to intestinal secretion. *Microbiol Rev* 1996;60:167–215.
51. Tamura K, Sakazaki R, Murase M, Kosako Y. Serotyping and categorisation of *Escherichia coli* strains isolated between 1958 and 1992 from diarrhoeal diseases in Asia. *J Med Microbiol* 1996;45:353–358.
52. MacDonald KL, Eidson M, Strohmeyer C, et al. A multistate outbreak of gastrointestinal illness caused by enterotoxigenic *Escherichia coli* in imported semisoft cheese. *J Infect Dis* 1985;151:716–720.
53. Taylor WR, Schell WL, Wells JG, et al. A foodborne outbreak of enterotoxigenic *Escherichia coli* diarrhea. *N Engl J Med* 1982;306:1093–1095.
54. Centers for Disease Control. Foodborne outbreaks of enterotoxigenic *Escherichia coli*—Rhode Island and

New Hampshire, 1993. *MMWR Morbid Mortal Wkly Rep* 1994;43:81, 87–89.
55. Roels T, Mellen J, Proctor M, et al. A large outbreak of enterotoxigenic *Escherichia coli* (ETEC) infection in Milwaukee. Abstract K149. 35th International Conference of Antimicrobial Agents and Chemotherapy. Washington, D.C.: American Society for Microbiology, 1995, p. 314.
56. Black RE, Merson MH, Rowe B, et al. Enterotoxigenic *Escherichia coli* diarrhoea: Acquired immunity and transmission in an endemic area. *Bull World Health Organ* 1981;59:263–268.
57. Agüero ME, Reyes L, Prado V, Ørskov I, Ørskov F, Cabello FC. Enterotoxigenic *Escherichia coli* in a population of infants with diarrhea in Chile. *J Clin Microbiol* 1985;22:576–581.
58. Echeverria P, Leksomboon U, Chaicumpa W, Seriwatana J, Tirapat C, Rowe B. Identification by DNA hybridisation of enterotoxigenic *Escherichia coli* in homes of children with diarrhoea. *Lancet* 1984;1: 63–65.
59. Blanco J, González EA, Bernádez I, Varela BR. Enterotoxigenic and enteropathogenic *Escherichia coli* in Galicia (North-west Spain). *Med Microbiol Immunol (Berl)* 1983;172:165–169.
60. DeWitt DE, Ostergaard D. Traveler's diarrhea: How to avoid "Montezuma's revenge"—and how to cope if you don't. *Postgrad Med* 1984;75:231–232.
61. Barry M. Traveler's diarrhea: New perspectives. *Infect Agents Dis* 1992;1:114–118.
62. Wolfe MS. Acute diarrhea associated with travel. *Am J Med* 1990;88(Suppl 6A):34S–37S.
63. Okhuysen PCX, Ericsson CD. Traveler's diarrhea: Prevention and treatment. *Med Clin North Am* 1992;76: 1357–1373.
64. Brown MR, DuPont HL, Sullivan PS. Effect of duration of exposure on diarrhea due to enterotoxigenic *Escherichia coli* in travelers from the United States to Mexico. *J Infect Dis* 1982;145:582.
65. Echeverria P, Blacklow NR, Sanford LB, Cukor GC. Travelers' diarrhea among American Peace Corps volunteers in rural Thailand. *J Infect Dis* 1981;143: 767–771.
66. Hyams KC, Bourgeois AL, Merrell BR, et al. Diarrheal disease during Operation Desert Shield. *N Engl J Med* 1991;325:1423–1428.
67. Willshaw GA, Cheasty T, Rowe B, Smith HR, Faithfull-Davies DN, Brooks TGJ. Isolation of enterotoxigenic *Escherichia coli* from British troops in Saudi Arabia. *Epidemiol Infect* 1995;115:455–463.
68. Wood LV, Wolfe WH, Ruiz-Palacios G, et al. An outbreak of gastroenteritis due to a heat-labile enterotoxin-producing strain of *Escherichia coli. Infect Immun* 1983;41:931–934.
69. Konowalchuk J, Speirs JI, Stavric S. Vero response to a cytotoxin of *Escherichia coli. Infect Immun* 1977;18: 775–779.
70. Tarr PI. *Escherichia coli* O157:H7: Clinical, diagnostic, and epidemiological aspects. *Clin Infect Dis* 1995; 20:1–10.
71. Calderwood SB, Acheson DWK, Keusch GT, et al. Proposed new nomenclature for SLT (VT) family. *ASM News* 1996;62:118–119.
72. Riley LW, Reims RS, Helgerson SD, et al. Hemorrhagic colitis associated with a rare *Escherichia coli* serotype. *N Engl J Med* 1983;308:681–685.
73. *Addressing emerging infectious disease threats: A prevention strategy for the United States.* Atlanta, GA: U.S. Department of Health and Human Services, 1994;p. 11.
74. Hughes JM, La Montagne JR. The challenges posed by emerging infectious diseases. *ASM News* 1994;60: 248–250.
75. Berkelman RL. Emerging infectious diseases in the United States, 1993. *J Infect Dis* 1994;170:272–277.
76. Karmali MA. Infection by verocytotoxin-producing *Escherichia coli. Clin Microbiol Rev* 1989;2:15–38.
77. Griffin PM, Tauxe RV. The epidemiology of infections caused by *Escherichia coli* O157:H7, other enterohemorrhagic *E. coli*, and the associated hemolytic uremic syndrome. *Epidemiol Rev* 1991;13:60–98.
78. Feng PF. *Escherichia coli* serotype O157:H7: Novel vehicles of infection and emergence of phenotypic variants. *Emerg Infect Dis* 1995;1:47–52.
79. Armstrong GL, Hollingsworth J, Morris JG Jr. Emerging foodborne pathogens: *Escherichia coli* O157:H7 as a model of entry of a new pathogen into the food supply of the developed world. *Epidemiol Rev* 1996; 18:29–51.
80. Ogorek CP, Fisher RS. Differentiation between Crohn's disease and ulcerative colitis. *Med Clin North Am* 1994;78:1249–1258.
81. Slutsker L, Ries AA, Greene KD, Wells JG, Hutwagner L, Griffin PM. *Escherichia coli* O157:H7 diarrhea in the United States: Clinical and epidemiologic features. *Ann Intern Med* 1997;126:505–513.
82. Karch H, Rüssmann H, Schmidt H, Schwarzkopf A, Heesemann J. Long-term shedding and clonal turnover of enterohemorrhagic *Escherichia coli* O157 in diarrheal diseases. *J Clin Microbiol* 1995;33:1602–1605.
83. Swerdlow DL, Griffin PM. Duration of faecal shedding of *Escherichia coli* O157:H7 among children in day-care centres. *Lancet* 1997;349:745–746.
84. Pai CH, Gordon R, Sims HV, Bryan LE. Sporadic cases of hemorrhagic colitis associated with *Escherichia coli* O157:H7. *Ann Intern Med* 1984;101: 738–742.
85. Centers for Disease Control. Isolation of *E. coli* O157:H7 from sporadic cases of hemorrhagic colitis—United States. *MMWR Morbid Mortal Wkly Rep* 1997;46:700–704.
86. Centers for Disease Control. Enhanced detection of sporadic *Escherichia coli* O157:H7 infections—New Jersey, July 1994. *MMWR Morbid Mortal Wkly Rep* 1995;44:417–418.
87. Bell BP, Goldoft M, Griffin PM, et al. A multistate outbreak of *Escherichia coli* O157:H7–associated bloody diarrhea and hemolytic uremic syndrome from hamburgers: The Washington experience. *JAMA* 1994; 272:1349–1353.
88. Izumiya H, Terajima J, Wada A, et al. Molecular typing of enterohemorrhagic *Escherichia coli* O157:H7 isolates in Japan by using pulsed-field gel electrophoresis. *J Clin Microbiol* 1997;35:1675–1680.
89. Gutierrez E. Is the Japanese O157:H7 *E. coli* epidemic over? *Lancet* 1996;348:1371.
90. Watanabe H, Wada A, Inagaki Y, Itoh K-I, Tamura K. Outbreaks of enterohemorrhagic *Escherichia coli* O157:H7 infection by two different genotype strains in Japan, 1996. *Lancet* 1996;348:831–832.
91. Swerdlow DL, Woodruff BA, Brady RC, et al. A waterborne outbreak in Missouri of *Escherichia coli*

O157:H7 associated with bloody diarrhea and death. *Ann Intern Med* 1992;117:812–819.
92. Morgan D, Newman CP, Hutchinson DN, Walker AM, Rowe B, Majid F. Verotoxin producing *Escherichia coli* O157 infections associated with the consumption of yoghurt. *Epidemiol Infect* 1993;111:181–187.
93. Upton P, Coia JE. Outbreak of *Escherichia coli* O157 infection associated with pasteurized milk supply. *Lancet* 1994;344:1015.
94. McCarthy M. *E. coli* O157:H7 outbreak in USA traced to apple juice. *Lancet* 1996;348:1299.
95. Keene WE, Sazie E, Kok J, et al. An outbreak of *Escherichia coli* O157:H7 infections traced to jerky made from deer meat. *JAMA* 1997;277:1229–1231.
96. Beutin L, Geier D, Zimmermann S, Karch H. Virulence markers of Shiga-like toxin–producing *Escherichia coli* strains originating from healthy domestic animals of different species. *J Clin Microbiol* 1995;33:631–635.
97. Kudva IT, Hatfield PG, Hovde CJ. *Escherichia coli* O157:H7 in microbial flora of sheep. *J Clin Microbiol* 1996;34:431–433.
98. Wallace JS, Cheasty T, Jones K. Isolation of vero cytotoxin–producing *Escherichia coli* O157 from wild birds. *J Appl Microbiol* 1997;82:399–404.
99. Besser TE, Hancock DD, Pritchett LC, McRae EM, Rice DH, Tarr PI. Duration of detection of fecal excretion of *Escherichia coli* O157:H7 in cattle. *J Infect Dis* 1997;175:726–729.
100. LeSaux N, Spika JS, Friesen B, et al. Ground beef consumption in noncommercial settings is a risk factor for sporadic *Escherichia coli* O157:H7 infection in Canada. *J Infect Dis* 1993;167:500–502.
101. Cieslak PR, Barrett TJ, Griffin PM, et al. *Escherichia coli* O157:H7 infection from a manured garden. *Lancet* 1993;342:367.
102. Keene WE, McAnulty JM, Hoesly FC, et al. A swimming-associated outbreak of hemorrhagic colitis caused by *Escherichia coli* O157:H7 and *Shigella sonnei*. *N Engl J Med* 1994;331:579–584.
103. Brewster DH, Brown MI, Robertson D, Houghton GL, Bimson J, Sharp JCM. An outbreak of *Escherichia coli* O157 associated with a children's paddling pool. *Epidemiol Infect* 1994;112:441–447.
104. Reida P, Wolff M, Pöhls H-W, et al. An outbreak due to enterohaemorrhagic *Escherichia coli* O157:H7 in a children's day care centre characterized by person-to-person transmission and environmental contamination. *Zentralbl Bakteriol* 1994;281:534–543.
105. Renwick SA, Wilson JB, Clarke RC, et al. Evidence of direct transmission of *Escherichia coli* O157:H7 infection between calves and a human. *J Infect Dis* 1993; 168:792–793.
106. Booth L, Rowe B. Possible occupational acquisition of *Escherichia coli* O157 infection. *Lancet* 1993;342: 1298–1299.
107. Burnens AP, Zbinden R, Kaempf L, Heinzer I, Nicolet J. A case of laboratory acquired infection with *Escherichia coli* O157:H7. *Zentralbl Bakteriol* 1993;279: 512–517.
108. Rao GG, Saunders BP, Masterton RG. Laboratory acquired verotoxin producing *Escherichia coli* (VTEC) infection [letter]. *J Hosp Infect* 1996;33:228–230.
109. Su C, Brandt LJ. *Escherichia coli* O157:H7 infection in humans. *Ann Intern Med* 1995;123:698–714.
110. Waters JR, Sharp JCM, Dev VK. Infection caused by *Escherichia coli* O157:H7 in Alberta, Canada, and in Scotland: A five-year review, 1987–1991. *Clin Infect Dis* 1994;19:834–843.
111. Sharp JCM, Coia JE, Curnow J, Reilly WJ. *Escherichia coli* O157 infections in Scotland. *J Med Microbiol* 1994;40:3–9.
112. Čobeljić M, Lepšanović Z, Velimirović S. Infrequent finding of verotoxin-producing *Escherichia coli* in diarrheal stools in Belgrade, Serbia. *Scand J Infect Dis* 1995;27:427–428.
113. Karmali MA, Petric M, Steele BT, Lim C. Sporadic cases of haemolytic-uraemic syndrome associated with faecal cytotoxin and cytotoxin-producing *Escherichia coli* in stools. *Lancet* 1983;1:619–620.
114. Boyce TG, Swerdlow DL, Griffin PM. *Escherichia coli* O157:H7 and the hemolytic-uremic syndrome. *N Engl J Med* 1995;333:364–368.
115. Ashkenazi S. Role of bacterial cytotoxins in hemolytic uremic syndrome and thrombotic thrombocytopenic purpura. *Annu Rev Med* 1993;44:11–18.
116. Siegler RL. Spectrum of extrarenal involvement in postdiarrheal hemolytic-uremic syndrome. *J Pediatr* 1994;125:511–518.
117. Taylor CM. Verocytotoxin-producing *Escherichia coli* and the haemolytic uraemic syndrome. *J Infect* 1995; 30:189–192.
118. Robson WLM, Leung AKC, Miller-Hughes DJ. Recurrent hemorrhagic colitis caused by *Escherichia coli* O157:H7. *Pediatr Infect Dis J* 1993;12:699–701.
119. Krishnan C, Fitzgerald VA, Dakin SJ, Behme RJ. Laboratory investigation of outbreak of hemorrhagic colitis caused by *Escherichia coli* O157:H7. *J Clin Microbiol* 1987;25:1043–1047.
120. Caprioli A, Luzzi I, Rosmini F, et al. Communitywide outbreak of hemolytic-uremic syndrome associated with non-O157 verocytotoxin-producing *Escherichia coli*. *J Infect Dis* 1994;169:208–211.
121. Centers for Disease Control. Outbreak of acute gastroenteritis attributable to *Escherichia coli* serotype O104:H21—Helena, Montana, 1994. *MMWR Morbid Mortal Wkly Rep* 1995;44:501–503.
122. Centers for Disease Control. Community outbreak of hemolytic uremic syndrome attributable to *Escherichia coli* O111:NM—South Australia, 1995. *MMWR Morbid Mortal Wkly Rep* 1995;44:550–554.
123. Banatvala N, Debeukelaer MM, Griffin PM, et al. Shiga-like toxin–producing *Escherichia coli* O111 and associated hemolytic-uremic syndrome: A family outbreak. *Pediatr Infect Dis J* 1996;15:1008–1011.
124. Tarr PI, Fouser LS, Stapleton AE, et al. Hemolytic-uremic syndrome in a six-year-old girl after a urinary tract infection with Shiga-toxin–producing *Escherichia coli* O103:H2. *N Engl J Med* 1996;335:635–638.
125. Ricotti GC, Buonomini MI, Merlitti A, Karch H, Luzzi I, Caprioli A. A fatal case of hemorrhagic colitis, thrombocytopenia, and renal failure associated with verocytotoxin-producing, non-O157 *Escherichia coli*. *Clin Infect Dis* 1994;19:815–816.
126. Lopez EI, Diaz M, Grinstein S, et al. Hemolytic uremic syndrome and diarrhea in Argentine children: The role of Shiga-like toxins. *J Infect Dis* 1989;160:469–475.
127. Mathewson JJ, Johnson PC, DuPont HL, et al. A newly recognized cause of travelers' diarrhea: Enteroadherent *Escherichia coli*. *J Infect Dis* 1985;151: 471–475.
128. Nataro JP, Kaper JB, Robins-Browne R, Prado V, Vial P, Levine MM. Patterns of adherence of diarrheagenic

Escherichia coli to HEp-2 cells. *Pediatr Infect Dis J* 1987;6:829–831.
129. Cravioto A, Tello A, Navarro A, et al. Association of *Escherichia coli* HEp-2 adherence patterns with type and duration of diarrhoea. *Lancet* 1991;337:262–264.
130. Mathewson JJ, Johnson PC, DuPont HL, Satterwhite TK, Winsor DK. Pathogenicity of enteroadherent *Escherichia coli* in adult volunteers. *J Infect Dis* 1986;154:524–527.
131. Nataro JP, Yikang D, Cookson S, et al. Heterogeneity of enteroaggregative *Escherichia coli* virulence demonstrated in volunteers. *J Infect Dis* 1995;171:465–468.
132. Scotland SM, Smith HR, Said B, Willshaw GA, Cheasty T, Rowe B. Identification of enteropathogenic *Escherichia coli* isolated in Britain as enteroaggregative or as members of a subclass of attaching and effacing *E. coli* not hybridizing with the EPEC adherence-factor probe. *J Med Microbiol* 1991;35:278–283.
133. Smith HR, Scotland SM, Willshaw GA, Rowe B, Cravioto A, Eslava C. Isolates of *Escherichia coli* O44:H18 of diverse origin are enteroaggregative. *J Infect Dis* 1994;170:1610–1613.
134. Čobeljić M, Miljković-Selimović B, Paunović-Todosijević D, et al. Enteroaggregative *Escherichia coli* associated with an outbreak of diarrhoea in a neonatal nursery ward. *Epidemiol Infect* 1996;117:11–16.
135. Mathewson JJ, Oberhelman RA, DuPont HL, de la Cabada FJ, Garibay EV. Enteroadherent *Escherichia coli* as a cause of diarrhea among children in Mexico. *J Clin Microbiol* 1987;25:1917–1919.
136. Bhan MK, Khoshoo V, Sommerfelt H, Raj P, Sazawal S, Srivastava R. Enteroaggregative *Escherichia coli* and *Salmonella* associated with nondysenteric persistent diarrhea. *Pediatr Infect Dis J* 1989;8:499–502.
137. Gíron JA, Jones T, Millán-Velasco F, et al. Diffuse-adhering *Escherichia coli* (DAEC) as a putative cause of diarrhea in Mayan children in Mexico. *J Infect Dis* 1991;163:507–513.
138. Wanke CA, Schorling JB, Barrett LJ, Desouza MA, Guerrant RL. Potential role of adherence traits of *Escherichia coli* in persistent diarrhea in an urban Brazilian slum. *Pediatr Infect Dis J* 1991;10:746–751.
139. Baqui AH, Sack R, Black RE, et al. Enteropathogens associated with acute and persistent diarrhea in Bangladeshi children <5 years of age. *J Infect Dis* 1992;166:792–796.
140. Gunzburg ST, Chang BJ, Elliott SJ, Burke V, Gracey M. Diffuse and enteroaggregative patterns of adherence of enteric *Escherichia coli* isolated from aboriginal children from the Kimberley region of western Australia. *J Infect Dis* 1993;167:755–758.
141. Mathewson JJ, Jiang ZD, Zumla A, et al. HEp-2 cell–adherent *Escherichia coli* in patients with human immunodeficiency virus–associated diarrhea. *J Infect Dis* 1995;171:1636–1639.
142. Cegielski JP, Msengi AE, Dukes CS, Levine MM. Pathogenic *Escherichia coli* in children with and without chronic diarrhea in Tanzania. *J Infect Dis* 1996; 174:675–676.
143. Cohen MB, Hawkins JA, Weckbach LS, Stanek JL, Levine MM, Heck JE. Colonization by enteroaggregative *Escherichia coli* in travelers with and without diarrhea. *J Clin Microbiol* 1993;31:351–353.
144. Morelli R, Baldassarri L, Falbo V, Donelli G, Caprioli A. Detection of enteroadherent *Escherichia coli* associated with diarrhoea in Italy. *J Med Microbiol* 1994; 41:399–404.
145. Forestier C, Meyer M, Favre-Bonte S, et al. Enteroadherent *Escherichia coli* and diarrhea in children: A prospective case-control study. *J Clin Microbiol* 1996; 34:2897–2903.
146. Johnson WM, Lior H. A new heat-labile cytolethal distending toxin (CLDT) produced by *Escherichia coli* isolates from clinical material. *Microb Pathog* 1988;4: 103–113.
147. Anderson JD, MacNab AJ, Gransden WR, Damm SMA, Johnson WM, Lior H. Gastroenteritis and encephalopathy associated with a strain of *Escherichia coli* O55:K59:H4 that produced a cytolethal distending toxin. *Pediatr Infect Dis J* 1987;6:1135–1136.
148. Bouzari S, Varghese A. Cytolethal distending toxin (CLDT) production by enteropathogenic *Escherichia coli* (EPEC). *FEMS Microbiol Lett* 1990;71:193–198.
149. Albert MJ, Faruque SM, Faruque ASG, et al. Controlled study of cytolethal distending toxin-producing *Escherichia coli* infections in Bangladeshi children. *J Clin Microbiol* 1996;34:717–719.
150. Caprioli A, Falbo V, Roda LG, Ruggeri FM, Zona C. Partial purification and characterization of an *Escherichia coli* toxic factor that induces morphological cell alterations. *Infect Immun* 1983;39:1300–1306.
151. Blanco J, Alonso MP, Gonzalez EA, Blanco M, Garabal JI. Virulence factors of bacteraemic *Escherichia coli* with particular reference to production of cytotoxic necrotising factor (CNF) by P-fimbriate strains. *J Med Microbiol* 1990;31:175–183.
152. Blanco J, Blanco M, Alonso MP, Blanco JE, González EA, Garabal JI. Characteristics of haemolytic *Escherichia coli* with particular reference to production of cytotoxic necrotizing factor type 1 (CNF1). *Res Microbiol* 1992;143:869–878.
153. Blanco J, González EA, Espinosa P, Blanco M, Garabal JI, Alonso MP. Enterotoxigenic and necrotizing *Escherichia coli* in human diarrhoea in Spain. *Eur J Epidemiol* 1992;8:548–552.
154. Ørskov I, Ørskov F. *Escherichia coli* in extra-intestinal infections. *J Hyg Camb* 1985;95:551–575.
155. Stamm WE, Hooton TM. Management of urinary tract infections in adults. *N Engl J Med* 1993;329:1328–1334.
156. Kunin CM. Urinary tract infections in females. *Clin Infect Dis* 1994;18:1–12.
157. Nicolle LE, Harding GKM, Preikaitis J, Ronald AR. The association of urinary tract infection with sexual intercourse. *J Infect Dis* 1982;146:579–583.
158. Foxman B, Zhang L, Tallman P, et al. Transmission of uropathogens between sex partners. *J Infect Dis* 1997; 175:989–992.
159. MacDonald P, Alexander D, Catz C, Edelman R. Summary of a workshop on maternal genitourinary infections and the outcome of pregnancy. *J Infect Dis* 1983; 147:596–605.
160. Ulleryd P, Lincoln K, Scheutz F, Sandberg T. Virulence characteristics of *Escherichia coli* in relation to host response in men with symptomatic urinary tract infection. *Clin Infect Dis* 1994;18:579–584.
161. Kuikka A, Sivonen A, Emelianova A, Valtonen VV. Prognostic factors associated with improved outcome of *Escherichia coli* in a Finnish university hospital. *Eur J Clin Microbiol Infect Dis* 1997;16:125–134.
162. Roberts FJ, Geere IW, Coldman A. A three-year study

of positive blood cultures, with emphasis on prognosis. *Rev Infect Dis* 1991;13:34–46.
163. Cockerill FR III, Hughes JG, Vetter EA, et al. Analysis of 281,797 consecutive blood cultures performed over an eight-year period: Trends in microorganisms isolated and the value of anaerobic culture of blood. *Clin Infect Dis* 1997;24:403–418.
164. Weinstein MP, Towns ML, Quartey SM, et al. The clinical significance of positive blood cultures in the 1990s: A prospective comprehensive evaluation of the microbiology, epidemiology, and outcome of bacteremia in adults. *Clin Infect Dis* 1997;24:584–602.
165. Grandsen WR, Eykyn SJ, Phillips I, Rowe B. Bacteremia due to *Escherichia coli:* A study of 861 episodes. *Rev Infect Dis* 1990;12:1008–1018.
166. Maslow JN, Mulligan ME, Adams KS, Justis JC, Arbeit RD. Bacterial adhesins and host factors: Role in the development and outcome of *Escherichia coli* bacteremia. *Clin Infect Dis* 1993;17:89–97.
167. Olesen B, Kolmos HJ, Ørskov F, Ørskov I, Gottschau A. Bacteraemia due to *Escherichia coli* in a Danish university hospital, 1986–1990. *Scand J Infect Dis* 1995;27:253–257.
168. Maslow JN, Mulligan ME, Arbeit RD. Recurrent *Escherichia coli* bacteremia. *J Clin Microbiol* 1994;32: 710–714.
169. Brauner A, Kaijser B, Kühn I. Recurrent *Escherichia coli* bacteraemia: Clinical characteristics and bacterial properties. *J Infect* 1994;28:49–57.
170. Bessesen MT, Wang E, Echeverria P, Blaser MJ. Enteroinvasive *Escherichia coli*: A cause of bacteremia in patients with AIDS. *J Clin Microbiol* 1991;29:2675–2677.
171. Rademaker CMA, Fluit AC, Fausze M, Rosenberg-Arska M, Glerum JH, Verhoef J. Enterovirulent *Escherichia coli* can cause bacteraemia in immunocompromised patients [letter]. *J Infect* 1993;26:226–228.
172. Levy I, Leibovici L, Drucker M, Samra Z, Konisberger H, Ashkenazi S. A prospective study of gram-negative bacteremia in children. *Pediatr Infect Dis J* 1996;15: 117–122.
173. Bingen E, Denamur E, Brahimi N, Elion J. Genotyping may provide rapid identification of *Escherichia coli* K1 organisms that cause neonatal menigitis. *Clin Infect Dis* 1996;22:152–156.
174. Silver RP, Aaronson W, Vaan WF. The K1 capsular polysaccharide of *Escherichia coli. Rev Infect Dis* 1988;10(Suppl):S282–S286.
175. Cross AS, Gemski P, Sadoff JC, Ørskov F, Ørskov I. The importance of the K1 capsule in invasive infections caused by *Escherichia coli. J Infect Dis* 1984; 149:184–193.
176. Christopher GW. *Escherichia coli* bacteremia, meningitis, and hemochromatosis. *Arch Intern Med* 1985; 145:1908.
177. Watanakunakorn C. *Escherichia coli* meningitis and septicemia associated with an epidural catheter [letter]. *Clin Infect Dis* 1995;21:713–714.
178. Johnson CC, Baldessarre J, Levison ME. Peritonitis: Update on pathophysiology, clinical manifestations, and management. *Clin Infect Dis* 1997;24:1035–1047.
179. Brook I, Frazier EH. Microbiological analysis of pancreatic abscess. *Clin Infect Dis* 1996;22:384–385.
180. Saxén H, Tarkka E, Hannikainen P, Nikku R, Rautio M, Siitonen A. *Escherichia coli* and appendicitis: Phenotypic characteristics of *E. coli* isolates from inflamed and noninflamed appendices. *Clin Infect Dis* 1996;23:1038–1042.
181. Berk SL, Neumann P, Holtsclaw S, Smith JK. *Escherichia coli* pneumonia in the elderly: With reference to the role of *E. coli* K1 capsular polysaccharide antigen. *Am J Med* 1982;72:899–902.
182. Jonas M, Cunha BA. Bacteremic *Escherichia coli* pneumonia. *Arch Intern Med* 1982;142:2157–2159.
183. Khaira HS, Allan A, Crowson MC. *Escherichia coli* infection of an abdominal aortic aneurysm. *J Infect* 1996;32:161–163.
184. Peacock SJ, Maxwell P, Stanton A, Jeffrey KJM. Two cases of infected atherosclerotic aneurysms and a comparison with infective endocarditis. *Eur J Clin Microbiol Infect Dis* 1995;14:1004–1008.
185. Miedema CJ, Kimpen JLL. Hematogenous dissemination of *Escherichia coli* to a preexistent subdural hematoma in a child. *Clin Infect Dis* 1996;23:662.
186. Rahav G, Levinger S, Frucht-Pery J. *Escherichia coli* endophthalmitis secondary to pyelonephritis: Another complication of diabetes? [letter]. *Clin Infect Dis* 1994;18:117–118.
187. Tseng C-Y, Liu PY-F, Shi Z-Y, et al. Endogenous endophthalmitis due to *Escherichia coli:* Case report and review. *Clin Infect Dis* 1996;22:1107–1108.
188. Tapsall JW, McIver CJ. Septicaemia caused by cysteine-requiring isolates of *Escherichia coli. J Med Microbiol* 1986;22:379–382.
189. March SB, Ratnam S. Sorbitol-MacConkey medium for detection of *Escherichia coli* O157:H7 associated with hemorrhagic colitis. *J Clin Microbiol* 1986;23: 869–872.
190. Zadik PM, Chapman PA, Siddons CA. Use of tellurite for the selection of verotoxigenic *Escherichia coli* O157. *J Med Microbiol* 1993;39:155–158.
191. Beutin L, Montenegro MA, Ørskov I, et al. Close association of verotoxin (Shiga-like toxin) production with enterohemolysin production in strains of *Escherichia coli. J Clin Microbiol* 1989;27:2559–2564.
192. Olsvik Ø, Popovic T, Skjerve E, et al. Magnetic separation techniques in diagnostic microbiology. *Clin Microbiol Rev* 1994;7:43–54.
193. Gicquelas KG, Baldini MM, Martinez J, et al. Practical and economical method for using biotynlated DNA probes with bacterial colony blots to identify diarrhea-causing *Escherichia coli. J Clin Microbiol* 1990;28: 2485–2490.
194. Abbott SL. Laboratory aspects of non-O157 toxigenic *E. coli. Clin Microbiol Newslett* 1997;19:105–108.
195. Domínguez A, Alcaide F, Pulido A, Ayats J, Pérez JL, Martín R. Use of commercial double-test tablet (Rosco PGUA/Indole) for screening *Escherichia coli. Diagn Microbiol Infect Dis* 1992;15:291–294.
196. Edberg SC, Kontnick CM. Comparison of β-glucuronidase–based substrate systems for identification of *Escherichia coli. J Clin Microbiol* 1986;24:368–371.
197. Thompson JS, Hodge DS, Borczyk AA. Rapid biochemical test to identify verocytotoxin-positive strains of *Escherichia coli* O157. *J Clin Microbiol* 1990;28: 2165–2168.
198. Miller JM, Rhoden DL. Preliminary evaluation of Biolog, a carbon source utilization method for bacterial identification. *J Clin Microbiol* 1991;29:1143–1147.
199. Pfaller MA, Sahm D, O'Hara C, et al. Comparison of

the AutoSCAN-W/A rapid bacterial identification system and the Vitek AutoMicrobic system for identification of gram-negative bacilli. *J Clin Microbiol* 1991; 29:1422–1428.
200. O'Hara CM, Rhoden DL, Miller JM. Reevaluation of the API 20E identification system versus conventional biochemicals for identification of members of the family *Enterobacteriaceae:* A new look at an old product. *J Clin Microbiol* 1992;30:123–125.
201. Kelly MT, Leicester C. Evaluation of the Autoscan Walkaway system for rapid identification and susceptibility testing of gram-negative bacilli. *J Clin Microbiol* 1992;30:1568–1571.
202. O'Hara CM, Miller JM. Evaluation of the autoSCAN-W/A system for rapid (2-hour) identification of members of the family *Enterobacteriaceae. J Clin Microbiol* 1992;30:1541–1543.
203. Visser MR, Bogaards L, Rozenberg-Arska M, Verhoef J. Comparison of the autoSCAN-W/A and Vitek Automicrobic systems for identification and susceptibility testing of bacteria. *Eur J Clin Microbiol Infect Dis* 1992;11:979–984.
204. York MK, Brooks GF, Fiss EH. Evaluation of the autoSCAN-W/A rapid systems for identification and susceptibility testing of gram-negative fermentative bacilli. *J Clin Microbiol* 1992;30:2903–2910.
205. O'Hara CM, Tenover FC, Miller JM. Parallel comparison of accuracy of API 20E, Vitek GNI, MicroScan Walk/Away Rapid ID, and Becton Dickinson Cobas Micro ID-E/NF for identification of members of the family *Enterobacteriaceae* and common gram-negative, non–glucose-fermenting bacilli. *J Clin Microbiol* 1993;31:3165–3169.
206. Kitch TT, Jacobs MR, Appelbaum PC. Evaluation of RapID onE for identification of 379 strains in the family *Enterobacteriaceae* and oxidase-negative, gram-negative nonfermenters. *J Clin Microbiol* 1994;32:931–934.
207. Holmes B, Costas M, Thaker T, Stevens M. Evaluation of two BBL Crystal systems for identification of some clinically important gram-negative bacteria. *J Clin Microbiol* 1994;32:2221–2224.
208. Wauters G, Boel A, Voorn GP, et al. Evaluation of a new identification system, Crystal Enteric/Non-Fermenter, for gram-negative bacilli. *J Clin Microbiol* 1995;33:845–849.
209. Robinson A, McCarter YS, Tetreault J. Comparison of Crystal Enteric/Nonfermenter system, API 20E system, and Vitek Automicrobic system for identification of gram-negative bacilli. *J Clin Microbiol* 1995;33: 364–370.
210. Rhoads S, Marinelli L, Imperatrice CA, Nachamkin I. Comparison of MicroScan WalkAway system and Vitek system for identification of gram-negative bacteria. *J Clin Microbiol* 1995;33:3044–3046.
211. McIver CJ, Tapsall JW. Assessment of conventional and commercial methods for identification of clinical isolates of cysteine-requiring strains of *Escherichia coli* and *Klebsiella* species. *J Clin Microbiol* 1990;28: 1947–1951.
212. Abbott SL, Hanson DF, Felland T, Connell S, Shum AH, Janda JM. *Escherichia coli* O157:H7 generates a unique biochemical profile on MicroScan conventional gram-negative identification panels. *J Clin Microbiol* 1994;32:823–824.
213. Haldane DJM, Damm MAS, Anderson JD. Improved biochemical screening procedure for small clinical laboratories for Vero (Shiga-like)-toxin–producing strains of *Escherichia coli* O157:H7. *J Clin Microbiol* 1986;24:652–653.
214. Brenner DJ, Davis BR, Steigerwalt AG, et al. Atypical biogroups of *Escherichia coli* found in clinical specimens and description of *Escherichia hermanni* sp. nov. *J Clin Microbiol* 1982;15:703–713.
215. Gunzer F, Böhm H, Rüssmann H, Bitzan M, Aleksić S, Karch H. Molecular detection of sorbitol-fermenting *Escherichia coli* O157 in patients with hemolytic-uremic syndrome. *J Clin Microbiol* 1992;30:1807–1810.
216. Field PI, Blom K, Hughes HJ, Helsel LO, Feng P, Swaminathan B. Molecular characterization of the gene encoding H antigen in *Escherichia coli* and development of a PCR–restriction fragment length polymorphism test for identification of *E. coli* O157:H7 and O157:NM. *J Clin Microbiol* 1997;35:1066–1070.
217. Farmer JJ III, Davis BR. H7 antiserum–sorbitol fermentation medium: A single tube screening medium for detecting *Escherichia coli* O157:H7 associated with hemorrhagic colitis. *J Clin Microbiol* 1985;22: 620–625.
218. Farmer JJ III, Davis BR, Hickman-Brenner FW, et al. Biochemical identification of new species and biogroups of *Enterobacteriaceae* isolated from clinical specimens. *J Clin Microbiol* 1985;21:46–76.
219. Toledo MRF, Trabulsi LR. Correlation between biochemical and serological characteristics of *Escherichia coli* and results of the Serény test. *J Clin Microbiol* 1983;17:419–421.
220. Burke V, Robinson J, Gracey M. Biotyping as a method of screening for enterotoxigenic *Escherichia coli. Pathology* 1987;19:80–83.
221. Albert MJ, Qadri F, Haque A, Bhuiyan NA. Bacterial clump formation at the surface of liquid culture as a rapid test for identification of enteroaggregative *Escherichia coli. J Clin Microbiol* 1993;31:1397–1399.
222. Aleksić S, Karch H, Bockemühl J. A biotyping scheme for Shiga-like (Vero) toxin–producing *Escherichia coli* O157 and a list of serological cross-reactions between O157 and other gram-negative bacteria. *Zentralbl Bakteriol* 1992;276:221–230.
223. Bitzan M, Ludwig K, Klemt M, König H, Büren J, Müller-Wiefel DE. The role of *Escherichia coli* O157 infections in the classical (enteropathic) haemolytic uraemic syndrome: Results of a central European, multicentre study. *Epidemiol Infect* 1993;110:183–196.
224. Hayes PS, Blkom K, Feng P, Lewis J, Strockbine NA, Swaminathan B. Isolation and characterization of a β-D-glucuronidase-producing strain of *Escherichia coli* serotype O157:H7 in the United States. *J Clin Microbiol* 1995;33:3347–3348.
225. Ojeda A, Prado V, Martinez J, et al. Sorbitol-negative phenotype among enterohemorrhagic *Escherichia coli* strains of different serotypes and from different sources. *J Clin Microbiol* 1995;33:2199–2201.
226. Crichton PB, Old DC. A biotyping scheme for the subspecific discrimination of *Escherichia coli. J Med Microbiol* 1982;15:233–242.
227. Ørskov F, Ørskov I. *Escherichia coli* serotyping and disease in man and animals. *Can J Microbiol* 1992;38: 699–704.
228. Jann K, Jann B. Polysaccharide antigens of *Escherichia coli. Rev Infect Dis* 1987;9(Suppl):S517–S526.

229. Bettelheim KA, Evangelidis H, Pearce JL, Sowers E, Strockbine NA. Isolation of a *Citrobacter freundii* strain which carries the *Escherichia coli* O157 antigen. *J Clin Microbiol* 1993;31:760–761.
230. Lior H, Borczyk AA. False positive identifications of *Escherichia coli* O157. *Lancet* 1987;2:333.
231. Ahmed R, Bopp C, Borczyk A, Katatiya S. Phage-typing scheme for *Escherichia coli* O157:H7. *J Infect Dis* 1987;155:806–809.
232. Krause U, Thomson-Carter FM, Pennington TH. Molecular epidemiology of *Escherichia coli* O157:H7 by pulsed-field gel electrophoresis and comparison with that of bacteriophage typing. *J Clin Microbiol* 1996; 34:959–961.
233. Selander RK, Caugant DA, Ochman H, et al. Methods for multilocus electrophoresis for bacterial population genetics and systematics. *J Appl Environ Microbiol* 1986;51:873–884.
234. Johnson JR, Moseley SL, Coyle MB, Stamm WE. Success of DNA fingerprinting after failure of biotyping, antimicrobial susceptibility testing, and plasmid analysis to reveal clonality of multiple blood and urine isolates from a patient with *Escherichia coli* urosepsis. *Diagn Microbiol Infect Dis* 1992;15:399–405.
235. Pacheco ABF, Guth BEC, Soares KCC, Nishimura L, DeAlemeida DF, Ferreira LCS. Random amplification of polymorphic DNA reveals serotype-specific clonal clusters among enterotoxigenic *Escherichia coli* strains isolated from humans. *J Clin Microbiol* 1997;35:1521–1525.
236. Mühldorfer I, Hacker J. Genetic aspects of *Escherichia coli* virulence. *Microb Pathog* 1994;16:171–181.
237. Janda JM, Desmond EP, Abbott SL. Animal and animal cell culture systems. In: Balows A, Hausler WJ, Hermann K, Isenberg HD, Shadomy HJ, eds. *Manual of clinical microbiology*. 5th ed. Washington, DC: ASM Press, 1991;137–146.
238. Oswald E, Pohl P, Jacquemin E, Lintermans P, Van Muylem K, O'Brien AD. Specific DNA probes to detect *Escherichia coli* strains producing cytotoxic necrotizing factor type 1 or type 2. *J Med Microbiol* 1994;40:428–434.
239. Tabouret M, DeRycke J. Detection of cytotoxic necrotizing factor (CNF) in extracts of *Escherichia coli* strains by enzyme-linked immunosorbent assay. *J Med Microbiol* 1990;32:73–81.
240. Czirok E, Semjem G, Steinruck H, et al. Comparison of rapid methods for detection of heat-labile (LT) and heat-stable (ST) enterotoxin in *Escherichia coli*. *J Med Microbiol* 1992;36:398–402.
241. Savarino SJ, Fasano A, Watson J, et al. Enteroaggregative *Escherichia coli* heat-stable enterotoxin 1 represents another subfamily of *E. coli* heat-stable toxin. *Proc Natl Acad Sci* 1993;90:3093–3097.
242. Tang P, Foubister V, Pucciarella MG, Finlay BB. Methods to study bacterial invasion. *J Microbiol Meth* 1993;18:227–240.
243. Donnenberg MS, Nataro JP. Methods for studying adhesion of diarrheagenic *Escherichia coli*. *Methods Enzymol* 1995;253:324–336.
244. Vial PA, Mathewson JJ, DuPont HL, Guers L, Levine MM. Comparison of two assay methods for patterns of adherence to HEp-2 cells of *Escherichia coli* from patients with diarrhea. *J Clin Microbiol* 1990;28: 882–885.
245. Haider K, Faruque SM, Albert MJ, Nahar S, Neogi PKB, Hossain A. Comparison of a modified adherence assay with existing assay methods for identification of enteroaggregative *Escherichia coli*. *J Clin Microbiol* 1992;30:1614–1616.
246. Scaletsky ICA, Silva MLM, Trabulsi LR. Distinctive patterns of adherence of enteropathogenic *Escherichia coli* to HeLa cells. *Infect Immun* 1984;45:534–536.
247. Jallat C, Livrelli V, Darfeuille-Michgaud A, Rich C, Joly B. *Escherichia coli* strains involved in diarrhea in France: High prevalence and heterogeneity of diffusely adhering strains. *J Clin Microbiol* 1993;31: 2031–2037.
248. Paul M, Tsukamoto T, Ghosh AR, et al. The significance of enteroaggregative *Escherichia coli* in the etiology of hospitalized diarrhoea in Calcutta, India and the demonstration of a new honey-combed pattern of aggregative adherence. *FEMS Microbiol Lett* 1994; 117:319–326.
249. Knutton S, Baldwin T, Williams PH, McNeish AS. Actin accumulation at sites of bacterial adhesion to tissue culture cells: Basis of a new diagnostic test for enteropathogenic and enterohemorrhagic *Escherichia coli*. *Infect Immun* 1989;57:1290–1298.
250. Law D. Virulence factors of enteropathogenic *Escherichia coli*. *J Med Microbiol* 1988;26:1–10.
251. Beutin L. The different hemolysins of *Escherichia coli*. *Med Microbiol Immunol* 1991;180:167–172.
252. Johnson JR. Virulence factors in *Escherichia coli* urinary tract infection. *Clin Microbiol Rev* 1991;4: 80–128.
253. Donnenberg MS, Kaper JB. Enteropathogenic *Escherichia coli*. *Infect Immun* 1992;60:3953–3961.
254. Hacker J. Role of fimbrial adhesins in the pathogenesis of *Escherichia coli* infections. *Can J Microbiol* 1992;38:720–727.
255. Gyles CL. *Escherichia coli* cytotoxins and enterotoxins. *Can J Microbiol* 1992;38:734–746.
256. Takeda Y, Kurazono H, Yamasaki S. Vero toxins (Shiga-like toxins) produced by enterohemorrhagic *Escherichia coli* (verocytotoxin-producing *E. coli*). *Microbiol Immunol* 1993;37:591–599.
257. Garcia M-I, Le Bouguénec C. Role of adhesion in pathogenicity of human uropathogenic and diarrhoeogenic *Escherichia coli*. *Bull Inst Pasteur* 1996; 94:201–236.
258. Dubreuil JD. *Escherichia coli* STb enterotoxin. *Microbiology* 1997;143:1783–1795.
259. McDaniel TK, Jarvis KG, Donnenberg MS, Kaper JB. A genetic locus of enterocyte effacement conserved among diverse enterobacterial pathogens. *Proc Natl Acad Sci* 1995;92:1664–1668.
260. Small PLC, Falkow S. Identification of regions on a 230-kilobase plasmid from enteroinvasive *Escherichia coli* that are required for entry into HEp-2 cells. *Infect Immun* 1988;56:225–229.
261. Waterman SR, Small PLC. Characterization of the acid resistance phenotype and *rpoS* alleles of Shiga-like toxin–producing *Escherichia coli*. *Infect Immun* 1996;64:2808–2811.
262. Agüero ME, Aron L, DeLuca AG, Timmis KN, Cabello FC. A plasmid-encoded outer membrane protein, TraT, enhances resistance of *Escherichia coli* to phagocytosis. *Infect Immun* 1984;46:740–746.
263. Timmis KN, Boulnois GJ, Bitter-Suermann D, Ca-

bello FC. Surface components of *Escherichia coli* that mediate resistance to the bactericidal activities of serum and phagocytes. *Curr Top Microbiol Immunol* 1985;118:197–218.
264. Devine DA, Roberts AP. K1, K5 and O antigens of *Escherichia coli* in relation to serum killing via the classical and alternative complement pathways. *J Med Microbiol* 1994;41:139–144.
265. Waters VL, Crosa JH. Colicin V virulence plasmids. *Microbiol Rev* 1991;55:437–450.
266. Nilius AM, Savage DC. Serum resistance encoded by colicin V plasmids in *Escherichia coli* and its relationship to the plasmid transfer system. *Infect Immun* 1984;43:947–953.
267. Montgomerie JZ, Bindereif A, Neilands JB, Kalmanson GM, Guze LB. Association of hydroxymate siderophore (aerobactin) with *Escherichia coli* isolated from patients with bacteremia. *Infect Immun* 1984;46:835–838.
268. Welch RA, Dellinger EP, Minshew B, Falkow S. Haemolysin contributes to virulence of extraintestinal *E. coli* infections. *Nature* 1981;294:665–667.
269. Jacoby GA, Medeiros AA. More extended-spectrum β-lactamases. *Antimicrob Agents Chemother* 1991;35: 1697–1704.
270. Sanders CC, Sanders Jr WE. β-Lactam resistance in gram-negative bacteria: Global trends and clinical impact. *Clin Infect Dis* 1992;15:824–839.
271. Jones RN, Pfaller MA, Allen SD, Gerlach EH, Fuchs PC, Aldredge KE. Antimicrobial activity of cefpirome: An update compared to five third-generation cephalosporins against nearly 6000 recent clinical isolates from five medical centers. *Diagn Microbiol Infect Dis* 1991;14:361–364.
272. Jones RN, Kehrberg EN, Erwin ME, Anderson SC, Fluoroquinolone Resistance Surveillance Group. Prevalence of important pathogens and antimicrobial activity of parenteral drugs at numerous medical centers in the United States. *Diagn Microbiol Infect Dis* 1994;19:203–215.
273. Thornsberry C. Susceptibility of clinical bacterial isolates to ciprofloxacin in the United States. *Infection* 1994;22(Suppl):S80–S89.
274. ASCP Susceptibility Testing Group. United States geographic bacteria susceptibility patterns: *Am J Clin Pathol* 1996;106:275–281.
275. Fuchs PC, Barry AL, Brown SD, the AST Surveillance Group. Survey of antimicrobial activity of four commonly used third-generation cephalosporins tested against recent bacterial isolates from ten American medical centers, and assessment of disk diffusion test performance. *Diagn Microbiol Infect Dis* 1996;24:213–219.
276. Threlfall EJ, Cheasty T, Graham A, Rowe B. High-level resistance to ciprofloxacin in *Escherichia coli. Lancet* 1997;349:403.
277. Lehn MN, Stöwer-Hoffmann J, Kott T, et al. Characterization of clinical isolates of *Escherichia coli* showing high levels of fluoroquinolone resistance. *J Clin Microbiol* 1996;34:597–602.
278. Sanders CC, Barry AL, Washington JA, et al. Detection of extended-spectrum β-lactamase–producing members of the family *Enterobacteriaceae* with Vitek ESBL test. *J Clin Microbiol* 1996;34:597–602.
279. O'Shaughnessy EM, Fahle GA, Witebsky FG. Correlation of in vitro susceptibility results for amoxicillin-clavulanate and ampicillin-sulbactam tested against *Escherichia coli. J Clin Microbiol* 1997;35:1902–1903.
280. McIver CJ, Tapsall JW. In vitro susceptibilities of clinical isolates of cysteine-requiring *Escherichia coli* to 12 antimicrobial agents. *Antimicrob Agents Chemother* 1991;35:995–997.
281. Johnson AP, Burns L, Woodford N, et al. Gentamicin resistance in clinical isolates of *Escherichia coli* encoded by genes of veterinary origin. *J Med Microbiol* 1994;40:221–226.

4

Other *Escherichia* Species

TAXONOMY

After enjoying the spotlight as the sole species within its genus, *Escherichia coli* was finally joined by a second species in 1973, some 54 years after the genus *Escherichia* was established. *E. coli* was never in danger of being upstaged by this new addition, however, since *Escherichia blattae,* isolated from the hindgut of the cockroach, *Blatta orientalis,*[1] is as rare as *E. coli* is common. *E. blattae* was gluconate-positive, unlike most *E. coli,* but otherwise it fit biochemically within the genus *Escherichia* and was placed there based on numerical identification.[1]

In the early 1980s, three more species were identified and rapidly added to the genus *Escherichia.* The first of these, *E. hermanii,* comprised the yellow-pigmented, cellobiose- and potassium cyanide–positive strains known as enteric group 11.[2] Although DNA–DNA homology studies found members of this group to be 84% to 91% related, clearly defining them as a species, the technique failed to identify a genus in which to place them. Although the type strain was 37% to 42% related to *E. coli,* it was also 35% to 50% related to four other genera in the *Enterobacteriaceae* (*Enterobacter, Klebsiella, Citrobacter,* and *Salmonella*). Phenotypic characteristics, by which these strains more closely resembled *Escherichia* than the other genera, were used to determine the genus into which they were placed. Also in 1982, Brenner and colleagues[3] introduced another species to the genus. Strains previously identified as enteric group 1, which were 75% related at 60°C, became *Escherichia vulneris.* Like its predecessor, *E. hermanii,* the type strain of *E. vulneris* showed relatedness to more than one genus. At 60°C these strains were 25% to 39% related to *Escherichia* and 24% to 35% related to *Enterobacter.* Again it fell to similarity of biochemical characteristics to determine the genus. Because *E. vulneris* strains gave negative responses on the Voges-Proskauer and citrate tests, reactions that are typically positive for *Enterobacter* species, the genus *Escherichia* was chosen. In 1985, strains in enteric group 10 joined the ranks of *Escherichia* as its fifth, and to date its last, addition.[4] Named *Escherichia fergusonii,* all members were highly related to each other, 90% to 97% at 60°C, and shared 49% to 63% homology with other *Escherichia* species. They differed from *E. coli* by positive fermentation reactions in adonitol and cellobiose and a negative reaction for D-sorbitol.

Table 1 lists all species of *Escherichia* other than *E. coli* along with their G+C ratios and American Type Culture Collection (ATCC) designations.

EPIDEMIOLOGY AND CLINICAL SIGNIFICANCE

E. blattae was first isolated from the hindgut of oriental cockroaches captured from various sites in England; whether this organism represented transient or resident intestinal flora was undetermined. However, isolation of *E. blattae* from the hindgut of cockroaches from other parts of the world[5] indicates that *E. blattae* is probably normal flora for this insect. Despite close cohabitation with cockroaches, there has

TABLE 1. Escherichia *species other than* E. coli

Nomenclature				
Current	Previous	G+C (% mol)	ATCC Number	Clinical Specimens/Significance
E. blattae	None	NA	—	No
E. hermanii	Enteric group 11	53–58	33650	Yes: wounds, questionable; other sites, opportunist
E. vulneris	Enteric group 1	58–59	33821	Yes: wounds, questionable; other sites, opportunist
E. fergusonii	Enteric group 10	NA	35469	Yes: feces, unknown; rare cause of UTI; other sites, opportunist

Data from references 1 through 4.
NA, not available; UTI, urinary tract infection.

been no transmission of this organism to humans.[1,5] Since cockroaches are capable of transmitting known enteric pathogens[1] to humans, *E. blattae* in all likelihood lacks virulence for humans and other animals. All three of the other *Escherichia* species covered in this chapter have been isolated from both human and nonhuman sources, including animals, foods, and the environment.[2–5] Animal sources include rabbit and bird for *E. vulneris*[3] and turkey, cow, swine, and horse for *E. fergusonii*.[4] Specific nonhuman sources have not been reported for *E. hermanii* other than three organisms listed as isolated from food.[5] Nevertheless, soil and water are likely habitats for both *E. hermanii* and *E. vulneris* because both organisms are associated with community-acquired wounds involving trauma, which are commonly colonized or infected with soil and water organisms.

Although isolated predominantly from wounds,[2,3,6–9] *E. vulneris* and *E. hermanii* are usually found in mixed infections[6,8,9] or in wounds in which no infection was noted.[7] Pien and coworkers[8] dismissed all of their isolates of *E. vulneris* (n=7) and *E. hermanii* (n=5) as colonizers after animal studies failed to demonstrate infection even when large numbers of organisms were inoculated into the peritoneum or soft tissue (10^7 and 10^6 organisms, respectively). In addition to wounds, *E. hermanii* has been isolated from sputum or lung, feces, blood, and spinal fluid.[2] Although its presence in blood and cerebrospinal fluid would indicate a pathogenic potential for this organism, there was no information on the patients from whom isolations were made. Cultures of *E. hermanii* submitted to our laboratory were most commonly isolated from feces (five isolates). One wound isolate was from a mixed infection, whereas a second wound yielded a pure culture. *E. hermanii* was also isolated in pure culture from the central venous catheter of a patient with leukemia.

E. vulneris has been isolated from sites other than wounds, including throat or sputum, blood, vagina, urine, feces, and lymph nodes.[3] Awsare and Lillo[10] ascribed a pathogenic role for *E. vulneris* in a case of urosepsis in an 86-year-old woman. The organism was isolated in pure culture from her blood and urine, and her infection resolved with appropriate therapy. Unarguably the cause of the urosepsis, this infection would still most appropriately be described as opportunistic given the patient's age and her history of type II diabetes and syncope. There was no comment on the source of the patient's illness. However, she had had cataract surgery 5 days previously and also had sustained an abrasion on her arm, which, although it showed no sign of infection, was a possible entry source. Neither the eyes nor the arm injury was cultured. A review of *E. vulneris* cultures received in our laboratory showed that 4 of 11 isolates were from blood. All but one of the patients with a blood isolate were compromised—one had a malignant tumor, another had postoperative peritonitis, and the third was an infant born prematurely. The four organisms isolated from wounds were from mixed infections, as were a lung and an ear isolate. The final strain was from feces in a patient with gastroenteritis.

Although it has been found in urine, blood, and an abdominal wound, *E. fergusonii* is most frequently isolated from feces. Funke and associates[11] presented a case in which *E. fergusonii* was isolated three times from the blood and once each from the gallbladder, feces, and a postoperative wound. The patient was a 69-year-old man with ascending cholangitis whose condition was further complicated by pancreatic carcinoma. The authors believed that the intestine was the source of the cholangiosepsis, but the patient died before it could be determined whether there was an animal or environmental source for the *E. fergusonii* isolate. It is not unreasonable that *E. fergusonii* is at least an opportunist, given this case and the reported isolations from urine and blood.[4] Patient histories on isolates in our culture collection would support this as well. *E. fergusonii* was isolated in pure culture from three patients—once from the blood of a bacteremic patient with severe combined immunodeficiency and twice from patients with urinary tract infections, one of whom had diarrhea as well. *E. fergusonii* was also present in the feces of two patients with bloody diarrhea and was isolated in large numbers from one of them. The last isolate was from a liver abscess and was in mixed culture with *E. coli,* suggesting the gut as a source for the infection. Whether *E. fergusonii* is a transient colonizer or a rare cause of diarrhea, the intestine appears to be the source of opportunistic infections involving this organism. Direct contact with food animals or consumption of their products (eg, meat, eggs, milk) is probably a source of infection for humans because isolations have been made from a variety of these animals.

LABORATORY IDENTIFICATION

Isolation

All species grow well on nonselective media such as chocolate, sheep blood, and heart infusion agars. *E. hermanii* usually appears as a yellow-pigmented colony, *E. vulneris* less often so. Neither *E. fergusonii* nor *E. blattae* colonies are yellow-pigmented.

Little information is available regarding the growth of *Escherichia* species on differential and selective agars, although it is probably similar to that of *E. coli.* Approximately half of all *E. hermanii* strains ferment lactose and/or sucrose, so their appearance on media containing those sugars is variable. Because *E. vulneris* strains are only occasionally found to utilize lactose or sucrose, colonies of this organism usually appear as nonfermenters (colorless) on differential and selective media. Neither *E. fergusonii* nor *E. blattae* uses either sugar, so both species grow as colorless colonies.

Identification

Conventional Systems

In case reports detailed previously,[7,9,11] API 20E (Vitek bioMérieux, Hazelwood, MO) easily and correctly identified strains of *E. vulneris, E. hermanii,* and *E. fergusonii.* In laboratory evaluations of commercial systems, both PASCO (PASCO Laboratories, Wheat Ridge, CO) and RapID onE (Innovative Diagnostic Systems, Norcross, GA) accurately identified all *Escherichia* species tested.[12,13] Twenty-nine strains of *Escherichia* (species unspecified) were identified in the former study, and one *E. vulneris,* one *E. fergusonii,* and two *E. hermanii* strains in the latter. In a comprehensive study, O'Hara and coworkers[14] tested four strains each of *E. hermanii* and *E. vulneris* by four commercial systems: API 20E, Vitek GNI card (Vitek bioMérieux), MicroScan Walk/Away Rapid ID (Dade MicroScan, West Sacramento, CA), and Cobas (Becton Dickinson Diagnostic Instrument Systems, Sparks, MD). In combination, the four systems misidentified these strains nine times, with Vitek and Cobas giving the best results and the MicroScan Walk/Away missing the most. No system correctly identified all eight strains. On additional testing, the MicroScan Walk/Away system misidentified one *E. hermanii* as *Klebsiella ozaenae* and one *E. vulneris* as *Enterobacter sakazakii;* Cobas misidentified one *E. vulneris* as *Yersinia pseudotuberculosis,* and Vitek GNI called *E. vulneris* an *E. hermanii.* API 20E identified all strains correctly with additional testing. There was no mention of *E. blattae* strains tested in any study. Given its absence in human clinical specimens and the rarity of available strains, it is unlikely this organism is included in the database of commonly used commercial systems.

Conventional Media

The four species of *Escherichia* considered here are easily separated from each other and from *E. coli* by conventional testing methods (Table 2).[5] Distinctive characteristics include positive results with potassium cyanide for *E. hermanii,* a positive malonate test for *E. vulneris,* fermentation of D-adonitol and D-arabitol for *E. fergusonii,* and fermentation of cellobiose for all three species. In addition, 98% of *E. hermanii* and about half of *E. vulneris* strains are yellow-pigmented. Pigment production can vary from an easily recognizable yellow color to a very pale yellow. Occasionally, to detect pigment it is necessary to scoop some growth and smear it on a white background (eg, filter paper) or to observe a colony picked up on a swab. Growth at 25°C sometimes also enhances pigment production. Biochemically, *E. blattae* is relatively unremarkable. It is distinguished from *E. coli* by negative tests for indol, *o*-nitrophenyl-β-D-galactopyranosidase (ONPG test), and fermentation of mannitol and sorbitol. Burgess and colleagues[1] reported two biotypes for *E. blattae* that were biochemically identical except for their citrate and malonate reactions, which were negative in biotype 1 and positive in biotype 2.

ANTIMICROBIAL SUSCEPTIBILITY

Limited susceptibility testing information is available for these *Escherichia* species, especially for *E. hermanii* and *E. fergusonii.* For *E. hermanii,* 31 of 32 strains were resistant to ampicillin, penicillin G, or carbenicillin by the Bauer-Kirby disk diffusion method.[2] Otherwise, all isolates of *E. hermanii* were reported to be susceptible to the following antimicrobials: gentamicin, kanamycin, tetracycline, chloramphenicol, cephalothin, cefazolin, cefoxitin, and ticarcillin.[2,9] *E. fergusonii* was the most resistant of the three species[4,11] and was reported to be susceptible to gentamicin, tobramycin, imipenem, ceftazidime, ceftriaxone, and trimethoprim-sulfamethoxazole and moderately susceptible to piperacillin. Susceptibility was variable for tetracycline, ampicillin, carbenicillin, and cephalothin. It was resistant to penicillin G and amoxacillin-clavulanate. Brenner and associates[3] reported that all strains of *E. vulneris* tested were susceptible to ampicillin, as did Kruczak and colleagues[7] and Awsare and Lillo[10] for their

TABLE 2. *Biochemical characteristics of* Escherichia *species*

	E. blattae	*E. coli*	*E. fergusonii*	*E. vulneris*	*E. hermanii*
Indol	−	+	+	−	+
ODC	+	V	+	−	+
KCN	−	−	−	−	+
Malonate	V	−	V	V	−
Acid from:					
Cellobiose	−	−	+	+	+
D-Adonitol	−	−	+	−	−
D-Arabitol	−	−	+	−	−
D-Sorbitol	−	+	−	−	−
Yellow pigment	−	−	−	V	+

Data from references 1 through 5.
ODC, ornithine decarboxylase; KCN, potassium cyanide; V, variable.

strains. However, the two strains tested by Dye and colleagues[6] gave minimal inhibitory concentrations of 8 and 4 μg/mL, indicating they would be refractory to oral but not to parenteral treatment with this drug. Whether this experience with ampicillin resistance was exceptional awaits more reports of testing with other strains. *E. vulneris* strains were otherwise susceptible to all aminoglycosides tested and to chloramphenicol, trimethoprim-sulfamethoxazole, tetracycline, ciprofloxacin, piperacillin, and a variety of second- and third-generation cephalosporins.

CONCLUSIONS

As with other rarely encountered *Enterobacteriaceae,* there are more unanswered questions than those for which we have answers. Perhaps the area that invites the most speculation is a possible role for *E. fergusonii* in diarrheal disease versus simple transient colonization of the gut. Further testing with *E. vulneris* and *E. hermanii* in wound models also should be done to lay to rest questions surrounding their possible role in these infections. Finally, as β-lactamases continue to increase among members of the *Enterobacteriaceae,* it will be necessary to watch these strains, particularly *E. vulneris,* for increased resistance to penicillins and cephalosporins.

REFERENCES

1. Burgess NRH, McDermott SN, Whiting J. Aerobic bacteria occurring in the hind-gut of the cockroach, *Blatta orientalis. J Hyg Camb* 1973;71:1–7.
2. Brenner DJ, Davis BR, Steigerwalt AG, et al. Atypical biogroups of *Escherichia coli* found in clinical specimens and description of *Escherichia hermanii* sp. nov. *J Clin Microbiol* 1982;15:703–713.
3. Brenner DJ, McWhorter AC, Knutson JKL, Steigerwalt AG. *Escherichia vulneris:* A new species of *Enterobacteriaceae* associated with human wounds. *J Clin Microbiol* 1982;15:1133–1140.
4. Farmer JJ III, Fanning GR, Davis BR, et al. *Escherichia fergusonii* and *Enterobacter taylorae,* two new species of *Enterobacteriaceae* isolated from clinical specimens. *J Clin Microbiol* 1985;21:77–81.
5. Farmer JJ III, Davis BR, Hickman-Brenner FW, et al. Biochemical identification of new species and biogroups of *Enterobacteriaceae* isolated from clinical specimens. *J Clin Microbiol* 1985;21:46–76.
6. Dye KR, Dall L, Yuhas J, Brockert JL. *Escherichia vulneris:* Isolation and treatment. *South Med J* 1984;77:1607–1608.
7. Kruczak PM, Cross A, Keiser JF, Almazan R. *Escherichia vulneris* from a human wound. *Clin Microbiol Newslett* 1985;7:5–6.
8. Pien FD, Shrum S, Swenson JM, Hill BC, Thornsberry C, Farmer JJ III. Colonization of human wounds by *Escherichia vulneris* and *Escherichia hermanii. J Clin Microbiol* 1985;22:283–285.
9. Berman M, Baron EJ. *Escherichia hermanii* wound infection. *Clin Microbiol Newslett* 1987;9:38–39.
10. Awsare SV, Lillo M. A case report of *Escherichia vulneris* urosepsis. *Rev Infect Dis* 1991;13:1247–1248.
11. Funke G, Hany A, Altwegg M. Isolation of *Escherichia fergusonii* from four different sites in a patient with pancreatic carcinoma and cholangiosepsis. *J Clin Microbiol* 1993;31:2201–2203.
12. Rhoden DL, Schable B, Smith PB. Evaluation of PASCO MIC-ID system for identifying gram-negative bacilli. *J Clin Microbiol* 1987;25:2362–2366.
13. Kitch TT, Jacobs MR, Appelbaum PC. Evaluation of RapID onE for identification of 379 strains in the family *Enterobacteriaceae* and oxidase-negative, gram-negative nonfermenters. *J Clin Microbiol* 1994;32:931–934.
14. O'Hara CM, Tenover FC, Miller JM. Parallel comparison of accuracy of API 20E, Vitek GNI, MicroScan Walk/Away Rapid ID, and Becton Dickinson Cobas Micro ID-F/NF for identification of members of the family *Enterobacteriaceae* and common gram-negative, non-glucose–fermenting bacilli. *J Clin Microbiol* 1993;31:3165–3169.

5

The Genus *Shigella*

TAXONOMY

The organism described by Shiga in 1898, and ultimately named *Shigella* in his honor in 1919, was actually discovered 10 years earlier. Chantemesse and Widal reported on strains similar to the Shiga bacillus isolated from dysenteric stools in 1888, as did Grigorieff in Russia in 1891.[1] Shiga's description stemmed from isolation of an organism from 34 of 36 patients with dysentery during the great epidemics in Japan that claimed 22,000 lives from 89,000 cases of disease. Castellani and Chalmers proposed the genus *Shigella* in 1919 with *Shigella dysenteriae,* the type species, corresponding to the organism described and named *Bacillus dysenteriae* by Shiga in 1898.[2] The species *Shigella flexneri* was also established in 1919.[2] It was named for Simon Flexner, who had described an organism similar to Shiga's bacillus isolated from a soldier with dysentery. *Shigella sonnei,* first described as *Bacillus sonnei* by Levine in 1920, was added to the genus in 1927 by Weldin.[2] The final member, *Shigella boydii,* was proposed 20 years later by Ewing in 1949.[2] Although *Shigella dispar* and *Shigella alkalescens* were considered members of the genus in the 1940s, they were excluded from *Shigella* in 1951 based on the recommendation of the International *Enterobacteriaceae* Subcommittee.[3] Because their biochemical reactions and somatic and capsular antigens more closely resembled those of *Escherichia,* they were moved to that genus and are known as *Escherichia coli* ADO (Alkalescens-Dispar organisms).

In 1973, Brenner and colleagues,[4] using DNA relatedness, conclusively confirmed that both shigellae and escherichiae belong in a single genus, a relationship long suspected based on serologic and biochemical similarities. In this landmark work, all shigellae, with the exception of *S. boydii* type 13 strains, were shown to be at least 80% related to each other and to *E. coli* at 60°C. Binding between members of a single species was not greater than that seen between various species of *Shigella* with two exceptions: *S. sonnei* showed 100% homology between the three strains tested, and *S. boydii* type 13 strains were only 68% to 75% related to other shigellae and 65% related to *E. coli.* In 1982, a more extensive study by Brenner and associates[5] confirmed the earlier finding that only *S. boydii* type 13, including aerogenic strains belonging to that type, were separable from other shigellae/escherichiae by DNA relatedness. Despite the fact that *Shigella* and *Escherichia* are actually a single genus, they remain taxonomically separated for clinical and epidemiologic reasons.

In 1982, Dodd and Jones[6] undertook a numerical taxonomic study of the genus *Shigella.* Based on 192 characteristics including morphology, biochemical reactions, and physiology, they concluded that it was possible to distinguish four species within the genus, a concept disputed by several studies in the 1970s.[6] However, they also concluded that considerable variation occurred within and between taxons and cautioned that the use of too few biochemicals would result in misidentifications. Table 1 includes nomen-

TABLE 1. *Nomenclature of the genus* Shigella

Designation	Subgroup	Number of Serotypes	ATCC Number of Type Strain
S. dysenteriae	A	1–13	13313
S. flexneri	B	1–6[a]	29903
S. boydii	C	1–18	8700
S. sonnei	D	None	29930

[a]Types 1 through 5 possess subtypes *a* and *b*.

clature and American Type Culture Collection (ATCC) numbers for the genus.

EPIDEMIOLOGY

Shigella are host-adapted to the human intestine. In animals, only rare isolations from dogs[2] and sporadic cases or outbreaks in primates have been reported.[7,8] Shigellosis in primates does not occur in the wild,[7] and cases in captive primates occur as a result of contact with humans.[7,8] A variety of primates have been documented to be susceptible to shigellosis, including Rhesus monkeys, tamarins, and marmosets.[8]

Although shigellosis is a worldwide problem, it is more prevalent in developing countries, with incidence rates several thousand times greater than those in the United States.[9] Most cases occur in children 6 months to 10 years of age. Infections in children younger than 6 months of age who are breastfed are uncommon; however, if contracted, the disease is severe in these infants.[9,10] Asymptomatic, transient carriers do exist.[10] Fecal shedding in patients who do not receive treatment can last 1 to 4 weeks; long-term carriage (>1 year) is known but rare.[11]

Spread of disease is primarily by person-to-person contact.[9,10,12] However, outbreaks do occur as a result of close contact (eg, day care centers, cruise ships, psychiatric institutions, Indian reservations), poor hygienic conditions, or contaminated food or water. Secondary transmission rates are high, especially for children. Approximately 60% of those younger than 1 year of age who are exposed to a primary case become ill; for older ages the rate is about 20%.[10] The low infectious dose of *Shigella,* 10 to 100 organisms,[9,12,13] accounts for much of the epidemiology seen with this genus; a few examples are given here. Simply swimming in contaminated water resulted in an outbreak with an 18% attack rate among persons who got water in their mouth[13]; the same strain of *S. sonnei* (determined by antibiograms, colicin, and phage typing) was isolated from the swimming area 1 month after the outbreak even though swimming had been banned. *Shigella* can survive in water at room temperature for up to 6 months.[10] Day care centers are a major setting for shigellosis, especially those caring for toddlers and infants.[14] Secondary transmission from these institutions to the community (frequently with antibiotic-resistant strains) is an additional concern aside from the outbreaks themselves.[15] In a retrospective review of cruise ship outbreaks occurring over a 7-year period, Koo and coworkers[16] reported that 4 of 31 episodes involved shigellae. *Shigella* accounted for a third of the bacterial outbreaks, and only enterotoxigenic *E. coli* caused more disease (n=5). Shigellosis transmitted sexually via oral-anal contact emerged as a problem in the 1970s.[17] Most cases, then and today, involve *S. flexneri* and subsequently there has been a rise in the incidence of infection with this subgroup in adult males.[12]

S. sonnei is the species most frequently encountered in industrialized countries, followed by *S. flexneri.* Isolations of *S. boydii* are relatively uncommon, and *S. dysenteriae* type 1 is seen only in developing countries or in persons traveling to areas where it is endemic.[9,10,12] In the United States most cases of *S. dysenteriae* type 1 are contracted in Mexico.[18] Even in areas of the world where

S. dysenteriae type 1 is endemic, except during epidemics of this organism, *S. flexneri* and *S. sonnei* account for most isolations.[19]

CLINICAL SIGNIFICANCE

Gastrointestinal Illness

Initial symptoms of shigellosis are fever and malaise, followed by watery diarrhea and abdominal pain.[9,10,12] Progression to classic dysentery with blood and mucus may occur, especially in patients with infections involving *S. dysenteriae* type 1 and *S. flexneri*.[9,12] DuPont and Pickering[10] estimated the incidence of blood at about 40% of cases and that of mucus at about 50%. In a study of shigellosis in children in India, Mathan and Mathan[20] reported two thirds had blood in their stools; in contrast, Halpern and associates[21] reported that only 26% of 116 adults in Israel presented with blood and mucus. As the disease proceeds, the volume of stool passed diminishes and is accompanied by tenesmus (the urge to defecate with little result and painful anal spasms).

It should be pointed out that not all cases of shigellosis, particularly disease caused by *S. sonnei,* present with the above symptomatology. Schulman[22] noted a number of cases of shigellosis with atypical symptoms during a community outbreak in New York. These included patients with fever but no gastrointestinal symptoms or with nausea and abdominal pain but no fever and scant diarrhea. Vomiting was an uncommon feature among patients in her practice, which included 267 cases, and the incidence of diminished appetite or the inability to eat a regular diet was minimal.

It was initially believed that early symptoms reflect small bowel involvement[10] and later symptoms are associated with colonic ulceration.[9,12] However, Butler and coinvestigators[23] showed that early in the disease the colon is unable to absorb water, sodium, and chloride, leading to diarrhea, and that only later, with colonic epithelial destruction, do stools become scant and bloody. Their results argue that the entire disease process results from the action of the organism on the colon only. Dehydration is rare in shigellosis,[12] and fluid loss usually does not exceed 30 mL/kg/d.[9] Anorexia, resulting in poor nutritional status, is a more prominent symptom and of greater concern than dehydration.[9,12,24] This is especially true in developing countries among malnourished children, in whom shigellosis may result in profound hypoproteinemia. Malnutrition is a major contributing factor to death in these patients, probably because it increases their susceptibility to secondary infections.[24,25] Respiratory symptoms, including a runny nose, cough, and chest pain, may accompany diarrheal symptoms.[10,18] Pneumonia is rare[9]; Halpern and associates[21] reported 2 cases of pneumonia among 116 adults, but Butler and coworkers[24] noted 11 cases among 37 fatal infections in children.

Sequelae of *Shigella* infection, although uncommon, are often fatal. Toxic megacolon and intestinal perforation are associated primarily with *S. dysenteriae* type 1 and *S. flexneri*.[9,12,26] Toxic megacolon, an acute nonobstructive dilation of the colon, has a high fatality rate and may be related to intense mucosal inflammation resulting in decreased colonic motility and distention, although the exact cause is unknown. Intestinal perforation is often difficult to detect and may be diagnosed only at autopsy.[26] It is found primarily in neonates and young children, particularly those who are malnourished. As with toxic megacolon, the exact cause is unclear; Bennish[26] indicated that diffuse small-vessel vasculitis, as opposed to mucosal ulceration, may be a factor.

Bacteremia

Sepsis associated with shigellosis is uncommon in immunocompetent persons in the United States because most disease is caused by *S. sonnei,* which is a generally less virulent species. *S. dysenteriae* type 1 is the most common cause of bacteremia, followed by *S. flexneri,* but any species may be involved. In Bangladesh, bacteremia developed in approximately 10% of children with shigellosis; either the infecting *Shigella* strain or a gram-negative rod was the cause.[26] The fatality rate for bacteremic children is twice that for nonbacteremic children and increases fivefold if sepsis is caused by a gram-negative rod.[11] Young, malnourished children whose condition results in decreased serum bactericidal activity are at greatest risk.[26] Hyponatremia and hypoglycemia, when they occur, are severe, with serum sodium levels of 110 to 120 mmol/L and blood glucose concentrations of less than 1.0 mmol/L.[26] At the International Centre for Diarrheal Disease Research, 25% of patients with shigellosis who die are hypoglycemic, probably because of the accompanying anorexia, which makes carbohydrate intake difficult.

Miscellaneous Infections

Hemolytic-uremic syndrome, with its triad of hemolytic anemia, thrombocytopenia, and acute renal failure, is a complication that occurs just as the patient appears to be recovering from the infection.[9,12] It is associated with *S. dysenteriae* type 1 and the production of Shiga toxin. Headache and nuchal rigidity are common in shigellosis, with 95% and 39% of patients, respectively, experiencing these symptoms.[27] Seizures, which may be caused by any species of *Shigella,*[9] occur primarily in children and are associated with elevated temperatures.[10,12] They are usually an early manifestation of shigellosis and in as many as 25% of cases may even precede gastrointestinal symptoms.[27] Cerebrospinal fluid appears normal in these patients, and cultures are negative. Although 12% to 45% of patients with shigellosis suffer convulsive episodes, approximately 42% develop encephalopathy.[27] Both focal and systemic central nervous system dysfunction occur but the latter is more common. Meningitis is extremely rare.[26]

Like other extraintestinal symptoms, urinary tract infections are rare; they may occur in patients who do not have gastrointestinal symptoms (40%) or positive stools (35%).[28] Both *S. flexneri* and *S. sonnei* have been isolated from urinary tract infections, but *S. flexneri* is more common; four of seven reported *S. sonnei* infections were asymptomatic.[28]

Reactive arthritis, osteomyelitis, and Reiter's syndrome are also sequelae of shigellosis. Patel and colleagues[29] reported on a patient with no history of gastrointestinal disease who had vertebral osteomyelitis caused by *S. boydii;* the source of the infection was undetermined. Eye involvement includes iritis and corneal ulceration.[9,19] Conjunctivitis occurs either as a symptom of Reiter's syndrome or alone.[30] Keratitis associated with *Shigella* has thus far occurred only in young children. In five of the seven reported cases keratitis was preceded by diarrhea, but shigellae were isolated from the stool in only three cases.[30] *S. sonnei* or *S. flexneri* were isolated from corneal scrapings or the conjunctival sac in five and two cases, respectively. Inoculation of the eye is believed to be caused by fecal contamination. Endophthalmitis has not been reported. Dermatologic manifestations, aside from diaper dermatitis or petechial or morbilliform rashes in children, are a curiosity. Stoll[31] reported on a homosexual patient with a penile furuncle with accompanying left inguinal adenopathy. Culture of the lesion yielded *S. flexneri.* No gastrointestinal symptoms were noted in the patient or his sexual partner, and both refused stool culture.

LABORATORY IDENTIFICATION

Isolation

Isolation of *Shigella* is usually straightforward because the severity of their disease forces patients to present early in the course of their illness, when the organism is easily isolated. There are some pitfalls, however, and these are discussed here. The specimen of choice is usually feces as opposed to a rectal swab, but field studies or surveys may require the use of swabs. Care should be taken to ensure that rectal, not anal, swabs are obtained because they yield a greater number of positive cultures. The use of both stool and a rectal swab, although impractical, yields the highest recovery rate. Adkins and Santiago[32] reported that, of 52 patients, 33 (63.5%) had a positive culture from both rectal swab and stool, 6 (11.5%) from stool only, and 13 (25%) by rectal swab alone.

If specimens are cultured within 2 to 4 hours of the time they are taken, transport medium is not required; preferably, all specimens should be held at 4°C while awaiting culture. If longer than 4 hours elapses, specimens should be placed in transport medium because shigellae are highly susceptible to the metabolic end products of normal flora and die-off occurs readily.[33] Buffered glycerol in saline (BGS) is superior to other media for the transport of shigellae. Wells and colleagues[34] found that, even at room temperature, BGS had an 85% recovery rate for *Shigella,* compared with 13% for Cary-Blair transport medium. Recovery improved to 88% and 63%, respectively, if samples were refrigerated. If longer than 3 days elapsed before the specimen arrived in the laboratory, recovery rates decreased significantly, even for samples transported in BGS.

Methylene blue wet mounts of fresh feces may reveal erythrocytes and polymorphonuclear leukocytes, usually occurring in sheets, but this test is not diagnostic for *Shigella*[20] and is not recommended on a routine basis.

Direct plating is the method of choice. Enrichment usually is not necessary for shigellae. Enrichment broths are intended for the isolation of salmonellae, and they eliminate normal flora *(E. coli),* to which shigellae are metabolically related. When enrichment is used, Hajna's gram-negative (GN) broth generally supports growth of shigellae better than other enrichments. Because the recommended incubation time of 6 to 8 hours does not readily fall into most laboratory schedules, overnight incubation may be used, but it should not exceed 16 to 18 hours. Because shigellae may be masked by competing normal flora on noninhibitory media or, conversely, may be too fastidious to grow on inhibitory agars, a variety of media are required for isolation. These include noninhibitory differential media such as MacConkey agar (MAC), eosin–methylene blue (EMB), or desoxycholate agar, on which shigellae appear as colorless, non–lactose-fermenting colonies; moderately selective media such as xylose-lysine-desoxycholate (XLD) or Hektoen enteric agar (HE); and selective media such as Salmonella-Shigella agar (SS) and desoxycholate citrate agar. *Shigella* colonies, again, present as colorless colonies on these media. Echeverria and coworkers[35] found that *S. flexneri* had a slightly higher recovery rate on MAC than on SS ($P<0.02$); *S. sonnei* grew equally well on MAC and SS, both of which were superior to HE ($P<0.01$). Teknaf enteric agar (MacConkey agar with 1 μg/mL of potassium tellurite added) did not support growth of *S. sonnei* ($P<0.001$) as well as MAC, SS, or HE did. Altwegg and associates[36] improved isolation rates for shigellae by adding xylose to MacConkey agar (MAC-X). The isolation rate for MAC-X was 91%, compared with 77% for HE, 86% for MAC, and 91% for XLD. Both XLD and MAC-X had a specificity of 78%, but XLD required longer incubation and was more difficult to read. As a result, these authors selected MAC-X as a testing medium for shigellae.

Colonies suspicious for *Shigella* should be screened with a combination of triple sugar iron (TSI) agar, lysine-iron agar (LIA), and, if possible, motility-indol-ornithine (MIO) agar.

Reactions of shigellae that may be confused with those of other *Enterobacteriaceae* in these media are presented in Table 2. Strains appearing to be shigellae in these media should be biochemically and serologically confirmed.

Rapid screening methods based on preenrichment in broth followed by latex agglutination are available. These tests eliminate the need for subculturing of broths that fail to agglutinate, thereby reducing workload. The Wellcolex Colour Salmonella and Shigella test (Murex Diagnostics, Norcross, GA), however, requires that shigellae be tested from direct or primary plating media only. In two clinical evaluations, it performed well for all *Shigella* species except *S. dysenteriae* type 2.[37,38] The Bactigen Salmonella-Shigella Latex Agglutination Slide test (Wampole Laboratories, Cranbury, NY) requires overnight incubation in GN enrichment broth. Results of clinical evaluations were less impressive with this sys-

TABLE 2. *Differentiation of shigellae from other fecal flora in TSI, LIA, and MIO agars*

Organism	TSI	LIA	MIO
Shigella spp.	K/A$^-$[a]	K/A$^-$	–/V/–[b]
Shigella sonnei	K/A$^-$	K/A$^-$	–/–/+
Providencia spp.	K/AG$^-$	K/A$^-$	+/+/–
Plesiomonas shigelloides[c]	K/A$^-$	K/K$^-$	+/+/+
Salmonella typhi	K/A$^-$	K/K$^-$	+^{w} or –/–/–
Typical *Escherichia coli*	A or K/AG$^-$	K/K$^-$	+/+/+
Escherichia coli ADO	K/A$^-$	K/A or K$^-$	–/V/+

[a] *S. flexneri* type 6 and *S. boydii* types 13 and 14 have aerogenic variants.
[b] *S. boydii* type 13 is ornithine-positive.
[c] *Plesiomonas* as well as various *Aeromonas* and *Vibrio* species that give K/A$^-$ reactions in TSI and LIA can be checked for oxidase from the LIA slant to separate them from shigellae.
TSI, triple sugar iron agar; LIA, lysine-iron agar; MIO, motility-indol-ornithine agar; K, alkaline; A, acid; G, gas; –, no H_2S produced; V, variable; ^{w}, weak; ADO, Alkalescens-Dispar organism.

tem; it missed two of four *S. sonnei* cultures in one study[39] and 3 of 29 in another.[40] Because of the numerous cross-reactions seen among the *Enterobacteriaceae,* all colonies giving agglutination reactions in either system should be confirmed biochemically.

Immunomagnetic separation (a technique involving monoclonal antibodies coated onto magnetic beads)[41] and polymerase chain reaction[42] assays for detection of *Shigella* have been developed but are not commercially available.

Identification

Commercial Systems

When a total of 77 strains of *Shigella* spp. were tested on eight commercial systems in separate studies, 15 (19.4%) failed to be identified as *Shigella;* 14 were misidentified at the genus level, and 1 was unidentified.[43–47] Of these eight systems, Crystal Enteric/Non-Fermenter (Becton Dickinson, Cockeysville, MD),[43] RapID onE (Innovative Diagnostic Systems, Norcross, GA),[44] Cathra Repliscan replicator system (Cathra, St. Paul, MN),[46] and PASCO MIC-ID (PASCO Laboratories, Wheat Ridge, CO)[47] all identified shigellae with a high degree of accuracy. API (Vitek bioMérieux, Hazelwood, MO) failed to identify shigellae to genus in 4 of 21 strains tested in one study[43] and initially missed 1 of 4 strains of *S. sonnei* in another[45] but gave a correct identification after additional testing. Vitek (Vitek bioMérieux) missed one each of *S. flexneri* and *S. sonnei,* and Cobas Micro ID-E/NF (Becton Dickinson Diagnostic Instrument Systems, Sparks, MD) missed two *S. flexneri.*[45] After additional testing, one *S. flexneri* was still incorrectly identified as *E. coli* ADO for both Vitek and Cobas, and Vitek reported the *S. sonnei* as unidentified. The MicroScan WalkAway Rapid ID (Dade MicroScan, West Sacramento, CA) misidentified one strain each of *S. boydii* and *S. sonnei* but after additional testing provided a correct identification.[45] This system also missed 3 of 10 *Shigella* spp. at the genus level in another study.[46]

Conventional Biochemicals

In the genus *Shigella* there is considerable biochemical variation between species and among types within a species. The most common species, *S. sonnei,* is the easiest to identify. It is positive on both the *o*-nitrophenyl-β-D-galactopyranosidase (ONPG) test and the ornithine decarboxylase test, a combined trait that is distinctive among the shigellae. It is also morphologically distinct. *S. sonnei* can usually be recognized easily on any plating medium (Fig. 1) by the presence of a distinctive mixture of smooth colonies (convex, regular edges, opaque) and rough colonies (larger, flat, irregular edges, translucent).

All *S. dysenteriae* types are mannitol-negative, whereas most other shigellae, with the exception of *S. flexneri* types 4 and 6 and

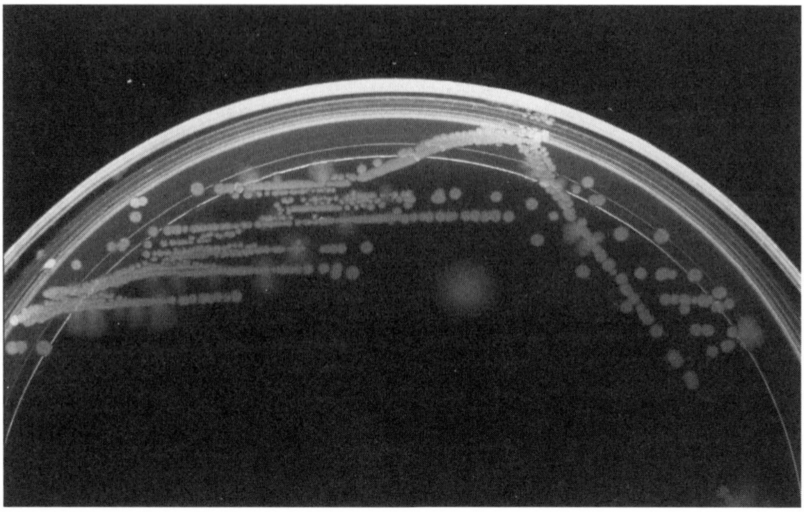

FIG. 1. Smooth and rough colony formation of *S. sonnei* on heart infusion agar. Large, flat, translucent colonies are rough, avirulent forms caused by the loss of a 180- to 200-kb plasmid.

S. boydii type 14, ferment mannitol. Other biochemicals helpful in separating types within species include dulcitol, xylose, D-rhamnose, and indol.[1] *S. flexneri* type 6 and *S. boydii* types 13 and 14 have aerogenic variants.[1,5,48]

Typical *E. coli* that are motile, produce gas from glucose, ferment lactose, and are lysine decarboxylase–positive are easily separable from *Shigella.* Inactive *E. coli* (including *E. coli* ADO) may be negative for any or all of these tests and may prove more difficult to differentiate. In these strains, sodium acetate and/or mucate may be helpful because a positive reaction may be obtained when other typical biochemicals are negative (Table 3). Among the shigellae, rare strains of *S. flexneri* type 4a are sodium acetate–positive and approximately 15% of *S. sonnei* may utilize mucate. Occasionally, strains may be indistinguishable biochemically; we report these as "*Escherichia/Shigella* group, unable to distinguish." Attempts should not be made to separate these strains on the basis of serology because of the cross-reactions that occur between these species (see Immunodiagnostics).

Biotyping

Only strains of *S. flexneri* type 6 have been biotyped[1] historically, and they are mentioned here only academically. Anaerogenic (non–gas-producing) strains of this serotype are re-

TABLE 3. *Separation of* Shigella *spp. and* E. coli

Characteristic	*Shigella* spp.	*S. sonnei*	*E. coli*	Inactive *E. coli*
Gas (from glucose)	–[a]	–	+	–
LDC	–	–	+	V
Motility	–	–	+	–
ONPG	–[b]	+	+	V
Mucate	–	V	+	V
Sodium acetate	–[c]	–	+	V

[a]*S. flexneri* type 6 and *S. boydii* types 13 and 14 are positive.
[b]*S. boydii* is rarely positive.
[c]*S. flexneri* type 4a is rarely positive.
LDC, lysine decarboxylase; ONPG, *o*-nitrophenyl-β-D-galactopyranosidase; –, <10% positive; +, >90% positive; V, 10%–90% positive.

ferred to as biotype Boyd 88. Aerogenic strains producing gas from both glucose and mannitol are termed biotype Manchester, whereas those producing gas from glucose only belong to the Newcastle biotype.

Immunodiagnostics

Serotyping of shigellae is based only on the presence of somatic cell surface antigens (all strains are nonmotile). Four subgroups are known: A, B, C, and D. All except subgroup D contain a varying number of serotypes, and subgroup B is broken down further into subserotypes (see Table 1). It is generally accepted that subgroup A *(S. dysenteriae)* contains 13 serotypes, numbered 1 through 13. Ansaruzzaman and colleagues[49] proposed that two provisional serotypes (strains characterized but not yet accepted and assigned a serotype number), E22383 and E23507, be promoted to serotype status as *S. dysenteriae* types 14 and 15, respectively. Subgroup B *(S. flexneri)* is by far the most complicated, serologically speaking. Although there are only six types, three group factors (3,4; 6; and 7,8) further divide these into 11 subtypes: 1a, 1b, 2a, 2b, 3a, 3b, 4a, 4b, 5a, 5b, and 6.[50] Additionally, strains that do not possess a type factor but do have group factors are designated as X (−:3,4) or Y (−:7,8) variants. For example, *S. flexneri* type 2a agglutinates in subgroup B, type 2, and group factor 3,4 sera. A strain that loses its type 2 antigen but still agglutinates in subgroup B and group factor 3,4 sera would be called *S. flexneri* X variant. Subgrouping and, if desired, typing may be performed with the use of commercial antisera, but complete serotyping is in general available only at large reference laboratories. Subgroup C *(S. boydii)* is the largest group with 18 known types. Subgroup D *(S. sonnei)* contains no types, although *S. sonnei* does produce both smooth (form I) and rough (form II) colonies. Genes encoding the lipopolysaccharide of *S. sonnei* reside on a 180- to 200-kb plasmid.[51] Loss of this unstable plasmid results in rough (form II), avirulent colonies.

Strains of *S. sonnei* do not cross-react significantly with other shigellae or *E. coli,* but some strains of *Plesiomonas shigelloides* may agglutinate in subgroup D serum. Other shigellae exhibit significant cross-reactions between types within a given subgroup, between members of different subgroups, or with *E. coli.* Further, the somatic antigens of 14 serotypes of *Shigella* share exact identity with various *E. coli* antigens[52]; for instance, the O antigen of *S. dysenteriae* type 2 is identical to that of *E. coli* type O112.[53]

Many isolates of shigellae, in particular *S. flexneri* type 6, are nonagglutinable when tested in homologous sera. In these strains, envelope antigens (somatic K antigens) mask the somatic antigen.[1] Gentle heating in a boiling water bath for 10 to 15 minutes inactivates the envelope antigen, and agglutination can then be performed. Care should be taken not to overheat the organisms because they readily become rough during heating. Because roughness can sometimes appear similar to agglutination in the homologous sera, normal rabbit antisera should be tested simultaneously. Most clinical laboratories must rely on commercial sera to confirm their identifications of shigellae. Because these organisms may be difficult to identify given (1) the few biochemicals included in some commercial systems, (2) the great variability in biochemical reactions between types, and (3) the biochemical similarity to *E. coli,* serotyping accuracy is important. Two published evaluations[54,55] indicated that many commercially available antisera are woefully lacking in their ability to accurately serogroup or type *Shigella* strains. In one study,[54] only 31 (42%) of 73 grouping and 30 (59%) of 51 typing sera were satisfactory. In addition to weak agglutination, negative reactions with homologous strains were noted with some sera. In the other survey,[55] only three of six companies (Denka Seiken, Accurate Chemical and Scientific Corp., Westbury, NY; Difco Laboratories, Detroit, MI; and Wellcolex) had a performance standard of 90% accuracy.

Typing

Epidemiologic typing of *Shigella* is used to trace strains during outbreaks or to define transmission within the general population. Serotyping is always the first epidemiologic

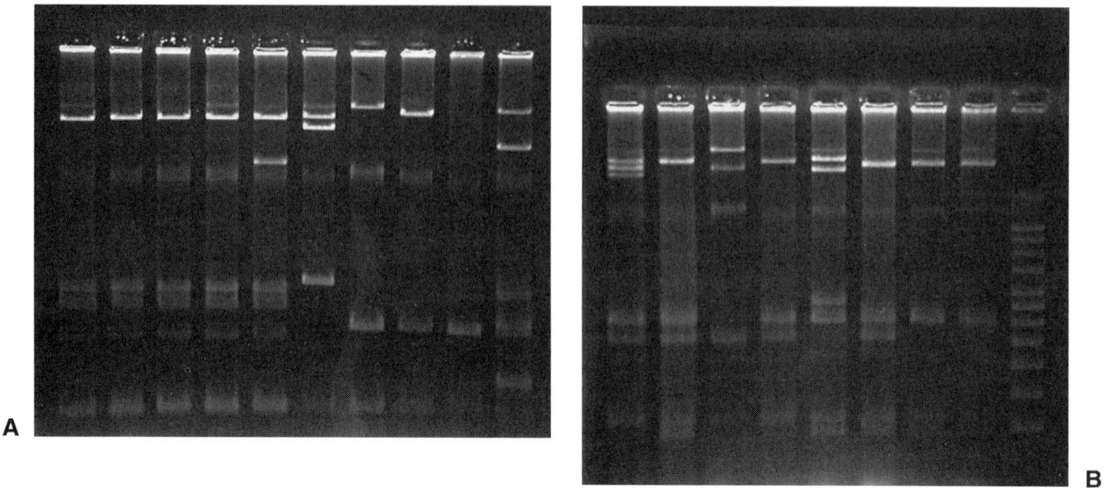

FIG. 2. A, Plasmid profiles linking a food handler to cases of shigellosis in two restaurants. From left to right, lane 1, patient who ate at restaurant 1; lanes 2, 3, and 5, patients who ate at restaurant 2; lane 3, food handler who worked at both restaurants; lanes 6 through 9, community controls of *S. sonnei* isolated during the same time frame; lane 10, molecular weight standards ranging from 120 to 1 Mda. **B,** Plasmid analysis of a case of laborlatory cross-contamination. Left to right, lanes 1 to 6, *S. sonnei* isolates from the previous month, none of which were related to the cross-contamination; lane 7, *S. sonnei* isolate from the bone graft; lane 8, isolate of *S. sonnei* from stool processed by an automatic inoculator just prior to processing the bone sample; lane 9, lambda ladder.

test performed, but the prevalence of *S. flexneri* and *S. sonnei* mandates use of other techniques to further define these strains. Techniques used include phage typing,[56] antibiograms,[57,58] plasmid fingerprinting,[57,58] ribotyping,[59,60] and pulsed-field gel electrophoresis.[57] Antibiograms have become less useful because of widespread resistance among strains. However, Yagupsky and coworkers[61] found antibiograms more useful than plasmid profiles or plasmid DNA restriction endonuclease analysis in following a prolonged epidemic of *S. sonnei.* Both Casalino and associates[58] and Litwin and coworkers[57] found plasmid fingerprinting a useful tool for tracing individual strains. Although Litwin and colleagues[57] noted that others have not found plasmid profiling useful, our laboratory has been able to track all *Shigella* outbreaks, except for those involving *S. flexneri* 2a, with this technique (unpublished data).

Fig. 2 illustrates the usefulness of plasmids epidemiologically. Fig. 2A is a plasmid gel (alkaline lysis extraction, 0.75% agarose electrophoresed 2.5 hours at 100 V) of an outbreak involving two restaurants located near each other with a potential common food handler. The food handler and patients who ate at either of the two restaurants all had the same profile, confirming the suspected link between cases. Fig. 2B represents a case of cross-contamination in a hospital laboratory. When checked for sterility before surgery, bone to be used in a graft yielded *S. sonnei.* Sample processing was performed on an automatic inoculator, and a stool sample, positive for *S. sonnei,* had been inoculated immediately before the bone sample. Plasmid analysis of the bone and stool isolates indicated that they were identical. Six other isolates of *S. sonnei* isolated that month had profiles that differed both from each other and from the bone isolate.

Litwin's group[57] also found pulsed-field gel electrophoresis to be an excellent means of tracing outbreak strains but less useful than plasmid analysis in determining the relations among sporadic community cases. Ribotyping has been used to trace sequence variations within *S. sonnei* strains collected from four continents over a period of 41 years.[59] Using ribotyping and a Shiga toxin probe, Strockbine and coinvestigators[60] determined that there had been little mingling among strains of *S. dysenteriae* type 1 isolated from Africa, Asia, North America, and Central America. In the same study, they showed that the epidemic of *S. dysenteriae* type 1 in Mexico in 1988 was caused by a strain that had been present in Central America since 1962.

PATHOGENESIS

Gastric acidity is an effective host defense mechanism against enteric pathogens. Generally, massive numbers of organisms must be ingested to allow for survival of sufficient bacteria to initiate infection in the intestine. From the small inoculating dose required to achieve shigellosis in human volunteers one would anticipate that *Shigella* are acid resistant. This was confirmed by Gorden and Small,[62] who found 44% of *S. flexneri* to be viable after 2 hours at pH 2.5, with 6% still recoverable after 5 hours. Acid resistance was fully expressed only in the stationary growth phase, and acid-exposed organisms lost their invasive ability. The authors proposed that excreted organisms, in response to a nonhuman environment, enter the stationary phase and thus become acid resistant. When ingested by a new host they readily survive gastric acidity. Lack of disease in the ileum may be explained by the need for several generations of growth (4 hours at pH 7.0 in vitro) before the shigellae regain their invasive capability.

Shigellae cause infection by invading the colonic mucosa.[12,63,64] Incapable of penetrating the microvillous brush border, these organisms enter via M cells and through the intercellular junctions of epithelial cells.[12,65–67] Entry is gained when the bacterium activates actin filaments in the host cell membrane, causing the formation of projections that eventually coalesce, phagocytizing the organism into a vacuole.[12,65,67] After uptake, the vacuole is quickly lysed (within 15–60 minutes) and the organism multiplies intracellularly in the rich medium of the cytoplasm.[64,67] During the multiplication phase, actin is again polymerized by the bacterium. It is accumulated at one end of the cell, forming an actin tail several microns in length that propels the organism, through protrusions, into adjacent cells.[65] As the tip of the protrusion penetrates an adjacent cell, a double membrane is formed from the membranes of the original and the newly invaded cell. This double vacuole is lysed and the process repeats itself in the adjoining cell, resulting in the intercellular spread of the organism and the formation of a lesion. Lesion formation, therefore, may ensue from the invasion of a single host cell by a single bacterium. *Shigella* may also move within the cell by an organelle-like motion in which the organisms bind to actin stress filaments by way of an actin-binding protein. This movement appears to be involved with intracellular multiplication and spread rather than intercellular spread.[67] After multiplication and invasion of an adjacent cell, the host cell dies by an unknown mechanism. In the in vitro macrophage-like cell line J774, cell death begins within 1 hour after infection, and all cells in the monolayer are dead within 4 hours.[64] A decrease in intracellular adenosine triphosphate (ATP) has been noted and may indicate interference with the host cell mitochondria, leading to its death.[64]

The survival of *Shigella* in the human host and the complex process of invasion require the functioning of both chromosomal genes and genes located on a 220-kb plasmid.[12,64,65,67] Chromosomally carried virulence factors include genes encoding for the synthesis of specific somatic antigens (eg, *rfb*) and an iron-binding siderophore, aerobactin (*iuc*A through *iuc*D), among other factors.[12] In addition, genes controlling regulation of plasmid-borne virulence factors are located on the chromosome (eg, *kcp*A, *vir*R).[12,67]

The 220-kb plasmid contains a 30- to 37-kb region[12,64,67] that, when cloned and transformed into *E. coli* K12, confers the ability to invade epithelial cells. Among the gene products encoded on the plasmid are invasion plasmid antigens, Ipa B, C, D, and A.[12,64,67] These genes are temperature regulated and are expressed only at 37°C; strains grown at

30°C are avirulent both in vivo and in vitro. Although the exact functions of the *ipa* gene products are yet to be worked out, they appear to trigger the formation of membrane projections by the host cell, which result in internalization of the bacterium. Strains with mutations in *ipa*B, *ipa*C, or *ipa*D adhere to the host cell but cannot induce actin polymerization, and the bacterium is not internalized.[65] Proteins encoded by *ipa* genes are released within 5 minutes after contact with cultured epithelial cells.[67] After entrance to the cell is effected, the lysis of the membrane vacuole occurs by means of a contact-dependent hemolysin encoded by the *ipa*B gene.

The plasmid also contains the loci for three additional genes (*mxi*C, *mxi*A, and *mxi*B) that are responsible for proteins involved in the transport and insertion of invasion plasmid antigens within the bacterial cell surface and the surrounding medium.[12,64,67] These genes, like the *ipa* genes, are temperature regulated. The last process of invasion, intracellular spread, can also be attributed to the plasmid. This includes both the formation of actin tails by polymerization and the lysis of membranes after penetration of adjacent cells. Actin tail formation is controlled by the gene *ics*A (identical to *vir*G), and a mutation at this site results in the inability of the bacterium to penetrate adjacent cells. The *ics*A gene encodes for a 120-kDa protein that is localized at the junction between the bacterium and the actin tail as well as throughout the tail itself.[65] Since *ics*A hydrolyzes ATP, it may provide energy for propulsion. Disruption of the double membrane surrounding organisms that have penetrated adjoining cells is carried out by the gene *ics*B.

Demonstration of invasion in the laboratory no longer requires an animal model. The Séreny test (instillation of shigellae into the cornea of a guinea pig eye, resulting in conjunctivitis) has, for the most part, been replaced by tissue culture assays.[68] *Shigella* invade a number of different epithelial cell lines of both human and nonhuman origin,[64,69] including Chinese hamster ovary (CHO), HEp-2, Madin-Darby canine kidney (MDCK), and pulmonary endothelial cell lines. Day and associates[68] have provided a detailed description of the in vitro invasion assay.

Shiga toxin is elaborated by *S. dysenteriae* type 1 and rare strains of *S. flexneri.* It plays a role in the development of microangiopathic lesions associated with hemorrhagic colitis and the symptoms of hemolytic-uremic syndrome.[12,70] It is a 70-kDa protein composed of one A subunit and five B subunits. The B subunits bind the toxin to the mammalian cell receptor, globotriaosylceramide. The A subunit must first be activated by a proteolytic process to form an A1 fragment, which is the active moiety. The A1 fragment removes adenine from a single specific adenosine in the 28S portion of the 60S ribosome, inhibiting protein synthesis and causing cell death. Neurologic symptoms, encountered only in rabbit and mouse models, are probably caused by the action of the toxin on brain and spinal cord vascular systems. Two enterotoxins, ShET1 and ShET2, have been described in all species of *Shigella.* They alter electrolyte transport, causing fluid secretion in rabbit ligated ileal loops, and neutralizing antibodies are formed in patients. However, no exact role has yet been ascribed for them in the initial watery phase of illness.[12]

ANTIMICROBIAL SUSCEPTIBILITY

Unlike salmonellosis, *Shigella* infections are usually treated. Arguments for treatment, even in milder infections, include shortened duration of disease and eradication of the organism in the stool to prevent further spread of the disease. Counterindications for treatment include cost (particularly in developing countries) and the development of resistance. The rapid emergence of resistant strains has been facilitated by the combination of two factors: the presence of antimicrobial resistance genes on a variety of plasmids, which promotes their spread, and the insertion of these genes into the chromosome, which maintains a stable population.[58]

Resistance to tetracycline in *S. flexneri* and *S. sonnei* is almost universal in developing countries.[58,71–73] Likewise, resistance to ampicillin occurs in 35% to 98% of *S. flexneri* and in 42% to 57% of *S. sonnei.*[72,73] In Somalia, all but 3 of 112 *S. flexneri* tested were multiply resistant.[58] In Thailand and in Israel, resistance to trimethoprim-sulfamethoxazole (SXT) is 88% and 29%, respectively, for *S. flexneri* and 72% and 83%, respectively, for *S. sonnei.*[72,73] Aleksić and associates[71] found resistance in *Shigella* spp. imported by travelers to developing nations to be 93.5% overall, compared with 62.7% in domestically acquired strains ($P<0.001$). Of 255 strains, 168 (66%) were resistant to three or more agents, usually some combination of ampicillin, chloramphenicol, streptomycin, sulfonamide, tetracycline, or SXT. All strains were susceptible to ofloxacin, enoxacin, and all second- and third-generation cephalosporins except cefaclor. However, 79 strains (31%) were resistant to ampicillin-sulbactam and 25 (10%) to piperacillin. In Israel, the study of Ashkenazi and coworkers[72] determined resistance to SXT in *S. sonnei* to be 2.7 times greater than in *S. flexneri* (83% versus 29%, respectively; $P<0.0001$). *S. sonnei* was also 1.6 times more resistant than *S. flexneri* to ampicillin alone and twice as resistant to the combination of ampicillin and SXT; percentage differences for both resistance rates were statistically significant. The authors speculated that the increase in *S. sonnei* strains in their country may have been caused in part by increased resistance in this species. In the United States, 20% of *S. flexneri* and 50% of *S. sonnei* are resistant to ampicillin.[74]

Most studies in which it has been tested show susceptibility to nalidixic acid (NA),[73,74] although resistance is increasing in *S. dysenteriae* type 1 strains in Bangladesh.[75] In other studies, as in the report by Aleksić and colleagues,[71] susceptibility to second- and third-generation cephalosporins remains.[73,75,76] These antimicrobials have not proved effective in treatment, however, because they do not significantly reduce symptoms and/or eradicate the organism from the stool.[73,75] In clinical trials, NA was as effective as norfloxacin in effecting a clinical response but eradicated only 70% of shigellae in stool, compared with 100% for norfloxacin. Pickering[74] indicated that ciprofloxacin was effective in adults even for strains resistant in vitro. For the time being, *Shigella* remain susceptible to quinolones.[71,73,77] They may be the drug of last resort in severely ill patients whose disease is caused by resistant strains, even in children, where a regimen of short duration (to avoid arthropathy) may save a life.

CONCLUSIONS

Any hope of eradicating this organism appears to rely on the development of an effective vaccine. Because the human intestine is the only reservoir for *Shigella,* such a vaccine could add this organism to the list of extinct species. Recent *S. flexneri* and *S. sonnei* LPS-conjugate vaccines have been developed. These vaccines trigger immunoglobulin G anti-LPS antibodies that, if leaked into the bowel lumen, would kill organisms at and within the mucosa.[11] Of the many attempted vaccines this one appears both safe and satisfactorily immunogenic, but its efficacy is as yet unproved. The rapid and widespread emergence of resistance to antimicrobials will be difficult to control without a vaccine, since treatment for shigellosis is usually required. Effective sanitary systems and provisions for safe drinking water in developing countries would go a long way toward eradicating this organism and many other enteric pathogens, but such infrastructure is extremely costly. Teaching good hygienic practices, particularly to care givers (eg, in day care centers) and food industry workers, would eliminate outbreaks of *Shigella* and other diarrheal agents and should be an achievable goal.

REFERENCES

1. Ewing EH. *Edwards and Ewing's identification of Enterobacteriaceae.* 4th ed. New York: Elsevier, 1986.
2. Rowe B, Gross RJ. *Shigella* Castellani and Chalmers 1919, 936.[AL] In: Krieg NR, Holt JG, eds. *Bergey's manual of systematic bacteriology.* Baltimore: Williams & Wilkins, 1984;423–427.
3. International *Enterobacteriaceae* Subcommittee. Nomenclature and classification of the dysentery bacillus. *Int Bull Bacteriol Nomen Taxon* 1951;1:150–167.
4. Brenner DJ, Fanning GR, Miklos GV, Steigerwalt AG.

Polynucleotide relatedness among *Shigella* species. *Int J Syst Bacteriol* 1973;23:1–7.
5. Brenner DJ, Steigerwalt AG, Walthen HG, Gross RJ, Rowe B. Confirmation of aerogenic strains of *Shigella boydii* 13 and further study of *Shigella* serotypes by DNA relatedness. *J Clin Microbiol* 1982;16:432–436.
6. Dodd CR, Jones D. A numerical taxonomic study of the genus *Shigella*. *J Gen Microbiol* 1982;128:1933–1957.
7. Watanabe H, Okamura N. The genus *Shigella*. In: Starr MPO, Stolp H, Trüper HG, Schlegel HG, eds. *The prokaryotes: A handbook on habitats, isolation and identification of bacteria*. Berlin: Springer-Verlag, 1981;2754–2759.
8. Cooper JE, Needham JR. An outbreak of shigellosis in laboratory marosets and tamarins (Family: Callithricidae). *J Hyg Camb* 1976;76:415–425.
9. Keusch GT, Bennish ML. Shigellosis: Recent progress, persisting problems and research issues. *Pediatr Infect Dis J* 1989;8:713–719.
10. DuPont HL, Pickering LK. *Infections of the gastrointestinal tract: Microbiology, pathophysiology, and clinical features*. New York: Plenum Medical Book Company, 1980.
11. Levine MM, DuPont HL, Khodabandelou M, Hornick RB. Long-term shigella-carrier state. *N Engl J Med* 1973;288:1169–1171.
12. Keusch GT, Acheson DWK. *Shigella* infection. In: Paradise LJ, Bendinelli M, Friedman H, eds. *Enteric infections and immunity*. New York: Plenum Press, 1996.
13. Rosenberg ML, Hazlet KK, Schaefer J, Wells JG, Pruneda RC. Shigellosis from swimming. *JAMA* 1976; 236:1849–1852.
14. Pickering LK. The day care center diarrhea dilemma. *Am J Public Health* 1986;76:623–624.
15. Johnson K, Boase J, Helgersen SD, Kobayashi JM. Shigellosis in day-care centers—Washington, 1983. *MMWR Morb Mortal Wkly Rep* 1984;33:250–251.
16. Koo D, Maloney K, Tauxe R. Epidemiology of diarrheal disease outbreaks on cruise ships, 1986 through 1993. *JAMA* 1996;275:545–547.
17. William DC, Felman YM, Marr JS, Shookhoff HB. Sexually transmitted enteric pathogens in male homosexual population. *N Y State J Med* 1977;2050–2052.
18. Parsonnet J, Greene KD, Gerber AR, Tauxe RV, Vallejo Aguilar OJ, Blake PA. *Shigella dysenteriae* type 1 infections in US travelers to Mexico, 1988. *Lancet* 1989;ii.
19. Taylor DN, Bodhidatta, Echeverria P. Epidemiologic aspects of shigellosis and other causes of dysentery in Thailand. *Rev Infect Dis* 1991;13(Suppl):S226–S230.
20. Mathan VI, Mathan MM. Intestinal manifestations of invasive diarrheas and their diagnosis. *Rev Infect Dis* 1991;13(Suppl):S311–S313.
21. Halpern Z, Dan M, Giladi M, Schwartz I, Sela O, Levo Y. Shigellosis in adults: Epidemiologic, clinical and laboratory features. *Medicine* 1989;68:210–217.
22. Schulman SK. Clinical experience during a large *Shigella sonnei* outbreak: When the "hoofbeat" were really "zebras." *Pediatr Infect Dis J* 1988;7:296–297.
23. Butler T, Speelman P, Kabir I, Banwell J. Colonic dysfunction during shigellosis. *J Infect Dis* 1986;154: 817–824.
24. Butler T, Dunn D, Dahms B, Islam M. Causes of death and the histopathologic finding in fatal shigellosis. *Pediatr Infect Dis J* 1989;8:767–772.
25. Bennish ML, Wojtyniak BJ. Mortality due to shigellosis: Community and hospital data. *Rev Infect Dis* 1991; 13(Suppl):S245–S251.
26. Bennish ML. Potentially lethal complications of shigellosis. *Rev Infect Dis* 1991;13(Suppl):S319–S324.
27. Mulligan K, Nelson S, Friedman HS, Andrews PI. Shigellosis-associated encephalopathy. *Pediatr Infect Dis J* 1992;11:889–890.
28. Papasian CJ, Enna-Kifer S, Garrison B. Symptomatic *Shigella sonnei* urinary tract infection. *J Clin Microbiol* 1995;33:2222–2223.
29. Patel R, Osman DR, Steckelberg JM, Dekutoski MB, Wilson WR. Vertebral osteomyelitis due to *Shigella boydii*. *Clin Infect Dis* 1996;22:863–864.
30. Tobias JD, Starke JR, Tosi MF. *Shigella* keratitis: A report of two cases and a review of the literature. *Pediatr Infect Dis J* 1987;6:79–81.
31. Stoll DM. Cutaneous shigellosis. *Arch Dermatol* 1986; 122:22.
32. Adkins HJ, Santiago LT. Increased recovery of enteric pathogens by use of both stool and rectal swab specimens. *J Clin Microbiol* 1987;25:158–159.
33. Hentges DJ. Inhibition of *Shigella flexneri* by the normal intestinal flora. *J Bacteriol* 1969;97:513–517.
34. Wells JG, Morris GK. Evaluation of transport methods for isolating *Shigella* spp. *J Clin Microbiol* 1981;13: 789–790.
35. Echeverria P, Sethabutr O, Pitarangsi C. Microbiology and diagnosis of infections with *Shigella* and enteroinvasive *Escherichia coli*. *Rev Infect Dis* 1991;13(Suppl): S220–S225.
36. Altwegg M, Buser J, von Graevenitz A. Stool cultures for *Shigella* spp: Improved specificity by using MacConkey agar with xylose. *Diagn Microbiol Infect Dis* 1996;24:121–124.
37. Bouvet PJM, Jeanjean S. Evaluation of two colored latex kits, the Wellcolex Colour Salmonella test and the Wellcolex Colour Shigella test for serological grouping of *Salmonella* and *Shigella* species. *J Clin Microbiol* 1992;30:2184–2186.
38. Kocka FE, Dorigan FD, Abbasi QT, Swiatlo AL, Hubbard-Shepard M. Clinical evaluations of the Wellcolex Colour Shigella and Salmonella test. *Diagn Microbiol Infect Dis* 1992;15:1–4.
39. Metzler J, Nachamkin I. Evaluation of a latex agglutination test for the detection of *Salmonella* and *Shigella* spp. by using broth enrichment. *J Clin Microbiol* 1988;26:2501–2504.
40. McGowan KL, Rubenstein MT. Use of a rapid latex agglutination test to detect salmonella and shigella antigens from gram-negative enrichment broth. *Am J Clin Pathol* 1989;92:679–682.
41. Islam D, Tzipori S, Islam M, Lindberg AA. Rapid detection of *Shigella dysenteriae* and *Shigella flexneri* in faeces by an immunomagnetic assay with monoclonal antibodies. *Eur J Clin Microbiol Infect Dis* 1993;12:25–32.
42. Frankel G, Riley L, Giron JA, et al. Detection of *Shigella* in feces using DNA amplification. *J Infect Dis* 1990;161:1252–1256.
43. Wauters G, Boel A, Voorn GP, et al. Evaluation of a new identification system, Crystal Enteric/Non-Fermenter, for gram-negative rods. *J Clin Microbiol* 1995;33:845–849.
44. Kitch TT, Jacobs MR, Appelbaum PC. Evaluation of RapID onE system for identification of 379 strains in the family *Enterobacteriaceae* and oxidase-negative,

gram-negative nonfermenters. *J Clin Microbiol* 1994; 32:931–934.
45. O'Hara CM, Tenover FC, Miller JM. Parallel comparison of accuracy of API 20E, Vitek GNI, MicroScan Walk/Away Rapid ID, and Becton Dickinson Cobas MicroID-E/NF for identification of members of the family *Enterobacteriaceae* and common gram-negative, non–glucose-fermenting bacilli. *J Clin Microbiol* 1993;31:3165–3169.
46. Kelly MT, Leicester C. Evaluation of the Autoscan Walkaway system for rapid identification and susceptibility testing of gram-negative bacilli. *J Clin Microbiol* 1992;30:1568–1571.
47. Rhoden DL, Schable B, Smith PB. Evaluation of PASCO MIC-ID system for identifying gram-negative bacilli. *J Clin Microbiol* 1987;25:2363–2366.
48. Toledo MRF, Silva RM, Trabulsi LR. Sach's "Enterobacterium A12" is an aerogenic variant of *Shigella boydii* 14. *Int J Syst Bacteriol* 1981;31:242–244.
49. Ansaruzzaman M, Kibriya KMG, Rahman A, et al. Detection of provisional serovars of *S. dysenteriae* and designation as *S. dysenteriae* serotypes 14 and 15. *J Clin Microbiol* 1995;33:1423–1425.
50. Brenner DJ. Recommendations on recent proposals for the classification of shigellae. *Int J Syst Bacteriol* 1984; 34:87–88.
51. Viret J, Bruderer U, Lang AB. Characterization of the *Shigella* serotype D *(S. sonnei)* O polysaccharide and the enterobacterial R1 lipopolysaccharide core by use of mouse monoclonal antibodies. *Infect Immun* 1992;60: 2741–2747.
52. Rowe B, Gross RJ, Guiney M. Antigenic relationships between *Escherichia coli* O antigens O149 to O163 and *Shigella* O antigens. *Int J Syst Bacteriol* 1976;26:76–78.
53. Lindberg AA, Kärnell A, Weintraub A. The lipopolysaccharide of *Shigella* bacteria as a virulence factor. *Rev Infect Dis* 1991;13(Suppl 4):S279–S284.
54. Evins GM, Gheesling LL, Tauxe RV. Quality of commercially produced *Shigella* serogrouping and serotyping antisera. *J Clin Microbiol* 1988;26:438–442.
55. Lefebvre J, Gosselin F, Ismaïl J, Lorange M, Lior H, Woodward D. Evaluation of commercial antisera for *Shigella* serogrouping. *J Clin Microbiol* 1995;33: 1997–2001.
56. Pruneda RC, Farmer III JJ. Bacteriophage typing of *Shigella sonnei*. *J Clin Microbiol* 1977;5:66–74.
57. Litwin CM, Leonard RB, Carroll KC, Drummond WK, Pavia AT. Characterization of endemic strains of *Shigella sonnei* by use of plasmid DNA analysis and pulsed-field gel electrophoresis to detect patterns of transmission. *J Infect Dis* 1997;175:864–870.
58. Casalino M, Nicoletti M, Salvia A, et al. Characterization of endemic *Shigella flexneri* strains in Somalia: Antimicrobial resistance, plasmid profiles, and serotype correlation. *J Clin Microbiol* 1994;32: 1179–1183.
59. Karalolis DKR, Lan R, Reeves PR. Sequence variation in *Shigella sonnei* (Sonnei), a pathogenic clone of *Escherichia coli*, over four continents and 41 years. *J Clin Microbiol* 1994;32:796–802.
60. Strockbine NA, Parsonnet J, Green K, Kiehlbauch JA, Wachsmuth IK. Molecular epidemiologic techniques in analysis of epidemic and endemic *Shigella dysenteriae* type 1 strains. *J Infect Dis* 1991;163:406–409.
61. Yagupsky P, Loeffelholz M, Bell K, Menegus MA. Use of multiple markers for investigation of an epidemic of *Shigella sonnei* infections in Monroe County, New York. *J Clin Microbiol* 1991;29:2850–2855.
62. Gorden J, Small PLC. Acid resistance in enteric bacteria. *Infect Immun* 1993;61:364–367.
63. Sansonetti PJ. Genetic and molecular basis of epithelial cell invasion by *Shigella* species. *Rev Infect Dis* 1991;13(Suppl 4):S285–S292.
64. Maurelli AT. *Shigella* inside and out: Lifestyles of the invasive and dysenteric. *ASM News* 1992;58:603–608.
65. Goldenberg MB, Sansonetti PJ. *Shigella* subversion of the cellular cytoskeleton: A strategy for epithelial colonization. *Infect Immun* 1993;61:4941–4946.
66. Sansonetti PJ, Arondel J, Cantey JR, Prévost MC, Huerre M. Infection of rabbit Peyer's patches by *Shigella flexneri:* Effect of adhesive or invasive bacterial phenotypes on follicle-associated epithelium. *Infect Immun* 1996;64:2752–2764.
67. Ménard R, Dehio C, Sansonetti PJ. Bacterial entry into epithelial cells: The paradigm of *Shigella*. *Trends Microbiol* 1996;4:220–225.
68. Day NP, Scotland SM, Rowe B. Comparison of an HEp-2 tissue culture test with the Séreny test for detection of enteroinvasiveness in *Shigella* spp. and *Escherichia coli*. *J Clin Microbiol* 1981;13:596–597.
69. Finlay BB, Falkow S. Comparison of the invasion strategies used by *Salmonella cholerae-suis, Shigella flexneri* and *Yersinia enterocolitica* to enter cultured animal cells: Endosome acidification is not required for bacterial invasion or intracellular replication. *Biochimie* 1988;70:1089–1099.
70. O'Brien AD, Holmes RK. Shiga and shiga-like toxins. *Microbiol Rev* 1987;51:206–220.
71. Aleksić S, Katz A, Aleksić V, Bockemühl J. Antibiotic resistance of shigella strains isolated in the Federal Republic of Germany 1989–1990. *Zentralbl Bakteriol* 1993;279:484–493.
72. Ashkenazi S, May-Zahav M, Dinari G, Gabbay U, Zilberberg R, Samra Z. Recent trends in the epidemiology of *Shigella* species in Israel. *Clin Infect Dis* 1993;17:897–899.
73. Lolekha S, Vibulbandhitkit S, Poonyarit P. Response to antimicrobial therapy for shigellosis in Thailand. *Rev Infect Dis* 1991;13(Suppl 4):S342–S346.
74. Pickering LK. Emerging antibiotic resistance in enteric bacterial pathogens. *Semin Pediatr Infect Dis* 1996;7: 272–280.
75. Gomez HF, Cleary TG. Drug-resistant *Salmonella, Shigella,* and diarrhea-associated *Escherichia coli. Pediatr Infect Dis J* 1996;7:212–222.
76. Fuchs PC, Barry AL, Brown SD, AST Surveillance Group. Survey of antimicrobial activity of four commonly used third-generation cephalosporins tested against recent bacterial isolates from ten American medical centers, and assessment of disk diffusion test performance. *Diagn Microbiol Infect Dis* 1996;24:213–219.
77. Thornsberry C. Susceptibility of clinical bacterial isolates to ciprofloxacin in the United States. *Infection* 1994;22(Suppl 2):S80–S89.

6
The Salmonellae

TAXONOMY

Although many genera within the *Enterobacteriaceae* have undergone protracted, laborious nomenclatural changes, *Salmonella* takes honors in this category. The genus name, *Salmonella*, was adopted in 1900 by Lignières in honor of Salmon, who isolated the organism now known as *Salmonella* serotype Choleraesuis, from pigs with hog cholera.[1] Although the epithet *Salmonella* has remained unchanged for 97 years, the species designation has been as volatile as the genus name has been constant, with at least 13 proposals or changes in 53 years. In the interest of brevity, and because many proposals have been similar in nature, only the more significant changes are covered here.

Initially, salmonellae were named for their clinical conditions and/or the animal host from which they were isolated (eg, the name "choleraesuis," from hog cholera).[2,3] This method of nomenclature was abandoned when it became apparent that many strains were not host specific or were not responsible for the disease for which they were named. As knowledge of the antigenic properties of *Salmonella* expanded, a schema was devised by Kauffmann and White.[3] *Salmonella* were then recognized by their specific somatic (O) and flagellar (H) antigen patterns. As new serotypes were discovered, they were named after the location from which they were isolated. Between 1960 and 1966, Kauffmann[3] divided the salmonellae into four subgenera; however, he stopped short of a formal proposal of nomenclature, and serotype designations were used as if they were species.

In 1963, Ewing proposed that salmonellae be limited to three species, a concept originally put forth by Borman in 1944.[3,4] The three species of *Salmonella* proposed by Ewing[4] were *S. choleraesuis, S. typhi,* and *S. enteritidis. S. choleraesuis* was retained because it was the type strain for the genus and the oldest valid name in the literature, *S. typhi* because of its clinical significance, and *S. enteritidis* because it had the third oldest standing after the other two. All other salmonellae were serotypes of *S. enteritidis.* At this time, Ewing[5] also proposed the genus *Arizona arizonae;* however, in 1969, the species designation was changed to *Arizona hinshawii.*[3] Previously, these strains, described by Caldwell and Ryerson in 1939, had been called *S. arizonae* until 1947 when the phrase Arizona group was coined.[4]

In 1973, *Salmonella* nomenclature took a huge step forward when Crosa and colleagues[6] analyzed 52 salmonella and arizona strains by DNA hybridization and concluded that they formed a single species with seven subgroups. Further, the level of DNA relatedness of these subgroups was consistent with the definition of a subspecies (all of which were distinguishable by biochemical tests). After this revelation, and in accordance with the *International Code of Nomenclature of Bacteria,* in 1982, the name of the type species, *S. choleraesuis,* was proposed as the only *Salmonella* species designation.[3] Subspecies (also called subgenera) names were as follows: I, *choleraesuis;* II, *salamae;* IIIa, *arizonae;* IIIb, *diarizonae;* IV, *houtenae;* V, *bongori;* VI, *indica.* All salmonellae then became serotypes of one of these subspecies (see next paragraph). The next landmark in *Salmonella* nomenclature came in 1987, when LeMinor and Popoff[3] recommended that the species name *S. choleraesuis* be replaced with *S. enterica.* This change avoided the inherent confusion arising from use of a species name that is also a serotype. Regretfully, this proposal was rejected by the Judicial Commission of the International Committee on Systematic Bacteriology in 1994,[7] leaving in place the nomenclature proposed previously in 1982 (ie, *S. choleraesuis* as the species name).

Therefore, the official name of the serotype *typhimurium* is *Salmonella choleraesuis* ssp. *choleraesuis* serotype Typhimurium. For convenience and clarity, however, serotypes are reported as *Salmonella* serotype Typhimurium or *Salmonella* Typhimurium. The serotype name is capitalized but not italicized.[2,8] Serotypes designated by an antigenic formula (all serotypes of subspecies II through VI discovered after 1966) are given as *Salmonella* followed by the subspecies number and the formula (eg, *Salmonella* IV, 48:z_4z_{32}:–).

In 1989, a proposal to elevate subspecies V *(bongori)* to the species level as *Salmonella bongori* was made by Reeves and associates.[9] This group used multilocus enzyme electrophoresis to identify *bongori* as a separate species; however, this finding was also supported by previous DNA relatedness data. The nomenclature of this genus is recapped in Table 1.

EPIDEMIOLOGY

While some serotypes of *Salmonella* have specific (eg, *S.* Gallinarum in poultry) or strong (eg, *S.* Dublin in cattle) associations with certain hosts, most are ubiquitous. They are capable of proliferating in a wide variety of domestic or wild, cold- or warm-blooded animals, where they may cause infections or simply reside. Although the natural habitat of *Salmonella* is the gastrointestinal tract, most can survive in the environment for prolonged periods. Once defecated into the environment by animals or humans, they can survive for weeks in water or for years in soil,[8] providing a ready reservoir for infection or reinfection. The ability of these organisms to grow at a wide range of temperatures (8°–45°C) and their rapid generation time[10] make *Salmonella* a formidable opponent. Although drying and freezing are generally a deterrent to survival, neither is a guarantee of sterility in the case of *Salmonella:* 20% of *S.* Typhimurium could be recovered from chow mein even after 9 months of storage at –25°C.[10]

TABLE 1. *Highlights in the evolution of* Salmonella *nomenclature*

Date	Proposal
1900	Lignierès establishes the genus *Salmonella*
—	Organisms named for host or clinical condition
—	Kauffmann-White antigenic schema expanded; serotypes used as species
1944	Borman introduces the concept of three species
1963	Ewing proposes *S. typhi, S. choleraesuis,* and *S. enteritidis* as the three species of *Salmonella*; all other salmonellae are serotypes of *S. enteritidis*
—	"Arizona group" named *Arizona hinshawii*
1960–1966	Kauffmann divides *Salmonella* into four subgenera
1973	Crosa et al. determine by DNA relatedness that all *Salmonella* belong to a single species
—	Six subgenera (including *Arizona*) are established
1982–1986	LeMinor et al. propose *S. choleraesuis* as type, and only, species for the genus, with six subgenera or subspecies
1987	LeMinor and Popoff propose *S. enterica* as the single species for *Salmonella*

Data from references 1, 3–7.

While *Salmonella* remains an important pathogen in developing countries of the world, it has also burgeoned in the last 20 years in industrialized countries. Changes in social and food practices have resulted in a rise of infections to the point that an estimated 1% of the population is diagnosed with *Salmonella* yearly.[11] Through the 1960s only 19% of cases of salmonellosis could be traced to commercial establishments. By 1982, this figure was 62%, representing a threefold increase, and infections acquired at home had increased from 12% to 22%.[10] The rise in salmonellosis is directly attributable to the demand for ready-to-eat, exotic, and unpasteurized foods and the increased volume or scale of food production.[11,12] The impact of salmonellosis on society is enormous in terms of both human suffering and financial losses. Although the mortality rate is relatively low (1% or less),[13] costs associated with deaths due to *Salmonella* infection are estimated at $372 million to $744 million in the United States alone.[14] Hospitalization costs range from $148 million to $157 million, physician care costs $30 million to $69 million, and infections not involving professional medical care cost $421 million to $424 million. This adds up to a grim $983 million to $1.4 billion dollars annually, not including losses to (1) commercial establishments that may be closed, temporarily or permanently; (2) manufacturers for foods removed from the market; (3) farmers for destroyed flocks or herds that are ill or the cause of infection; or (4) the public for regulatory agencies. Effective public health monitoring of *Salmonella* infections using serotyping does lead to interruption of transmission and reduction of the number of infections. For example, in finding and removing a contaminated chocolate product, a rapid public health response saved an estimated 5 deaths, 185 to 230 hospitalizations, and 29,000 to 36,000 cases (figures projected if the product had been left on the market for consumption).[14] Similarly, an outbreak caused by *S.* Heidelberg–contaminated cheese would have been 74% larger, with more than 23,000 additional infections and an added $3 million loss without public health intervention.[15]

Tracking the source of infections in an outbreak can be difficult because of delays between onset of infection and notification of public health officials. Further, the incubation period of 24 to 72 hours between ingestion and onset of symptoms can confuse patients, who often associate the last meal eaten with the cause of their illness. Incubation periods may be affected by the infective dose, which varies greatly in salmonellosis.[10,16] The accepted infective dose for salmonellae is believed to be between 10^4 and 10^8 organisms, figures that were derived from human volunteer studies. Blaser and Newman,[16] however, found that doses ingested in a natural setting were frequently lower; in 6 of 11 outbreaks they reviewed, the infective dose was 10^3 organisms or less. Doses as low as 100 to 250 and 100 to 500 organisms have been associated with contaminated chocolate and cheese, respectively.[16] Higher infective doses generally correlate with very high attack rates and short incubation periods.[16,17] Mintz and colleagues,[17] in studying dose-response effects in an outbreak of *S.* Enteritidis contaminating hollandaise sauce, found that as the dose increased the incubation time decreased from 37 to 17.5 hours ($P=0.006$). The severity of symptoms concomitantly increased with dose ($P=0.0007$), as indicated by significant increases in vomiting, body aches, and weight loss.

Salmonella remains present in the stool even after cessation of symptoms. The median duration of excretion is 5 weeks, and 90% of adults and older children are culture-negative by 9 weeks.[18] The duration of carriage increases in children younger than 5 years of age, to a median of 7 weeks of excretion with only 40% culture-negative at 5 months. It is not unusual for children in this last group to shed *Salmonella* in their stool for up to 1 year. Permanent carriers or persistent excretors are rare, fewer than 1% of those infected.[18] Shedding may be intermittent, although testing of stools, as opposed to rectal swabs, improves isolation; it may be that intermittent shedding is actually an artifact caused by testing variables. Asymptomatic carriers involved in sensitive professions (eg, food handlers, care givers) are rarely involved in the transmission of salmonellae.[18–20] In a review of 396 outbreaks of *Salmonella,* only 2 (0.5%) were attributable to asymptomatic excretors.[18] Although food handlers and care givers should never be allowed to work while they are symptomatic, there is little evidence to justify prevention of asymptomatic persons with formed stools from returning to work.[19] A reasonable approach in employment practices may prevent workers from hiding illnesses in order to continue to work.

Waterborne infections with nontyphoidal *Salmonella* are rare in comparison with foodborne disease, but they are not unheard of. Foodborne infections may be caused by almost any of the known serotypes of salmonellae. Serotypes such as *S.* Typhi, *S.* Paratyphi A and C, and *S.* Sendai, although host-adapted to humans only, cause infection by person-to-person transmission or, more often, by fecal contamination of foods. Some host-adapted serotypes (eg, *S.* Gallinarum in poultry) have yet to cause disease in humans; others, when contracted by humans, often cause bacteremia and other serious infections (eg, *S.* Choleraesuis in swine). Most serotypes, however, are indiscriminate and fail to associate with any particular source. They may be transmitted by humans, domestic poultry and animals raised for consumption or as pets, fish, wild birds and mammals (especially rodents), reptiles, and houseflies. Reptiles have been recognized increasingly as a source of human infection as they have gained in popularity as pets.[21,22] Very young children and infants (we have seen children as young as 10 days) are particularly susceptible to infections from these sources even though they themselves may not come into direct contact with the reptile. *Salmonella* may be transmitted on the hands of siblings or parents or by inanimate objects, as when the same sink is used to wash a reptile (or cage) and then a baby. Their young age puts these patients at risk for severe complications such as bacteremia and meningitis.[22] Reptile-associated infections often involve unusual serotypes, particularly those belonging to subgenera III, II, and IV; in contrast, 99.5% of salmonellosis from other sources is caused by serotypes in subgenus I.[23] Aleksić and associates[23] found that, of 90,201 salmonellae serotyped over a 15-year period, 403 (0.45%) belonged to subgenera II through VI or *S. bongori* (formerly *Salmonella* subgenus V). Of these 403 isolates, 215 were from humans, including two *S. bongori* and one subgenus VI isolate, both of which are extraordinarily rare in humans; it was not indicated whether these isolates were from symptomatic patients. In general, the range of symptoms of these unusual serotypes was the same as that observed for subgenus I serotypes. However, males were almost two times more likely to be infected with these serotypes than females, in contrast to the normal ratio of 1:1.

CLINICAL SIGNIFICANCE

Clinical disease caused by *Salmonella* is usually divided into four syndromes: gastroenteritis, enteric fever, septicemia with or without focalized infection, and asymptomatic infection or carrier state.[13,24,25] Most *Salmonella* infections fall into the category of gastroenteritis, which has been well characterized and is not described in detail here. Diarrheal stools range from a mildly watery consistency to cholera-like, and they may be positive for occult blood or visibly bloody (sometimes grossly so) with mucus. Usually the diarrhea subsides within 3 to 4 days. Fever, up to 39°C (102°F), occurs in about 50% of cases but is usually of short duration (2 days). High or persistent fever usually indicates septicemia[25]; about 1% to 4% of patients have a transient bacteremia. Abdominal pain ranges from mild to severe and, if mesenteric lymph nodes are invaded (mesenteric lymphadentitis), can mimic appendicitis.[26] Development of irritable bowel syndrome, characterized by diarrhea and the urgency to defecate, after *Salmonella* gastroenteritis has been described. McKendrick and Read[27] monitored 38 patients for 1 year after their recovery from salmonellosis and found that 12 patients (31%, mostly women >40 years of age) developed irritable bowel syndrome.

Enteric fever is usually caused by *S.* Typhi or *S.* Paratyphi A or C, but other serotypes can be involved (see Chapter 7). Patients in whom *Salmonella* septicemia develops frequently have underlying illnesses, especially diseases of the reticuloendothelial system such as lymphoma, leukemia, sickle cell disease, cancer, systemic lupus erythematosus, or other collagen-vascular disorders. The frequency of *Salmonella* bacteremia in cancer patients has been reported to be 35%, and in patients with the acquired immunodeficiency syndrome (AIDS), it is as high as 78%.[28] Unlike the general population, 90% of whom have gastrointestinal symptoms before or during their sepsis, only 20% of AIDS patients with salmonellosis present with diarrheal complaints.[28] Extraintestinal infections, including septicemia, occur in 3% to 16% of cases of salmonellosis.[28–31] Wilkins and Roberts[29] reported that blood (34%) and urine (32%) accounted for two thirds of their isolates from extraintestinal sites. In 7 (11%) of their 65 patients, sepsis was associated with a culture-positive focal infection, including chest, spine, meninges, and a blood clot. Twenty-seven percent of their 62 urine isolates were believed to be the result of contamination during collection. Twenty-five patients had cystitis and one had pyelonephritis; half of these experienced diarrhea. Ramos and colleagues,[32] in examining 28 cases of bacteriuria, found 21 (75%) patients were symptomatic with cystitis (n=16), pyelonephritis (n=3), or renal abscess (n=2). All of the patients in their series were severely compromised or had urologic abnormalities, or both. Of the remainder of the extraintestinal isolates of Wilkins and Roberts,[29] 44 were from pus and included patients with cholecystitis, peritonitis, rectal abscesses, and infections of the spleen, scrotum, or Bartholin's glands. Eleven additional cases had localized abscesses only, and three infants were colonized after vertical transmission during birth. A further nine isolates each were obtained from cerebrospinal fluid (in patients of meningitis, four of whom died) and bone (in patients with osteomyelitis-arthritis).

The most common metastatic infections of *Salmonella* are osteomyelitis and cholecystitis.[25] Between 51% and 77% of patients with sickle cell disease develop *Salmonella* osteomyelitis.[33] The source for these infections is generally believed to be asymptomatic carriage in the stool or gallbladder, but Wright and coworkers[33] could not document this carriage in their study of 55 patients with sickle cell disease. In studying 32 episodes of osteomyelitis and 30 of sepsis, they found sterile avascular necrosis of the bone to be a risk factor for development of *Salmonella* osteomyelitis. They proposed that *Salmonella* may be carried asymptomatically within leukocytes responding to the inflammation caused by dead marrow within areas of sterile avascular necrosis and thereby seeded in these vulnerable areas of bone. They also noted none of the patients with osteomyelitis died, probably because of the known association between sickle cell disease and *Salmonella* osteomyelitis which resulted in early intervention. However, almost a quarter of those with sepsis died, usually within 2 days of onset. *Salmonella* osteomyelitis is also associated with trauma from accident or surgery and with malignancies.[13] It usually occurs in the long bones or spine, but 12 cases of osteomyelitis of the skull have been reported.[34] Overt symptoms included headache, seizures, and focal neurologic signs, but one patient with an epidural abscess exhibited only local swelling. Despite the fact that an intracranial abscess was present in nine patients, fever and leukocytosis were not common findings. Recommended treatment included both surgical debridement and prolonged antimicrobial therapy and, if carried out, led to a good prognosis.[34]

Salmonella meningitis occurs primarily in children, especially infants and neonates. The incidence noted in the literature is between 1.5% and 6%.[31] The course of the disease is rapid, and even with effective treatment there is a high rate of sequelae.[25,31] Eradication of the organism from the cerebrospinal fluid is difficult, and 64% of cases represent treatment failures or relapses; treatment with ampicillin alone is not recommended.[25] Huang[35] reported a rare case of meningitis complicated by brain infarctions in a 3-month-old girl. Seizures reappeared at 8 months, and at 1 year the child had the gross motor skills of a child at the 4- to 5-month level.

Salmonella mycotic aneurysm of the aorta is rare in most parts of the world, accounting for between 0.8% and 3.4% of all aortic aneurysms.[36,37] In addition to their patient, Meerkin and colleagues[36] found only 13 cases in the English language literature between 1950 and 1993. Before 1970, *Salmonella* was the predominant cause of aortic aneurysms in the Western world, but it has since been supplanted by *Staphylococcus*.[37] The data of Chan and associates[37] and Wang and colleagues,[38] who reported a combined total of 16 cases in 5 years in Taiwan, indicate that *Salmonella* remains a primary contributor of this disease in developing countries. It was always fatal in the past, and the mortality rate is still high at 40%. Preexisting atherosclerosis is a strong risk factor.[36,37] Fever, back pain, a palpable mass, and relapsing bacteremia are all indications of an intravascular lesion.[36,37] Early recognition and surgical intervention, coupled with antimicrobial therapy, improve the prognosis.

LABORATORY IDENTIFICATION

Isolation

The relative merits of traditional enteric differential media—MacConkey (MAC), xylose-lysine-desoxycholate (XLD), Hektoen (HE), and Salmonella-Shigella (SS) agar—for the isolation of salmonellae have been well debated over the years. These media are based on the principle that this genus does not ferment lactose (with the exception of subgenus III organisms, which do not commonly do so on isolation media) or sucrose and appears as colorless colonies. On media with an H_2S mechanism (HE, XLD, SS), *Salmonella* colonies have a black center, indicating that H_2S has been produced. All of these media can be used for the isolation of other enteric pathogens simultaneously and, as such, are multipurpose media. Some agars, however, have been designed to select specifically for *Salmonella*. The best known of the older generation of this type of media are brilliant green (BG) and bismuth sulfite (BS) agars, although neither is used extensively in clinical laboratories in the United States. BG is inhibitory to *S.* Typhi; other salmonellae appear as whitish, opaque colonies that turn the surrounding medium brilliant red. Most other flora are inhibited, and those that grow and ferment lactose/sucrose appear as yellow-green colonies with a yellow-green zone around the colony. BS is less inhibitory than BG and relies not on the differential fermentation of a carbohydrate but on the ability of *Salmonella* to produce H_2S. *Salmonella* appear as dark black colonies with a shiny black halo (which may require a full 24 hours to form) in the surrounding medium (Fig. 1; Color Plate 12). It is excellent for the recovery of *S.* Typhi and rare strains of salmonellae that have acquired the ability to ferment lactose or sucrose.[39] Although it is a boiled medium and easy to prepare, its short shelf life (4 days) discourages its use.

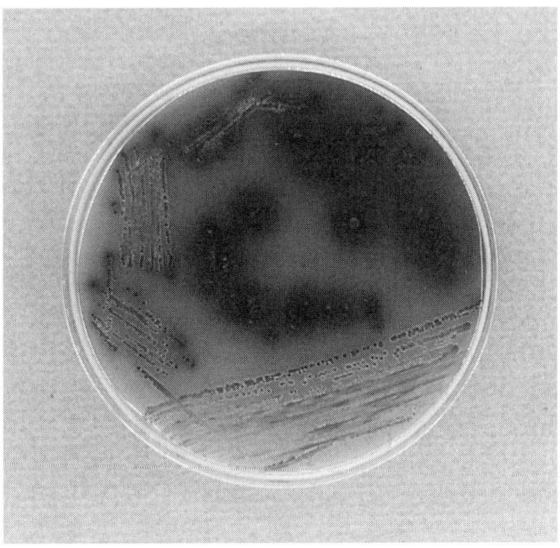

FIG. 1. *Salmonella* on bismuth sulfite agar. See also Color Plate 12.

TABLE 2. *Specific agars used to isolate* Salmonella

Medium	Differential Characteristics	Inhibitory Agent/ Condition	Colony Appearance		Isolates *S. typhi/paratyphi* A	Commercially Available
			Salmonella	Others		
Rambach	Propylene glycol (PG), β-D-galactoside (gal)	Bile salts	Red (PG+, gal–)	Colorless (PG–, gal–), blue (PG–, gal+), violet (PG+, gal+)	No	Yes: Merck, Gene Trac
MSRV	Swarming motility	↑ Osmotic pressure, malachite green, incubation at 42°C	Swarming	Inhibited, if present + or – swarming	No	Yes: Difco
NBGL	H_2S, glycerol, lactose	Novobiocin, brilliant green	H_2S+, black	H_2S inhibited from fermentation of glycerol, lactose; colonies colorless	No	No
XLT 4	H_2S, lysine	Tergitol 4	Black	Inhibited, white, yellow if present	No	Yes: Difco
SM-ID	Glucuronate (glu), β-D-galactoside (gal)	Bile salts, brilliant green	Pink (glu+, gal–)	Colorless (glu–, gal–), blue (glu–, gal+), violet (glu+, gal+)	Yes	Yes: bioMérieux
SS-NOV	H_2S, lactose	Bile salts, citrate, novobiocin	Colorless, black center	Pink	NT	No
BS	H_2S	Bismuth, brilliant green	Black with black halo	Black-brown, no halo, green, colorless	Excellent	Yes: Difco
BG	Lactose, sucrose	Brilliant green	Whitish, red halo	Yellow-green, green halo	No	Yes: Difco

MSRV, modified, semisolid Rappaport-Vassailidis; NBGL, novobiocin-brilliant green-glycerol-lactose; XLT 4, xylose-lysine-tergitol 4; SM-ID, Salmonella isolation and identification; SS-NOV, Salmonella-Shigella-novobiocin; BS, bismuth sulfite; BG, brilliant green; NT, not tested.

Because no single medium has ever supported the growth of all members of this genus, the search for better methods has continued. The mid-1980s saw the introduction of a number of agars and screening techniques specific for salmonellae. These media include the Rambach (RA), modified semisolid Rappaport-Vassailidis (MSRV), novobiocin-brilliant green-glycerol-lactose (NBGL), *Salmonella* isolation and identification (SM-ID), xylose-lysine-tergitol 4 (XLT 4), and Novobiocin-Salmonella-Shigella (NOV-SS) agars.[39–51] Table 2 lists the relevant characteristics of these media. Data from comparative studies (Table 3) indicate that these agars have not escaped the pitfalls of earlier media and must be used in combination with other agars or enrichment broths to achieve maximum yield of *Salmonella.* Although RA (Fig. 2; Color Plate 13) showed promise in early studies,[40–43] it and SM-ID are less sensitive than their counterparts.[44–47] Overall, MSRV was rated the most sensitive and had excellent specificity.[46–49] In two studies, MSRV raised the isolation rate for salmonellae by 22% and 16%.[48,49] However, MSRV is not without disadvantages: its semisolid consistency (spillage) and condensate formation in the lid (dripping) present hazards to the microbiologist handling the plate; it does not detect nonmotile salmonellae or *S.* Typhi; and it requires incubation at 42°C for maximum inhibitory effect and a full 24-hour incubation for maximum sensitivity.[45,47,50]

Most comparative media studies included HE or SS as a standard or control, and these two agars fared relatively well in terms of sensitivity in comparison with the newer media. However, their specificity was dismal, owing to the large numbers of colonies that required workup to rule out *Salmonella.* Specificity for these media can be improved by use of a screening method that detects C_8-esterase. C_8-esterase, produced by *Salmonella,* cleaves 4-methylumbelliferone, which is conjugated to an eight-carbon ester, caprylate (MUCAP). Release of the umbelliferone compound results in fluorescence, which can be detected at 365 nm by a Wood's lamp. Laboratories using MUCAP to screen for *Salmonella* colonies report sensitivity rates of 95% to 100% and specificity rates that vary between 80% and 99%.[52–56] C_8-esterase in-

TABLE 3. *Sensitivity, specificity, and numbers of* Salmonella *recovered by direct or selenite enrichment culture on newer* Salmonella*-specific agars*

Direct Plating				Enrichment	
Study	Medium	Sensitivity (%)	Specificity (%)	Direct (#)	Selenite (#)
Ruiz et al.,[47] 81 isolates	SS	69	99[a]	56	72
	MSRV	95	98	77	81
	NBGL	79	92	64	72
	RA	57	97[a]	46	71
	SM-ID	54	92	44	71
Dusch & Altwegg,[46] 32 isolates	MSRV	100	ND	32	32
	NBGL	53	100	17	28
	RA	69	91	29	31
	SM-ID	59	84	19	31
	XLT 4	88	100	28	31
	HE	75	84	24	31
Ruiz et al.,[45] 88 isolates	SM-ID	39	44	29	82
	BS	55	21	41	80
	HE	37	34	27	NT
	SS	65	19	48	81
	NBGL	78	61	58	80

[a]Colonies screened with C_8-esterase.

SS, Salmonella-Shigella; MSRV, modified semisolid Rappaport-Vassailidis; NBGL, novobiocin-brilliant green-glycerol-lactose; RA, Rambach; SM-ID, *Salmonella* isolation and identification; XLT 4, xylose-lysine-tergitol 4; HE, Hektoen enteric; BS, bismuth sulfite; ND, not done; NT, not tested.

A

B

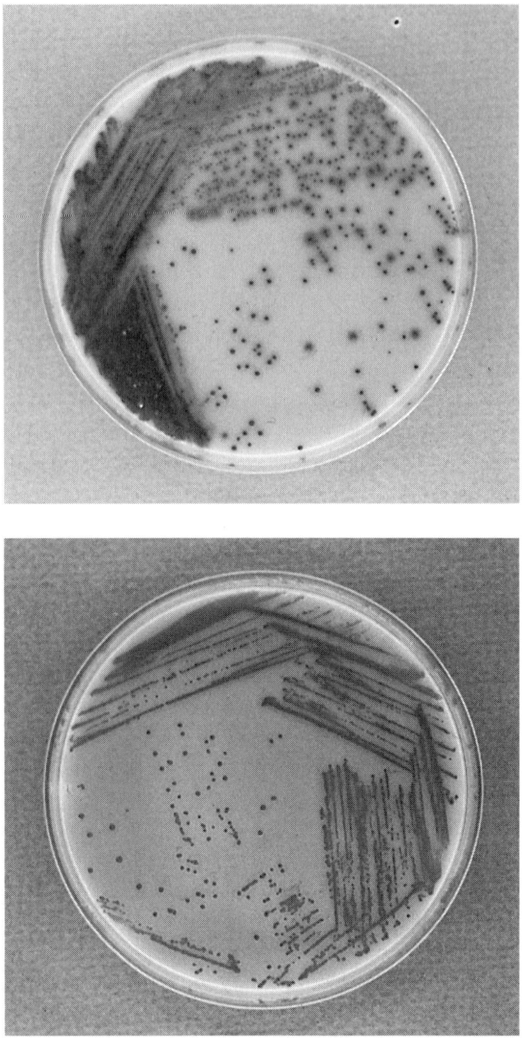

FIG. 2. A, *Citrobacter* on Rambach agar. **B,** *Salmonella* on Rambach agar. See also Color Plate 13.

creased the positive predictive value of SS from 17% to 98.8% by eliminating false-positive results,[45] and Aguirre and coworkers[52] found 70% of suspicious colonies could have been discarded without further workup. It cannot be used on BS, SM-ID, MSRV, or NBGL agars but works effectively on HE, SS, MAC, and RA.[45,53] The majority of false-positive findings were attributed to *Pseudomonas* and *Acinetobacter* species.[41,55,56] The Wood's lamp can be used to check for the presence of autofluorescing colonies before application of the reagent. The presence of L-pyrrolidonyl peptidase (PYR) is a useful test for separating *Salmonella* (which are negative) from H_2S-producing *Citrobacter* species (which are positive).[57] However, other H_2S-positive members of the *Enterobacteriaceae* (*Budivicia, Leminorella, Pragia, Trabulsiella,* and *Edwardsiella tarda*) are also PYR-negative. A colored-latex agglutination assay that can be performed as a screening test on colonies on direct plates or from enrichment broths is also available. The Wellcolex Colour Salmonella and Shigella test (Murex Diagnostics, Norcross, GA) detects *Salmonella* groups A, B, C, D, E, and G, and the Vi antigen. In three studies, sensitivity ranged from 87% to 100% with specificities of 98% or 99%.[58–60] Use of monoclonal antibody in an enzyme-linked immunosorbent assay (ELISA),[61] DNA probes,[62] or polymerase chain reaction[63–65] is impractical at this time because reagents are not commercially available.

Highest numbers of salmonellae are cultured when enrichment broths such as selenite F or Hajna's gram-negative broth are used. Only MSRV gave isolation rates as high as, or higher than, enrichment.[46,47] Selenite F with cysteine excels by far over other enrichments in the recovery of *S.* Typhi. Tetrathionate enrichment does not recover *S.* Typhi or *S.* Paratyphi C, and the broth must be used within 24 hours once the iodine has been added.

Traditional screening methods vary and include the use of a combination of triple sugar iron (TSI) or Kligler's agar, lysine-iron (LIA) agar, motility-indol-ornithine (MIO) agar, urea, indol, or tryptophan, the latter two by conventional or strip tests. Table 4 describes the reactions of the *Enterobacteriaceae* most often isolated from enteric plating media that must be differentiated from *Salmonella.*

Biochemical Methods

Commercial Systems

By and large, commercial systems do not appear to be unduly challenged by the identi-

TABLE 4. *Separation of* Salmonella *from other enteric flora by commonly used screening media*

Organism	TSI	LIA	MIO	Urea	Indol	TDA
Nontyphoidal *Salmonella* (including subgenera III)	K/AG⁺ K/AG⁻	K/K⁺	+/–/+	–	–	–
H_2S^+ *Proteus* spp.	K/AG⁺	K/A⁺	+/–/+ +/+/–	+ 4 h	–	+
H_2S^+ *Citrobacter* spp.	K/AG⁺ A/AG⁺/⁻	K/A⁺	+/–/V	+ 24 h	–	–
Edwardsiella tarda	K/AG⁺	K/K⁺	+/+/+	–	+	–
Escherichia coli	K/AG⁻	K/K⁻	+/+[a]/+	–	+[a]	–
Hafnia alvei	K/AG⁻	K/K⁻	+/–/+	–	–	–

[a]Many strains negative.
TSI, triple sugar iron agar; LIA, lysine-iron agar; MIO, motility-indol-ornithine agar; TDA, tryptophan deaminase agar; K, alkaline; A, acid; G, gas; ⁺, H_2S-positive; ⁻, H_2S-negative; V, variable.

fication of *Salmonella.* Atypical strains of subgenus I caused the greatest number of errors in seven studies[66–72] evaluating eight commercial systems. The systems evaluated were PASCO-ID (PASCO Laboratories, Wheat Ridge, CO), Autoscan Walkaway (Dade MicroScan, West Sacramento, CA), Cathra Repliscan replicator (Cathra, St. Paul, MN), API 20E (Vitek bioMérieux, Hazelwood, MO), Cobas Micro ID-E/NF (Becton Dickinson Diagnostic Instrument Systems, Sparks, MD), RapID onE (Innovative Diagnostic Systems, Norcross GA), and Crystal Enteric/Non-Fermenter ID (Becton Dickinson, Cockeysville, MD). All of the errors were at the genus level (10 of 183 cases) or the strain was unidentified (3 of 183).[66–69,72] Three *S.* Gallinarum strains were unidentified because of database error.[66] All *Salmonella* identifications should include TSI and motility medium in addition to any commercial system; many misidentifications or mixed cultures can be averted with their inclusion.

Conventional Media

As with commercial systems, salmonellae are usually easily identified using a minimal number of conventional media (Table 5). Production of H_2S easily separates nontyphoidal *Salmonella* from other *Enterobacteriaceae* with the exception of some species of *Citrobacter, Proteus,* and *Edwardsiella* and several rare genera (see Table 5). Except for *Tra-*

TABLE 5. *Biochemical reactions separating nontyphoidal* Salmonella *from other* Enterobacteriaceae

Organism	H_2S	Indol	Urea	Citrate	VP	KCN
Nontyphoidal *Salmonella*	+[a]	–	–	+	–	–[b]
H_2S^+ *Citrobacter* spp.	+	–	+	+	–	+
H_2S^+ *Proteus* spp.	+	V	+	V	V	+
Edwardsiella tarda	+	+	–	–	–	–
Leminorella grimontii	+	–	–	+	–	–
Leminorella richardii	+	–	–	–	–	–
Trabulsiella guamensis	+	V	–	V	–	+
Budvicia aquatica	+	–	V	–	–	–
Pragia fontium	+	–	–	V	–	–
Escherichia coli[c]	–	+	–	–	–	–
Hafnia alvei[c]	–		–	–	+	+

[a]Some strains negative.
[b]Subgenera IV strains positive.
[c]Often submitted as H_2S^- *Salmonella* for serotyping.
VP, Voges-Proskauer; KCN, potassium cyanide; +, ≥90%; V, 10%–90%; –, ≤10%.

bulsiella guamensis, all of the other H_2S-producing members of the *Enterobacteriaceae* are lysine- and ornithine decarboxylase–negative.[73] Other characteristics that separate H_2S-positive *Citrobacter* species from *Salmonella* are positive reactions in urea and potassium cyanide (KCN) tests; the H_2S-positive *Proteus* species rapidly produce urease and phenylalanine deaminase. *Edwardsiella tarda* strains are indol-positive and do not ferment a number of common carbohydrates (L-arabinose, D-rhamnose, xylose, trehalose) that are readily fermented by salmonellae. On the tests for citrate, production of gas from glucose or H_2S in TSI, and motility, about 4% to 5% of salmonellae are negative; likewise, about 2% fail to produce lysine decarboxylase. These strains may be very difficult to separate from urea-negative *Citrobacter* species or indol-negative *E. coli.* A positive PYR reaction may help identify a strain as *Citrobacter,* and a negative C_8-esterase result would indicate *E. coli. S.* Choleraesuis strains differ from other subgenus I serotypes by their negative reactions in tests for citrate, H_2S, and fermentation of dulcitol, arabinose, and trehalose. H_2S-positive variants of *S.* Choleraesuis are referred to as var. *kunzendorf* and are more predominant in the United States than *S.* Choleraesuis. *S.* Paratyphi A strains are lysine decarboxylase– and H_2S-negative, although we have seen strains that produce weak H_2S or are weakly positive in Möeller's lysine decarboxylase broth. Serotypes of subgenera II through VI and *S. bongori* can be separated from those of subgenus I by the biochemical characteristics given in Table 6.

Immunodiagnostics

The Kauffmann-White schema contains the antigenic composition of more than 2000 *Salmonella* serotypes and is exceeded in complexity and numbers of serotypes only by that of *E. coli* (>10,000 serotypes). *Salmonella* are broken down into groups based on which O somatic or cell surface antigens (lipopolysaccharide) are present. Between 95% and 97% of commonly isolated serotypes belong to groups A, B, C_1, C_2, D, and E_1 through E_4, although the schema contains 44 groups. Groups are designated by capital letters (groups after X are designated by their antigen number), but the actual somatic antigenic factors are given numeral designations. These factors occur singly (21, group L) or in complexes (9,12, group D). Generally, in clinical laboratories, identification of a group is sufficient. However, in reference laboratories where the strain is taken to serotype, specific antigens within a complex must be identified. This ensures that the agglutination reaction is not caused by cross-reactivity of a nonspecific antigen. For instance, groups A (2,12), B (4,5,12), and D (9,12) all contain the common antigen 12, which may cause a cross-reaction. Therefore, serum absorbed to contain only factor 2 (group A) or 9 (group D), for example, would be tested to ensure the group identity.

Typing is achieved by defining the specific H flagellar (protein) antigens present. These antigens are designated by either small letters or numbers; they also occur singly ("i" or "d") or as complexes ("e,h" or "1,2"). Again, as with somatic factors, a specific antigen within

TABLE 6. *Biochemicals that separate* Salmonella *subgenera and* Salmonella bongori

Organism	Dulcitol	ONPG	Malonate	KCN	Sorbitol
Salmonella choleraesuis subgenera					
I choleraesuis (ATCC 13313)	A	-	-	-	A
II salamae (ATCC 6959)	A	V	+	–	A
IIIa arizonae (ATCC 13314)	–	+	+	–	A
IIIb diarizonae	–	+	+	–	A
IV houtenae (NCTC 10401)	–	–	–	+	A
VI indica	V	V	–	–	–
Salmonella bongori (CIP 8233)	A	+	–	+	A

ONPG, *o*-nitrophenyl-β-D-galactopyranoside; KCN, potassium cyanide; A, acid ≥90%; V, 10%–90%; –, ≤10%; ATCC, American Type Culture Collection; NCTC, National Culture Type Collection; CIP, Collection Institut Pasteur.

a complex must be identified. For example, for complexes 1,2; 1,5; 1,6; and 1,7, factor 2, 5, 6, or 7 would be tested; testing for factor 1 would not be done since it is common to all complexes.

To complicate matters further, *Salmonella* may produce one (monophasic; gm:–) or two (diphasic; i:1,2) sets of flagella, but only one type is expressed by a cell at any given time. For example, in a culture of *S.* Typhimurium, which produces flagella designated i:1,2, some cells will express "i," others "1,2," but no single cell will produce both simultaneously. Both phases must be demonstrated in order to serotype the strain, but often only one phase can be demonstrated. The phase that is present is then suppressed by immobilizing the strain; for example, if only "i" is demonstrated, the strain is grown in the presence of "i" antisera. This causes a genetic shift, allowing expression of the second phase—in this case, "1,2"—which restores motility to the organism. Thus, serotyping of a strain such as *S.* Typhimurium would be accomplished by agglutination in group B serum; agglutination in absorbed, somatic single-factor 4, 5, or 27 sera; agglutination in flagellar factors "i" and "1,2"; and agglutination in absorbed, flagellar single-factor 2 serum. If one of the flagellar phases ("i" or "1,2") was not present, the strain would be suppressed to demonstrate the missing phase.

Most laboratories use a saline somatic antigen made by suspension of heavy overnight growth in 0.85% saline. Reference laboratories commonly use alcohol antigens made by subjecting the organism to absolute alcohol, heat (60°C), and phenol. This treatment renders the flagella nonagglutinable (somatic antisera also contain flagellar antibodies of the strain used to make the serum) and makes the antigen suspension more specific. Testing is carried out by a slide agglutination technique. Flagellar testing is done with the use of a formalinized broth suspension, which renders the somatic antigen nonagglutinable and increases the specificity of this antigen. A tube agglutination assay (a single dilution of 1:1000 or 1:4000) is used. Suppressions usually are carried out in motility deeps, but we use a filter paper strip method. A 1-inch by 0.25-inch piece of sterile filter paper, impregnated with the appropriate antiserum, is placed on an agar plate; one end of the strip is inoculated with the culture, and the plate is incubated overnight. Monophasic organisms remain at the site of inoculation; diphasic organisms that have undergone a genetic shift move away from the site of inoculation down the entire length of the strip, indicating that the second phase is present. The entire strip is placed in broth with a sterile forceps, incubated for 4 hours, formalinized, and then tested for the appropriate phase.

Typing

Occasionally serotyping alone may be sufficient to identify an outbreak or cluster of cases because of the serotype's rarity (eg, an outbreak caused by *S.* Saphra). With common serotypes, additional epidemiologic test methods, such as phage typing,[74–76] plasmid analysis,[77–79] pulsed-field gel electrophoresis (PFGE),[78,80] ribotyping,[77–81] probes for insertion sequences,[77,78,80] and outer membrane protein analysis,[79] must be employed. Phage typing schemes for a variety of serotypes have been developed[74] but are probably most important for *S.* Enteritidis[75] and *S.* Typhimurium[76] because of the frequency of their isolation and their association with outbreaks. Epidemics of *S.* Enteritidis associated with eggs and poultry in the United Kingdom, in Europe, and on both the east and west coasts of the United States have been tracked by phage typing. *S.* Enteritidis phage type 4 (SE PT4) was predominant in Europe and the western United States,[82,83] whereas SE PT8 created epidemics along the eastern coast of the United States.[84] *S.* Typhimurium var. Copenhagen DT 104 (definitive type is used instead of phage type) has caused major outbreaks in Europe and is second only to SE PT4 in terms of frequency of isolation.[85] It has also appeared in the United States, both sporadically and in an outbreak.[86] Contact with ill farm animals and consumption of poultry, pork sausages, or meat paste have been linked to DT 104 infections, but the organism has been isolated from a wide variety of both domestic and wild animals.[85,86] These strains are associated with increased virulence and are resistant to five antimicrobials: ampicillin, chloramphenicol, sulfonamides, streptomycin, and tetracycline. Some strains are resistant to trimethoprim and/or ciprofloxacin as well.

Molecular typing is used extensively with the salmonellae to trace an outbreak back to its source. Plasmid analysis is a very useful tool[77–79] although many serotypes, especially those less commonly encountered, may not possess them. Still other serotypes (eg, *S.* Dublin, *S.* Typhimurium, *S.* Enteritidis) possess virulence plasmids common to most strains of that serotype, and these cannot be used as identifying profiles. Ribotyping, the comparison of conserved sequences of ribosomal RNA using enzyme-restricted DNA and a universal 16S and 23S RNA probe, has been used successfully with some serotypes.[81,87] However, both ribotyping and probes used to detect the insertion sequence IS*200* have been found less useful than other molecular techniques.[77,78,80] PFGE is probably the most commonly used molecular technique at this time. However, it has several disadvantages, including the long time required for completion (4 days) and the fact that highly clonal serotypes such as *S.* Enteritidis fail to be differentiated by this technique.[88] A PFGE gel of *S.* Typhimurium var. Copenhagen isolates from an outbreak involving contaminated homemade cheese vended door-to-door is shown in Fig. 3.

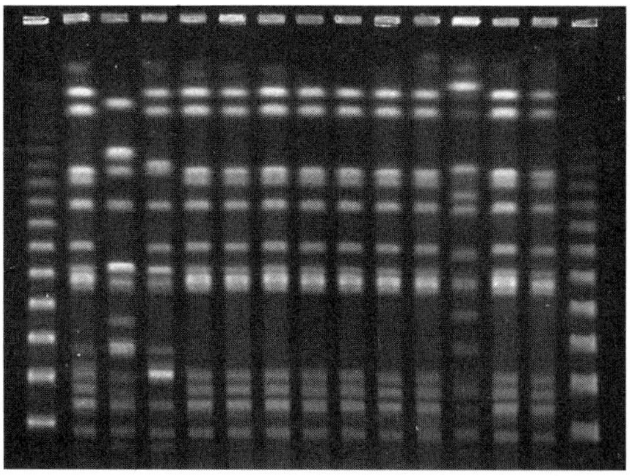

FIG. 3. Pulsed-field gel electrophoresis gel of isolates from an outbreak of *S.* Typhimurium var. Copenhagen involving contaminated homemade cheese vended door-to-door. Lane 14, cheese isolate; lanes 5–11 and 13, patients who consumed the contaminated cheese; lanes 2, 3, 4, and 12, other patients with same serotype but not involved in the outbreak; lanes 1 and 15, lambda ladders.

PATHOGENESIS

Unlike enteric pathogens such as *E. coli* and *Yersinia enterocolitica,* all salmonellae are generally regarded as virulent, whether they are recovered from diarrheal episodes or associated with the carrier state. Based on this assumption, all isolates of those serotypes that are associated with human infections inherently have the capacity to incite disease. However, both the type and degree of disease produced can differ because pathogenicity depends on a number of interrelated factors. These include strain virulence, initial dose, route of infection, and overall host susceptibility,[89] each of which can play a critical role in the spectrum of *Salmonella* infections encountered (Fig. 4). Several types of infection can result as a direct consequence of the ingestion of salmonellae. The most common clinical manifestation of salmonellosis is gastroenteritis, which can also be described as early infection. Gastroenteritis sometimes progresses into later disease, such as enteric fever, nontyphoidal bacteremia in AIDS patients, or septicemia in apparently healthy persons. In these settings (early infection, late infection), different virulence factors may be operative in selected serotypes that control both the type and extent of disease that ensues. As with other classic enteric pathogens, these virulence determinants are located on both the bacterial chromosome and plasmids. It is estimated that at least 150 to 300 of the more than 3000 genes located on the *Salmonella* chromosome play a role in pathogenic-

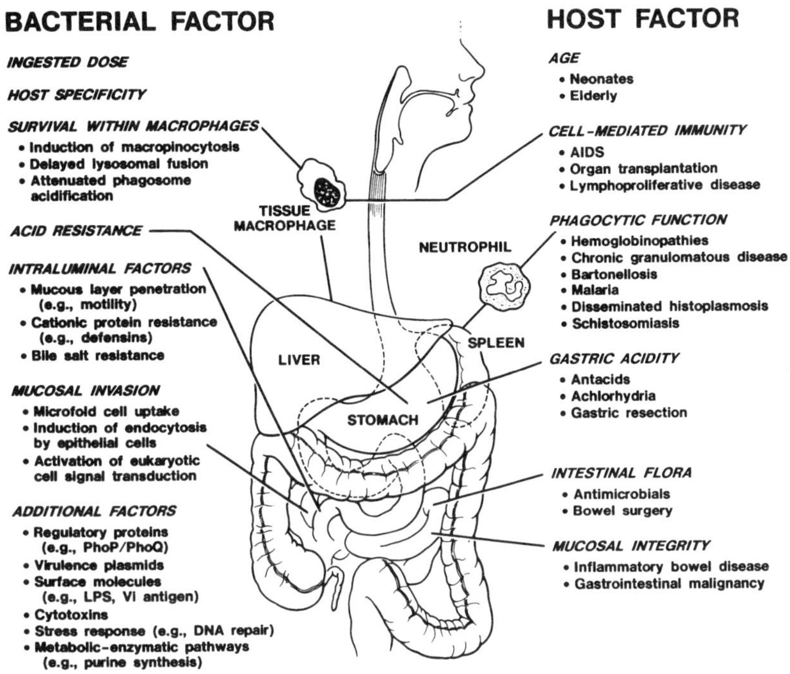

FIG. 4. Pathogenic bacterial properties and host factors conferring increased susceptibility to *Salmonella* infection. (Reprinted with permission from Pegues DA, Hohlmann EL, Miller SI. *Salmonella* including *S. typhi.* In: Blaser MJ, Smith PD, Ravdin JI, Greenberg HB, Guerrant RL, eds. *Infections of the gastrointestinal tract.* New York: Raven Press, 1995;785–809.)

ity.[89] Because not all of these factors have been investigated in every serotype, generalities must be drawn regarding their role in pathogenesis (Table 7).

On ingestion, salmonellae must overcome several physiologic and immunologic barriers to establish localized infection of the gut. Foster and Hall[90] found that when *S.* Typhimurium was shifted from an almost neutral pH to extremely acidic conditions, few if any bacteria survived. However, if a single doubling dilution in mildly acidic conditions was allowed, more than 1000 times as many organisms survived. Development of acid resistance after exposure to a stepwise change in pH in the external environment (from pH 7.6 to pH 5.5 to pH 3.3) was termed the acid tolerance response by Foster and Hall.[91] The response is induced in bacteria exposed to pH values between 5.5 and 6.0; subsequent ability to survive at low pH is the result of maintenance of internal pH homeostasis by the microorganism. The acid tolerance response could conceivably play a role in the ability of salmonellae to withstand acidic conditions in the stomach and to survive the acidic environment of the phagosome, once ingested.

After salmonellae have survived the effects of gastric acidity, they migrate to the small intestine, where they attach to the terminal ileum, the preferred site of infection. This process requires attachment to epithelial cells/M cells in order for infection to take place. One series of bacterial determinants thought to play a role in this scenario is the fimbriae. *S.* Typhimurium and *S.* Enteritidis are the most extensively investigated strains in regard to these structures. In both serotypes, type 1 (mannose-sensitive), type 3 (hemagglutination of tannic acid–treated cells), and GVVPQ (thin fimbriae, aggregative) pili have been isolated.[92–95] Bäumler and Heffron[95] have described a new type of fimbriae in *S.* Typhimurium. This gene, *lpf* (long

TABLE 7. *Selected virulence factors associated with* Salmonella *pathogenicity*

Factor or Trait	Phenotype/ Genotype	Location	Proposed Functions	Primary Stage of Infection
Acid tolerance	ATR	Chromosome	Resistance to gastric acidity, phagocytosis inhibition	Early/late
Colonization	Type 1	Chromosome	Gastrointestinal epithelium attachment	Early
	Type 3	Chromosome	Attachment	Early
	GVVPQ	Chromosome	Attachment	Early
	Type 4	Chromosome	Lymphoid follicle (Peyer's patches) attachment	Early
Invasion	*inv/spa*	Chromosome	Invasion of M cells, macrophages, epithelial cells	Early
Virulence plasmid	*spv*	Plasmid	Enhances multiplication at extraintestinal sites	Late
Serum resistance	*tra*T, *rsk, rck;* LPS	Plasmid/ chromosome	Resistance to serum killing; resistance to complement killing	Late

LPS, lipopolysaccharide.

polar fimbriae; 10–20 μm), encodes for fimbriae that morphologically resemble type 4 pili but genetically do not appear to share homologous sequences. The *lpf* operon appears to be involved in attachment of salmonellae to murine Peyer's patches, supporting its role in pathogenesis.[96] Organisms with mutations in this operon fail to destroy M cells, a normal consequence of attachment and invasion of these cells by salmonellae. Although the evidence is inconclusive, both type 1 fimbriae and the bacterial flagella also appear to play roles in this attachment process in at least some serotypes.[94,95]

After attachment, infection of the gut by salmonellae occurs by means of a rapid invasion of adsorptive epithelial cells, M cells, and macrophages.[96,97] At least six different genetic loci appear to be involved in the process.[98] One of these loci involves a 40-kb region of chromosomal DNA that represents pathogenicity island SPI 1.[99] SPI 1 codes for invasion sequences, surface presentation of antigen (*spa,* similar to that found for *Shigella flexneri*), and a type III secretory system.[100] A second *Salmonella* pathogenicity island, SPI 2, has also been detected.[101] Its function is presently unknown, although it also codes for a type III secretory system. The invasion genotype of salmonellae correlates in vitro with the ability to penetrate and replicate in a variety of cell lines, including HEp-2, HeLa, Vero, Henle 407, Caco-2, macrophages, and T and B cells.[102–106] In vivo, a preferred target of this invasion process appears to be macrophages. Invasive *S.* Typhimurium replicate within macrophages and subsequently reside within large vacuoles.[106] Within hours, virulent salmonellae induce host cell membrane ruffling, cytotoxic reactions, and apoptosis.[107] This response presumably permits infection, bacterial survival, and escape from the host's immune response on systemic dissemination.

Probably the best studied factor involved in *Salmonella* pathogenicity is the virulence plasmid.[108–111] Virulence plasmids are high-molecular-weight (50–100 kb) plasmids that are present in only a few *Salmonella* serotypes (Table 8). Although the size of these extrachromosomal elements varies from serotype to serotype, an 8-kb region encoding for five genes is consistently present in all virulence plasmids.[109] This region has been designated *spv* (*s*almonella *p*lasmid *v*irulence). Recent studies have found little evidence that the *spv* genes play a prominent role in early infections such as gastroenteritis. Rather, in

TABLE 8. Salmonella *serotypes containing virulence plasmids*

Group 1[a]	Group 2[b]
S. Dublin (cattle)	*S.* Typhimurium
S. Choleraesuis (swine)	*S.* Enteritidis
S. Gallinarum-pullorum (poultry)	Most other serotypes
S. Abortusovis (sheep)	

[a]Host-adapted. All wild-type strains may carry plasmids encoding for a variety of factors.
[b]Broad host range. Virulence plasmid carriage variable; plasmids encoding other factors may be present.

those serotypes where the virulence plasmid is only sporadically present (eg, *S.* Typhimurium), a higher percentage of blood isolates (76%) carry this marker than do those originating from feces (42%). This suggests that *spv* plays a featured role in late infections such as septicemia.[112] It was thought that the spread of salmonellae to extraintestinal sites (eg, mesenteric lymph nodes, liver, spleen) was promoted more efficiently by serotypes with *spv* than by virulence plasmid–negative isolates, but this, too, appears not to be the case. Currently, the best evidence suggests that *spv* genes function by enhancing certain serotypes to multiply in extraintestinal tissues preferentially over strains lacking the virulence plasmids.[110] When *spv*$^+$ *S.* Dublin strains infect their natural host (calves) instead of mice (not a host to study enteritis), the animals develop severe diarrhea and become moribund.[113] Mutant strains (*spv* regulator defective) cause only a mild enteritis and no signs of systemic disease, suggesting that these virulence genes play a role in both intestinal and extraintestinal proliferation.

Several other factors may be active in *Salmonella* virulence. Resistance to complement-mediated lysis of serum is specified on at least four different loci outside the *spv* region on the virulence plasmid.[109] These genes are *tra*T, *rsk* (*r*esistance to *s*erum *k*illing), *rck* (*r*esistance to *c*omplement *k*illing), and a fourth partially characterized locus.[109,114,115] In addition to these sites, lipopolysaccharide genes on the bacterial chromosome also play a major role in serum resistance. Many common serotypes of *Salmonella* contain DNA that codes for a 29-kDa protein with enterotoxin activity.[116] However, the enterotoxin is difficult to detect in cell-free supernatants, and detection often requires cell lysis by treatment with polymixin B. Its role in pathogenesis is unclear. L-forms of *Salmonella* produce a 32-kDa protein that is cytotoxic for macrophages.[117] This molecule may function intracellularly by permitting replicated salmonellae to lyse macrophages and escape immune surveillance.

ANTIMICROBIAL SUSCEPTIBILITY

Treatment of uncomplicated gastroenteritis caused by nontyphoidal salmonellae is a matter of some contention. People with milder infections usually do not seek treatment, and those who present to physicians are usually quite ill and in great discomfort and expect antibiotics to relieve their condition. However, antimicrobial agents should be administered only to those patients at risk for a poor outcome.[10,118–121] Such patients include those who are at the extremes of age; those who are immunodeficient or receiving immunosuppressive therapy; and those who have gastric hypochlorhydria, inflammatory bowel disease, heart disease, diabetes, or renal or rheumatoid disease.[118] Limiting treatment avoids the risk of emerging resistance in *Salmonella* strains after antimicrobial exposure. Furthermore, antibiotic treatment is known to prolong excretion in some patients.[10,118,120–122] This can have an extremely detrimental effect on food handlers and care givers, who may be prevented from returning to work by health professionals or employers who fear the consequences of employing carriers. Elimination of either convalescent or chronic carriage is difficult at best. Oral ciprofloxacin has been used, but convalescent carriers have relapsed, with bacteria detected in their stools 2 or more weeks later.[118] An expert panel of the British Society for the Study of Infection recommended that stools be cultured 3 weeks after treatment.[118]

Emerging resistance in salmonellae after exposure to antibiotics, whether in a clinical, veterinary, or environmental setting (45% of all antibiotics are used in animal feeds[10]), has reached an alarmingly high rate. The Centers for Disease Control tested a random cross-section of salmonellae taken nationwide at three 5-year intervals (1979–1980, 1984–1985, and 1989–1990) for antimicrobial susceptibilities.[119] Of the 12 antimicrobials tested, 6 (streptomycin, tetracycline, ampicillin, sulfisoxazole, kanamycin, and gentamicin) showed increased resistance between the second and third sampling periods. Trimethoprim-sulfamethoxazole (SXT) and cephalothin resistance remained approximately the same during this interval. In addition to increased resistance, a number of other interesting changes between the first and third sampling periods were noted and are summarized in Table 9. Increased resistance also has been reported from the United Kingdom, especially for *S.* Typhimurium: 39% of isolates were resistant to four or more antimicrobials,[123] and this was especially true among strains of definitive type DT 104 (see Typing). Multidrug resistance patterns included ampicillin and SXT. Because of increased resistance to ampicillin and tetracycline, third-generation cephalosporins and fluoroquinolones are being used increasingly (see Pickering,[121] Table 9). At this time, resistance to these agents is minimal, but resistant strains are beginning to emerge. *S.* Typhimurium strains with mutations in the *gyr*A gene associated with ciprofloxacin resistance have been reported.[121] A novel extend-spectrum β-lactamase, PER-1, unrelated to the classic SHV or TEM families of β-lactamases, has been reported in Turkey and confers resistance to the newer cephalosporins.[124] Hohl and colleagues[125] reported that all 37 *Salmonella* tested were susceptible to ciprofloxacin, ceftriaxone, and SXT, but the 50% minimal inhibitory concentration (MIC_{50}) for amoxacillin-clavulanate was 16 g/mL, indicating resistance. In testing 70 *Salmonella* against four third-generation cephalosporins (cefotaxime, cetriaxone, ceftizoxime, and cetazidime), Fuchs and coworkers[126] found that 97% of strains were susceptible to each drug. Thornsberry[127] reported that all 903 *Salmonella* isolates tested were susceptible to ciprofloxacin.

TABLE 9. *Changes in isolation rates and trends in antimicrobial use and resistance in* Salmonella *between 1979 and 1990 (in %)*

Parameter	1979–1980 (n=378)	1989–1990 (n=484)
Patient age		
<1 y	22	14
>60 y	17	28
Isolates from blood		
Total	1	11
Resistant	0	50
Isolates from stool		
Total	95	86
Resistant	17	30
Isolates of *S.* Enteritidis		
Total	6	25
Resistant	13	10
Isolates of *S.* Typhimurium		
Total	42	21
Resistant	14	47
Treatment		
Amoxicillin/ampicillin	59[a]	32
Trimethoprim-sulfamethoxazole	27[a]	23
Ciprofloxacin	0[a]	22
Third-generation cephalosporins	0[a]	12

From Lee et al.[119]
[a]Information not available for 1979–1980, data from 1984–1985.

CONCLUSIONS

Until *Campylobacter jejuni* ssp. *jejuni* arrived on the scene, nontyphoidal *Salmonella* was the prevalent cause of gastroenteritis in most areas of most industrialized nations. Salmonellosis is on the rise in the United States and other Western countries because societal and lifestyle changes have encouraged or demanded food preparation measures, from farm to table, that incur increased risk of *Salmonella* contamination. To improve the situation now existing will require concentration on good hygiene and elimination of risky practices all the way from hoof or field to consumption. The 1997 President's Food Safety Initiative is aimed at decreasing infections caused by foodborne agents and is a step in the right direction. It is also necessary to continue funding public health efforts (1) to track and prevent outbreaks and (2) to understand the dynamics of sporadic infections in order to reduce them. Because resistance is easily acquired by *Salmonella,* especially via plasmids, limited exposure to antibiotics in both patients and animals is crucial. However, the biggest cure for foodborne diseases lies in education for everyone from farmer to consumer and all handlers in between. *Salmonella,* as an inhabitant of the gastrointestinal tract, is unlikely to ever be eliminated, but its damage can be reduced and controlled if everyone understands the dynamics of how this organism is transmitted.

REFERENCES

1. LeMinor L. Genus III: *Salmonella* Lignières 1900, 389[AL] In: Krieg NR, Holt JG, eds. *Bergey's manual of systematic bacteriology.* Baltimore: Williams & Wilkins, 1984;427–458.
2. Old DC. Nomenclature in *Salmonella. J Med Microbiol* 1992;37:361–363.
3. LeMinor L, Popoff MY. Request for an opinion: Designation of *Salmonella enterica* sp. nov., nom. rev., as the type and only species of the genus *Salmonella. Int J Syst Bacteriol* 1987;37:465–468.
4. Ewing EH, Edwards PR. *Identification of the* Enterobacteriaceae. 3rd ed. Minneapolis: Burgess Publishing Company, 1972.
5. Ewing EH. An outline of nomenclature for the family *Enterobacteriaceae. Int Bull Bacteriol Nomen Taxon* 1963;13:95–110.
6. Crosa JH, Brenner DJ, Ewing EH, Falkow S. Molecular relationship among the *Salmonelleae. J Bacteriol* 1973;115:307–315.
7. Wayne LG. Actions of the Judicial Commission of the International Committee on Systematic Bacteriology on requests for opinions published between January 1985 and July 1993. *Int J Syst Bacteriol* 1994;44:177–178.
8. LeMinor L. The genus *Salmonella*. In: Starr MP, Stolp H, Trüper HG, Schlegel HG, eds. *The prokaryotes: A handbook on habitats, isolation and identification of bacteria.* Berlin: Springer-Verlag, 1981;2760–2774.
9. Reeves MW, Evins GM, Heiba AA, Plikaytis BD, Farmer JJ III. Clonal nature of *Salmonella typhi* and its genetic relatedness to other salmonellae as shown by multilocus enzyme electrophoresis, and proposal of *Salmonella bongori* comb. nov. *J Clin Microbiol* 1989; 27:313–320.
10. Guthrie RK. *Salmonella*. Boca Raton: CRC Press, 1992.
11. Blaser MJ. How safe is our food? *N Engl J Med* 1996;334:1324–1325.
12. Waltner-Toews D. Food, sex and salmonella: The risks of environmental intimacy. *Can J Public Health* 1991; 82:219–220.
13. Pegues DA, Hohmann EL, Miller SI. *Salmonella* including *S. typhi.* In: Blaser MJ, Smith PD, Ravdin JI, Greenberg HB, Guerrant RL, eds. *Infections of the gastrointestinal tract.* New York: Raven Press, 1995.
14. Rubino JR. The economic impact of human *Salmonella* infection. *Clin Microbiol Newslett* 1997;19: 25–29.
15. Cohen ML, Fontaine RE, Pollard RA, Von Allmen SD, Vernon TM, Gangarosa EJ. An assessment of patient-related economic costs in an outbreak of salmonellosis. *N Engl J Med* 1978;299:459–460.
16. Blaser MJ, Newman LS. A review of human salmonellosis: I. Infective dose. *Rev Infect Dis* 1982;4: 1096–1106.
17. Mintz ED, Cartter ML, Hadler JL, Wassell JT, Zingeser JA, Tauxe RV. Dose-response effects in an outbreak of *Salmonella enteritidis. Epidemiol Infect* 1994;112:13–23.
18. Buchwald DS, Blaser MJ. A review of human salmonellosis: II. Duration of excretion following infection with nontyphi *Salmonella. Rev Infect Dis* 1984;6: 345–356.
19. Cruickshank JG, Humphrey TJ. The carrier food-handler and non-typhoid salmonellosis. *Epidemiol Infect* 1987;98;223–230.
20. Tauxe RV, Hassan LF, Findeisen KO, Sharrar RG, Blake P. Salmonellosis in nurses: Lack of transmission to patients. *J Infect Dis* 1988;157:370–373.
21. Mermin J, Hoar B, Angulo FJ. Iguanas and *Salmonella* Marina infection in children: A reflection of the increasing incidence of reptile-associated salmonellosis in the United States. *Pediatrics* 1997;99:399–402.
22. Weinstein JW, Seltzer EG, Nelson RS, et al. Reptile-associated salmonellosis—Selected states, 1994–1995. *MMWR Morb Mortal Wkly Rep* 1995;44:347–350.
23. Aleksić S, Heinzerling F, Bockemühl J. Human infection caused by salmonellae of subspecies II to VI in Germany, 1977–1992. *Zentralbl Bakteriol* 1996;283: 391–398.
24. DuPont HL, Pickering LK. *Infections of the gastrointestinal tract: Microbiology, pathophysiology, and clinical features.* New York: Plenum Medical Book Company, 1980.

25. Goldberg MB, Rubin RH. The spectrum of *Salmonella* infection. In: Moellering RC, Gorbach SL, eds. *Infectious diarrhea. Infect Dis Clin North Am* 1988;2: 571–599.
26. Garcia-Corbeira P, Ramos JM, Aguado JM, Soriano F. Six cases in which mesenteric lymphadenitis due to non-typhi *Salmonella* caused an appendicitis-like syndrome. *Clin Infect Dis* 1995;21:232–234.
27. McKendrick MW, Read NW. Irritable-bowel syndrome—Post salmonella infection. *J Infect* 1994;29:1–3.
28. Sperber SJ, Schleupner CJ. Salmonellosis during infection with human immunodeficiency virus. *Rev Infect Dis* 1987;9:925–934.
29. Wilkins EDL, Roberts C. Extraintestinal salmonellosis. *Epidemiol Infect* 1988;100:361–368.
30. Yang CH, Tseng HH, Chen KJ, Liu JD. *Salmonella* infections: A retrospective 10-year analysis of 134 cases in a regional hospital in Taiwan. *Scand J Infect Dis* 1996;28:171–175.
31. Davis RC. *Salmonella* sepsis in infancy. *Am J Dis Child* 1981;135:1096–1099.
32. Ramos JM, Aguado JM, Garcia-Corbeira P, Alés, Sorian F. Clinical spectrum of urinary tract infections due to nontyphoidal *Salmonella* species. *Clin Infect Dis* 1996;23:388–390.
33. Wright J, Thomas P, Serjeant GR. Septicemia caused by salmonella infection: An overlooked complication of sickle cell disease. *J Pediatr* 1997;130:394–399.
34. Kamarulzaman A, Briggs RJS, Fabinyi G, Richards MJ. Skull osteomyelitis due to *Salmonella* species: Two case reports and review. *Clin Infect Dis* 1996;22: 638–641.
35. Huang LT. *Salmonella* meningitis complicated by brain infarctions. *Clin Infect Dis* 1996;22:194–195.
36. Meerkin D, Yinnon Am, Munter RG, Shemesh O, Hiller N, Abraham AS. *Salmonella* mycotic aneurysm of the aortic arch: Case report and review. *Clin Infect Dis* 1995;21:523–528.
37. Chan P, Tsai CW, Huang JJ, Chuang YC, Hung JS. Salmonellosis and mycotic aneurysm of the aorta: A report of 10 cases. *J Infect* 1995;30:129–133.
38. Wang JH, Liu YC, Yen MY, et al. Mycotic aneurysm due to non-typhi *Salmonella:* Report of 16 cases. *Clin Infect Dis* 1996;23:734–747.
39. Ruiz JR, Núñez ML, Díaz J, Sempere MA, Gómez J, Usera MA. Note: Comparison of media for the isolation of lactose-positive *Salmonella. J Appl Bacteriol* 1996;81:571–574.
40. Rambach A. New plate medium for facilitated differentiation of *Salmonella* spp. from *Proteus* spp. and other enteric bacteria. *Appl Environ Microbiol* 1990; 56:301–303.
41. Freydiere AM, Gille Y. Detection of salmonellae by using Rambach agar and by a C8 esterase spot test. *J Clin Microbiol* 1991;29:2357–2359.
42. Grunewald R, Henderson RW, Yappow S. Use of Rambach propylene glycol–containing agar for identification of *Salmonella* spp. *J Clin Microbiol* 1991;29: 2354–2356.
43. Heinzmann WR. Evaluation of Rambach agar for rapid detection of *Salmonella* in human feces. *Eur J Clin Microbiol Infect Dis* 1993;12:306–307.
44. Pignato S, Giammanco G, Giammanco G. Rambach agar and SM-ID medium sensitivity for presumptive identification of *Salmonella* subspecies I–VI. *J Med Microbiol* 1995;43:668–671.
45. Ruiz J, Núñez ML, Díaz J, Lorente I, Pérez J, Gómez J. Comparison of five plating media for isolation of *Salmonella* species from human stools. *J Clin Microbiol* 1996;34:686–688.
46. Dusch H, Altwegg M. Evaluation of five new plating media for isolation of *Salmonella* species. *J Clin Microbiol* 1995;33:802–804.
47. Ruiz J, Núñez ML, Lorente I, Pérez J, Simarro E, Gómez J. Performance of six culture media for isolation of *Salmonella* species from stool samples. *Eur J Clin Micriobiol Infect Dis* 1996;15:922–926.
48. Goossens H, Wauters G, De Boeck M, Janssens M, Butzler JP. Semisolid selective-motility enrichment medium for isolation of salmonellae from fecal specimens. *J Clin Microbiol* 1984;19:940–941.
49. Aspinall ST, Hindle MA, Hutchinson DN. Improved isolation of salmonellae from faeces using a semisolid Rappaport-Vassailidis medium. *Eur J Clin Microbiol Infect Dis* 1992;11:936–938.
50. Hine EA, Steffen EK, Wagner JE. New semisolid agar for the detection of motile salmonellae. *J Clin Microbiol* 1988;26:875–878.
51. Núñez ML, Díaz J, Sempere MA, Gomez J, Ruiz J. Novobiocin-Salmonella-Shigella agar for isolation of *Salmonella* species. *Eur J Clin Microbiol Infect Dis* 1996;15:96–97.
52. Aguirre PM, Cacho JB, Folgueira L, López M, García J, Velasco AC. Rapid fluorescence method for screening *Salmonella* spp. from enteric differential agars. *J Clin Microbiol* 1990;28:148–149.
53. Olsson M, Syk A, Wollin R. Identification of salmonellae with the 4-methylumbelliferyl caprilate fluorescence test. *J Clin Microbiol* 1991;29:2631–2632.
54. Ruiz J, Sempere MA, Varela MC, Gomez J. Modification of the methodology of stool culture for *Salmonella* detection. *J Clin Microbiol* 1992;30:525–526.
55. Manafi M, Sommer R. Comparison of three rapid screening methods for *Salmonella* spp.: "MUCAP test, Microscreen^R Latex and Rambach Agar." *Letts Appl Microbiol* 1992;14:163–166.
56. Muñoz P, Díaz MD, Cercenado E, Rodríguez-Créixems M, Berenguer J, Bouza E. Rapid screening of *Salmonella* species from stool cultures. *Am J Clin Pathol* 1993;100:404–406.
57. Inoue K, Miki K, Tamura K, Sakazaki R. Evaluation of L-pyrrolidonyl peptidase paper strip test for differentiation of members of the family *Enterobacteriaceae,* particularly *Salmonella* spp. *J Clin Microbiol* 1996;34: 1811–1812.
58. Bouvet PJM, Jeanjean S. Evaluation of two colored latex kits, the Wellcolex Colour *Salmonella* test and the Wellcolex Colour Shigella test, for serological grouping of *Salmonella* and Shigella species. *J Clin Microbiol* 1992;30:2184–2186.
59. Rohner P, Dharan S, Auckenthaler R. Evaluation of the Wellcolex Colour *Salmonella* test for detection of *Salmonella* spp. in enrichment broths. *J Clin Microbiol* 1992;30:3274–3276.
60. Kocka FE, Dorigan FD, Abbasi QT, Swiatlo AL, Hubbard-Shepard M. Clinical evaluation of the Wellcolex Colour Shigella and Salmonella tests. *Diagn Microbiol Infect Dis* 1992;15:1–4.
61. Kerr S, Ball HJ, Mackie DP, Pollock DA, Finlay DA. Diagnostic application of monoclonal antibodies to outer membrane protein for rapid detection of salmonella. *J Appl Bacteriol* 1992;72:302–308.

62. Scholl DR, Kaufmann C, Jollick JD, York CK, Goodrum GR, Charache P. Clinical application of novel sample processing technology for the identification of salmonellae by using DNA probes. *J Clin Microbiol* 1990;28:237–241.
63. Widjojoatmodjo MN, Fluit AC, Torensma R, Verdonk GPHT, Verhoef J. The magnetic immuno polymerase chain reaction assay for direct detection of salmonellae in fecal samples. *J Clin Microbiol* 1992;30:3195–3199.
64. Stone GG, Oberst RD, Hays MP, McVey S, Chengappa MM. Detection of *Salmonella* serovars from clinical samples by enrichment broth cultivation-PCR procedure. *J Clin Microbiol* 1994;32:1742–1749.
65. Chiu CH, Ou JT. Rapid identification of *Salmonella* serovars in feces by specific detection of virulence genes, *invA* and *spvC,* by an enrichment broth culture-multiplex PCR combination assay. *J Clin Microbiol* 1996;34:2619–2622.
66. Rhoden DL, Schable B, Smith PB. Evaluation of PASCO MIC-ID System for identifying gram-negative bacilli. *J Clin Microbiol* 1987;25:2363–2366.
67. O'Hara CM, Miller JM. Evaluation of the autoSCAN-W/A System for rapid (2-hour) identification of members of the family *Enterobacteriaceae. J Clin Microbiol* 1992;30:1541–1543.
68. Kelly MT, Leicester C. Evaluation of the Autoscan Walkaway system for rapid identification and susceptibility testing of gram-negative bacilli. *J Clin Microbiol* 1992;30:1568–1571.
69. O'Hara CM, Tenover FC, Miller JM. Parallel comparison of accuracy of API 20E, Vitek GNI, MicroScan Walk/Away Rapid ID, and Becton Dickinson Cobas Micro ID-E/NF for identification of members of the family *Enterobacteriaceae* and common gram-negative, non–glucose-fermenting bacilli. *J Clin Microbiol* 1993;31:3165–3169.
70. Kitch TT, Jacobs MR, Appelbaum PC. Evaluation of RapID onE System for identification of 379 strains in the family *Enterobacteriaceae* and oxidase-negative gram-negative nonfermenters. *J Clin Microbiol* 1994; 32:931–934.
71. Rhoads S, Marinelli L, Imperatrice CA, Nachamkin I. Comparison of MicroScan WalkAway System and Vitek System for identification of gram-negative bacteria. *J Clin Microbiol* 1995;33:3044–3046.
72. Wauters G, Boel A, Voorn GP, et al. Evaluation of a new identification system, Crystal Enteric/Non-fermenter, for gram-negative bacilli. *J Clin Microbiol* 1995;33:845–849.
73. Farmer JJ III. *Enterobacteriaceae:* Introduction and identification. In: Murray PR, Baron EJ, Pfaller MA, Tenover FC, Yolken RH, eds. *Manual of Clinical Microbiology.* 6th ed. Washington, DC: ASM Press 1995;438–449.
74. Castro D, Moriñigo MA, Martinez-Manzanares E, et al. Development and application of a new scheme of phages for typing and differentiating *Salmonella* strains from different sources. *J Clin Microbiol* 1992; 30:1418–1423.
75. Ward LR, deSa JDH, Rowe B. A phage-typing scheme for *Salmonella enteritidis. Epidemiol Infect* 1987;99: 291–294.
76. Barker R, Old DC. Phage type/biotype groups of *Salmonella typhimurium* in Scotland 1974–1976: Variation during spread of epidemic clones. *J Hyg Camb* 1980;84:115–125.
77. Millemann Y, Lesage MC, Chaslus-Dancla E, Lafont JP. Value of plasmid profiling, ribotyping, and detection of IS*200* for tracing avian isolates of *Salmonella typhimurium* and *S. enteritidis. J Clin Microbiol* 1995; 33:173–179.
78. Olsen JE, Brown DJ, Baggesen DL, Bisgaard M. Biochemical and molecular characterization of *Salmonella enterica* serovar *berta,* and comparison of methods for typing. *Epidemiol Infect* 1992;108:243–260.
79. Liebisch B, Schwarz S. Evaluation and comparison of molecular techniques for epidemiological typing of *Salmonella enterica* subsp. *enterica* serovar *dublin. J Clin Microbiol* 1996;34:641–646.
80. Baquar N, Burnens A, Stanley J. Comparative evaluation of molecular typing of strains from a national epidemic due to *Salmonella brandenburg* by rRNA gene and IS*200* probes and pulsed-field gel electrophoresis. *J Clin Microbiol* 1994;32:1876–1880.
81. Esteban E, Snipes K, Hird D, Kasten R, Kinde H. Use of ribotyping for characterization of *Salmonella* serotypes. *J Clin Microbiol* 1993;31:233–237.
82. Coyle EF, Ribiero CD, Howard AJ, Palmer SR, Jones HI, Ward L. *Salmonella enteritidis* phage type 4 infection: Association with hen's eggs. *Lancet* 1988;2: 1295–1297.
83. Boyce TG, Koo D, Swerdlow DL, et al. Recurrent outbreaks of *Salmonella enteritidis* infections in a Texas restaurant: Phage type 4 arrives in the United States. *Epidemiol Infect* 1996;117:29–34.
84. Hickman-Brenner FW, Stubbs AD, Farmer JJ III. Phage typing of *Salmonella enteritidis* in the United States. *J Clin Microbiol* 1991;29:2817–2823.
85. Centers for Disease Control. Multidrug-resistant *Salmonella* serotype Typhimurium—United States, 1996. *MMWR Morbid Mortal Wkly Rep* 1997;46: 308–310.
86. Wall PG, Threllfall EJ, Ward LR, Rowe B. Multiresistant *Salmonella typhimurium* DT104 in cats: A public health risk. *Lancet* 1996;348:471.
87. Altwegg M, Hickman-Brenner FW, Farmer JJ III. Ribosomal RNA gene restriction patterns provide increased sensitivity for typing *Salmonella typhi* strains. *J Infect Dis* 1989;160:145–149.
88. Olsen JE, Skov MN, Threlfall EJ, Brown DJ. Clonal lines of *Salmonella enterica* serotype Enteritidis documented by IS*200*-, ribo-, pulsed-field gel electrophoresis and RFLP typing. *J Med Microbiol* 1994;40:15–22.
89. Groisman EA, Fields PI, Heffron F. Molecular biology of *Salmonella* pathogenesis. In: Iglewski BH, Clark VL, eds. *Molecular basis of bacterial pathogenesis.* San Diego: Academic Press, 1990;251–272.
90. Foster JW, Hall HK. Adaptive acidification tolerance response of *Salmonella typhimurium. J Bacteriol* 1990;172:771–778.
91. Foster JW, Hall HK. Inducible pH homeostasis and the acid tolerance response of *Salmonella typhimurium. J Bacteriol* 1991;173:5129–5135.
92. Collinson SK, Emödy L, Müller K-H, Trust TJ, Kay WW. Purification and characterization of thin, aggregative fimbriae from *Salmonella enteritidis. J Bacteriol* 1991;173:1773–1781.
93. Müller K-H, Collinson SK, Trust TJ, Kay WW. Type 1 fimbriae of *Salmonella enteritidis. J Bacteriol* 1991; 173:4765–4772.
94. Lockman, HA, Curtiss R III. Virulence of non-type 1 fimbriated and nonfimbriated nonflagellated *Salmo-*

nella typhimurium mutants in murine typhoid fever. *Infect Immun* 1992;60:491–496.
95. Bäumler AJ, Heffron F. Identification and sequence analysis of lpfABCDE, a putative fimbrial operon of *Salmonella typhimurium. J Bacteriol* 1995;177: 2087–2097.
96. Bäumler AJ, Tsolis RM, Heffron F. The *lpf* fimbrial operon mediates adhesion of *Salmonella typhimurium* to murine Peyer's patches. *Proc Natl Acad Sci U S A* 1996;93:279–283.
97. Finlay BB, Leung KY, Rosenshine I, Garcia-del Portillo F. *Salmonella* interactions with epithelial cells. *ASM News* 1992;58:486–489.
98. Betts J, Finlay BB. Identification of *Salmonella typhimurium* invasiveness loci. *Can J Microbiol* 1993; 38:852–857.
99. Mills DM, Bajaj V, Lee CA. A 40 kb chromosomal fragment encoding *Salmonella typhimurium* invasion genes is absent from the corresponding region of the *Escherichia coli* K-12 chromosome. *Mol Microbiol* 1995;15:749–759.
100. Mecsas J, Strauss EJ. Molecular mechanisms of bacterial virulence: Type III secretion and pathogenicity islands. *Emerg Infect Dis* 1996;2:271–288.
101. Shea J, Hensel M, Gleeson C, Holden DW. Identification of a virulence locus encoding a second type III secretion system for *Salmonella typhimurium. Proc Natl Acad Sci U S A* 1996;93:2593–2597.
102. Barrow PA, Lovell MA. Invasion of Vero cells by *Salmonella* species. *J Med Microbiol* 1989;28:59–67.
103. Gahring LC, Heffron F, Finlay BB, Falkow S. Invasion and replication of *Salmonella typhimurium* in animal cells. *Infect Immun* 1990;58:443–448.
104. Douce GR, Amin II, Stephen J. Invasion of HEp-2 cells by strains of *Salmonella typhimurium* of different virulence in relation to gastroenteritis. *J Med Microbiol* 1991;35:349–357.
105. Verjans GM, Ringrose JH, van Alphen L, Feltkamp TEW, Kusters JG. Entrance and survival of *Salmonella typhimurium* and *Yersinia enterocolitica* within human B- and T-cell lines. *Infect Immun* 1994;62:2229–2235.
106. Lingren SW, Stoljilkovic I, Heffron F. Macrophage killing is an essential virulence mechanism of *Salmonella typhimurium. Proc Natl Acad Sci U S A* 1996;93: 4197–4201.
107. Monack DM, Raupach B, Hromockyj AE, Falkow S. *Salmonella typhimurium* invasion induces apoptosis in infected macrophages. *Proc Natl Acad Sci U S A* 1996;93:9833–9838.
108. Gulig PA. Virulence plasmids of *Salmonella typhimurium* and other salmonellae. *Microb Pathog* 1990;8:3–11.
109. Gulig PA, Danbara H, Guiney DG, Lax AJ, Norel F, Rhen M. Molecular analysis of *spv* virulence genes of the salmonella virulence plasmid. *Mol Microbiol* 1993;7:825–830.
110. Guiney DG, Fang FC, Krause M, Libby S. Plasmid-mediated virulence genes in non-typhoid *Salmonella* serovars. *FEMS Microbiol Lett* 1994;124:1–10.
111. Guiney DG, Fang FC, Krause M, Libby S, Buchmeier NA, Fierer J. Biology and clinical significance of virulence plasmids in *Salmonella* serovars. *Clin Infect Dis* 1995;21(Suppl):S146–S151.
112. Fierer J, Krause M, Tauxe R, Guiney D. *Salmonella typhimurium* bacteremia: Association with the virulence plasmid. *J Infect Dis* 1992;166:639–642.
113. Libby SJ, Adams LG, Ficht TA, et al. The *spv* genes on the *Salmonella dublin* virulence plasmid are required for severe enteritis and systemic infection in the natural host. *Infect Immun* 1997;65:1786–1792.
114. Vandenbosch JL, Rabert DK, Kurlandsky DR, Jones GW. Sequence analysis of *rsk,* a portion of the 95-kilobase plasmid of *Salmonella typhimurium* associated with resistance to the bactericidal activity of serum. *Infect Immun* 1989;57:850–857.
115. Hefferman EJ, Harwood J, Fierer J, Guiney D. The *Salmonella typhimurium* virulence plasmid complement resistance gene *rck* is homologous to a family of virulence-related outer membrane protein genes, including *pagC* and *ail. Infect Immun* 1992;174:84–91.
116. Prager R, Fruth A, Tschäpe H. *Salmonella* enterotoxin *(stn)* gene is prevalent among strains of *Salmonella enterica,* but not among *Salmonella bongori* and other *Enterobacteriaceae. FEMS Immunol Microbiol* 1995; 12:47–50.
117. Kita E, Kamikaidou N, Nakano A, Kashiba S. Isolation of a cytotoxin from L-form *Salmonella typhimurium. FEMS Microbiol Lett* 1993;109:179–184.
118. Farthing M, Feldman R, Finch R, et al. The management of infective gastroenteritis in adults: A consensus statement by an expert panel convened by the British Society for the Study of Infection. *J Infect* 1996;33: 143–152.
119. Lee LA, Puhr ND, Maloney K, Bean NH, Tauxe RV. Increase in antimicrobial-resistant *Salmonella* infections in the United States, 1989–1990. *J Infect Dis* 1994;170:128–134.
120. Gomez HF, Cleary TG. Drug-resistant *Salmonella, Shigella,* and diarrhea-associated *Escherichia coli. Pediatr Infect Dis* 1996;7:212–222.
121. Pickering LK. Emerging antibiotic resistance in enteric bacterial pathogens. *Pediatr Infect Dis* 1996;7: 272–280.
122. Aserkoff B, Bennett JV. Effect of therapy in acute salmonellosis on salmonellae in feces. *N Engl J Med* 1969;281:636–640.
123. Rahal K, Wang F, Schindler J, et al. Reports on surveillance of antimicrobial resistance in individual countries. *Clin Infect Dis* 1997;24(Suppl):S169–S175.
124. Vahaboglu H, Dodanli S, Eroglu C, et al. Characterization of multiple-antibiotic-resistant *Salmonella typhimurium* strains: Molecular epidemiology of PER-1 producing isolates and evidence for nosocomial plasmid exchange by a clone. *J Clin Microbiol* 1996;34: 2942–2946.
125. Hohl P, Lüthy-Hottenstein J, Zollinger-Iten J, Altwegg M. In vitro activities of fleroxacin, cefetamet, ciprofloxacin, ceftriaxone, trimethoprim-sulfamethoxazole, and amoxacillin-clavulanic acid against rare members of the family *Enterobacteriaceae* primarily of human (clinical) origin. *Antimicrob Agents Chemother* 1990;34:1605–1608.
126. Fuchs PC, Barry AL, Brown SD, AST Surveillance Group. Survey of antimicrobial activity of four commonly used third generation cephalosporins tested against recent bacterial isolates from ten American medical centers, and assessment of disk diffusion test performance. *Diagn Microbiol Infect Dis* 1996;24: 213–219.
127. Thornsberry C. Susceptibility of clinical bacterial isolates to ciprofloxacin in the United States. *Infection* 1994;22(Suppl):S80–S89.

7

Salmonella Typhi and *Salmonella* Paratyphi A, B, and C

TAXONOMY

Typhoid fever was defined as a disease in 1829, but the responsible agent was not discovered until 55 years later, when Gaffkey detected the organism for the first time in the spleens of typhoid patients.[1] However, Schroeter is credited with the isolation of *Salmonella* Typhi, an organism he called *Bacillus typhi,* in 1886.[2] Although Lignières established the genus *Salmonella* in 1900, *Bacillus typhi* was not transferred to *Salmonella* as *S.* Typhi until 1930. In 1902, an organism causing a typhoid-like disease was detected and called *Bacterium paratyphi* typus A, the name "paratyphi" meaning "alongside typhi."[2] *Salmonella* Paratyphi B was also described in 1902; it was called *Bacillus paratyphi* typus B and was later known as *Salmonella schottmulleri.* Twenty-five years later, Hirschfeld isolated an agent he called paratyphoid C bacillus, which ultimately was named *Salmonella hirschfeldii* in his honor. In 1934, it was renamed *Salmonella* Paratyphi C.

Because all *Salmonella* form a single species,[3] called *Salmonella choleraesuis,*[4] *S.* Typhi and *S.* Paratyphi A, B, and C are actually serotypes of *S. choleraesuis* ssp. *choleraesuis* (see Chapter 6). As with other serotypes, they are referred to as *Salmonella* serotype Typhi or *Salmonella* Typhi for convenience and because of their clinical significance.

EPIDEMIOLOGY

S. Typhi and *S.* Paratyphi A and C are host-adapted to humans only. *S.* Paratyphi B has a D-tartrate–positive variant, *Salmonella* Java, that can cause enteritis in animals as well as humans. Although these organisms are restricted to humans, fecal excretion and subsequent contamination of food, and especially of water, permits their widespread transmission. *S.* Typhi is endemic in economically impoverished countries with poor water and sanitary systems. The World Health Organization estimated that there were 12.5 million cases worldwide in 1980, with 7 million infections in Southeast Asia alone.[5] In Western countries such as the United States, rates have dropped from more than 35,000 cases in 1920 to 500 in 1991.[5] In industrialized areas the incidence is now 0.24 to 3.7 cases per 100,000 population, compared with 500 cases per 100,000 in the developing world excluding China.[1]

Mortality rates differ considerably also, probably because of differences in access to health care rather than host or organism virulence factors. Death rates as high as 12% to 32% have been reported from Indonesia, India, and Nigeria,[5] whereas the incidence of mortality in the United States is 1.3%.

The epidemiology of *S.* Typhi in the United States is relatively simple (Table 1). Foreign travel is a major source of infection, with 74% of cases in 1984 acquired in this manner.[1] Forty-five percent of cases were associated with travel to Mexico, 20% and 15%, respectively, with travel to Asia and India.[6] Of domestic (U.S.) typhoid cases, almost one fourth represent known or newly discovered carriers, another quarter are associated with outbreaks involving contaminated food or water, 6% are

TABLE 1. *Epidemiology of Salmonella Typhi in the United States*

Parameter	*S.* Typhi
Morbidity	0.24–3.7/ 100,000 population
Mortality	1.3%
How acquired	
Foreign travel	74% of cases
Mexico	
Asia	
Indian subcontinent	
South America	
Central America	
Domestic cases	
Established carrier	8%
New carrier	15%
Outbreaks	24%
Contaminated food or water	23%
Household contact	6%
Laboratory-acquired	4%
Incubation period	7–21 d (average 14 d)
Duration of symptoms	2–4 wk, untreated
Carrier state	
Convalescent (90 d)	10%
Chronic	3%–5%

acquired by person-to-person contact, and 4% are laboratory-acquired.[6] In a retrospective report on the characteristics of typhoid fever in children in a metropolitan area of the United States, Misra and colleagues[7] noted the following: 43 (78%) of 55 patients had traveled outside the country; 61% (38/43) to India or Pakistan, 34% to Central or South America, and 5% to Africa. Of those who had not traveled, five had a household contact who was a typhoid carrier, three had had contact with a person recently arrived from an endemic area, and four had no discernible source.

Large outbreaks are rare in the United States. In 1981, 80 cases were reported over a 9-week period in persons who had consumed barbacoa, a meat dish made from steamed bovine heads.[8] In 1989, 43 culture-confirmed and 24 probable cases were linked to orange juice consumption at a restaurant.[9] Two persons from this outbreak underwent surgery for intestinal perforation. In neither outbreak was the source of contamination determined, although in both situations serious food-handling breaches were identified.

Carrier rates for temporary excretors (approximately 90 days) range from 7% to 20%; about 2% to 5% of persons who have typhoid fever become chronic carriers.[10] Women tend to become carriers more often than men. As with nontyphoidal salmonellae, when proper culture collection (feces as opposed to swabs) and laboratory techniques are used, excretors are found to shed fairly constant numbers of organisms on a regular basis rather than intermittently.[10] Also, as with nontyphoidal salmonellae, the actual infecting dose is often 10^4 organisms or less, although ranges determined by volunteer studies are higher at 10^4 to 10^7 organisms.[10]

The incubation period is prolonged. With *S.* Typhi it ranges from 7 to 21 days (average 14 days); times as long as 60 days have been reported.[5,10,11] Therefore, new cases related to an outbreak often continue to occur even after a source has been identified and transmission interrupted.

CLINICAL SIGNIFICANCE

Cardinal features of enteric fever, also called typhoid or paratyphoid fever, include prolonged fever, abdominal pain, hepatomegaly and/or splenomegaly, and rose spots. However, these and other symptoms (Table 2) are variable. Although patients are usually acutely ill, some present with milder disease, especially early in the course of the illness. In one outbreak report it was believed that only 20% of those infected had a typical clinical illness.[10] Diarrhea is irregularly present; if present, it occurs early. Diarrhea occurs in 33% to 50% of cases[12] but may be more common in patients with the acquired immunodeficiency syndrome (AIDS) and in children younger than 1 year of age.[5] Roy and associates,[12] studying 42 Bangladeshi patients with blood culture–positive enteric fever, reported that 100% had diarrhea which began about 3.5 days after the fever. All stools were watery (98%) and contained polymorphonuclear leukocytes (70%) or monocytes (30%). Only 7% of stools were visibly bloody, but 29% were guaiac test–positive. Leukocyte counts varied from 400 to 16,500 cells/mm^3. In the report of Misra and colleagues[7] regarding

TABLE 2. *Characteristics of untreated enteric fever*

Time	Symptoms	Cultures
Week 1	Early symptoms Diarrhea (33%–50%) Fever (99%) Abdominal pain (20%–40%) Muscle aches (59%) Cough/sore throat (20%–40%) Malaise (90%) Anorexia (80%) Headache (34%) Anemia	Blood: usually negative Stool: transiently positive Bone marrow: negative Urine: negative
Week 2	Established disease Bacteremia Leukocytosis Hepatosplenomegaly (50%) Rose spots (30%–60%) Neuropsychiatric manifestations: confusion, psychosis (5%–10%)	Blood: 80%–90% positive Stool: 80% positive Bone marrow: 80%–90% positive Urine: 25% positive Rose spots: 60% positive (punch biopsy)
Week 3 to 5	Late stage/convalescence: complications Relapse (3%–15%) Intestinal hemorrhage (3%–20%) Intestinal perforation (3%) Cholecystitis[a] Osteomyelitis[a] Focal infections: pericarditis/myocarditis, orchitis, soft tissue infections (spleen and liver)[a]	Blood: usually negative unless relapse Stool: 50% positive Bone marrow and rose spots: decreasing positive numbers Urine: 10% positive

Data from Pegues et al.[5] and Goldberg & Rubin.[11]
[a]Most common complications; can occur years after initial infection.

typhoid fever in children in the United States, 77% had diarrhea; 54% had positive stool cultures and 20% were both stool- and blood culture-positive.

Intestinal or ileal perforation can occur if areas of hyperplasia within Peyer's patches develop into necrotic ulcers.[13] The incidence varies geographically but is about 3% worldwide, with a mortality rate of approximately 40%.[13] Mortality rates increase to as high as 83% if the delay between onset and treatment is longer than 96 hours. Survival rates are higher with surgical intervention than if only antimicrobials are administered (74% versus 30%, respectively).[13] Free air in the peritoneum on radiography is insensitive for the detection of perforation.[13] High remittent fever, abdominal pain, and change in bowel habits are typical symptoms but are not diagnostic for perforation.

Fever, initially low-grade (37.2°C [99°F]), increases stepwise for 1 to 2 weeks, then slowly declines. It may be preceded by a variety of nonspecific, flu-like symptoms including malaise, body aches, lack of appetite, cough, and sore throat. About 50% of patients with enteric fever develop hepatomegaly or splenomegaly, or both. Rose spots occur in 10% to 30% of patients but are not specific for enteric fever.[11] They appear as small (2–4 mm) pink macules, and punch biopsies of these lesions are frequently culture-positive for *S.* Typhi. Neuropsychiatric symptoms, generally psychosis and confusion, occur in the second or third week of disease[5,11] in 5% to 10% of patients. Seizures are rare, and cerebrospinal fluid is normal. Van de Wetering and coworkers[14] reported on a rare case of cerebral edema complicating enteric fever. The patient had been receiving antimicrobial therapy for 10 days at the time the edema developed; he survived with no residual sequelae after administration of mannitol and dexamethasone reduced the swelling in his brain.

Relapse occurs in about 3% to 15% of patients, usually during the third or fourth week of disease, and according to Goldberg and Rubin[11] is not altered by therapy. Although the list of complications is long,[5,11] most are infrequently encountered; osteomyelitis, cholecystitis, and soft tissue abscesses, especially of the spleen and liver, occur most often. Several reports have highlighted some of the more unusual aspects of enteric fever, including hair loss,[15] sternal osteomyelitis in an immunocompetent patient,[16] and bacteriuria.[17] Disease begins to resolve after about 3 to 4 weeks without antibiotic therapy, but convalescence can be prolonged, with weakness, weight loss, and general debilitation lasting for months.

LABORATORY IDENTIFICATION

Isolation

Isolation of *S.* Typhi and paratyphoid organisms from blood is variable depending on the stage of disease (see Table 2) and whether antimicrobial therapy has been initiated. It is further hampered by the low numbers of organisms usually present (for *S.* Typhi, <15/mL of blood).[5] Use of centrifuged blood to concentrate mononuclear cells and platelets does not increase culture sensitivity.[5] However, use of a medium with 10% bile (Oxgall) in distilled water does enhance isolation.[18] When either whole or clotted blood was added to Oxgall medium, almost 15% more *S.* Typhi and *S.* Paratyphi A were recovered than when trypticase soy broth (TSB) was inoculated with whole blood.[18] Escamilla and colleagues[18] showed that, although a blood clot used for inoculum was as sensitive as whole blood, the use of whole blood with bile medium resulted in isolation 1 to 3 days faster than either clot with bile medium or whole blood with TSB.

Bone marrow culture should be considered if other sites are negative for *S.* Typhi. Sensitivity of bone marrow culture may be greater (90%) than blood, stool, or urine.[5,11] Also, growth of *S.* Typhi from cultures of bone marrow or punch biopsy material from rose spots is unaffected (ie, not inhibited) by antimicrobial therapy.

Isolation of *S.* Typhi from stool is equally as variable as from the blood and likewise changes with the stage of disease (see Table 2). Feces are preferable to rectal swabs. Although direct plating should always be carried out with routine enteric isolation media, maximum yield is obtained with enrichment (particularly with selenite F cysteine) followed by plating to enteric media. If urine is cultured, both direct plating onto enteric isolation media and enrichment (1 mL of urine per 10 mL of enrichment broth) should be used. Of the specialized media developed for isolation of *Salmonella,* bismuth sulfite achieves the best recovery of *S.* Typhi and paratyphoid organisms (see Chapter 6). Use of duodenal string capsules in conjunction with blood and bone marrow culture increased the positivity rate from 84% (from both blood and bone marrow) to 98% (from all three sites) in one study.[11]

In conventional screening media *S.* Typhi appear similar to other *Salmonella,* except that in triple sugar iron (TSI) agar it shows no gas and a very weak H_2S reaction. The H_2S reaction can appear as a ring of blackening at the interface of the butt and slant or as a dot in the stab (at the butt/slant interface), or it can be negative altogether (Fig. 1; Color Plate 14). If motility-indol-ornithine (MIO) medium is used to screen for *S.* Typhi, strains may appear weakly motile or non-

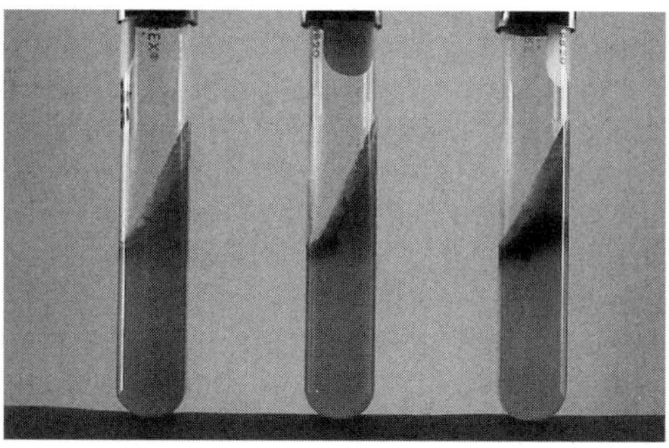

FIG. 1. Production of H_2S by *S.* Typhi in TSI. See also Color Plate 14.

TABLE 3. *Separation of* Salmonella *Typhi and* Salmonella *Paratyphi A from other subgenus I salmonellae*

Organism	TSI	LDC	ODC	Citrate	Gas from Glucose	Acid from Xylose
S. Typhi	K/A^w	+	–	–	–	V
S. Paratyphi A	K/AG^-	–	+	V	+	–
Other salmonellae (subgenus I)	K/AG^+	+	+	+	+	+

TSI, triple sugar iron agar; LDC, lysine decarboxylase; ODC, ornithine decarboxylase; K, alkaline; A, acid; G, gas; w, weak reaction; +, ≥90% positive; V, 10%–90% positive; –, ≤10% positive.

motile. *S.* Paratyphi A cultures are negative for H_2S in TSI (although rare weak positive strains occur) and are lysine decarboxylase–negative, so they appear negative (yellow) in the butt with lysine-iron agar (LIA) screening medium.

Identification

For commercial systems, see Chapter 6. With conventional media, *S.* Typhi is biochemically distinctive and easily separated from other *Salmonella* (H_2S weak-positive, anaerogenic, ornithine decarboxylase- and citrate-negative) and from members of the *Enterobacteriaceae. S.* Paratyphi A also has differential biochemical characteristics (negative for H_2S, citrate, and lysine decarboxylase) that set it apart. However, *S.* Paratyphi B and C biochemically resemble most other subgenera I members of the genus *Salmonella* and must be distinguished from them serologically. Table 3 lists biochemicals useful in identifying *S.* Typhi and *S.* Paratyphi A.

Immunodiagnostics

In the Kauffmann-White schema (see Chapter 6), *S.* Typhi belongs to group D and is monophasic, possessing only flagellar factor "d." *S.* Typhi and some strains of *S.* Paratyphi C possess a capsule, commonly referred to as the Vi (virulence) antigen, which envelops the somatic or cell surface antigen. To serologically identify these strains, a saline-prepared antigen (AG) is split and one portion is heated to 100°C (occasionally longer is required). Both heated and unheated AGs are tested with Vi, group D (group C_1, if Paratyphi C), and normal rabbit antisera. The unheated AG should agglutinate in the Vi serum but not in the group D serum, whereas the reverse reaction is expected with the heated AG (negative in Vi, positive in group D). Along with the correct biochemical profile, this serologic pattern is sufficient to diagnose *S.* Typhi, making flagellar typing unnecessary. Some strains of *S.* Typhi do not possess Vi AG (but are nonetheless virulent), and those cultures require flagellar typing for complete identification. Because the capsule is spontaneously and easily lost, cultures may contain both Vi-positive and Vi-negative cells, in which case agglutination occurs in both Vi and group D sera without heating the AG. With some strains it is impossible to remove the Vi AG regardless of the duration of heating. Flagellar typing must also be performed to correctly identify these strains.

Typing

Phage typing has been, and often still is, the primary method used to type strains of *S.* Typhi (Fig. 2).[19] However, 4% to 5% of strains are untypable, and a small number of types (E_1, M_1) account for most strains seen.[20] Multilocus enzyme electrophoresis of 26 strains of *S.* Typhi yielded an identical isoenzyme pattern for all strains,[20] and therefore this method cannot be considered a useful epidemiologic tool for typing. However, it and other methods (eg, analysis of outer membrane proteins) indicate that *S.* Typhi is clonal.[20–22] Selander and coworkers[23] found only two clones, one restricted to West Africa

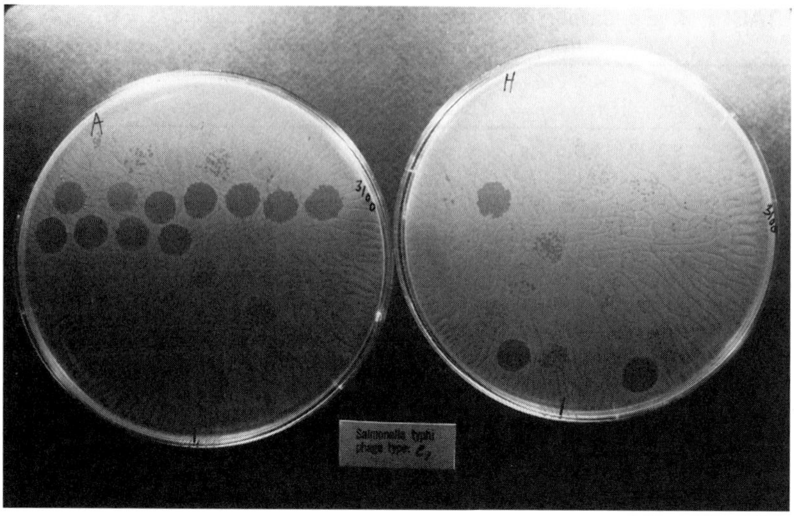

FIG. 2. Phage typing of *S.* Typhi, type C_1 shown.

and the other of worldwide distribution. Use of plasmids to characterize isolates is also unhelpful because they are lacking (only 6% to 33% positive) in most strains.[24] Insertion sequence typing (IS*200*) is unproductive as well.[25] Other molecular methods, such as ribotyping and pulsed-field gel electrophoresis, are able to discriminate among isolates of *S.* Typhi despite its clonal nature.[25,26]

PATHOGENESIS

S. Typhi and *S.* Paratyphi A, B, and C are invasive pathogens. To reach the small intestine, the site of infection, they must first survive gastric acidity. This is achieved by an acid tolerance response (see Chapter 6) which is possessed by all *Salmonella,* including typhoid strains. As with other *Salmonella,* M (microfold) cells overlying Peyer's patches are a preferential target for typhoid and paratyphoid bacilli, although other intestinal epithelial cells can be induced to phagocytize the bacterial cell through bacteria-mediated endocytosis.[5] Regardless of the method, after uptake the organism interacts with macrophages and lymphocytes or other lymphoid tissue in the submucosa. Survival within macrophages (see Chapter 6) allows spread from the intestine to the mesenteric lymph nodes, followed by passage through the thoracic duct to the systemic circulation, infecting the liver and biliary tree.[11] Intact motility,[27] iron transport systems (enterochelin),[22] and the Vi antigen (which prevents binding by complement C3, reducing complement killing) are all additional, important virulence factors in typhoidal strains. For a more detailed description of *Salmonella* pathogenesis, see Chapter 6.

ANTIMICROBIAL SUSCEPTIBILITY

Although *S.* Typhi strains resistant to ampicillin (A), chloramphenicol (C), and tetracycline (T) were noted as early as 1967, epidemics with such multidrug-resistant (MDR) strains first occurred in 1972 in Mexico, India, Vietnam, Korea, and Peru.[28,29] Resistance to CSSuT (chloramphenicol, streptomycin [S], sulfonamide [Su], and tetracycline) or resistance to CSuT plus aminoglycosides, borne on plasmids of the H1 incompatibility group, caused widespread morbidity and mortality. Such strains became endemic in Southeast Asia so rapidly that, by 1989, 80% of *S.* Typhi strains carried a 98-MDa plasmid encoding resistance to ACGSuTTm (ACSuT plus gentamicin [G] and trimethoprim [Tm]) and cephazolin.[28] Likewise, in Pakistan, 90% of *S.* Typhi became resistant to ACSSuTTm. In India, by

1990, MDR *S.* Typhi strains with H1 incompatibility group plasmids of 110 to 120 MDa became predominant.[28] Immigrant workers from these countries introduced multidrug resistance to countries of the Persian Gulf, so that today 30% to 40% of *S.* Typhi isolated from this region are resistant as well.[28,29] Although Africa, South America, and Central America continue to suffer large outbreaks, most *S.* Typhi strains remain susceptible to one or more of the first-line drugs used in treatment (ie, chloramphenicol or ampicillin). In the United Kingdom, chloramphenicol remained the drug of choice for treatment of enteric fever until 1990, when soaring resistance to this drug forced a change to fluoroquinolones.[29] Resistance rose from 1.5% in the 4-year interval between 1986 and 1989 to 26% in 1990–1994.[29] In 1990, 20% of strains were also resistant to ampicillin and tetracycline, as well as chloramphenicol. By 1995, resistance to chloramphenicol had jumped to 34% and almost all strains were resistant to ampicillin and trimethoprim. *S.* Typhi containing plasmids bearing resistance to ACSSuTTm and belonging to incompatibility group H1 were isolated from patients returning from India and Pakistan. In the United States, resistance in *S.* Typhi remains at about 2% to 3%, with strains resistant primarily to chloramphenicol and ampicillin.[30] Ling and associates[31] reported that 88% of *S.* Typhi isolated in Hong Kong are susceptible to 21 antimicrobial agents, including 6 cephalosporins, aztreonam, imipenem, 4 aminoglycosides, 4 fluoroquinolones, trimethoprim, sulfamethoxazole, tetracycline, and chloramphenicol. Chloramphenicol resistance occurred in only 8 of 182 strains, 4 of which were MDR. One strain was resistant to nine antibiotics: ACSuTTm, cephalothin, cefuroxime, cefamandole, and nalidixic acid. Most resistance was carried on 106-, 116-, or 211-kb plasmids that belonged not only to incompatibility group H but also to I1 and K. Four of five ampicillin-resistant strains produced a β-lactamase with an isoelectric point of 5.5.

Although ciprofloxacin has become the drug of choice in countries with high percentages of MDR *S.* Typhi, it has not been immune to emerging drug resistance either. In 1992, a ciprofloxacin-resistant strain was reported in the United Kingdom.[32] Also resistant to ACTm via a plasmid, resistance to ciprofloxacin in this strain (and all subsequent strains so far) was found to be chromosomally encoded. It was isolated from a 1-year-old child with enteric fever who failed to respond to ciprofloxacin therapy. In 1994 and 1995, nine ciprofloxacin-resistant strains were identified. They were MDR as well and were isolated from patients who had traveled to the Indian subcontinent.[32] In 1991, 1992, and 1995, strains that were resistant only to ciprofloxacin were isolated; resistance was chromosomally mediated in all cases. However, in two of these patients a second *S.* Typhi strain resistant to ACTm (but ciprofloxacin-susceptible) was cultured simultaneously with the ciprofloxacin-resistant strain from blood.[32]

Plasmid-borne multidrug resistance is associated with greater virulence in terms of both increased severity of disease and frequency of complications.[28,33] Jesudason and colleagues[33] reported on an epidemic caused by an MDR strain of *S.* Typhi that lasted approximately 2 years. During this time a chloramphenicol-susceptible strain of *S.* Typhi, endemic before the outbreak, remained constant without increasing or decreasing in prevalence. The authors proposed that the increase in MDR strains may have been caused by virulence properties present on the plasmid that conferred the multidrug resistance.

In countries such as the United States with a low prevalence of drug resistance, chloramphenicol, ampicillin, and trimethoprim-sulfamethoxazole may still be used successfully for treatment of enteric fever. In areas of the world with high levels of MDR *S.* Typhi, treatment, which must be initiated before antimicrobial susceptibility results are known, should comprise either a fluoroquinolone or an expanded-spectrum cephalosporin. Because of the severity of disease associated with MDR *S.* Typhi, fluoroquinolone therapy is recommended even in children; the attendant risk of arthropathy from these drugs is the lesser of two evils. Although cefotaxime, ceftriaxone, and cefoperazone are efficacious against MDR *S.* Typhi, a fluoroquinolone yields better results than an expanded-spectrum cephalosporin.[31,34] Significant differences in regard to cure ($P < 0.01$) were found between ofloxacin (given 5 days) and ceftriaxone (given 8 days).[31] Although other antimicrobials may show good activity in vitro, many are ineffective in vivo; for example, aminoglycosides are ineffective because of their poor ability to kill intracellular organisms. Even with effective therapy, fever may continue for as long as 1 week after cessation of bacteremia.[34] Relapse of symptoms also may occur despite appropriate treatment.

CONCLUSIONS

An effective vaccine holds the greatest promise for eliminating enteric fever. Over the years a number of vaccines have been developed, but none has enough efficacy to recommend its widespread use, particularly for travelers to endemic areas of the world.[5] The older, heat/phenol-inactivated vaccine for *S.* Typhi was only 51% to 77% effective. It was followed by an acetone-inactivated vaccine with rates of effectiveness reported at 55% to 88%[11] and 79% to 94%.[5] However, this vaccine is expensive and has many side effects. Ty21a, a live-attenuated oral vaccine, has an efficacy of 43% to 96%[5,11] in persons living in endemic areas and has virtually no side effects. However, in a small study of travelers to endemic areas it did not show any efficacy, and it is not recommended in children younger than 6 years of age because of the potential for developing infection. Because it is live, refrigeration of the capsules is required during the 7 days needed for administration (one capsule every other day); this is a major drawback in developing areas of the world, where refrigeration may not be available. Finally, a Vi capsular polysaccharide vaccine administered intramuscularly has undergone trials and has shown an effective rate of 75%.[11] It is stable and may be kept at room temperature, advantages over the oral vaccine. Adverse reactions occur but do not compare with those of the acetone-inactivated vaccine. The phenol-inactivated and oral vaccines both require boosters every 5 years; the Vi capsule vaccine showed immunity for 17 months[11] and will undoubtedly require a booster as well. For now, the oral vaccine offers the greatest protection (without major side effects) for those traveling to developing countries for a prolonged period. However, the need for effective vaccines with minimal adverse reactions and lasting immunity, not only for travelers but for those in living in endemic areas, is of paramount importance to reduce the morbidity and mortality associated with this disease.

Vaccines alone cannot eliminate this disease. In order for vaccines to have a major impact, drinking water and sanitation systems in developing countries must also improve. World Health Organization workers estimate from field observation data that even a remedy as simple as construction of latrines diminishes the rate of transmission by 50%.[10] They also assert that when sanitation is introduced with health education and other changes that raise the standard of living, the cumulative effects result in a steady decline in infections.

Perhaps the most frightening aspect of typhoid fever today is the rapid resistance that has developed in strains in endemic areas, so that all first-line drugs used in treatment are virtually useless. Indiscriminate use of the drugs remaining for treatment of these infections will ultimately result in our inability to treat a disease that absolutely requires antibiotic intervention.

REFERENCES

1. Edelman R, Levine MM. Summary of an international workshop on typhoid fever. *Rev Infect Dis* 1986;8: 329–349.
2. LeMinor L. *Salmonella* Lignières 1900, 389[AL]. In: Kreig NR, Holt JG, eds. *Bergey's manual of systematic bacteriology*. Baltimore: Williams & Wilkins, 1984;427–458.
3. Crosa JH, Brenner DJ, Ewing EH, Falkow S. Molecular relationship among the *Salmonellae*. *J Bacteriol* 1973; 115:307–315.
4. LeMinor L, Popoff MY. Request for an opinion: Designation of *Salmonella enterica* sp. nov., nom. rev., as the

type and only species of the genus *Salmonella*. *Int J Syst Bacteriol* 1987;37:465–468.
5. Pegues DA, Hohlmann EL, Miller SI. *Salmonella* including *S. typhi*. In: Blaser MJ, Smith PD, Ravdin JI, Greenberg HB, Guerrant RL, eds. *Infections of the gastrointestinal tract*. New York: Raven Press, 1995; 785–809.
6. Taylor DN, Pollard RA, Blake PA. Typhoid in the United States and the risk to the international traveler. *J Infect Dis* 1983;148:599–602.
7. Misra S, Diaz PS, Rowley AH. Characteristics of typhoid fever in children and adolescents in a major metropolitan area in the United States. *Clin Infect Dis* 1997; 24:998–1000.
8. Taylor JP, Shandera WX, Betz TG, et al. Typhoid fever in San Antonio, Texas: An outbreak traced to a continuing source. *J Infect Dis* 1984;149:553–557.
9. Birkhead GS, Morse DL, Levine WC, et al. Typhoid fever at a resort hotel in New York: A large outbreak with an unusual vehicle. *J Infect Dis* 1993;167:1228–1232.
10. World Health Organization. Dynamics of acute bacterial diseases. Part II: Typhoid fever. *Bull World Health Organ* 1978;56:45–63.
11. Goldberg MB, Rubin RH. The spectrum of *Salmonella* infection. *Infect Dis Clin North Am* 1988;2:571–598.
12. Roy SK, Speelman P, Butler T, Nath S, Rahman H, Stoll BJ. Diarrhea associated with typhoid fever. *J Infect Dis* 1985;151:1138–1143.
13. van Basten JP, Stockenbrügger R. Typhoid perforation. *Trop Geogr Med* 1994;46:336–339.
14. van de Wetering J, Visser LG, van Buchem MA, van der Hoeven JG. A case of typhoid fever complicated by unexpected diffuse cerebral edema. *Clin Infect Dis* 1995; 21:1057–1058.
15. Haefeli WE, Ott R, Bircher AJ, Burnens AP. Hair loss following typhoid fever: A forgotten phenomenon. *Clin Infect Dis* 1995;20:723–724.
16. Weiss K, Laverdière M. *Salmonella typhi* osteomyelitis in an immunocompetent patient. *Clin Microbiol Newslett* 1995;17:135–136.
17. Mathai E, John TJ, Rani M, et al. Significance of *Salmonella typhi* bacteriuria. *J Clin Microbiol* 1995;33: 1791–1792.
18. Escamilla J, Florez-Ugarte H, Kilpatrick ME. Evaluation of blood clot cultures of *Salmonella typhi, Salmonella paratyphi-A,* and *Brucella melitensis*. *J Clin Microbiol* 1986;24:388–390.
19. Anderson ES, Fraser A. The influence of the factors determining Vi-type specificity in *Salmonella typhi* on the adaptation of Vi-phage II. *J Gen Bacteriol* 1995;13: 519–532.
20. Reeves MW, Evins GM, Heiba AA, Plikaytis BD, Farmer JJ III. Clonal nature of *Salmonella typhi* and its genetic relatedness to other salmonellae as shown by multilocus enzyme electrophoresis, and proposal of *Salmonella bongori* comb. nov. *J Clin Microbiol* 1989;27: 313–320.
21. Franco A, Gonzalez C, Levine O, et al. Further consideration of the clonal nature of *Salmonella typhi:* Evaluation of molecular and clinical characteristics of strains from Indonesia and Peru. *J Clin Microbiol* 1992;30: 2187–2190.
22. Faundez G, Aron L, Cabello FC. Chromosomal DNA, iron-transport systems, outer membrane proteins, and enterotoxin (heat labile) production in *Salmonella typhi* strains. *J Clin Microbiol* 1990;28:894–897.
23. Selander RK, Beltran P, Smith NH, et al. Evolutionary genetic relationships of clones of *Salmonella* serovars that cause human typhoid and other enteric fevers. *Infect Immun* 1990;58:2262–2275.
24. Finch MJ, Franco A, Gotuzzo E, et al. Plasmids in *Salmonella typhi* in Lima, Peru, 1987–1988: Epidemiology and lack of association with severity of illness or clinical complications. *Am J Trop Med Hyg* 1992;47:390–396.
25. Navarro F, Llovet T, Echeita MA, et al. Molecular typing of *Salmonella enterica* serovar Typhi. *J Clin Microbiol* 1996;34:2831–2834.
26. Fica AE, Prat-Miranda S, Fernandez-Ricci A, D'Ottone K, Cabello FC. Epidemic typhoid in Chile: Analysis by molecular and conventional methods of *Salmonella typhi* strain diversity in epidemic (1977 and 1981) and nonepidemic (1990) years. *J Clin Microbiol* 1996;34: 1701–1707.
27. Liu SL, Ezaki T, Miura H, Matsui K, Yabuuchi E. Intact motility as a *Salmonella typhi* invasion-related factor. *Infect Immun* 1988;56:1967–1973.
28. Mizra SH, Beeching NJ, Hart CA. Multi-drug resistant typhoid: A global problem. *J Med Microbiol* 1996;44: 317–319.
29. Rowe B, Ward LR, Threlfall EJ. Multi-drug resistant *Salmonella typhi:* A worldwide epidemic. *Clin Infect Dis* 1997;24(Suppl):S106–S109.
30. Pickering LK. Emerging antibiotic resistance in enteric bacterial pathogens. *Pediatr Infect Dis J* 1996;7: 272–280.
31. Ling JM, Lo NWS, Ho YM, et al. Emerging resistance in *Salmonella enterica* serotype Typhi in Hong Kong. *Int J Antimicrob Agents* 1996;7:161–166.
32. Rowe B, Ward LR, Threlfall EJ. Ciprofloxacin-resistant *Salmonella typhi* in the UK. *Lancet* 1995;346:1302.
33. Jesudason MV, John R, John TJ. The concurrent prevalence of chloramphenicol-sensitive and multidrug-resistant *Salmonella typhi* in Vellore, S. India. *Epidemiol Infect* 1996;116:225–227.
34. Gomez HF, Cleary TG. Drug-resistant *Salmonella, Shigella,* and diarrhea-associated *Escherichia coli*. *Pediatr Infect Dis J* 1996;7:212–222.

8

The Genus *Klebsiella*

TAXONOMY

Klebsiella is among the oldest genera in the *Enterobacteriaceae.* Established in 1885,[1,2] its four most commonly encountered species had all been described by the turn of the century. This list included *Klebsiella pneumoniae* (Friedländer's pneumobacillus), *Klebsiella oxytoca, Klebsiella ozaenae,* and *Klebsiella rhinoscleromatis,* although neither *K. oxytoca* nor *K. ozaenae* was placed in the genus *Klebsiella* until later (Table 1).

For the most part, the klebsiellae have escaped the tumultuous taxonomic changes that other *Enterobacteriaceae* have endured. However, for almost 70 years microbiologists were unable to separate *K. pneumoniae* from *Aerobacter aerogenes (Enterobacter aerogenes).*[1–4] This conflict should have been resolved by the introduction of Möeller's decarboxylase tests and the subsequent transfer of *A. aerogenes* to the new genus *Enterobacter* in 1960. Instead, in 1960 Cowan and associates[5] categorized *Klebsiella* into six groups based on fimbriation and 13 biochemical tests. The most notable aspects of this proposal were the inclusion of nonmotile strains of *A. aerogenes* as *Klebsiella aerogenes* and the introduction of a new species with two subspecies, *Klebsiella edwardsii* ssp. *edwardsii* and *K. edwardsii* ssp. *atlantae. K. pneumoniae, K. ozaenae,* and *K. rhinoscleromatis* comprised the remaining three categories. The next landmark in *Klebsiella* nomenclature occurred in 1971, when Bascomb and colleagues[6] again proposed six taxa although along somewhat different lines than those of Cowan and associates.[5] Using numerical taxonomy as a basis, they also retained *K. pneumoniae, K. ozaenae,* and *K. rhinoscleromatis* as separate groups. However, *K. aerogenes, K. edwardsii,* and *K. oxytoca* were placed in a single taxon, and a new group, which remained unnamed, was established. But by far the biggest change was the inclusion of *E. aerogenes* as a motile species of *Klebsiella* named *Klebsiella mobilis.* Subsequently, Brenner and colleagues[4] published DNA hybridization data indicating that *E. aerogenes* was only 51% to 56% related to *Klebsiella* at 60°C. Relatedness dropped further to 16% to 22% at the more stringent temperature of 75°C, thus establishing that neither motile nor nonmotile strains of *E. aerogenes* should be placed in the genus *Klebsiella.* Additionally, the DNA data showed that *K. ozaenae, K. edwardsii,* and *K. rhinoscleromatis* could not be separated from *K. pneumoniae.* However, since *K. ozaenae* and *K. rhinoscleromatis* are associated with distinct disease syndromes, each was retained as a subspecies of *K. pneumoniae.*[1,2] For convenience, both are referred to as if they had species status.[7] The name *K. edwardsii* is still in use in the United Kingdom but not in the United States. Of the four species proposed before the turn of the century, *K. oxytoca* was the last to be confirmed as a species. Until 1977, it was usually dismissed as an indol-positive variant of *K. pneumoniae.* Although it was given species rank in 1956 by Lautrop,[1] it was only in 1977 that molecular techniques firmly established *K. oxytoca* as a legitimate species.[8]

In 1981 two additional species gained admittance to the genus *Klebsiella.* Although genetically distinct, they are phenotypically diffi-

TABLE 1. *Nomenclature of the genus* Klebsiella

Current Designation	Previous Designation	ATCC Number	G+C Ratio (mol%)
Klebsiella pneumoniae subspecies *pneumoniae*	Friedländer's bacillus	13883	56–58
Klebsiella pneumoniae subspecies *ozaenae*		11296	NA
Klebsiella pneumoniae subspecies *rhinoscleromatis*		13384	NA
Klebsiella oxytoca	*K. pneumoniae* indol+	13182	55–58
Klebsiella planticola	Group K, *K. trevisanii*	33531	54–55
Klebsiella terrigena	Group L	33257	56
Klebsiella ornithinolytica	Enteric group 47, NIH 12	NIH 90-72	57–58

From Grimont et al.,[1] Ørskov,[2] and Sakazaki et al.[14]
NA, not available; NIH, National Institute of Health, Tokyo.

cult to separate from *K. pneumoniae.* Originally called group K and group L,[9] both species consist primarily of environmental isolates. Izard and coworkers[9] named group L *K. terrigena* after finding that all but 2 of 20 strains were 87% to 100% related to each other but only 48% to 63% related to *K. pneumoniae.* Two studies confirmed group K to be a unique, homogeneous group, the first by Naemura and colleagues[10] using numerical taxonomy and another by Woodward and associates[11] using DNA–DNA hybridization. In 1981, after further corroboration, Bagley and coworkers[12] named group K *K. planticola.* The species *K. trevisanii,* named in 1983, was found to be a subjective synonym of *K. planticola* in 1986[13] and is not recognized. The last organism to be proposed as a new species of *Klebsiella* was named *K. ornithinolytica* in 1989 by Sakazaki and associates.[14] These strains were previously referred to as enteric group 47 by the Centers for Disease Control, Atlanta, Georgia, and as NIH group 12 by the National Institute of Health, Tokyo, Japan. This is the only species of *Klebsiella* that decarboxylates ornithine. A recap of the nomenclature, ATCC strain numbers, and G+C ratios of the genus *Klebsiella* can be found in Table 1.

EPIDEMIOLOGY

The inherent difficulties in recognizing and separating *Klebsiella* species, both in the past and the present, have somewhat clouded the interpretation of the epidemiologic data regarding this genus. Although this appears to be especially true for environmental studies,[1,10,11] it is also relevant in clinical situations. Of note, for many years medical laboratories identified *K. oxytoca* as indol-positive *K. pneumoniae.* Though they share a similar epidemiology, some studies suggest *K. oxytoca* may be the more virulent, making the distinction more important.[15] Further, isolates of environmental *Klebsiella* (ie, *K. terrigena* and *K. planticola*) which have been obtained from humans are easily misidentified as *K. pneumoniae.* Neither species can be separated from *K. pneumoniae* by usual identification techniques used in clinical laboratories.

Between 1980 and 1989, the number of *Klebsiella* nosocomial infections declined.[16] Nonetheless, it has remained the fourth and fifth most common agent of nosocomial pneumonia and urinary tract infection (UTI), respectively. The numbers of hospital-acquired bacteremias and wound infections also remained significant. Spencer,[17] in a 1-day point-prevalence-of-infection analysis of 10,038 patients in intensive care units (ICUs), found that 8.1% of infections were caused by *Klebsiella* species.

It is generally agreed that hospitalized patients, as opposed to the hospital environment, serve as the reservoir for nosocomial *Klebsiella* infections (Fig. 1).[1,15,18] Lucet and associates[19] found that colonization was a prerequisite for acquisition of extended-spectrum β-lactamase (ESBL)–producing *Klebsiella* in an ICU setting. Risk of infection

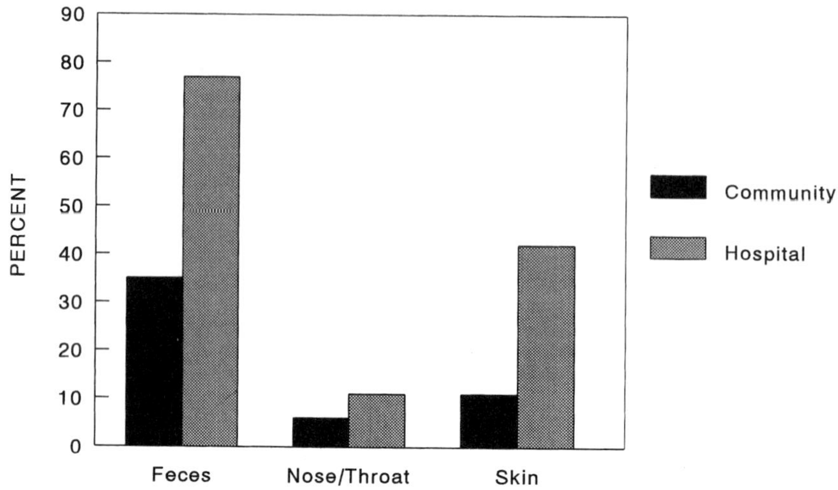

FIG. 1. Carriage rates of klebsiellae in the community versus hospital settings. (From Montgomerie[18] and Smith et al.[65])

rose with time of hospitalization, increasing sixfold by the fourth week of stay. The gastrointestinal tract was the first site colonized, and after colonization infection developed in 22 of 62 patients. Pulsed-field gel electrophoresis (PFGE) indicated that the majority of infections were caused by a single strain. Besides the bowel, common carriage sites include the skin and nasopharynx.[15,18] However, because carriage rates are higher for feces, it is probably the most significant source of patient infections. Approximately one third of people carry *Klebsiella* in their stool, but the rate increases with hospitalization and subsequent use of antimicrobials.[1,18] Patients receiving antibiotics have a carriage rate three times greater and carry more *Klebsiella* per gram of feces than those not receiving such treatment. Neonates appear to be an exception; during outbreaks, colonization rates as high as 90% to 100% have been detected among neonates not given antimicrobial therapy. *Klebsiella* persist longer in infant feces as well, up to 1 month or more. Babies receiving antibiotics have had *Klebsiella* isolated from their feces for up to 1 year. In adults, persistence, although variable, seldom exceeds days or weeks. Fecal carriage of *Klebsiella* in hospital personnel has not been studied extensively but does not appear to be a significant source for hospital-acquired infections.[18]

One percent to 6% of healthy persons carry *Klebsiella* in the nose or throat. Studies on hospital patients indicate carriage rates at these sites of 2% to 9%.[18] Similar rates were found in two surveys of moderately ill (9%) and dying (11%) patients. Besides use of antimicrobials, alcoholism, tracheal intubation, hypertension, acidosis, and increased blood urea nitrogen and increased leukocytes enhance the odds of nasopharyngeal colonization. Carriage rates of 50% or more have been described in premature infants and neonates in ICUs. Unlike fecal carriage in infants, respiratory tract colonization is higher when antibiotics are used.[18] Nurses and other hospital staff may have *Klebsiella* in their nares, but no nosocomial significance has been attributed to this carriage.

Unlike the nose and throat, skin carriage rates have not been well documented. However, in one study 11% of patients admitted to the hospital had *Klebsiella* on their hands.[18] This rate rose to 42% after 14 days of hospitalization, and carriage was associated with antimicrobial use. Hospital personnel frequently carry *Klebsiella* on their hands, even if only transiently. Seventeen percent of nurses in an ICU acquired *Klebsiella* on their hands after handling patients.[18] In one neonatal outbreak 30% of nurses had *Klebsiella* on their hands, and on a urology ward two thirds of personnel carrying antibiotic-resistant bacteria had *Klebsiella* on their hands.

Outbreaks in neonatal wards are of particular concern. Risk factors associated with *Klebsiella* infection in one study[20] included low birth weight (50% of that of control infants), length of gestation (<32 weeks), a lower Apgar score at 5 minutes (score of 6, versus 9 for controls) and male gender. Colonized, noninfected infants served as the reservoir for infections; invasive (bacteremic) strains were more effective at colonizing than noninvasive strains.

Hospital surfaces are not a common source of *Klebsiella*. However, sporadic cases and outbreaks have been related to contaminated respiratory equipment, urinary catheters, and intravenous (IV) solutions and other fluids.[15,18] Vegetables and various foods, especially those with a high sugar content, also harbor *Klebsiella* and may be a source of bowel colonization.

In nonhumans, *Klebsiella* infections can cause serious financial losses, particularly in cases of metritis in horses and mastitis in cows, in addition to suffering in the animals.[1,21] Squirrel monkeys and lemurs in zoos have died from *Klebsiella* infection; other animals affected include rhesus monkeys, guinea pigs, and muskrats.[1]

Klebsiella, including *K. pneumoniae* sensu stricto, have a wide distribution in the environment. However, isolates of *K. pneumoniae* from seeds, vegetables, bark, and so on utilize 5-ketogluconate, gentisate, and *m*-hydroxybenzoate, whereas those involved in human and animal infections do not.[1,10] *Klebsiella* can readily be isolated from wood and pulp mill products and wastewater, vegetables (especially root vegetables), other plant material, and water.[1,10,11] Knittel and colleagues[22] found *Klebsiella* to exceed 10^3 organisms per gram of surface on potatoes or lettuce even when room temperature incubation was used.

K. rhinoscleromatis is the causative agent of rhinoscleroma (see Respiratory Infections). Although an ancient disease[23] present in central Europe for many years, it is now thought to be endemic primarily in tropical and subtropical areas worldwide.[23,24] It is not found in animals or in the noses of healthy humans,[25] but Krasilnikov and coworkers[26] recognized mild and subclinical infections in humans during a 12-year study of the focal incidence of rhinoscleroma. Affected persons usually have low socioeconomic status, live in rural areas, and have poor nutrition, sanitation, and hygiene.[23,24] The disease occurs more often in women than men (1.3:1 ratio) and generally afflicts those in their teens and 20s, although a patient may be symptomatic for many years before actually seeking medical attention.[23,27] Transmission of the disease is from person to person, probably via airborne secretions, but close and prolonged contact, usually within a household, is required.[23,26] Paul and associates,[24] however, reported on a patient with human immunodeficiency virus infection who had limited exposure before developing the disease. Presumably his immune status contributed to acquisition, which was believed to have occurred during travel to tropical areas on just four occasions over an 11-year period.

CLINICAL SIGNIFICANCE

Central Nervous System

Although *Klebsiella* infections involving the central nervous system are relatively infrequent, attendant mortality rates are high, ranging from 38% to 91%.[28] Old age and diabetes mellitus, in particular, are frequently associated factors factors in these infections. Tang and Chen[28] reviewed 42 cases of *K. pneumoniae* meningitis diagnosed at their institution over a 10-year period (1981–1991). The infection rate more than doubled in the last 5 years, rising to 16% from 7% in the first 6 ycars. Most infections were community acquired, but approximately one third were nosocomial; 12 of 13 patients acquired their infection after a neurosurgical procedure. Men were three times more commonly in-

TABLE 2. *Factors associated with mortality in* Klebsiella *infections*

Central Nervous System Infections	Bacteremia Infections
Age >60	Alcoholism/cirrhosis
Diabetes mellitus	Diabetes mellitus
Stage II mental status (coma/delirium)	Cancer/malignancy
Mean lactate levels 19.0±7.6 mmol/L	Hypotension
	Serum urea 127.2 mmol/L
	Serum bilirubin 84.1 mmol/L

From Krasilnikov et al.,[26] Montgomerie & Ota,[32] and Garcia de la Torre et al.[33]

fected than women, which the authors noted had been reported previously.[28] Twenty-three patients (54.8%) had a positive blood culture, including four whose cerebrospinal fluid was culture-negative but who had clinical signs suggestive of meningitis. Three factors were significantly associated with death (Table 2); age older than 60 years ($P<0.02$), diabetes mellitus ($P<0.05$), and stage II neurologic status (coma or delirium, $P<0.02$). Lactate levels, established for 25 (59%) of 42 patients, were always elevated. Acidosis was a significant finding. Mean lactate values were significantly different between patients who survived and those who died (13.3±4.3 versus 19.0±7.6 mmol/L, respectively).

Holder and Halkias[29] reported an unusual case of *K. pneumoniae* meningitis that relapsed five times during a 7-month period and ultimately resulted in the death of the patient. The authors believed that the inability of the patient to mount an immunologic response due to underlying acquired immunodeficiency was the major contributing factor to his relapsing meningitis. Although all strains isolated had similar antibiograms, definitive epidemiologic typing was not performed on all of the isolates.

Cases of *K. ozaenae* cerebral abscess and meningitis are rarer than *K. pneumoniae* central nervous system infections. In 1987, Strampfer and colleagues[30] published a case report on a 72-year-old woman with diabetes mellitus and heart disease who developed a cerebral abscess. She presented with a chronically draining right ear infection that yielded *K. ozaenae;* her left ear, for which she had a history of chronic otitis media, was not cultured. Because the abscess occurred in the left temporal lobe, it was assumed the left ear was colonized and that there was contiguous spread from the ear. *K. ozaenae* was cultured from the abscess. Despite surgery and treatment with chloramphenicol, ceftazidime, and ampicillin, the infection was fatal. Tang and Chen[31] reported on two cases of *K. ozaenae* meningitis and reviewed two cases from the literature. All four patients from whom *K. ozaenae* was isolated from the cerebrospinal fluid were older than 50 years of age and debilitated; two had hyperglycemia or diabetes mellitus. Nasopharyngeal disease, including sinusitis and otitis media, carcinoma, and atrophic rhinitis (AR) treated with surgery, was present in three of the four patients. Blood cultures were also positive in two patients, including one who died.

Bacteremia

Despite the declining incidence of nosocomial *K. pneumoniae* infections reported by the National Nosocomial Infections Surveillance System (NNIS),[16] *Klebsiella* bacteremia remains a serious concern. Common characteristics of *Klebsiella* bacteremia taken from four studies published between 1980 and 1995 are presented in Table 3.[32–35] Hospital-acquired infections were more common in all four reports than community-acquired infections. Generally, acquisition rates reflected the type of institution, with teaching or large medical facilities having significantly more infections than community hospitals. The portal of entry for infections varied. When source could be determined, the urinary tract was more often implicated in sepsis than other sites. However, Feldman and coinvestigators[34] reported the respiratory tract to be 3.5 times more com-

TABLE 3. *Characteristics of* Klebsiella *bacteremia*

Characteristic	Montgomerie & Ota[32]	Garcia de la Torre et al.[33]	Korvick et al.[35]	Feldman et al.[34]
No. of patients	41	100	230	47
How acquired				
Community (%)	2	23	47	47
Nosocomial (%)	98	77	53	53
Male/female ratio	22/19	—	—	31/16
Entry portal (%)				
Urinary tract	27	31	28	17
Gastrointestinal	24	15[a]	21[b]	4
Respiratory	15	15	10	59
Surgical	0	8	5	0
IV line	20	4	5	0
Other	4	0	5	16
Unknown	10	27	26	4
Underlying conditions	Alcoholism/cirrhosis, age >50 y, premature birth	Neoplastic disease, cirrhosis, diabetes mellitus, age >60 y	Alcoholism, diabetes mellitus, cancer/malignancy, respiratory disease	Alcoholism, diabetes mellitus, neoplastic disease
Predisposing conditions	Prior antimicrobial therapy, catheters, surgery	Indwelling urinary catheter, surgery, prior antimicrobial therapy	Hypotension, prior antimicrobial therapy	
Mortality (%)	37	25	28	55
Laboratory findings (%)				
Monomicrobic		88	74	
Polymicrobic		12	26	
K. pneumoniae		75	82	
K. oxytoca		25	15	
K. ozaenae			<1	

[a]Includes biliary tract and peritoneum.
[b]Includes biliary tract and abdomen.

mon as a source than the urinary tract. Alcoholism/cirrhosis was the major underlying condition reported in all four studies. Diabetes mellitus, cancer/malignancy, and increased age were also frequent contributory underlying factors. Patients with indwelling catheters and those who underwent surgery and/or were on antimicrobial therapy were at highest risk for infection. Twenty-five percent to 55% of infections were fatal. In two studies,[34,35] hypotension was significantly associated with mortality (see Table 2). Feldman and colleagues[34] also found that the levels of serum urea (127.2 mmol/L) and serum bilirubin (84.1 mmol/L) in those who died were significantly higher than in those who survived. Most infections (>74%) were monomicrobial. When species were identified, *K. pneumoniae* accounted for 75% to 100% in three studies and *K. oxytoca* 15% to 25% in two of the studies; *K. ozaenae* was isolated from one patient.

In considering a source for outbreaks or clusters of *Klebsiella* bacteremia, contaminated IV solutions should always be given consideration.[18,36] The report of Goetz and associates[36] describing *K. pneumoniae* bacteremia in a liver transplant unit exemplifies such a cluster. Five episodes of sepsis occurred in four patients within the unit over a 1-month period. All had midline or central catheters that were flushed with normal saline from a 100-mL bag equipped with IV tubing and a stopcock. At the time the outbreak was discovered, two saline bags, the IV tubing, and the stopcock were culture-positive. All samples from the blood and catheter tips of the four patients yielded *K. pneumoniae,* as did the catheter tip of a patient without bacteremia. PFGE analysis of the strains revealed that they were identical; three other, unrelated strains gave different profiles from those of the outbreak.

Cases of *Klebsiella* bacteremia found in conjunction with rare underlying conditions also occasionally appear in the literature. In 1981, Murray and colleagues[37] reported on two patients with Hansen's disease who developed *K. ozaenae* sepsis. One patient, a 74-year-old man with an extensive list of underlying illnesses including alcoholism, died. The second patient had severe Hansen's disease, and *K. ozaenae* was isolated from a leg ulcer as well as from blood culture. This patient recovered from his infection, although at follow-up 6 and 12 months later his nares continued to be colonized.

Klebsiella sepsis associated with toxic epidermal necrosis (TEN) also has been reported. TEN, a dermatologic condition resembling severe scalding, usually occurs in adults and is primarily caused by exposure to drugs. Picard and coinvestigators[38] described a case in a 4-month-old male infant with cystic fibrosis in whom TEN developed simultaneously with fever, thrombocytopenia, and a positive *K. pneumoniae* blood culture; the infection was fatal. They also summarized three other cases of TEN in infants with sepsis (all male, 6 weeks old). *K. pneumoniae* was isolated from two of these three babies.

Although *K. planticola* is known to colonize rather than cause disease in humans, it has been isolated in pure culture from two patients with sepsis.[39] Both infections occurred postoperatively in patients with heart disease. Neither water nor any hospital surface was implicated, and fecal carriage was assumed to be the source of these infections.

Respiratory Infections

Klebsiella lower respiratory tract infections are usually characterized by fever and chills and, typically, by leukocytosis.[15] The sputum is described as resembling red-currant jelly.[15,40] In pneumonia, upper lobe involvement is more common, although lower lobes also may be involved. Cavitation is generally unilateral.[15] Although suggestive, the bulging fissure sign and radiologic data (sharp infiltrate margins) once thought diagnostic of *Klebsiella* infection are found in pneumonias with other causes as well.[15,40,41]

Klebsiella is usually considered to be the most common cause of community-acquired pneumonia (CAP), but this view has been contested. In his 1990 review, Carpenter[40] concluded that large public hospitals see fewer than three cases of *Klebsiella* CAP per year, with pneumococcal, hemophilus, and anaerobic CAPs all being more common. Carpenter believed that much of the earlier *Klebsiella* data relied on inaccurate or inconclusive techniques. Diagnosis of *Klebsiella* pulmonary infections should be based on positive blood and/or pleural fluid cultures. In other specimens, including transtracheal aspirates and sputa which meet criteria based on the ratio of polymorphonuclear cells to oropharyngeal cells, *Klebsiella* must be present in pure or predominant culture to be diagnostic.[40]

Despite Carpenter's assertions,[40] which were based on data from U.S. hospitals, *Klebsiella* still appears to be a prominent cause of CAP in other parts of the world, particularly in alcoholic patients.[41,42] Feldman and associates[41] found that 38 (32%) of 119 patients with CAP from whom an organism was isolated had *K. pneumoniae;* the diagnosis was based on a positive blood culture in 29 of the 38 cases. Likewise, Jong and colleagues[42] found 11 patients over a 3-year period with *K. pneumoniae* pneumonia and bacteremia. All of the patients in this report died of their infection; in all cases there was both a short onset (~48 hours) and a short survival time (~36 hours).

Lung abscesses are a fairly frequent complication.[40] However, pulmonary gangrene, a condition in which segments or an entire lobe of the lung become necrotic and slough off, is a rare event in patients with *Klebsiella* pneumonia.[39,43]

Klebsiellae are also involved in two uncommon upper respiratory diseases, rhinoscleroma (RS) and atrophic rhinitis (AR). RS caused by *K. rhinoscleromatis* is a slowly progressive disease; patients may have symptoms for as long as 5 to 10 years before presenting for treatment.[23,27,44] Immunologic tolerance and fibrotic tissue formation probably contribute to the chronicity of RS by slowing the spread of the disease.[23,27] Whether it is cause or effect is unknown, but the cellular immunity of RS patients is skewed, with increased CD8 and decreased CD4 lymphocyte production.[23,24] The disease, which is difficult to cure, has a tendency to relapse and is considered to have three stages.[23,44] The catarrhal-atrophic or exudative stage, lasting weeks or months, is characterized by a foul-smelling, sometimes bloody discharge with some crusting of the nasal mucosa. The cause is frequently undeterminable at this stage because biopsy does not yet show cells typical of RS. The diagnosis is easiest to ascertain in the second stage, when Russell bodies and Mikulicz's cells harboring gram-negative rods are apparent in biopsy. At this time cultures should yield *K. rhinoscleromatis.* This stage, termed granulomatous or proliferative, is also the phase during which inflammatory masses form, giving rise to deformities and possibly to obstructed nasal passages. Spread to other facial and upper respiratory structures, including the trachea and larynx (causing hoarseness and dysphagia), and invasion of cartilage and bone occur in this stage. Patients' blood counts and chemistry panels are generally unremarkable,[25,44,45] with the possible exception of iron deficiency. Lack of iron weakens the integrity of the epithelium and may be a predisposing factor in RS.[23] There is a notable lack of fever or cervical lymphadenopathy; dense fibrotic tissue formation may block the spread of the disease to, and subsequently by, the lymphatic system.[23,27] However, in one rare case of RS multiple bilateral lymph nodes were infected in a 15-year-old girl.[27] The third stage is referred to as the sclerotic or fibrotic phase. Deformity and fibrotic tissue formation are more pronounced at this time, and again, as in the first stage, *K. rhinoscleromatis* is infrequently isolated. Therapy for RS requires prolonged administration of an antimicrobial agent and, in severe cases, surgery using a carbon dioxide laser.[23,44]

Like RS, AR is a chronic disease of the upper respiratory tract but it is generally restricted to the nose. Because isolation of *K. ozaenae* is common in AR, the disease is also referred to as ozena.[46] Symptoms include an atrophied mucosa (but with no submucosal involvement) and a foul-smelling discharge that forms crusts as it dries; bleeding is not uncommon. Using ciliated nasal epithelial cells and a culture of *K. ozaenae* to study the pathogenesis of AR, Ferguson and colleagues[46] concluded that the bacterium inhibited the beating activity of cilia, which normally cleanse tissue. Electron microscopy showed the cilia to be clumped by an amorphous material that the authors assumed was capsular polysaccharide (CPS). It is also assumed that damage or injury to the mucus blanket predisposes for colonization by *K. ozaenae.* Speculation that nasal surgery, which dries the mucosa, may be a common precipitating event for AR arises since many patients undergo such surgery before development of their disease.[46] Prolonged antimicrobial therapy clears the infection. Intranasal instillation of antimicrobials physically decreases the bacterial load by its irrigating activity and may be a better method of treatment.[46]

Urinary Tract Infections

Klebsiella, particularly *K. pneumoniae,* is recognized as a frequent cause of UTIs. In the NNIS report for 1986–1989, *K. pneumoniae* was the fifth most commonly reported agent of nosocomial UTI.[16] In nosocomial infections, catheterization is believed to be an important factor in the incidence of disease and spread of strains.[18] However, a study of more than 11,000 urine isolates showed essentially no difference in isolation rates between clean-catch and catheterized specimens.[47] This was true not only for *K. pneumoniae* but for all common etiologic agents of UTI. This study occurred over a 3-year period and rates were determined yearly; isolates were from residents in 50 different nursing homes. During this time *Klebsiella* was isolated from 6% to 8% of both clean-catch and catheterized specimens.

Davis,[48] using an in vivo rat bladder model, showed that the surface interactions occurring in UTIs are similar among the five most common gram-negative agents. These steps include attachment, exfoliation of the epithelium, and strand formation.

The spectrum of *Klebsiella* UTIs ranges from asymptomatic bacteriuria to cystitis to pyelonephritis.[49] Podschun and coworkers[50] compared virulence characteristics among isolates from these three categories and isolates from feces. Characteristics examined included production of urease, type 1 and type 3 pili, production of aerobactin and enterochelin, capsular type, and serum resistance. Of these factors, only type 1 piliation was significantly associated with any particular type of UTI; it was more strongly associated with pyelonephritis than with asymptomatic bacteriuria or cystitis. There was no significance in any other virulence factor between groups. The authors concluded that, in addition to the urovirulence characteristics of the infecting bacterium, the host contributes significantly to the infectious process. Patients with diabetes, poor immune status, or structural abnormalities are more prone to infection and to more severe disease.

Significant bacteriuria has been ascribed infrequently to both *K. ozaenae*[51] and *K. terrigena.*[52]

Gastroenteritis

The ability of *Klebsiella* to cause gastroenteritis is unclear, but in general colonization is probably a more accurate assessment. Karpas and Boman[53] presented seven cases of enteritis they believed were attributable to *K. pneumoniae.* All but one of their patients was 55 years of age or older with severe underlying impairments. Only one case report was given in detail, and it involved an alcoholic with severe malnutrition. Both blood and feces yielded *K. pneumoniae* type 2. The bowel showed extensive destruction, with hemorrhage and perforation of the midileum.

Olarte and colleagues,[54] in their study of an outbreak of diarrhea in infants, concluded that the *Klebsiella* isolated was not involved. They fed strains of the predominant serotype involved in the outbreak (type 19) to volunteers. Volunteers receiving 1.3×10^{10} organisms per milliliter failed to develop diarrhea, and only two excreted *Klebsiella* capsule type 19 in their feces.

Oliver and associates,[55] however, showed compelling data indicating a role for *K. oxytoca* in antibiotic-associated nonpseudomembranous colitis. Seven of 13 patients with disease, ranging from mild right-sided colitis to diffuse ulcerative colitis as determined by colonoscopy or rectosigmoidoscopy, had *K. oxytoca* isolated in pure culture or as the predominant flora from mucosal biopsies. All seven were being treated with ampicillin or a derivative, and all recovered with withdrawal of the antimicrobial.

Other Infections

Although a variety of infections are presented in this section, these patients in general share the same predisposing conditions (eg, older age, alcoholism, diabetes mellitus). In one such group of patients in Southeast Asia, those with pyogenic liver abscess (PLA), the number of infections has increased dramatically in the last 10 years. In a retrospective review of 146 cases of culture-confirmed PLA in their institution, Chang and Chou[56] found that 78% were caused by *K. pneumoniae;* all were community-acquired. Of their patients, more than 90% had a solitary abscess, 50% had septicemia, and 95% had monomicrobial infections, all statistically significant findings in comparison with patients with PLA caused by other agents. Their study, along with one by Han,[57] found that no more than 30% of the infections were associated with the biliary tract or portal system, both usual sources for PLA. Septic or endogenous endophthalmitis, which results in blindness, is a complication of PLA.[57–59] Underlying diabetes mellitus contributes to the increased risk of metastasis of *K. pneumoniae* to the eye.[57] Chou and Kuo[58] reported that endogenous endophthalmitis developed in 11 (3.1%) of 352 of patients with PLA. All had blood and/or liver abscesses, and three had vitreous fluid that was culture-positive for *K. pneumoniae.* Nagase and coworkers[59] recorded a case of a 64-year-old woman with diabetes mellitus who developed bilateral endophthalmitis due to *K. pneumoniae.* This diagnosis led to the discovery of an unknown liver abscess. *Klebsiella* subsequently spread via the bloodstream to the lung, causing pulmonary emboli, and to the brain, where multiple abscesses developed in the basal ganglia. Whereas all of these infections involved *K. pneumoniae,* Chowdhury and Stein[60] reported on a patient with PLA who had pulmonary emboli due to *K. ozaenae.* The source for this community-acquired infection was not ascertained. The patient, a nonalcoholic, healthy 69-year-old woman, had recently been diagnosed with type II diabetes mellitus.

Two unusual infections, a primary gas-containing mediastinal abscess[61] and peritonitis occurring after injection sclerotherapy,[62] have been attributed to *K. pneumoniae.* The patient with the mediastinal abscess had underlying diabetes and had been treated for a gas-containing hepatic abscess 2 years earlier. The patient with peritonitis was an alcoholic. It was surmised that his oropharnyx was colonized with *K. pneumoniae.* His bloodstream became seeded during treatment of bleeding esophageal varices, with subsequent spread to his peritoneum.

K. pneumoniae has also been associated with acute cholecystitis complicated by bacteremia.[63] In 78 patients with positive blood cultures, a single isolate was present 87% of the time. The most common organism was *Escherichia coli* (56%), followed by *K. pneumoniae* (12%). These two agents were also the most common isolates from bile.

Pontell and colleagues,[64] in a review of 20 cases of retropharyngeal abscesses, found *K. pneumoniae* to be the second most common agent in this type of infection. Seventy percent of their cases occurred in adults, and 35% of the abscesses were either secondary to an upper respiratory infection or of unknown origin.

Burn patients are at particular risk for *Klebsiella* infection. Of 58 acute burn patients,[65] *Klebsiella* was isolated 14 times from burn sites, 23 times from urine or Foley catheters, and once from blood. One patient died as a result of *Klebsiella* infection. The source for *Klebsiella* was the gastrointestinal tract, where it was isolated on 157 occasions in 45 of the 58 patients. Although these patients were admitted with *Pseudomonas aeruginosa, Klebsiella* colonization occurred only after admission.

Finally, two unusual manifestations of *K. pneumoniae* bear mention. The first is a case of crepitant myonecrosis in a 78-year-old diabetic man.[66] The infection which began on the buttock spread to the right gluteal muscle with necrosis of the soft tissue and perinephritic abscess formation. Ultimately, blood, urine, muscle, and kidney were all positive for *K. pneumoniae.* To our knowledge, this was only the second reported case of *Klebsiella* myonecrosis, and both infections were fatal. The last case involves septic arthritis caused by a seashell spine that punctured the knee of an 8-year-old girl.[67] Although *K. pneumoniae* is known to spread from the pulmonary system and urinary tract via the bloodstream, we are unaware of any other case of posttraumatic infection caused by *K. pneumoniae.*

LABORATORY IDENTIFICATION

Isolation

Excepting perhaps *K. ozaenae* and *K. rhinoscleromatis,* klebsiellae grow readily on virtually all commonly used laboratory media, including most differential and selective enteric agars. All species, again with the exception of *K. ozaenae* and *K. rhinoscleromatis,* normally ferment both lactose and sucrose. Appearing as fermenters on differential media, they are indistinguishable from other *Enterobacteriaceae* involved in nosocomial or opportunistic infections. For this reason, a number of specialized media have been developed to isolate *Klebsiella* during epidemiologic investigations of nosocomial outbreaks as well as from environmental sources.[1,18,68,69]

On any medium, *Klebsiella* appear as convex, moist colonies, because they are prodigious producers of CPSs. Often colonies take on a mucoid aspect and may almost ooze across the agar; at other times, they are sticky. Although it is generally considered to be characteristic of *Klebsiella* species, other genera also may appear mucoid, including some that commonly have capsules (eg, *Escherichia*) and some for which a capsule is not common (eg, nontyphoidal salmonellae). Although most strains of *Klebsiella,* as mentioned, grow well on blood or other nondifferential media, approximately 1.5% of *Klebsiella* isolated from UTIs are cysteine-dependent strains. These strains grow as dwarf colonies, between 0.5 and 1.0 mm in diameter, on regular laboratory media.[70]

Biochemical Methods

Commercial Systems

Of the clinically relevant species, *K. ozaenae* and *K. rhinoscleromatis* do the most poorly in commercial systems.[71] This is not surprising since (1) they sometimes grow poorly and (2) they can be difficult to separate even with conventional media. API 20E (API, Vitek bioMérieux, Hazelwood, MO) and MicroScan Walk/Away Rapid ID (Dade MicroScan, West Sacramento, CA) misidentified two and three of four *K. rhinoscleromatis* and five and three of five *K. ozaenae* strains, respectively.[71] However, in that same study, Cobas Micro ID-E/NF (Becton Dickinson Diagnostic Instrument Systems, Sparks, MD) correctly identified all nine strains, and the Vitek GNI (Vitek bioMérieux) missed only one strain of *K. ozaenae.*

Even with *K. pneumoniae* and *K. oxytoca,* commercial systems did not perform as well as expected.[71–77] In many studies, however, data were presented for *Klebsiella* as a group, without any indication of which or how many of each species were tested. Correct identification to the genus level varied from 71% to 100%. RapID onE (Innovative Diagnostic Systems, Norcross, GA) identified all strains tested (20 *K. pneumoniae* and 10 *K. oxytoca*), but this did not include any *K. ozaenae* or *K. rhinoscleromatis.*[75] For *K. pneumoniae,* identification to the genus level ranged from 83% to 100%. In the only study where such information was provided, *K. pneumoniae* was misidentified as *Kluyvera ascorbata, E. coli,* or *E. aerogenes* or yielded a code for no identification.[77]

K. terrigena, which cannot be separated from *K. pneumoniae* or *K. planticola* by routine biochemical tests found in most commercial systems, is identified by API 20EC (Vitek bioMérieux).[78] In this system, *K. terrigena* generated numerical codes 1777771 and 1777671 almost exclusively. However, supplemental testing for growth at 10° and 44.5°C, which separates *K. pneumoniae* and *K. terrigena* (see later discussion), was also performed to confirm identification.

Conventional Media

K. pneumoniae and *K. oxytoca* typically are easily differentiated from other genera and from those species within their genus that are most commonly associated with human infections (Table 4). They themselves, however, are distinguishable by essentially only one characteristic, indol, with *K. pneumoniae* giving a negative and *K. oxytoca* a positive reaction. Separation of *K. ozaenae* and *K. rhinoscleromatis,* which may be difficult, relies primarily on reactions in two tests, malonate and *o*-nitrophenyl-β-D-galactopyranoside (ONPG)[79] (see Table 4). *K. terrigena* and *K. planticola* are indistinguishable from *K. pneumoniae* and from each other by conventional biochemical tests. Growth tests at

TABLE 4. *Biochemical characteristics of the genus* Klebsiella

Species	Indol	VP	CIT	ODC	MAL	Urea
K. pneumoniae	–	+	+	–	+	+
K. oxytoca	+[a]	+	+	–	+	+
K. rhinoscleromatis	–	–[a]	–	–	+	–
K. ozaenae	–	–[a]	V	–	–[a]	V
K. planticola	V	+	+	–	+[a]	+
K. terrigena	–	+	+	–	+	+
K. ornithinolytica	–	+	+	+[a]	+	+

VP, Voges-Proskauer; CIT, citrate; ODC, ornithine decarboxylase; MAL, malonate.
[a]Indicates significant reactions.

TABLE 5. *Characteristics helpful in separating unusual* Klebsiella *spp. from* K. pneumoniae *and* K. oxytoca

Characteristic	*K. planticola*	*K. terrigena*	*K. ornithinolytica*	*K. pneumoniae*	*K. oxytoca*
Growth 10°C	+	+	+	–	–
Growth 44°C	–	–	NA	+	+
ODC	–	–	+	–	–
D-Melezitose fermentation	–	+	+	–	–
Carbon assimilation					
DL-3-Hydroxy-butyrate	+	+	+	V	–
m-Hydroxy-benzoate	–	+	–	V	+
Pectinase	–	–	–	–	+

From references 1, 9, 12, 13, 77, and 78.
ODC, ornithine decarboxylase; NA, not available.

two temperatures, 10° and 44.5°C, can separate *K. pneumoniae* from the other two species, but neither incubation temperature is readily available in clinical laboratories. Both *K. terrigena* and *K. planticola* grow at 10°C but not at 44.5°, whereas the reverse (no growth at 10°, growth at 44.5°C) is true for *K. pneumoniae.*[1,9,12] Reactions that separate *K. planticola* and *K. terrigena* from each other and from other *Klebsiella* species are listed in Table 5. Monnet and Freney[80] devised a schema for the separation of *Klebsiella* species based on four-carbon assimilation tests, indol production, and ornithine decarboxylase (ODC) activity; these reactions are also included in Table 5. *K. ornithinolytica* strains resemble *K. oxytoca* in routine biochemical tests except for their positive ODC reaction and growth at 4°C. They can be further differentiated by carbon assimilation tests (see Table 5).

Biotyping

A number of biotyping systems have been proposed over the years,[1] but no one system has been adopted universally. Tests most commonly used in these schemes included any combination of the following: urea, citrate, malonate, indol, Voges-Proskauer, mucate, and fermentation of dulcitol, sorbose, or D-melezitose. All together 21 different assays, not including carbon assimilation tests, were employed, with any given schema using anywhere from 6 to 11 tests. These biotyping methods generated fairly large numbers of biogroups, two of them containing as many as 26 to 29 biotypes. Probably the most helpful subtyping system employs carbon assimilation tests that discriminate between strains of *K. pneumoniae* found in clinical versus environmental specimens.[1,10] These tests include 5-ketogluconate,[1] *m*-hydroxybenzoate, and gentisate,[10] which are utilized by environmental but not by clinical *K. pneumoniae* strains.

Bagley and coinvestigators[12] reported three biogroups within *K. planticola* that were separable by fermentation of adonitol and utilization of D-tartrate and L-arginine. However, their biogroup 3 contained only two strains.

Biotyping is generally considered to be a more powerful epidemiologic method for *Klebsiella* species when it is used in conjunction with capsular typing (see Immunodiagnostics).[1] Both are available only at a limited number of large reference laboratories.

Immunodiagnostics

Serotyping is most often accomplished by means of antisera directed at somatic or flagellar cellular components; however, for *Klebsiella* species, the capsule is the preferential target. There are 77 known capsular antigens (K antigens); they are numbered K1 through K82, several K types having been found to be identical.[81] Besides the historic quellung reaction, a number of typing methods have been adapted for use in *Klebsiella* K typing. They include immunofluorescence, double-diffusion gel precipitation, countercurrent immunoelectrophoresis, *Staphylococ-*

cus coagglutination, and latex agglutination.[1,82] Immunofluorescence and coagglutination assays are reported to be more sensitive than either the quellung reaction or latex agglutination, resulting in fewer untypable strains.[82] The quellung reaction is the oldest[83] and perhaps still the most commonly used technique. In this method, precipitation of the capsule can be visualized microscopically under a cover slip after equal volumes of suspended organism and antiserum are mixed. The precipitated capsule is always compared with an India ink capsule preparation to which no antibody has been added. In poorly encapsulated strains, CPS production can be enhanced on media with a high carbohydrate to nitrogen ratio (eg, Worfel-Ferguson medium).[83,84]

Capsular antigens K1 through K6 are the most common types, with K1 through K5 being more virulent.[1,85] K1 and K2 are considered especially virulent. There also appears to be a correlation with site of isolation and capsular type, at least for K2. In urine and sputum specimens, which comprised almost 51% of *Klebsiella* isolates, Riser and Noone[86] found K types 1, 3, and 4 to be most common in sputum and 2, 9, and 10 in urine. Podschun and colleagues[50] also found K2 to be most common in urine. Commercial antisera are available for types K1 through K6,[82,83] but complete typing is available in only a very limited number of laboratories.

Testing for the lipopolysaccharide (LPS) or somatic (O) antigen of *Klebsiella* is problematic because of the capsule. The capsule, which is immunogenic, usually is not completely removed by boiling; therefore, the production of specific *Klebsiella* somatic antisera relies on capsule-deficient mutants. To eliminate the need for such difficult-to-prepare antisera, two groups of researchers developed inhibition- or competitive-based enzyme-linked immunosorbent assays (ELISAs) using O antigen that is released into surrounding media.[87,88] There are 8 recognized O antigens, although 12 were originally described; the accepted types are O1, O2ab, O2ac, O3, O4, O5, O7, and O12.[88] Trautman and coworkers[88] tested 82 *Klebsiella,* including five different species, and found the following O types, given in order of frequency, to be the most common: O3, O2ab, O1/2ab, and O2ac. Antisera raised to either the CPS or LPS can be used to type all species of *Klebsiella.*

Typing

Because *Klebsiella* is a common nosocomial agent[16] and serologic typing is not readily accessible, other typing methods have been sought. Bacteriocin typing, developed in the 1960s and 1970s, has had limited success because of poor reproducibility.[1,89,90] By using a streak-and-point method, Bauernfeind and associates[91] eliminated much of the instability inherent in bacteriocin typing and improved reproducibility considerably. Several bacteriophage systems have evolved over the years, but they too suffer from lack of reproducibility.[92] Molecular typing methods, including plasmid analysis,[93] ribosomal RNA gene restriction or ribotyping,[1] PFGE,[19,94,95] and random amplified polymorphic DNA (RAPD) analysis,[94] have been used successfully to track strains epidemiologically. Gori and colleagues[94] found that PFGE and RAPD gave concordant results and believed that both were useful techniques; PFGE yielded more subclonal variants, but RAPD was more efficient.

PATHOGENESIS

In *Klebsiella,* as with most bacteria capable of producing disease, virulence is both multifactorial and complex. CPS,[85,96–99] LPS,[96,97] iron-scavenging systems,[50,100–103] and adhesins[50,101,104–108] all contribute to pathogenicity. In addition, strains producing a variety of toxins[109–115] have been isolated.

Several studies[85,96,97] have shown that both the CPS and LPS of *Klebsiella* are effective at resisting phagocytosis and complement-mediated killing by serum. Phagocytosis of encapsulated strains requires opsonization of specific anti-capsular antibody. But antibody to CPS, a relatively poorly immunogenic substance compared with LPS, is not rapidly produced, and this retards phagocytosis.[85]

Further, the CPS can mask LPS, preventing activation of the serum protein C3b, which initiates complement-mediated killing.[96] In strains where both LPS and CPS are exposed, C3b is activated, with subsequent attachment to the LPS. However, this occurs at the end of the long LPS chains, far removed from the cell membrane. Formation of the C5b–9 membrane attack complex is either not promoted or occurs too far away for insertion into the membrane. The role of CPS and LPS in resistance to phagocytosis is twofold. First, CPS blocks attachment of specific anti-LPS immunoglobulin G and/or C3b which serve as ligands for phagocytic binding, thus blocking phagocytosis. Second, both CPS and LPS are hydrophilic and produce a negative charge on the bacterial cell surface, which also impedes phagocytosis. Third, although the capsule may not prevent formation of immune complexes with the LPS, the size of the capsule can determine whether such complexes are exposed to phagocytosis.[97] Knowledge regarding bacterial growth conditions (especially subinhibitory concentrations of antimicrobials) that affect capsule production would prove valuable. Alteration of these conditions could reduce the capsule, allowing penetration of antibodies beneath the CPS.

The extremely mucoid, viscous appearance of certain strains of *Klebsiella* is not the result of the CPS. The material is of an unknown nature, but it is not related to the hyperproduction of CPS or colanic acid (as in *E. coli*).[99] Loss of this plasmid-mediated substance results in significant loss of virulence. It is presumed this material also acts as a barrier to antibody.

Kabha and associates[98] found that certain *Klebsiella* possess Man-α2/3-Man or similar sequences in their CPS. Such sequences render bacteria, especially those in serum-poor environments such as the respiratory tract, susceptible to macrophages possessing a mannose receptor. Capsular types without the mannose sequence, which escape phagocytosis, multiply uninhibited and produce infection if other virulence factors are expressed. Studies done by Broug-Holub and colleagues[100] found that alveolar macrophages play a role in containing and clearing *Klebsiella* during pneumonia. All mice treated to eliminate alveolar macrophages and then challenged intratracheally with 100 cfu of *Klebsiella* died, but 100% of control mice survived. There was a 20- to 27-fold increase in numbers of *Klebsiella* in the lung. Failure to clear *Klebsiella* from the lung was not caused by inhibition of polymorphonuclear leukocytes (PMNs), which were recruited in large numbers to the infected lungs. It appeared that the PMNs, which were functional, could not adequately clear the lungs in the absence of the alveolar macrophages.

Siderophore systems which allow bacteria to complete for iron in tissue and blood, where the supply is restricted, are a requisite for pathogenicity. The catechol iron chelator enterobactin (enterochelin), common among *Enterobacteriaceae,* was found in all but one *Klebsiella* strain in two studies of urinary tract isolates.[50,101] In contrast to strains examined by Tarkkanen and colleagues,[101] Podschun and coworkers[50] found five strains that produced aerobactin, a more effective iron chelator associated with virulence. All five isolates were from patients with pyelonephritis. Aerobactin production also has been detected in bacteremic isolates.[102,103] In these strains it was found to be encoded on plasmids ranging from 180 to 185 kb in size.

Klebsiella are known to produce several fimbrial and nonfimbrial adhesins. Multiple nonfimbrial adhesion factors have been described,[104–106] with three distinct patterns of adherence to the human intestine-407 cell line. Diffuse adherence was characterized by the scattering of bacteria over the cell surface, aggregative adherence by attachment to both the tissue cells and the glass surface, and localized adherence by discreet clumps of bacteria on cells. A 29-kDa plasmid-encoded surface protein is responsible for diffuse adherence,[104] which was the least common pattern in this study. Aggregative adherence accounted for 57% of the adherence found in the 30 strains of *K. pneumoniae* studied.[106] Visually, by electron microscopy, this pattern is similar to that seen with enteroaggregrative *E. coli,* but strains did not hybridize with a 1-kb probe to the plasmid pCDV432 of *E. coli.* Strains with aggregative adherence were isolated primarily from UTIs and wounds. Thirty percent of *K. pneumoniae* showed localized adherence.[106] These were bacteremic and respiratory isolates from pneumonia that were resistant to multiple β-lactams, aminoglycosides, and other antimicrobials. β-Lactamases carried by these bacteria are known to be plasmid mediated, and studies are underway to determine whether localized adherence is carried in this manner. If so, it would account for the fact these strains have been involved in numerous outbreaks and have spread readily between hospitals.[106] The 29-kDa nonfimbrial adhesin designated CF29k, which produces diffuse adherence, resides on a plasmid that also mediates mucoid colony formation and aerobactin, an iron-scavenging system.[105] The two plasmids that mediate CF29k and KPF 28 (a fimbrial adhesin) also encode for the ESBL TEM-5 and the β-lactamase SHV-4, respectively, suggesting a link between ability to adhere to the intestine and antibiotic resistance.[107] Di Martino and colleagues[107] further found that strains with plasmids encoding other β-lactamases (TEM-1, TEM-3, TEM-8, TEM-24, and TEM-16) also adhered to Caco-2 tissue cells. None expressed CF29k or KPF 28, suggesting the presence of other, as yet unidentified adhesions.

Three fimbriae produced by *Klebsiella,* types 1, 3, and 6, mediate adherence in UTIs. Würker and colleagues[108] presented convincing evidence that strains possessing both type 1 and 3 fimbriae were significantly ($P<0.05$) more capable of adhering to buccal, tracheal, pulmonary, or uroepithelial cells than strains with type 1 or type 3 alone. Tarkkanen and associates,[101] on the other hand, questioned the role of type 1 fimbria in UTI. In their study, type 1 fimbriated strains failed to bind to epithelial cells in human urine sediment even though fimbriae were present on 85% of the 39 strains investigated. They found that type 3 fimbriae were universally present and bound efficiently to urinary tract epithelial cells. From these findings and the knowledge that type 3 fimbriae bind to type V collagen and renal extracellular matrix, they concluded that type 3 fimbriae played a role in lower UTIs, particularly with regard to persistence of *Klebsiella* in catheter-associated bacteriurias. In contrast, Podschun and coworkers,[50] in a study of 146 strains involved in UTIs, found strains with type 1 fimbriae to be significantly associated with pyelonephritis rather than with lower UTIs ($P<0.05$). The distribution of type 3 fimbriae, highest in patients with cystitis (80%), was not significantly different in other types of infection (asymptomatic bacteriuria, 67%; pyelonephritis, 58%). Since these authors also found that type 3 fimbriae were expressed more frequently in fecal isolates than in UTI isolates, they concluded that these structures were not associated with UTI.

A variety of toxins have also been reported in strains of *K. pneumoniae* and *K. oxytoca.* Klipstein and Engert[109] reported on the presence of a heat- and acid-stable (ST) enterotoxin with a molecular weight of 5000 in strains of *K. pneumoniae* isolated from patients with tropical sprue. In 1986, Koo and Stein[110] presented data on a heat-labile enterotoxin elaborated in encapsulated *K. pneumoniae* isolates from 15 burn patients. Crude toxin preparations produced from all 15 isolates contained a toxin of about 26 kDa that was inactivated at 100° and 121°C. They surmised that the capsule provided resistance to phagocytosis while the toxin caused tissue damage and aided in invasion of the bloodstream. In 1989, Guarino[111] further characterized the ST of Klipstein and Engert.[109] Although it shared a mechanism of action similar to that of *E. coli* STa in that it activated the guanylate cyclase–cyclic guanosine monophosphate (cGMP) system, its amino acid sequence varied from that of *E. coli. K. oxytoca* has been shown to produce a cytotoxin active in ampicillin-associated hemorrhagic colitis infections.[112–115] It appears to be a very small toxin with a molecular weight of 217 Da[114] that is heat labile[113,114] and chromosomally mediated.[114] It induces fluid accumulation in both the ileum and the colon in rabbit ileal loops. Mucosal damage, however, occurs only in the ileum despite the fact that five times as much toxin is required to induce a response in the ileum as in the colon.[115] Although 10^8 *Klebsiella* per gram of feces is required to produce damage, studies on *K. oxytoca* in hemorrhagic colitis infections found 10^7 cells per gram of feces during the acute stage and 10^5 cells during convalescence. Therefore, abnormal proliferation of *Klebsiella* in the gut, such as would be permitted in the presence of antimicrobial therapy, could result in sufficient numbers of bacteria to produce the amount of toxin necessary for disease.

ANTIMICROBIAL SUSCEPTIBILITY

Historically, *Klebsiella* strains have exhibited resistance to penicillins, especially ampicillin and carbenicillin, via the penicillinase SHV-1. Still today, surveys indicate, SHV-1 is the most prevalent β-lactamase in clinical isolates, with 11% to 73% strains harboring this enzyme.[116] However, 25 years ago a second β-lactamase, TEM-1, appeared in *Klebsiella,* and many strains began to carry it alone or in combination with SHV-1.[106,116] Between 1984 and 1986, strains of *Enterobacteriaceae* emerged in Europe, the United States, and eventually worldwide that contained ESBLs.[116] These enzymes are a highly successful bacterial response to the extended-spectrum β-lactam antibiotics which were produced to counter the first family of β-lactamases (TEM-1, TEM-2, SHV-1). They essentially arose as mutations of TEM-1, TEM-2, and SHV-1, and more than 30 documented or putative ESBLs are known.[117] *K. pneumoniae* has the dubious distinction of being the most common species harboring ESBLs.[116,117] These enzymes hydrolyze newer cephalosporins and monobactams (aztreonam) and also confer resistance to β-lactamase inhibitor/β-lactam antibiotic combinations (eg, amoxicillin/clavulanate); resistance rates to such combination drugs range from 8% to 34% in *Klebsiella* species.[116] In the United States, the incidence of ESBL *Klebsiella* pneumoniae is about 8% to 11%.[118] Tsakris and associates[119] in Greece found that 60% of 48 *Klebsiella* pneumoniae were resistant to third-generation cephalosporins and aztreonam via the ESBL SHV-5. Two of these strains also possessed class C cephalosporinases, and 32 strains (66%) were resistant to ampicillin/sulbactam via the β-lactamase TEM-1. This report emphasizes not only the high rate but also the diversity of β-lactamases that may be present in one institution. In France, ESBLs in *K. pneumoniae* increased from less than 1% in 1985 to 11% in 1988[116] and rose to 15% in 1990.[120] Sirot and colleagues[120] and Livrelli and associates[106] reported that with the rise in ESBLs there has been a concomitant increase in aminoglycoside resistance. Such trends are alarming because therapy for *Klebsiella* generally involves use of a β-lactam antibiotic or aminoglycoside, or both.[34,35] French and coworkers[121] found a *K. pneumoniae* strain that produced five times the normal amount of the ESBL SHV-5, resulting in resistance to all β-lactamase inhibitor/β-lactam drug combinations tested. This suggests that such drugs may not be reliable in treating infections with organisms possessing ESBLs.

Detection of resistance due to ESBL can be difficult. Some ESBLs confer low levels of resistance or are selective for certain β-lactams, and organisms possessing these enzymes may not appear resistant by National Committee for Clinical Laboratory Standards breakpoints.[122] The presence of ESBLs can be detected by a double-disk diffusion assay which detects a synergistic response between a disk containing clavulanate (amoxicillin/clavulanate) and a disk containing a third-generation cephalosporin (ceftazidime, cefotaxime) or aztreonam. Disks of each are placed 30 mm apart, and an extension of the zone around any of the disks toward the amoxicillin/clavulanate disk is interpreted as synergy. Any strain showing full or intermediate resistance or synergy to any of the drugs should be considered resistant.[120] Sirot and coinvestigators[120] found that 50% of resistant strains would have been misclassified as susceptible by routine testing. Jacoby and Han[122] found ceftazidime or aztreonam disks to be superior to cefotaxime disks in detection of ESBLs.

Resistance to antimicrobials other than β-lactams was an unusual finding among more than 11,000 strains tested in multiple studies.[78,123–127] Livrelli and associates,[106] however, found 11 strains resistant to aminoglycosides, 14 to tetracycline, and 9 to chloramphenicol out of 30 strains of *K. pneumoniae.* Of 16 *K. oxytoca* strains, 9 were resistant to aminoglycosides, 5 to chloramphenicol, and 4 to tetracycline. Occasional strains of both species were resistant to nalidixic acid and sulfonamides. In the United States, susceptibility to ciprofloxacin ranged from 94% for *K. pneumoniae* to 98% for *K. oxytoca* and 99% for *K. ozaenae.*[123] No ciprofloxacin resistance was found among 5 strains each of *K. ornithinolytica, K. ozaenae,* and *K. planticola* and 4 strains of *K. rhinoscleromatis,*[124] nor in 10 strains of *K. terrigena.*[78]

Out of 80 strains of *K. terrigena* tested,[78,126,127] resistance was typical of *Klebsiella* in general and occurred primarily with penicillins, cephalothin, cefoxitin, ceftazidime, or amoxicillin/clavulanate. Variable resistance was noted for trimethoprim-sulfamethoxazole (SXT), with 6 of 10 strains resistant in one study[78] and 50 of 50 susceptible in another.[127] Likewise, both intermediate resistance[78] and susceptibility[126,127] were reported for doxycycline and tetracycline. No resistance to any aminoglycoside was found. Very few strains of *K. planticola, K. ornithinolytica,* or *K. rhinoscleromatis* were tested,[125,126] but resistance was reported only for amoxicillin/clavulanate. Cefetamet, ceftriaxone, SXT, cefotaxime, cefoxitin, doxycycline, and all aminoglycosides were active against these strains in vitro.

In the study of Sirot and colleagues,[120] strains carrying ESBL β-lactamases were highest in the surgical ICU ward (30%), but rates in the medical ICU, surgical, medical, and pediatric wards ranged from 13% to 18%. Resistance by specimen source in order of decreasing frequency was 27% for the respiratory tract, 23% for blood, 14% for miscellaneous sites, and 11% for urine.

Korvick and coworkers[35] studied monotherapy versus treatment with an antibiotic combination in 230 cases of *Klebsiella* bacteremia. They concluded that treatment with a β-lactamase or aminoglycoside alone was acceptable in less severely ill patients (ie, those who were immunocompetent, had an urinary tract portal, and were mentally alert or with normal vital signs). Patients with hypotension required both a β-lactam and an aminoglycoside for a successful outcome. However, Feldman and colleagues[34] found that, among patients surviving longer than 48 hours (no hypotension), none of nine receiving combination therapy but about half of those receiving only a β-lactam or an aminoglycoside died.

CONCLUSIONS

Probably the most serious challenge facing the medical community is the rapidly rising resistance seen in this species. Both increased awareness of and better methods for detection of β-lactamases are needed. Hospitals armed with accurate information regarding the numbers of resistant strains will be better able to curtail indiscriminate and/or inappropriate use of antimicrobials. The commonplace practice of prescribing amoxicillin/clavulanate in the community has added to the problem, because strains found in community-acquired infections have also shown increased resistance.[116] Recent drops in *Klebsiella* nosocomial infection rates[16] are promising, but continued efforts are necessary to ensure that the decline continues. Redoubled efforts to increase our knowledge of the capsular and other defense mechanisms of these strains could result in a vaccine[128] that could save many lives. Meanwhile, support is needed for experiments to find and study therapeutic agents such as sodium salicylate or bismuth salts which decrease the bacterial capsule, rendering strains susceptible to phagocytosis and normal serum killing.[129]

REFERENCES

1. Grimont F, Grimont PAD, Richard C. The genus *Klebsiella.* In: Starr MP, Stolp H, Truper HG, Balows A, Schlegel HG, eds. *The prokaryotes: A handbook on habitats, isolation, and identification of bacteria.* 2nd ed. Berlin: Springer-Verlag, 1991;2775–2796.
2. Ørskov, I. The genus *Klebsiella* Trevisan 1885. In: Krieg NR, Holt JG, eds. *Bergey's manual of systematic bacteriology.* Vol 1. Baltimore: Williams & Wilkins, 1984;461–465.
3. Edwards PR, Ewing WH. *Identification of the Enterobacteriaceae.* 2nd ed. Minneapolis: Burgess Publishing Co., 1962.
4. Brenner DJ, Steigerwalt AG, Fanning GR. Differentiation of *Enterobacter aerogenes* from *Klebsiellae* by deoxyribonucleic acid reassociation. *Int J Syst Bacteriol* 1972;22:193–200.
5. Cowan ST, Steel KJ, Shaw C. A classification of the *Klebsiella* group. *J Gen Microbiol* 1960;23:601–612.
6. Bascomb S, Lapage SP, Willcox WR, Curtis MA. Numerical classification of the tribe *Klebsiellae. J Gen Microbiol* 1971;66:279–295.
7. Farmer JJ III, Davis BR, Hickman-Brenner FW, et al. Biochemical identification of new species and biogroups of *Enterobacteriaceae* isolated from clinical specimens. *J Clin Microbiol* 1985;21:46–76.
8. Brenner DJ, Farmer JJ III, Hickman FW, Asbury MA, Steigerwalt AG. Taxonomic and nomenclatural changes in *Enterobacteriaceae.* Atlanta: Centers for Disease Control, 1977.
9. Izard D, Ferragut C, Gavini F, Kersters K, De Ley J, Leclerc H. *Klebsiella terrigena,* a new species from soil and water. *Int J Syst Bacteriol* 1981;31:116–127.
10. Naemura LG, Bagley ST, Seidler RJ, Kaper JB, Colwell RR. Numerical taxonomy of *Klebsiella* pneumoniae strains isolated from clinical and nonclinical sources. *Curr Microbiol* 1979;2:175–180.
11. Woodward BW, Carter M, Seidler RJ. Most nonclinical *Klebsiella* strains are not *K. pneumoniae* sensu stricto. *Curr Microbiol* 1979;2:181–185.
12. Bagley ST, Seidler RL, Brenner DJ. *Klebsiella planticola* sp. nov.: A new species of *Enterobacteriaceae* found primarily in nonclinical environments. *Curr Microbiol* 1981;6:105–109.
13. Gavini F, Izard D, Grimont PAD, Beji A, Agernon E, Leclerc H. Priority of *Klebsiella planticola* Bagley, Seidler, and Brenner 1982 over *Klebsiella trevisanii* Ferragut, Izard, Gavini, Kersters, DeLey, and Leclerc 1983. *Int J Syst Bacteriol* 1986;36:486–488.
14. Sakazaki R, Tamura K, Kosako Y, Yoshizaki E. *Klebsiella ornithinolytica* sp. nov., formerly known as ornithine-positive *Klebsiella oxytoca. Curr Microbiol* 1989;18:201–206.
15. Stratton CW. *Klebsiella. Infect Control* 1984;5: 343–347.
16. Schaberg DR, Culver DH, Gaynes RP. Major trends in the microbial etiology of nosocomial infection. *Am J Med* 1991;91(Suppl 3B):72S-3B74S.
17. Spencer RC. Predominant pathogens found in the European prevalence of infection in intensive care study. *Eur J Clin Microbiol Infect Dis* 1996;15:281–285.
18. Montgomerie JZ. Epidemiology of *Klebsiella* and hospital-associated infections. *Rev Infect Dis* 1979;1: 736–753.
19. Lucet JC, Chevret S, Decré D, et al. Outbreak of multiply resistant *Enterobacteriaceae* in an intensive care unit: Epidemiology and risk factors for acquisition. *Clin Infect Dis* 1996;22:430–436.
20. Fryklund B, Tullus K, Burnam LG. Epidemiology and attack index of gram-negative bacteria causing invasive infection in three special-care neonatal units and risk factors for infection. *Infection* 1995;23:76–80.
21. Kikuchi N, Kagota C, Nomura T, Hiramune T, Takahashi T, Yanagawa R. Plasmid profiles of *Klebsiella pneumoniae* isolated from bovine mastitis. *Vet Microbiol* 1995;47:9–15.
22. Knittel MD, Seidler RJ, Eby C, Cabe LM. Colonization of the botanical environment by *Klebsiella* isolates of pathogenic origin. *Appl Environ Microbiol* 1977;34:557–563.
23. Andraca R, Edson RS, Kern EB. Rhinoscleroma: A growing concern in the United States? Mayo Clinic experience. *Mayo Clin Proc* 1993;68:1151–1157.
24. Paul C, Pialoux G, Dupont B, et al. Infection due to *Klebsiella rhinoscleromatis* in two patients infected with human immunodeficiency virus. *Clin Infect Dis* 1993;16:441–442.
25. Altmann G, Ostfield E, Zohar S, Theodor E. Rhinoscleroma. *Isr J Med Sci* 1977;13:62–64.
26. Krasilnikov AP, Izraitel NA, Krylov IA. Focal incidence of scleroma. *J Hyg Epidemiol Microbiol Immunol* 1971;15:243–257.
27. Gera N, Bhatia A, Tickoo SK. Cytomorphic features of scleromatous lymphadenitis on fine-needle aspiration cytology: A case report. *Diagn Cytopathol* 1995;13: 320–321.
28. Tang L, Chen S. *Klebsiella pneumoniae* meningitis: Prognostic factors. *Scand J Infect Dis* 1994;26: 95–102.
29. Holder CD, Halkias D. Case report: Relapsing, bacteremic *Klebsiella pneumoniae* meningitis in an AIDS patient. *Am J Med Sci* 1988;295:55–59.
30. Strampfer MJ, Schoch PE, Cunha BA. Cerebral abscess by *Klebsiella ozaenae. J Clin Microbiol* 1987; 25:1553–1554.
31. Tang L, Chen S. *Klebsiella ozaenae* meningitis: Report of two cases and review of the literature. *Infection* 1994;22:58–61.
32. Montgomerie JZ, Ota JK. *Klebsiella* bacteremia. *Arch Intern Med* 1980;140:525–527.
33. García de la Torre M, Romero-Vivas J, Martinez-Beltrán J, Guerrero A, Meseguer M, Bouza E. *Klebsiella* bacteremia: An analysis of 100 episodes. *Rev Infect Dis* 1985;7:143–150.
34. Feldman C, Smith C, Levy H, Ginsburg P, Miller SD, Koornhof HJ. *Klebsiella pneumoniae* bacteraemia at an urban general hospital. *J Infect* 1990;20:21–31.
35. Korvick JA, Bryan CS, Farber B, et al. Prospective observational study of *Klebsiella* bacteremia in 230 patients: Outcome for antibiotic combinations versus monotherapy. *Antimicrob Agents Chemother* 1992; 36:2639–2644.
36. Goetz AM, Rihs JD, Chow JW, Singh N, Muder RR. An outbreak of infusion-related *Klebsiella pneumoniae* bacteremia in a liver transplantation unit. *Clin Infect Dis* 1995;21:1501–1503.
37. Murray KA, Clements BH, Keas SE. *Klebsiella ozaenae* septicemia associated with Hansen's disease. *J Clin Microbiol* 1981;14:703–705.

38. Picard E, Gillis D, Klapholz L, Schreiber L, Engelhard D. Toxic epidermal necrolysis associated with *Klebsiella pneumoniae* sepsis. *Pediatr Dermatol* 1994;11: 331–334.
39. Freney J, Gavini R, Alexandre H, et al. Nosocomial infection and colonization by *Klebsiella trevisanii. J Clin Microbiol* 1986;23:948–950.
40. Carpenter JL. *Klebsiella* pulmonary infections: Occurrence at one medical center and review. *Rev Infect Dis* 1990;12:672–682.
41. Feldman C, Ross S, Mahomed AG, Omar J, Smith C. The aetiology of severe community-acquired pneumonia and its impact on initial, empiric, antimicrobial chemotherapy. *Respir Med* 1995;89:187–192.
42. Jong GM, Hsuie TR, Chen CR, Chang HY, Chen CW. Rapidly fatal outcome of bacteremic *Klebsiella pneumoniae* pneumonia in alcoholics. *Chest* 1995;107: 214–217.
43. Penner C, Maycher B, Long R. Pulmonary gangrene a complication of bacterial pneumonia. *Chest* 1994;105: 567–573.
44. Avery RK, Salman SD, Baker AS. Rhinoscleroma treated with ciprofloxacin: A case report. *Laryngoscope* 1995;105:854–856.
45. Maloway MS, Chester B, Allerhand J. Isolation and microbiologic differentiation of *Klebsiella rhinoscleromatis* and *Klebsiella ozaenae* in cases of chronic rhinitis. *Am J Clin Pathol* 1972;58:550–553.
46. Ferguson JL, McCaffrey TV, Kern EB, Martin WJ III. Effect of *Klebsiella ozaenae* on ciliary activity in vitro: Implications in the pathogenesis of atrophic rhinitis. *Otolaryngol Head Neck Surg* 1990;102:207–211.
47. Flournoy DJ. Incidence of bacteria from clean-catch and catheterized urine of nursing home residents. *J Okla State Med Assoc* 1993;86:550–552.
48. Davis CP. Bladder cytopathic effects associated with gram-negative bacterial genera. *Curr Microbiol* 1981; 6:37–42.
49. Stamm WE, Hooton TM. Management of urinary tract infections in adults. *N Engl J Med* 1993;329: 1328–1334.
50. Podschun R, Sievers D, Fischer A, Ullmann U. Serotypes, hemagglutinins, siderophore synthesis, and serum resistance of *Klebsiella* isolates causing urinary tract infections. *J Infect Dis* 1993;168:1415–1421.
51. Goldstein EJC, Lewis RP, Martin WJ, Edelstein PH. Infections caused by *Klebsiella ozaenae:* A changing disease spectrum. *J Clin Microbiol* 1978;8:413–418.
52. Podschun R, Ullmann U. Incidence of *Klebsiella planticola* among clinical *Klebsiella* isolates. *Med Microbiol Lett* 1994;3:90–95.
53. Karpas CM, Boman I. The significance of *Klebsiella enteritis. Am J Clin Pathol* 1966;46:632–640.
54. Olarte J, Ferguson WW, Henderson ND, Torregrosa L. *Klebsiella* strains isolated from diarrheal infants. *Am J Dis Child* 1961:101:763–770.
55. Oliver JM, Dorval ED, Loulergue J, Bacq Y, Duval G, Metman EH. *Klebsiella oxytoca* and antibiotic-associated non-pseudomembranous colitis. *Gastroenterology* 1991;100:A605.
56. Chang FG, Chou MY. Comparison of pyogenic liver abscesses caused by *Klebsiella pneumoniae* and non–*Klebsiella pneumoniae* pathogens. *J Formos Med Assoc* 1995;94:232–237.
57. Han SB. Review of hepatic abscess from *Klebsiella pneumoniae* in association with diabetes mellitus and septic endophthalmitis. *West J Med* 1995;162:220–224.
58. Chou FF, Kuo HK. Endogenous endophthalmitis associated with pyogenic hepatic abscess. *J Am Coll Surg* 1996;182:33–36.
59. Nagase T, Wada S, Naramura R, et al. Magnetic resonance imaging of multiple brain abscesses of the bilateral basal ganglia. *Intern Med* 1995;34:554–558.
60. Chowdhury P, Stein DS. Pyogenic hepatic abscess and septic pulmonary emboli associated with *Klebsiella ozaenae* bacteremia. *South Med J* 1992;85:638–641.
61. Lee TY, Ko SF, Cheng YF, Wang YL, Chen WY. Primary gas-containing mediastinal abscess in a diabetic patient. *Am J Emerg Med* 1995;13:427–429.
62. Trotter JF, Peterson WL. Injection sclerotherapy as a cause of *Klebsiella pneumoniae* peritonitis. *J Clin Gastroenterol* 1992;15:329–331.
63. Kuo CH, Changchien CS, Chen JJ, Tai DI, Chiou SS, Lee CM. Septic acute cholecystitis. *Scand J Gastroenterol* 1995;30:272–275.
64. Pontell J, Har-El G, Lucente FE. Retropharyngeal abscess: Clinical review. *Ear Nose Throat J* 1995;74: 701–704.
65. Smith RF, Dayton SL, Chipps DD, Blasi D. Intestinal carriage of *Klebsiella* and *Pseudomonas* in burned children and their comparative role in nosocomial infection. *Health Lab Sci* 1973;10:173–179.
66. Bruno-Murtha LA, Sedghivaziri MA, Arbeit RD. Crepitant myonecrosis caused by *Klebsiella pneumoniae* in an immunocompromised diabetic patient. *J Infect Dis* 1990;162:1415–1416.
67. Ritter MS, Mroch H, Burns MJ. Soaring suppurative sea shells from the sea shore: *Pseudomonas aeruginosa* and *Klebsiella pneumoniae* septic arthritis after a marine shell injury. *Pediatr Emerg Care* 1993;9:289–291.
68. Bruce SK, Schick DG, Tanaka L, Jimenez EM, Montgomerie JZ. Selective medium for isolation of *Klebsiella pneumoniae. J Clin Microbiol* 1981;13: 1114–1116.
69. Thom BT. *Klebsiella* in feces. *Lancet* 1970;2:1033.
70. McIver CJ, Tapsall JW. Further studies of clinical isolates of cysteine-requiring *Escherichia coli* and *Klebsiella* and possible mechanisms for their selection in vivo. *J Med Microbiol* 1993;39:382–387.
71. O'Hara CM, Tenover FC, Miller JM. Parallel comparison of accuracy of API 20E, Vitek GNI, MicroScan Walk/Away Rapid ID, and Becton Dickinson Cobas Micro ID-E/NF for identification of members of the family *Enterobacteriaceae* and common gram-negative, non–glucose-fermenting bacilli. *J Clin Microbiol* 1993;31:3165–3169.
72. Rhoden DL, Schable B, Smith PB. Evaluation of PASCO MIC-ID system of identifying gram-negative bacilli. *J Clin Microbiol* 1987;25:2363–2366.
73. O'Hara CM, Miller JM. Evaluation of the AutoSCAN-W/A system for rapid (2 hour) identification of members of the family *Enterobacteriaceae. J Clin Microbiol* 1992;30:1541–1543.
74. Kelly MT, Leicester C. Evaluation of the Autoscan Walkaway system for rapid identification and susceptibility testing of gram-negative bacilli. *J Clin Microbiol* 1992;30:1568–1571.
75. Kitch TT, Jacobs MR, Appelbaum PC. Evaluation of RapID onE system for identification of 379 strains in the family *Enterobacteriaceae* and oxidase-negative

gram-negative nonfermenters. *J Clin Microbiol* 1994; 32:931–934.
76. Wauters G, Boel A, Vorn GP, et al. Evaluation of a new identification system, Crystal Enteric/Non-Fermenter, for gram-negative bacilli. *J Clin Microbiol* 1995;33: 845–849.
77. Rhoads S, Marinelli L, Imperatrice CA, Nachamkin I. Comparison of MicroScan WalkAway system and Vitek system for identification of gram-negative bacteria. *J Clin Microbiol* 1995;33:3044–3046.
78. Podschun R, Ullmann U. Isolation of *Klebsiella terrigena* from clinical specimens. *Eur J Clin Microbiol Infect Dis* 1992;11:349–352.
79. Farmer JJ III. *Enterobacteriaceae:* Introduction and identification. In: Murray PR, Baron EJ, Pfaller MA, Tenover FC, Yolken RH, eds. *Manual of clinical microbiology.* 6th ed. Washington DC: American Society for Microbiology 1995;438–449.
80. Monnet D, Freney J. Method for differentiating *Klebsiella planticola* and *Klebsiella terrigena* from other *Klebsiella. J Clin Microbiol* 1994;32:1121–1122.
81. Ørskov I, Fife-Asbury MA. New *Klebsiella* antigen K 82 and the deletion of five of those previously assigned. *Int J Syst Bacteriol* 1977;27:386–387.
82. Smith SM, Digori JT, Eng RHK. Epidemiology of *Klebsiella* antibiotic resistance and serotypes. *J Clin Microbiol* 1982;16:868–873.
83. Ewing EH. *Edwards and Ewing's identification of* Enterobacteriaceae. 3rd ed. Minneapolis: Burgess Publishing Co., 1972.
84. Pruzzo C, Satta G. Capsular antigenic variations by lysogenic conversion in *Klebsiella pneumoniae:* Relationship with virulence. *Curr Microbiol* 1988;16: 259–263.
85. Simoons-Smit AM, Verweij-Van Vaught AMJJ, Maclaren DM. The role of K antigens as virulence factors in *Klebsiella.* 1986;21:133–137.
86. Riser E, Noone P. *Klebsiella* capsular type versus site of isolation. *J Clin Pathol* 1981;34:552–555.
87. Albertí S, Hernández-Allés S, Gil J, et al. Development of an enzyme-linked immunosorbent assay method for typing and quantitation of *Klebsiella pneumoniae* lipopolysaccharide: Application to serotype O1. *J Clin Microbiol* 1993;31:1379–1381.
88. Trautman M, Cross AS, Reich G, Held H, Podschun R, Marre R. Evaluation of a competitive ELISA method for the determination of *Klebsiella* O antigens. *J Med Microbiol* 1996;44:44–51.
89. Buffenmyer CL, Rycheck RR, Yee RB. Bacteriocin (Klebocin) sensitivity typing of *Klebsiella. J Clin Microbiol* 1976;4:239–244.
90. Ichhpujani RL, Bhatia R, Aggarwal P, Anand BR, Chowdhury ANR. Klebocin type and diarrhoeagenicity of *Klebsiella pneumoniae. Indian J Med Res* 1983;77:605–607.
91. Bauernfeind A, Petermüller C, Schneider R. Bacteriocins as tools in analysis of nosocomial *Klebsiella pneumoniae* infections. *J Clin Microbiol* 1981;14:15–19.
92. Gaston MA, Ayling-Smith BA, Pitt TL. New bacteriophage typing scheme for subdivision of the frequent capsular serotypes of *Klebsiella* spp. *J Clin Microbiol* 1987;25:1228–1232.
93. Hartstein AI, Morthland VH, Rouke JW, Sykes R, Rashad AL. Plasmid DNA analysis, biotyping, and antimicrobic susceptibility as subtyping tests for *Klebsiella pneumoniae* and *Klebsiella oxytoca. Diagn Microbiol Infect Dis* 1993;16:35–41.
94. Gori A, Espinasse F, Deplano A, Nonhoff C, Nicolas MH, Struelens MJ. Comparison of pulsed-field gel electrophoresis and randomly amplified DNA polymorphism analysis for typing extended-spectrum-β-lactamase–producing *Klebsiella pneumoniae. J Clin Microbiol* 1996;34:2448–2453.
95. Prodinger WM, Fille M, Bauernfeind A, et al. Molecular epidemiology of *Klebsiella pneumoniae* producing SHV-5 β-lactamase: Parallel outbreaks due to multiple plasmid transfer. *J Clin Microbiol* 1996;34:564–568.
96. Merino S, Camprubí S, Albertí S, Benedí V, Tomás JM. Mechanisms of *Klebsiella pneumoniae* resistance to complement-mediated killing. *Infect Immun* 1992; 60:2529–2535.
97. Williams P, Lambert PA, Brown MRW. Penetration of immunogloblins through the *Klebsiella* capsule and their effect on cell-surface hydrophobicity. *J Med Microbiol* 1988;26:29–35.
98. Kabha K, Nissimov L, Athamna A, et al. Relationships among capsular structure, phagocytosis, and mouse virulence in *Klebsiella pneumoniae. Infect Immun* 1995;63:847–852.
99. Podschun R, Ullmann U. *Klebsiella* capsular type K7 in relation to toxicity, susceptibility to phagocytosis and resistance to serum. *J Med Microbiol* 1992;36: 250–254.
100. Broug-Holub E, Toews GB, Van Iwaarden JF, et al. Alveolar macrophages are required for protective pulmonary defense in murine *Klebsiella* pneumonia: Elimination of alveolar macrophages increases neutrophil recruitment but decreases bacterial clearance and survival. *Infect Immun* 1997;65:1139–1146.
101. Tarkkanen A, Allen BL, Williams PH, et al. Fimbriation, capsulation, and iron-scavenging systems of *Klebsiella* strains associated with human urinary tract infection. *Infect Immun* 1992;60:1187–1192.
102. Nassif X, Fournier J, Arondel J, Sansonetti PJ. Mucoid phenotype of *Klebsiella pneumoniae* is a plasmid-encoded virulence factor. *Infect Immun* 1989;57:546–552.
103. Vernet V, Philippon A, Madoulet C, Vistelle R, Jaussaud R, Chippaux C. Virulence factors (aerobactin and mucoid phenotype) in *Klebsiella pneumoniae* and *Escherichia coli* blood culture isolates. *FEMS Microbiol Lett* 1995;130:51–57.
104. Darfeuille-Michaud A, Jallat C, Aubel, et al. R-plasmid–encoded adhesive factor in *Klebsiella pneumoniae* strains responsible for human nosocomial infections. *Infect Immun* 1992;60:44–55.
105. Favre-Bonte, S, Darfeuille-Michaud A, Forestier C. Aggregative adherence of *Klebsiella pneumoniae* to human intestine-407 cells. *Infect Immun* 1995;63: 1318–1328.
106. Livrelli V, De Champs C, Di Martino P, Darfeuille-Michaud A, Forestier C, Joly B. Adhesive properties and antibiotic resistance of *Klebsiella, Enterobacter,* and *Serratia* clinical isolates involved in nosocomial infections. *J Clin Microbiol* 1996;34:1963–1969.
107. Di Martino P, Sirot D, Joly B, Rich C, Darfeuille-Michaud A. Relationship between adhesion to intestinal Caco-2 cells and multidrug resistance in *Klebsiella pneumoniae* clinical isolates. *J Clin Microbiol* 1997; 35:1499–1503.
108. Würker M, Beuth J, Ko HL, Przondo-Mordarska A,

Pulverer G. Type of fimbriation determines adherence of *Klebsiella* bacteria to human epithelial cells. *Zentralbl Bakteriol* 1990;274:239–245.
109. Klipstein FA, Engert RF. Purification and properties of *Klebsiella pneumoniae* heat-stable toxin. *Infect Immun* 1976;13:373–381.
110. Koo FCW, Stein MD. Detection of an extracellular toxin produced by burn isolates of *Klebsiella pneumoniae. Curr Microbiol* 1986;13:23–27.
111. Guarino A, Guandalini S, Alessio M, et al. Characteristics and mechanism of action of a heat-stable enterotoxin produced by *Klebsiella pneumoniae* from infants with secretory diarrhea. *Pediatr Res* 1989;25:514–518.
112. Minami J, Okabe A, Shiode J, Hayashi H. Production of a unique cytotoxin by *Klebsiella pneumoniae. Microbial Pathog* 1989;7:203–211.
113. Higaki M, Chida T, Takano H, Nakaya R. Cytotoxic component(s) of *Klebsiella oxytoca* on HEp-2 cells. *Microbiol Immunol* 1990;34:147–151.
114. Minami J, Saito S, Yoshida T, Uemara T, Okabe A. Biological activities and chemical composition of a cytotoxin of *Klebsiella oxytoca. J Gen Microbiol* 1992; 138:1921–1927.
115. Minami J, Katayama S, Matsushita O, Sakamoto H, Okabe A. Enterotoxic activity of *Klebsiella oxytoca* cytotoxin in rabbit intestinal loops. *Infect Immun* 1994; 62:172–177.
116. Sanders CC, Sanders WE Jr. β-Lactam resistance in gram-negative bacteria: Global trends and clinical impact. *Clin Infect Dis* 1992;15:824–839.
117. Jacoby GA, Medeiros AA. More extended-spectrum β-lactamases. *Antimicrob Agents Chemother* 1991;35: 1697–1704.
118. Chartrand SA, Thompson KJ, Sanders CC. Antibiotic-resistant, gram-negative bacillary infections. *Pediatr Infect Dis* 1996;7:187–203.
119. Tsakris A, Tzouvelekis LS, Douboyas T, Tzelepi E. Diversity of β-lactamases among *Klebsiella pneumoniae* strains isolated in a university hospital in Greece. *Eur J Epidemiol* 1997;13:103–107.
120. Sirot DL, Goldstein FW, Soussy CJ, et al. Resistance to cefotaxime and seven other β-lactams in members of the family *Enterobacteriaceae:* A 3 year study in France. *Antimicrob Agents Chemother* 1992;36: 1677–1681.
121. French GL, Shannon KP, Simmons N. Hospital outbreak of *Klebsiella pneumoniae* resistant to broad-spectrum cephalosporins and β-lactam–β-lactamase inhibitor combinations by hyperproduction of SHV-5 β-lactamase. *J Clin Microbiol* 1996;34:358–363.
122. Jacoby GA, Han P. Detection of extended-spectrum β-lactamases in clinical isolates of *Klebsiella pneumoniae* and *Escherichia coli. J Clin Microbiol* 1996; 34:908–911.
123. Thornsberry C. Susceptibility of clinical bacterial isolates to ciprofloxacin in the United States. *Infection* 1994;22(Suppl):S80–S89.
124. Jones RN, Pfaller MA, Allen SD, Gerlack EH, Fuchs PC, Alderidge KE. Antimicrobial activity of cefpirome An update compared to five third-generation cephalosporins against nearly 6000 recent clinical isolates from five medical centers. *Diagn Microbiol Infect Dis* 1991;14:361–364.
125. Hohl P, Lüthy-Hottenstein J, Zollinger-Iten J, Altwegg M. In vitro activities of fleroxacin, cefetamet, ciprofloxacin, ceftriaxone, trimethoprim-sulfamethoxazole, and amoxicillin-clavulanic acid against rare members of the family *Enterobacteriaceae* primarily of human (clinical) origin. *Antimicrob Agents Chemother* 1990; 34:1605–1608.
126. Freney J, Husson MO, Gavini F, et al. Susceptibilities to antibiotics and antiseptics of new species of the family *Enterobacteriaceae. Antimicrob Agents Chemother* 1988;32:873–876.
127. Podschun R. Isolation of *Klebsiella terrigena* from human feces: Biochemical reactions, capsule types, and antibiotic sensitivity. *Zentralbl Bakteriol* 1991; 275:73–78.
128. Cross AS, Sadoff JC, Furer E, Cryz SJ Jr. *Escherichia coli* and *Klebsiella* vaccines and immunotherapy. *Infect Dis Clin North Am* 1990;4:271–282.
129. Salo RJ, Domenico P, Tomás JM, et al. Salicylate-enhanced exposure of *Klebsiella pneumoniae* subcapsular components. *Infection* 1995;6:371–377.

9

The Genus *Serratia*

TAXONOMY

Historical accounts, dating as far back as the second century, of phenomena wherein foodstuffs took on the appearance of blood, are now attributed to red-pigmented bacteria. Although a number of such bacteria may have accounted for these phenomena, only the genus *Serratia* has been recovered from a documented investigation of such an event.

Certainly the most celebrated appearance of blood on food occurred in the year 1264, at the Mass of Bolsena, with the presence of a red substance on sacramental bread.[1] It was this apparent miracle that ultimately led to the adoption of communion for parishioners, a rite previously reserved for the clergy only. "Blood" sightings continued throughout the centuries until 1819, when the discovery of blood which covered polenta and spread to yet other foods led to civic upheaval among the villagers of Legarno, Italy.[1] Panic was so great that priests were called on to expunge the supernatural forces at work, and this ultimately resulted in the formation of a commission to investigate incidences in the village. Vincenzo Sette, a surgeon and district physician who led the investigation, concluded that the phenomena were the result of a fungus, *Zoagalactina imetrofa,* which readily grew in the moist, warm environment found in village homes.[1] However, Bartolomeo Bizio, a pharmacist who had disagreed with an earlier conclusion by Father Pietro Melo that the blood was an end product of fermentation, also took exception to the results found by the Sette's commission. Conducting his own independent studies in 1879, Bizio agreed with Sette regarding the necessity of damp, warm conditions for growth but named the seeds he found *Serratia marcescens.* Despite the proposal of numerous other epithets for this organism over the years, *S. marcescens* endured and was formally accepted in 1923.[2]

Over time, the number of species included within the genus *Serratia* has waxed and waned. In 1974, *Bergey's Manual of Systematic Bacteriology*[3] contained only a single species, *S. marcescens,* with Voges-Proskauer–negative variants contained in a subspecies designated *kiliensis.*[3] Farmer[4] currently recognizes eight species, retaining *Serratia liquefaciens* as the *S. liquefaciens* group because its members are not separable by routine biochemical tests. Grimont and Grimont,[2] however, proposed 10 species within the genus *Serratia,* splitting the *S. liquefaciens* group into three species, *S. liquefaciens* sensu stricto, *Serratia grimesii,* and *Serratia proteamaculans.*

S. liquefaciens is the second oldest species of serratiae. Originally placed in the genus *Aerobacter* in 1931, it was transferred to *Enterobacter* when *Aerobacter* was rejected in 1963 and was finally moved to *Serratia* in 1971.[5] The transfer of strains of *Erwinia proteamaculans,* which were biochemically similar to *S. liquefaciens,* to the genus *Serratia* resulted in a controversial proposal to replace *liquefaciens* with the older epithet *proteamaculans.*[6] The taxonomic storm surrounding this heterologous group (containing seven types) subsided after DNA hybridiz ies demonstrated that *S. prot* a distinct species from third species, *S. grimesii,* also emerged from this study. Two of the original seven biotypes, RQ and Adc, approached definition of a species but were included in *S. proteamaculans* and *S. grimesii,* respectively. Strains of biotype RQ were then elevated to subspecies status as *S. proteamaculans* ssp *quinovara.*[5] The three isolates of Adc included in *S. grimesii* remain as such, without being designated subspecies, pending the isolation of further strains for study.[5]

Serratia rubidaea, the third described member of the genus,[2] was originally named *Bacterium rubidaeum* in 1940 and was proposed as a species of *Serratia* in 1973.[8] Although it was briefly referred to as *Serratia marinorubra,* the older of the two epithets, *rubidaea,* had priority when both names were made subjective synonyms.[2] Composed primarily of water isolates, *Serratia plymuthica* was confirmed as belonging to *Serratia* as a species distinct from *S. liquefaciens* and was added to the genus in 1978.[9] That same year, the designation *Serratia odorifera* was proposed for an unnamed cluster of strains of human origin. These strains were similar to a previously described but unidentified isolate known as *Serratia* strain 38, which had been recovered from mushrooms.[9] *S. odorifera,* named for its distinctive odor, contains two biogroups, 1 and 2. In 1979, the designation *Serratia fonticola* was assigned to a group of strains that were first described in 1965 and referred to as *Citrobacter*-like or class C bacteria.[10] Phenotypically atypical from the other serratiae, *S. fonticola*'s inclusion in the genus was never fully embraced. In 1990, it was unquestionably placed within the psychrophilic *Serratia* species (see Epidemiology) by a study comparing a sequence of more than 1400 nucleotides of 16S ribosomal RNA (rRNA) in each of the 10 species of serratiae.[11] In 1979, yet another species, *Serratia ficaria,* was added after strains fitting the description of serratiae were discovered in figs and fig wasps.[12] *Serratia entomophila,* the last species to be placed in this genus, causes amber disease in the New Zealand grass grub, *Costelytra zealandica.*[13] Table 1 recaps the nomenclature of this genus along with American Type Culture Collection (ATCC) numbers, G+C ratios, and other pertinent information.

EPIDEMIOLOGY

Serratiae are water, soil, and plant organisms. Colonization and infection with *Serratia* in reptiles, insects, and mammals, includ-

TABLE 1. *Members of the genus* Serratia

Current Designation	Previous Designation	ATCC Number	G+C Ratio (mol%)	Isolation from Humans
S. marcescens		13880	57–60	Commonly
S. liquefacie… sensu …	*Aerobacter liquefaciens, Enterobacter liquefaciens, Serratia proteamaculans*	27592	53–54	Occasionally
S…	*Erwinia proteamaculans, Serratia liquefaciens*	19323	NA	Rarely
	…*rratia liquefaciens*	14460	NA	Rarely
	…*terium rubidaeum,* …*ratia marinorubra*	27593	53–58	Rarely
	…*cter*-like …class C	29844	48–52	Rarely
		183	53–56	Rarely
		33105	59	Rarely
		33077	54	Rarely
		43705	57–58	No

…e species.

ing humans, arises from their contact with these environments. Water is the primary ecologic niche of *S. plymuthica*[2,9] and *S. fonticola,*[2,10] and most serratiae also can be isolated from this source. In one study, 75% of 150 strains isolated from river water were *S. marcescens.*[2] Likewise, a variety of species of serratiae can be isolated from numerous plants. The most notable exception is *S. ficaria,* which is specifically associated with figs and the fig wasp, *Blastocenes psenes.*[12] The wasps transmit *S. ficaria* as they pollinate trees. Figs, which are sterile before wasp contact, provide limited support of *S. ficaria,* and despite the presence of 10 to 300 cfu per fig, spoilage of the fruit does not occur.[2] Grimont and Grimont,[2] dividing plants surveyed for serratiae into three groups, found the isolation rates for two of them, grasses and tree/shrub leaves, to be relatively low at 24% and 8%, respectively. Eight species were identified, but *S. liquefaciens* and *S. proteamaculans* predominated. In the third group, wet plants (eg, vegetables, mushrooms, mosses), 54% of samples yielded one or more *Serratia* species, with counts of 2000 to 20,000 cfu per plant. In a study on the microbial flora of vegetable salads, Wright and colleagues[14] found that 11% to 29% harbored *S. marcescens, S. liquefaciens,* and *S. rubidaea.* Approximately one third of the salads were positive for serratiae, and counts ranged as high as 10^6 cfu. Although these results suggest that vegetables may serve as one source for the colonization of hospital patients with *Serratia,* no study has established that the same strains are involved.

In domesticated animals, serratiae have been isolated from horses, rabbits, cows, pigs, and poultry.[2,15] Isolations from wild animals include deer, water buffalo, red fox, a variety of rodents, sparrows, and other birds.[2,15] *Serratia* are pathogenic in mice, rats, guinea pigs, dogs, and turtles,[15] and chronic infections in reptiles have been well documented.[2] Both lethal and nonlethal associations have been noted with insects. Only *S. rubidaea,* which does not produce chitinase, a trait necessary for virulence, is seldom found in insects.[2] Thus far, the species *S. entomophila* has been isolated exclusively (with the exception of one water isolation) from grass grubs, for which it is lethal.[13] It causes amber disease, which is named for the amber discoloration that occurs in the gut of the grub; the infected grub larvae stop feeding and die within 4 to 6 weeks. This organism has been used successfully in New Zealand to exterminate grass grubs, an agricultural pest that destroys thousands of acres of pastureland at a cost of millions of dollars annually.[13]

Except for infections of the eyes, serratiae in humans are almost exclusively acquired nosocomially, and colonization or infection only rarely involves a species other than *S. marcescens.*

Now a well established pathogen, at one time *S. marcescens* was believed to be completely innocuous. Because its pigment provided a convenient marker, it was used in experiments to study transmission of bacteria by means of aerosols. These studies, all inconceivable today, included the following: (1) gargling a culture of *Serratia* to prove that speech is as effective as coughing or sneezing in spreading germs; (2) painting cultures of *Serratia* on gums and teeth to prove that transient bacteremia occurs after dental extraction; (3) swabbing *Serratia* on the perineum of patients with indwelling catheters to prove that catheters serve as a means of entry for bacteria into the urinary tract.[15] The usefulness of *S. marcescens* in the study of aerosols did not escape the notice of the military either. *S. marcescens* was used in germ warfare population-vulnerability experiments in three sites in the United States.[15] The strain selected for use, *S. marcescens* 8UK, was red pigmented and was easily trackable with a rare profile: biotype A6, serotype O8:H3, phage type 678.[16] Fortunately, this distinctive profile was not found in any of 100 U.S. outbreaks studied by the Centers for Disease Control. Only 7 of 3000 strains of *S. marcescens* that were serotyped belonged to serotype O8:H3, and 20 of 2062 strains were biotype A6.[16] No isolate possessed all of the markers that identified the strain used by the military.

S. marcescens was first recognized and reported as a nosocomial pathogen in 1951; a second report followed the next year. According to the National Nosocomial Infections Surveillance System (NNIS), *S. marcescens* currently accounts for 1% of hospital-acquired urinary tract infections (UTIs), wound infections, and bloodstream infections and 4% of nosocomial pneumonias.[17] Hospital equipment, intravenous (IV) fluids, and other solutions are instrumental in nosocomial outbreaks, whereas hand-to-hand transmission is the primary mode of spread in hospital cross-infections.[15,18] From data obtained in 23 hospital outbreak investigations, Farmer and associates[18] found four distinct patterns of *S. marcescens* infection: (1) relatively few isolations at any given time; (2) outbreaks traceable to a common or point source, in which the number of patients infected or colonized depends on their exposure to the reservoir; (3) transmission from one patient to another, especially from those with indwelling catheters, on the hands of hospital personnel; and (4) pseudo-outbreaks caused by contamination of specimens from an exogenous source. Antimicrobial resistance of strains varies from pattern to pattern. Isolates from the first two patterns tend to be sensitive to all antibiotics or, at most, resistant to a single drug. In strains transmitted person-to-person, as more toxic drugs are added to the regimen, resistance eventually escalates from a few drugs to all antimicrobials used within the institution.

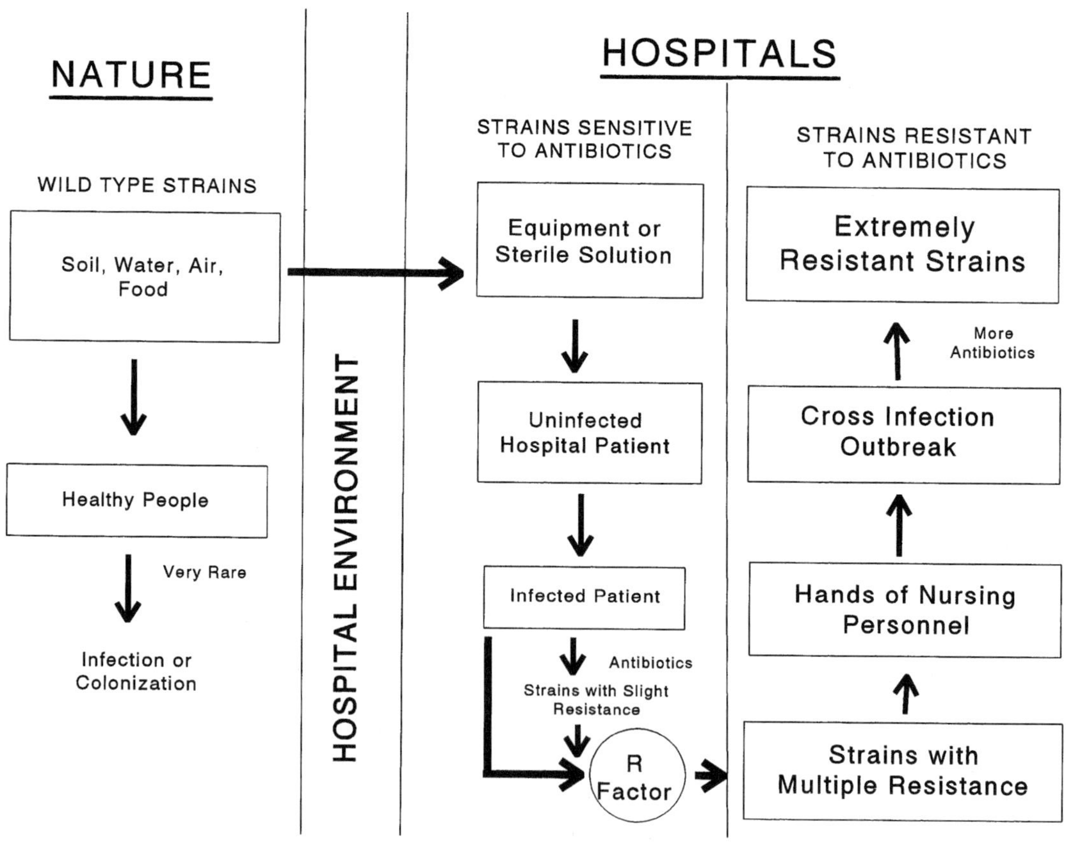

FIG. 1. The dynamics of intrahospital transmission of infection, development of antimicrobial resistance with antibiotic use, and emergence of resistant strains. (With permission from Farmer JJ III, Davis BR, Hickman FW. Detection of *Serratia* outbreaks in hospital. *Lancet* 1976;2:455–459.)

The dynamics of intrahospital transmission of infection, development of antimicrobial resistance with antibiotic use, and emergence of resistant strains is shown in Fig. 1.

The list of agents and fomites involved in common source outbreaks is seemingly unlimited, with a moist environment the only prerequisite. Some of the objects and solutions found to be culpable in *Serratia* outbreaks are listed in Table 2. Even antiseptics are not immune to serratiae; *S. marcescens*–contaminated benzalkonium chloride was ultimately traced as the source of 10 cases of septic arthritis.[19] The outbreak strain survived in a 1:100 dilution of this antiseptic and grew to more than 10^5 cfu/mL in γ-myristyl picolinium chloride–preserved vials of methylprednisolone that had been disinfected with the benzalkonium chloride.[20] In another example of its tenacity, *S. marcescens* survived on materials leached from autoclaved transfusion bags that had been cooled with nonsterile water and held under oxygen-restricted conditions.[21] Organisms present on the outside of the bags ultimately contaminated the blood and resulted in a number of transfusion-related infections.[21]

Infections transmitted from person to person are by far the most common[2,15,18] and the most difficult to control. The primary reservoir is patients, usually those with UTIs associated with indwelling catheters or colonized or infected respiratory tracts[2,15,22]; spread occurs primarily via the hands of hospital personnel. Patients most susceptible to infection are those undergoing invasive procedures, especially procedures involving the respiratory tract, and those who are immunocompromised or receiving antimicrobial therapy. Although the gastrointestinal tract is not generally a significant source of infection in adults, it is a common reservoir in neonates and children.[15,18,23] In one neonatal outbreak of septicemia and meningitis, 68% of the babies who remained asymptomatic carried *S. marcescens* in their gastrointestinal or nasopharyngeal tract.[24] Although point source outbreaks are readily detected as numbers of isolations rise, hand-to-hand transmission is often insidious. It may occur over a long period and may subside and peak several times before infection control measures become effective. Although policies mandating strict hand washing and aseptic technique and removal of catheters whenever possible are valuable in interrupting transmission, cohorting and isolation of patients with *Serratia* is probably the most effective measure.[22,24]

Community-acquired infections of serratiae are rare.[22,25] Out of 44 cases of *Serratia* bacteremia, Watanakunakorn[25] found that 31 (70%) had been contracted within 3 to 43 days of hospitalization, compared with 5 (11%) from nursing homes and 8 (18%) from the community. Two types of infections typically are acquired outside of the hospital environment: contact lens–induced acute red eye (CLARE)[26] and endocarditis secondary to IV drug abuse.[27] CLARE is an inflammatory reaction causing pain, corneal infiltration, excessive tearing, light sensitivity, and, as the name implies, reddening of the conjunctiva.[26] *S. marcescens* has been isolated from 24% of lenses of persons with CLARE. Epityping based on serotype, biotype, and ribotype

TABLE 2. *List of fomites/agents involved in* Serratia *nosocomial infections*

Vehicle	Source
Hands of hospital personnel	Colonized or infected patients' catheters, hand lotions, adhesive tape
Respiratory equipment	Ultrasonic nebulizers, respirators, intermittent positive-pressure breathing machines, inhalation medications, bronchoscopes
Solutions/lotions	Intravenous, irrigation fluids, disinfectants, contact lens solutions
Blood, plasma derivatives	
Miscellaneous medical devices	Electrocardiogram leads, scalp vein needles, arterial pressure monitors, presurgical shaving brushes and soaps, unfilled transfusion bags
Hospital environment	Sinks, flowers and plants, food

From references 15, 18, 22, 45, and 46.

could not identify a particular strain associated with CLARE infection.[26] Patterns of CLARE and hospital-acquired isolates were different, suggesting that different strains are involved in these two groups of infections. Numbers were insufficient to be statistically significant but suggested that contamination of the lenses or eyes by hands or fingers was not the source for these infections. IV drug abusers also contract *S. marcescens* infections in the community, presumably from contaminated paraphernalia and/or drugs.[27] These patients present to the hospital with endocarditis, and most are apparently infected with multiple strains of *S. marcescens* (determined by serotype) which have varying antibiograms. The mortality rate in patients with left-sided disease, which predominates, was reported to be as high as 70%.

CLINICAL SIGNIFICANCE

Bacteremia

Serratia bacteremia is relatively uncommon. Between 1986 and 1989, *S. marcescens* accounted for 367 (1%) of bloodstream infections reported to the NNIS.[17] In 1989, two reviews, one from a large cancer center[28] and the other from a 750-bed teaching hospital,[25] showed infection rates for *Serratia* bacteremia of 7 and 1.8 per 10,000 admissions, respectively. Likewise, in a 1991 review, Roberts and coworkers[29] found only 43 *Serratia* isolates (2.1%) among 1972 positive blood cultures. However, the mortality rate of *Serratia* (53.8%) exceeded that for all other gram-positive and gram-negative organisms involved in bacteremia in this study, including *Pseudomonas aeruginosa.* Other reviews reported mortality rates ranging from 33% to 52%.[25,28,30] Factors associated with poor outcome include septic shock,[28,30] presence and extent of pneumonia,[28] underlying rapidly or ultimately fatal condition,[25] low platelet count (<50,000/mm^3),[28,29] and total serum bilirubin higher than 18 μmol/L.[25] All of the above factors were statistically significant. Most patients present with symptoms typical of any gram-negative sepsis with fever the most common symptom.[28]

The portal of entry is frequently undetermined; when it can be established, the most common source is the respiratory tract, from which 18% to 30% of bacteremic episodes are seeded.[25,28,29] After the respiratory tract there is considerable difference between reviews regarding source. The genitourinary tract was the portal of entry in 3.8% to 13.6% of infections[25,28,29]; IV lines, in 6.8% and 9.0%[25,28]; skin and soft tissue, in 18%[28]; and wounds, in 9.0%.[25] Watanakunakorn[25] included two cases of cholangitis and one case each of osteomyelitis, ruptured diverticulitis, and premature ruptured membranes as sources of bacteremia.

Serratia bacteremia is predominantly monomicrobic[22]; Roberts and colleagues[29] and Watanakunakorn[25] respectively reported that 60% and 79% of blood cultures yielded pure growth. As mentioned, *S. marcescens* is the species isolated in the overwhelming number of cases. From 162 cases of bacteremia, *S. marcescens* was identified in 130 instances (80%), *S. liquefaciens* 16 times (9.8%), and *S. rubidaea* once; in 15 cases, no species identification was assigned. There has been one report of *S. liquefaciens* septicemia acquired through transfusion.[31] A previously healthy 21-year-old patient died within 48 hours after receiving the transfusion; a source of contamination for the blood was never found. Except for *S. liquefaciens,* only a smattering of reports of septicemia caused by species other than *S. marcescens* are found in the literature. The majority of *S. plymuthica* infections are reported from Spain, including all but one case of sepsis.[32–36] Of seven cases of *S. plymuthica* bacteremia, six were nosocomially acquired. The community-acquired case occurred in an immunocompetent 50-year-old woman with cholangitis whose sepsis developed 3 days after she underwent an endoscopic procedure.[32] The remaining five patients all had compromising conditions such as oncologic disorders or stroke,[34] alcoholic cirrhosis,[35] or heart disease.[36] No portal of entry was clearly established in any of these cases. Although sources were surmised for each infection, appropriate cultures either were not obtained or were found to be negative. All of these infections responded very rapidly to antimicrobial therapy without apparent need for further intervention (eg, removal of IV or central venous lines). The fact that the infections resolved so readily in these immunocompromised patients indicates a limited pathogenic potential for this species. Bacteremia caused by other species (three cases of *S. odorifera* biogroup 1, both community and hospital acquired; two of *S. rubidaea;* and one of *S. ficaria*) follow a similar profile.[37–42] Like their counterparts with *S. plymuthica* sepsis, patients were compromised, had typical sources for their infections (UTI, invasive procedures, catheterization), and were cured with antibiotic treatment. The exception was a fatal case of sepsis in which *S. rubidaea* became rapidly disseminated.[40] The patient had positive sputum, chest tube, and blood cultures after surgery for pericarditis; all of these sources were sterile at the time of surgery.[40] *S. rubidaea* was isolated from lung tissue, heart tissue, peritoneal fluid, and other specimens at autopsy. Later, the patient's corneas, cultured before preparation for transplant, also yielded *S. rubidaea.*

Central Nervous System

S. marcescens meningitis is thankfully rare. It occurs primarily in neonates and infants,[43] and its mortality rate (45%) exceeds that of other, more common gram-negative causative agents.[44] Surviving infants usually have neurologic deficits. Premature birth, respiratory distress syndrome, and necrotizing enterocolitis are all associated with increased risk of infection.[44] As with adults, catheters and previous antibiotic therapy are also predisposing factors. In a retrospective review of 10 cases of *S. marcescens* meningitis from three outbreak clusters, Campbell and associates[44] found that 9 infants were septic when meningitis developed; 3 had normal cerebrospinal fluid (CSF) parameters and negative CSF cultures, and 2 others had few leukocytes present in the CSF. Eight patients were thrombocytopenic, either initially or rapidly after onset; leukocytosis also occurred, but to a lesser extent. Clinically, most were lethargic and had apnea or tachypnea, and half were hypotensive. Half the patients were already being treated for an *S. marcescens* infection at the time meningitis developed, prompting the authors to recommend meningitic doses of antibiotics in septic infants. Complete, daily evaluations of CSF was also recommended to monitor for ventriculitis, a common progression of *S. marcescens* meningitis. Cases of meningitis in adults are rare. An unusual, and fatal, case of *S. marcescens* meningitis in an immunocompetent adult who had undergone ear surgery was reported in 1991; it was the second report with the ear as a source.[43]

Miscellaneous Infections

Because colonization of the nasopharynx with *S. marcescens* is relatively common in hospitalized patients,[22] subsequent respiratory infections are inevitable. Accordingly, the NNIS report found that 4% of nosocomially acquired pneumonias are caused by serratiae.[17] Lung abscesses, bronchiectasis, pneumonia, and empyema have all been documented.[15] Invasive procedures, particularly in the presence of antimicrobial therapy, coincide with these infections. This can be illustrated by two clusters of pneumonia, one traced to a contaminated bronchoscope[45] and the other associated with electrocardiogram bulbs used in cardiac surgery.[46] Nosocomial pneumonias attributed to *S. proteamaculans* ssp. *quinovora*[47] and *S. plymuthica*[48] have also been reported in two compromised patients. The case of *S. plymuthica* pneumonia in a homeless, alcoholic, 43-year-old man resulted from an obstructing abscess on the floor of the mouth that spread from a dental abscess.[48] Cultures to determine the cause of the pneumonia, which probably resulted from inhalation of pus from the abscess, yielded large numbers of *S. plymuthica* in pure culture. Approximately 1 L of pus was drained from the lung at one time; although the pneumonia was the precipitating factor, the patient ultimately died of multiorgan failure.

Musculoskeletal infections, including cellulitis,[49] septic bursitis,[50] septic arthritis,[19] and osteomyelitis,[51] have been reported for serratiae. All are rare infections and occur primarily in compromised patients.[19,49,50] *Serratia* eye infections include CLARE[26] (see Epidemiology) and corneal ulcers,[52] both associated with contact lens wear, and rare cases of endophthalmitis in adults and infants.[53] The only species reported thus far from ocular infections has been *S. marcescens.* Surgical wounds are another category of infection from which the serratiae are isolated. The NNIS reports that 1% of nosocomial wounds are caused by serratiae. Community-acquired wound infections have also been reported, including a case of *S. ficaria* in mixed culture in a leg ulcer.[54] Although three other agents were present in the wound, *S. ficaria* was isolated in heavy numbers and was believed to be involved in the infection.

LABORATORY IDENTIFICATION

Isolation

Serratia grow readily on routine laboratory media (heart infusion, trypticase soy, nutrient agars) and produce moist, opaque to white colonies. All species grow well between 15° and 30°C; only *S. marcescens* and *S. rubidaea* do not grow at 4°C. Other species grow at 4°C within 24 hours,[9,10,12] except the *S. liquefaciens* group and *S. entomophila,* which are usually delayed at this temperature and require 2 to 4 days for growth.[5,13] *S. liquefaciens* group organisms are, however, more biochemically active at 30°C than at 35° to 37°C.[5,8,55]

In their extraordinary study of blood culture isolation methods, Cockerill and coinvestigators[56] found that the Isolator tube (Wampole Laboratories, Cranbury, NJ) recovered 67 cultures of *S. marcescens,* compared with 38 for the Septi-Chek bottle (Becton Dickinson Diagnostic Instrument Systems, Sparks, MD); the difference was statistically significant. A small number of strains were isolated only in nonvented trypticase soy broth (anaerobically).

Isolation of serratiae from nonsterile body sites, foods, or environmental habitats, either in the hospital or in the community, may require specialized media. Two such media are deoxyribonuclease-toluidine blue-cephalothin agar (DTC), formulated by Farmer and coworkers,[57] and the caprylate-thallous (CT) agar of Starr and colleagues.[58] On DTC, serratiae colonies are defined by a reddish halo surrounding the colonies within 48 hours; non-*Serratia* either do not grow or produce no halo.[57] A small percentage of *S. marcescens* (2%–18%), *S. liquefaciens* group (5%), and *S. rubidaea* (1%) strains are negative for deoxyribonuclease (DNase–) and therefore are missed on DTC medium. All strains of *S. fonticola* are DNase– and do not produce a halo even when growth is present. In CT medium, the caprylate is used as a carbon source while the thallous salts inhibit other opportunistic pathogens or environmental bacteria, especially *P. aeruginosa.*[58] As with DTC, CT agar does not detect *S. fonticola,*[2] and in studies in which *S. plymuthica* is the targeted organism the thallous salts should be reduced to 0.012%.[58] CT is a highly selective medium, and only rare genera other than *Serratia* (eg, *Pseudomonas, Enterobacter, Proteus*) grow larger than pinpoint colonies.[2,58] Although CT agar is very selective for testing of human specimens from nonsterile sites, it is not as effective for food or environmental samples.[2]

Biochemical Identification

Commercial Systems

The serratiae presented no challenge to any of the commercial systems evaluated in seven major studies.[59–64] Of 239 *Serratia* isolates, 13 (5%) were identified to genus only or were misidentified as another species of *Serratia;* only 3 (1%) were incorrectly identified. Where species were specifically indicated, Crystal Enteric/Non-Fermenter (Becton Dickinson, Cockeysville, MD) misidentified a single isolate of *S. marcescens,* and identified a second strain to genus only, out of 26 strains tested.[64] Two strains of *S. rubidaea* were emisidentified as *Klebsiella ozaenae* and *Enterobacter gergoviae* by PASCO MIC-ID (PASCO Laboratories, Wheat Ridge, CO) and AutoSCAN-Walk/Away (Dade MicroScan, West Sacramento, CA), respectively.[58,59] *S. fonticola* was identified to genus only by Cobas Micro ID-E/NF (Becton Dickinson Diagnostic Instrument Systems, Sparks, MD) and reportedly was not included in the database of RapID onE (Innovative Diagnostic Systems, Norcross, GA).[61,62] Separation of species within the *S. liquefaciens* group requires carbon source utilization tests. These tests can be made in house by adding neutral carbon sources to M70, a minimal medium base.[2] Alternatively, testing can be performed with the use of API 50 CH, API 50 AO, and API 50 AA strips (Vitek bioMérieux, Hazelwood, MO), which contain carbohydrates, organic acids, or amino acids, respectively.[2]

Conventional Media

S. marcescens, the most common clinically encountered *Serratia,* can be speciated with the use of four tests: fermentation broth for arabinose and rhamnose (both negative) and lysine and ornithine decarboxylase broths (both positive).[4] For *S. marcescens* and *S. rubidaea,* biochemical testing should be performed at 35° to 37°C, although pigment production may require an incubation temperature of 25° to 30°C. For biochemical separation of other serratiae, especially *S. liquefaciens* group and *S. plymuthica,* testing at 30°C is preferable because more assays yield positive results at this temperature.[8,25,55] However, because incubation temperatures other than 35°C may not be available in all clinical laboratories, the *Serratia* biochemical profiles of Farmer[4] are based on 36°C incubation. Biochemicals useful in separating all members of the genus *Serratia* are given in Table 3. Several features in this table are of note. *S. entomophila,* a species not encountered in humans, is quite inactive biochemically, a characteristic that does not change even when incubation at 30°C is used.[13] *S. fonticola,* rarely found in clinical specimens, fails to produce DNase, lipase, or gelatinase, all hallmark traits of the genus *Serratia.*[4] Lack of these characteristics cast doubt on the inclusion of this organism within the genus until recently (see Taxonomy). The *S. liquefaciens* group can be separated from other

TABLE 3. *Differential biochemical characteristics for the genus* Serratia

Species or Group	LDC	ODC	Acid from ARAB	RHAM	SORB	XYL
S. marcescens	+	+	–	–	+	–
S. marcescens biogroup 1	V	V	–	–	+	–
S. liquefaciens group[a]	+	+	+	V	+	+
S. rubidaea	V	–	+	–	–	+
S. odorifera biogroup 1[b]	+	+	+	+	+	+
S. odorifera biogroup 2[b]	+	–	+	+	+	+
S. plymuthica[a]	–	–	+	–	V	+
S. ficaria[a]	–	–	+	V	+	+
S. entomophila	–	–	–	–	–	V
S. fonticola[c]	+	+	+	V	+	V

From Farmer[4] and Grimont et al.[13]
[a]Reactions given are for 35°–37°C; these species may be more biochemically active at 22°–25°C.
[b]*S. odorifera* biogroup 1 strains ferment sucrose; biogroup 2 strains do not.
[c]*S. fonticola* do not produce DNase, lipase, or gelatinase; all other serratiae do.
LDC, lysine decarboxylase; ODC, ornithine decarboxylase; ARAB, arabinose; RHAM, D-rhamnose; SORB, sorbitol; XYL, xylose; +, ≥90% positive; –, ≥90% negative; V, 10%–90% positive.

TABLE 4. *Identification of members of the* S. liquefaciens *group*

Species or Group	Growth on		Arginine Dihydrolase	Esculin Hydrolysis
	D-Malate	*m*-Tartrate		
S. liquefaciens	+	+	–	+
S. proteamaculans ssp. *proteamaculans*	–	–	–	+
S. proteamaculans ssp. *quinovora*	V	–	–	–
S. grimesii	V	V	+	+

From Grimont & Grimont[2] and Grimont et al.[5]
+, positive ≥90%; –, negative ≥90%; V, variable reactions.

cold-tolerant serratiae and from *S. rubidaea* by positive tests for lysine and ornithine decarboxylase and failure to ferment cellobiose. Again, members within the *S. liquefaciens* group can be separated only by carbon source utilization assays, and those useful for this purpose are presented in Table 4.

Two other distinguishing characteristics are associated with *Serratia,* although neither is found in all species within the genus. The first and most renowned is the production of prodigiosin, a nondiffusable red pigment (Fig. 2; Color Plate 15). Only some strains of *S. marcescens, S. rubidaea,* or *S. plymuthica* produce this pigment.[2,55] *S. marcescens* biotype A4 strains produce a diffusible pink pigment called pyrimine that may be misinterpreted as weak production of prodigiosin; however, no strains of this biotype produce prodigiosin.[2] The second characteristic, al-

FIG. 2. Unusual pigment production by *S. rubidaea.* See also Color Plate 15.

though not as well known, is immediately evident when encountered. Virtually any strain of *S. odorifera* or *S. ficaria,* and some strains of *S. rubidaea,* produce a pungent odor described as potato-like, although this description hardly does it justice. The odor is the result of alkyl-methoxypyrazine production.[2]

Biotyping

Biotyping is an important tool in the epidemiology of nosocomial *S. marcescens* isolates. Profiles attained by conventional biochemical methods do not show sufficient variability to allow for biotyping. Carbon assimilation or carbon source utilization assays, however, can detect small quantities of acid that may go undetected by fermentation.[2] Grimont and Grimont[65] devised a biotyping scheme based on an eight-carbon source panel plus a tetrathionate reductase assay and pigment production. The carbon sources used included benzoate, DL-carnitine, *m*-erythritol, 3-hydroxybenzoate, lactose, D-quinate, and trigonelline; they are added to a minimal medium. Incubation usually requires 7 to 14 days (minimum 4 days) at 30°C (66). API enzymatic test strips (see Commercial Systems) can also be used and require only 1 to 4 days' incubation.[2] Grimont and Grimont[2] recognize 18 biotypes of *S. marcescens.* Biotypes of *S. proteamaculans, S. grimesii, S. rubidaea, S. odorifera,*[2] and *S. entomophila*[13] can also be defined by carbon assimilation schemes. Nonpigmented biotypes A5, A3a, A3b, A3c, and A3d of *S. marcescens* are found almost exclusively in hospitalized patients. Several recent studies have used biotyping to examine hospital strains of *S. marcescens.*[66–69] Of these, only one[68] found bacteriocin typing preferable to biotyping, and this was based on the speed with which a bacteriocin type could be produced. Sifuentes-Osornio and associates[66] and Bale and colleagues[67] found biotyping to be extremely useful in determining whether isolates in their institutions were sporadic or epidemic. Bale and coworkers[67] further showed that 23 of 26 patients retained the same biotype regardless of the anatomic site from which it was isolated or the length of time between isolations (1 day to 5 months). Reproducibility was 100%, both within and between runs, in their institution.

Immunodiagnostics

There are 24 recognized somatic (O) and 26 flagellar (H) antigens of *S. marcescens.*[70] Cross-reactions among O groups is extensive, especially between O14 and several other O groups common in clinical specimens, O6, O1, and O21. Gaston and Pitt[70] were the first to recognize that the cross-agglutination reactions seen during serotyping were actually caused by microcapsules present in multiple serotypes rather than being simple crosses resulting from lipopolysaccharide (LPS) alone. After extensive immunoblot tests and cross-absorptions, they established 17 somatic types with corresponding LPS-specific sera.[70] A dot enzyme immunoassay (EIA) is recommended over slide agglutination because, although it is more time consuming (3–4 hours),[70] its results are more accurate.[71] Aucken and associates[72] identified 14 distinct acidic polysaccharides or capsules. These are readily detected by the quellung reaction (see Chapter 8), but strains produced capsules only when grown on Kings A agar as opposed to nutrient agar. Capsule production was best when incubated at 30°C overnight. H typing can be determined by immobilization assays done with specific sera since *S. marcescens* produces only one flagellar phase.[2,70] Serotyping schemes have been devised for *S. ficaria* and *S. rubidaea* as well.[2] Antisera for *S. marcescens* is commercially available from Denka Seiken (Accurate Chemical and Scientific Corp., Westbury, NY) but is not approved by the U.S. Food and Drug Administration.

Typing

A variety of typing methods have been employed with *S. marcescens,* including bacteriocin typing[68] and phage typing,[2] but producer strains and phages are not readily available.

Multilocus enzyme electrophoresis[73] and electrophoresis of periplasmic proteins[74] correlated well with biotyping and are alternative epidemiologic methods. Applicable molecular methods used successfully to track strains include plasmid analysis,[73] which is limited by the fact that only 25% of strains contain plasmids; ribotyping[2,69]; pulsed-field gel electrophoresis[68,75,76]; and random amplified polymerase chain reaction.[77]

PATHOGENESIS

Although characteristics associated with pathogenesis in other bacteria have been found in *Serratia,* virulence has not been well worked out for this genus. *Serratia* may express any or all of three types of fimbriation: type 1 mannose-sensitive hemagglutinins (MSHA), type 3 mannose-resistant *Klebsiella*-like fimbriae, and mannose-resistant *Proteus*-like hemagglutinins.[78] Parment and colleagues,[78] however, found that although 46 of 58 *S. marcescens* isolates produced at least one type of fimbriae there was no correlation between hemagglutinin type and source of isolate (UTI, wound, contact lens) or adhesion to urinary epithelial cells. As with a number of other bacteria, *S. marcescens* may mediate adhesion to tissues by means of a hydrophobic cell surface, because many strains have been shown to be hydrophobic.[79] In strains of *S. marcescens* that possess MSHAs and are hydrophobic (factors associated with bacterial clearance from the blood as well as tissue adherence), only MSHAs correlate with the host's ability to remove bacteria from the bloodstream.[79] Most strains of *S. marcescens* exhibit delayed serum sensitivity (cells killed within several hours of exposure to serum), with the remaining strains equally divided between rapid serum sensitivity and serum resistance.[2] In those strains ultimately susceptible to killing by serum, an alternative human complement pathway is involved; strains rapidly killed by serum activate the classic complement pathway.[2]

Strains of *S. marcescens* have also been found to produce what was originally believed to be a cell-bound hemolysin.[80,81] Schiebel and associates[82] were able to show that this hemolysin was actually excreted (with a short half-life of 3 minutes) into the surrounding medium in addition to being cell bound. Genes encoding this hemolysin are iron-regulated, and expression is increased threefold in iron-limited medium.[82] Only early log-phase bacteria produce the hemolysin, which is not calcium dependent.[80–82] Marre and coworkers[81] found that hemolysin-positive strains of *S. marcescens* contributed to renal colonization in an experimental rat pyelonephritis model. The hemolysin was also found to trigger an inflammatory response, measured by increased leukocytes in the urine and a thickening of the bladder wall in this same model.

Serratia primarily acquire iron under iron-limited conditions by means of siderophores, usually enterobactin and, rarely, aerobactin.[2] Létoffé and associates[83] identified an extracellular heme-binding protein that delivers free heme to a specific receptor on the outer membrane.

Extracellular proteases produced by serratiae are responsible for dermal hemorrhage, ocular damage, and experimental pneumonia.[2,84] Lyerly and Kreger[84] demonstrated that purified protease caused lung damage in guinea pigs and mice that was similar to that produced by *S. marcescens.* Further, vaccination of animals against the protease protected against disease, and protease was actively produced by *Serratia* strains in animals with pneumonia.

Carbonell and colleagues[85] found that *S. marcescens* strains are cytotoxic for Vero cells. The heat-labile cytotoxin produces a cytopathic effect similar to that of the Shiga toxins of *Escherichia coli* on this cell line. Cytotoxicity did not appear to correlate with production of hemolysin or nuclease; these enzymes are not readily produced at 37°C, whereas cytotoxicity was maximal at this temperature. Likewise, proteinases were ruled out because their production is reduced in the presence of glucose, whereas strains grown in a glucose-containing medium had high cytotoxicity.

ANTIMICROBIAL SUSCEPTIBILITY

The serratiae are among the most resistant bacteria encountered in the hospital setting. Berkowitz and Metchock,[86] in a study of cephalosporin-resistant bacteria colonizing pediatric patients, encountered three strains of *Serratia* that were resistant to ampicillin/sulbactam, piperacillin, cefazolin, cefotetan, cefotaxime, ceftriaxone, ceftazidime, and ceftizoxime; two of the three strains were also resistant to aztreonam. Strains that are initially susceptible rapidly become resistant during antimicrobial therapy.[15,87–89] Both dosage and duration, not simply use of an antibiotic, are risk factors for development of resistance.[15] Resistance is mediated by transposons, plasmids, and chromosomally.[90,91] Use of a single antibiotic may select or induce resistance to multiple drugs,[15,91] and such use may not only confer resistance to a new antimicrobial agent but increase resistance to an already resistant one.[15] Serratiae are resistant to cephalothin,[15,89] tetracycline,[15,92] most penicillins except piperacillin and ticarcillin,[93] chloramphenicol, and amoxicillin/clavulanate.[94] Resistance to aminoglycosides is variable, with as many as 30% of strains resistant to tobramycin and netilmicin.[95] In one report from Argentina,[90] 90% of *S. marcescens* were resistant to gentamicin, and resistance to amikacin varied between 31% and 70% during an 8-year period. Seventy-five percent of *S. marcescens* carried three enzymes AAC(3)-V, AAC(6′)-I, and APH(3′)-I, which resulted in resistance to gentamicin, tobramycin, netilmicin, and amikacin. In a study of antimicrobial resistance in U.S. hospitals,[93] amikacin was more active, with 9 of 14 laboratories reporting 99% to 100% susceptibility and the remaining 3 laboratories 93% to 96% susceptibility. Gentamicin was 98% to 100% susceptible in 8 of 14 laboratories, 96% susceptible in 3, and only 89% to 90% susceptible in 4. Lortholary and coworkers[96] reported a 50% minimal inhibitory concentration (MIC_{50}) of 1 μg/mL for both gentamicin and amikacin but the 90% MIC was 8 μg/mL, moving gentamicin into the resistant category. Ninety percent or more of serratiae are susceptible to ciprofloxacin.[93,97,98] Trimethoprim-sulfamethoxazole is generally effective against serratiae, but susceptibility percentages as low as 85% have been reported.[93,98] Aztreonam and imipenem remain potent against most strains of *S. marcescens,* although resistance is slowly eroding the efficacy of these powerful antibiotics. Sirot and colleagues[89] reported 12% resistance to aztreonam, and a Chicago hospital[93] reported 17% of *S. marcescens* resistant to imipenem. Although such strains are rare at this time, carbapenem-hydrolyzing β-lactamases have been found in strains in Japan and England.[91] In Japan, the metallo-β-lactamase IMP-1 is encoded by the bla_{IMP} gene and carried on the chromosome,[99] and London strains carry the group 2f serine β-lactamase, Sme-1. Thus far Sme-1 has been found on the chromosome only in *Serratia,* but the presence of a metallo-β-lactamase on a transposon residing in a plasmid in *Klebsiella* may hasten spread of this enzyme among the *Enterobacteriaceae.*[91] Two of the English strains also carried Amp-C–type enzymes, which attack penicillins and cephalosporins. Extended-spectrum β-lactamases (ESBLs) have been responsible for an increase of about 20% in resistance to new cephalosporins in France and are twice as likely to occur in teaching hospitals as in smaller institutions.[87] Tzouvelekis and colleagues[100] found that when the plasmid-mediated ESBL SHV-5 was present the MIC for cepirome doubled, although strains remained susceptible to this cephalosporin. Susceptibilities to various cephalosporins are reported in Table 5. In general, other clinically encountered species of *Serratia* had susceptibility profiles similar to *S. marcescens* except that they appeared more susceptible to aminoglycosides.[93,97,98]

CONCLUSIONS

The serratiae present the medical community with three distinct challenges. First, given that serratiae are ubiquitous organisms, thriving with little encouragement (ie, moisture and warmth), hospitals must make a concentrated

TABLE 5. *Activity of newer cephalosporins against* Serratia *(percent susceptible)*

Drug	Jones et al.[101] (129 Smar/10 Ssp)	Sirot et al.[89] (336 Smar)	ASCP Group[93] (1604 Smar)	Fuchs et al.[102] (92 Ssp)	Jones et al.[103] (102 Smar)
Cefotaxime	98/100	—	89–100[a]	89	—
Ceftriaxone	99/100	—	86–100[b]	89	90
Cefoperazone	98/100	—	—	—	—
Ceftizoxime	100/100	61	90	87	—
Ceftazidime	100/100	92	86–100[c]	97	96
Cefpirome	100/100	—	—	—	—

From references 89, 93, 101–103.
[a]Results from 5 hospitals.
[b]Results from 6 hospitals.
[c]Results from 12 hospitals.
Smar, *S. marcescens;* Ssp, *Serratia* species.

effort to reduce their transmission. This includes transmission from equipment to patients and from one patient to another via the hands of hospital personnel. These efforts, particularly hand washing and cohorting of patients known to be colonized, must be stressed continuously as a method to prevent, rather than remedy, outbreaks. The second challenge involves the need for a better understanding of how the serratiae produce disease. With this knowledge, for example, vaccines or drugs aimed at bacterial attachment or protease production could be devised to reduce the risk of disease. The last and certainly the most critical challenge is control of antimicrobial use in hospitals. The evolvement of ESBLs and carbapenem-hydrolyzing β-lactamases, even if they are not extremely common as yet, is a warning that should not be ignored. Hospital-mandated restrictions and recommendations calling for the judicious use of antimicrobials in the treatment of all infections may be the best method for preventing further development of resistance by the serratiae.

REFERENCES

1. Cullen JC. The Miracle of Bolsena. *ASM News* 1994; 60:187–190.
2. Grimont F, Grimont PAD. The genus *Serratia.* In: Starr MP, Stolp H, Trüper HG, Schlegel HG, eds. *The Prokaryotes: A handbook on habitats, isolation and identification of bacteria.* Berlin: Springer-Verlag, 1981;2822–2848.
3. Sakazaki R. Genus IX: *Serratia* Bizio 1823,288. In: Buchanan RE, Gibbons NE, eds. *Bergey's manual of systematic bacteriology.* 8th ed. Baltimore: Williams & Wilkins, 1974;326.
4. Farmer JJ III. *Enterobacteriaceae:* Introduction and identification. In: Murray PR, Baron EJ, Pfaller MA, Tenover FC, Yolken RH, eds. *Manual of clinical microbiology.* 6th ed. Washington, DC: American Society for Microbiology, 1995;438–449.
5. Grimont PAD, Grimont F, Irino K. Biochemical characterization of *Serratia liquefaciens* sensu stricto, *Serratia proteamaculans,* and *Serratia grimesii* sp. nov. *Curr Microbiol* 1982;7:69–74.
6. Grimont PAD, Grimont F, Starr MP. Comment on the request to the judicial commission to conserve the specific epithet *Serratia liquefaciens* over the specific epithet *proteamaculans* in the name of the organism currently known as *Serratia liquefaciens. Int J Syst Bacteriol* 1981;31:211–212.
7. Grimont PAD, Irino K, Grimont F. The *Serratia liquefaciens–S. proteamaculans–S. grimesii* complex: DNA relatedness. *Current Microbiol* 1982;7:63–68.
8. Ewing WH, Davis BR, Fife MA, Lessel EF. Biochemical characterization of *Serratia liquefaciens* (Grimes and Hennerty) Bascomb et al. (formerly *Enterobacter liquefaciens*) and *Serratia rubidaea* (Stapp) comb. nov. and designation of type and neotype strains. *Int J Syst Bacteriol* 1973;23:217–225.
9. Grimont PAD, Grimont F, Richard C, Davis BR, Steigerwalt AG, Brenner DJ. Deoxyribonucleic acid relatedness between *Serratia plymuthica* and other *Serratia* species, with a description of *Serratia odorifera* sp. nov. (type strain: ICPB 3995). *Int J Syst Bacteriol* 1978;28:453–463.
10. Gavini F, Ferragut C, Izard D, et al. *Serratia fonticola,* a new species from water. *Int J Syst Bacteriol* 1979; 29:92–101.
11. Dauga C, Grimont F, Grimont PAD. Nucleotide sequences of 16S rRNA from ten *Serratia* species. *Res Microbiol* 1990;141:1139–1149.
12. Grimont PAD, Grimont F, Starr MP. *Serratia ficaria* sp. nov., a bacterial species associated with Smyrna figs and the fig wasp *Blastophaga psenes. Curr Microbiol* 1979;2:277–282.
13. Grimont PAD, Jackson TA, Ageron E, Noonan MJ. *Serratia entomophila* sp. nov. associated with amber disease in the New Zealand grass grub *Costelytra zealandica. Int J Syst Bacteriol* 1988;38:1–6.
14. Wright C, Kominos SD, Yee RB. *Enterobacteriaceae* and *Pseudomonas aeruginosa* recovered from veg-

etable salads. *Appl Environ Microbiol* 1976;31: 453–454.

15. Yu VL. *Serratia marcescens* historical perspective and clinical review. *New Engl J Med* 1979;300:887–893.
16. Farmer JJ III, Davis BR, Grimont PAD, Grimont F. Source of American *Serratia. Lancet* 1977;2:459–460.
17. Schaberg DR, Culver DH, Gaynes RP. Major trends in the microbial etiology of nosocomial infection. *Am J Med* 1991;91(Suppl 3B):72S–75S.
18. Farmer JJ III, Davis BR, Hickman FW. Detection of *Serratia* outbreaks in hospital. *Lancet* 1976;2:455–459.
19. Nakashima AK, McCarthy M, Martone WJ, Anderson RL. Epidemic septic arthritis caused by *Serratia marcescens* and associated with a benzalkonium chloride antiseptic. *J Clin Microbiol* 1987;25:1014–1018.
20. Nakashima AK, Highsmith AK, Martone WJ. Survival of *Serratia marcescens* in benzalkonium chloride and in multiple-dose medication vials: Relationship to epidemic septic arthritis. *J Clin Microbiol* 1987;25: 1019–1021.
21. Yannelli B, Schoch PE, Cunha BA. *Serratia marcescens. Clin Microbiol Newslett* 1987;9:157–160.
22. Szewzyk U, Szewzyk R, Stenströn. Growth and survival of *Serratia marcescens* under aerobic and anaerobic conditions in the presence of materials from blood bags. *J Clin Microbiol* 1993;31:1826–1830.
23. McGeer A, Low DE, Penner J, Ng J, Goldman C, Simor AE. Use of molecular typing to study the epidemiology of *Serratia marcescens. J Clin Microbiol* 1990;28:55–58.
24. Zaidi M, Sifuentes J, Bobadilla M, Moncada D, Ponce de León S. Epidemic of *Serratia marcescens* bacteremia and meningitis in a neonatal unit in Mexico City. *Infect Control Hosp Epidemiol* 1989;10:14–20.
25. Watanakunakorn C. *Serratia* bacteremia: A review of 44 episodes. *Scand J Infect Dis* 1989;21:477–483.
26. Hune EBH, Willcox MDP, Sweeney DF, Holden BA. An examination of the clonal variants of *Serratia marcescens* that infect the eye during contact lens wear. *J Med Microbiol* 1996;45:127–132.
27. Cooper R, Mills J. *Serratia* endocarditis. *Arch Intern Med* 1980;140:199–202.
28. Saito H, Elting L, Bodey GP, Berkey P. *Serratia* bacteremia: Review of 118 cases. *Rev Infect Dis* 1989;11: 912–920.
29. Roberts FJ, Geere IW, Coldman A. A three-year study of positive blood cultures, with emphasis on prognosis. *Rev Infect Dis* 1991;13:34–46.
30. Arribas JR, Dominguez A, Folgueira MD, et al. Prognostic factors in *Serratia* bacteremia. *Rev Infect Dis* 1990;12:563–564.
31. Woodfield DG. Transfusion acquired *Serratia liquefaciens* septicemia. *N Z Med J* 1991;104:141.
32. Reina J, Borrell N, Llompart I. Community-acquired bacteremia caused by *Serratia plymuthica* case report and review of the literature. *Diagn Microbiol Infect Dis* 1992;15:449–452.
33. Ramos JM, Fernández-Roblas R, Gadea I, Cuenca-Estrella M, Soriano F. Nosocomially acquired bacteremia caused by *Serratia plymuthica. Clin Microbiol Newslett* 1995;17:156–157.
34. Carrerro P, Garrote JA, Pacheco S, García AI, Gil R, Carabajosa SG. Report of six cases of human infection by *Serratia plymuthica. J Clin Microbiol* 1995; 33:275–276.
35. Horowitz HW, Nadelman RB, Van Horn KG, Weekes SE, Goyburu L, Wormser GP. *Serratia plymuthica* sepsis associated with infection of central venous catheter. *J Clin Microbiol* 1987;25:1562–1563.
36. Domingo D, Limia A, Alarcón T, Sanz JC, Del Rey MC, López-Brea M. Nosocomial septicemia caused by *Serratia plymuthica. J Clin Microbiol* 1994;32: 575–577.
37. Chmel H. *Serratia odorifera* biogroup 1 causing an invasive human infection. *J Clin Microbiol* 1988;26: 1244–1245.
38. Mermel LA, Spiegel CA. Nosocomial sepsis due to *Serratia odorifera* biovar 1. *Clin Infect Dis* 1992;14: 208–210.
39. Glustein JZ, Rudensky B, Abrahamov A. Catheter-associated sepsis caused by *Serratia odorifera* biovar 1 in an adolescent patient. *Eur J Clin Microbiol Infect Dis* 1994;13:183–184.
40. Libby KJ. Infection with *Serratia rubidaea. Clin Microbiol Newslett* 1980;2:4.
41. Darbas H, Jean-Pierre H, Paillisson J. Case report and review of septicemia due to *Serratia ficaria. J Clin Microbiol* 1994;32:2285–2288.
42. Ursua PR, Unzaga MJ, Melera P, Iturburu I, Expeleta C, Cisterna R. *Serratia rubidaea* as an invasive pathogen. *J Clin Microbiol* 1996;34:216–217.
43. Theccanat G, Hirschfield L, Isenberg H. *Serratia marcescens* meningitis. *J Clin Microbiol* 1991;29:822–823.
44. Campbell JR, Diacovo T, Baker CJ. *Serratia marcescens* meningitis in neonates. *Pediatr Infect Dis J* 1992; 11:881–886.
45. Vandenbrouck-Grauls CMJE, Baars ACM, Visser MR, Hulstaert PF, Verhoef J. An outbreak of *Serratia marcescens* traced to a contaminated bronchoscope. *J Hosp Infect* 1993;23:263–270.
46. Sokalski SJ, Jewell MA, Asmus-Shillington AC, Mulcahy J, Segreti J. An outbreak of *Serratia marcescens* in 14 adult cardiac surgical patients associated with 12-lead electrocardiogram bulbs. *Arch Intern Med* 1992; 152:841–844.
47. Bollet C, Grimont P, Gainnier M, Geissler A, Sainty J, De Micco P. Fatal pneumonia due to *Serratia proteamaculans* subsp. *quinovora. J Clin Microbiol* 1993;31: 444–445.
48. Boyd RS, DeMaio J. *Serratia plymuthica* nosocomial pneumonia in an immunocompromised patient. *Lab Medicine* 1991;22:861–863.
49. Bornstein PF, Ditto AM, Noskin GA. *Serratia marcescens* cellulitis in a patient on hemodialysis. *Am J Nephrol* 1992;12:374–376.
50. Vartian CV, Septimus EJ. Septic bursitis caused by a gram-negative bacilli. *J Infect Dis* 1989;160:908.
51. Zbinden R, Blass R. *Serratia plymuthica* osteomyelitis following a motorcycle accident. *J Clin Microbiol* 1988;26:1409–1410.
52. Lass JH, Haaf J, Foster C, Belcher C. Visual outcome in eight cases of *Serratia marcescens* keratitis. *Am J Ophthalmol* 1981;92:384–390.
53. Al Hazzaa SAF, Tabbara KF, Gammon JA. Pink hypopyon: A sign of *Serratia marcescens* endophthalmitis. *Brit J Ophthalmol* 1992;76:764–765.
54. Pien FD, Farmer JJ III. *Serratia ficaria* isolated from a leg ulcer. *South Med J* 1983;76:1591–1592.
55. Grimont PAD, Grimont F. Genus VIII: *Serratia* Bizio 1823, 288^{AL} In: Krieg NR, Holt JG, eds. *Bergey's*

manual of systematic bacteriology. Baltimore: Williams & Wilkins, 1984;477–484.
56. Cockerill FR III, Hughes JG, Vetter EA, et al. Analysis of 281,797 consecutive blood cultures performed over an eight-year period: Trends in microorganisms isolated and the value of anaerobic culture of blood. *Clin Infect Dis* 1997;24:403–418.
57. Farmer JJ III, Silva F, Williams DR. Isolation of *Serratia marcescens* on deoxyribonuclease-toluidine blue-cephalothin agar. *Appl Microbiol* 1973;25: 151–152.
58. Starr MP, Grimont PAD, Grimont F, Starr PB. Caprylate-thallous agar medium for selectively isolating *Serratia* and its utility in the clinical laboratory. *J Clin Microbiol* 1976;4:270–276.
59. Rhoden DL, Schable B, Smith PB. Evaluation of PASCO MIC-ID system for identifying gram-negative bacilli. *J Clin Microbiol* 1987;25:2363–2366.
60. O'Hara CM, Miller JM. Evaluation of the autoSCAN-W/A system for rapid (2-hour) identification of members of the family *Enterobacteriaceae. J Clin Microbiol* 1992;30:1541–1543.
61. Kelly MT, Leicester C. Evaluation of the Autoscan Walkaway system for rapid identification and susceptibility testing of gram-negative bacilli. *J Clin Microbiol* 1992;30:1568–1571.
62. O'Hara CM, Tenover FC, Miller JM. Parallel comparison of accuracy of API 20E, Vitek GNI, MicroScan Walk/Away Rapid ID, and Becton Dickinson Cobas Micro ID-E/NF for identification of members of the family *Enterobacteriaceae* and common gram-negative, non–glucose-fermenting bacilli. *J Clin Microbiol* 1993;31:3165–3169.
63. Kitch TT, Jacobs MR, Appelbaum PC. Evaluation of RapID onE system for identification of 379 strains in the family *Enterobacteriaceae* and oxidase-negative, gram-negative nonfermenters. *J Clin Microbiol* 1994; 32:931–934.
64. Wauters G, Boel A, Voorn GP, et al. Evaluation of a new identification system, Crystal Enteric/Non-Fermenter, for gram-negative bacilli. *J Clin Microbiol* 1995;33:845–849.
65. Grimont PAD, Grimont F. Biotyping of *Serratia marcescens* and its use in epidemiological studies. *J Clin Microbiol* 1978;8:73–83.
66. Sifuentes-Osornio J, Ruiz-Palacios GM, Gröschel DHM. Analysis of epidemiologic markers of nosocomial *Serratia marcescens* isolates with special reference to the Grimont biotyping system. *J Clin Microbiol* 1986;23:230–234.
67. Bale M, Sanford M, Hollis R, Pfaller MA. Application of a biotyping system and DNA restriction fragment analysis to the study of *Serratia marcescens* from hospitalized patients. *Diagn Microbiol Infect Dis* 1993; 16:1–7.
68. Traub WH, Eiden A, Leonhard B, Bauer D. Phenotypic variation of clinical *Serratia marcescens* isolates repeatedly recovered from individual patients. *Zentralbl Bakteriol* 1996;284:124–135.
69. Alsonso R, Aucken HM, Perez-Diaz JC, Cookson BD, Baquero F, Pitt TL. Comparison of serotype, biotype and bacteriocin type with rDNA RFLP patterns for the type identification of *Serratia marcescens. Epidemiol Infect* 1993;111:99–107.
70. Gaston MA, Pitt TL. O-Antigen specificities of the serotype strains of *Serratia marcescens. J Clin Microbiol* 1989;27:2697–2701.
71. Gaston MA, Pitt TL. Improved O-serotyping method for *Serratia marcescens. J Clin Microbiol* 1989;27: 2702–2705.
72. Aucken HM, Wilkinson SG, Pitt TL. Identification of capsular antigens in *Serratia marcescens. J Clin Microbiol* 1997;35:59–63.
73. Gargallo-Viola D. Enzyme polymorphism, prodigiosin production, and plasmid fingerprints in clinical and naturally occurring isolates of *Serratia marcescens. J Clin Microbiol* 1989;27:860–868.
74. Gargallo-Viola D, López D. Numerical analysis of electrophoretic periplasmic protein patterns, a possible marker system for epidemiologic studies. *J Clin Microbiol* 1990;28:136–139.
75. Lehmacher A, Olszewski M, Wolter M, et al. Pulsed-field gel electrophoresis and typing of nosocomial *Serratia marcescens* isolates. *Med Microbiol Lett* 1996;5:245–252.
76. Shi Z, Lui PY, Lau Y, Lin Y, Hu B. Use of pulsed-field gel electrophoresis to investigate an outbreak of *Serratia marcescens. J Clin Microbiol* 1997;35:325–327.
77. Debast SB, Melchers WJG, Voss A, Hoogkamp-Korstanje JAA, Meis JFGM. Epidemiological survey of an outbreak of multiresistant *Serratia marcescens* by PCR-fingerprinting. *Infection* 1995;23:267–271.
78. Parment PA, Svanborg-Ede C, Chaknis MJ. Hemagglutination (fimbriae) and hydrophobicity in adherence of *Serratia marcescens* to urinary tract epithelium and contact lenses. *Curr Microbiol* 1992;25:113–118.
79. Rumelt S, Metzger Z, Kariv N, Rosenberg M. Clearance of *Serratia marcescens* from blood in mice: Role of hydrophobic versus mannose-sensitive interactions. *Infect Immun* 1988;56:1167–1170.
80. Braun V, Günther H, Neuss B, Tautz C. Hemolytic activity of *Serratia marcescens. Arch Microbiol* 1985; 141:371–376.
81. Marre R, Hacker J, Braun V. The cell-bound hemolysin of *Serratia marcescens* contributes to uropathogenicity. *Microb Pathog* 1989;7:153–156.
82. Scheibel E, Schwarz H, Braun V. Subcellular location and unique secretion of the hemolysin of *Serratia marcescens. J Biol Chem* 1989;264:16311–16320.
83. Létoffé S, Ghigo JM, Wandersman C. Iron acquisition from heme and hemoglobin by a *Serratia marcescens* extracellular protein. *Proc Natl Acad Sci* 1994;91: 9876–9880.
84. Lyerly DM, Kreger AS. Importance of *Serratia* protease in the pathogenesis of experimental *Serratia marcescens* pneumonia. *Infect Immun* 1983;40: 113–119.
85. Carbonell GV, Fonseca BAL, Figueiredo LTM, Darini ALC, Yanaguita RM. Culture conditions affect cytotoxin production by *Serratia marcescens. FEMS Immunol Med Microbiol* 1996;16:299–307.
86. Berkowitz FE, Metchock B. Third generation cephalothin-resistant gram-negative bacilli in the feces of hospitalized children. *Pediatr Infect Dis J* 1995;14: 97–100.
87. Sanders CC, Sanders WE Jr. β-lactam resistance in gram-negative bacteria: Global trends and clinical impact. *Clin Infect Dis* 1992;15:824–839.
88. Hindler JA. Antimicrobial susceptibility testing of gram-negative bacteria: Meeting the challenge of in-

creasing resistance and decreasing budget. *Clin Microbiol Newslett* 1995;17:77–80.
89. Sirot DL, Goldstein FW, Soussy CJ, et al. Resistance to cefotaxime and seven other β-lactams in members of the family *Enterobacteriaceae:* A 3-year survey in France. *Antimicrob Agents Chemother* 1992;36:1677–1681.
90. Garcia DC, Woloj GM, Piñeiro S, Sordelli, Kaufman S. An 8-year study of resistance to amikacin in gram-negative bacilli isolates from patients with nosocomial infection at one hospital in Argentina. *J Med Microbiol* 1995;42:283–290.
91. Rasmussen BA, Bush K. Carbapenem-hydrolyzing β-lactamases. *Antimicrob Agents Chemother* 1997;41: 223–232.
92. Klein NC, Cunha BA. Tetracyclines. *Med Clin North Am* 1995;79:789–801.
93. 1995 ASCP Susceptibility Testing Group. United States geographic bacteria susceptibility patterns. *Am J Clin Path* 1996;106:275–281.
94. Freney J, Husson MO, Gavini F, et al. Susceptibilities to antibiotics and antiseptics of new species of the family *Enterobacteriaceae. Antimicrob Agents Chemother* 1988;32:873–876.
95. Pancoast SJ. Aminoglycoside antibiotics in clinical use. *Med Clin North Am* 1988;72:581–612.
96. Lortholary O, Tod M, Cohen Y, Petitjean O. Aminoglycosides. *Med Clin North Am* 1995;79:761–789.
97. Thornsberry C. Susceptibility of clinical bacterial isolates to ciprofloxacin in the United States. *Infection* 1994;22:S80–S89.
98. Hohl P, Lüthy-Hottenstein, Zollinger-Iten J, Altwegg M. In vitro activities of fleroxacin, cefetamet, ciprofloxacin, ceftriaxone, trimethoprim-sulfamethoxazole, and amoxacillin-clavulanic acid against rare members of the family *Enterobacteriaceae* primarily of human (clinical) origin. *Antimicrob Agents Chemother* 1990; 34:1605–1608.
99. Osano E, Arakawa Y, Wacharotayankun R, et al. Molecular characterization of an enterobacterial metallo-β-lactamase found in a clinical isolate of *Serratia marcescens* that shows imipenem resistance. *Antimicrob Agents Chemother* 1994;38:71–78.
100. Tzouvelekis LS, Tzelepi E, Mentis AF, Legakis NJ. In vitro activity of cefpirome against selected clinical enterobacterial isolates with β-lactamase–mediated resistance. *Infection* 1995;23:384–388.
101. Jones RN, Pfaller MA, Allen SD, Gerlach EH, Fuchs PC, Alderidge KE. Antimicrobial activity of cefpirome: An update compared to five third-generation cephalosporins against nearly 6000 recent clinical isolates from five medical centers. *Diagn Microbiol Infect Dis* 1991;14:361–364.
102. Fuchs PC, Barry AL, Brown SD, AST Surveillance Group. Survey of antimicrobial activity of four commonly used third generation cephalosporins tested against recent bacterial isolates from ten American medical centers, and assessment of disk diffusion test performance. *Diagn Microbiol Infect Dis* 1996;24: 213–219.
103. Jones RN, Kehrberg EN, Erwin ME, Anderson SC, Fluoroquinolone Resistance Surveillance Group. Prevalence of important pathogens and antimicrobial activity of parenteral drugs at numerous medical centers in the United States: I. Study on the threat of emerging resistances: Real or perceived? *Diagn Microbiol Infect Dis* 1994;19:203–215.

10

The Genus *Enterobacter*

TAXONOMY

Historical

The genus *Enterobacter* was first proposed by Hormaeche and Edwards[1] in 1960 to resolve taxonomic inconsistencies arising from strains formerly assigned to the genus *Aerobacter*. The genus *Aerobacter* had previously contained both motile and nonmotile forms, many of which could not be phenotypically separated at one time from *Klebsiella pneumoniae*. The subsequent development of biochemical tests, most notably Moeller's amino acid decarboxylase reactions, led to useful characteristics which clearly helped define each genera. True members of the klebsiellae were nonmotile (MOT–) and ornithine decarboxylase–negative (ODC–), whereas *Aerobacter* isolates were MOT+ and ODC+. Because of the inclusion of multiple genera under the single epithet, *Aerobacter*, the genus name *Enterobacter* was proposed by Hormaeche and Edwards as a replacement to avoid confusion.[1]

Although the ancestry of the genus *Enterobacter* is nebulous and confusing in the older medical literature, it is clear that strains that currently reside within this genus were isolated as early as the late 19th century.[2] These strains were placed in a number of different taxa, including *Bacillus aerogenes*, *"Bacterium cloacae," "Cloaca cloacae,"* and *Aerobacter*. By the middle to late 1950s, two groups of *Aerobacter* were distinguished based on arginine dihydrolase (ADH) activity and the fermentation of glycerol and *m*-inositol.[3] These two motile groups, previously designated "Cloaca A" and "Cloaca B" by Hormaeche and Munilla,[4] were renamed *Aerobacter cloacae* and *Aerobacter aerogenes*, respectively, by Hormaeche and Edwards in 1958.[3] When the genus *Enterobacter* was proposed to replace the genus *Aerobacter* by the same authors in 1960, *Enterobacter cloacae* (type species) and *Enterobacter aerogenes* became the first two species assigned to the new genus.[5] In 1972, a third species, *Enterobacter agglomerans*, whose members had formerly resided in the phytopathogenic genus *Erwinia*, was added to the genus *Enterobacter*.[6]

Between 1970 and 1980, considerable flux in the number of species assigned to the genus *Enterobacter* occurred. Examples of these changes are shown in Table 1. Several groups, including *Serratia liquefaciens* and *Hafnia alvei*, were included as species in the genus *Enterobacter* before their taxonomic reassignment to other genera based on molecular methodologies. Most of the changes that occurred during this decade were the result of the initial misplacement of species in *Enterobacter* based on phenotypic (numerical taxonomic) and morphologic properties. Most of these studies predated the use of molecular methods such as DNA–DNA hybridization as routine techniques in systematic investigations. By 1980, only three species were commonly recognized as members of the genus *Enterobacter* (see Table 1).

With the advent of DNA–DNA hybridization as a standard tool in taxonomic studies, a more precise species definition for inclusion in the genus *Enterobacter* unfolded. With the

TABLE 1. *Taxonomic changes in the genus* Enterobacter *(1970–1985) as shown in the* Manual of Clinical Microbiology[a]

First Edition, 1970 (Four Species)	Second Edition, 1974 (Four Species)	Third Edition, 1980 (Three Species)	Fourth Edition, 1985 (Five Species)
E. cloacae	*E. cloacae*	*E. cloacae*	*E. cloacae*
E. aerogenes	*E. aerogenes*	*E. aerogenes*	*E. aerogenes*
E. hafniae	*E. hafniae*	*E. agglomerans*	*E. agglomerans*
E. liquefaciens	*E. agglomerans*		*E. gergoviae*
			E. sakazakii

[a]ASM Press, Washington, DC.

use of DNA reassociation kinetics as a gold standard, 10 *E. aerogenes* isolates were found to be highly related (83%–100%) to labeled DNA from a reference strain of *E. aerogenes.*[7] However, in the same study, *E. aerogenes* was found to be less related to an *E. cloacae* isolate (44%) than to *Klebsiella* species (47%–64%). Unlike the high degree of homology found among *E. aerogenes* strains,[7] two distinct groups within *E. cloacae* were detected using similar methods.[8] These two groups were phenotypically distinguishable based on the presence or absence of yellow pigmentation. Nonpigmented strains were found to be highly related to each other (59%–100%) but much less so to yellow-pigmented *E. cloacae* (35%–41%). In a reciprocal fashion, yellow-pigmented strains were 83% to 100% related to each other but only 47% to 54% related to nonpigmented *E. cloacae,* indicating that significant heterogeneity existed within *E. cloacae.* These findings by Steigerwalt and associates[8] fueled later investigations of yellow-pigmented *E. cloacae,* with the result that this group was elevated to species rank within *Enterobacter* as *Enterobacter sakazakii.*[9] Later, independent hybridization studies by Izard and colleagues[10] confirmed that all *E. sakazakii* strains were highly related with no other close relationships within members of the enterobacteria detected. Since 1980, 11 new or reassigned taxa have been added to the genus *Enterobacter.*[2] Several of these species were isolated from trees, including *Enterobacter nimipressuralis* (elms, wetwood disease), *Enterobacter pyrinus* (pear, brown leaf spot disease), and *Enterobacter cancerogenus* (poplars). The latest addition to this genus is *Enterobacter kobei,* formerly referred to as atypical *E. cloacae.*[18] These strains bear features strikingly similar to those of Centers for Disease Control (CDC) enteric group 69. Current species in the genus *Enterobacter* are listed in Table 2.

Taxonomic Problems and Issues

Probably no single group of bacteria in the family *Enterobacteriaceae* has more unresolved taxonomic problems than does the genus *Enterobacter.* These issues range from correct taxonomic designations for species to the known existence of multiple hybridization groups (genomospecies) within previously established taxa (Table 3). Many of these issues are complex in nature and probably will not be settled in the near future.

Changes in species designation for several established enterobacterial groups have been proposed. Von Graevenitz proposed that *Enterobacter intermedium* be changed to *Enterobacter intermedius,* making use of the new Latin ending for the adjective.[19] *Enterobacter cancerogenus* is the present designation[14] for a group of bacteria that have also been known as *Enterobacter taylorae* or CDC enteric group 19.[20] In 1988, Dickey and Zumoff[21] proposed the transfer of *Erwinia cancerogena* to the genus *Enterobacter* as *E. cancerogenus* based on biochemical characteristics and a similarity matrix program. In 1989, Grimont and Ageron[14] showed by DNA hybridization (S1 nuclease method) that four

TABLE 2. *Current species composition within the genus* Enterobacter

Species	Type Strain	Isolated from Clinical Material	Synonyms	Ref.
E. cloacae	ATCC 13047	Yes	*Aerobacter cloacae*	1
E. aerogenes	ATCC 13048	Yes	*Aerobacter aerogenes, Klebsiella mobilis*	1, 2
E. gergoviae	ATCC 33028	Yes		11
E. sakazakii	ATCC 29544	Yes		9
E. intermedium	ATCC 33110	Yes		12
E. amnigenus	ATCC 33072	Yes		13
E. cancerogenus	ATCC 33241	Yes	*Enterobacter taylorae,* CDC enteric group 19	14
E. asburiae	ATCC 35953	Yes	CDC enteric group 17	15
E. nimipressuralis	ATCC 9912	No	*Erwinia nimipressuralis*	15
E. dissolvens	ATCC 23373	No	*Erwinia dissolvens*	15
E. agglomerans	ATCC 27155	Yes	*Erwinia lathyri,* Herbicola-Agglomerans group	6
E. hormaechei	ATCC 49162	Yes	CDC enteric group 75	16
E. pyrinus	ATCC 49851	No	*Erwinia pyrinus*	17
E. kobei	JCM 8580	Yes	*E. cloacae* (VP–)	18

ATCC, American Type Culture Collection; JCM, Japanese Collection of Microorganisms, RIKEN; VP, Voges-Proskauer.

E. taylorae strains were 74% to 81% related to *E. cancerogenus* NCPPB 2176; both groups additionally had identical carbon source utilization patterns. Therefore, based on the priority of the epithet "cancerogena" in the literature, *E. taylorae* must be considered a junior subjective synonym of *E. cancerogenus.* Discrepant biochemical reactions between these two groups previously noted by others were subsequently resolved by Schønheyder and coworkers.[22] These collective results suggest that *E. cancerogenus* is the valid name for this species.

A second unresolved issue concerns the correct taxonomic designation for *E. aerogenes*. Early studies by Brenner and coinvestigators[7] showed that *E. aerogenes* 1627-66 was more closely related to *Klebsiella* (47%–61%) than to *E. cloacae* (44%). This report was followed by data from Izard and associates,[23] who demonstrated that 14 *E. aerogenes* strains were 53% related to a strain of *K. pneumoniae* but only 39% related to an *E. cloacae* isolate. The authors went on to propose that *E. aerogenes* be transferred to the genus *Klebsiella* as *Klebsiella mobilis,*[23] and others have supported this change.[2] However, caution should be exercised, because it is known that klebsiellae are more highly related to each other than are members of the genus *Enterobacter.*[7] At present, *E. aerogenes* is still the preferred name for this group.

The single most complicated issue regarding the genus *Enterobacter* concerns strains that are currently identified as *E. cloacae.* Reports indicate that this species is genetically heterogenous at the DNA level.[2,14] At least five distinct DNA groups are presently recognized within *E. cloacae,* one of which (DNA group IV) contains three subgroups within it.[2] One of the most troubling aspects of these findings involves the location of reference

TABLE 3. *Current taxonomic issues in the genus* Enterobacter

Problem	Issue to Be Resolved	Ref.
E. cancerogenus	Correct species designation	14, 21, 22
E. aerogenes	Correct genus assignment	23
E. cloacae	Genetically heterogenous at the DNA level	2, 14
E. agglomerans	Genetically heterogenous at the DNA level	24–27
E. intermedius	Nomenclature change	18

TABLE 4. *DNA groups in* Enterobacter cloacae

DNA Grouping According to			
Grimont & Ageron, 1989[14]	Grimont & Grimont, 1991[2]	Strains[a]	Reference Cultures
DNA group I	DNA group I	3	*E. cloacae* ATCC 13047, *E. dissolvens* ATCC 23373
DNA group II	DNA group II	7	*E. cloacae* ATCC 29941
DNA group III	DNA group III	15	*E. hormaechei* ATCC 49162, *E. cloacae* CUETM 77-121
DNA group IV	DNA group IVa–c	17	*E. asburiae* ATCC 35953, *E. cloacae* 54-81
DNA group V			
DNA group VI	DNA group V	3	*E. cloacae* 30-81

[a]Number of *E. cloacae* strains analyzed by Grimont & Grimont.[2]
ATCC, American Type Culture Collection; CUETM, Collection de l'Unité d'Eco-Toxicologie Microbienne (Lille, France).

stains in different DNA groups within *E. cloacae* (Table 4). The type strain of *E. cloacae* (ATCC 13047) resides in DNA group I. This group is infrequently isolated from clinical material. If the type strain is maintained, the isolation of *E. cloacae* would almost completely disappear from clinical specimens. Besides *E. cloacae,* other type strains of enterobacters reside within the phenospecies. These include *Enterobacter dissolvens* (ATCC 23373), *Enterobacter hormaechei* (ATCC 49162), and *Enterobacter asburiae* (ATCC 35953). These results, if corroborated, would require the redefinition of some species *(E. hormaechei, E. asburiae)* and the possible elimination of others *(E. dissolvens).*

E. agglomerans is in a similar taxonomic quagmire. Brenner and associates[24] found this species to be genetically heterogenous, containing at least 18 different hybridization groups among the 124 strains studied. Most hybridization groups were not closely related to other genera in the family *Enterobacteriaceae,* with the exceptions of DNA group VII (56% related to *Enterobacter*) and DNA groups II through VI (41% to 47% related to *Erwinia*). Only some of these groups could be separated biochemically. Despite these pioneering efforts, this study was handicapped in several aspects because of the limited number and sources of strains and the failure to include the type strain of *E. agglomerans* (ATCC 27155^T) as part of the taxonomic analysis. In 1986, Lind and Ursing[25] found that 52 (60%) of 86 clinical isolates of *E. agglomerans* formed a tight genetic cluster (≥89% related). They also demonstrated that *Erwinia herbicola* and *Erwinia milletiae* were 92% to 99% related to a reference strain of *E. agglomerans,* suggesting that these species were identical. This latter observation was confirmed by Beji and colleagues[26] with the use of DNA–DNA hybridization. The authors further showed that a highly related group of *E. agglomerans* strains (including ATCC 27155^T) were identical to DNA group XIII of Brenner and colleagues.[26] *E. agglomerans* was transferred to a new genus *(Pantoea)* in 1989 along with a second species, *Pantoea dispersa,* formerly referred to as DNA group II.[27]

EPIDEMIOLOGY

Most of our present knowledge concerning the environmental distribution of enterobacters stems from studies dating from the mid-1950s to the late 1970s. These collective reports indicate that the genus *Enterobacter* is widely dispersed in nature.[28–30] Plants (including cultivatable crops) and soil are often found to harbor *Enterobacter* species, as are those ecologic niches intimately associated with both habitats, such as seedlings and the microbial rhizosphere.[28,31] Enterobacters also can be isolated from a number of vertebrate and invertebrate species, including farm animals, primates, fish, and insects.[28] Most often they are recovered from animals as commensals, although in rare instances they are involved in infectious processes.[28] Because of their frequent occurrence in both plants and animals, it is not surprising that they have been isolated on many different occasions from foods. Common consumables found to contain *Enterobacter* species include meats, dairy products, poultry, and vegetable salads.[28,29]

Since 1980, only a limited number of studies have addressed the incidence of *Enterobacter* in environmental samples. Table 5 lists some of the sites from which *Enterobacter* species have been isolated. Both beef and pork products are common reservoirs for members of the *Enterobacteriaceae,* including *Enterobacter.*[32–35] Stiles and Ng[32] reported that *E. cloacae* and *E. agglomerans* were among the seven enteric bacteria most commonly recovered from a variety of retail meats including ground beef, pork sausage, pork loin, pork chops, and beef steaks. Three *Enterobacter* species (*E. cloacae, E. aerogenes,* and *E. agglomerans*) were present at varying frequencies (2.6% to 31.2%) in ground beef (low- and high-fat content varieties), packaged pork sausage, and vacuum-packed beef trim and primal cuts just before retail handling.[33] *E. agglomerans* was the predominant *Enterobacter* species identified in all of the meat products. All three species were also isolated from surfaces in the packing plant that had been in contact with these meat products. In other studies *Enterobacter* species were identified as common microbial contaminants of fresh ground beef *(E. aerogenes)* and Vienna sausages.[34,35] In addition, these organisms have frequently been identified in a variety of poultry products, including chicken meats[37] and chicken rinses.[34] Zeitoun and colleagues[37] found *E. cloacae* to be the fourth most common enteric microorganism isolated from six chicken legs immediately after slaughter (11%). In fish, histamine-producing strains of *E. cloacae* and *E. aerogenes* were detected in tuna caught in the Pacific Ocean.[39] Next to *Morganella, Enterobacter* species *(E. cloacae, E. aerogenes, E. agglomerans)* were the second most common gram-negative facultatively anaerobic bacilli identified.

TABLE 5. *Environmental sources of* Enterobacter *species*

Source	Ref.
Soil	29
Foods	
Beef	31–33
Powdered milk	34
Powdered formula	49, 50
Poultry	32, 35
Milk and dairy products	36
Seafoods	
Scallops	32
Tuna	37
Medical equipment	
Heart-lung machine	38
Vacuum suction apparatus	39
Blood gas analyzer	39, 41
Hemodialyzers	40
Equipment arm	38
Catheters	42–45
Perfusion lines	43
Rectal thermometers	46
Blenders	47, 48
Desk	46
Scale	46
Shelf	41
Sinks	38, 41
Siphons	46
Spot lamp	41
Toilet	41

By far the most important epidemiologic trend noted with *Enterobacter* concerns its emergence as a major nosocomial pathogen over the past 15 years. In the latest published report by the National Nosocomial Infections Surveillance System (NNIS), *Enterobacter* was one of several genera whose relative incidence in causing nosocomial infections is on the rise.[51] In the NNIS survey, *Enterobacter* species were the sixth leading cause of hospital-acquired infections; among cases of nosocomially transmitted pneumonias, *Enterobacter* ranked third (11%). Patients in intensive care units (ICUs) such as neonatal, cardiac, or burn units, who are constantly monitored and are prone to receiving numerous intravenous (IV) solutions, are most susceptible to development of *Enterobacter* infections. Although some surveillance investigations have failed to identify endemic foci within the hospital environment responsible for such illnesses,[52,53] many studies have documented instances in which contaminated devices or instrumentation led to colonization or infection with *Enterobacter* species. In a 1982 review by John and associates,[53] a number of products associated with *E. cloacae* outbreaks were highlighted, including IV solutions, pressure monitoring devices, gauze, platelets, and ice. Products found to be contaminated with enterobacters have included cardioplegic solutions,[54] parenteral nutrition solutions,[55] distilled water,[56] and blood products.[57] Of particular interest is the frequent isolation of *E. agglomerans,* primarily an environmental organism, from blood and IV solutions.[57,58] Probably the single greatest outbreak of *Enterobacter* infection in the United States occurred from mid-1970 through the spring of 1971. During that time, at least 378 patients nationwide developed septicemia caused by either *E. cloacae* or *E. aerogenes.*[58] The source of all of these infections was a contaminated Elastomer liner in threaded caps of bottles containing parenteral solutions produced by a single manufacturer. The outbreak became apparent when the incidence of sepsis caused by these organisms increased dramatically and in a patient population not generally at high risk for *Enterobacter* infection (ie, with no predisposing factors or other sites of infection). Later reports have most often linked *Enterobacter* infections to contaminated lines such as vascular,[44,47] central venous,[43,46] tracheal catheters,[42] and total parenteral nutrition lines.[55]

Another alarming tendency is *Enterobacter* infection or colonization in neonates as a result of contaminated medical devices or products. Lacey and Want[41] described five cases of neonatal bacteremia caused by *E. cloacae* over a 3-month period. Although these illnesses initially were suspected to be the result of cross-contamination, further epidemiologic investigations associated them with a contaminated inlet port and cover of a blood gas analyzer. Acolet and associates[43] also found a blood gas analyzer to be the major reservoir for cases of *E. cloacae* septicemia or colonization in neonates in an ICU in London, England. Over a 26-month period, 15 (36%) of 42 *E. cloacae* isolates recovered from infants were epidemiologically tied to an identical strain recovered from the washer and effluent fluid of a blood gas machine. Sporadic cases of septicemia in adults, caused by a variety of gram-negative bacteria including *Enterobacter,* have been tied to contaminated O-rings from hemodialyzers,[42] pressurized canisters,[54] and respirator humidifiers.[56] Contaminated infant formula has been a common source of life-threatening infections in neonates and infants.[49,50] Most of these infections were attributed to *E. sakazakii* rather than *E. cloacae.* In 1989, Simmons and coworkers[49] described four cases of *E. sakazakii* infection or colonization in neonates in a 20-bed ICU over a 6-week period. Illnesses in these infants ranged from bloody diarrhea to sepsis. Both the powdered infant formula and the blender used to mix the formula were found to be contaminated with *Enterobacter.* The powdered infant formula yielded low numbers of *Enterobacter* species by the most probable number (MPN) method (*E. cloacae,* 48 cfu/g; *E. sakazakii,* 8 cfu/g), but the blender, cleaned only by a tap water rinse, was heavily seeded with *E. sakazakii. E. sakazakii* also was shown to be a persistent colonizer (2 to 8 weeks) of the respiratory and gastrointestinal tracts of severely ill infants in an ICU.[59] Although the organism was not recovered from invasive sites (blood), several infants colonized with *E. sakazakii* died with signs of sepsis. No environmental source for the outbreak could be identified.

CLINICAL SIGNIFICANCE

Enterobacter is an increasingly important human pathogen, particularly in the hospital setting.[30,60] Its emergence as a common nosocomial pathogen led one reviewer to remark that enterobacters are opportunists poised to flourish in the 21st century.[61] An NNIS study[51] found that *Enterobacter* accounted for 5% to 11% of all nosocomially transmitted blood, wound, respiratory tract, and urinary tract infections. Many of these illnesses arose in persons who were previously colonized at various anatomic sites with enterobacteria and who were undergoing medical treatments such as chemotherapy, surgery, hyperalimentation, and catheterization. Flynn and associates[52] found that 23% to 28% of patients preparing for cardiac surgery or angioplasty were rectally colonized with either *E. cloacae* or *E. aerogenes.* Within 72 hours after their operative procedure, *Enterobacter* colonization rates had increased in surgical patients to 49%, presumably because of the postsurgical administration of antibiotics. Patients undergoing angioplasty procedures did not receive antibiotics and showed no increase in colonization rates. John and colleagues[53] described six burn patients with *E. cloacae* septicemia. The wounds of all six patients were found to have been colonized with *E. cloacae* before their septic episode. In a review of patients receiving long-term care, Nicolle and coinvestigators[62] found that between 14% and 19% of all nursing home residents have oropharyngeal colonization with *Enterobacter*. Such colonization is presumably a prerequisite step for the 1% to 11% *Enterobacter*-associated pneumonia rate observed in such patients. These epidemiologic data indicate that frequent colonization of mucosal surfaces by *Enterobacter* species produces serious endemic foci for subsequent endogenously acquired infections or for outbreaks related to cross-contamination.

Bacteremia

Enterobacter is responsible for between 4% and 12% of all cases of gram-negative bacteremia.[30,53,63] All age groups are affected. In adults *Enterobacter* is the third or fourth leading cause of sepsis after *Escherichia coli, Klebsiella* spp., and *Pseudomonas aeruginosa.* A large number of clinical investigations detailing the epidemiology and clinical spectrum of *Enterobacter* bacteremia in adults have been published,[61] some of which are listed in Table 6. *Enterobacter* sepsis most commonly occurs as a nosocomially acquired bacteremia in older men. Fever, the symptom most often associated with such infections, typically occurs simultaneously with the onset of bacteremia.[61] Most patients (92%–98%) develop fever during their septic crisis, although in one study the frequency was reported to be much lower (61%).[69] Other symptoms often associated with *Enterobacter* bacteremia are leukocytosis (58%–68%), septic shock (9%–34%), and thrombocytopenia (31%–56%). In addition, some studies have identified tachypnea (32%) and tachycardia (29%) as common findings and oliguria (12%) and metabolic acidosis (10%) as less frequently reported complications.[65] Disseminated intravascular coagulation is an infrequent complication of *Enterobacter* sepsis, occurring in only 1% to 6% of all cases.[63,68,69] None of the symptoms mentioned is useful in distinguishing septicemia caused by *Enterobacter* spp. from bacteremias caused by other members of the *Enterobacteriaceae.* The annual incidence (prevalence) of *Enterobacter* septicemia varies from 0.3 to 1.8 episodes per 1000 admissions; it is higher in general and tertiary medical centers than in community hospitals. Almost 100% of persons who develop *Enterobacter* septicemia have one or more underlying diseases,[63,65–68,70] although Haddy and colleagues[69] found that 9% of patients in their community-based study (7 persons) had no known defect. Other conditions linked to *Enterobacter* bacteremia and not listed in Table 6 include hepatobiliary disease,[63,65,66] pulmonary problems,[63,65,66] renal failure,[65,66,68] cerebrovascular accidents,[66] transplantation,[67] and burns or trauma.[67,68,70] The respiratory and genitourinary tracts are the two most commonly cited portals of entry from which *Enterobacter* septicemia develops. Other frequent sites include surgical wounds/skin[63,65] and intravascular catheters.[66] A large number of investigations have reported a high frequency (19%–72%) of "site unknown" as the portal of entry.[63,65–69] This indicates that in many cases *Enterobacter* bacteremia develops without previous microbiologically-confirmed colonization.

The fatality rate associated with *Enterobacter* sepsis varies from 20% to 50%. How-

TABLE 6. Enterobacter *bacteremia*

Characteristic	Bouza et al.[63] (1985)	Burchard et al.[64] (1986)	Fung et al.[65] (1988)	Watanakunakorn et al.[66] (1989)	Chow et al.[67] (1991)	Bodey et al.[68] (1991)	Haddy et al.[69] (1991)
Location of study	Spain	USA	China	USA	USA	USA	USA
Study date	1977–1983	1980–1984	1985	1980–1986	Unspecified	1972–1986	1985–1987
Hospital type	General	General	General Veterans	Community	Multicenter	Tertiary-care center	Community hospitals
No. patients/episodes	147	63	50	55	129	281	75
Prevalence (per 1000 admissions)	1.25	NA	1.13	0.30	NA	1.8	0.30
Age in years (median/mean [range])	56 (1–87)	61	56 (17–82)	65 (1 d–85)	59 (17–95)	49 (1–83)	64 (10 d–98)
Male/female ratio	1.3	2.9	3.0	1.9	2.8	1.3	1.5
Nosocomial (%)	76	NA	68	72	84	74	80
Portal of entry[a] (%)	Respiratory (18%)	Respiratory (47%)	Genitourinary (27%)	Respiratory (34%)	Abdominal (39%)	Respiratory (8%)	GI/biliary (19%), Genitourinary (19%)
Underlying disease[a] (%)	Cardiovascular (20%)	Gastrointestinal (37%)	Cancer (32%)	Diabetes mellitus (42%)	Cancer (22%)	Cancer (100%)	Cancer (23%)
Mortality rate (%)	42	35	46	69	20	23	29

[a]Most common site.
NA, not available.

ever, a rate as high as 69% has been published.[66] The high mortality rate reported in this community-based survey may be related to several factors, namely inclusion of postmortem blood cultures in the database, the lungs as the most common portal of septicemic infection, and patient demographics (median age, 65 years). Risk factors associated with *Enterobacter* bacteremia include urinary, arterial, or central venous catheterization[63,64,67,70]; surgical procedures[63,64,67]; recent antimicrobial therapy[63,64]; endotracheal tubes[63,70]; and stay in an ICU.[63,67] Factors associated with a poor prognosis in patients with *Enterobacter* bacteremia include septic shock,[63,65,68] ultimately or rapidly fatal underlying disease,[63,66,69] inappropriate or incomplete chemotherapy,[63,67] and respiratory disease, including pneumonia.[66,68] Other conditions less often seen but associated with an extremely grave prognosis include metabolic acidosis[65] and sepsis without fever.[67]

Antimicrobial therapy plays a critical role especially in patient care and outcome. Bouza and associates[63] found that only 67% of 46 septic persons received correct antimicrobial chemotherapy. The mortality rate associated with adequate treatment was significantly lower than that associated with inadequate or no chemotherapy (29% versus 80%, respectively). Other studies have demonstrated similar findings, indicating that rapid and appropriate antimicrobial intervention increases therapeutic response and decreases mortality. In most of these clinical settings, persons receiving appropriate chemoprophylaxis have higher survival rates (77%–84%) than those with inadequate or no antimicrobial therapy (29%–45%).[63,65–67] Only one study found no difference in mortality rate after 3 days of chemotherapy between patients receiving appropriate or inappropriate regimens (54% versus 56%, respectively).[66] The reasons for such observed differences are not clear. Although some studies detected no difference in patient outcome with monotherapy (aminoglycoside or β-lactam) versus two antibiotics,[63] Bodey and coworkers[68] found that single therapy with aminoglycosides (59% response) was inferior to β-lactams alone (75%–85% response). The prior administration of antibiotics, particularly third-generation cephalosporins, before development of *Enterobacter* bacteremia significantly increased the isolation rate (69% versus 20%) of multiply-resistant *Enterobacter* strains from initial positive blood cultures, compared with no β-lactams.[67] Emergence of drug-resistant strains in patients undergoing antimicrobial treatment has been reported to occur in 4% to 6% of cases.[67,68]

E. cloacae is the *Enterobacter* species most frequently isolated from blood (Fig. 1). Together with *E. aerogenes,* it is responsible

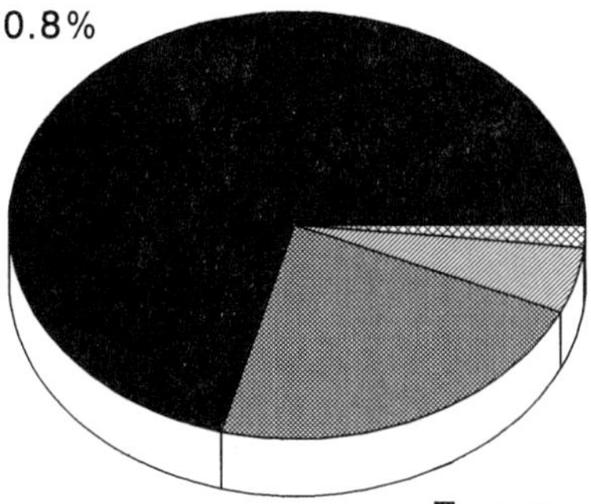

FIG. 1. Species distribution of *Enterobacter* isolates associated with 789 episodes of bacteremia. Other species include *E. sakazakii* (n=7), *E. amnigenus* (n=2), *E. gergoviae* (n=1), and *E. cancerogenus* (*"E. taylorae";* n=1). From references 44, 45, 64–70.

for more than 90% of all episodes of *Enterobacter* bacteremia. Although most cases of *Enterobacter* septicemia are monomicrobic infections, between 18% and 34% are polymicrobic in nature.[44,45,63,66] The bacteria most commonly associated with polymicrobic sepsis due to *Enterobacter* are *E. coli, Klebsiella,* coagulase-negative staphylococci, and enterococci. Although most of the less frequently encountered *Enterobacter* species have been associated with blood infections, little information is available regarding the epidemiology and clinical significance of such isolations. Rubinstein and colleagues[71] reported on two cases of *E. cancerogenus* (formerly *E. taylorae*) in two 75-year-old men. In one case, *E. cancerogenus* bacteremia, pneumonia, and cholangitis occurred in a man initially admitted for cholestatic jaundice. In a second instance, *E. cancerogenus* was isolated from the blood of a man admitted for a coronary artery bypass graft and a left ventricular aneurysmectomy. Both patients died during hospitalization, and *E. cancerogenus* was thought to play a contributory role to each case. The two men shared other common risk factors for enterobacterial infection, including presurgical chemotherapy with a cephalosporin (no activity against *E. cancerogenus*) and medical instrumentation procedures. A third case of *E. cancerogenus* bacteremia was described in a 30-year-old woman who suffered multiple trauma to her head, vertebrae, and long bones.[72] *Enterobacter amnigenus* was recovered from the blood and a catheter of a immunocompromised 52-year-old male who had undergone heart transplantation.[73] Because this strain was sporadically isolated from bacteremic specimens also containing *P. aeruginosa,* its clinical significance remains questionable. *E. intermedium* has also been isolated from blood, but no clinical information is available.[74] Similar circumstances surround the recovery of both *Enterobacter gergoviae* and *E. asburiae* from blood.[11,15] In 1991, a case of *E. sakazakii* bacteremia in an adult was reported by Hawkins and colleagues.[75] An outbreak of *E. hormaechei* bacteremia in five premature infants was reported from The Hospital of the University of Pennsylvania.[76] All five infants survived their septic episodes. Cultures from isolettes and a doorknob were positive for *E. hormaechei.*

Bacteremia caused by *Enterobacter* species in pediatric patients is similar in most ways to that observed in adults.[44,45,77] Most bacteremias (55%–67%) are nosocomial infections and occur in children with underlying health problems, including liver disease, cancer, other immunodeficient conditions, bronchopulmonary dysplasia, neurologic deficits, and short bowel syndrome.[44,45,77] In the study of Andresen and associates,[46] 3 of 4 children who developed septicemia caused by *E. aerogenes* had short bowel syndrome. Risk factors associated with *Enterobacter* sepsis in children include preceding antibiotic therapy and catheterization, particularly with a central venous catheter.[44,45] The overall mortality rate in pediatric patients is 6% to 18%.[44,45,77] This is lower than the rate observed in adults, perhaps partially because of less serious underlying disorders in children. Although the prevalence of *Enterobacter* bacteremia is low (0.4–0.8 episodes per 1000 admissions), data suggest that the frequency of *Enterobacter* septicemia is on the rise.[44,45,77,78] Between 1989 and 1992, *Enterobacter* replaced *Klebsiella* as the most commonly isolated gram-negative rod at one children's hospital.[44] During those years, it accounted for 33% of all gram-negative nosocomial bacteremias. Similar conclusions were drawn by Gürses,[78] who found *Enterobacter* to be the most common nosocomial pathogen involved in septicemia in neonates and children in two institutions in Turkey. These findings present an ominous warning regarding a possible dramatic increase in the frequency of *Enterobacter* sepsis on a global basis.

Endocarditis is a rare infectious sequela of *Enterobacter* bacteremia.[61] Tunkel and collaborators[79] reviewed 18 cases of *Enterobacter* endocarditis reported since 1949. As with bacteremia, most patients were male, and 67% of all those with endocarditis had underlying cardiac anomalies (eg, rheumatic heart disease, prosthetic valves). The mitral valve was the site most commonly infected. The overall mortality rate was 44%, although this value was substantially lower (23%) for cases described since 1973.

Central Nervous System

A rare but severe sequela of *Enterobacter* bacteremia is central nervous system (CNS) disease, especially meningitis. *Enterobacter* meningitis occurs in both children and adults, although it is a far more serious complication in the former group. At least 21 cases of *Enterobacter* meningitis in pediatric patients have been described in the literature since 1979.[80–82] Most (67%) occurred in neonates who were predisposed to developing systemic diseases because of prematurity, low birth weight (<2500 g), or prolonged rupture of membranes before delivery.[80,81,83] In some instances, however, *E. sakazakii* meningitis developed in normal infants who received large infectious doses of bacteria through contaminated formula.[81] The typical clinical picture observed in children presenting with *Enterobacter* infection is an ill-appearing infant with one or more nonspecific symptoms suggestive of possible meningitis associated with gram-negative organisms. Such symptomatology commonly includes poor feeding, lethargy, irritability, jaundice, apnea, and fever. Other early manifestations of meningitis are a bulging or full anterior fontanelle, hypertonia, twitching, convulsions, and seizures.[80–83] In many cases (about 40%), *Enterobacter* meningitis is accompanied by secondary lesions in the brain.[80] These hypodense areas, as detected by computed tomography, often mimic brain abscess formation with ring enhancement.[80–83] However, these lesions usually represent sterile cerebral infarctions consisting of liquefied white matter. Laboratory analysis of cerebrospinal fluid from infected infants generates a typical profile indicative of meningitis,[84] namely pleocytosis (leukocytes >10/mm^3), elevated protein (>45 mg/dL), and depressed glucose concentration (one half of the serum glucose level or less). In many instances, both the blood and the cerebrospinal fluid are positive for *Enterobacter,* although either specimen may be negative. The most common species associated with neonatal meningitis is *E. sakazakii,*[80,82,85] but cases of *E. cloacae* meningitis have also been described.[84] A portal of entry for such illnesses is usually not apparent.[83] However, one series of three Icelandic infants who developed *Enterobacter* meningitis had their infections epidemiologically traced to powdered milk.[80] Such CNS disease also can arise endogenously, because a 24-day-old infant was unequivocably shown through molecular epidemiologic techniques to have developed his meningitis from prior colonization of the gastrointestinal tract.[84] Despite aggressive medical intervention including multiple antibiotic regimens, the case fatality rate in infants remains exceedingly high. Most children who survive *Enterobacter*-associated meningitis (94%) have long-term neurologic deficits, in many instances attributable to secondary cerebral infarcts. Commonly noted problems include impaired visual and hearing responses, quadriplegia, and developmental delays. Meis and associates[86] used meropenem to cure a brain abscess in a 7-year-old boy with underlying lymphoblastic leukemia, without apparent residual sequelae.

In contrast to pediatric infections, *Enterobacter*-associated meningitis in adults is a less serious disease. Wolff and colleagues[87] reported on 13 cases of *Enterobacter* meningitis occurring at their medical center and summarized an additional 33 cases reported since 1980. Over a 9-year period, *Enterobacter* was the second most common gram-negative organism isolated from cerebrospinal fluid at one teaching hospital.[87] Most patients with meningitis present with fever, obtundation, and/or decreased mental status. Most have a recent history of trauma or neurologic procedures. The most common species associated with adult cases are *E. cloacae* and *E. aerogenes.* Approximately half (54%) of such patients have had *Enterobacter* isolated from sites other than cerebrospinal fluid or blood before or simultaneous with the onset of meningitis. Such sites include sputum, wounds, pleural fluid, catheters, prenasal sinus, and a subgaleal collection. Wolff and colleagues[87] also reviewed various therapeutic modalities used in the treatment of adult cases of *Enterobacter* meningitis. They found that trimethoprim-sulfamethoxazole, given alone or in combination with a β-lactam agent, was more likely to achieve a clinical cure without development of resistance than was a β-lactam agent given alone or in combination with intraventricular administration of an aminoglycoside. The authors suggested that, although trimethoprim-sulfamethoxazole is not considered a first-line agent in the treatment of bacterial meningitis, it may be a valuable alternative, particularly in cases in which in vitro resistance to β-lactam agents is initially detected in infecting strains.[87] Similarities and differences in the presentation of *Enterobacter* meningitis in children and adults are presented in Table 7.

Pulmonary System

There is conflicting evidence regarding the significance of the isolation of *Enterobacter* species from respiratory tract specimens. In the landmark review by John and associates,[53] only two cases of *E. cloacae* pneumonia could be found. Crane and Lerner[88] later stated that, although the incidence of community-acquired *Enterobacter* pneumonia was very low, the rate of nosocomially transmitted infections (ie, in hospitals, nursing homes) was much higher (5%–14.3%). The most recent NNIS study, published in 1991, lists *Enterobacter* species as the cause of 11% of all nosocomially acquired pneumonias.[51] A 3-month investigation conducted at a Veterans Administration Medical Center in 1991 found respiratory secretions to be the second most common anatomic site from which *E. aerogenes* was recovered.[44] Most isolates in this study (67%) were found to represent infection. Characteristics of persons colonized or infected with *E. aerogenes* included a mean age of 65 years; recent hospital exposure (88%); underlying chronic disease such as cardiorespiratory or urologic disease, diabetes, or alcoholism (91%); and prior invasive procedures (79%). Füssle and coworkers[89] reported on 38 mechanically ventilated patients with pulmonary infections caused by *E. cloacae* who were admitted to surgical or neurosurgical units at the University of Giessen. Using a number of criteria to establish infection (eg, fever, leukocytosis $>10^5$/mL, *E. cloacae* from tracheobronchial secretions, chest radiography results), all 38 persons were judged to have pulmonary disease caused by *Enterobacter*. The median age of these patients was 46 years, and the male/female ratio was 3.2:1. All had underlying health problems such as multiple trauma/head injuries (58%), postoperative complications (34%), and peritonitis (8%). Treatment with imipenem or ciprofloxacin was successful in 86% to 94% of these cases; those who initially received cefotaxime and/or tobramycin experienced a high failure rate (>70%) because of the development of drug resistance.

Glustein and Kaplan[90] reported an unusual complication of pulmonary infection with *Enterobacter* in a 7-week-old infant. The child, born prematurely and underweight (890 g),

TABLE 7. *Clinical presentation of* Enterobacter *meningitis in children and adults*

Characteristic	Adults	Children
Age (range)	21–69 y	3 d–5 wk
Presenting symptoms	Fever, altered mental status	Poor feeding, lethargy, irritability
Predisposing conditions	Craniotomy, craniofacial trauma	Prematurity, low birth weight
Portal of entry	Unknown	Unknown
Most common species	*E. cloacae, E. aerogenes*	*E. sakazakii*
Secondary sequelae	Infrequent	Common
Mortality	16%	50%

From references 80–83 and 87.

had respiratory distress syndrome requiring intubation and ventilation. At 7 weeks of age, pneumonia with pleural effusion developed and was treated with gentamicin and imipenem. During resolution of the pulmonary disease (14 days of treatment), chest radiographs detected formation of a pneumatocele. Development of this cyst-like lesion required an additional 4 weeks of chemotherapy to obtain marked clinical improvement. A second interesting presentation concerned a 54-year-old man with a retropharyngeal abscess caused by *E. cloacae* and *Candida albicans.*[91] His condition deteriorated despite incision, drainage, and treatment of the initial abscess, with subsequent development into a deep neck abscess and further extension into the right pleural and pericardial cavities. By day 18 of hospitalization, abscess formation in the right pleural cavity due to *E. cloacae* was observed. Resolution of the descending necrotizing mediastinitis required drainage, extensive irrigation, antimicrobial chemotherapy (amikacin), and prolonged hospitalization (52 days).

Wounds

Although not commonly considered wound pathogens, *Enterobacter* species can be involved in polymicrobic infections of the skin, muscle, subcutaneous tissues, and fascia. Early studies recorded a prominent role for enterobacteria in infections in burn patients. Various investigators found colonization rates in such patients to range from 10% to 50%.[53] Colonization often led to infection of necrotic tissues, in many cases ending in fatal septicemic episodes. However, the role of *Enterobacter* species in such illnesses now appears to be on the decline. A 1995 microbiologic survey of more than 100 persons with or without a history of IV drug abuse isolated *Enterobacter* from 2% to 3% of all skin and soft tissue abscess specimens.[92] More invasive wound disease, such as necrotizing fasciitis or synergistic necrotizing cellulitis, often involves mixed infections with strict anaerobes such as *Bacteroides* or *Peptostreptococcus.*[93] In these illnesses, facultative bacteria such as *Enterobacter, Klebsiella,* or *E. coli* play an accessory role in the rapid spread of a painful and necrotic process. *E. aerogenes* has also been isolated from pancreatic abscesses in conjunction with anaerobes.[94]

The key to understanding what role the genus *Enterobacter* plays in wound microbiology stems from reports of strains recovered in pure culture from infections. There are few such citations in the literature. A case of crepitant cellulitis due to *E. aerogenes* in a 78-year-old diabetic man was reported by Michowitz and colleagues.[95] Subcutaneous crepitation of the entire left leg was observed, and radiographic studies indicated extensive subcutaneous gas formation. A Gram stain of the wound yielded purulent material (many leukocytes) and gram-negative bacilli; only *E. aerogenes* was isolated. A milder case of *Enterobacter* cellulitis was reported in 1992 by Ganelin and Ellis.[96] The illness arose in a 58-year-old doctor who developed a hematoma of the right toe after playing tennis. The initial trauma did not respond to drainage and debridement, and culture of the infected area grew *E. cloacae* as the predominant flora (with a trace amount of staphylococci). Treatment with imipenem on day 5 of the illness brought about rapid resolution of the symptomatology. *E. cloacae* was isolated as the sole pathogen in a case of suppurative pyomyositis in a 26-year-old man who received a blunt trauma to the right gluteal muscles.[97] The patient presented with increasing pain in the right buttocks area, fever to 40°C (104°F), and a leukocyte count of 15,000/mm^3. A computed tomography–guided needle aspirate of the abscess cavity showed polymorphonuclear infiltration and muscle necrosis on histologic analysis. *E. cloacae* was recovered from muscle tissue. *E. cloacae* also was reported as the cause of fatal gas gangrene in a 61-year-old man with pancytopenia.[98] A final unusual skin manifestation of *E. cloacae* infection, formation of hemorrhagic bullae in conjunction with sepsis, was reported by Livingston and colleagues.[99] Hemorrhagic bullae, which are more commonly associated with septicemia caused by *P. aeruginosa* or members of the family *Vibrionaceae,* erupted on the right leg of a lethargic 75-year-old man with chronic cardiac problems and a 5-month history of bullous pemphigoid. The latter condition had been controlled with a topical corticosteroid preparation and prednisone. *E. cloacae* was isolated from blood and bullae cultures, and diverticulosis was suspected as the original source of the infection. The patient was treated with ciprofloxacin, to which he responded favorably, although he continued to have other medical problems. In 1997, Abbott and Janda[72] described four *E. cancerogenus* wound infections in persons who had sustained severe trauma or crush injuries. All strains were recovered in pure culture. Infections ranged from mild cases of cellulitis to infected hematomas and chronic draining processes.

Urinary Tract Infection

Enterobacter is one of several genera in the family *Enterobacteriaceae (Citrobacter, Morganella)* that are occasionally associated with urinary tract infections (UTIs) in adults and children.[53,100] Prevalence studies conducted in the United States[101] and Spain[102] estimate the overall frequency of enterobacters in positive urine specimens to be between 1.7% and 1.9%. Probably most people with UTIs caused by *Enterobacter* species have underlying urologic anomalies or have recently undergone one of a variety of medical procedures. A 2-year, multicenter study from Europe, the Middle East, Asia, and South America found *Enterobacter* to be the cause of 5% of all serious nosocomially acquired upper UTIs in adults[103]; ceftazidime, either alone or in combination with tobramycin, eradicated all of these infections. Stamm and associates[104] studied women with recurrent UTIs. During intervals when patients were not receiving antimicrobial chemotherapy, *Enterobacter* species caused 5% of asymptomatic bacteriurias, 4% of all cases of acute pyelonephritis, and 0.2% of acute cystitis infections. Catheter-associated bacteriuria infections in adults frequently yield *Klebsiella* or *Enterobacter* (15%–21%) in varying concentrations.[105] Despite evidence that *Enterobacter* can be an important urinary tract pathogen, there is a paucity of reports documenting this fact. However, Reina and associates, in 1989,[106] described a case of UTI caused by *E. cancerogenus ("E. taylorae").* The patient, a 70-year-old man with urinary lithiasis, was seen in an emergency room with complaints of dysuria and hematuria. A urine culture grew *E. cancerogenus* at more than 10^5 cfu/mL. He was treated with trimethoprim-sulfamethoxazole and released. Although individual case reports have not been published, both *E. gergoviae* and *E. asburiae* have been recovered on occasion from urine.[11,15]

Bone and Joint Disease

Several reports have been published that describe *Enterobacter* infections of the bone. A case of osteomyelitis involving the leg occurred in an 18-year-old man who sustained a serious fracture to the left femur in an automobile accident.[107] *E. cancerogenus ("E. taylorae")* was isolated from the bone. Despite therapy with multiple antibiotics (cefotaxime, tobramycin), the infection failed to resolve and chronic osteomyelitis developed. A second case of *E. cancerogenus*–associated osteomyelitis was reported in a 52-year-old man who sustained a major trauma to his tibia (open fracture). The infection required 14 months to resolve completely.[108] Three cases of vertebral osteomyelitis caused by *E. cloacae* have been published lately.[109–111] These involve the L1–L2 vertebrae,[109] the D8–D9 dorsal vertebral bodies,[110] and the L4–L5 region.[111] In only one report,[104] that of a 28-year-old man with human immunodeficiency virus infection, was a potential source for the bone infection identified (IV self-inoculation). Diagnosis in most cases of *Enterobacter*-associated osteomyelitis is achieved through radiologic studies and needle biopsy of areas exhibiting enhanced uptake. All three patients recovered after appropriate medical treatment. *E. cloacae* also caused hemorrhagic bursitis of the right knee in an elderly woman.[112] Bursitis developed after she fell on her knees in the street. Her infection was cured by surgical drainage and chemotherapy which included cefotaxime and netilmicin.

Miscellaneous Infections

A number of other diseases have occasionally been associated with infections caused by the genus *Enterobacter. Enterobacter* species have been isolated from the bile (7%–37%) of persons with common bile duct stones, biliary strictures, or a biliary endoprosthesis[61,113]; approximately 4% of these patients with acute cholangitis were positive for *Enterobacter* in their blood. One European investigation found *Enterobacter* to be the third most common microbe (after *Enterococcus* and *E. coli*) recovered from 25 biliary stents removed during endoscopy from 17 patients.[114] *E. cloacae* was isolated from ascites in a woman with peritonitis caused by obstruction of the small bowel.[115] In rare instances, *Enterobacter* species are involved in ocular disease.[61,116,117] Infections such as corneal ulceration[116] and endophthalmitis[117] typically appear in elderly persons undergoing cataract surgery. These ocular infections can be extremely difficult to treat and can result in serious permanent complications.[117]

Sanders and Sanders[61] described a number of clinical syndromes mimicked by *Enterobacter* species that traditionally were associated with other pathogens. Such syndromes include ecthyma gangrenosum and acute purpura fulminans.

LABORATORY IDENTIFICATION

Isolation

Because their recovery from environmental sources and the gastrointestinal contents of humans has no apparent significance, development of enrichment techniques and selective media designed for the specific isolation of *Enterobacter* from such samples has not been attempted. The majority of *Enterobacter* isolates are both sucrose-positive (with the exception of *E. cancerogenus*) and lactose-positive (with the exceptions of *E. cancerogenus* and *E. hormaechei*). Therefore most species, such as *E. cloacae* and *E. aerogenes,* appear as pink (lactose-positive) or yellow (sucrose-positive) colonies on standard selective media containing these sugars (eg, MacConkey agar, Hektoen enteric agar). On nonselective media such as trypticase soy or nutrient agars, *E. sakazakii* and many isolates of *E. agglomerans* produce a nondiffusible yellow pigment. The intensity of this pigment varies depending on the temperature of incubation (stronger at 25°C than at 35°C) and the growth medium. More esoteric media, such as corn oil lipase agar, enhance pigment production by these two groups when incubated at 25°C.

Identification

Commercial Systems

Enterobacter is one of the most difficult groups to correctly identify to species using commercial automated, semiautomated, or manual identification systems.[118–130] Cumulative results on the accuracy of several such systems are listed in Table 8. Commercial systems in general do not identify species currently residing in the genus *Enterobacter* to a satisfactory degree (ie, to >90% accuracy). An exception to this rule may be the RapID onE system (Innovative Diagnostics Systems, Norcross, GA), which correctly identified 57 (95%) of 60 isolates belonging to one of nine *Enterobacter* species (*E. hormaechei* is not included in the RapID onE database). However, only one major study working with this system has been published,[125] and conclusions regarding its overall accuracy await further rigorous challenges. Most manual systems (API 20E, API BioMérieux, Hazelwood, MO; Crystal E/NF, Becton Dickinson, Cockeysville, MD; RapID onE) appear to be better equipped to identify *Enterobacter* species accurately than fully automated panels such as the MicroScan WalkAway and the Vitek units.

TABLE 8. *Identification of* Enterobacter *species by commercial systems*

System	No. Tested	*E. cloacae*		*E. aerogenes*		*E. agglomerans*		*E. sakazakii*		Other Species[a]	
		CI[b]	Inc[c]	CI	Inc	CI	Inc	CI	Inc	CI	Inc
MicroScan W/A	269	89[d]	11	87	13	73	27	67	33	60	40
API 20E	259	85	15	90	10	74	26	67	33	91	9
Vitek	235	86	16	100	0	50	50	100	0	72	28
Crystal ID E/NF	151	87	13	91	9	63	37	100	0	100	0
Biolog GN	80	32	68	28	72	77	23	47	53	75	25
RapID onE	60	100	0	100	0	100	0	100	0	82	18
Cobas	35	50	50	100	0	100	0	67	33	36	64

From references 118–130.
[a]Includes *E. cancerogenus, E. asburiae, E. gergoviae,* and *E. hormaechei.*
[b]CI, correct identification (includes those correct with low probability).
[c]Inc, incorrect (includes those with no identification).
[d]Percent of strains falling into each category.

One explanation for these differences may be the longer incubation periods used by most manual systems, which allows for induction of substrate-specific enzymes. However, two newer systems from Europe (ATB Rapid ID 32E, ATB ID 32E, bioMérieux, France) performed poorly in the accurate identification of 118 *E. cloacae* strains in one Norwegian study.[131] The Biolog system, which can be read manually or with an automatic reader, fared poorly in identification of *Enterobacter* species, with the exception of *E. agglomerans.* Holmes and colleagues[126] found that only 7% to 20% of *E. aerogenes* strains were correctly identified after 24-hour incubation on the Biolog GN (Biolog, Hayward, CA) system, whether the panel was read manually or automatically. Miller and Rhoden[118] noted that many *Enterobacter* species were simply too active on this panel, leading to misidentifications. The two major automated systems (MicroScan, Vitek) exhibited very little difference in their overall accuracy of identification of *Enterobacter* species. The Vitek system was superior to MicroScan in identifying *E. aerogenes* and *E. sakazakii,* and MicroScan was better for *E. agglomerans.* Kelly and Leicester[121] found that the WalkAway system produced a disproportionate number of identification errors with *E. aerogenes* and *E. cloacae* in regard to their overall frequency in the test sample population.[121] Most errors generated by both systems resulted in the assignment of an isolate to two or more phenotypically similar groups or in misidentification as another *Enterobacter* species or one of several genera in the *Enterobacteriaceae* such as *Klebsiella, Citrobacter,* or *Serratia.*[119,123,128,130] Both automated and manual systems correctly identify most *E. gergoviae, E. asburiae,* and *E. cancerogenus* isolates, although one study[124] noted specific problems regarding *E. asburiae* (Vitek) and *E. cancerogenus* (MicroScan).

Conventional Media

Because of the taxonomic uncertainties that exist within this group (see Taxonomy), it is currently impossible to construct accurate defining criteria for inclusion in the genus *Enterobacter.* Until systematic issues are resolved (eg, where such species as *E. aerogenes* and the *E. agglomerans* complex should reside), the genus will continue to be defined in imprecise terms. Table 9 lists some of the major biochemical features that have traditionally been associated with inclusion in the genus *Enterobacter.* For many of these tests, one or more *Enterobacter* species deviate from the ideal phenotype for the genus. Besides the properties listed in Table 9, most *Enterobacter* isolates are indole-negative and *o*-nitrophenyl-β-D-galactopyranoside (ONPG)–positive, reduce nitrates, and ferment D-mannitol, maltose, trehalose, and cellobiose.[132]

TABLE 9. *Present defining biochemical reactions for the genus* Enterobacter[a]

Test	Phenotype[b]	Exceptions[c]
Voges-Proskauer	+	*E. asburiae, E. kobei*
LDC	–	*E. aerogenes, E. gergoviae*
ODC	+	*E. amnigenus* biogroup 1
Citrate utilization	+	*E. intermedium, E. amnigenus* biogroup 1
Acid from		
L-Arabinose	+	
L-Rhamnose	+	*E. asburiae*
Raffinose	+	*E. asburiae, E. cancerogenus, E. hormaechei*
D-Xylose	+	
Lipase	–	
Gelatinase	–	

[a]Clinically relevant species or biogroups (excluding *E. agglomerans* complex).
[b]Reactions at 24–48 h.
[c]>10% of strains deviating from the phenotype.
LDC, lysine decarboxylase; ODC, ornithine decarboxylase.

A majority of *Enterobacter* isolates can be identified to species with the use of a small number of additional biochemical characteristics including amino acid decarboxylases and sugar fermentation tests (Table 10). *E. intermedium,* which is phenotypically similar to *E. cloacae,* can be separated from the latter species by its failure to elaborate ADH, a positive methyl red test, and failure to grow at 41°C.[74,132]

Three *Enterobacter* species have not been isolated from clinical material to date. Two of these, *E. nimipressuralis* and *E. dissolvens,* phenotypically resemble *E. cloacae. E. nimipressuralis* can be separated biochemically from *E. cloacae* based on its negative sucrose and raffinose reactions. *E. dissolvens* and *E. cloacae* cannot be separated phenotypically. Whether *E. dissolvens* should remain a distinct species or be grouped together with *E. cloacae* remains to be determined.[15] The third species, *E. pyrinus,* resembles *E. gergoviae.* It can be easily distinguished from all other enterobacters by its ability to produce H_2S on triple sugar iron slants.[17]

Biogroups and Genomospecies

Because of the increasing association between *Enterobacter* species and nosocomial

TABLE 10. *Identification of* Enterobacter *isolates to species*

Organism	Decarboxylase Group[a]	Production of Acid: Sucrose	D-Sorbitol	Melibiose	Dulcitol
E. amnigenus[b]	1	+	–	+	–
E. asburiae	1	+	+	–	–
E. cancerogenus	1	–	–	–	–
E. cloacae, E. kobei[c]	1	+	+	+	V
E. hormaechei	1	+	–	–	+
E. sakazakii	1	+	–	+	–
E. aerogenes	2	+	+	+	–
E. gergoviae	2	+	–	+	–
E. intermedium	3	V	+	+	+
E. agglomerans	4	V	V	V	V

[a]Lysine decarboxylase (LDC), arginine dihydrolase (ADH), and ornithine decarboxylase (ODC) pattern: group 1 (–,+/V,+/V), group 2 (+,–,+), group 3 (–,–,+), group 4 (–,–,–).
[b]Biogroup 1 strains only.
[c]*E. kobei* is Voges-Proskauer–negative.
+, >85% of strains positive; –, >85% of strains negative; V, variable (15%–85%) positive.

Color Plates

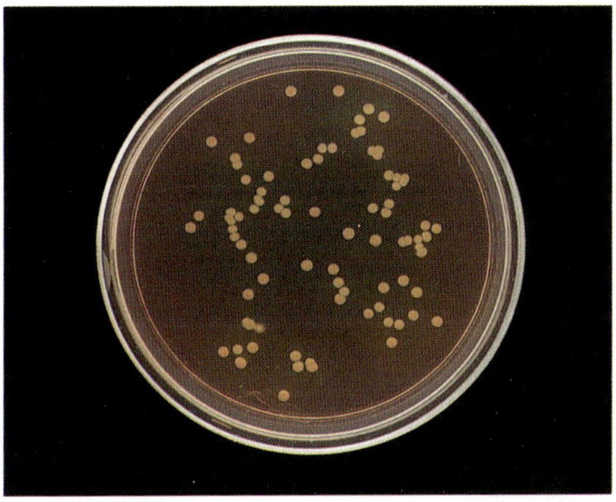

A

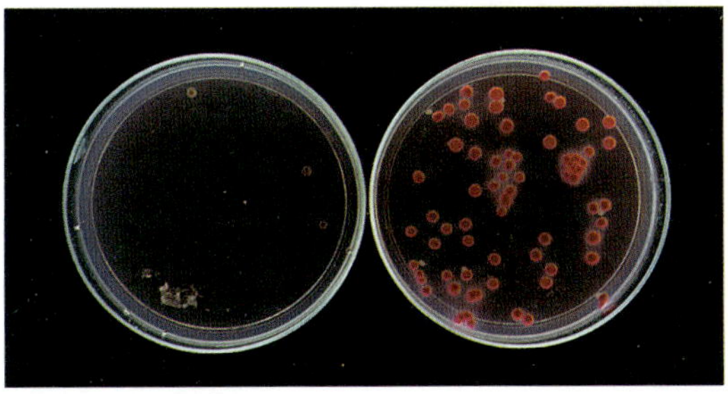

B

Color Plate 1. A, *E. coli* O157:H7 on SMAC displaying sorbitol-negative colonies. **B,** CT-SMAC showing inhibition of normal *E. coli* flora with several sorbitol-negative O157:H7 colonies *(left)*; SMAC alone *(right)*.

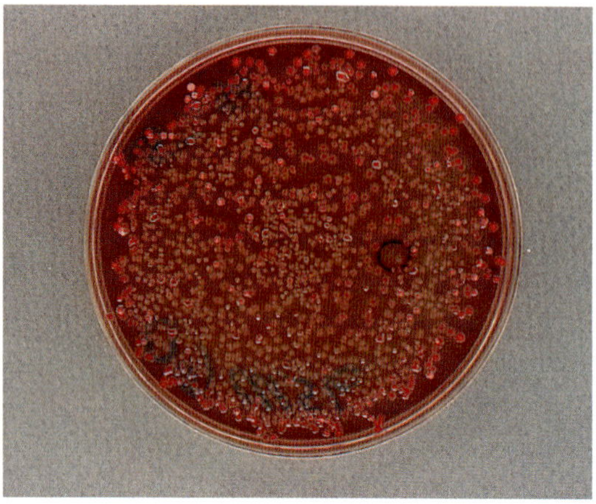

A

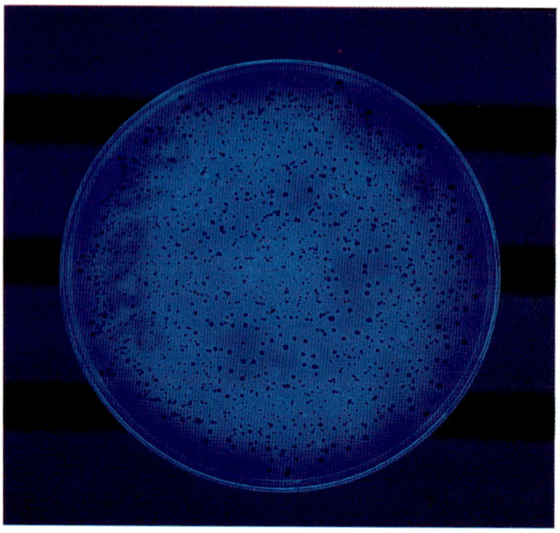

B

Color Plate 2. A, SMAC agar with β-D-glucuronide without fluoresence; circle marks *E. coli* O157:H7 colony. **B,** SMAC agar with β-D-glucuronide; O157:H7 colony masked by fluorescence of fecal *E. coli.*

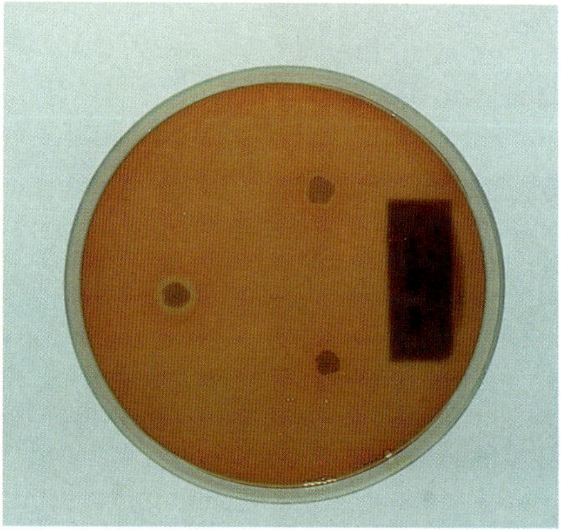

Color Plate 3. Blood agar plate with washed sheep RBCs and $CaCl_2$; enterohemolysin-positive, Stx+ *E. coli (top)*; α-hemolytic Stx– *E. coli (left)*; nonhemolytic Stx– *E. coli (bottom).*

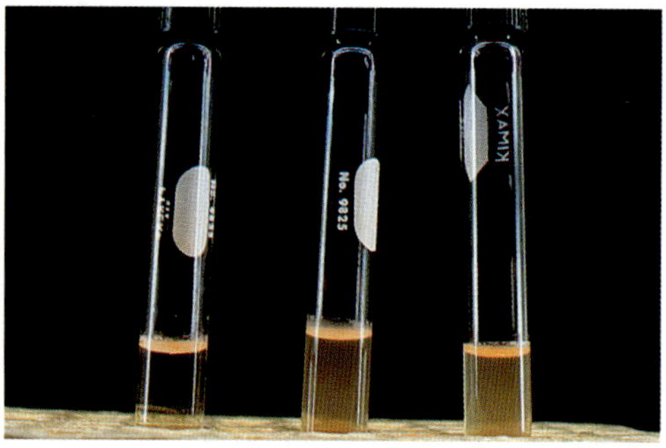

Color Plate 4. Pellicle formation by three EAggEC strains in Mueller-Hinton broth; strongest reaction is indicated at far left (with autoagglutination); weakest reaction at right.

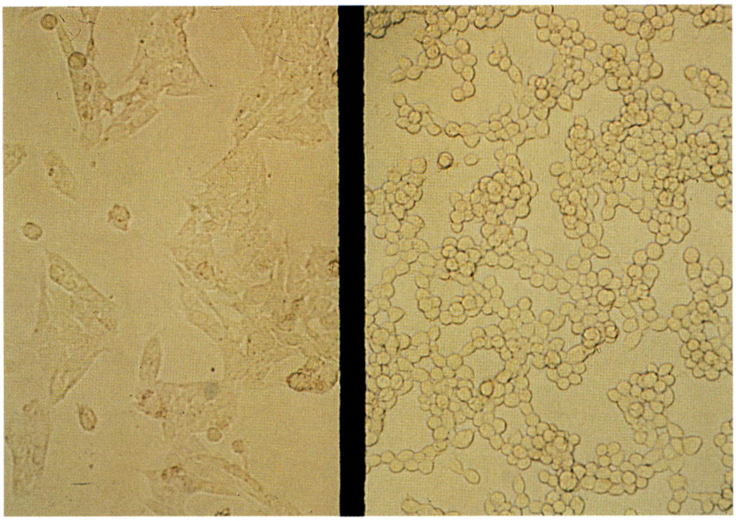

Color Plate 5. Normal Y1 cell sheet *(left)*; Y1 cells exposed to heat-labile (LT) enterotoxin for 24 hours *(right)*.

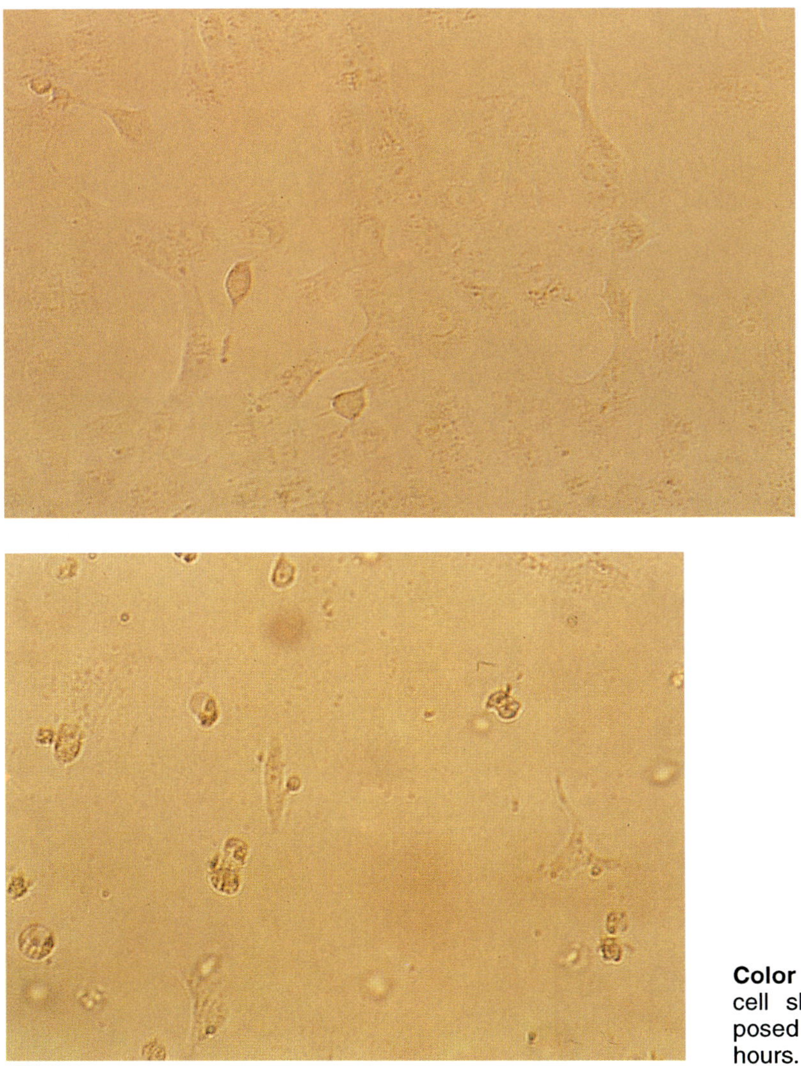

A

B

Color Plate 6. A, Normal Vero cell sheet. **B,** Vero cells exposed to Shiga toxin (Stx) for 48 hours.

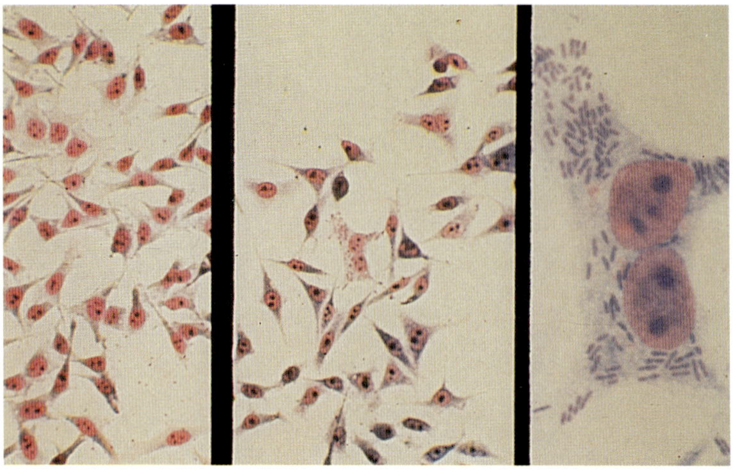

Color Plate 7. Uninoculated HEp-2 cell sheet *(left)*; HEp-2 invaded by EIEC, ×20 *(center)*; closeup view of invaded HEp-2 cells showing multiplication of cells intracellularly *(right)*.

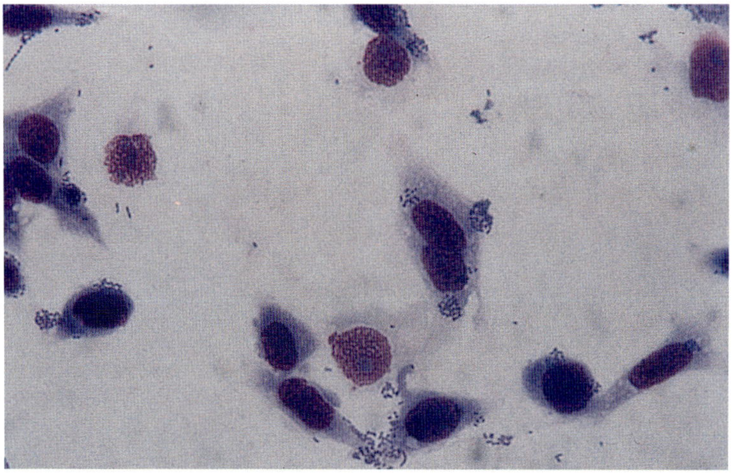

Color Plate 8. EPEC showing localized adherence to HEp-2 cells.

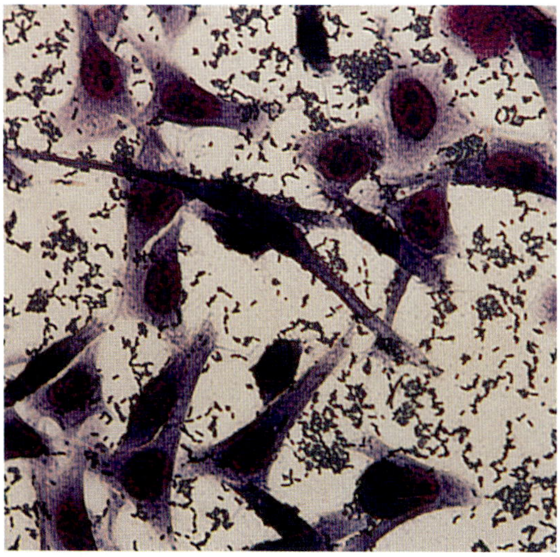

Color Plate 9. EAggEC showing aggregative adherence to HEp-2 cells.

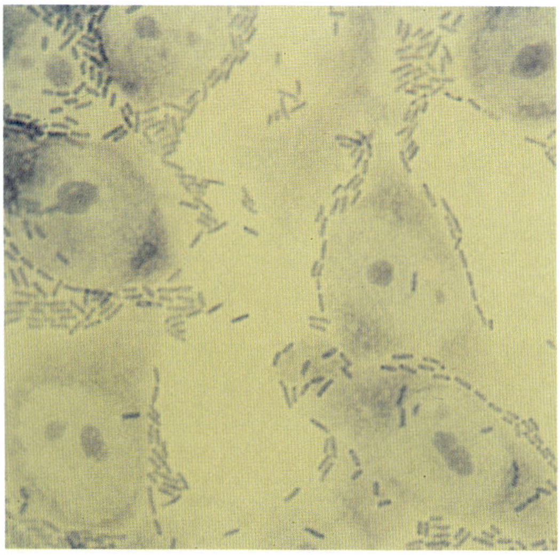

Color Plate 10. DAEC showing diffuse adherence to HEp-2 cells.

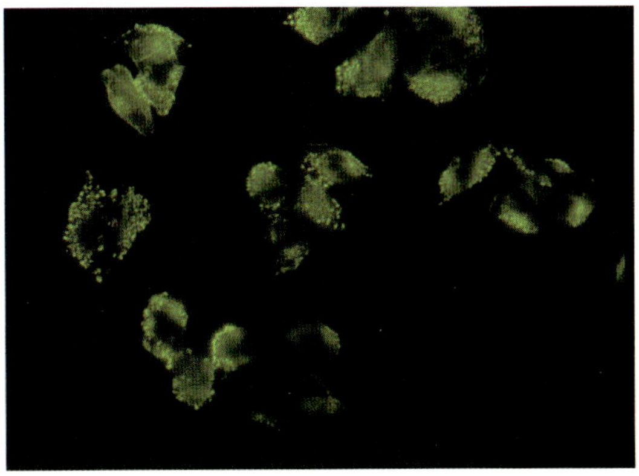

Color Plate 11. Fluorescent actin stain (FAS) of EPEC-infected HEp-2 cells showing bundles of actin filaments.

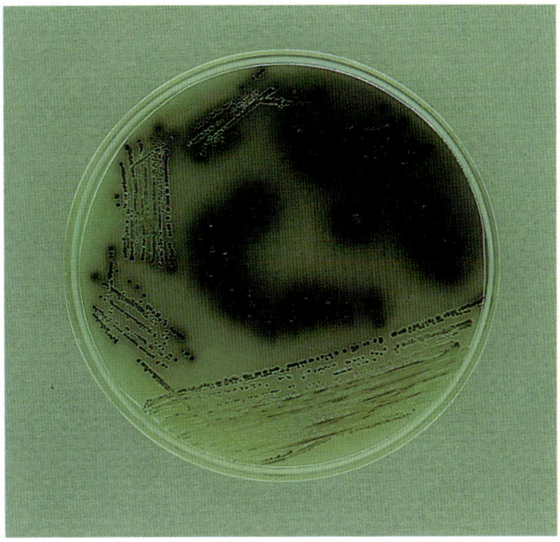

Color Plate 12. *Salmonella* on bismuth sulfite agar.

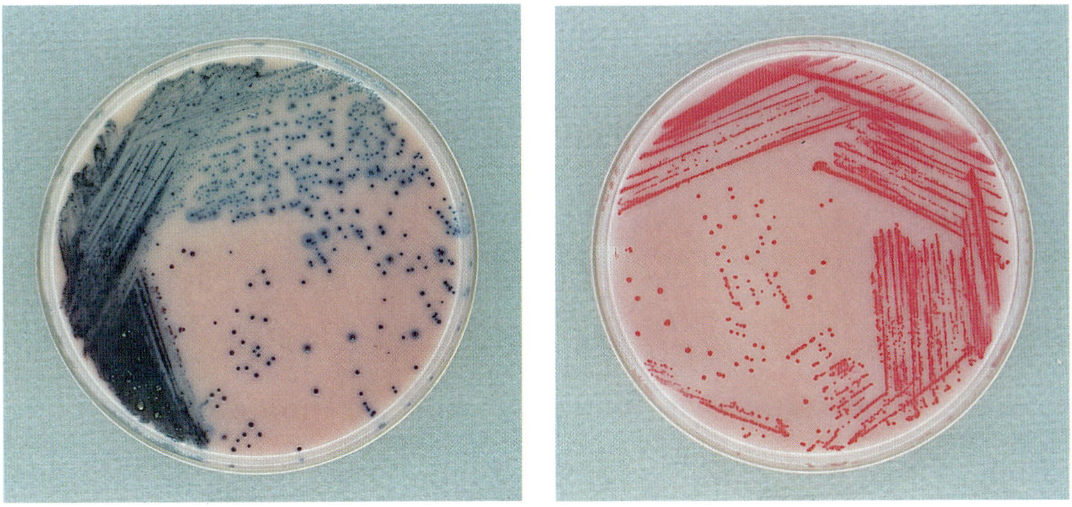

A

B

Color Plate 13. A, *Citrobacter* on Rambach agar. **B**, *Salmonella* on Rambach agar.

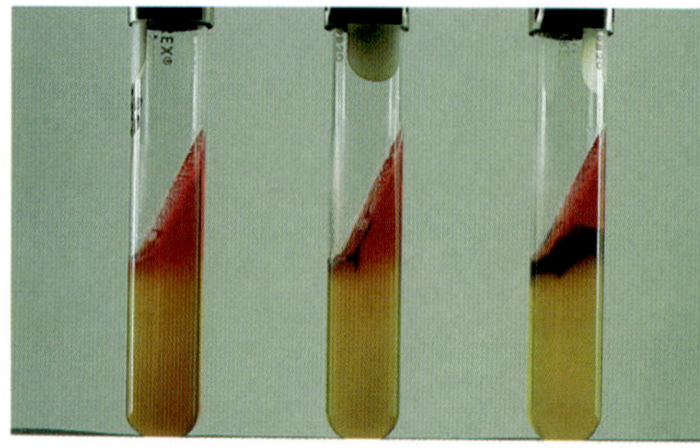

Color Plate 14. Production of H_2S by *S.* Typhi in TSI.

Color Plate 15. Unusual pigment production by *S. rubidea.*

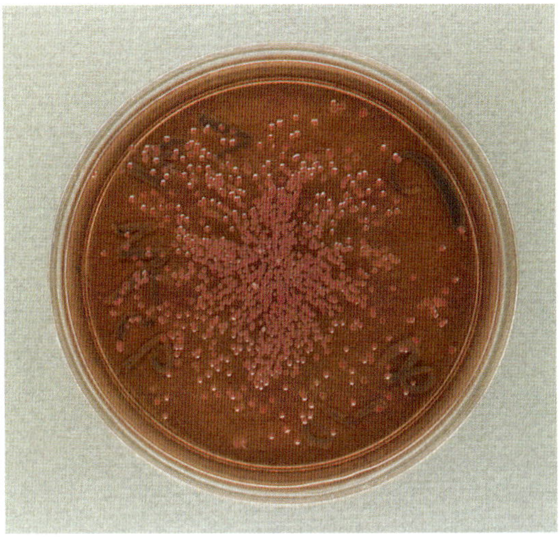

Color Plate 16. *Citrobacter* on CIN agar.

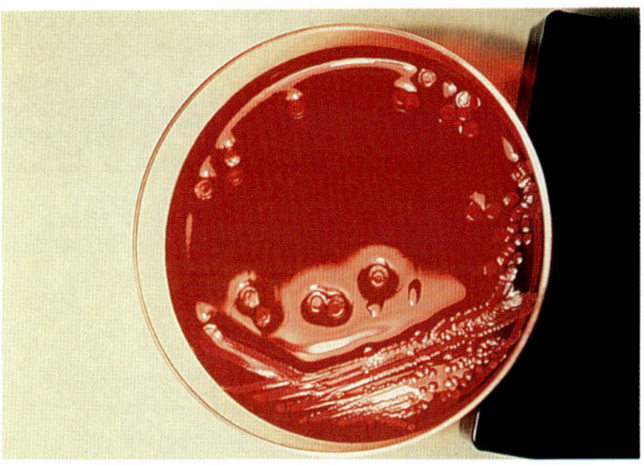

Color Plate 17. Pectinase production by colonies of *Y. enterocolitica* causing pitting (dissolving) on pectin agar.

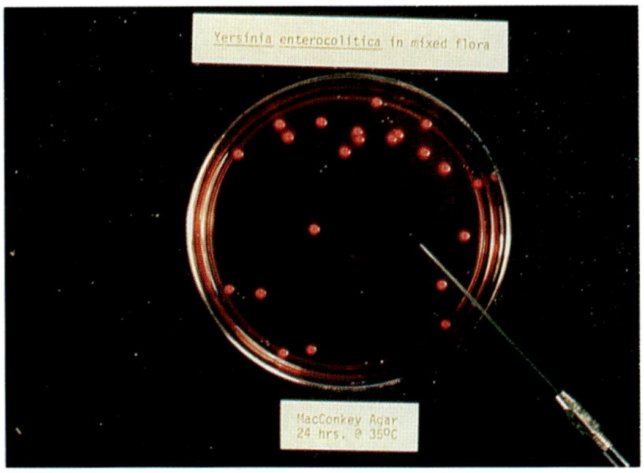

Color Plate 18. Pinpoint colonies of *Y. enterocolitica* on MacConkey agar.

Color Plate 19. Darkening of TSI slant by *Morganella.*

Color Plate 20. Production of reddish-brown pigment on DL-tryptophan slants by *Morganella* after prolonged incubation.

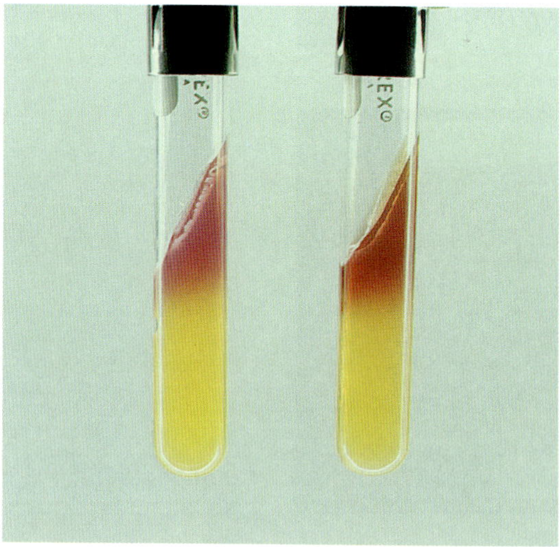

Color Plate 21. Deamination of lysine (red slant) by *Providencia (right tube).*

infections, biotyping schemes have been developed as a first approximation of strain relatedness in outbreak situations. The most common system used to assess strain identity has been the API 20E strip (Vitek bioMérieux, Hazelwood, MO). Gaston and associates[133] analyzed API 20E profiles for 790 isolates of *E. cloacae* not related to outbreaks. Thirty-six profiles were found, with one biotype (API profile 3305573) accounting for two thirds of all isolates. Together with a second, less common profile (3305773), it accounted for almost 80% of all isolates. These results have generally been confirmed in other, less extensive investigations.[40,44,55,131,134] Biochemical properties on the API 20E strip found to most often exhibit strain-to-strain variability include ADH and gelatinase activity and fermentation of inositol. Because the API 20E strip yields two predominating biogroups, biotyping of *E. cloacae* isolates by this method has often been unsuccessful. However, Haertl and Bandlow[45] found 11 of 12 *E. cloacae* strains in a neonatal ICU outbreak to be linked by an unusual API 20E profile (1305573). These ADH-negative isolates were subsequently confirmed to be part of the same genetic cluster by other molecular typing techniques. De Champs and colleagues[135] used both the API 20E and API 50 CH galleries to biotype 23 imipenem-resistant and imipenem-susceptible *E. aerogenes* strains. Although the API 20E gallery could identify only two different biotypes, the 50 CH system recognized five groups based on dulcitol, inulin, 5-ketogluconate, and citrate reactions. The Vitek system has also been used to biotype 62 *E. aerogenes* isolates.[44] Although 12 patterns could be distinguished, two profiles accounted for 80% of all strains. As with the API 20E system, the overall predominance of a few biogroups makes these galleries impractical under most situations for biotyping of strains.

Old[136] proposed a biotyping system for *E. cloacae* using seven differential tests on conventional media. Of 110 isolates typed, 86 (78%) resided in biotype 4, again emphasizing the lack of discriminatory value that most phenotypic typing schemes have. A biotyping system for most genomic group 3 strains (phenotypically *E. cloacae*) was proposed by Grimont and Grimont.[2] This scheme relied on the differential utilization by *E. cloacae* of eight carbohydrates: D-melibiose, D-raffinose, D-arabitol, fucose, methyl-α-D-galactoside, sorbitol, adonitol, and 3-methylglucose. Based on these tests, seven biogroups (labeled 3a through 3g) were recognized, with most blood isolates of *E. cloacae* falling into groups 3b, 3d, or 3f. In addition to *E. cloacae* biotypes, two biogroups of *E. amnigenus* are recognized.[132] Biogroup 1 strains have been recovered from a number of clinical sources such as blood, wounds, feces, the respiratory tract, and a catheter.[73,132] The chief differences between these two groups are listed in Table 11. Two biogroups of *E. kobei* are also recognized.[18] One biogroup is L-pyrrolidonyl aminopeptidase–positive and 5-ketogluconate–negative, and the second exhibits the opposite reactions.

Genomic groups (genomospecies) that currently reside within the *E. cloacae* complex cannot be separated biochemically with any certainty. Studies conducted at the Institut Pasteur indicate that individual groups are partially separable with the aid of carbon source utilization assays.[2] Production of a gluconate dehydrogenase is a unique characteristic associated with genomic group 3. Most isolates of genomic group 4 are malonate-negative; other genomospecies are positive for this trait. Strains contained in genomovar 2 are unable to use xylose as a sole carbon source, but the remaining genomic groups can. Despite these accomplish-

TABLE 11. *Biogroups of* Enterobacter amnigenus

Characteristic	*E. amnigenus* Biogroup 1	*E. amnigenus* Biogroup 2
Sources	Water, humans	Water
Pathogenicity	?	–
Acid from		
Sucrose	+	–
D-Sorbitol	–	+
Raffinose	+	–

ments, it does not appear at present that a straightforward identification scheme to separate members of the *E. cloacae* complex will become achievable in the foreseeable future.

Immunodiagnostics

A serogrouping scheme based on the presence of unique somatic antigens was devised for *E. cloacae* in 1983.[137] Twenty-eight serogroups were originally established after antigenic analysis of more than 300 *E. cloacae* isolates, with serogroups O:3 (21%) and O:8 (13%) predominating. This scheme has since been expanded to include two additional serogroups, for a total of 30.[30] Serogrouping has a greater discriminatory value than biotyping in determining strain relatedness, and at least 85% of all *E. cloacae* are typable.[30] Serogrouping of *E. cloacae* isolates were shown to be useful in tracing outbreaks of β-lactam–resistant *E. cloacae* in a cardiothoracic/vascular surgery unit and in establishing the environmental source (blood gas analyzer) of an outbreak in a neonatal ICU.[40,43] However, serogrouping has several disadvantages. Some outbreaks are caused by nontypable strains,[55,131] and another 10% of isolates autoagglutinate in saline. Therefore, serogrouping has distinct advantages over biotyping but may not be able to stand alone as an epidemiologic typing system in outbreaks due to common serogroups (O:3, O:8, O:13), or when isolates are not typable or spontaneously autoagglutinate.

Typing

Several protocols have been proposed to epidemiologically fingerprint *Enterobacter* strains associated with nosocomial outbreaks. These protocols fall into two major categories, namely traditional assays and newer molecular methods. The Central Public Health Laboratory in London, England, has taken a leadership role in the development of more traditional approaches to typing *Enterobacter* species, particularly *E. cloacae.* In addition to biotyping and serotyping schemes, early studies found that *E. cloacae* isolates could be typed according to their susceptibility to bacteriocins. A 1982 German publication found that only 79% of more than 300 *E. cloacae* isolates were susceptible to 1 or more of 16 different bacteriocins[138]; 52 provisional bacteriocin types were defined in this study. A subsequent report by Bauernfeind and Petermüller[139] achieved a higher rate of typability (>96%) using eight producer strains, which were subsequently deposited with the American Type Culture Collection in Bethesda, Maryland. Bacteriocins produced in this study also showed activity against isolates of other *Enterobacter* species *(E. aerogenes, E. agglomerans, E. sakazakii)* and some enteric bacilli *(Klebsiella, C. freundii, S.* Typhimurium*).* They differed from bacteriocins previously described[138] in their susceptibility to trypsinization (50 μg/mL) and their sedimentation characteristics. Bacteriocin typing has also been applied to *E. aerogenes;* 70 (92%) of 76 isolates could be assigned to 1 of 15 sensitivity patterns based on bacteriocins secreted by one of five producer strains.[140] A phage typing scheme has been constructed for *E. cloacae.* Gaston[141] developed a set of 25 bacteriophages (including some *Klebsiella* bacteriophages) that was able to type 76 (83%) of 92 field isolates. Although this system is able to type serologically rough strains and those that fail to react in specific somatic antisera, it is less reproducible than biotyping, and small variations in the susceptibility of sequential isolates of the same strain to a standard set of bacteriophages has been noted.

A number of molecular methods have been used as aids in determining strain relatedness among clusters of *Enterobacter* isolates associated with human infections. Of this group, plasmid analysis is by far the most widely employed technique.[56] Most strains of *E. cloacae* harbor extrachromosomal elements ranging in molecular mass from 3 to 150 MDa. Because of the frequent plasmid carriage by *Enterobacter* species, plasmid analysis is an extremely useful fingerprinting tool.[40,44,45,56,142] However, Clark and associates[143] cautioned that, because of plasmid instability, alternative methods may be necessary to confirm primary results. Gaston and Warner[144] used multilocus enzyme electrophoresis (MEE) to analyze the electrophoretic mobility of 12 different cellular enzymes for 62 *E. cloacae* strains. Each enzyme generated between two and five different electromorphs. A subset of four enzymes (lactate dehydrogenase, 6-phosphogluconate dehydrogenase, glutamate dehydrogenase, and an unidentified enzyme band) were found useful in typing most *E. cloacae* strains that belonged to serogroups other than O:3/O:8. Although MEE is a valuable technique, it is relatively difficult to develop, technically demanding, and labor intensive. Ribotyping, a common epidemiologic technique used with other gram-negative bacteria, has also been found to be highly discriminatory in distinguishing strain relatedness in *Enterobacter*-associated outbreaks or incidents.[145] It can offer advantages over other molecular strategies (eg, restriction enzyme analysis) in both specificity and ease of gel interpretation (fewer bands).

Over the past several years, a number of investigations have traced outbreaks of disease or monitored the incidence of nosocomial *Enterobacter* infections using various epidemiologic tools (Table 12). All of these studies have employed a combination of traditional methods and molecular techniques (two to five tests per study) to fingerprint epidemiologically related clusters of *Enterobacter* isolates. Antibiograms, although not as sensitive as molecular tests, have been used in almost all investigations as a red flag to alert in-house staff regarding possible outbreaks of drug-resistant *Enterobacter*. Of the remaining classic typing procedures, phage typing appears to be the most discriminatory, based on both the number of phage patterns observed and the frequency distribution of *E. cloacae* isolates within each group. Its major drawbacks include its limited availability (as for serotyping) and the previously noted minor fluctuations in phage patterns detected among multiple isolates of the same strain.[145] In virtually all epidemiologic investigations using a combination of techniques, a molecular method has been found to be superior in determining strain relatedness. Polymerase chain reaction amplification of variable-length regions of genomic DNA between repetitive extragenic palindromic (REP) and enterobacterial repetitive intergenic consensus (ERIC) sequences has been used successfully used to type outbreaks of *E. aerogenes* infections.[44,142] Other methods successfully applied in epidemiologic investigations include ribotyping and pulsed-field gel electrophoresis (see Table 12).

However, the use of molecular technology to fingerprint *Enterobacter* isolates should not be regarded as foolproof. Garaizar and colleagues[145] found that, although ribotyping was slightly more discriminatory than a combination of traditional tests (biotyping, serology, phage typing), it offered little information in addition to that provided by these classic methods. In addition, in 6 (13%) of 45 instances, ribotyping failed to agree with strain designations assigned to individual isolates by the other three methods, suggesting that even a single molecular technique has a finite capability in strain differentiation. An unusual observation was reported by Davin-Regli and associates in France.[148] An outbreak of fluoroquinolone-resistant enterobacteria were identified as *E. cloacae* by both API 20E and random amplification of polymorphic DNA. However, when these strains were subsequently subjected to DNA hybridization, they were genetically identified as *E. hormaechei*. At present, therefore, it seems necessary to use more than one technique to determine strain relatedness among *Enterobacter* isolates linked to outbreak situations.

PATHOGENICITY

Few studies have investigated potential virulence factors associated with the genus *Enterobacter* in human infections. The lack of significant research in this area is probably a reflection of the predominant role of these organisms as nosocomial pathogens. Despite this paucity of information, several factors have been described (Table 13). Many *Enterobacter* species cause the in vitro hemagglutination of

TABLE 12. *Epidemiologic typing of* Enterobacter *species in selected surveys*[a]

Study	Location	Organism	No. Isolates	Traditional Methods				Molecular Methods		Best Assay
				Biotype	Serotype	Antibiogram[b]	Phage	Plasmid	Other	
Andersen et al.[40]	Norway	*E. cloacae*	15	X	X	X	X	X		Plasmid
Acolet et al.[43]	England	*E. cloacae*	42	X	X		X		Rbt	Ribotype
Georghiou et al.[44]	USA	*E. aerogenes*	43	X		X		X	REA, PCR	PCR
Haertl & Bandlow[45]	Germany	*E. cloacae*	62	X		X		X	REA, PFGE	REA/PFGE
Poilane et al.[48]	France	*E. cloacae*	33	X		X			Rbt	Ribotype
Verschraegen et al.[55]	Belgium	*E. cloacae*	11		X		X			Phage
Wang et al.[56]	Republic of China	*E. cloacae*	17			X		X		Plasmid
Neuwirth et al.[142]	France	*E. aerogenes*	20			X		X	PCR, PFGE	PFGE
Clark et al.[143]	USA	*E. sakazakii*	40			X		X	REA, Rbt, MEE	Ribotype/ MEE
McConkey et al.[146]	Ireland	*E. cloacae*	33	X	X	X	X	X		Plasmid
Verweij et al.[147]	France	*E. cloacae*	38	X					PCR	PCR

[a]The symbol "X" indicates the technique used in study.
[b]Includes all antibiotic-associated assays.
Rbt, ribotyping; REA, restriction endonuclease analysis; PCR, polymerase chain reaction; PFGE, pulsed-field gel electrophoresis; MEE, multilocus enzyme electrophoresis.

TABLE 13. *Virulence properties associated with the genus* Enterobacter

Gene Product	Phenotype Frequency	Disease Association	Species	Biologic Functions
MSHA/MRHA	75%	Unknown	*E. amnigenus, E. cloacae, E. sakazakii, E. gergoviae, E. intermedium, E. aerogenes*	Cellular adhesion
Aerobactin	71%	Septicemia/meningitis	*E. cloacae*	Iron acquisition
α-Hemolysin	<1%	Unknown	*E. cloacae*	Cellular cytotoxicity
SLT	<1%	HUS	*E. cloacae*	Cellular cytotoxicity
OmpX	?	Septicemia	*E. cloacae*	Tissue invasion, resistance to complement lysis, drug resistance

From references 149, 150, 153, 154, and 157.
MSHA, mannose-sensitive hemagglutination; MRHA, mannose-resistant hemagglutination; SLT, Shiga-like toxin; HUS, hemolytic-uremic syndrome.

erythrocytes obtained from a number of vertebrate species. These hemagglutinins are either mannose-sensitive (MSHA) or mannose-resistant (MRHA), and their phenotypic expression strongly correlates with production of type 1 or type 3 fimbriae, respectively.[149] Only *E. gergoviae* consistently expresses MRHA. Most *E. cloacae* isolates produce a hydroxyamate-type siderophore, aerobactin,[150] which is commonly associated with microbial species causing invasive disease.[151] The *E. cloacae* aerobactin-mediated iron uptake system is both genetically and structurally distinct from that of *E. coli* (pColV-K30), although a 72-kDa aerobactin-associated protein expressed in *E. coli* immunologically cross-reacts with antisera raised against the pColV-K30 74-kDa aerobactin receptor.[152]

Several other pathogenic properties have been identified in random isolates of *E. cloacae.* Prada and Beutin[153] identified a hemolytic strain of *E. cloacae* in the feces of an 11-month-old boy. *Bam*H1-digested DNA from this strain reacted with an *E. coli* α-hemolysin–specific probe. Paton and Paton[154] described the isolation of a Shiga-like toxin (SLT)–producing strain of *E. cloacae* from an infant with hemolytic-uremic syndrome. This strain reacted with *slt*-II– but not *slt*-I–specific DNA probes. French pressure-treated cell extracts yielded SLT activity detectable on Vero cells. Because this strain was simultaneously isolated from feces along with another SLT-producing organism (*E. coli* OR:H9), its clinical significance remains unknown. Probably the most intriguing virulence factor identified so far in *E. cloacae* is the outer membrane protein, OmpX.[155] This 17-kDa protein, encoded by a chromosomal gene, causes decreased porin production (OmpF, OmpC) in *E. coli* transformants, leading to increased antimicrobial resistance to several β-lactam compounds.[155,156] OmpX also exhibits 50% to 60% amino acid sequence homology with an invasion protein (Ail) of *Yersinia enterocolitica* and other factors (serum resistance, Rsk; survival in macrophages, PagC) in *Salmonella* Typhimurium.[157] Although no differences in invasive characteristics in organ culture were seen between a wild-type strain carrying the *ompX* gene and a deletion mutant, a 10-fold higher invasive rate was found when a hyperproducing OmpX strain (*E. cloacae* JS101[pJSO4]) was compared with an OmpX-negative strain. One postulate suggests that upregulation of OmpX expression could be controlled in vivo by use of antibiotics such as cefamandole. In such a scenario, cephalosporin chemotherapy could cause hyperproduction of OmpX by *E. cloacae* strains residing in the gut, leading to increased invasive capabilities.

ANTIMICROBIAL SUSCEPTIBILITY

The single most alarming trend regarding the genus *Enterobacter* is the increasing incidence of serious, life-threatening infections caused by strains resistant to multiple antibiotics. In the latest NNIS report,[51] *Enterobacter* species were the third most common cause of gram-negative nosocomial infections. This distressing fact is exacerbated by the increasing tendency of *E. cloacae* and *E. aerogenes* to gain resistance to many commonly administered antibiotics, including third-generation cephalosporins. Such resistance often arises de novo in patients receiving empiric therapy for systemic infections.[158] Later, these drug-resistant enterobacters, primarily residing in the gut, may emerge to produce severe or fulminant illnesses in debilitated persons.[159]

The general susceptibility profile for the genus is listed in Table 14; some variation exists from species to species. Of major importance is the increasing incidence of drug resistance noted in the genus. In a 1994 survey of 43 medical centers throughout the United States, Jones and colleagues[160] found that 31% (range 5%–58%) of all *E. cloacae* produced a type I cephalosporinase, as determined by their in vitro susceptibility to ceftazidime. This study went on to show that enterobacters were the fourth most prevalent group (9.2%) producing class I cephalosporinases. A 3-year French study (1988–1990) found the annual rate of cefotaxime and ceftazidime resistance in *Enterobacter* species to range between 20% and 29%.[168] Similar findings on the increasing incidence of resistance to third-generation cephalosporins in the most common *Enterobacter* species (*E. cloacae, E. aerogenes,* and *E. agglomerans*) have been reported by other groups.[162] Cephalosporin resistance in *Enterobacter* species is caused by constitutive expression of a chromosomal β-lactamase (Bush group 1) or similar activity carried on a plasmid.[61,169–171] Plasmid-mediated β-lactamase activity most commonly involves extrachromosomal elements harboring homologous sequences to either the TEM or SHV families.[169] Both chromosomal and plasmid-mediated extended-spectrum β-lactamase (ESBL) activity impart general resistance to most β-lactam antibiotics except imipenem. Most plasmid-mediated ESBL activity has been shown to involve either *E. aerogenes* or *E. cloacae,*[142] although such properties have been reported in China in a strain of *E. gergoviae* carrying a 60-kb plasmid with SHV-2–type sequences.[172]

Coupled with this increasing trend toward antibiotic resistance in *Enterobacter* species are the numerous reports documenting the emergence of drug resistance in these organisms during the course of antimicrobial chemotherapy. Quinn and colleagues[173] described in 1987 two patients with serious *E. cloacae* disease (bacteremia, meningitis) whose infections relapsed after therapy with ceftazidime and cefotaxime, respectively. Each posttherapy isolate of the original strain had become resistant to third-generation cephalosporins, including their initial drug regimens. Laboratory analysis of pretherapy isolates in

TABLE 14. *General susceptibility pattern of the genus* Enterobacter

Susceptible[a]	Variable	Resistant
Aminoglycosides Carbapenems	β-Lactamase inhibitor combinations Cephalosporins (second-generation)	Penicillins/aminopenicillins Cephalosporins (first-generation)
Cephalosporins[b] (fourth-generation) Fluoroquinolones	Cephalosporins (third-generation) Monobactams Extended-spectrum penicillins Trimethoprim-sulfamethoxazole	Tetracyclines Chloramphenicol

From references 160–167.
[a]<5% resistance in group as a whole.
[b]Cefepime.

each case found both strains to produce inducible β-lactamases, whereas both posttherapy strains constitutively expressed this enzymatic activity. These observations have been supported by a number of larger clinical investigations. Chow and associates[67] found that multiresistant bacteremic isolates of *E. cloacae* were much more likely to occur in patients previously treated with third-generation cephalosporins than with other antimicrobial agents. Modi and coworkers[159] found that use of cefuroxime to control an outbreak of *S. epidermidis* in a neonatal ICU resulted in the appearance of cefotaxime-resistant *E. cloacae* in the same unit 3 months later. In this same facility, an outbreak of serious *E. cloacae* disease (ventriculitis, septicemia) occurred in which five of six outbreak strains were susceptible to cefotaxime initially in vitro. All five persons harboring these strains were subsequently treated with cefotaxime, and in four instances (80%) susceptibility to this agent was lost during treatment. Füssle and coinvestigators[89] had similar findings in an analysis of ICU patients with pulmonary infections caused by *E. cloacae.* In 18 (47%) of 38 persons treated with cefotaxime and tobramycin, resistance to cefotaxime was detected within 6 days. These *E. cloacae* isolates developed resistance to cefotaxime in a stepwise fashion with an initial minimal inhibitory concentration (MIC) of 1 μg/mL increasing to 32 μg/mL at day 3 and more than 64 μg/mL at day 6. Yu and colleagues,[174] using the spiral plater technique, found that drug resistance in *Enterobacter* emerges faster when isolates are exposed to third-generation cephalosporins and aztreonam than to other agents such as imipenem and ciprofloxacin. Although not all series have shown such a correlation,[44] the overwhelming body of evidence strongly supports an association between drug resistance in *Enterobacter* and use of third-generation cephalosporins.

While rare, resistance to imipenem in *Enterobacter* has been documented.[175,176] In most instances, imipenem-resistance in *E. cloacae* has been associated with elevated expression of chromosomal β-lactamase and permeability changes in the outer membrane.[171,177,178] Imipenem-resistant *E. aerogenes* was described in a 67-year-old leukemic patient who underwent several rounds of chemotherapy including ceftazidime/netilmicin and imipenem/ciprofloxacin. The imipenem-resistant isolate was found to constitutively express β-lactamase activity, and structural analysis of the cell wall indicated the absence of a 39-kDa outer membrane protein and high-molecular-weight lipopolysaccharide.[179] Introduction of a plasmid carrying a wild-type *ampD* allele into this imipenem-resistant isolate restored full susceptibility to β-lactam antibiotics.

TABLE 15. *Susceptibility of major* Enterobacter *species to select antimicrobial agents*

	E. aerogenes		*E. agglomerans*		*E. cloacae*	
Compound	MIC_{50}[a]	MIC_{90}[a]	MIC_{50}	MIC_{90}	MIC_{50}	MIC_{90}
Cephalothin	>128	>128	32	>128	>128	>128
Cefoperazone	0.25	128	2.0	128	1.0	>32
Cefotaxime	0.125	64	0.5	32	0.5	>32
Ceftriaxone	0.25	16	0.25	32	0.5	>32
Ceftazidime	0.25	32	0.25	4.0	1.0	>16
Ceftizoxime	0.25	32	0.25	>32	0.25	>32
Cefepime	2.0	4.0	0.02	NA	2.0	16.0
Ciprofloxacin	0.5	1.0	0.5	1.0	0.5	1.0
Ofloxacin	0.12	0.25	0.06	0.25	0.12	0.25
Norfloxacin	0.125	0.25	0.06	2.0	0.06	0.5
Aztreonam	NA	16.0	NA	0.5	NA	8.0
Imipenem	1.0	2.0	0.5	0.5	1.0	4.0

From references 161, 163, 164–167, 180, and 181.

[a]Maximum reported MIC (μg/mL) from reported studies; number of isolates tested, source of strains, and conditions of testing vary from study to study.

NA, not available.

Table 15 lists relative MIC values for some antimicrobial agents commonly used to treat *E. cloacae, E. aerogenes,* and *E. agglomerans* infections. Overall, *E. cloacae* and *E. aerogenes* are the two most resistant species, although MIC_{50} and MIC_{90} values for some agents differ significantly from study to study. Reasons for most of these discrepancies are not always evident. MIC_{50} values for many third-generation cephalosporins are well within the susceptible range (<8 μg/mL), but MIC_{90} values for these agents fall into the resistant category (cefotaxime, ceftazidime, ceftriaxone). Of the less frequently encountered species, *E. cancerogenus* appears to be one of the more resistant species based on MIC_{50} and MIC_{90} values, although only a limited number of strains have been tested.[161,165,182] Most of these less frequently isolated species do have lower MIC values to most antimicrobial agents used to treat *Enterobacter* infections than the more common species associated with systemic infections (see Table 15).

CONCLUSIONS

The genus *Enterobacter* will continue to increase in medical significance for the foreseeable future owing to its role in nosocomial infections. Although hospital-acquired infections predominate, several case studies suggest that members of this group may cause gastroenteritis[154,157]; however, the frequency and relative significance of such infections remains to be determined. Furthermore, evidence indicates that SLTs are unstable in *E. cloacae,* making the original observation a potentially spurious finding.[183] Community-acquired infections have been largely overlooked, and very limited data exist in this area. The changing taxonomy of the genus *Enterobacter* has expanded the number of recognized species without a concomitant increase in our knowledge of the pathogenicities, disease spectra, and resistance patterns of newly described species. Of paramount concern is the needed redefinition of species assigned to the genus *Enterobacter,* which, when completed, should allow for better species identification and association of pathogenic types with disease. Finally, the increasing level of antimicrobial resistance seen in the major species of this genus is a sign of future problems in the treatment of serious *Enterobacter* infections. If resistance to imipenem and the fluoroquinolones begins to appear, physicians will be left with a dwindling armamentarium of antimicrobial agents to treat infections caused by strains with ESBL activity.

REFERENCES

1. Hormaeche E, Edwards PR. A proposed genus *Enterobacter. Int Bull Bacteriol Nomencl Taxon* 1960;10: 71–74.
2. Grimont F, Grimont PAD. The genus *Enterobacter.* In: Balows A, Trüper HG, Dworkin M, Harder W, Schleifer K-H, eds *The Prokaryotes: A handbook on the biology of bacteria ecophysiology, isolation, identification, applications.* 2nd ed. Berlin: Springer-Verlag, 1991;2797–2815.
3. Hormaeche E, Edwards PR. Observations on the genus *Aerobacter* with a description of two species. *Int Bull Bacteriol Nomencl Taxon* 1958;8:111–115.
4. Hormaeche E, Munilla M. Biochemical tests for differentiation of *Klebsiella* and *Cloacae. Int Bull Bacteriol Nomencl Taxon* 1957;7:1–20.
5. Hormaeche E, Edwards PR. Proposal for the rejection of the generic name *Cloaca* Castellani and Chalmers, and proposal of *Enterobacter* as a generic name with designation of type species and of its type culture. With request for an opinion. *Int Bull Bacteriol Nomencl Taxon* 1960;10:75–76.
6. Ewing WH, Fife MA. *Enterobacter agglomerans* (Beijerinck) comb. nov. (the Herbicola-Lathyri bacteria). *Int J Syst Bacteriol* 1972;22:4–11.
7. Brenner DJ, Steigerwalt AG, Fanning GR. Differentiation of *Enterobacter aerogenes* from klebsiellae by deoxyribonucleic acid reassociation. *Int J Syst Bacteriol* 1972;22:193–200.
8. Steigerwalt AG, Fanning GR, Fife-Asbury MA, Brenner DJ. DNA relatedness among species of *Enterobacter* and *Serratia. Can J Microbiol* 1976;22:121–137.
9. Farmer JJ III, Asbury MA, Hickman FW, Brenner DJ. *Enterobacter sakazakii:* A new species of *"Enterobacteriaceae"* isolated from clinical specimens. *Int J Syst Bacteriol* 1980;30:569–584.
10. Izard D, Richard C, Leclerc H. DNA relatedness between *Enterobacter sakazakii* and other members of the genus *Enterobacter. Ann Microbiol* 1983;134: 241–244.
11. Brenner DJ, Richard C, Steigerwalt AG, Asbury MA, Mandel M. *Enterobacter gergoviae* sp. nov.: A new species of *Enterobacteriaceae* found in clinical specimens and the environment. *Int J Syst Bacteriol* 1980; 30:1–6.
12. Izard D, Gavini F, Leclerc H. Polynucleotide sequence relatedness and genome size among *Enterobacter in-*

termedium sp. nov. and the species *Enterobacter cloacae* and *Klebsiella pneumoniae. Zentralbl Bakteriol* 1980;1:51–60.

13. Izard D, Gavini F, Trinel PA, Leclerc H. Deoxyribonucleic acid relatedness between *Enterobacter cloacae* and *Enterobacter amnigenus* sp. nov. *Int J Syst Bacteriol* 1981;31:35–42.
14. Grimont PAD, Ageron E. *Enterobacter cancerogenus* (Uroĉević, 1966) Dickey and Zumoff 1988, a senior subjective synonym of *Enterobacter taylorae* Farmer et al. (1985). *Res Microbiol* 1989;140:459–465.
15. Brenner DJ, McWhorter AC, Kai A, Steigerwalt AG, Farmer JJ III. *Enterobacter asburiae* sp. nov., a new species found in clinical specimens, and reassignment of *Erwinia dissolvens* and *Erwinia nimipressuralis* to the genus *Enterobacter* as *Enterobacter dissolvens* comb. nov. and *Enterobacter nimipressuralis* comb. nov. *J Clin Microbiol* 1986;23:1114–1120.
16. O'Hara CM, Steigerwalt AG, Hill BC, Farmer JJ III, Fanning GR, Brenner DJ. *Enterobacter hormaechei,* a new species of the family *Enterobacteriaceae* formerly known as enteric group 75. *J Clin Microbiol* 1989;27:2046–2049.
17. Chung YR, Brenner DJ, Steigerwalt AG, Kim BS, Kim HT, Cho KY. *Enterobacter pyrinus* sp. nov., an organism associated with brown leaf spot disease of pear trees. *Int J Syst Bacteriol* 1993;43:157–161.
18. Kosako Y, Tamura K, Sakazaki R, Miki K. *Enterobacter kobei* sp. nov., a new species of the family *Enterobacteriaceae* resembling *Enterobacter cloacae. Curr Microbiol* 1996;33:261–265.
19. von Graevenitz A. Revised nomenclature of *Campylobacter laridis, Enterobacter intermedium,* and *"Flavobacterium branchiophila." Int J Syst Bacteriol* 1990; 40:211.
20. Farmer JJ III, Fanning GR, Davis BR, et al. *Escherichia fergusonii* and *Enterobacter taylorae,* two new species of *Enterobacteriaceae* isolated from clinical specimens. *J Clin Microbiol* 1985;21:77–81.
21. Dickey RS, Zumoff CH. Emended description of *Enterobacter cancerogenus* comb. nov. (formerly *Erwinia cancerogena*). *Int J Syst Bacteriol* 1988;38:371–374.
22. Schønheyder HC, Jensen KT, Frederiksen W. Taxonomic notes: Synonymy of *Enterobacter cancerogenus* (Uroĉević 1966) Dickey and Zumoff 1988 and *Enterobacter taylorae* Farmer et al. 1985 and resolution of an ambiguity in the biochemical profile. *Int J Syst Bacteriol* 1994;44:586–587.
23. Izard D, Gavini F, Trinel PA, Krubwa F, Leclerc H. Contribution of DNA–DNA hybridization to the transfer of *Enterobacter aerogenes* to the genus *Klebsiella* as *K. mobilis. Zentralbl Bakteriol* 1980;1:257–263.
24. Brenner DJ, Fanning GR, Knutson JKL, Steigerwalt AG, Krichevsky MI. Attempts to classify Herbicola-group–*Enterobacter agglomerans* strains by deoxyribonucleic acid hybridization and phenotypic tests. *Int J Syst Bacteriol* 1984;34:45–55.
25. Lind E, Ursing J. Clinical strains of *Enterobacter agglomerans* (synonyms: *Erwinia herbicola, Erwinia milletiae*) identified by DNA–DNA hybridization. *Acta Path Microbiol Immunol Scand B* 1986;94: 205–213.
26. Beji A, Mergaert J, Gavini F, et al. Subjective synonymy of *Erwinia herbicola, Erwinia milletiae, Enterobacter agglomerans* and redefinition of the taxon by genotypic and phenotypic data. *Int J Syst Bacteriol* 1988;38:77–88.
27. Gavini F, Mergaert J, Beji A, et al. Transfer of *Enterobacter agglomerans* (Beijerinck 1888) Ewing and Fife 1972 to *Pantoea* gen. nov. as *Pantoea agglomerans* comb. nov. and description of *Pantoea dispersa* sp. nov. *Int J Syst Bacteriol* 1989;39:337–345.
28. Brenner DJ. The genus *Enterobacter.* In: Starr MP, Stolp H, Trüper HG, Balows A, Schlegel HG, eds. *The prokaryotes: A handbook on habitats, isolation, and identification of bacteria.* Berlin: Springer-Verlag, 1981;1173–1180.
29. Stiles ME. Less recognized or presumptive pathogenic bacteria. In: Doyle MP, ed. *Foodborne bacterial pathogens.* New York: Marcel Dekker, 1989;673–720.
30. Gaston MA. *Enterobacter:* An emerging nosocomial pathogen. *J Hosp Infect* 1988;11:197–208.
31. Hinton DM, Watson CW. *Enterobacter cloacae* is an endophytic symbiont of corn. *Mycopathologia* 1995; 129:117–125.
32. Stiles ME, Ng L-K. Biochemical characteristics and identification of *Enterobacteriaceae* isolated from meats. *Appl Environ Microbiol* 1981;41:639–645.
33. Stiles ME, Ng L-K. *Enterobacteriaceae* associated with meats and meat handling. *Appl Environ Microbiol* 1981;41:867–872.
34. Cox NA, Bailey JS. *Enterobacteriaceae* identification from stock cultures and high moisture foods with a four hour system (API Rapid E). *J Food Protect* 1986; 49:605–607.
35. von Holy A, Holzapfel WH, Dykes GA. Bacterial populations associated with Vienna sausage packaging. *Food Microbiol* 1992;9:45–53.
36. Muytjens HL, Roelofs-Willemse H, Jaspar GHJ. Quality of powdered substitutes for breast milk with regards to members of the family *Enterobacteriaceae. J Clin Microbiol* 1988;26:743–746.
37. Zeitoun AAM, Debevere JM, Mossel DAA. Significance of *Enterobacteriaceae* as index organisms for hygiene on fresh untreated poultry, poultry treated with lactic acid and poultry stored in a modified atmosphere. *Food Microbiol* 1994;11:169–176.
38. Wessels D, Jooste PJ, Mostert JF. Psychrotrophic, proteolytic and lipolytic properties of *Enterobacteriaceae* isolated from milk and dairy products. *Int J Food Microbiol* 1989; 9:79–83.
39. López-Sabater EI, Rodríguez-Jerez JJ, Roig-Sagués AX, Mora-Ventura MAT. Bacteriological quality of tuna fish *(Thunnus thynnus)* destined for canning: Effect of tuna handling on presence of histidine decarboxylase bacteria and histamine level. *J Food Protect* 1994;57:318–323.
40. Andersen BM, Sørlie D, Hotvedt R, et al. Multiply beta-lactam resistant *Enterobacter cloacae* infections linked to the environmental flora in a unit for cardiothoracic and vascular surgery. *Scand J Infect Dis* 1989;21:181–191.
41. Lacey SL, Want SV. An outbreak of *Enterobacter cloacae* associated with contamination of a blood gas machine. *J Infect* 1995;30:223–226.
42. Flaherty JP, Garcia-Houchins S, Chudy R, Arnow PM. An outbreak of gram-negative bacteremia traced to contaminated O-rings in reprocessed dialyzers. *Ann Intern Med* 1993;119:1072–1078.
43. Acolet D, Ahmet Z, Houang E, Hurley R, Kaufmann

ME. *Enterobacter cloacae* in a neonatal intensive care unit: Account of an outbreak and its relationship to use of third generation cephalosporins. *J Hosp Infect* 1994;28:273–286.
44. Georghiou PR, Hamill RJ, Wright CE, et al. Molecular epidemiology of infections due to Enterobacter aerogenes: Identification of hospital outbreak–associated strains by molecular techniques. *Clin Infect Dis* 1995; 20:84–94.
45. Haertl R, Bandlow G. Epidemiological fingerprinting of *Enterobacter cloacae* by small-fragment restriction endonuclease analysis and pulsed-field gel electrophoresis of genomic restriction fragments. *J Clin Microbiol* 1993;31:128–133.
46. Andresen J, Asmar BI, Dajani AS. Increasing *Enterobacter* bacteremia in pediatric patients. *Pediatr Infect Dis J* 1994;13:787–792.
47. Gallagher PG. *Enterobacter* bacteremia in pediatric patients. *Rev Infect Dis* 1990;12:808–812.
48. Poilane I, Cruaud P, Lachassinne E, et al. *Enterobacter cloacae* cross-colonization in neonates demonstrated by ribotyping. *Eur J Clin Microbiol Infect Dis* 1993; 12:820–826.
49. Simmons BP, Gelfand MS, Haas M, Metts L, Ferguson J. *Enterobacter sakazakii* infections in neonates associated with intrinsic contamination of a powdered infant formula. *Infect Control Hosp Epidemiol* 1989;10: 398–401.
50. Noriega FR, Kotloff KL, Martin MA, Schwalbe RS. Nosocomial bacteremia caused by *Enterobacter sakazakii* and *Leuconostoc mesenteroides* resulting from extrinsic contamination of infant formula. *Pediatr Infect Dis J* 1990;9:447–449.
51. Schaberg DR, Culver DH, Gaynes RP. Major trends in the microbial etiology of nosocomial infection. *Am J Med* 1991;91(Suppl 3B):S72–S75.
52. Flynn DM, Weinstein RA, Nathan C, Gaston MA, Kabins SA. Patient's endogenous flora as the source of "nosocomial" *Enterobacter* in cardiac surgery. *J Infect Dis* 1987;156:363–368.
53. John JF, Sharbaugh RJ, Bannister ER. *Enterobacter cloacae:* Bacteremia, epidemiology, and antibiotic resistance. *Rev Infect Dis* 1982;4:13–28.
54. Talbot GH, Miller DE, Doorley M, et al. *Enterobacter cloacae*–contaminated cardioplegic solution: Discovery and eradication of a pharmacy reservoir. *Am J Infect Control* 1984;12:239–244.
55. Verschraegen G, Claeys G, Delanghe M, Pattyn P. Serotyping and phage typing to identify *Enterobacter cloacae* contaminating total parenteral nutrition. *Eur J Clin Microbiol Infect Dis* 1988;7:306–307.
56. Wang C-C, Chu M-L, Ho L-J, Hwang R-C. Analysis of plasmid pattern in paediatric intensive care unit outbreaks of nosocomial infection due to *Enterobacter cloacae. J Hosp Infect* 1991;19:33–40.
57. Stenhouse MAE. *Enterobacter agglomerans* as a contaminant of blood [letter]. *Transfusion* 1992;32:86.
58. Maki DG, Rhame FS, Mackel DC, Bennett JV. Nationwide epidemic of septicemia caused by contaminated intravenous products. I: Epidemiologic and clinical features. *Am J Med* 1976;60:471–485.
59. Arseni A, Malamou-Ladas E, Koutsia C, Xanthou M, Trikka E. Outbreak of colonization of neonates with *Enterobacter sakazakii. J Hosp Infect* 1987;9: 143–150.
60. Chow JW, Yu VL, Shales DM. Epidemiologic perspectives on *Enterobacter* for the infection control professional. *Am J Infect Control* 1994;22:195–201.
61. Sanders Jr WE, Sanders CC. *Enterobacter* spp.: Pathogens poised to flourish at the turn of the century. *Clin Microbiol Rev* 1997;10:220–241.
62. Nicolle LE, Strausbaugh LJ, Garibaldi RA. Infections and antibiotic resistance in nursing homes. *Clin Microbiol Rev* 1996;9:1–17.
63. Bouza E, García de la Torre M, Erice A, Loza E, María Diaz-Borrego J, Buzón L. *Enterobacter* bacteremia: An analysis of 50 episodes. *Arch Intern Med* 1985; 145:1024–1027.
64. Burchard KW, Barrall DT, Reed M, Slotman GJ. *Enterobacter* bacteremia in surgical patients. *Surgery* 1986;100:857–861.
65. Fung C-P, Wang L-S, Juang Y-C, Liu C-Y, Cheng D-L. *Enterobacter cloacae* bacteremia: Clinical analysis of 41 cases. *Chin Med J (Taipei)* 1988;42:297–304.
66. Watanakunakorn C, Weber J. *Enterobacter* bacteremia: A review of 58 episodes. *Scand J Infect Dis* 1989; 21:1–8.
67. Chow JW, Fine MJ, Shlaes DM, et al. *Enterobacter* bacteremia: Clinical features and emergence of antibiotic resistance during therapy. *Ann Intern Med* 1991; 115:585–590.
68. Bodey GP, Elting LS, Rodriguez S. Bacteremia caused by *Enterobacter:* 15 years experience in a cancer hospital. *Rev Infect Dis* 1991;13:550–558.
69. Haddy RI, Cecil ML, Norris LL, Markert RJ. *Enterobacter* bacteremia in the community hospital. *J Fam Pract* 1991;32:601–606.
70. Weischer M, Kolmos HJ. Retrospective 6-year study of enterobacter bacteraemia in a Danish university hospital. *J Hosp Infect* 1992;20:15–24.
71. Rubinstein EM, Klevjer-Anderson P, Smith CA, Drouin MT, Patterson JE. *Enterobacter taylorae,* a new opportunistic pathogen: Report of four cases. *J Clin Microbiol* 1993;31:249–254.
72. Abbott SL, Janda JM. *Enterobacter cancerogenus ("Enterobacter taylorae")* infections associated with severe trauma or crush injuries. *Am J Clin Pathol* 1997;107:359–361.
73. Bollet C, Elkouby A, Pietri P, de Micco P. Isolation of *Enterobacter amnigenus* from a heart transplant recipient. *Eur J Clin Microbiol Infect Dis* 1991;10: 1071–1073.
74. Prats G, Richard C, Mirelis B, Lopez P. Human isolates of *Enterobacter intermedium. Zentralbl Bakteriol* 1987;266:422–424.
75. Hawkins RE, Lissner CR, Sanford JP. *Enterobacter sakazakii* bacteremia in an adult. *South Med J* 1991; 84:793–795.
76. Wenger PN, Tokars JI, Brennan P, et al. An outbreak of *Enterobacter hormaechei* infection and colonization in an intensive care nursery. *Clin Infect Dis* 1997;24: 1243–1244.
77. Bonadio WA, Margolis D, Tovar M. *Enterobacter cloacae* bacteremia in children: A review of 30 cases in 12 years. *Clin Pediatr* 1991;30:310–313.
78. Gürses N. *Enterobacter* septicemia in neonates [letter]. *Pediatr Infect Dis J* 1995;14:638.
79. Tunkel AR, Fisch MJ, Schlein A, Scheld WM. *Enterobacter* endocarditis. *Scand J Infect Dis* 1992;24: 233–240.

80. Willis J, Robinson JE. *Enterobacter sakazakii* meningitis in neonates. *Pediatr Infect Dis J* 1988;7:196–199.
81. Biering G, Karlsson S, Clark NC, Jónsdóttir KE, Lúdvígsson P, Steingrímsson Ø. Three cases of neonatal meningitis caused by *Enterobacter sakazakii* in powdered milk. *J Clin Microbiol* 1989;27:2054–2056.
82. Gallagher PG, Ball WS. Cerebral infarctions due to CNS infection with *Enterobacter sakazakii. Pediatr Radiol* 1991;21:135–136.
83. Muytjens HL, Zanen HC, Sonderkamp HJ, Kollée LA, Wachsmuth K, Farmer JJ III. Analysis of eight cases of neonatal meningitis and sepsis due to *Enterobacter sakazakii. J Clin Microbiol* 1983;18:115–120.
84. Lambert-Zechovsky N, Bingen E, Denamur E, et al. Molecular analysis provides evidence for the endogenous origin of bacteremia and meningitis due to *Enterobacter cloacae* in an infant. *Clin Infect Dis* 1992;15:30–32.
85. Nazarowec-White M, Farber JM. *Enterobacter sakazakii:* A review. *Int J Food Microbiol* 1997;34: 103–113.
86. Meis JFGM, Groot-Loonen J, Hoogkamp-Korstanje JAA. A brain abscess due to multiply-resistant *Enterobacter cloacae* successfully treated with meropenem [correspondence]. *Clin Infect Dis* 1995;20:1567.
87. Wolff MA, Young CL, Ramphal R. Antibiotic therapy for *Enterobacter* meningitis: A retrospective review of 13 episodes and review of the literature. *Clin Infect Dis* 1993;16:772–777.
88. Crane LR, Lerner AM. Gram-negative bacillary pneumonias. In: Pennington JE, ed. *Respiratory infections: Diagnosis and management.* New York: Raven Press, 1983;227–250.
89. Füssle R, Biscoping J, Behr R, Sziegoleit A. Development of resistance by *Enterobacter cloacae* during therapy of pulmonary infections in intensive care patients. *Clin Invest* 1994;72:1015–1019.
90. Glustein JZ, Kaplan M. *Enterobacter cloacae* causing pneumatocele in a neonate. *Acta Paediatr* 1994;83: 990–991.
91. Takao M, Ido M, Hamaguchi K, Chikusa H, Namikawa S, Kusagawa M. Descending necrotizing mediastinitis secondary to a retropharyngeal abscess. *Eur Respir J* 1994;7:1716–1718.
92. Summanen PH, Talan DA, Strong C, et al. Bacteriology of skin and soft-tissue infections: Comparison of infections in intravenous drug users and individuals with no history of intravenous drug use. *Clin Infect Dis* 1995;20(Suppl):S279–S282.
93. Kahn RM, Goldstein EJC. Common bacterial skin infections: Diagnostic clues and therapeutic options. *Postgrad Med* 1993;93:175–182.
94. Brook I, Frazier EH. Microbiological analysis of pancreatic abscess. *Clin Infect Dis* 1996;22:384–385.
95. Michowitz M, Berger SA, Nagar H, Chayen D. Crepitant cellulitis due to *Enterobacter aerogenes. Israel J Med Sci* 1985;21:546–547.
96. Ganelin RS, Ellis M. Cellulitis caused by *Enterobacter cloacae* [letter]. *J Infect* 1992;24: 218–219.
97. Falasca GF, Reginato AJ. The spectrum of myositis and rhabdomyolysis associated with bacterial infection. *J Rheumatol* 1994;21:1932–1937.
98. Fata F, Chittivelu S, Tessler S, Kupfer Y. Gas gangrene of the arm due to *Enterobacter cloacae* in a neutropenic patient. *South Med J* 1996;89:1095–1096.
99. Livingston W, Grossman ME, Garvey G. Hemorrhagic bullae in association with *Enterobacter cloacae* septicemia. *J Am Acad Dermatol* 1992;27:637–638.
100. Shapiro ED. Infections of the urinary tract. *Pediatr Infect Dis J* 1992;11:165–168.
101. Markowitz SM. Urinary tract specimens. In: Dalton HC, Nottebart HC Jr, eds. *Interpretive medical microbiology.* New York: Churchill Livingstone, 1986;523–592.
102. Muñoz P, Cercenado E, Rodríguez–Créixems M, Díaz MD, Vicente T, Bouza E. The CLED agar option in urine culture routine: A prospective and comparative evaluation. *Diagn Microbiol Infect Dis* 1992;15: 287–290.
103. Rubinstein E, Lode H, Grassi C, and an antibiotic study group. Ceftazidime monotherapy vs. ceftriaxone/tobramycin for serious hospital-acquired gram-negative infections. *Clin Infect Dis* 1995;20:1217–1228.
104. Stamm WE, McKevitt M, Roberts PL, White NJ. Natural history of recurrent urinary tract infections in women. *Rev Infect Dis* 1991;13:77–84.
105. Stark RP, Maki DG. Bacteriuria in the catheterized patient: What quantitative level of bacteriuria is relevant? *N Engl J Med* 1984;311:560–564.
106. Reina J, Salva F, Gil J, Alomar P. Urinary tract infection caused by *Enterobacter taylorae. J Clin Microbiol* 1989;27:2877.
107. Westblom TU, Coggins ME. Osteomyelitis caused by *Enterobacter taylorae,* formerly enteric group 19. *J Clin Microbiol* 1987;2432–2433.
108. Krag ML, Schønheyder HC. *Enterobacter cancerogenus* (syn. *Enterobacter taylorae*) from a case of traumatic tibial osteomyelitis. *Clin Microbiol Newslett* 1997;19:31–32.
109. Chassagne P, Mejjad O, Daragon A, Lecomte R, Le Loet X, Deshayes P. Spondylodiscite à *Enterobacter cloacae* traitée par le céfixime. *Presse Med* 1990;19: 673–674.
110. Solans R, Simeon P, Cuenca R, Fonollosa V, Bago J, Vilardell M. Infectious discitis caused by *Enterobacter cloacae. Ann Rheum Dis* 1992;51:906–907.
111. Marce S, Antoine J-F, Schaeverbeke T, Vernhes J-P, Bannwarth B, Dehais J. *Enterobacter cloacae* vertebral infection in a heroin addict with HIV infection [Matters Arising]. *Ann Rheum Dis* 1993;52:695–696.
112. Quint L, Fantin B, Belmatoug N, Carbon C. Bursite hémorragique infectée par *Enterobacter cloacae. Presse Med* 1991;20:1787.
113. van der Hazel SJ, Speelman P, Tytgat GNJ, Dankert J, van Leeuwen DJ. Role of antibiotics in the treatment and prevention of acute and recurrent cholangitis. *Clin Infect Dis* 1994;19:279–286.
114. Molinari G, Pugliese V, Schito GC, Guzmán CA. Bacteria involved in the blockage of biliary stents and their susceptibility to antibacterial agents. *Eur J Clin Microbiol Infect Dis* 1996;15:88–92.
115. Scheider DM, King PD, Miedema BW. Ascites and secondary bacterial peritonitis associated with small bowel obstruction. *Am J Gastroenterol* 1994;89: 1238–1240.
116. Gross ND, Meyer RF. *Enterobacter cloacae* ulceration in a failed corneal graft: A case report. *Br J Ophthalmol* 1985;69:542–544.
117. Milewski SA, Klevjer-Anderson P. Endophthalmitis caused by *Enterobacter cloacae. Ann Ophthalmol* 1993;25:309–311.

118. Miller JM, Rhoden DL. Preliminary evaluation of Biolog, a carbon source utilization method for bacterial identification. *J Clin Microbiol* 1991;29:1143–1147.
119. Pfaller MA, Sahm D, Ciaglia C, et al. Comparison of the AutoSCAN-W/A rapid bacterial identification system and the Vitek AutoMicrobic system for identification of gram-negative bacilli. *J Clin Microbiol* 1991; 29:1422–1428.
120. O'Hara CM, Rhoden DL, Miller JM. Reevaluation of the API 20E identification system versus conventional biochemicals for identification of members of the family *Enterobacteriaceae:* A new look at an old product. *J Clin Microbiol* 1992;30:123–125.
121. Kelly MT, Leicester C. Evaluation of the Autoscan Walkaway system for rapid identification and susceptibility testing of gram-negative bacilli. *J Clin Microbiol* 1992;30:1568–1571.
122. Visser MR, Bogaards L, Rozenberg-Arska M, Verhoef J. Comparison of the autoSCAN-W/A and vitek Automicrobic systems for identification and susceptibility testing of bacteria. *Eur J Clin Microbiol Infect Dis* 1992;11:979–984.
123. York MK, Brooks GF, Fiss EH. Evaluation of the autoSCAN-W/A rapid system for identification and susceptibility testing of gram-negative fermentative bacilli. *J Clin Microbiol* 1992;30:2903–2910.
124. O'Hara CM, Tenover FC, Miller JM. Parallel comparison of accuracy of API 20E, Vitek GNI, MicroScan Walk/Away Rapid ID, and Becton Dickinson Cobas Micro ID-E/NF for identification of members of the family *Enterobacteriaceae* and common gram-negative, non–glucose-fermenting bacilli. *J Clin Microbiol* 1993;31:3165–3169.
125. Kitch TT, Jacobs MR, Appelbaum PC. Evaluation of RapID onE system for identification of 379 strains in the family *Enterobacteriaceae* and oxidase-negative, gram-negative nonfermenters. *J Clin Microbiol* 1994; 32:931–934.
126. Holmes B, Costas M, Ganner M, On SLW, Stevens M. Evaluation of Biolog system for identification of some gram-negative bacteria of clinical importance. *J Clin Microbiol* 1994;32:1970–1975.
127. Holmes B, Costas M, Thaker T, Stevens M. Evaluation of two BBL Crystal systems for identification of some clinically important gram-negative bacteria. *J Clin Microbiol* 1994;32:2221–2224.
128. Robinson A, McCarter YS, Tetreault J. Comparison of Crystal Enteric/Non-Fermenter system, API 20E system, and Vitek Automicrobic system for identification of gram-negative bacilli. *J Clin Microbiol* 1995;33: 364–370.
129. Wauters G, Boel A, Voorn GP, et al. Evaluation of a new identification system, Crystal Enteric/Non-Fermenter, for gram-negative bacilli. *J Clin Microbiol* 1995;33:845–849.
130. Rhoads S, Marinelli L, Imperatrice CA, Nachamkin I. Comparison of MicroScan WalkAway system and Vitek system for identification of gram-negative bacteria. *J Clin Microbiol* 1995;33:3044–3046.
131. Andersen BM. Biochemical profiles and serotypes of nosocomial *Enterobacter cloacae* strains in Northern Norway: Biochemical identification problems with commercial test systems. *Infection* 1995;23:339–343.
132. Farmer JJ III, Davis BR, Hickman-Brenner FW, et al. Biochemical identification of new species and biogroups of *Enterobacteriaceae* isolated from clinical specimens. *J Clin Microbiol* 1985;21:46–76.
133. Gaston MA, Crees-Morris JA, Pitt TL. Serotypes and biochemical profiles of British hospital strains of *Enterobacter cloacae* in relation to site of infection and antibiotic susceptibility. *J Hosp Infect* 1987;10:17–27.
134. Bingen E, Denamur E, Lambert-Zechovsky, Brahimi N, El Lakany M, Elion J. Rapid genotyping shows the absence of cross-contamination in *Enterobacter cloacae* nosocomial infections. *J Hosp Infect* 1992;21:95–101.
135. de Champs C, Henquell C, Guelon D, Sirot D, Gazuy N, Sirot J. Clinical and bacteriological study of nosocomial infections due to *Enterobacter aerogenes* resistant to imipenem. *J Clin Microbiol* 1993;31:123–127.
136. Old D. Biotyping of *Enterobacter cloacae. J Clin Pathol* 1982;35:875–878.
137. Gaston MA, Bucher C, Pitt TL. O serotyping scheme for *Enterobacter cloacae. J Clin Microbiol* 1983;18: 1079–1083.
138. Traub WH, Spohr M, Blech R. Bacteriocin typing of clinical isolates of *Enterobacter cloacae. J Clin Microbiol* 1982;16:885–889.
139. Bauernfeind A, Petermüller C. Typing of *Enterobacter* spp. by bacteriocin susceptibility and its use in epidemiological analysis. *J Clin Microbiol* 1984;20:70–73.
140. Gaston MA, Strickland MA, Ayling-Smith BA, Pitt TL. Epidemiological typing of *Enterobacter aerogenes. J Clin Microbiol* 1989;27:564–565.
141. Gaston MA. Isolation and selection of bacteriophage-typing set for *Enterobacter cloacae. J Med Microbiol* 1987;285–290.
142. Neuwirth C, Siebor E, Lopez J, Pechinot A, Kazmierczak A. Outbreak of TEM-24–producing *Enterobacter aerogenes* in an intensive care unit and dissemination of the extended-spectrum β-lactamase to other members of the family *Enterobacteriaceae. J Clin Microbiol* 1996;34:76–79.
143. Clark NC, Hill BC, O'Hara CM, Steingrimsson O, Cooksey RC. Epidemiologic typing of *Enterobacter sakazakii* in two neonatal nosocomial outbreaks. *Diagn Microbiol Infect Dis* 1990;13:467–472.
144. Gaston MA, Warner M. Electrophoretic typing of *Enterobacter cloacae* with a limited set of enzyme stains. *Epidemiol Infect* 1989;103:255–264.
145. Garaizar J, Kaufmann ME, Pitt TL. Comparison of ribotyping with conventional methods for the type identification of *Enterobacter cloacae. J Clin Microbiol* 1991;29:1303–1307.
146. McConkey SJ, Coleman DC, Falkiner FR, McCann SR, Daly PA. *Enterobacter cloacae* in a haematology/oncology ward—first impressions. *J Hosp Infect* 1989;14:277–284.
147. Verweij PE, Van Belkum A, Melchers WJG, Voss A, Hoogkamp-Korstanje JAA, Meis JFGM. Interrepeat fingerprinting of third-generation cephalosporin-resistant *Enterobacter cloacae* isolated during an outbreak in a neonatal intensive care unit. *Infect Control Hosp Epidemiol* 1995;16:25–29.
148. Davin-Regli A, Bosi C, Charrel R, et al. A nosocomial outbreak due to *Enterobacter cloacae* strains with the *E. hormaechei* genotype in patients treated with fluoroquinolones. *J Clin Microbiol* 1997;35:1008–1010.
149. Adegbola RA, Old DC. Fimbrial haemagglutinins in *Enterobacter* species. *J Gen Microbiol* 1983;129: 2175–2180.

150. Martinez JL, Cercenado E, Baquero F, Pérez-Díaz JC, Delgado-Iribarren A. Incidence of aerobactin production in gram-negative hospital isolates. *FEMS Microbiol Lett* 1987;43:351–353.
151. Payne SM. Iron and virulence in the family *Enterobacteriaceae. Crit Rev Microbiol* 1988;16:81–111.
152. Crosa LM, Wolf MK, Actis LA, Sanders-Loehr J, Crosa JH. New aerobactin-mediated iron uptake system in a septicemia-causing strain of *Enterobacter cloacae. J Bacteriol* 1988;170:5539–5544.
153. Prada J, Beutin L. Detection of *Escherichia coli* α-haemolysin genes and their expression in a human faecal strain of *Enterobacter cloacae. FEMS Microbiol Lett* 1991;79:111–114.
154. Paton AW, Paton JC. *Enterobacter cloacae* producing a Shiga-like toxin II–related cytotoxin associated with a case of hemolytic-uremic syndrome. *J Clin Microbiol* 1996;34:463–465.
155. Stoorvogel J, van Bussel MJAWM, Tommassen J, van de Klundert JAM. Molecular characterization of an *Enterobacter cloacae* outer membrane protein (OmpX). *J Bacteriol* 1991;173:156–160.
156. Stoorvogel J, van Bussel JAWM, van de Klundert JAM. Cloning of a beta-lactam resistance determinant of *Enterobacter cloacae* affecting outer membrane proteins of *Enterobacteriaceae. FEMS Microbiol Lett* 1987;48:277–281.
157. de Kort G, Bolton A, Martin G, Stephen J, van de Klundert JAM. Invasion of rabbit ileal tissue by *Enterobacter cloacae* varies with the concentration of OmpX in the outer membrane. *Infect Immun* 1994;62:4722–4726.
158. Conus P, Francioli P. Relationship between ceftriaxone use and resistance of *Enterobacter* species. *J Clin Pharm Therpeu* 1992;17:303–305.
159. Modi N, Damjanovic V, Cooke RWI. Outbreak of cephalosporin resistant *Enterobacter cloacae* infection in a neonatal intensive care unit. *Arch Dis Child* 1987;62:148–151.
160. Jones RN, Kehrberg EN, Erwin ME, Anderson SC, and the Fluoroquinolone Resistance Surveillance Group. Prevalence of important pathogens and antimicrobial activity of parenteral drugs at numerous medical centers in the United States: I. Study on the threat of emerging resistances: Real or perceived? *Diagn Microbiol Infect Dis* 1994;19:203–215.
161. Muytjens HL, van der Ros-van de Repe J. Comparative in vitro susceptibilities of eight *Enterobacter* species, with special reference to *Enterobacter sakazakii. Antimicrob Agents Chemother* 1986;29:367–370.
162. Jones RN, Pfaller MA, Allen SD, Gerlach EH, Fuchs PC, Aldridge KE. Antimicrobial activity of cefpirome: An update compared to five third-generation cephalosporins against nearly 6000 recent clinical isolates from five medical centers. *Diagn Microbiol Infect Dis* 1991;14:361–364.
163. Thornsberry C. Susceptibility of clinical bacterial isolates to ciprofloxacin in the United States. *Infection* 1994;22(Suppl):S80–S89.
164. Chin N-X, Gu J-W, Yu K-W, Zhang Y-X, Neu HC. In vitro activity of sparfloxacin. *Antimicrob Agents Chemother* 1991;35:567–571.
165. Freney J, Husson MO, Gavini F, et al. Susceptibilities to antibiotics and antiseptics of new species of the family *Enterobacteriaceae. Antimicrob Agents Chemother* 1988;32:873–876.
166. Lipman B, Neu HC. Imipenem: A new carbapenem. *Med Clin North Am* 1988;72:567–579.
167. Neu HC. Aztreonam: The first monobactam. *Med Clin North Am* 1988;72:555–566.
168. Sirot DL, Goldstein FW, Soussy CJ, et al. Resistance to cefotaxime and seven other β-lactams in members of the family *Enterobacteriaceae:* A 3-year survey in France. *Antimicrob Agents Chemother* 1992;36:1677–1681.
169. Jacoby GA, Archer GL. New mechanisms of bacterial resistance to antimicrobial agents. *N Engl J Med* 1991;324:601–612.
170. Jacoby GA, Medeiros AA. More extended-spectrum β-lactamases. *Antimicrob Agents Chemother* 1991;35:1697–1704.
171. Sanders CC, Sanders Jr WE. β-Lactam resistance in gram-negative bacteria: Global trends and clinical impact. *Clin Infect Dis* 1992;15:824–839.
172. Cheng Y, Chen M. Extended-spectrum β-lactamases in clinical isolates of *Enterobacter gergoviae* and *Escherichia coli* in China. *Antimicrob Agents Chemother* 1994;38:2838–2842.
173. Quinn JP, DiVincenzo CA, Foster J. Emergence of resistance to ceftazidime during therapy for *Enterobacter cloacae. J Infect Dis* 1987;155:942–947.
174. Yu CM, Chow JW, Yu VL. Quantitative comparison in vitro of mutational antibiotic resistance of *Enterobacter* spp. using a spiral plater. *J Antimicrob Chemother* 1996;37:233–242.
175. Thomson KS, Sanders CC, Chmel H. Imipenem resistance in *Enterobacter. Eur J Clin Microbiol Infect Dis* 1993;12:610–613.
176. Rasmussen BA, Bush K. Carbapenem-hydrolyzing β-lactamases. *Antimicrob Agents Chemother* 1997;41:223–232.
177. Lee EH, Nicolas MH, Kitzis MD, Pialoux G, Collatz E, Gutmann L. Association of two resistance mechanisms in a clinical isolate of *Enterobacter cloacae* with high-level resistance to imipenem. *Antimicrob Agents Chemother* 1991;35:1093–1098.
178. Raimondi A, Traverso A, Nikaido H. Imipenem- and meropenem-resistant mutants of *Enterobacter cloacae* and *Proteus rettgeri* lack porins. *Antimicrob Agents Chemother* 1991;35:1174–1180.
179. Tzouvelekis LS, Tzelepi E, Kaufmann ME, Mentis AF. Consecutive mutations leading to the emergence in vivo of imipenem resistance in a clinical strain of *Enterobacter aerogenes. J Med Microbiol* 1994;40:403–407.
180. Kessler RE, Bies M, Buck RE, et al. Comparison of a new cephalosporin, BMY 28142, with other broad-spectrum β-lactam antibiotics. *Antimicrob Agents Chemother* 1985;27:207–216.
181. Johnson CC, Livornese L, Gold MJ, Pitsakis PG, Taylor S, Levison ME. Activity of cefepime against ceftazidime-resistant gram-negative bacilli using low and high inocula. *J Antimicrob Chemother* 1995;35:765–773.
182. Cormican MG, Jones RN. Antimicrobial activity of cefotaxime tested against infrequently isolated pathogenic species (unusual pathogens). *Diagn Microbiol Infect Dis* 1995;22:43–48.
183. Paton JC, Paton AW. Instability of Shiga toxin type 2 gene in *Enterobacter cloacae* [correspondence]. *J Clin Microbiol* 1997;35:1917.

11
The Genus *Pantoea*

The genus *Pantoea* is one of several recently established genera in the family *Enterobacteriaceae.* Its genesis was the result of taxonomic problems associated with members of the *Enterobacter agglomerans* complex (see Chapter 9). The *E. agglomerans* complex had long been recognized as a catchall group for human, animal, plant, and environmental isolates that shared similar phenotypic features. However, a number of lines of evidence suggested that *E. agglomerans* was in fact a phenospecies composed of genetically distinct species that phenotypically resembled each other. Progress on characterizing these genomovars was hampered by the difficulty in determining the correct generic assignment for each taxa and identification of biochemical properties useful in their recognition.

Don Brenner and colleagues[1] provided the first major taxonomic assessment of the *E. agglomerans* complex using DNA reassociation kinetics. Of 124 *E. agglomerans* strains analyzed by DNA–DNA hybridization, at least 14 taxons were identified, including one cluster of 34 isolates that did not fit into any of 13 established hybridization groups (HGs). With the use of a dichotomic scheme, only 6 of 10 HGs that contained three or more strains were separable by phenotypic methods. Some of these groups, such as HG XI, were subsequently renamed *Leclercia adecarboxylata* (*"Escherichia adecarboxylata"*),[2] and other unclustered isolates were assigned to *Rahnella aquatilis.*[3]

The seminal observations of Brenner and colleagues[1] led to the first concerted efforts to address some of the major taxonomic issues inherent in the *E. agglomerans–Erwinia herbicola* group. A Swedish investigation by Lind and Ursing[4] found that 52 (60%) of 86 *E. agglomerans* isolates were highly related to the type strain, American Type Culture Collection (ATCC) 27155 or National Culture Type Collection (NCTC) 9381. This was followed by another European survey that characterized 54 *E. agglomerans* isolates by DNA relatedness and by protein electrophoretic profiles.[5] DNA–DNA hybridization results indicated that 24 of these isolates, regardless of protein profile, were highly related to ATCC 27155 (referred to as DNA HG 27155). This group of bacteria also included the type strains of *Erwinia milletiae* and *Erwinia herbicola* and two strains from Brenner's HG XIII (ATCC 12287 and International Collection of Phytopathogenic Bacteria [ICPB] EH103). The authors concluded that DNA HGs 27715 and XIII and the 52 strains from Lind and Ursing's study[4] constituted a single, homogeneous species.

In 1989, Gavini and coworkers[6] identified 10 strains of *E. herbicola* and *E. agglomerans* that were highly related by DNA–DNA hybridization. These 10 strains exhibited relative binding ratios of 75% to 100% to the reference strain, *E. herbicola* ATCC 14589. Although the 14589 group was determined to belong to the *Enterobacteriaceae,* it was not highly related to other members of this family, exhibiting only 4% *(Leminorella)* to 37% *(Escherichia blattae)* relatedness to other established genera, including *Enterobacter* species (15%–24% relatedness). The nearest neighbor to the 14589 group was found to be HG 27155 of Beji and colleagues,[5] with which it exhibited 41% to 53% homology. The reference strain for Brenner's HG III (Centers for Disease Control and Prevention [CDC] 1429-71) was found to reside within this group because it was 85% homologous with ATCC 14589. Based on these observations, Gavini concluded that HGs 14589 and 27155 should coexist as separate species in a new genus named *Pantoea,* meaning "all sorts and sources" (ie, from diverse geographical samples). HG 14589 was named *Pantoea dispersa,* and HG 27155, which contained the type strain of *E. agglomerans,* was renamed *Pantoea agglomerans.*

Over the past 5 years, several other species have been added to the genus *Pantoea.* In 1992, Kageyama and colleagues[7] proposed three new species for inclusion, *Pantoea punctata, Pantoea citrea,* and *Pantoea terrea.* These three species were isolated from fruit and soil samples and produced 2,5-diketo-D-gluconic acid from D-glucose. Interspecies relatedness ranged from 30% to 45%, whereas lower degrees of homology were noted with *P. dispersa* (9%–18%) and *P. agglomerans* (12%–13%). Mergaert and colleagues[8] proposed the transfer of *Erwinia ananas ("Erwinia uredovora")* and *Erwinia stewartii* to the genus *Pantoea* as *Pantoea ananas* and *Pantoea stewartii,* respectively. *P. ananas* contains five strains assigned to Brenner's HG VI, indicating that these two groups are identical. Current composition of the genus *Pantoea* is presented in Table 1.

Despite many improvements in the taxonomy of the *E. agglomerans–E. herbicola–Pantoea* group, there are still enormous problems to deal with. It is presently not clear whether all current *Pantoea* species are sufficiently related to each other to remain in a single genus. Another problem concerns "bridge isolates" such as the reference strain for Brenner's HG V (CDC 3482-71). This strain has been found to be only 58% to 62% related (with 9% divergence) to *P. agglomerans* ATCC 27155 (Brenner's HG XIII) despite producing electropherograms almost identical to those of other strains that meet criteria (≥70% homology) for inclusion within this species.[4,5] This has led to proposals ranging from inclusion of HG V within the genomospecies *P. agglomerans* (thus lowering genetic criteria for inclusion) to provisional classification of HG V as a *Pantoea* species.[5,8] In addition, there is an exceedingly high number of *E. agglomerans–E. herbicola* isolates that remain ungrouped (unclustered), implying that they represent undescribed taxa or species. Such strains, which constitute a significant percentage of the total isolates analyzed in some studies,[1] indicate that there is still tremendous genetic diversity within this group. Where Brenner's HGs I, II, IV, V, VII–X, and XII fit into this taxonomic quagmire remains to be determined. Much work needs to be done.

Although members of the *E. agglomerans–E. herbicola* group are widely dispersed

TABLE 1. *Current composition of the genus* Pantoea

Species	HG	G+C (mol%)	Isolated from Human Material	Type or Reference Strain	Synonyms	Ref.
P. agglomerans	XIII	55.1–56.8	Yes	ATCC 27155	*Erwinia agglomerans, Erwinia herbicola*	6
P. dispersa	III	56.5–60.6	Yes	ATCC 14589		6
P. punctata	—	50.0–50.3	No	ATCC 31626		7
P. citrea	—	49.7	No	ATCC 31623		7
P. terrea	—	51.9	No	ATCC 31628		7
P. ananas	VI	55.0	Yes	ATCC 33244	*Erwinia ananas, Erwinia uredovora*	8
P. stewartii	—	55.1–56.4	No	ATCC 8199	*Erwinia stewartii*	8
Unnamed	V	—	Yes	CDC 3482-71		1,4,5

HG, hybridization group of Brenner et al.[1]

throughout the plant and animal kingdoms, most environmental studies predate the revised taxonomy of this group. Therefore, it is difficult to determine whether early ecologic studies are representative for all new species and redesignated taxa. In addition to human infections, some *Pantoea* species are thought to be phytopathogens involved in pineapple rot *(P. ananas),* bacterial wilt of corn *(P. stewartii),* leaf spot disease of foxtail and pearl millet *(P. stewartii),* stalk and leaf necrosis of onions *(P. agglomerans),* and galls on *Wisteria* and other plants.[6,8] Table 2 lists sources from which *Pantoea* species have been isolated.

Acceptance of the revised taxonomy regarding the *E. agglomerans–E. herbicola* group generally has been lacking. Most case reports continue to refer to these bacteria simply as *E. agglomerans* without further biochemical characterization.[11] Until the newer taxonomy is adopted, a clear understanding of the roles individual *Pantoea* species play in human infections will not be available.

In 1989, Gavini and colleagues defined the genus *Pantoea* as gram-negative, nonencapsulated rods that produce smooth, convex colonies on nutrient agar with or without yellow pigmentation.[6] All *Pantoea* species are facultatively anaerobic and oxidase-negative, and most are motile by means of peritrichous flagella. Acid is produced from the fermentation of various pentose and hexose sugars such as D-xylose, D-galactose, and D-mannose. Most studies, including older reports in the literature, have consistently described members of this group as negative for ornithine decarboxylase (ODC), lysine decarboxylase (LDC), and arginine dihydrolase (ADH) activities. However, Gavini and associates[6] found some *P. agglomerans* strains that were ODC+. This observation has not been substantiated by other investigations in which a large number of genetically characterized *P. agglomerans* strains have been studied.[10] Most *Pantoea* strains are Voges Proskauer– and β-galactosidase–positive. Urea is not hydrolyzed.

The salient features for current nomenspecies in the genus *Pantoea* are listed in Table 3. These characteristics are extremely tenuous for all groups except *P. agglomerans* at present, due to the limited number of strains tested and the use of API 20E and API CH50E systems (Vitek bioMérieux, Hazelwood, MO) to construct databases rather than relying on conventional assays. Most *Pantoea* isolates recovered from clinical specimens produce a yellow pigment, distinguishing this group from all other *Enterobacteriaceae* except *Enterobacter sakazakii* and some *Escherichia* species.[8] *E. sakazakii* is, however, ODC+ and ADH+, which clearly separates these organisms from *Pantoea* species. Gas production, used as a marker for distinct biogroups within *E. agglomerans* in early surveys, has been found to be an unreliable characteristic in

TABLE 2. *Sources of* Pantoea *species*

Species	Human Sources	Nonhuman Sources
P. agglomerans	Leg wounds, knee exudate, sputum, trachea, spinal fluid, urine, conjunctiva, peritoneal dialysis	Water, plants, carrot juice, mango
P. dispersa	Blood, leg wound	Intravenous fluid cap, roses, hibiscus, sorghum, snail vine, seeds
P. ananas	Blood, wounds, finger wound	Environment (unspecified), banana, cereal rusts, pineapple, soil
P. stewartii		Foxtail millet, pearl millet, corn, cluster bean, corn flea beetle, beetle
P. punctata		Mandarin orange
P. citrea		Mandarin orange, pineapple
P. terrea		Soil

From references 1, 4–10, and 12.

TABLE 3. *Differential characteristics of* Pantoea *species*

Characteristic	*P. agglomerans*	*P. dispersa*	*P. ananas*	*P. stewartii*	*P. citrea*	*P. punctata*	*P. terrea*
Yellow pigment	+	V	+	+	−	−	−
Motility	+	+	+	V	−	−	+
Indole	−	−	+	V	−	−	−
Esculin	+	−	V	V	−	V	+
ONPG[a]	+	+	+	+	+	−	−
Gelatinase[b]	+	+	V	−	−	−	−
Acid from							
Salicin	+	−	+	V	V	−	+
L-Rhamnose	+	+	V	V	−	−	−
D-Mannitol	+	+	NA	NA	+	−	−
α-Methylglucoside	NA	NA	+	−	−	−	−

[a]*o*-Nitrophenyl-β-D-galactopyranoside.
[b]Activity weak and detected only after extended incubations (>48 h).
V, variable (15%–85% positive); NA, data not available.

more recent investigations.[10] Some aerogenic isolates have been found to become anaerogenic over time. Although most commercial systems have not yet been evaluated for their ability to correctly identify *Pantoea* species, one system (API 20E) has been found to be unsatisfactory for such purposes.[6,8]

Lindh and associates[10] described four biotypes of *P. agglomerans* based on acid production from α-methylglucoside and citrate utilization. They also noted that, on subculture, nonpigmented segregants arose from yellow-pigmented colonies. These nonpigmented varieties differed from the original pigmented colonies by their inability to ferment certain carbohydrates such as salicin and maltose. Two subspecies of *P. stewartii* are recognized.[8] Their separation is based on differences in indole production, citrate utilization, and fermentation of cellobiose, lactose, maltose, arbutin, salicin, and esculin hydrolysis. These differences at present do not appear to have any major ecologic significance.

Very little additional information is available on the genus *Pantoea.* Cha and associates[12] have tentatively identified the gene in *P. citrea* responsible for pink disease of pineapple. The gene, designated *pnk,* codes for a D-glucose dehydrogenase. Mutations in *pnk* abolish the ability of *P. citrea* to cause pink disease. Pathogenic factors in other species have not been identified. For further general background on clinical, cultural, biochemical, and antimicrobial susceptibility studies on these organisms, see the discussion of *E. agglomerans* in Chapter 10.

REFERENCES

1. Brenner DJ, Fanning GR, Knutson JKL, Steigerwalt AG, Krichevsky MI. Attempts to classify *Herbicola* group-*Enterobacter agglomerans* strains by deoxyribonucleic acid hybridization and phenotypic tests. *Int J Syst Bacteriol* 1984;34:45–55.
2. Tamura K, Sakazaki R, Kosako Y, Yoshizaki E. *Leclercia adecarboxylata* gen. nov., comb. nov., formerly known as *Escherichia adecarboxylata. Curr Microbiol* 1986;13:179–184.
3. Grimont F, Grimont PAD. Chapter 148. The genus *Enterobacter.* In: Balows A, Trüper HG, Harder W, Schleifer K-H, eds. *The prokaryotes: A handbook on the biology of bacteria ecophysiology, isolation, identification, applications.* 2nd ed. Berlin: Springer-Verlag, 1991;2797–2815.
4. Lind E, Ursing J. Clinical strains of *Enterobacter agglomerans* (synonyms: *Erwinia herbicola, Erwinia milletiae*) identified by DNA–DNA hybridization. *Acta Pathol Microbiol Immunol Scand B* 1986;94:205–213.
5. Beji A, Mergaert J, Gavini F, et al. Subjective synonymy of *Erwinina herbicola, Erwinia milletiae,* and *Enterobacter agglomerans* and redefinition of the taxon by genotypic and phenotypic data. *Int J Syst Bacteriol* 1988;38:77–88.
6. Gavini F, Mergaert J, Beji A, et al. Transfer of *Enterobacter agglomerans* (Beijerinck 1888) Ewing and Fife 1972 to *Pantoea* gen. nov. as *Pantoea agglomerans* comb. nov. and description of *Pantoea dispersa* sp. nov. *Int J Syst Bacteriol* 1989;39:337–345.
7. Kageyama B, Nakae M, Yagi S, Sonoyama T. *Pantoea punctata* sp. nov., *Pantoea citrea* sp. nov., and *Pantoea terrea* sp. nov. isolated from fruit and soil samples. *Int J Syst Bacteriol* 1992;42:203–210.

8. Mergaert J, Verdonck L, Kersters K. Transfer of *Erwinia ananas* (synonym, *Erwinia uredovora*) and *Erwinia stewartii* to the genus *Pantoea* emend. as *Pantoea ananas* (Serrano 1928) comb. nov. and *Pantoea stewartii* (Smith 1898) comb. nov., respectively, and description of *Pantoea stewartii* subsp. *indologenes* subsp. nov. *Int J Syst Bacteriol* 1993;43:162–173.
9. Verdonck L, Mergaert J, Rijckaert C, Swings J, Kersters K, De Ley J. Genus *Erwinia:* Numerical analysis of phenotypic features. *Int J Syst Bacteriol* 1987;37:4–18.
10. Lindh E, Kjaeldgaard P, Frederiksen W, Ursing J. Phenotypical properties of *Enterobacter agglomerans (Pantoea agglomerans)* from human, animal and plant sources. *Acta Pathol Microbiol Immunol Scand* 1991; 99:347–352.
11. Astagneau P, Gottot S, Gobin Y, et al. Nosocomial outbreak of *Enterobacter agglomerans* pseudobacteraemia associated with non-sterile blood collection tubes. *J Hosp Infect* 1994;27:73–79.
12. Cha J-S, Pujol C, Kado CI. Isolation and characterization of the gene in *Pantoea citrea* responsible for the pink disease of pineapple. *Phytopathology* 1995;85:1156.

12

The Genus *Citrobacter*

TAXONOMY

The genus *Citrobacter* has undergone extensive taxonomic revisions since its original description. In 1932, Werkman and Gillen[1] described a group of gram-negative bacteria that utilized sodium citrate and produced trimethylene glycol from glycerol (citrobacters). Over the next four decades, organisms we now know to be true citrobacters were placed in a number of other genera (eg, *Escherichia, Salmonella*) or assigned to poorly defined clusters such as the Bethesda-Ballerup group. Studies conducted during this period were complicated by the lack of complete biochemical characterization and the failure to preserve type or reference strains of respective species or groups. By the late 1960s and early 1970s, three major groups of biochemically defined citrobacters were recognized, namely *Citrobacter freundii,* which dated back to the original work of Braak in 1928[2]; *Citrobacter koseri,* described by Frederiksen in 1970[3]; and *Citrobacter amalonaticus,* an organism that was originally placed in the genus *Levinea* by Young and coworkers.[4] In 1972, Ewing and Davis[5] described *Citrobacter diversus,* which retrospectively appeared to be phenotypically similar to *C. koseri.*

In 1974, the first major taxonomic investigations on the genus *Citrobacter* using molecular techniques were published by the Walter Reed Army Institute of Research.[6] Results of these pioneering studies were subsequently summarized by Farmer[7] and indicated that three distinct relatedness groups, *C. freundii* (indole-negative, H_2S-positive), *C. diversus* (indole- and malonate-positive, H_2S-negative), and *C. amalonaticus* (*"Levinea amalonatica";* indole-positive, H_2S-, and malonate-negative), existed within the genus. Crosa and associates[6] also clearly showed that members of the genus *Levinea* were no more closely related to each other than they were to members of the genus *Citrobacter,* and the proposal to transfer *L. amalonatica* to *C. amalonaticus* was made.[7] Crosa's study[6] found significant genetic heterogeneity within *C. freundii;* ^{32}P-labeled DNA from selected *C. freundii* isolates exhibited only 43% to 100% relatedness (binding) to reference strains at 60°C (with ΔT_m ranging from 0° to 13.7°C). These observations suggested the possible existence of multiple taxa (species) within *C. freundii.*

In 1993, an international collaborative effort resulted in the proposal to recognize a number of new species (genomospecies) within the genus *Citrobacter.*[8] Based on DNA–DNA reassociation kinetics, 11 distinct genomospecies were established. Of the three currently recognized species within the genus, only *C. koseri ("C. diversus")* was found to be genetically homogeneous. The phenospecies *C. freundii* was found to be composed of eight genomospecies, including the type species, *C. freundii* sensu stricto. Four of the seven new genomospecies within the *C. freundii* complex were named *(Citrobacter youngae, Citrobacter braakii, Citrobacter werkmanii, Citrobacter sedlakii)* with genomospecies 9, 10, and 11 remaining unnamed due to the limited number of strains available for analysis. Subsequent investigations on genomospecies 9 led to the proposal to name this group *Citrobacter ro-*

TABLE 1. *Current taxonomic composition of the genus* Citrobacter

Genomospecies	Species Designation: Current	Species Designation: Previous	Type Strain	Isolated from Clinical Material
1	*C. freundii*		ATCC 8090	Yes
2	*C. koseri*	*C. diversus, L. malonatica*	ATCC 27028	Yes
3	*C. amalonaticus*	*L. amalonatica*	ATCC 25407	Yes
4	*C. farmeri*	*C. amalonaticus* (biogroup 1)	ATCC 51112	Yes
5	*C. youngae*		ATCC 29935	Yes
6	*C. braakii*		ATCC 51113	Yes
7	*C. werkmanii*		ATCC 51114	Yes
8	*C. sedlakii*		ATCC 51115	Yes
9	*C. rodentium*		ATCC 51116	No
10	Unnamed *("C. gillenii")*		ATCC 51117	Yes
11	Unnamed *("C. murliniae")*		ATCC 51118	Yes

dentium (biotype 4280), the causative agent of murine colonic hyperplasia.[9] A previously recognized biogroup of *C. amalonaticus* (biogroup 1) was found by DNA hybridization to deserve species status, and the name *Citrobacter farmeri* was proposed.[8] In addition to these taxonomic proposals, the correct name for *C. diversus ("C. koseri")* was challenged.[10] Although the most commonly used name for this group over the past 20 years has been *C. diversus,* Frederiksen[10] argued that the name *C. diversus* should be rejected because these strains deviated in at least eight phenotypic characteristics from the original *C. diversum* strains described by Werkman and Gillen.[1] In 1994, the Judicial Commission of the International Committee on Systematic Bacteriology ruled that *C. diversus* is a nomen dubium and therefore *C. koseri* is the correct name for this group of organisms.[11] The current taxonomic composition of the genus *Citrobacter* is listed in Table 1.

EPIDEMIOLOGY

Very little information is available on the distribution of *Citrobacter* in nature. In part this is a result of the inexact characterization of *Citrobacter*-like organisms before 1970, which makes interpretation of results difficult. Since the 1970s, only a limited number of investigations have focused on the environmental distribution of these organisms. *Citrobacter* species can be recovered from river water, sewage, soil, foods, and a variety of animals including mammals, birds, amphibians, and reptiles.[12–14] *Citrobacter* has also been linked to the marine environment.[15] In a retrospective survey of more than 8000 strains by Sedlák,[13] 14% originated from foodstuffs and 5% from either water or sewage. In addition to being part of the normal microbial fecal flora of humans, citrobacters have been recovered from the intestinal tract of dogs, cats, horses, cows, birds, and tortoises.[16] *Citrobacter* species are increasingly implicated as important fish pathogens. They have been recovered from various organs of moribund fish, including trout, bass, and channel catfish, and from inhabitants of tropical aquaria.[16]

Most *Citrobacter* infections are nosocomially acquired, and a seasonal variation in the incidence of human infections has not been noted. In a 1991 survey of participating laboratories in the National Nosocomial Infections Surveillance System of the Centers for Disease Control and Prevention (CDC), between 1% and 2% of all nosocomially acquired bacteremias, pneumonias, and wound and urinary tract infections (UTIs) were caused by *Citrobacter.*[17] Most nosocomial infections occur by horizontal transmission, although vertical transmission from mother to offspring has been reported in one instance,[18] and colonization of neonatal fecal flora with *C. freundii* has been linked to contaminated infant formula.[19] Hospital-acquired sporadic infections or outbreaks have been postulated

TABLE 2. *Distribution of recently described* Citrobacter *species*

Species	Human	Animal	Environmental
C. farmeri	Wounds, blood, feces		Soil
C. youngae	Stool, blood, wounds, urine	Unspecified[a]	
C. braakii	Stool, urine, wounds, respiratory tract	Unspecified	Food
C. werkmanii	Stool, blood	Turkey	Soil
C. sedlakii	Stool, urine, blood, wounds, CSF		Water, melon pulp

Data from Brenner et al.,[8] Miki et al.,[23] and Aldová et al.[24]
[a]Animals cited as source but species not given.
CSF, cerebrospinal fluid

to involve hospital personnel through umbilicus-hand or fecal-hand transmission. During such outbreaks, between 8% and 27% of the infants from whom cultures are obtained may be asymptomatically colonized with *C. koseri* in the feces, umbilicus, or nares.[20,21] Medical personnel are often colonized with *Citrobacter* in their gastrointestinal tract, and their hands are often positive as a result of contact with colonized or infected infants. Sampling within hospitals for inanimate or environmental reservoirs containing *Citrobacter* has usually yielded negative results, although one case of *C. freundii* bacteremia was traced to contaminated O-rings from a reprocessed dialyzer.[22] Only a few reports on the distribution of the new *Citrobacter* species have been published since the taxonomic revisions of Brenner and colleagues.[8] These reports are summarized in Table 2.

CLINICAL SIGNIFICANCE

Members of the genus *Citrobacter* are commonly isolated from clinical specimens and are involved in a variety of human infections. In a Veterans Administration Medical Center study, Hodges and coworkers[25] found that more than 80% of all *Citrobacter* isolates recovered from 83 patients with possible infection were clinically significant or possibly significant (indeterminate category). Most infections involved the urinary tract. More than half of the patients with bacteriuria had underlying conditions predisposing them to infection, such as anatomic defects, urethral catheterization, or both.[25] Other sites often found to contain citrobacters included the respiratory tract (pneumonia, respiratory distress), soft tissue or skin infections (surgical, stasis ulcers), and blood (bacteremia). More than two thirds of all *Citrobacter* infections were hospital acquired.

Central Nervous System

The most prominent diseases linked to members of the genus *Citrobacter* involve infections of the central nervous system (CNS; Table 3). As early as 1960, it was recognized that citrobacters had the capacity to cause life-threatening infections such as meningitis in neonates. Subsequent studies have borne out this association: *Citrobacter* has been found to account for more than 1% of all cases of neonatal meningitis reported to the CDC's Bacterial Meningitis and Meningococcemia Surveillance System.[27] The disease involves significant morbidity, and mortality rates approaching 50% have been reported.[26]

Citrobacter meningitis is found exclusively in infants younger than 2 months of age. A single species, *C. koseri* (formerly *C. diversus*), accounts for more than 90% of all reported cases of *Citrobacter* meningitis[27] and is the only species linked to reported outbreaks of neonatal meningitis. Symptoms in infected infants may range from rather vague, generalized complaints such as irritability, lethargy, and poor feeding habits to more pathognomonic signs of bacterial meningitis including fever, vomiting, nuchal rigidity, seizures or convulsions, and bulging fontanelles.[20,21,39,40] Two distinct, age-related populations of infants presenting with *Citrobacter* meningitis have been described.[27] The larger group (>80%) com-

TABLE 3. *Clinical presentations associated with* Citrobacter *infection*

System/Site	Clinical Manifestations	Infection Type	Major Target Population	Associated Risk Factors	Incidence	Mortality	Ref.
CNS	Meningitis	Primary	Infants (<2 mo)	None	1.3%	34%–50%	26–28
	Brain abscess	Primary	Infants (<2 mo)	Preceding meningitis	1.0%	38%	27, 28
	Ventriculitis	Primary	Neonates, adults	Accompanies meningitis/ brain abscess	Rare	Unknown	29, 30
Cardiovascular	Septicemia	Secondary	Adults (>60 y)	Gallbladder disease, cancer; catheterization; recent antimicrobial therapy	0.7%	28%–48%	31, 32
	Endocarditis, pericarditis	Secondary	Adults	Unknown	Rare	33%	26, 33
Intraabdominal	Peritonitis	Secondary	Adults (>60 y)	Spontaneous or surgical disruption of gastrointestinal tract	Unknown	37%–44%	26, 34
	Cholangitis, cholecystitis	Primary	Adults (>60 y)	None	Unknown	37%	34
Genitourinary	Urinary tract infection	Secondary	Adults (>65 y)	Urinary tract abnormalities, indwelling catheters, cystoscopy	5%	0%	26
Bone and joint	Osteomyelitis, spondylodiscitis, arthritis	Secondary	Adults (>55 y)	Vascular compromise; trauma	Uncommon	0%	26, 35, 36
Respiratory	Pneumonia, pneumonitis, bronchitis	Secondary	Adults	Intubation, chronic pulmonary disease	Uncommon	Unknown	26
Wounds	Cellulitis, soft tissue infections (pyomyositis), abscesses	Secondary	Adults	Surgery, trauma, catheterization	5%	Unknown	26, 37
Gastrointestinal	Enteritis	Primary	Children (<3 y)	None	Unknown	0%	38

prises neonates with a mean age at onset of meningitis of 7.5 days; the smaller cluster consists of older, postneonatal infants, with a mean age at onset exceeding 30 days. A higher mortality rate (39% versus 10%) has been associated with the former group.[27] Only a third of these infected infants are simultaneously bacteremic with *Citrobacter.*

An additional sequela of *C. koseri* CNS disease is the development of brain abscesses in infants presenting with *Citrobacter* meningitis. Retrospective computed tomography studies indicate that this occurs in more than 75% of infants with *Citrobacter* meningitis. This disease progression is a unique feature of *C. koseri* because other gram-negative pathogens producing such sequelae (ie, brain abscesses) do so at a much lower rate (≤10%). In addition to an increased mortality rate, more than 90% of the survivors are mentally impaired. Significant long-term sequelae associated with *C. koseri* meningitis and brain abscess formation include abnormal visual responses, motor skill impairment, developmental delays, hydrocephalus, porencephaly, and hemiparesis.[21,40,41] Joaquin and coworkers[42] described an unusual case of neonatal meningitis and bilateral cerebellar abscess formation caused by *C. freundii.* On rare occasions, *C. koseri* ventriculitis and brain abscess formation have been described in adults.[30,43] Two adult cases of *Citrobacter* meningitis developed subsequent to transphenoid surgery and ventriculostomy in diabetic patients.[43] In the first instance, *C. koseri* and *Klebsiella oxytoca* were isolated from the blood and cerebrospinal fluid (CSF) of a 46-year-old man after surgery to remove a pituitary tumor. He was treated with gentamicin, chloramphenicol, and penicillin and survived. A second, fatal episode of *Citrobacter* meningitis was caused by a strain resistant to multiple drugs (including third-generation cephalosporins) of *C. freundii* and occurred in a 64-year-old woman with a cerebellar hematoma causing obstructive hydrocephalus. Both illnesses were thought to have arisen by direct invasion after the neurosurgical procedure.

The risk factors that play a major role in the development of neonatal meningitis are not well appreciated. During outbreaks, *Citrobacter* colonization rates can run as high as 27% in nursery infants,[20] whereas in nonoutbreak situations levels fall lower than 1%.[21] A disease/colonization ratio of 4% has been proposed for *Citrobacter,*[40] yet many infants with CNS disease have no apparent focus of infection. Some studies have linked high colonization rates to intensive care therapy, gavage feeding, and prenatal intrauterine monitoring.[40] Other reports have found statistically significant associations between meningitis/abscess formation and male gender, low birth weight (<2500 g), and care by house pediatricians[21]; many of these infants also received prenatal intrauterine monitoring via scalp electrodes. However, these findings have not always been observed by other investigators, and many full-term infants develop CNS disease without any recognizable risk factors. One report documented in utero transmission of *C. koseri* from an expectant mother with chorioamnionitis to her infant, who later developed sepsis and presumed meningitis.[44] Therefore, full-term infants may well develop nosocomial infections from contaminated or colonized medical personnel or, less frequently, from the mother.

Bacteremia

The genus *Citrobacter* accounts for 0.3% to 0.7% of all cases of bacteremia, and an annual rate of 1.2 to 4.8 cases of septicemia per 10,000 admissions/discharges has been projected.[29,32,45] In a study of more than 17,000 blood isolates belonging to the family *Enterobacteriaceae* submitted to the Statens Seruminstitut (Denmark) or to the Public Health Laboratory Service Surveillance Centre (United Kingdom), *Citrobacter* accounted for 1.3% to 1.5% of all enteric isolates involved in sepsis.[2] Sepsis is most commonly seen in adults, typically occurring in the elderly or those suffering from serious underlying medical complications such as leukemia or solid tumors.[32] A 12-year survey of 45 cases of *Citrobacter* bacteremia in a Taiwanese hospital found that 49% of patients had underlying malignancies and an additional 22% had hepatobiliary disease.[46] Fever was the single most common symptom associated with septicemia. Other prominent clinical features were chills (78%), leukocytosis (65%), hypotension (33%), oliguria (33%), and an altered mental status (18%) in one Southeast Asian study.[46] Most investigations have found that fewer than 20% of patients studied had signs of hypotension or shock during their septic crisis. On occasion, *C. freundii* bacteremia can mimic typhoid (enteric) fever.[47] One report from India found *Citrobacter* to be the most common cause of gram-negative sepsis in an analysis of 136 episodes of neonatal bacteremia.[48] These neonates typically were preterm and had low birth weights[49]; case fatality rates exceeded 60%. Cultures collected from infected infants often indicate multisystem disease, with umbilicus stumps, eye swabs, pustules, urine, and catheter tips often yielding citrobacters. *C. freundii* and *C. koseri* are the two species responsible for virtually all cases of adult bacteremia, although the relative percentages of each species vary depending on the patient population, medical facility, and diagnostic services offered.[31,32,34] One major retrospective study of more than 280,000 blood cultures drawn at the Mayo Clinic between 1984 and 1992 found *C. freundii* and *C. koseri* to account for 0.6% and 0.2% of all positive bacterial cultures, respectively.[50]

Citrobacter bacteremia is frequently polymicrobic in nature and is more often nosocomially acquired (65%–77%) than community transmitted. Drelichman and Band[31] found 35% of *Citrobacter* bacteremias to contain another copathogen (as opposed to only 3% of *Escherichia coli* bacteremias). Common copathogens included *E. coli, Klebsiella pneumoniae, Bacteroides fragilis,* and enterococci.[46] Patients particularly predisposed to development of nosocomial infections included those undergoing invasive procedures (eg, catheterization, cystoscopy) and those receiving antimicrobial chemotherapy within 10 days before their septic episode. The single most common portal of entry for *Citrobacter* bacteremia is the urinary tract (31%–38% of all such illnesses).[31,32] Other common sources of sepsis are the gastrointestinal and biliary tracts.[31,34] In the study by Drelichman and Band,[31] *C. koseri* septicemia more often originated from the genitourinary tract, whereas persons with gallbladder or gastrointestinal disease more commonly presented with *C. freundii* bacteremia. A high mortality rate (48%–50%) has been associated with *Citrobacter* sepsis in several studies,[26,31] and higher fatality rates have been associated with polymicrobic sepsis (54% versus 14%) and nosocomially acquired infections (54% versus 29%). Samonis and colleagues[32] studied 65 episodes of monomicrobic bacteremia in cancer patients and found several clinical characteristics to be signs of an unfavorable outcome of the septic crisis. These prognostic features included isolation of *Citrobacter* from blood after the initiation of appropriate therapy and the presence of pneumonia, hemorrhage, or shock. No differences in response rate were observed related to the species of *Citrobacter* isolated, the source of infection, or the patient's age, sex, race, or underlying disease status.[31,32]

Urinary Tract Infections

The urinary tract is the most common anatomic site that citrobacters are isolated from in the clinical laboratory. Most strains are clinically significant (60%–65%) and are present as pure cultures (60%–75%).[26,51] Early studies conducted by Fields and coinvestigators[51] found that more than 15% of all *Citrobacter* UTIs involve the upper urinary tract. More than 95% of all adults presenting with *Citrobacter* UTI have one or more preexisting urologic complications, such as neoplasias (eg, prostatic cancer), indwelling catheters, or diabetes mellitus, which may predispose them to infection.[26,51] On rare occasions, more serious urinary tract problems, such as cystitis cystica or perinephric abscesses due to *Citrobacter,* may develop.[26] A rare complication of UTI in diabetic patients is emphysematous pyelonephritis. Fischer and coworkers[52] described a case of emphysematous pyelonephritis caused by *C. freundii* in a 56-year-old woman with an 11-year history of diabetes mellitus. She initially presented with fever and pain in her right flank. *C. freundii* was recovered from her urine and from gluteal and perinephric abscesses. Resolution of this life-threatening infection required a nephrectomy in addition to antibiotic therapy.

Other Infections

Citrobacter species are commonly isolated from superficial wounds; the clinical significance of this fact is questionable because the organisms are recovered from such sites as part of a mixed microbial flora. In most instances citrobacters are opportunistic pathogens. Murphey and coworkers[53] highlighted *C. koseri* as a wound pathogen associated with freshwater and saltwater injuries. *C. freundii* was recovered along with other aquatic bacteria from a patient with a catfish-related injury of the right arm and thigh. *C. freundii* pyomyositis was described in a 72-year-old man with an infected left thigh[37] and in a 45-year-old man with human immunodeficiency virus infection, cirrhosis, diabetes, and chronic leg ulcers.[54] Spondylitis due to *C. koseri* has been reported on a couple of occasions.[36,55]

The role of citrobacters in respiratory tract disease is also speculative in most instances because the organisms usually are recovered in mixed culture. Lipsky and associates[26] found that only one of five sputum cultures containing *Citrobacter* yielded pure growth; only three of these isolates were thought to be involved as secondary pathogens. In a larger, dated study of 43 respiratory isolates, only four strains (<10%) were judged to be clinically significant, and only one (3%) represented a primary infection.[51]

Citrobacters may play a role as etiologic agents of infectious gastroenteritis. Long recognized as normal inhabitants of the gastrointestinal tract, they have on occasion been implicated as causes of sporadic diarrhea in infants.[56] One study from Italy found a significant association between the presence of *Citrobacter* in feces and the diarrheal state.[38] *Citrobacter* was recovered more than twice as often from patients as from controls (14% versus 6.5%). Affected children had mild to severe watery diarrhea causing dehydration and required hospitalization; toxigenic citrobacters were recovered from their stools. In 1995, a German investigation implicated verotoxigenic *C. freundii* as the etiologic agent in a diarrheal disease outbreak associated with hemolytic-uremic syndrome.[57] The outbreak was linked to the consumption of sandwiches made with green butter by nursery school children. Several other clinical conditions infrequently associated with *Citrobacter* are listed in Table 2.

LABORATORY IDENTIFICATION

Isolation

Citrobacter species grow on a wide array of common laboratory media, including selective and differential agars designed for the isolation of gram-negative aerobic and facultatively anaerobic bacilli. Their frequent isolation from contaminated body sites (eg, sputum, wounds) indicates that recovery usually is not a problem (Fig. 1). On fecal specimen workups, some strains of *Citrobacter* can be recovered on virtually any medium used for the isolation of enteric pathogens. Although many citrobacters grow on bismuth sulfite agar, they to not possess the normal colonial morphology of *Salmonella* (black with metallic sheen) but rather appear as brownish-black colonies without a sheen. On Salmonella-Shigella (SS) agar, most citrobacters are lactose-positive and are readily distinguishable from both *Salmonella* and *Shigella;* formation of H_2S is most likely inhibited by acid production from carbohydrate metabolism. Occasionally, lactose-negative H_2S-positive strains of *Citrobacter* may grow on SS and are morphologically indistinguishable from *Salmonella.* When *Citrobacter* species are actively sought from fecal specimens, their isolation can pose a serious problem because their ability to ferment lactose can mask their presence on such media as

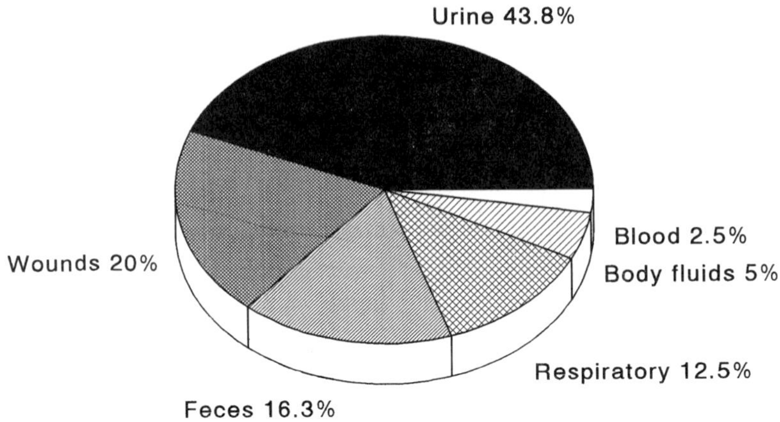

FIG. 1. Anatomic distribution of 80 unselected *Citrobacter* isolates recovered from seven different medical centers.

MacConkey and xylose-lysine-desoxycholate (XLD) agars. *Citrobacter* reportedly can be isolated from feces on media containing 0.5% L-tyrosine; citrobacters are one of the few genera in the *Enterobacteriaceae* besides the *Morganella-Proteus-Providencia* group that can clear tyrosine,[58] possibly by elaboration of a tyrosine phenol lyase.[2,59] One other attractive alternative for isolation of citrobacters is the use of cefsulodin-irgasan-novobiocin (CIN) agar, a medium originally formulated for the recovery of *Yersinia enterocolitica.* Many strains of *Citrobacter* grow extremely well on CIN agar. After 24 hours of incubation, *Citrobacter* colonies are identical in appearance to those of *Yersinia,* exhibiting a classic "bull's-eye" morphology with a red center and a translucent edge (Fig. 2; Color Plate 16).

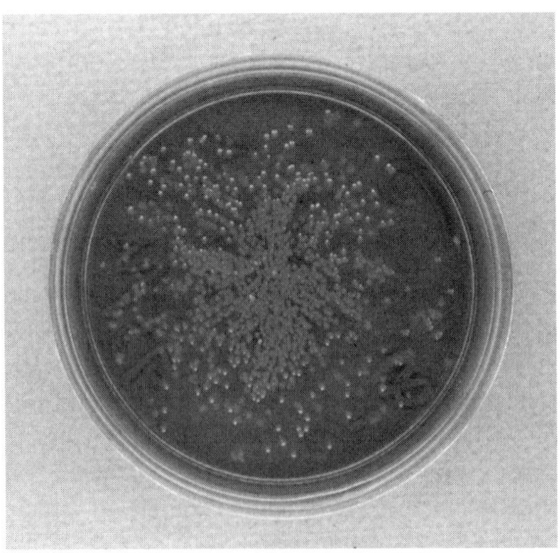

FIG. 2. *Citrobacter* on CIN agar. Also see Color Plate 16.

Biochemical Methods

Commercial Systems

Most laboratories use manual, semiautomated, or automated commercial test kit systems to biochemically identify gram-negative rods. Cumulative results on the ability of several of these systems to accurately identify *Citrobacter* isolates to species are listed in Table 4. Assuming a minimum accuracy of 90% or better for an acceptable species identification, most systems have problems identifying one or more *Citrobacter* species. Problems encountered in *Citrobacter* identification vary from company to company. Many isolates also require supplemental conventional biochemical tests in addition to test kit results for an accurate species identification. York and colleagues[64] found in an analysis of more than 400 gram-negative bacilli tested on the autoSCAN-W/A system (Dade MicroScan, West Sacramento, CA) that more than 50% of strains required additional biochemical tests; 32 of these

TABLE 4. *Identification of* Citrobacter *species by commercial systems*

		C. freundii Complex				*C. koseri*				*C. amalonaticus*			
System	Time[a]	No.[b]	CI[c]	CLP	Inc[d]	No.	CI	CLP	Inc	No.	CI	CLP	Inc
Biolog GN	24 h	22	100[e]	0	0	19	100	0	0	18	100	0	0
AutoSCAN W/A	2 h	71	76	7	17	39	92	0	8	28	89	4	7
Cathra Repliscan	18–24 h	10	100	0	0	15	100	0	0	NT	—	—	—
Vitek AMS	4–18 h	58	91	3	5	29	100	0	0	17	82	12	6
API 20E	21–24 h	74	95	0	5	58	78	0	22	27	59	0	41
Cobas Micro ID-E/NF	21 h	10	80	0	20	5	80	0	20	5	80	0	20
RapID onE	4 h	12	100	0	0	10	100	0	0	10	90	0	10
Crystal ID E/NF	18–20 h	38	97	0	3	29	97	0	3	8	88	0	12

Data from references 60–71.
[a]Time of final identification.
[b]Number of strains tested.
[c]Includes strains requiring supplemental tests for identification.
[d]Includes unidentified strains and those requiring prolonged incubations.
[e]Percent correctly identified.
CI, correct identification to species; CLP, correct (low probability); Inc, incorrect; NT, not tested.

isolates were citrobacters that could not be assigned unambiguously to either *C. koseri* or *C. amalonaticus* without malonate, adonitol, and potassium cyanide (KCN) results. Other systems, such as the API 20E (Vitek bioMérieux, Hazelwood, MO), do not separate *C. koseri/C. amalonaticus* in their database.[70] Some systems, such as the Biolog GN microplate (Biolog, Hayward, CA), are much more accurate in identifying *Citrobacter* species at 24 hours (100%) than at 4 hours (27%) using an automated reading device.[60,68] Most systems incorrectly identify citrobacters as either *E. coli, Escherichia vulneris, Enterobacter cloacae, Enterobacter amnigenus,* or *K. pneumoniae.* One study performed by the CDC in 1995, after the taxonomic revisions within the genus, suggested that many commercial systems should be able to identify these newly described species with only minor adjustments to their current algorithms.[72]

Conventional Tests

The genus *Citrobacter* can easily be differentiated from most other genera in the family *Enterobacteriaceae* on the basis of several key biochemical reactions. Most citrobacters produce L-pyrrolidonyl peptidase (PYR); this helps distinguish them from *Salmonella,* which are negative for this trait.[73] Almost all citrobacters except *C. farmeri* utilize sodium citrate (Simmons' citrate agar) and are methyl red–positive, Voges-Proskauer–negative, and lysine decarboxylase–negative. Biochemical separation of the three traditional species (groups) can be accomplished according to the phenotypic characteristics listed in Table 5. Although almost all *C. koseri* strains produce H_2S in cysteine-based media,[74] only *C. freundii* complex members are positive on triple sugar iron (TSI) slants. *C. koseri* strains can easily be differentiated from *C. amalonaticus* and *C. freundii* by their ability to ferment D-arabitol and D-adonitol. A number of other tests have been found useful in separating these groups, including growth in KCN broth, esculin hydrolysis, and salicin fermentation[2,12]; approximately 15% of *C. amalonaticus* strains are gelatinase-positive.[2] *C. farmeri,* formerly a biogroup of *C. amalonaticus,*[75] ferments sucrose, melibiose, and raffinose, which allows for its easy separation from classic *C. amalonaticus* strains (sensu stricto).

Because commercial systems are unable to separate members of the *C. freundii* complex, identification of isolates to species requires the use of conventional tests. A large number of biochemical properties are useful for separating members of the *C. freundii* complex (Table 6). Although the data presented in Table 6 represent cumulative results at 7 days, more than 90% of all positive results for each species occur within the first 48 hours. Some tests (eg,

TABLE 5. *Distinguishing biochemical characteristics of major groups within the genus* Citrobacter

Test	*C. freundii* Complex	*C. koseri*	*C. amalonaticus* Group[a]
Indole	–	+	+
H_2S production (TSI)	+	–	–
Acid from			
Adonitol	–	+	–
D-Arabitol	–	+	–
Malonate	V	+	–
Cephalothin (30 μg)	R	S	S
Carbenicillin (100 μg)	S	R	R
C_8-Esterase[b]	–	–	+
β-Glucosidase[b]	–	+	–

Adapted from Farmer[7] and Janda et al.[74]; API-ZYM data from unpublished material.
[a]Includes *C. farmeri.*
[b]Based on API-ZYM results.
+, >90% positive; –, >90% negative; V, 10%–90% positive; TSI, triple sugar iron agar; S, susceptible; R, resistant.

malonate) must have final readings at 48 hours, since prolonged incubations may produce uninterpretable results. Other useful tests, such as acid production from salicin and α-methyl-D-glucoside fermentation as proposed by Brenner and colleagues,[8] require longer incubation periods (>72 hours) for positive reactions to appear, making their use impractical in most laboratory situations. Carbon source utilization tests (BioMérieux, La Balme les Grottes, France) are also capable of separating members of the *C. freundii* complex strains into their respective genomospecies.[8]

In one study,[74] the use of a battery of conventional biochemical tests allowed for easy assignment of more than 85% of all *Citrobacter* strains to specific genomospecies (Fig. 3). Approximately 50% of all *Citrobacter* isolates of clinical origin were identified as either *C. freundii* sensu stricto or *C. koseri;* these two species were responsible for UTIs and life-threatening infections (bacteremia). Of the newly recognized groups within the *C. freundii* complex, both *C. youngae* (24%) and *C. braakii* (14%) were commonly isolated from clinical material and were more common than *C. amalonaticus. C. youngae, C. werkmanii,* and *C. sedlakii* have all been isolated from blood, indicating that these newly designated species have the capacity to cause serious human disease.[8,74]

Biotyping

Biotyping is a technique used primarily for epidemiologic purposes. Two biotypes of *C.*

TABLE 6. *Biochemical identification of species within the* Citrobacter freundii *complex*

Test	*C. freundii*	*C. youngae*	*C. braakii*	*C. werkmanii*	*C. sedlakii*
Indole	–	–	–	–	+
Citrate	+	+	+	+	+
Urea	+	+	+	+	+
ODC	–	–	+	–	+
Acid from					
Sucrose	+	V	–	V	–
Dulcitol	V	+	V	–	+
Melibiose	+	–	+	–	+
Malonate	–	V	–	+	+

Adapted from Janda et al.[74]
+, ≥90% positive; –, ≤10% positive; V, 10%–90% positive; ODC, ornithine decarboxylase.

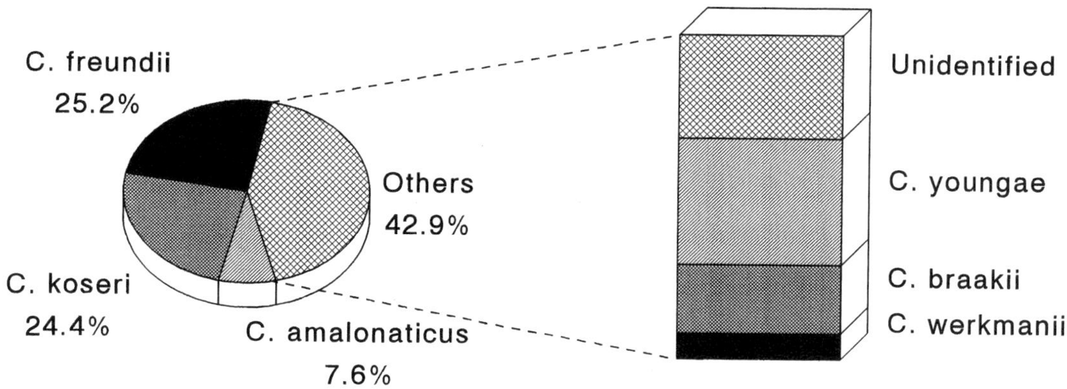

FIG. 3. Relative frequency of *Citrobacter* species in clinical material. Expanded pie slice indicates actual species identification of "other" species. Data on 238 isolates from Janda et al.[74]

freundii were distinguished in the older literature based on the ability of citrobacters to produce H_2S on TSI slants.[26] For *C. koseri,* a major cause of nosocomial outbreaks of neonatal meningitis, Richard and associates[76] described six distinct biotypes based on the fermentation of four sugars (Table 7). In this study of 49 *C. koseri* strains (referred to as *L. amalonatica*), biotypes a and c through f were detected, with biotypes a and c predominating. Most U.S. and U.K. studies have implicated *C. koseri* biotypes d[77,78] and e[20,21] as the cause of reported outbreaks. Gross and Rowe[79] determined the biotype distribution for 517 strains of *C. koseri* and found biotype d (48%) to be the most common, followed by biotypes a (25%), e (16%), c (10%), and b (0.7%); no biotype f strains have been reported to cause outbreaks, although Šourek and Aldová[80] noted that 1.9% of *C. koseri* strains surveyed belonged to biotype f. Of 41 CSF isolates of *C. koseri*, biotype d (46%) again predominated, suggesting that this biotype does not have enhanced pathogenicity for humans[79]; biotype e (37%), however, was found more often associated with cases of meningitis, which suggests the possibility of enhanced virulence. Šourek and Aldová[80] have also used this scheme to biotype strains of *C. amalonaticus* and *C. farmeri* (formerly *C. amalonaticus* biotype 1). Among *C. amalonaticus* strains, biotype b (68%) was the most common; all *C. farmeri* isolates were found to be biotype e.

Immunodiagnostics

A number of different antigenic typing systems have been developed for *Citrobacter* species. Early studies recognized consider-

TABLE 7. *Biotypes of* Citrobacter koseri

	Carbohydrate Fermentation			
Biotype	Dulcitol	Sucrose	Sorbose	Rhamnose
a	–	–	(+)[a]	+
b	–	–	+	+
c	+	–	(+)	+
d	+	+	(+)	+
e	–	+	(+)	+
f	–	+	(+)	–

Reprinted from Richard et al.[76] with permission of Institut Pasteur.
[a]Delayed reactions >48 h.

able antigenic diversity in *C. freundii,* with 32 to 48 different O (somatic) and 87 to 90 H (flagellar) types.[7] Chart and coinvestigators[81] identified five major lipopolysaccharide (LPS) patterns within the *C. freundii* complex based on the electrophoretic migration of O polysaccharide side chains in sodium dodecyl sulfate–polyacrylamide gel electrophoresis (SDS-PAGE) gels. Many *C. freundii* strains exhibit serologic cross-reactivity with *Salmonella* antisera, and some strains possess the Vi antigen of *Salmonella* Typhi.[2] This cross-reactivity with *Salmonella* may be caused by structurally identical O-specific polysaccharide chains, which in some cases differ only in the presence of O-acetyl groups.[82] Two strains of *C. freundii* recovered from meat and one isolated from the intestinal tract of a 10-week-old infant who died of sudden infant death syndrome were shown to have somatic antigens similar or identical to those of *E. coli* O157.[81,83] Both strains from meat belonged to LPS type A, an electrophoretic chemotype whose members are known to cross-react with O157.[81] Šourek and Aldová[80] serologically classified *C. koseri* into 9 serovars, *C. amalonaticus* into 20 serovars, and *C. farmeri* into 4 serovars. To date, most of these typing schemes have been used sparingly, the most widely used being that of Gross and Rowe.[84] This scheme recognizes 17 different *C. koseri* O antigens and has been used in a number of epidemiologic investigations.[85–87] Serogroups O1 and O2 have been linked to most outbreaks of *C. koseri* meningitis in the United States and United Kingdom, and they are the two serogroups most commonly detected among *C. koseri* strains from diverse sources, accounting for 60% of all typable strains[79]; among CSF isolates only, serogroups O2 (39%) and O3 (32%) predominate.

Typing

Although many different molecular techniques have been used to fingerprint *C. koseri,* no single assay has sufficient discriminatory power to conclusively type all isolates. Most studies have relied on a combination of conventional (biochemical, serologic) and molecular techniques to achieve maximum sensitivity. Plasmid analysis is a technique frequently employed in determining relatedness between *C. koseri* isolates involved in outbreak situations. Most *C. koseri* strains carry small plasmids of 10 MDa or less, although extrachromosomal elements as large as 50 MDa have been reported.[78] Several outbreaks of meningitis have been epidemiologically linked to *C. koseri* strains possessing single or multiple plasmids in the 4- to 8-MDa range[20,78,87]; however, background (control) isolates often harbor identical elements.[20] Morris and colleagues[87] found 17 different plasmid patterns among 63 strains studied; two plasmid profiles (types I and II) accounted for more than 50% of all strains, indicating that plasmid analysis is not as highly discriminatory as one would like. Particular plasmid patterns could not be linked to other phenotypic traits (biotype) or serotype. This study also investigated the use of chromosomal digests (restriction endonuclease analysis [REA]) as a means of typing isolates. Using the restriction enzyme *Hind*III, related strains were shown by REA to be identical. Morgan and coinvestigators[41] found that of 11 restriction endonucleases tested, *Sac*I digests produced the best discriminatory patterns. Meningitis-related isolates were identical and could easily be differentiated from 11 nonrelated strains tested by this method. REA is expensive and time-consuming when a substantial number of isolates are involved, and, because high-molecular-weight bands may be difficult to resolve visually, it does not always provide sufficient analytic detail.

Several other procedures have been studied as possible methods to fingerprint *C. koseri* strains or detect genetic relatedness. Kline and colleagues[88] surveyed outer membrane proteins (OMPs) of *C. koseri* as possible epidemiologic markers. Of seven OMPs studied, two (35K and 40K) were universally present, and the others varied from strain to strain. This strain-to-strain variation in five OMPs allowed for 11 different electrophoretic OMP profiles to be distinguished among 41 isolates. Related strains generated identical OMP profiles, including minor bands. Other studies have examined multilocus enzyme analysis as a tool in strain discrimination and determination of genetic relatedness.[89] Of 20 enzymatic loci screened, 16 were found to be polymorphic, yielding 16 different electrophoretic types among 42 different isolates; 14 unrelated isolates recovered from CSF produced 8 different electrophoretic types. Repetitive extragenic palindromic (REP) elements present in *C. koseri* and common in many enteric organisms have been amplified by polymerase chain reaction (PCR) to generate complex genomic DNA band patterns.[44,90] Using PCR-REP technology, highly related strains were found to be similar on visual inspection and yielded similarity coefficients of 0.952 or greater; fewer than 3% of more than 200 comparisons between 21 nonrelated isolates produced similar values.[90] A case of neonatal bacteremia and septic shock caused by *C. koseri* infection was traced to the patient's mother, who had developed chorioamnionitis. Two molecular techniques (ribotyping and pulsed-field gel electrophoresis) provided unequivocal evidence of strain relatedness between the maternal and infant isolates.[91]

PATHOGENESIS

Pathogenetic studies on *C. koseri* have focused on determining what virulence characteristics play key roles in the development of neonatal meningitis. Most strains of *C. koseri,* regardless of site of isolation, are piliated and resistant to lysis in 50% pooled human serum.[92] Piliation appears to be mannose-sensitive (type I) in *C. koseri*.[93] Some studies have noted weak hemolysis of human O blood cells on prolonged incubation, but this is not a consistently observed feature in most studies.[92] Kline and associates[92] noted no differences in biotype, piliation, resistance to complement-mediated lysis, or hemolysin production between *C. koseri* strains originating from CSF and non-CSF sources. They did, however, note a striking difference in the expression of a minor, 32-kDa OMP. More than 80% of all *C. koseri* strains recovered from patients with meningitis expressed the 32-kDa protein, whereas fewer than 20% of all non-CSF isolates possessed the same OMP. Further support for the potential pathogenic significance of this OMP comes from the work of Li and coworkers,[89] who demonstrated its presence in genotypically distinct clones from diverse geographic regions. Such widespread dissemination in divergent strains associated with neonatal meningitis lends further credence to the possible pathogenic significance of this protein.

Two animal models have been established to study *C. koseri* meningitis. The infant rat model involves the simultaneous inoculation of the test strain by intraperitoneal and intranasal routes; infection leads to meningitis with abscess formation.[94] A second model involves the intracranial inoculation of about 3×10^3 *C. koseri* cells into 2-day-old mice.[95] CNS disease progresses from a suppurative ventriculitis in the early phases (48 hours) to ventriculomegaly accompanied by bilateral ventricular abscesses 72 to 96 hours after infection. At 6 days, mice exhibit periventricular abscesses, glial scarring, and extensive cortical damage. Mortality in this model correlates with the presence of bacteremia. The severity of the infection depends on not only the strain and the inoculum but also the litter and age of mice. Despite these models, investigators have been unable to link virulence in either system with presence or absence of the 32-kDa OMP, plasmids, attachment or invasion of HEp-2 cells, or the site from which citrobacters were initially recovered.[93,95]

Toxigenic activity has been detected in selected members of the *C. freundii* complex, potentially explaining their sporadic association with bacterial gastroenteritis. Guarino and colleagues[38] described a low-molecular-weight (<10,000), heat-stable, methanol-soluble factor in three strains of *C. freundii* that produced fluid accumulation in suckling mice; the enterotoxigenic activity was stable at acid or neutral pH but was inactivated under alkaline conditions (pH>8). The active agent was purified to relative homogeneity and was found to be an 18-amino-acid protein identical to the STa (ST Ia) enterotoxin of *E. coli*.[96] In 1993, Schmidt and colleagues[97] described seven strains belonging to the *C. freundii* complex (two from feces, five from beef) that reacted with SLT-II gene sequences of *E. coli* O157:H7. These strains produced weak cytotoxic activity against Vero cells, which was partially neutralized by antisera raised against an SLT-II variant, SLT-IIvhc. PCR amplification of one *C. freundii* isolate with four sets of SLT-II–derived oligonucleotide primers yielded amplicons consistent with the presence of a single chromosomal copy of the SLT-II gene. Despite these findings, five of these seven strains rapidly became nontoxigenic on subculture; loss of activity correlated with loss of the *slt* gene. A 1995 outbreak of diarrheal disease was attributed to *slt*-II–producing *C. freundii*.[57] The significance of these collective observations remains unclear.

Another possible mode by which citrobacters produce gastrointestinal or systemic illnesses is by an invasive mechanism. Although invasive strains of *C. freundii* have been mentioned in the literature, hard evidence supporting such a mechanism has been lacking. Two isolates of *C. freundii* recovered from UTIs were shown to invade T24 and 5637 cell lines (bladder) at relatively high rates, whereas other monolayers such as INT407 (intestine), HCT-8 (ileocecum), and A-498 (kidney) were found to be less permissive.[98] In T24 cells, *C. freundii* invasion was blocked by inhibitors of protein synthesis, microtubules (colchicine), and coated-pit formation (monodansylcadaverine) but not by agents interfering with microfilament formation (eg, cytochalasin D). *C. freundii* uptake into T24 cells was also blocked in the presence of monensin, an inhibitor of endosome acidification. Invasion in other cell lines such as INT407 and 5637 was both microfilament and microtubule dependent, implying that at least two different cell-specific pathways existed for their uptake. *C. rodentium* biotype 4280 causes murine colonic hyperplasia and produces attaching and effacing (AE) lesions in the mouse colon similar to those induced by enteropathogenic *E. coli* (EPEC).[99] Biotype 4280 exhibits sequence homology with the *eae* gene of EPEC, a locus partially responsible for the characteristic AE lesions seen in the course of the disease. By creation of an internal deletion in the open reading frame containing the *eae* gene of *C. rodentium,* the resulting mutant was unable to colonize the mouse colon[100]; but plasmid-cured isolates of biotype 4280 with intact chromosomal *eae* genes were able to colonize as well as the wild-type strain. Four probes constructed to a large (35-kb) region of DNA encoding the locus for enterocyte effacement (LEE) in EPEC hybridized to the AE-positive biotype 4280 strain but not to two AE-negative isolates.[101] However, *C. rodentium* biotype 4280 is a pathogen of rodents only, and no other *Citrobacter* species has been found to contain *eae*-like sequences.

Of less well characterized virulence factors, a number of *Citrobacter* isolates have been found to produce aerobactin by bioassay.[102] Of 35 strains tested, 11.4% were positive for this hydroxamate siderophore. Khashe and Janda[103] found that all *Citrobacter* species produce an enterobactin-like siderophore but only *C. koseri* elaborates an aerobactin-like compound.

ANTIMICROBIAL SUSCEPTIBILITY

To date, extensive investigations on the susceptibility of *Citrobacter* species to a variety of antimicrobial agents have not been performed, no doubt partly because of previous evidence that citrobacters generate fairly consistent susceptibility profiles to most commonly administered antibiotics. Studies dating back to the early 1980s, however, recognized that *C. koseri, C. amalonaticus,* and *C. freundii* did differ in their susceptibility to ampicillin, carbenicillin, and cephalothin, as determined by disk diffusion.[7,26] These differences have been useful in the identification of isolates to species.[7]

Overall, members of the *C. freundii* complex are inherently more resistant to a greater number of antibiotics than either *C. koseri* or *C. amalonaticus* (Table 8). Lew and associates[34] compared 90% minimal inhibitory concentration (MIC_{90}) values for more than 100 *C. freundii* and *C. koseri* strains subjected to 13 antibiotics over a 5-year period. For virtually every antimicrobial agent tested, *C. freundii* MIC_{90} values for each drug were equal to or greater than those of *C. koseri,* with the exception of carbenicillin and gentamicin. This study also noted significant trends in increasing *C. freundii* resistance to cefaperazone, cefamandole, gentamicin, and tetracycline during the period of epidemiologic surveillance. For *C. koseri,* similar increases were documented for cephalothin, cefoxitin, and chloramphenicol. Two 1994 studies further stressed the alarming trend of increasing resistance within the *C. freundii* complex. In a large study from 43 medical centers,[104] increasing resistance to third-generation cephalosporins was seen in *C. freundii,* and more than 25% of all isolates tested were found to be resistant to ceftazidime. This study also found considerable resistance (~10%) to quinolones, a fact confirmed by the report of Thornsberry,[105] in which 8% to 10% of more than 2000 isolates of *C. freundii* were found to be resistant to ciprofloxacin (MIC range <0.5 to >64). Resistance to cephalosporins appears to be mediated by both plasmid and chromosomal factors. Many *C. freundii* harbor plasmids that carry one of the β-lactamase genes of the TEM family (TEM-1), which can inactivate such agents as ampicillin, carbenicillin, piperacillin, and first-, second-, and third-generation cephalosporins.[106] An additional operative mechanism in *C. freundii* is a chromosomally located inducible β-lactamase gene analogous to the *ampC* gene of *E. coli.*[100] In certain strains, this activity can become constitutively expressed at such levels that hypersecretion of β-lactamase activity leads to inactivation of many third-generation cephalosporins, including cefoxitin, cefotetan, and moxalactam.[106] Most strains of *Citrobacter* are currently susceptible to amikacin, although one Argentinian study found annual amikacin resistance rates in *C. freundii* as high as 21%[107]; some strains of *C. freundii* produce acetyltransferases or nucleotidyltransferases, which inactivate aminoglycosides.[106]

Two studies have examined the reliability of *Citrobacter* susceptibility results on commer-

TABLE 8. *Antimicrobial susceptibility of* Citrobacter *species*

Antimicrobial Agent	Susceptible (%) *C. freundii* Complex	Susceptible (%) *C. koseri*
Ampicillin	5–8[a]	0–1
Carbenicillin	50–87	0–1
Ticarcillin/clavulanic acid	68	NA
Cephalothin	0–19	88–96
Cefazolin	11	NA
Cefuroxime	68	NA
Ceftriaxone	75	NA
Ceftazidime	75	NA
Gentamicin	81–100	83–100
Tobramycin	100	100
Amikacin	89	100
Ofloxacin	90	NA
Ciprofloxacin	90–92	99
Chloramphenicol	76–97	98–100
Trimethoprim-sulfamethoxazole	38–100	99–100
Tetracycline	65–100	80–98
Imipenem	100	NA

From references 26, 31, 34, 79, and 105.
[a]Range in percent susceptibility reported for specific agent.
NA, not available.

cial systems. Kelly and Leicester[63] found an unusually high number of *C. koseri* susceptibility disagreements on the Walkaway system when compared with the relative frequency of that species in test samples. *C. freundii* produced only 3% discrepancies on the same system; specific errors were not identified. York and colleagues,[64] in a more detailed investigation, found that *Citrobacter* species were involved in about 30% of the very major errors in susceptibility to ceftriaxone reported by the MicroScan Walkaway (Dade MicroScan, West Sacramento, CA) system (eg, organisms reported as susceptible, which were resistant by microdilution). Other antimicrobials producing very major errors included piperacillin and ceftazidime.

CONCLUSIONS

Because of the redefinition of species within the *C. freundii* complex, most clinical and epidemiologic data are now outdated, including in vitro susceptibility studies. The reasons for the predilection of *C. koseri* to cause CNS disease remain largely a mystery.

REFERENCES

1. Werkman CH, Gillen GF. Bacteria producing trimethylene glycol. *J Bacteriol* 1932;23:167–182.
2. Frederiksen W, Søgaard P. The genus *Citrobacter*. In: Balows A, Trüper HG, Dworkin M, Harder W, Schleifer K-H, eds. *The prokaryotes: A handbook on the biology of bacteria ecophysiology, isolation, identification, applications.* 2nd ed. Berlin: Springer-Verlag, 1991;2744–2753.
3. Frederiksen W. *Citrobacter koseri* (n. sp.) a new species within the genus *Citrobacter*, with a comment on the taxonomic position of *Citrobacter intermedium* (Werkman and Gillen). *Publ Fac Sci Univ J E Purkyne Brno* 1970;47:89–94.
4. Young VM, Kenton DM, Hobbs BJ, Moody MR. *Levinea*, a new genus of the family *Enterobacteriaceae. Int J Syst Bacteriol* 1971;21:58–63.
5. Ewing WH, Davis BR. Biochemical characterization of *Citrobacter diversus* (Burkey) Werkman and Gillen and designation of the neotype strain. *Int J Syst Bacteriol* 1972;22:12–18.
6. Crosa JH, Steigerwalt AG, Fanning GR, Brenner DJ. Polynucleotide sequence divergence in the genus *Citrobacter. J Gen Microbiol* 1974;83:271–282.
7. Farmer JJ III. The genus *Citrobacter*. In: Starr MP, Stolp H, Trüper HG, Schlegel HG, eds. *The Prokaryotes: A handbook on habitats, isolation, and identification of bacteria*. Berlin: Springer-Verlag, 1981; 1140–1147.
8. Brenner DJ, Grimont PAD, Steigerwalt AG, Fanning GR, Ageron E, Riddle CF. Classification of citrobacteria by DNA hybridization: Designation of *Citrobacter farmeri* sp. nov., *Citrobacter youngae* sp. nov., *Citrobacter braakii* sp. nov., *Citrobacter werkmanii* sp. nov., *Citrobacter sedlakii* sp. nov., and three unnamed *Citrobacter* genomospecies. *Int J Syst Bacteriol* 1993; 43:645–658.
9. Schauer DB, Zabel BA, Pedraza IF, O'Hara CM, Steigerwalt AG, Brenner DJ. Genetic and biochemical characterization of *Citrobacter rodentium* sp. nov. *J Clin Microbiol* 1995;33:2064–2068.
10. Frederiksen W. Correct names of the species *Citrobacter koseri, Levinea malonatica,* and *Citrobacter diversus:* Request for an opinion. *Int J Syst Bacteriol* 1990;40:107–108.
11. Wayne LG. Actions of the Judicial Commission of the International Committee on Systematic Bacteriology on requests for opinions published between January 1985 and July 1993. *Int J Syst Bacteriol* 1994;44: 177–178.
12. Sakazaki R. Genus IV. *Citrobacter* Werkman and Gillen 1932, 173^{AL}. In: Krieg NR, Holt JG, eds. *Bergey's manual of systematic bacteriology.* Baltimore: Williams & Wilkins, 1984;458–461.
13. Sedlák J. Present knowledge and aspects of *Citrobacter. Curr Top Microbiol Immunol* 1973;62:41–59.
14. Goldstein EJC, Agyare EO, Vagvolgyi AE, Halpern M. Aerobic bacterial oral flora of garter snakes: Development of normal flora and pathogenic potential for snakes and humans. *J Clin Microbiol* 1981;13:954–956.
15. Kueh CSW, Kutarski P, Brunton M. Contaminated marine wounds: The risk of acquiring acute bacterial infection from marine recreational beaches. *J Appl Bacteriol* 1992;73:412–420.
16. Toranzo AE, Cutrín JM, Roberson BS, Núñez S, Abell JM, Hetrick FM, Baya AM. Comparison of the taxonomy, serology, drug resistance transfer, and virulence of *Citrobacter freundii* strains from mammals and poikilothermic hosts. *Appl Environ Microbiol* 1994; 60:1789–1797.
17. Schaberg DR, Culver DH, Gaynes RP. Major trends in the microbial etiology of nosocomial infection. *Am J Med* 1991;91:3B72S–3B75S.
18. Finn A, Talbot GH, Anday E, Skros M, Provencher M, Hoegg C. Vertical transmission of *Citrobacter diversus* from mother to infant. *Pediatr Infect Dis J* 1988; 7:293–294.
19. Thurm V, Gericke B. Identification of infant food as a vehicle in a nosocomial outbreak of *Citrobacter freundii:* Epidemiological subtyping by alloenzyme, whole-cell protein and antibiotic resistance. *J Appl Bacteriol* 1994;76:553–558.
20. Williams WW, Mariano J, Spurrier M, et al. Nosocomial meningitis due to *Citrobacter diversus* in neonates: New aspects of the epidemiology. *J Infect Dis* 1984;150:229–235.
21. Lin F-YC, Devoe WF, Morrison C, et al. Outbreak of neonatal *Citrobacter diversus* meningitis in a suburban hospital. *Pediatr Infect Dis J* 1987;6:50–55.
22. Flaherty JP, Garcia-Houchins S, Chudy S, Arnow PM. An outbreak of gram-negative bacteremia traced to contaminated O-rings in reprocessed dialyzers. *Ann Intern Med* 1993;119:1072–1078.
23. Miki K, Tamura K, Sakazaki R, Kosako Y. Re-specia-

tion of the original reference strains of serovars in the *Citrobacter freundii* (Bethesda-Ballerup group) antigenic scheme of West and Edwards. *Microbiol Immunol* 1996;40:915–921.
24. Aldová E, Schindler J, Šourek J, Nemec A, Urbášková P. Detection and identification of *Citrobacter sedlakii* in the Czech republic. *Zentralbl Bakteriol* 1997;285: 389–396.
25. Hodges GR, Degener CE, Barnes WG. Clinical significance of *Citrobacter* isolates. *Am J Clin Pathol* 1978;70:37–40.
26. Lipsky BA, Hook EW III, Smith AA, Plorde JJ. *Citrobacter* infections in humans: Experience at the Seattle Veterans Administration Medical Center and a review of the literature. *Rev Infect Dis* 1980;2:746–760.
27. Graham DR, Band JD. *Citrobacter diversus* brain abscess and meningitis in neonates. *JAMA* 1981;245: 1923–1925.
28. Kline MW. *Citrobacter* meningitis and brain abscess in infancy: Epidemiology, pathogenesis, and treatment. *J Pediatr* 1988;113:430–434.
29. Eppes SC, Woods CR, Mayer AS, Klein JD. Recurring ventriculitis due to *Citrobacter diversus:* Clinical and bacteriologic analysis. *Clin Infect Dis* 1993;17: 437–440.
30. Booth LV, Palmer JD, Pateman J, Tuck AC. *Citrobacter diversus* ventriculitis and brain abscesses in an adult. *J Infect* 1993;26:207–209.
31. Drelichman V, Band JD. Bacteremias due to *Citrobacter diversus* and *Citrobacter freundii. Arch Intern Med* 1985;145:1808–1810.
32. Samonis G, Anaissie E, Elting L, Bodey GP. Review of *Citrobacter* bacteremia in cancer patients over a sixteen-year period. *Eur J Clin Microbiol Infect Dis* 1991;10:479–485.
33. Farr RW, Khakoo RA, Maxwell LP, Hill RC. *Citrobacter* pericarditis secondary to a subphrenic abscess. *Clin Infect Dis* 1994;18:838–839.
34. Lew PD, Baker AS, Kunz LJ, Moellering RC Jr. Intra-abdominal *Citrobacter* infections: Association with biliary or upper gastrointestinal source. *Surgery* 1984; 95:398–403.
35. Bruehl CL, Listernick R. *Citrobacter freundii* septic arthritis. *J Paediatr Child Health* 1992;28:402–403.
36. Sotto A, Bernard JC, Brunschwig C, Blin D, Lopez FM, Combe B. Bifocal cervical spondylodiscitis due to *Citrobacter diversus. Infection* 1994;22:423–424.
37. Fincher R-ME, Jackson MW, Fischer AQ. Case report. *Citrobacter freundii:* A newly reported cause of pyomyositis. *Am J Med Sci* 1990;299:331–333.
38. Guarino A, Capano G, Malamisura B, Alessio M, Guandalini S, Rubino A. Production of *Escherichia coli* STa-like heat-stable enterotoxin by *Citrobacter freundii* isolated from humans. *J Clin Microbiol* 1987; 25:110–114.
39. Parry MF, Hutchinson JH, Brown NA, Wu C-H, Estreller L. Gram-negative sepsis in neonates: A nursery outbreak due to hand carriage of *Citrobacter diversus. Pediatrics* 1980;65:1105–1109.
40. Graham DR, Anderson RL, Ariel FE, Ehrenkranz NJ, Rowe B, Boer HR, Dixon RE. Epidemic nosocomial meningitis due to *Citrobacter diversus* in neonates. *J Infect Dis* 1980;144:203–209.
41. Morgan MG, Stuart C, Leanord AT, Enright M, Cole GF. *Citrobacter diversus* brain abscess: Case report and molecular epidemiology. *J Med Microbiol* 1992; 36:273–278.
42. Joaquin A, Khan S, Russell N, Al Fayez N. Neonatal meningitis and bilateral cerebellar abscesses due to *Citrobacter freundii. Pediatr Neurosurg* 1991;17: 23–24.
43. Tang L-M, Chen S-T, Lui T-N. *Citrobacter* meningitis in adults. *Clin Neurol Neurosurg* 1994;96:52–57.
44. Harvey BS, Koeuth T, Versalovic J, Woods CR, Lupski JR. Vertical transmission of *Citrobacter diversus* documented by DNA fingerprinting. *Infect Control Hosp Epidemiol* 1995;16:564–569.
45. Roberts FJ. A review of positive blood cultures: Identification and source of microorganisms and patterns of sensitivity to antibiotics. *Rev Infect Dis* 1980;2: 329–339.
46. Shih C-C, Chen Y-C, Chang S-C, Luh K-T, Hsieh W-C. Bacteremia due to *Citrobacter* species: Significance of primary intraabdominal infection. *Clin Infect Dis* 1996;23:543–549.
47. Flegg PJ, Mandal BK. *Citrobacter freundii* bacteraemia presenting as typhoid fever. *J Infect* 1989;18: 171–173.
48. Sugandhi RP, Beena VK, Shivananda PG, Baliga M. *Citrobacter* sepsis in infants. *Ind J Pediatr* 1992;59: 309–312.
49. Christo GG, Mathai J, Nalini B, Baliga M, Venkatesh A. Neonatal *Citrobacter* sepsis: Clinical and epidemiological aspects. *Ind J Pediatr* 1990;57:781–784.
50. Cockerill FR III, Hughes JG, Vetter EA, et al. Analysis of 281,797 consecutive blood cultures performed over an eight-year period: Trends in microorganisms isolated and the value of anaerobic culture of blood. *Clin Infect Dis* 1997;24:403–418.
51. Fields BN, Uwaydah MM, Kunz LJ, Swartz MN. The so-called "paracolon" bacteria. *Am J Med* 1967;42: 89–106.
52. Fischer C, Kallerhoff M, Weidner W, Ringert RH. *Citrobacter* emphysematous pyelonephritis in a tuberculous kidney caused by *Citrobacter:* A case report in a diabetic patient. *Ann Urol* (Paris) 1996;30:108–111.
53. Murphey DK, Septimus EJ, Waagner DC. Catfish-related injury and infection: Report of two cases and review of the literature. *Clin Infect Dis* 1992;14: 689–693.
54. Widrow CA, Kellie SM, Saltzman BR, Mathur-Wagh U. Pyomyositis in patients with the human immunodeficiency virus: An unusual form of disseminated bacterial infection. *Am J Med* 1991;91:129–136.
55. Müller GF, Keller H. Spondylitis durch *Citrobacter diversus. Schweiz Med Wochenschr* 1992;122:1173–1177.
56. Guerrant RL, Dickens MD, Wenzel RP, Kapikian AZ. Toxigenic bacterial diarrhea: Nursery outbreak involving multiple bacterial strains. *J Pediatr* 1976;89: 885–891.
57. Tschäpe H, Prager R, Streckel W, Fruth A, Tietze E, Böhme G. Verotoxinogenic *Citrobacter freundii* associated with severe gastroenteritis and cases of haemolytic uraemic syndrome in a nursery school: green butter as the infection source. *Epidemiol Infect* 1995;114:441–450.
58. Müller HE. Occurrence and clinical significance of tyrosine clearing positive *Citrobacter freundii* in human faecal specimens. *Zentralbl Bakteriol* 1986;262: 412–416.

59. Chen H, Gollnick P, Phillips RS. Site-directed mutagenesis of His343→Ala in Citrobacter freundii tyrosine phenol-lyase: Effects on the kinetic mechanism and rate determining step. *Eur J Biochem* 1995;229:540–549.
60. Miller JM, Rhoden DL. Preliminary evaluation of Biolog, a carbon source utilization method for bacterial identification. *J Clin Microbiol* 1991;29:1143–1147.
61. Pfaller MA, Sahm D, O'Hara C, Ciaglia C, Yu M, Yamane N, Scharnweber G, Rhoden D. Comparison of the AutoSCAN-W/A rapid bacterial identification system and the Vitek AutoMicrobic system for identification of gram-negative bacilli. *J Clin Microbiol* 1991;29:1422–1428.
62. O'Hara CM, Rhoden DL, Miller JM. Reevaluation of the API 20E identification system versus conventional biochemicals for identification of members of the family *Enterobacteriaceae:* A new look at an old product. *J Clin Microbiol* 1992;30:123–125.
63. Kelly MT, Leicester C. Evaluation of the Autoscan Walkaway system for rapid identification and susceptibility testing of gram-negative bacilli. *J Clin Microbiol* 1992;30:1568–1571.
64. York MK, Brooks GF, Fiss EH. Evaluation of the autoSCAN-W/A rapid system for identification and susceptibility testing of gram-negative fermentative bacilli. *J Clin Microbiol* 1992;30:2903–2910.
65. Visser MR, Bogaards L, Rozenberg-Arska, Verhoef J. Comparison of the autoSCAN-W/A and Vitek Automicrobic systems for identification and susceptibility testing of bacteria. *Eur J Clin Microbiol Infect Dis* 1992;11:979–984.
66. O'Hara CM, Tenover FC, Miller JM. Parallel comparison of accuracy of API 20E, Vitek GNI, MicroScan Walk/Away Rapid ID, and Becton Dickinson Cobas Micro ID-E/NF for identification of members of the family *Enterobacteriaceae* and common gram-negative, non–glucose-fermenting bacilli. *J Clin Microbiol* 1993;31:3165–3169.
67. Kitch TT, Jacobs MR, Appelbaum PC. Evaluation of RapID onE system for identification of 379 strains in the family *Enterobacteriaceae* and oxidase-negative, gram-negative nonfermenters. *J Clin Microbiol* 1994;32:931–934.
68. Holmes B, Costas M, Ganner M, On SLW, Stevens M. Evaluation of Biolog system for identification of some gram-negative bacteria of clinical importance. *J Clin Microbiol* 1994;32:1970–1975.
69. Holmes B, Costas M, Thaker T, Stevens M. Evaluation of two BBL Crystal systems for identification of some clinically important gram-negative bacteria. *J Clin Microbiol* 1994;32:2221–2224.
70. Robinson A, McCarter, Tetreault J. Comparison of Crystal Enteric/Nonfermenter system, API 20E system, and Vitek Automicrobic system for identification of gram-negative bacilli. *J Clin Microbiol* 1995;33:364–370.
71. Wauters G, Boel A, Voorn GP, Verhaegen J, Meunier F, Janssens M, Verbist L. Evaluation of a new identification system, Crystal Enteric/Non-Fermenter, for gram-negative bacilli. *J Clin Microbiol* 1995;33:845–849.
72. O'Hara CM, Roman SB, Miller JM. Ability of commercial identification systems to identify newly recognized species of *Citrobacter. J Clin Microbiol* 1995;33:242–245.
73. Inoue K, Miki K, Tamura K, Sakazaki R. Evaluation of L-pyrrolidonyl peptidase paper strip test for differentiation of members of the family *Enterobacteriaceae,* particularly *Salmonella* spp. *J Clin Microbiol* 1996;34:1811–1812.
74. Janda JM, Abbott SL, Cheung WKW, Hanson DF. Biochemical identification of citrobacteria in the clinical laboratory. *J Clin Microbiol* 1994;32:1850–1854.
75. Farmer JJ III, Davis BR, Hickman-Brenner FW, et al. Biochemical identification of new species and biogroups of *Enterobacteriaceae* isolated from clinical specimens. *J Clin Microbiol* 1985;21:46–76.
76. Richard C, Brisou B, Lioult J. Étude taxonomique de *"Levinea,"* nouveau genre de la famille des Entérobacterériès. *Ann Inst Pasteur* 1972;122:1137–1146.
77. Graham DR, Anderson RL, Ariel FE, Ehrenkranz NJ, Rowe B, Boer HR, Dixon RE. Epidemic nosocomial meningitis due to *Citrobacter diversus* in neonates. *J Infect Dis* 1981;144:203–209.
78. Goering RV, Ehrenkranz J, Sanders CC, Sanders WE Jr. Long term epidemiological analysis of *Citrobacter diversus* in a neonatal intensive care unit. *Pediatr Infect Dis J* 1992;11:99–104.
79. Gross RJ, Rowe B. *Citrobacter koseri* (syn. *C. diversus*): Biotype, serogroup and drug resistance patterns of 517 strains. *J Hyg Camb* 1983;90:233–239.
80. Šourek J, Aldová E. Importance of serological tests in the diagnosis of *Citrobacter diversus* and *Citrobacter amalonaticus. Syst Appl Microbiol* 1988;11:60–66.
81. Chart H, Willshaw GA, Cheasty T, Rowe B. Structure and antigenic properties of *Citrobacter freundii* lipolysaccharides. *J Appl Bacteriol* 1993;74:583–587.
82. Kocharova NA, Knirel YA, Stanislavsky ES, et al. Structural and serological studies of lipopolysaccharides of *Citrobacter* O35 and O38 antigenically related to *Salmonella. FEMS Microbiol Immunol* 1996;13:1–8.
83. Bettelheim KA, Evangelidis H, Pearce JL, Sowers E, Strockbine NA. Isolation of a *Citrobacter freundii* strain which carries the *Escherichia coli* O157 antigen. *J Clin Microbiol* 1993;31:760–761.
84. Gross RJ, Rowe B. *Citrobacter koseri.* I: An extended antigenic scheme for *Citrobacter koseri* (syn. *C. diversus, Levinea malonatica*). *J Hyg Camb* 1975;75:121–127.
85. Gross RJ, Rowe B, Sechter I, Cahan D, Altmamm G. Antigenic scheme for *Citrobacter koseri* (syn. *C. diversus, Levinea malonatica*): Three new antigens recognized in strains from Israel. *J Hyg* 1981;86:111–115.
86. Sechter I, Altmann G, Cahan D, Braunstein I, Gerichter CB. Antigenic scheme of *Citrobacter koseri. J Clin Microbiol* 1981;14:620–622.
87. Morris JG Jr, Lin F-YC, Morrison CB, et al. Molecular epidemiology of neonatal meningitis due to *Citrobacter diversus:* A study of isolates from hospitals in Maryland. *J Infect Dis* 1986;154:409–414.
88. Kline MW, Mason EO Jr, Kaplan SL. Epidemiologic marker system for *Citrobacter diversus* using outer membrane protein profiles. *J Clin Microbiol* 1989;27:1793–1796.
89. Li J, Musser JM, Beltran P, Kline MW, Selander RK. Genotypic heterogeneity of strains of *Citrobacter diversus* expressing a 32-kilodalton outer membrane protein associated with neonatal meningitis. *J Clin Microbiol* 1990;28:1760–1763.
90. Woods CR Jr, Versalovic J, Koeuth T, Lupski JR.

Analysis of relationships among isolates of *Citrobacter diversus* by using DNA fingerprints generated by repetitive sequence-based primers in the polymerase chain reaction. *J Clin Microbiol* 1992;30:2921–2929.
91. Papasian CJ, Kinney J, Coffman S, Hollis RJ, Pfaller MA. Transmission of *Citrobacter koseri* from mother to infant documented by ribotyping and pulsed-field gel electrophoresis. *Diagn Microbiol Infect Dis* 1996; 26:63–67.
92. Kline MW, Mason EO Jr, Kaplan SL. Characterization of *Citrobacter diversus* strains causing neonatal meningitis. *J Infect Dis* 1988;157:101–105.
93. Woods CR Jr, Mason EO Jr, Kaplan SL. Interaction of *Citrobacter* diversus strains with HEp-2 epithelial and human umbilical vein endothelial cells. *J Infect Dis* 1992;166:1035–1044.
94. Kline MW, Kaplan SL, Hawkins EP, Mason EO Jr. Pathogenesis of brain abscess formation in an infant rat model of *Citrobacter diversus* bacteremia and meningitis. *J Infect Dis* 1988;157:106–112.
95. Soriano AL, Russell RG, Johnson D, Lagos R, Sechter I, Morris JG Jr. Pathophysiology of *Citrobacter diversus* neonatal meningitis: Comparative studies in an infant mouse model. *Infect Immun* 1991;59:1352–1358.
96. Guarino A, Giannella R, Thompson MR. *Citrobacter freundii* produces an 18-amino-acid heat-stable enterotoxin identical to the 18-amino-acid *Escherichia coli* heat-stable enterotoxin (ST Ia). *Infect Immun* 1989; 57:649–652.
97. Schmidt H, Montag M. Bockemühl J, Heesemann J, Karch H. Shiga-like toxin II-related cytotoxins in *Citrobacter freundii* strains from humans and beef samples. *Infect Immun* 1993;61:534–543.
98. Oelschlaeger TA, Guerry P, Kopecko DJ. Unusual microtubule-dependent endocytosis mechanisms triggered by *Campylobacter jejuni* and *Citrobacter freundii. Proc Natl Acad Sci U S A* 1993;90:6884–6888.
99. Schauer DB, Falkow S. Attaching and effacing locus of a *Citrobacter freundii* biotype that causes transmissible murine colonic hyperplasia. *Infect Immun* 1993; 61:2486–2492.
100. Schauer DB, Falkow S. The *eae* gene of *Citrobacter freundii* biotype 4280 is necessary for colonization in transmissable murine colonic hyperplasia. *Infect Immun* 1993;61:4654–4661.
101. McDaniel TK, Jarvis KG, Donnenberg MS, Kaper JB. A genetic locus of enterocyte effacement conserved among diverse enterobacterial pathogens. *Proc Natl Acad Sci* 1995;92:1664–1668.
102. Martinez JL, Cercenado E, Baquero F, Peréz-Díaz JC, Delgado-Iribarren A. Incidence of aerobactin production in gram-negative hospital isolates. *FEMS Microbiol Lett* 1987;43:351–353.
103. Khashe S, Janda JM. Iron utilization studies in *Citrobacter* species. *FEMS Microbiol Lett* 1996;137: 141–146.
104. Jones RN, Kehrberg EN, Erwin ME, Anderson SC, and the Fluoroquinolone Resistance Surveillance Group. Prevalence of important pathogens and antimicrobial activity of parenteral drugs at numerous medical centers in the United States: I. Study on the threat of emerging resistances: Real or perceived? *Diagn Microbiol Infect Dis* 1994;19:203–215.
105. Thornsberry C. Susceptibility of clinical bacterial isolates to ciprofloxacin in the United States. *Infection* 1994;22(Suppl. 2).
106. Jacoby GA, Archer GL. New mechanisms of bacterial resistance to antimicrobial agents. *N Engl J Med* 1991;324:601–612.
107. Garcia DC, Woloj GM, Piñeiro S, Sordelli DO, Kaufman S. An 8-year study of resistance in gram-negative bacilli isolates from patients with nosocomial infection at one hospital in Argentina. *J Med Microbiol* 1995;42:283–290.

13

The Genus *Yersinia*

TAXONOMY

In 1944, Van Loghem[1] proposed a new genus, *Yersinia,* to honor A.J.E. Yersin, the discoverer of the plague bacillus. The new genus was created to house two zoonotic pathogens, *Yersinia pestis* and *Yersinia pseudotuberculosis* (*"Pasteurella rodentium"*), previously placed in the genus *Pasteurella.* Both of these species differed significantly in morphologic, cultural, antigenic, and biochemical properties from the then type species of *Pasteurella.* Van Loghem reasoned that enough differences existed between the two groups to justify creation of the genus *Yersinia.* Other historical synonyms of *Y. pestis* and *Y. pseudotuberculosis,* which date from the late 1890s through the 1940s, can be found in the reviews of Bottone[2] and Bercovier and Mollaret.[3]

By the mid-1950s three major groups of yersiniae had been defined. In addition to the zoonotic species already mentioned, a third group was described by Schleifstein and Coleman[4] in 1939. They reported on five human isolates of an unidentified bacterium isolated from gastrointestinal specimens (feces, ulcer), a granuloma of the chin, and a glanders-like infection of the left cheek. In 1943, these same authors proposed the name *Bacterium enterocoliticum* for this unidentified group of microorganisms.[5] The group was later referred to as *Yersinia enterocolitica* by Frederiksen in 1964.[6] Unlike *Y. pestis* and *Y. pseudotuberculosis, Y. enterocolitica* was found to be an extremely "loose" species phenotypically. Various biotyping schemes based on indole production, esculin hydrolysis, and carbohydrate fermentation were developed in the late 1960s and early 1970s in recognition of this biochemical heterogeneity. The most notable of these schemes were those of Niléhn,[7] Wauters (Ph.D. thesis), and Knapp and Thal.[8] Almost concurrent to the development of these biotyping schemes, reports began to surface regarding the isolation of biochemically aberrant *Y. enterocolitica.* These strains, often referred to as "rhamnose-positive," "melibiose-positive," or "sucrose-negative" variants of *Y. enterocolitica* or *Y. enterocolitica*-like bacteria, became increasingly common as more laboratories attempted to isolate and identify yersiniae. By the early 1970s, it was increasingly unclear whether these variants represented unusual varieties of *Y. enterocolitica* or new, unrecognized taxa.[9]

A numerical taxonomy study using 111 tests was conducted on 116 *Yersinia* strains by Stevens and Mair in 1973.[10] Based on the analysis of test results performed at both 25°C and 35°C, two major groups of *Y. enterocolitica* strains could be observed. The first group, which consisted of typical *Y. enterocolitica* strains, contained three subgroups based on isolate source (humans, chinchillas) and biochemical properties (xylose fermentation). The second cluster was composed solely of strains recovered from hares. These latter strains differed in several ways from typical isolates; for example, they showed growth in the presence of 3% to 4% NaCl, and they failed to ferment either trehalose or xylose. All *Y. pestis* and *Y. pseudotuberculosis* strains were found to be contained in a single cluster. This study pointed to possible genetic heterogeneity within *Y. enterocolitica* and was followed 3 years later by a study from the Centers for Disease Control (CDC) employing DNA hybridization.[11] In that study of *Y. enterocolitica* and *Y. pseudotuberculosis* strains, both species were found to exhibit a core-level relatedness (10%–25%) to other members of the family *Enterobacteriaceae;* less relatedness was observed between these same two groups and *Pasteurella multocida* (6%–12%). Within *Y. enterocolitica,* four potential DNA relatedness groups were detected by hybridization. These four groups phenotypically corresponded to "typical" *Y. enterocolitica,* two atypical "rhamnose-positive" groups, and one atypical "sucrose-negative" cluster.[11] Typical *Y. enterocolitica* strains contained biogroups 1, 2, and 3 of Knapp and Thal,[8] while biogroup 4, which previously had contained atypical strains, harbored the other 3 DNA relatedness groups. *Y. pseudotuberculosis* was determined to be an extremely homogeneous species.

In 1980, a series of seminal publications by an international group of collaborators helped redefine *Y. enterocolitica* and *Y. enterocolitica*–like organisms.[12] *Y. enterocolitica* sensu stricto was found to contain several biogroups, including the biochemically inactive biogroup 5 of Stevens and Mair[10] that was recovered from hares (Fig. 1). Also found to reside within *Y. enterocolitica* were a number of isolates expressing unusual biochemical properties, such as citrate-positive or urease-negative strains.[13] Among previously recognized atypical *Y. enterocolitica* groups,[11] two rhamnose-positive species were named. *Yersinia intermedia* was the name chosen for a group of rhamnose-positive yersiniae that produced acid from melibiose and raffinose fermentation.[14] *Y. intermedia* was only 60% related to *Y. enterocolitica* at 60°C, with 11% to 12% divergence in sequences. The second new rhamnose-positive species was designated *Yersinia frederiksenii.*[15] Unlike *Y. intermedia, Y. frederiksenii* is negative for meli-

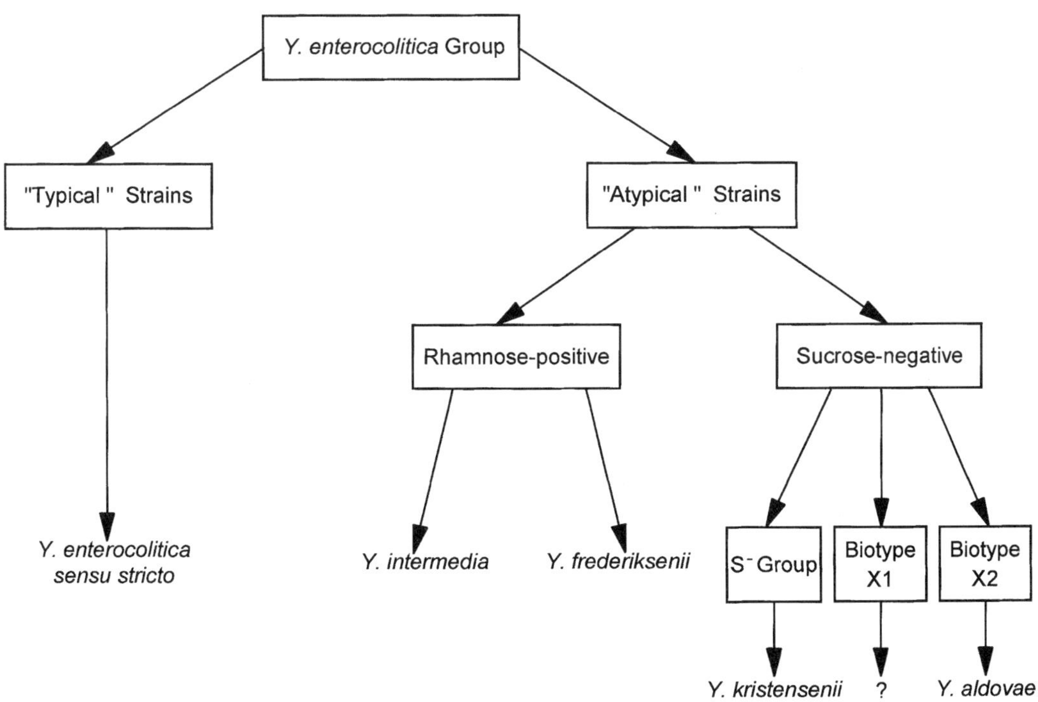

FIG. 1 Taxonomic evolution of species within *Yersinia enterocolitica* group.

biose and raffinose fermentation. Intraspecies hybridization results indicate that the degree of homology between *Y. frederiksenii* and *Y. enterocolitica* (55%–60%) is similar to that between *Y. intermedia* and *Y. enterocolitica.* The final species named by this international group of scientists was a sucrose-negative biotype originally described by Bercovier and colleagues.[16] The name *Yersinia kristensenii* was proposed for this genotypically distinct taxa.[17] *Y. kristensenii* shares a significantly higher homology (59%–74%) at 60°C with *Y. enterocolitica* than either of the two rhamnose-positive species. This degree of relatedness bordered at the level of species identity (>70% homology), suggesting that these sucrose-negative strains may in fact belong within *Y. enterocolitica.* However, the thermal stability of such heteroduplexes indicated an intraspecies divergence of 7% to 11.5%, suggesting that these were distinct groups. Further support for this concept came from hybridization studies conducted under more stringent conditions (75°C). Intraspecies relatedness values between *Y. kristensenii* and *Y. enterocolitica* were significantly reduced (24%–58%), indicating that these are legitimately distinct taxa. Brenner and associates[12] and Bercovier and collaborators[17] also detected two other sucrose-negative *Y. enterocolitica*–like groups that appeared to merit species status. These groups, called biogroups X1[16] and X2, differed in biochemical properties from *Y. kristensenii* based on L-sorbose, L-rhamnose, and D-cellobiose fermentation and ornithine decarboxylase (ODC) activity.[13] In 1984, the name *Yersinia aldovae* was proposed for biogroup X2.[18]

Several other *Yersinia* species have been identified in addition to this main cluster of *Y. enterocolitica*–like organisms. In a systematic investigation of yersiniae, Brenner noted that four strains of the red mouth bacterium were 30% related (15% divergence) to *Y. enterocolitica.*[11] Further analysis indicated that on a genetic basis the red mouth bacterium was most closely related to *Yersinia* and *Serratia* species compared with other *Enterobacteriaceae.* Because the red mouth bacterium had a GC content and biochemical features more closely related to *Yersinia* than *Serratia,* the name *Yersinia ruckeri* was chosen.[19] In 1987, the species name *Yersinia rohdei* was given to a group of indole-negative yersiniae primarily isolated from dog feces.[20] A year later, two groups of strains resembling biogroup 3 of *Y. enterocolitica* were given species status.[21] These species differed from biogroup 3 strains in their ability to degrade mucate and their failure to produce acetylmethylcarbinol (Voges-Proskauer test). The epithets *Yersinia mollaretii* and *Yersinia bercovieri* were chosen for biogroups 3A and 3B, respectively, to honor two microbiologists who had contributed significantly to *Yersinia* taxonomy.[21] Current composition of the genus *Yersinia* is listed in Table 1.

Although there have been many improvements in the taxonomy of yersiniae over the past two decades, there are still a number of problematic areas. *Y. pestis* and *Y. pseudotuberculosis* are more than 90% related at the DNA level.[23] Because of this, *Y. pestis* is actually a subspecies of *Y. pseudotuberculosis* based on priority of recognition and description (see Table 1). However, from a public health and laboratory standpoint, these two groups should still be reported as separate species to avoid confusion. The correct taxonomic position of *Y. ruckeri* is still a matter of some conjecture. However, 16S sequence data appear to support its inclusion in the genus *Yersinia.*[22] *Y. frederiksenii* is known to contain at least three DNA relatedness groups, all of which appear to deserve species status.[15] Analysis of the cellular enzymes of *Y. frederiksenii* by multilocus enzyme electrophoresis supports this conclusion.[24] However, because these DNA relatedness groups cannot be separated biochemically, they remain within *Y. frederiksenii,* making this group a phenospecies composed of at least three genomospecies at present. Other named taxa, such as *Yersinia philomiragia,*[25] originally grouped with yersiniae, have been demonstrated by hybridization not to be true *Yersinia* species.[26] *Y. philomiragia* has been renamed *Francisella philomiragia.*[27] A final point con-

TABLE 1. *Current composition of the genus* Yersinia

Species	Isolated from Humans	Pathogenic for Humans	G+C (mol%)	Subline[a]	Biogroups[b]	Type Strain	Synonyms
Y. pseudotuberculosis	Yes	Yes	46.5	2	1	ATCC 29833	*Y. pseudotuberculosis* ssp. *pseudotuberculosis*
Y. pestis	Yes	Yes	46.0	2	3	ATCC 19428	*Y. pseudotuberculosis* ssp. *pestis*
Y. enterocolitica	Yes	Yes	48.5	1	5	ATCC 9610	
Y. intermedia	Yes	No	48.5	3	8	ATCC 29909	
Y. frederiksenii	Yes	No	48.0	2	4	ATCC 33641	
Y. kristensenii	Yes	No	48.5	3	1	ATCC 33638	
Y. aldovae	No	No	48.0	3	1	ATCC 35236	Biogroup X2
Y. mollaretii	Yes	No	50.0	3	1	ATCC 43969	*Y. enterocolitica* biogroup 3A
Y. bercovieri	Yes	No	50.0	3	1	ATCC 43970	*Y. enterocolitica* biogroup 3B
Y. rohdei	Yes	No	49.1	4	2	ATCC 43380	
Y. ruckeri	Yes	No	47.5–48.5	5	1	ATCC 29473	Red mouth bacterium

[a]Based on the data of Ibrahim et al.[22]
[b]Biotypes or biovars.

cerns the large number of groups or individual strains that either appear to represent new unnamed taxa (biogroup X1) or cannot be placed within any recognized taxa at present.[12,13,17] Some of these latter strains appear to be "bridge" organisms, that is, microbes that biochemically resemble one species but at the DNA level are almost equidistantly related genetically to two distinct species. What to do with such strains is presently unresolved.

EPIDEMIOLOGY

Yersinia species are widely distributed in the microbial biosphere. They inhabit a number of naturally occurring ecosystems, including the aquatic environment (marine and freshwater), soil, the alimentary tract of vertebrates, and products of the biologic food chain (Table 2). Yersiniae have been isolated from almost every geographic region of the world where they have been actively sought. Of 11 current nomenspecies, four are inherently pathogenic for humans or animals (see Table 2). Some groups (eg, *Y. kristensenii*) have been reputed to cause disease in humans but definitive data establishing them as true pathogens are lacking.

As pointed out by Bercovier and colleagues,[23] although *Y. pestis* and *Y. pseudotuberculosis* share some genetic and ecologic similarities, they are strikingly different in epidemiology and in the type of disease presentations associated with each agent. *Y. pestis* is the causative agent of plague and the

TABLE 2. *Environmental distribution of yersiniae*

Species	Natural Ecosystems	Pathogenic Hosts
Y. pseudotuberculosis	Lagomorphs, birds	Rodents, humans
Y. pestis	Rodents	Rodents, humans, cats
Y. enterocolitica	Pigs, rodents	Chinchillas, hares, monkeys, humans
Y. intermedia	Aquatic	None identified
Y. frederiksenii	Aquatic	None identified
Y. kristensenii	Terrestrial	None identified
Y. aldovae	Aquatic	None identified
Y. ruckeri	Aquatic	Rainbow trout
Y. rohdei	Aquatic	None identified
Y. mollaretii	Terrestrial	None identified
Y. bercovieri	Terrestrial	None identified

notorious Black Death that brought frightening morbidity and mortality to Europe during medieval times.[28] Today *Y. pestis* is responsible for epizootic plague in a number of rodent species, including ground squirrels, chipmunks, and prairie dogs.[3,29] It has been isolated from more than 100 different species of rodents in their natural setting.[3] Domestic dogs and ferrets, coyotes, raccoons, and skunks appear to be highly resistant to infection.[30] Human infections are transmitted by diseased animals or carriers through the bite of a flea. Once infected, *Y. pestis* multiplies in the flea until the stomach is essentially blocked by bacteria; plague bacilli are then regurgitated, after which they are introduced into the human host during a blood meal. In the human, the injected microorganisms migrate from the bloodstream to lymphatics, where they produce inflammation and enlargement of regional lymph nodes (buboes) near the infected flea bite. Such localized infections can lead to septicemia and secondary pneumonia if left untreated. During the past 60 years, endemic cases of plague in the United States have been restricted to random infections involving direct or indirect contact with contaminated rodents.[29] Almost all of these cases have been located in the southwestern section of the United States. In a review of plague cases occurring in the United States between 1970 and 1994, Perry and Fetherston[30] found squirrels (44%) to be the most common vector, followed by rabbits (7%) and prairie dogs (6%); the source of infection for a third of all plague cases could not be determined. In addition to this classic route of transmission (rodent to flea to human), both bubonic and pneumonic plague can be acquired by intimate contact with infected cats. Such illnesses result from cat scratches (bubonic) or inhalation of aerosolized droplets from pneumonic-infected cats held in close proximity.[31–33]

The lack of large numbers of bubonic or pneumonic plague cases worldwide should not be interpreted as a disappearance of plague or the lack of a need for public health measures geared toward its potential emergence. These facts were recently brought to the forefront when a major outbreak of bubonic and pneumonic plague was reported from India in 1994.[34] Over a 1-month period, several hundred cases of plague were detected first in the Bir district, Mararashstra state (bubonic) and then subsequently in the city of Surat, Gujarat state (pneumonic). These areas had been naive to human plague infection for almost 30 years.[35] The outbreak followed a 3-year increase in the plague seropositivity rate in rats in Bangalore, as reported by India's Plague Surveillance Unit.[36] Precautions were immediately instituted to prevent further spread of this outbreak through global travel.

Y. pseudotuberculosis is found in a wide range of animals as either a primary pathogen or a commensal. Occasionally it is also recovered from foods and potable water sources.[37,38] In Europe and the United States, birds, rodents, and pigs appear to be the major reservoirs of infection.[2,37] Among species of birds, turkeys yielded the highest isolation rate of *Y. pseudotuberculosis* in one British survey, while other North American investigations have documented its presence in fowl.[2] A German study of isolates recovered from animals over a 10-year period (1983–1993) found hares (25.3%) and ducks (7.9%) to be the two most common species associated with *Y. pseudotuberculosis* infection.[39] Among 90 European brown hare carcasses examined over an 18-month period, 13.3% yielded *Y. pseudotuberculosis* on direct culture of internal organs.[40] In Japan, the reservoir of infection appears to be markedly different; swine, and not birds or hares, appear to be the predominant host for *Y. pseudotuberculosis.*[37]

Infection or colonization with *Y. pseudotuberculosis,* in contrast to *Y. pestis,* results from oral ingestion of yersiniae rather than from the bite of an infected flea.[3] Besides epizootic outbreaks of disease in rodents and fowl, *Y. pseudotuberculosis* has been shown to cause disease in goats and primates (monkeys, humans). Cattle, swine, dogs, cats, and rats have all been found to be carriers of *Y. pseudotuberculosis* on occasion.[37] Illnesses in animals typically range from chronic diarrhea and mesenteric adenitis to fatal episodes of septicemia in disseminated cases.[3] Human disease with *Y. pseudotuberculosis* closely parallels that observed in animals (mesenteric adenitis, septicemia).

Many different environmental sources serve as reservoirs for or transiently contain *Y. enterocolitica*. Chief among these ecologic niches is the alimentary or gastrointestinal tract of animals. *Y. enterocolitica* has been isolated from numerous species of domesticated or wild animals; these reports are discussed in detail in the review of Hurvell.[41] Among this pool, swine, small rodents (field mice, rats, voles), and dogs are the most commonly cited carriers of *Y. enterocolitica*. In a study of 695 serogroup O:3 *Y. enterocolitica* strains isolated from human and nonhuman sources in Japan, 75% of isolates were recovered from pigs.[42] In an epidemiologic investigation conducted by the New York State Department of Health in 1981, pigs (5.5%) and cows (5.2%) were the leading domestic animals found to be positive for *Y. enterocolitica*. *Y. enterocolitica* infections in chinchillas, monkeys, and hares range from diarrhea and wasting disease (weight loss, anorexia), to abscess formation in internal organs (liver, spleen) and septicemia.[40,41] Other reported sources yielding *Y. enterocolitica* include prawns,[38] river and surface water samples,[42,43] frogs,[44] soil,[16] earthworms,[16] crops,[16] birds,[16,41] flies,[42] and animal feed.[42]

Because *Y. enterocolitica* is so widely dispersed in the animal kingdom, it is not surprising that many foods have been demonstrated to contain these bacteria, including meats, dairy products, seafoods, and vegetables.[45] Several food surveillance investigations have reported high incidences of *Y. enterocolitica* in raw meat samples, particularly pork, destined for human consumption.[46,47] Logue[47] found that 89% of all samples of fresh (raw) meat contained yersiniae, 68% harboring *Y. enterocolitica*. Among such meats, beef (88%), lamb (71%), pork (71%), and chicken (55%) had the highest contamination rates.[47] *Y. enterocolitica* was found, in order of decreasing incidence, in raw meat, deli meat, cooked meat, and prepackaged cooked meat. Quantitative cultures indicated that in most instances (~80%) yersiniae were present in low numbers (<500 cfu/g), particularly in cooked meats (<5 cfu/g). However, some poultry, pork, sausage, beef steak, and minced meat samples contained yersiniae in numbers up to 5×10^5 cfu/g.

Despite their prevalence in environmental samples, the epidemiology of *Y. enterocolitica* is extremely complicated. There are significant differences in the prevalence of *Y. enterocolitica* as a human pathogen, which appear to be related to thermal gradients.[48] Colder climates, such as those in Europe, Canada, northern sections of the United States, and Japan, have higher rates of *Y. enterocolitica* infection than do warmer locales. Swaminathan and colleagues[49] found in a review of *Y. enterocolitica* strains recovered from humans, animals, and the environment between 1974 and 1977 that Belgium, Canada, Hungary, and Sweden had the highest isolation rates (>1000 isolates). This fact may be linked to the ability of yersiniae to multiply at low temperatures, potentially enhancing their chance of infectivity. In addition to a geographic tropism for disease production, is the fact that both pathogenic and nonpathogenic strains (the latter often referred to as environmental strains) exist within *Y. enterocolitica* sensu stricto. At the cellular level, these two groups are distinguished on the basis of somatic antigens, biochemical reactions (biotypes), and phenotypic markers of pathogenicity (see Biotyping and Pathogenesis). Most pathogenic strains for humans fall into one of five serogroups, namely, O:3, O:8, O:9, O:5,27, and O:1,2,3. These serogroups are also geographically restricted: most European infections are caused by serogroups O:3 and O:9, whereas O:8 predominates in the United States, O:3 and O:8 in Canada, and O:3 in Japan.[50] Furthermore, outbreaks of disease due to serogroup O:8 are common, but those involving O:3 and O:9 are rare. Reasons for these differences are not readily apparent.

Most strains of *Y. enterocolitica* recovered from environmental samples, foods, or animals, including those detected by enrichment techniques, fall into nonpathogenic serogroups and biogroups. Because person-to-person transmission of yersiniosis is an extremely rare event, the source of most human infections remained a mystery until the last 10 years. Data now suggest that swine are a common source of serogroup O:3 infections.[50] This supposition is based on several lines of evidence. First, pigs have been found to be a major reservoir for *Y. enterocolitica* in various regions of the globe.[45–47,50,51] Second, only swine routinely carry pathogenic strains of *Y. enterocolitica* in their caecum, as defined by serologic and phenotypic markers.[50] Such strains are indistinguishable from those isolated from humans. Third, a number of well-controlled studies, including one consisting of 40 persons from Belgium (which has the highest incidence of yersiniosis in Europe), have epidemiologically linked *Y. enterocolitica* infection to consumption of raw pork.[52] Areas of the world in which small amounts of pork are consumed (eg, the Middle East) have low rates of infection with *Y. enterocolitica*. Over the past decade, O:3 has emerged as a leading cause of yersiniosis in the United States on both coasts,[53–55] and infections in infants and children have been tied to ingestion of raw pork (chitterlings) in some instances.[56] These collective lines of evidence provide a compelling scenario for the role of raw pork in O:3 infections on a worldwide basis. The source of O:8 infections, now also reported from Japan,[57,58] still remains an enigma. Some epidemiologic data, including the isolation of serogroup O:8 from mice and voles in Japan,[58] suggest that rodents are a potential

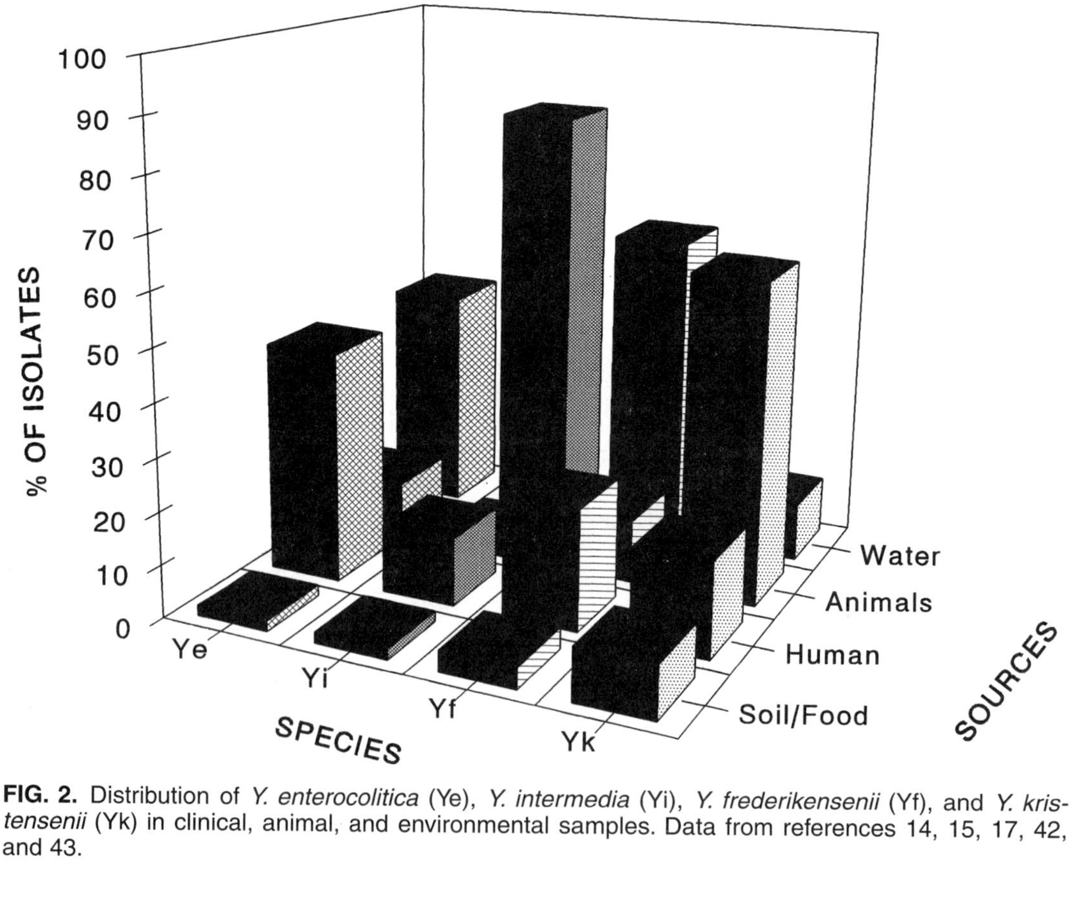

FIG. 2. Distribution of *Y. enterocolitica* (Ye), *Y. intermedia* (Yi), *Y. frederikensenii* (Yf), and *Y. kristensenii* (Yk) in clinical, animal, and environmental samples. Data from references 14, 15, 17, 42, and 43.

source of yersiniosis.[59] Why serogroup "shifts" have occurred in some areas (United States, Japan) and not in others (Europe) is unclear.

The distribution of some *Y. enterocolitica*–like species is illustrated in Fig. 2. Unlike *Y. enterocolitica,* these species tend to predominate in nonhuman sources and are infrequently isolated from clinical material. They are common inhabitants of water, soil, the intestinal tracts of animals, and foods.[14,15,17,38,40,47,50] Although Fig. 2 highlights these differences in distribution between human and nonhuman sources, the data presented may not be an accurate reflection of the primary habitat of each taxa. For instance, one study of sucrose-negative (S^-) strains (later named *Y. kristensenii*) found 70% of soil samples to be positive for this biogroup, suggesting that soil is the predominant ecosystem where *Y. kristensenii* may be found.[16] However, strains used in the taxonomic study naming this group were recovered primarily from animals.[17] Even when these other species are recovered from clinical material (eg, *Y. intermedia* from blood), their role as bona fide pathogens has not been unquestionably established, and little clinical information is available on most of these extraintestinal isolates. Some of the sources from which other *Yersinia* species have been isolated are listed in Table 3.

Outbreaks of disease caused by yersiniae are relatively rare, but their clinical significance and public health impact can be quite dramatic. Besides the 1994 outbreak of pneumonic and bubonic plague in India caused by *Y. pestis,*[34–36] both *Y. pseudotuberculosis* and *Y. enterocolitica* have been documented to cause outbreaks of disease. Infections with these latter two species, unlike many other enteric bacteria *(Serratia, Klebsiella, Enterobacter),* are almost exclusively community acquired and usually involve the consumption of contaminated foods rather than person-to-person transmission.

Y. pseudotuberculosis has been noted to cause rare outbreaks of disease in Europe[61] and in Japan.[37] Often the vehicle of infection cannot be identified.[61] In a review of 12 outbreaks occurring in Japan between 1977 and 1986, Tsubokura and colleagues[37] were able to identify a potential source of infection in 8; the suspected vehicles included water (3 outbreaks), barbecue (2 outbreaks, same restaurant), and a school lunch, sandwiches, and vegetable juice (1 outbreak each). Most *Y. pseudotuberculosis* outbreaks affect primarily young children, who present with a variety of symptoms including gastrointestinal complaints and pseudoappendicitis.[37,61]

The importance of *Y. enterocolitica* as an enteric pathogen went basically unnoticed until the first large-scale foodborne outbreak of yersiniosis in 1976.[62] More than 200 children attending five schools in Holland Patent, an upper New York state community, became in-

TABLE 3. *Reported sources from which uncommon* Yersinia *species have been isolated*

Species	Human Sources	Nonhuman Sources
Y. intermedia	Blood, feces, eye, urine, wound, abscess, knee fluid	Water/rivers, sewage, fish, oysters/shrimp, snails, rodents, milk/cream, meats, poultry, pigs/cows/hares, soil, earthworms
Y. frederiksenii	Feces, blood, sputum	Water, sewage, fish, cattle, rodents, cows/pigs, dogs, hares, milk, sandwiches, meats, soil
Y. kristensenii	Feces, eye	Soil, earthworms, birds, hares/rabbits, water, crops, vegetables, meats, pigs, foods
Y. aldovae		Water, fish, swimming pools
Y. ruckeri		Rainbow trout, Chinook salmon, steelhead trout, sockeye trout
Y. rohdei	Feces	Dogs, water
Y. mollaretii	Feces	Soil, raw vegetables, water, meat, hares
Y. bercovieri	Feces	Soil, raw vegetables, water, hares, coypu, pigs/dogs

Data from references 14–21, 38, 40, 42–47, 50, and 60.

fected with *Y. enterocolitica.*[63,64] Sixteen children underwent appendectomies for what was presumed initially to be acute appendicitis but was in fact a *Y. enterocolitica* mesenteric lymphadenitis infection (pseudoappendicitis). Chocolate milk was epidemiologically implicated as the source of infection. Numerous other large-scale outbreaks of yersiniosis were reported later from the United States, Canada, Japan, and Europe.[48] In the United States, dairy products have often been implicated as the vehicle of infection. Morse and coworkers[65] described an outbreak of yersiniosis in a rural summer camp in New York state in 1981. At least 159 children and camp staff were affected, including the head cook (index case), who apparently transmitted the bacterium to others through his contaminated hands while preparing powdered milk and turkey chow mein. One year later, a major three-state outbreak of *Y. enterocolitica* infections in the southern United States was linked to the consumption of contaminated pasteurized milk.[66] A retrospective analysis at one site (Greenwood, MS) projected that 800 or more persons were infected. Other foods incriminated as sources of yersiniosis outbreaks include tofu, spring water, and bean sprouts.[48] Until the late 1980s, most U.S. outbreaks were caused by serogroup O:8 (in Europe and Japan, by serogroup O:3). However, with the changing seroepidemiology of *Y. enterocolitica* infections in the United States, outbreaks caused by O:3 (chitterlings) have been published.[56] Rarely, nongastrointestinal outbreaks of disease (eg, pharyngitis) caused by *Y. enterocolitica* have also been reported.[67]

CLINICAL SIGNIFICANCE

Despite their infrequent occurrence in clinical specimens relative to other *Enterobacteriaceae,* yersiniae have been documented to cause a variety of intestinal and extraintestinal infections in humans. These illnesses range from pseudoappendicular syndromes to septicemia, pharyngitis, and infections of the joints and bones (arthritis, osteomyelitis). Of the three major species associated with disease, *Y. enterocolitica* is the preeminent pathogen causing the majority of human infections and the widest variation in clinical manifestations.

Gastrointestinal Syndromes

Y. enterocolitica causes a number of distinct gastrointestinal syndromes ranging in severity from innocuous infections to more pronounced illnesses requiring surgical intervention. The type of infection that develops depends in part on the host's susceptibility, the infectious dose, and the serotype of the infecting strain, with serogroup O:8 tending to cause more severe disease than O:3. Bottone[64] highlighted the pathologic progression from mild to severe enteric disease caused by *Y. enterocolitica* in the typical host (Fig. 3). After initial colonization of the gastrointestinal tract, yersiniae migrate through the lamina propria into the mucosa, where they replicate, producing a mild, localized infection such as an enteritis or enterocolitis. On occasion yersiniae may proceed to the Peyer's patches of the terminal ileum, where they multiply in mononuclear cells, producing terminal ileitis, a relatively infrequent complication. A more common route of invasive enteric disease involves the translocation of *Y. enterocolitica* through the mucosa, submucosa, and muscularis externa via the lymphatics to mesenteric lymph nodes.[64] Proliferation within these lymph nodes results in mesenteric lymphadenitis, a clinical condition that is often mistaken for acute appendicitis. What factors arrest disease progression or allow it to proceed in susceptible persons is not completely understood.

Enteritis (enterocolitis) is the gastrointestinal complaint most often associated with *Y. enterocolitica* infection. Most cases of *Y. enterocolitica* enteritis occur in young children (<5 years), with some prospective and retrospective investigations reporting median ages of infection as low as 6 months,[55] 8 months,[68] or 24 months.[69] Hoogkamp-Korstanje and associates[70] evaluated 84 cases of *Y. enterocolitica* enteritis in pediatric patients submitted to

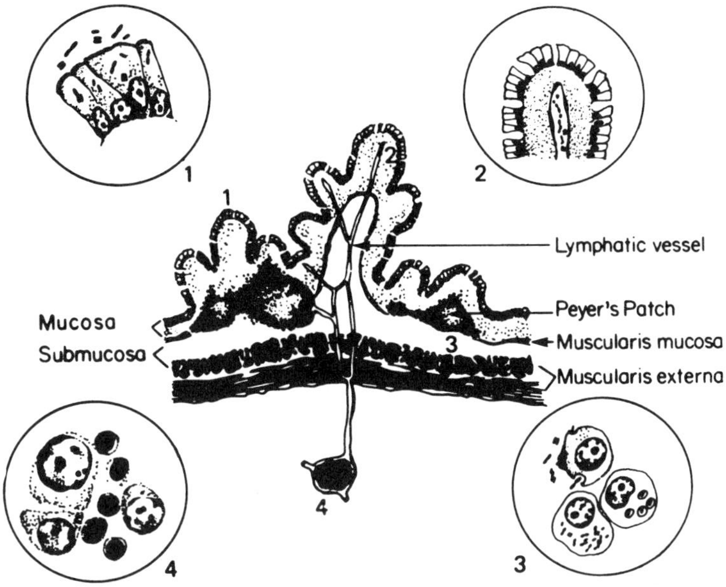

FIG. 3. Pathologic progression of enteric disease produced by *Y. enterocolitica:* 1, adherence to gastrointestinal epithelium; 2, passage through lamina propria; 3, reticuloendothelial tissues of Peyer's patches; 4, mesenteric lymph nodes. From Bottone[64] with permission of Marcel Dekker, Inc.

the Public Health Laboratory of the Netherlands between 1982 and 1991. More than 70% of these infections occurred in children younger than 5 years of age. Less often, *Y. enterocolitica* enteritis was observed in older children (6–15 years).

Children with *Y. enterocolitica* enteritis present with a triad of symptoms including diarrhea, low-grade fever, and abdominal pain.[48] Surveys and outbreak investigations indicate that between 68% and 100% of people with *Yersinia* enteritis have diarrhea.[55,66,68–70] In most instances the diarrhea is nonbloody; one Canadian study of 181 pediatric cases of yersiniosis found that most children experienced 5 to 10 episodes per day of diarrhea during the first week of illness.[69] In some instances, the enteritis may be severe enough to require hospitalization. Naqvi and coinvestigators[68] found that 7 (39%) of 18 children hospitalized for *Yersinia* gastroenteritis were admitted with dehydration as either a primary or a secondary cause. Other studies have reported similar frequencies of dehydration (30%) warranting hospitalization.[70] Although secretory (watery) diarrhea predominates in *Y. enterocolitica*–associated enteritis, the number of persons with macroscopically bloody stools varies from study to study, ranging from 20% to 46%.[55,66,68,69] Cover and Aber[48] estimated that up to 25% of patients with yersiniosis have gross blood evident in their stools.

One of the most intriguing aspects of *Y. enterocolitica* gastroenteritis is its duration. In a study of 84 children with enteritis, Hoogkamp-Korstanje and coworkers[70] observed that 22% of diarrheal cases resolved after 7 days; in 55%, symptoms persisted for 2 to 4 weeks, and in 23%, for 2 to 12 months. Marks and colleagues[69] found the mean duration of *Y. enterocolitica* enteritis (without treatment) to be 14 days (range 1–46 days). In some instances, people carried yersiniae in their feces for up to 97 days after their initial onset of symptoms (79 days after cessation of diarrhea). A CDC survey of *Y. enterocolitica* serogroup O:3 gastroenteritis associated with black pediatric patients from six different hospitals estimated the mean duration of diarrhea to be 10 days or longer.[55] A Norwegian investigation found that, of 448 children and adults serologically confirmed to have *Y. enterocolitica* O:3 or O:9 infection, only 200 (45%) had diarrhea and 184 abdominal pain (41%) on first admission.[71] Those with more severe complaints (ie, bloody diarrhea, vomiting) were usually younger than 10 years of age. During the follow-up period, 38 (8%) of these patients were readmitted to hospitals for chronic abdominal pain and/or diarrhea. Collectively, these data indicate that *Y. enterocolitica* enteritis can have a protracted course in a high percentage of cases. Secondary spread of yersiniosis has also been documented. Both Marks[69] and Naqvi[68] and their colleagues estimated that one third of all close contacts or symptomatic family members developed diarrhea. In the former investigation, these results were confirmed by a positive fecal culture and serology (agglutination titers ≥1:200).

Y. enterocolitica enteritis is also associated with low-grade fever in 33% to 94% of patients.[55,66,68–70] A 4-year study of *Y. enterocolitica* enteritis in the St. Louis area found that 77% of affected children had rectal temperatures higher than 38°C, 48% higher than 38.8°C. High fevers (38.8°–40.5°C) were particularly associated with young infants (<3 months).[68] Abdominal pain, typically localized to the right lower quadrant, is a common and prominent finding. Other clinical or laboratory features associated with *Y. enterocolitica* enteritis include vomiting (39%–50%), sore throat (22%), rash (22%), weight loss (18%), and leukocytosis.[55,66,68–70]

The incidence of yersiniosis varies dramatically in different geographic regions of the world. It is highest in Europe, Canada, and northern sections of the United States. In some studies, including those from Canada and Italy, the incidence of yersiniosis has ranged from 1.2% to 2.9%.[68,69,72] Other investigations from the United States[55,73] and Ethiopia[74] have reported lower rates (0.2%–0.9%). Many epidemiologic surveys on the infectious etiology of bacterial gastroenteritis, particularly from the Middle East and South America, have failed to uncover a single case of *Y. enterocolitica* enteritis.[75,76] These differences in incidence probably result from a multiplicity of factors such as microbiologic techniques (selective media, enrichment), geographic locale (thermal gradients), socioeconomic factors (ethnic foods), and animal and environmental reservoirs. In similar fashion, the peak seasonal incidence of yersiniosis has been reported to be both summer/fall[69] and winter.[68] Such discrepancies in the seasonality of yersiniosis may not only be due to the above described factors but also to the changing seroepidemiology of *Y. enterocolitica* gastroenteritis, as witnessed in the United States. Our past experience with yersiniae in California found the peak to occur between late fall to early spring, but since the emergence of serogroup O:3, this seasonal pattern may be somewhat changed.

The most common gastrointestinal complication of *Y. enterocolitica* enteritis is mesenteric lymphadenitis. Unlike enterocolitis, mesenteric lymphadenitis is seen predominantly in older children (>6 years) and is manifested by symptoms that include fever, acute abdominal pain, tenderness in the lower right quadrant, and leukocytosis.[48,70,77] Because of the nature of these symptoms, the disease is often mistaken clinically for acute appendicitis ("pseudoappendicitis"). In an analysis of 363 children who underwent appendectomies for abdominal pain, *Y. enterocolitica* was obtained from surgical (appendix) and/or anatomic (feces) specimens in 5 (1.4%), only 1 of whom had pathologic evidence of acute appendicitis.[77] In the remaining four children, the appendix was normal but enlarged lymph nodes were noted. Hoogkamp-Korstanje and colleagues[70] found that 20 (18%) of 114 children with enteric *Y. enterocolitica* infection developed pseudoappendicular syndromes ("acute abdomen") accompanied by abdominal pain and tenderness in the right lower quadrant. Thirteen underwent laparotomy, and in nine instances mesenteric lymphadenitis with terminal ileitis and a normal appendix were found; in the other four cases, both mesenteric lymphadenitis and an inflamed appendix were detected. Ileitis by itself is an uncommon complication of enteric infection with *Y. enterocolitica.*[70,71] Hoogkamp-Korstanje and colleagues[70] described two cases of terminal ileitis in teenagers with chronic and relapsing diarrhea with abdominal pain, fever, and weight loss; bloody diarrhea was occasionally present. These symptoms persisted longer than 10 months. An endoscopic examination revealed the intestinal mucosa to be swollen and edematous, with nonspecific lymphocytic infiltrates in the submucosa and muscular layers.[70] Traditionally, severe enteric diseases (ie, mesenteric adenitis, ileitis) caused by *Y. enterocolitica* have been associated with invasive serogroups such as O:8, O:5,27, and O:1,2,3. However, a number of reports have linked cases of pseudoappendicitis to infection with serogroup O:3.[77] Monno and associates[78] described a case of mesenteric lymphadenitis in a 13-year-old boy with β-thalessemia major. He had previously received multiple transfusions and was treated with deferoxamine. His symptoms included fever, diarrhea, and abdominal pain in the right iliac fossa. A laparotomy and appendectomy were performed, and *Y. enterocolitica* O:3 was isolated from edematous mesenteric lymph nodes.

Mesenteric adenitis and terminal ileitis are the two chief syndromes associated with *Y. pseudotuberculosis* infection.[61,79] These syndromes typically occur in older children with no prior history of gastroenteritis. *Y. pseudotuberculosis* is infrequently recovered from fecal specimens in such cases, and positive results are most often obtained after cold enrichment only. Such features help to distinguish enteric *Y. pseudotuberculosis* infections from those caused by *Y. enterocolitica.*[2,79]

El-Maraghi and Mair[80] reviewed the histopathology of 70 patients (age range 4–33 years) with serologically confirmed *Y. pseudotuberculosis* infection. In 69, acute appendicitis was the preoperative diagnosis. Symptoms in these persons included fever and colicky abdominal pain localized to the right iliac fossa with tenderness and muscle guarding. Gross pathologic findings revealed mesenteric lymphadenitis in all but one case. In most instances, gross pathology indicated a normal or mildly congested appendix and an edematous terminal ileum with a congested serosa, mucosal ulcerations, and hypertrophy in the Peyer's patches.[80] Histologic sections of mesenteric lymph nodes indicated lymphoid hyperplasia with prominent cortical follicles and large, active germinal centers in 90% of cases. Analysis of the frequency of granuloma formation in both the ileum and the appendix supports the hypothesis that the terminal ileum is the most likely portal of entry for *Y. pseudotuberculosis* enteric infections. Intestinal *Y. pseudotuberculosis* infections have also been linked to several syndromes of unknown infectious etiology. Sato and associates[81] recovered *Y. pseudotuberculosis* from gastrointestinal samples of 12 children (in eight cases, by cold enrichment culture only) whose symptoms included prolonged fever (>5 days) and a macropapular erythematous, "scarlet fever–like" rash accompanied by diarrhea (92%), periungual desquamation (92%), and a strawberry tongue (83%). Two of these children met the strict definition of Kawasaki syndrome, and another was diagnosed initially as having Izumi fever, a relapsing febrile illness causing a biphasic rash, abdominal pain resembling appendicitis, desquamation, and strawberry tongue, which was first described in 1929. Follow-up serologic investigations of 29 people involved in several outbreaks of Izumi fever clearly linked those illnesses to *Y. pseudotuberculosis* infection.[82] Tertti and coworkers[61] also noted a high rate (>50%) of postinfectious complications primarily associated with gastrointestinal *Y. pseudotuberculosis* infections. These sequelae included erythema nodosum, arthritis, iritis, and nephritis. One Finnish study of an outbreak of *Y. pseudotuberculosis* Ia infection found 26% of infected children to be without symptoms.[83] The authors concluded that *Y. pseudotuberculosis* is capable of inducing a strong immune response in persons who are otherwise without overt symptomatology.

Septicemia

Septicemia is an uncommon yet life-threatening complication of *Y. enterocolitica* infection. Bacteremic disease is most often observed in older persons, although any age group can be affected. Two publications have retrospectively analyzed cumulative data on 143 cases of *Y. enterocolitica* bacteremia. Gayraud and associates[84] reviewed 43 cases of septicemia reported to the French national registry between 1985 and 1991. The mean age was 61 years (range 3–89 years), and 77% had underlying disease or medical conditions predisposing them to potential infection. In 1995, Jensen and colleagues[85] published data on the largest series of bacteremic episodes to date, involving 100 cases of septicemia from 44 Danish hospitals over a 19-year period. Again, *Yersinia* sepsis was seen predominantly in persons of advanced age (mean 76 years). Fifty-three percent of all patients had factors predisposing them to infection, and of these 43% had multiple underlying abnormalities. Not all reports have generated identical findings. Foberg and associates[86] described seven cases of *Y. enterocolitica* septicemia in elderly persons (mean age 73 years); only two (29%) had underlying disease. Reasons for these differences are not readily apparent, although the small number of cases in the latter report may be a contributing factor.

The frequency of septicemia due to *Y. enterocolitica* is extremely low in the general population. Several major surveys of sepsis caused by gram-negative bacilli have failed to uncover a single case of bacteremia caused by yersiniae. However, in certain defined patient populations, *Y. enterocolitica* is an important pathogen of the blood. At least four distinct settings are recognizable, including patients with certain underlying medical complications, those with iron overload conditions, those with transfusion-related sepsis, and neonates (<3 months) with bacteremic disease (Table 4). By far the largest group subject to *Y. enterocolitica* septicemia comprises patients who are immunocompromised or have underlying metabolic disorders. Immunocompromised conditions associated with *Yersinia* bacteremia include cancer (tumors, hematologic malignancies), liver disease (cirrhosis, hepatitis, alcohol abuse), and corticosteroid therapy.[48,84,85] *Y. enterocolitica* bacteremia in patients with the acquired immunodeficiency syndrome (AIDS) has been described on rare occasions.[87] Several metabolic diseases (eg, diabetes mellitus, hemochromatosis) also have been noted to predispose patients to *Y. enterocolitica* septicemia.[84,86,88] Fatality rates from such cases range from 7.5% to 15%.[84,85]

Conditions leading to iron excess result in a significantly higher risk of developing bacteremia. This unique relationship between iron levels in the host's serum and tissues and the microbe's need for the essential micronutrient for growth is driven by the inability of several pathogenic serogroups of *Y. enterocolitica* (O:3, O:5,27, and O:9) to synthesize siderophores, a class of iron-sequestering compounds (see Pathogenesis). Such siderophore-negative serogroups compete ineffectively against the host's own iron-sequestering systems for available free iron. Under condi-

TABLE 4. *Major presentations of* Yersinia enterocolitica *bacteremia*

Setting or Condition	Affected Population	Underlying Diseases	Risk Factors	Mortality (%)
Immunocompromised, metabolic disorders	Adults ≥60 y	Cancer, liver disease, diabetes mellitus	No specific factors	10–25
Iron overload	All age groups	Hemochromatosis, β-thalassemia, anemia, renal disease	Iron dextran transfusion, deferoxamine therapy, blood transfusions, iron intoxication, hemodialysis	<10
Transfusion-sepsis	All age groups	β-thalassemia, anemia	Contaminated whole blood or red cell concentrate transfusions	59–70
Children	Infants ≤3 mo	None	None	<10

tions favoring iron overload, yersiniae are better able to compete for free iron by scavenging the micronutrient from iron-chelating compounds or by using siderophores produced by heterologous bacteria. Boelaert and associates[89] identified nine hemodialysis patients who developed sepsis caused by *Y. enterocolitica* O:3 or O:9 after iron dextran transfusions. In all nine patients, iron overload was demonstrated by increased serum ferritin levels (629–8000 ng/mL, normal range 20–300 ng/mL); intense iron staining of bone marrow (five cases), hepatocytes, and Kupffer cells with Prussian blue, and/or attenuated liver values. During a 5-year period, *Yersinia* bacteremia developed in 5 (0.9%) of 539 hemodialysis patients with iron overload (serum ferritin >500 ng/mL).[89] Human leukocyte antigen (HLA) typing was performed on eight of the nine patients, and four were found to carry "hemochromatosis alleles" (HLA-A3, HLA-B7, or HLA-B14). Hemochromatosis is a metabolic disorder in which ingested iron is excessively absorbed by the body and abnormally distributed in tissues as hemosiderin (hemosiderosis) or through saturation of iron-binding proteins (transferrin). Vadillo and coinvestigators[88] reported three cases of liver abscess caused by *Y. enterocolitica* O:3, one of which was accompanied by positive blood cultures. In each instance, primary hemochromatosis was diagnosed within 24 hours after an exploratory procedure to determine the cause of the liver abscess. A review of past cases of liver abscesses caused by *Y. enterocolitica* indicates that positive blood cultures occurred in 60% of all cases in which *Yersinia* was concomitantly isolated from abscess material. In an equal percentage of cases (n=24), an iron overload condition existed as a predisposing factor to infection. Such conditions included primary hemochromatosis (n=19), Bantu hemosiderosis (n=2), thalassemia major (n=2), and long-term iron therapy (n=1). A corollary to this observation is the association between the use of deferoxamine, an iron chelator used to treat iron overload or iron toxemia, and *Yersinia* septicemia. It is theorized that when deferoxamine complexes with iron in vivo, it provides the micronutrient in a utilizable form to *Y. enterocolitica.* Various studies have described cases of *Y. enterocolitica* septicemia in people with blood dyscrasias (eg, β-thalassemia, sideroblastic anemia) who manifested signs of sepsis 8 to 48 hours after administration of deferoxamine to relieve iron overload situations.[90]

Although most iron overload states result from permanent or recurring medical conditions, a surprising number of cases of accidental iron poisoning have resulted in *Yersinia* septicemia. A 1982 report described two instances of *Y. enterocolitica* O:3 septicemia in infant girls who accidentally consumed large amounts of either ferrocarbonate capsules or a ferrous fumarate mixture.[91] Both patients had signs of sepsis (high fever, diarrhea, vomiting, general deterioration) 24 hours after the accidental overdose. In one infant, *Y. enterocolitica* O:3 was isolated from both blood and stool, indicating that both children were probably colonized in the gut with yersiniae before the overdose occurred. In 1989, Milteer and colleagues[92] published two case reports on related 13-month-old infants who ingested an unknown amount of ferrous sulfate. Deferoxamine therapy was administered to both infants on admission. Approximately 36 to 48 hours later, clinical signs of sepsis appeared, including bloody diarrhea. Although *Y. enterocolitica* was isolated from the blood and stool of one of the infants, the other child was suspected of having yersiniosis based on the similar medical history and clinical course. Both infants had several predisposing factors for development of *Y. enterocolitica* bacteremia (eg, iron overload, deferoxamine therapy). They were each treated with cefotaxime and recovered completely.

A third scenario by which *Y. enterocolitica* can gain access to the bloodstream is through transfusion-related sepsis.[93] In a review by Wagner and colleagues,[94] *Y. enterocolitica* was found to be the leading cause (51%) of bacteremia resulting from contaminated red blood cell (RBC) transfusions. These case histories have been extensively examined in reviews by Jacobs and associates,[95] Wagner and coworkers,[94] and Bottone.[96] Together they present a remarkably similar picture regarding the epidemiology surrounding *Y. enterocolitica* transfusion-associated sepsis. All of these reports implicated heterologous whole or red cell concentrate transfusions as the source of primary infection, except for one case of septicemia involving an autologous transfusion in a 15-year-old boy undergoing surgery for idiopathic thoracic scoliosis.[97] Retrospective examination of donation records for these transfused blood products indicated that almost all were stored for 16 to 40 days; in one case, an asymptomatic infection in a 45-year-old woman receiving oral norfloxacin was traced to a 7-day-old leukocyte-depleted erythrocyte concentrate transfusion.[95] Traceback investigations indicated that at least 50% of the persons who had donated the contaminated blood had no recent history of illness and were healthy at the time of donation.[95,98,99] This suggests that blood products are initially seeded with small concentrations of yersiniae; as they are stored for prolonged periods, the bacteria are able to replicate because of their ability to multiply under conditions of refrigeration (4°C). Also, as blood products age, cells begin to lyse, liberating iron-containing compounds (hemoglobin), which can be metabolized by siderophore-negative serogroups of *Y. enterocolitica,* further favoring growth. One study found that the concentration of *Y. enterocolitica* present in a residual portion of a 31-day-old RBC pack transfusion contained 3.2×10^8 cfu/mL of bacteria and 4970 ng/mL of endotoxin.[100] Transfusion-associated bacteremia can easily be differentiated from bacteremia associated with iron overload in that signs of sepsis (eg, fever, hypotension) often develop within 2 hours and in some instances in as little as 10 minutes after the transfusion has begun.[99] To date, almost all transfusion-related cases are associated with *Y. enterocolitica* siderophore-negative serogroups O:3, O:9, and O:5,27, and most of these infections have been reported from Europe.[94,96] Although early reports documented high mortality (~59%) in such patients,[95,101] more recent investigations suggest a declining fatality rate, possibly because of better antimicrobial agents and physician awareness regarding this syndrome.[94] A number of proposals have been suggested to reduce the mortality risk further, including shorter shelf lives for blood products, leukodepletion, screening of donors for histories of diarrhea, and testing of RBC units that are 25 or more days old for the presence of bacteria and endotoxin. None of these proposals has met with universal approval.

An unusual presentation of yersiniosis involves septicemia in very young children. In 1981, Shapiro[102] published two cases of *Y. enterocolitica* bacteremia in 6-week-old boys admitted to Children's Hospital of Pittsburgh with chief complaints of diarrhea, fever, dehydration, and lethargy. Blood cultures drawn on admission from both infants grew *Y. enterocolitica;* both illnesses responded to appropriate chemotherapy, although one child had a protracted clinical course as a result of secondary complications. Neither child had any known underlying deficits. Since this study, other reports have described similar findings. Antonio-Santiago and associates[103] reported a case of *Y. enterocolitica* septicemia in a 7-month-old black infant with a prolonged fever (39.4°C) of unexplained origin. One week after admission, an elevated erythrocyte sedimentation rate (125 mm/h) and leukocyte count (34,500 cells/mm^3) were noted, suggesting possible infection. *Y. enterocolitica* was isolated from blood cultures drawn on the 13th day of hospitalization. Again, no underlying diseases were detected. In 1993, Naqvi and associates[68] defined this syndrome in pediatric patients presenting with *Y. enterocolitica* enteritis. Of 45 such children, 18 required hospitalization. Seventy-six percent of hospitalized children were younger than 3 months of age. Four of these infants, representing 28% of all children younger than 3 months, developed sepsis. In contrast to the previous association of risk factors with *Y. enterocolitica* septicemia in children,[104] none of these four infants had thalassemia or sickle cell disease. The authors concluded that age was a defining risk factor for infection and suggested that development of *Y. enterocolitica* septicemia after gastroenteritis in children younger than 3 months of age mirrored a similar pattern of disease found in young infants with *Salmonella* enteritis.[68] These observations have been supported by other studies describing apparently healthy young infants in whom *Y. enterocolitica* septicemia developed after initial gastrointestinal infection[53,105–108]; cumulative results are depicted in Table 5.

Rabson and coinvestigators,[109] in their classic study of *Y. enterocolitica* bacteremia in 13 patients, defined two major syndromes associated with sepsis. One presentation, called the acute septicemic form, was seen in persons complaining of abdominal pain, diarrhea, and vomiting. This acute form has also been referred to as a "typhoid fever–like" condition. However, unlike typhoid fever, the acute form typically presents with early diarrhea and hepatomegaly and without a rash. The second syndrome was described as a subacute localizing form. This form was observed in older men who had severe abdominal pain that was maximal in the right hypochondrium and aggravated by pressure over the liver[109]; the absence of diarrhea or vomiting was a cardinal feature of this form. All of these patients had been ill for prolonged periods and had underlying cirrhosis of the liver and/or hemochromatosis. Although the two syndromes apparently were distinct in this study,[109] they can overlap.[86] In the study of Rabson and associates,[109] the two groups were almost equally represented, although it is clear from later studies that the acute septicemic form predominates in the general population. In one survey of 100 cases of *Y. enterocolitica* bacteremia, only one patient with the subacute localizing form was detected.[85] Most people with the acute form have fever, an elevated erythrocyte sedimentation rate, and leukocytosis.[84,86,110] Other common findings (30%–45%) include diarrhea, nausea, vomiting, and abdominal pain.[84,85] One of the remarkable attributes of *Y. enterocolitica* bacteremia is the number of postbacteremic sequelae associated with infection.[48,85] One large study found 54% of patients with *Y. enterocolitica* bacteremia had secondary complications.[85] These complications include cardiovascular problems, cutaneous manifestations (eg, bullous lesions), pulmonary disease, abscesses, arthritis, intracranial extensions, and suppurative lymphadenitis.[48,84–86,111–113] Chitterlings have been implicated as the source of infection in at least two instances in the United States,[106,113] although the source in most cases remains unknown.

The most common extraintestinal site from which *Y. pseudotuberculosis* is isolated is the blood. Ljunberg and coworkers[114] reviewed the medical literature regarding 54 published cases of *Y. pseudotuberculosis* septicemia. The pat-

TABLE 5. Yersinia enterocolitica *bacteremia in healthy infants*

Case	Sex/Age	Birth Weight (kg)	Diarrhea	Temperature (°C)	Leukocytes/ mm³	Therapy	Outcome	Ref.
1	M/6 wk	2.4	+	39	12,000	Amp/Km	Survived	102
2	M/6 wk	3.8	+(B)[a]	36.8	11,600	Gm	Survived	102
3	M/8 wk	—	+	38.6	11,800	Gm	Survived	107
4	M/7 mo	—	+(B)	39.4	34,500	Gm	Survived[b]	103
5	M/7 wk	—	+(B)	37.2	30,300	Amp/Km	Survived	105
6	M/3 wk	3.0	+(B)	39.6	12,100	Amp/Gm	Survived	105
7	F/5 wk	3.5	+(B)	39.8	25,400	Amp/Km	Survived	105
8	M/10 wk	3.2	+(B)	38.5	27,900	Amp/Cfx	Survived	105
9	M/2 wk	3.2	+	38.5	7,100	Cfx/Cfz/Gm	Survived	106
10	F/4 mo	—	+(B)	39.5	18,100	Cfx	Survived	108

[a](B), bloody diarrhea either macroscopically or occult (guaiac-positive).
[b]Recovered spontaneously but was still treated with gentamicin.
Amp, ampicillin; Km, kanamycin; Gm, gentamicin; Cfx, ceftriaxone; Cfz, ceftizoxime.

tern of disease presentation, susceptible groups, and underlying diseases associated with *Y. pseudotuberculosis* infection were similar to those observed for *Y. enterocolitica*. Hemochromatosis/hemosiderosis (17%) was the most common underlying disease associated with sepsis, followed by diabetes (15%), hepatic cirrhosis (13%), and alcohol abuse (7%); in almost one third of all cases, no underlying illness was apparent. Frequent signs of sepsis in infected individuals included fever (67%), hepatomegaly (33%), icterus (26%), diarrhea (26%), abdominal pain or tenderness (26%), and anorexia, vomiting, nausea, and malaise (22%). Mortality rates approached 28% in treated cases and 100% for untreated patients.[114] Bacteremia with associated hepatic disease has also been described. Farrer and colleagues[115] reported a case of septicemia in a 50-year-old man who had multiple small hepatic abscesses due to *Y. pseudotuberculosis* diagnosed by computed tomography (CT). The patient may have acquired his infection several months earlier while hunting turkey. Like *Y. enterocolitica, Y. pseudotuberculosis* can cause secondary complications on occasion. Two cases of bloody diarrhea caused by *Y. pseudotuberculosis* in an elderly woman and a young man apparently triggered renal failure as a postinfectious sequela.[116] Both recovered after chemotherapy and medical intervention (hemodialysis, transfusion). Each infection was diagnosed serologically.

Y. pestis can cause septicemia without lymphadenopathy in 25% of plague cases, based on epidemiologic data dating from the early 1980s.[30] Presenting symptoms are typical of those associated with gram-negative bacillary infections (ie, fever, chills, headache, malaise, and gastrointestinal problems).[30] Reported fatality rates range from 30% to 50%.

Cardiovascular Complications

Endocarditis and pericarditis are infrequent cardiac anomalies resulting from *Y. enterocolitica* septicemia. Giamarellou and coauthors[117] published a review of the medical literature pertaining to *Y. enterocolitica* endocarditis. Of 12 known cases, only 8 were reported in sufficient detail to summarize. Seven of the eight cases of *Y. enterocolitica*–associated endocarditis involved older men (mean age 65 years). Three of these patients had underlying illnesses (rheumatic heart disease), and two had prior prosthetic valve replacements. Most presented with signs of sepsis (fever, chills, malaise, diarrhea), although two had focal neurologic symptoms suggestive of a possible underlying endocarditis. Vegetations were detected by transthoracic echocardiography in three instances and by transesophageal echocardiography in one case. The mitral valve was affected in five instances, the aortic valve in one, and both valves in two persons. The overall mortality rate was 38%. The authors suggested that a third-generation cephalosporin plus an aminoglycoside was a possible first-choice regimen and that a quinolone, alone or in combination with a β-lactam, was an alternative therapy.[117] Pericarditis caused by *Y. enterocolitica* O:9 was described in a 48-year-old man who, 3 weeks after a traumatic insult, developed fever, a pericardial friction rub, and pericarditis with dyspnea.[118] Surgical drainage recovered 750 mL of serohematic fluid which yielded *Y. enterocolitica*. He was treated with perfloxacin and tobramycin and recovered completely.

Another unusual cardiovascular complication of yersiniosis is mycotic aneurysm. Plotkin and O'Rourke[119] described a mycotic aneurysm of the left internal carotid artery in a 53-year-old diabetic man. He was treated initially for presumed sepsis but returned at 11 and 21 days after release from the hospital with a tender swelling of the left side of the neck. An angiogram revealed an aneurysm of the left carotid artery. Antimicrobial therapy plus surgical excision and a jugular vein graft replacement brought about a favorable outcome. A 1994 case report highlighted an abdominal aneurysm caused by *Y. enterocolitica* in a 57-year-old man with arteriosclerotic cardiovascular disease.[120] Although resection and graft replacement were successful, the man died of postoperative complications.

Pharyngitis

Pharyngitis is a recognized complication of *Y. enterocolitica* infection. Early studies dating from the mid-1970s described cases of *Yersinia* enterocolitis accompanied by signs of acute exudative pharyngitis. However, yersiniae were not specifically sought in throat cultures from these investigations. The first direct evidence linking *Y. enterocolitica* to bacterial pharyngitis came to light during a multistate investigation of *Yersinia* enterocolitis associated with the consumption of contaminated pasteurized milk.[66] In 1983, Tacket and colleagues[121] reported that 14 (8%) of 172 patients from this outbreak presented with acute pharyngitis without enteric disease. All were adults (median age 38.5 years, range 18–83 years) and had positive throat cultures for *Y. enterocolitica*. The predominant symptoms were sore throat (100%) and fever (93%, range 37.8°–40°C); miscellaneous findings included joint pain (21%), nausea (14%), and abdominal pain (14%). Seven (88%) of eight persons evaluated had enlarged cervical lymph nodes; leukocyte counts ranged from 10,000 to 20,000 cells/mm^3 (mean 16,000 cells/mm^3) for four adults with blood workups. In addition to positive throat cultures, eight patients demonstrated significant convalescent antibody titers (range 1:256–1:8192) to the epidemic strain. Several years later, a familial outbreak of *Y. enterocolitica* was reported.[67] The wife, who had a 1-week history of cramping abdominal pain, became acutely ill with massive exudative tonsillitis and cervical lymphadenopathy. On her second day of hospitalization she became hypotensive and died. Pathology specimens demonstrated ulcerative lesions in the colon and massive mesenteric adenopathy.[67] *Y. enterocolitica* was isolated from premortem blood and stool specimens (no throat culture was taken). Her husband presented with a sore throat without enteritis. His throat, blood, and stool cultures were all positive for *Y. enterocolitica*. He was treated with gentamicin and trimethoprim-sulfamethoxazole and recovered. Both of these adults had been healthy before their bouts of disease caused by *Yersinia*. In 1988, Blinkhorn and Marino[122] published a case of *Y. enterocolitica* pharyngitis that developed into a lateral pharyngeal abscess in an adult with diabetes. The patient was hospitalized because of breathing difficulties and the inability to swallow solids. A CT scan indicated a multiloculated collection in the right neck extending to the right parapharyngeal space, with deviation of the trachea to the left and marked narrowing of the upper airway.[122] The abscess was drained, and *Y. enterocolitica* O:8 was isolated. Seroconversion was demonstrated with agglutinating titers against the infecting strain of 1:2048 (acute) and 1:256 (convalescent).

There are several noteworthy points regarding *Y. enterocolitica* pharyngitis. Because symptoms often suggest a viral pharyngitis, streptococcal infection, or infectious mononucleosis, yersiniosis may not be suspected. When throat cultures fail to reveal a pathogen or symptoms persist after penicillin therapy, *Y. enterocolitica* should be considered as part of the differential diagnosis. Even when throat cultures are taken, *Y. enterocolitica* may not be present in large numbers. In the study of 14 cases of *Y. enterocolitica* pharyngitis, Tacket and coauthors[121] found that most isolates failed to yield heavy growth on MacConkey (MAC) plates incubated at 37°C. Finally, *Y. enterocolitica* pharyngitis can lead to more severe disease that includes difficulty in swallowing and breathing. In some instances fulminant disease has resulted. Such sequelae point to the serious nature of *Yersinia* pharyngitis and a need for rapid and appropriate treatment.

Arthritis

Two distinct types of arthritic processes have been associated with *Y. enterocolitica* infection. Reactive arthritis, the more common form, is characterized by inflammation of the sacroiliac and peripheral joints and the spine.[123] The disease is endemic in Scandinavian countries, especially Finland, and is less frequently encountered in other regions of the world.[124] Typically, between three and eight joints are involved, with tenderness and swelling being the chief medical complaints.[123] A prodromal period consisting of fever, diarrhea, headache, or malaise sometimes occurs a week before the onset of disease. Reactive arthritis is most often an asymmetric polyarthropathy, which helps to distinguish it from rheumatoid arthritis.[124] It is most often diagnosed serologically. People at risk for development of reactive arthritis include those with immunologic diseases (eg, psoriasis, inflammatory bowel disease) and carriers of the HLA-B27 allele, the histocompatibility antigen associated with ankylosing spondylitis. Borg and colleagues[124] looked at the sera of 86 persons with a clinical history suggestive of reactive arthritis. Twelve (14%) were serologically diagnosed as having *Yersinia*-associated reactive arthritis. The most common serogroups involved in infection were O:3 and O:9. None of the patients had underlying disease, although eight (75%) were positive for HLA-B27. The joints most commonly affected were the knees and feet; 33% of patients developed chronic joint disease over a 4- to 21-month period.[124] A review found 25 additional cases of reactive arthritis due to yersiniae in the United Kingdom between 1979 and 1990.[124] On occasion, *Y. pseudotuberculosis* has also been demonstrated serologically to cause reactive arthritis, but the incidence is extremely low compared with that induced by *Y. enterocolitica.*[124]

The second, less commonly observed arthritic condition is septic arthritis. Unlike reactive arthritis, this complication arises as a result of sepsis and the seeding of bloodborne bacteria to various skeletal bones, producing septic arthritis and/or osteomyelitis. The disease is not linked to HLA-B27, although affected persons often have underlying problems such as osteoarthritis, liver disease, thalassemia, and diabetes mellitus, conditions known to predispose individuals to *Y. enterocolitica* septicemia. Crowe and associates[125] reviewed 24 cases of *Y. enterocolitica* septic arthritis, which occurred in the United Kingdom between 1970 and 1994. The overall mortality rate was 17%. Most patients received combination therapy (a third-generation cephalosporin plus an aminoglycoside); one was treated successfully with ciprofloxacin alone.[125]

Wound Infections

Yersiniae are rarely recovered from wound specimens. Even so, they have been documented to cause infections of both cutaneous and soft tissue structures. Krogstad and colleagues[126] described three superficial *Y. enterocolitica* wound infections in young children. The first two patients originally presented with pimples or blisters on the extremities. Both children were treated initially by excision and drainage of the wound site and administration of a first-generation cephalosporin. These innocuous skin manifestations failed to heal, developing into bullae, which then progressed into more advanced forms, including granulation tissue (abscess) in one instance and a raised lesion with friable tissue and a purulent exudate in the other.[126] Pathogenic *Y. enterocolitica* O:4 was isolated from each child. Further treatment including a change in antimicrobial therapy (trimethoprim-sulfamethoxazole, tetracycline) brought about rapid improvement. A third child developed cellulitis after sustaining a minor head trauma. Again, under a similar chemotherapeutic regimen, this superficial abrasion progressed into two indurated erythematous lesions with lymphangitic streaks. The infection continued to cause the child problems over a period of several days. When *Y. enterocolitica* O:8 was finally isolated from purulent material, therapy was changed (gentamicin) and the lesions began to heal within 72 hours. Although the sources of these illnesses were unknown, water was the likely vehicle of infection in two instances.[126]

Kahn and associates[127] reported an infection of the psoas muscle caused by *Y. enterocolitica.* The patient, a 71-year-old man, presented with fever, right lower abdominal pain, and hip flexion and pain on hip extension.[127] A CT scan of the pelvis revealed a low-density region of the right psoas muscle. This area was aspirated and cultured, and *Y. enterocolitica* was recovered in pure culture. *Y. enterocolitica* can also cause infections in deep muscle, and several cases of pyomyositis have been described. One case report from New York City involved a 62-year-old diabetic man with a 1-month history of pain and swelling in his left calf.[128] After the patient failed to respond to initial supportive therapy, an ultrasound of the calf was performed, revealing a semisolid mass. On surgical intervention an abscess was detected that tracked down to the gastrocnemius muscle. *Y. enterocolitica* was isolated in pure culture from purulent material. A second case of pyomyositis was described in a 77-year-old man with a palpable mass of the left anterior thigh.[129] This tumor-like mass was removed and found to contain pus harboring *Y. enterocolitica* O:9. A traumatic injury to the left quadriceps tendon sustained 55 years earlier was not ruled out by the authors as a possible source for this infection, despite the delayed onset of symptoms.

Pulmonary Disease

There have been fewer than a dozen medical reports implicating *Y. enterocolitica* as a respiratory pathogen. Most of these pulmonary infections involved cases of pneumonia and/or empyema. Cropp and colleagues[130] summarized the results of six pulmonary infections due to *Y. enterocolitica* that occurred between 1976 and 1984. All of these patients were adult men (mean age 57 years, range 43–66 years); 67% had evidence of underlying liver disease or ethanol abuse. Pulmonary lesions ranged from interstitial pneumonia with nodular infiltrates to cavitary lesions, the right upper lung being the site most commonly infected. In four of the six cases, both sputum and blood cultures were positive for *Y. enterocolitica.* Other rare foci associated with *Y. enterocolitica* pulmonary infection include the gastrointestinal tract, the joints, and the ribs.[130] All of these patients recovered after receiving prolonged courses of chemotherapeutic agents (10 days to 8 weeks). Since this review, several additional reports have surfaced. Sodervik and colleagues[131] described four cases of pulmonary disease and sepsis caused by *Y. enterocolitica* O:3. Three of these patients had micronodular interstitial pneumonia, and two additionally presented with diarrheal disease (positive stool cultures for O:3). More recently, Kane and Reuman[132] reported a case of pneumonia and empyema in a 5-month-old girl who had previously had recurrent bouts of otitis media. Initial chest roentgenograms detected a right lower lobe infiltrate and a pleural effusion. Gross pus obtained via thoracentesis contained 10 to 25 RBCs and 25 to 50 neutrophils per high-powered field. *Y. enterocolitica* was recovered from this purulent material. Therapy included intravenous cefuroxime (16 days) followed by oral trimethoprim-sulfamethoxazole (5 days).

There is some conjecture regarding the route of disease progression in cases of pulmonary infection due to *Y. enterocolitica.* Cropp and coworkers[130] stated that five of the six pulmonary infections reviewed were probably primary processes with bacteremia developing secondary to respiratory disease. This hypothesis contrasts with that of Ettensohn and Roberts,[133] who described what they believed was only the second report of a primary pneumonic process caused by *Y. enterocolitica,* in a 43-year-old alcoholic male. In light of the normal routes of infection of yersiniae, it seems more likely that this second hypothesis is correct and that most pulmonary infections result from secondary spread of *Y. enterocolitica* from the blood to the lungs.

Empyema has also been associated with *Y. enterocolitica* infection. Some cases of empyema involve frank respiratory disease,[132] whereas other pyogenic infections accompany overt sepsis.[134] Clarridge and associates[135] detailed a case of *Y. enterocolitica* sepsis and empyema in a 58-year-old man with alcoholic liver disease who developed a pleural effusion on the left side. Thoracentesis yielded 4100

RBCs and 4200 leukocytes/mm^3. Both pleural fluid and blood were positive for an amikacin-susceptible strain of *Y. enterocolitica*. The patient was treated with this aminoglycoside and subsequently defervesced. In addition to these illnesses, hilar involvement (lymphadenopathy, calcifications) has been linked to pulmonary disease and other extraintestinal sequelae (erythema nodosum) associated with severe *Y. enterocolitica* infections.[133,136]

Y. pestis also can cause primary pneumonia (pneumonic plague). This respiratory infection, which is spread by droplets, progresses from a flu-like illness to a fulminant pneumonia within a couple of days.[30] The disease has an extraordinarily high mortality rate.

Hepatic Disease

Infection of the liver is another consequence of dissemination of yersiniae by the hematogenous route. Persons prone to development of hepatic abscesses include those with iron overload conditions such as primary hemachromatosis, diabetes mellitus, and hemosiderosis.[88] Infections occur primarily in middle-aged men; only a couple of cases of *Yersinia*-associated liver disease have been reported in people younger than 37 years of age. Nemoto and colleagues[137] and Vadillo and coauthors[88] both reviewed the existing literature regarding *Y. enterocolitica* and liver disease. Their results are summarized in Table 6. Confirmation of hepatic disease due to yersiniae is most often achieved by the recovery of *Y. enterocolitica* from the blood, liver, or both. In some instances, yersiniae (viable, nonviable) are not detected in liver granulomas apparently induced by systemic *Y. enterocolitica* infection.[137,138] Reasons why hepatic cultures are sterile in such cases are unknown. Although cumulative data indicate that fatality rates associated with *Y. enterocolitica* sepsis and hepatic disease are high, reports since 1986 suggest a much lower rate (<20%), possibly owing to the availability of better chemotherapeutic agents. Liver disease has also been reported on occasion in association with *Y. pseudotuberculosis* septicemia. Farrer and companions[115] reported a case of sepsis in a 50-year-old man who had gone turkey hunting before the onset of his disease. A CT scan of the abdomen demonstrated multiple low-density lesions. Open liver biopsy yielded *Y. pseudotuberculosis*. He was given ampicillin and gentamicin and was discharged from the hospital approximately 1 month after admission.

Erythema Nodosum

Erythema nodosum is a skin eruption characterized by inflammation of subcutaneous adipose tissue resulting in the appearance of painful nodular lesions. These lesions typically appear on the legs and arms but can extend to the feet, hands, and neck.[139] They can be triggered by many dif-

TABLE 6. Yersinia enterocolitica *and liver disease*

Character	Nemoto et al.,[137] 1992	Vadillo et al.,[88] 1994[a]
No. cases	29	38
Patient demographics		
Age (range)	3 mo–75 y	3 mo–75 y
Age (mean)	57 y	55 y
Hepatic abscesses (multiple)	82%	85%
Culture		
Liver+/blood–	28%	31%
Liver+/blood+	38%	58%
Mortality rate	59%	45%
Predominant serotypes	O:3 (64%), O:5 (22%)[b]	Not stated

[a]Cases reported overlap.
[b]Probably serogroup O:5,27 strains.

ferent infectious diseases, including yersiniosis. The most common illnesses associated with *Yersinia*-induced erythema nodosum are arthritis in HLA-B27–negative patients[136,139,140] and enteric infections with severe abdominal manifestations (eg, appendicitis).[137] Diagnosis is achieved primarily through serologic tests to detect agglutinating antibodies to *Y. enterocolitica* serogroups O:3 and O:9. A few cases of erythema nodosum after *Y. pseudotuberculosis* infection have also been reported.[139]

Chronic Infections

One of the hallmarks of yersiniosis is the development of chronic illness after primary *Y. enterocolitica* infection. Chronic yersiniosis can develop from virtually any of the panorama of clinical manifestations associated with *Y. enterocolitica* infection. Hoogkamp-Korstanje and associates[141] described 10 *Y. enterocolitica* primary infections, ranging from enteritis to lymphadenopathy, that developed into chronic illnesses 1 to 8 years later. Although most of these chronic conditions occurred at the same site as the primary infection (ie, ileitis to chronic ileitis, lymphadenopathy to chronic lymphadenopathy), unusual secondary sequelae have been reported. In one patient, hepatosplenomegaly and acute pleurisy developed 18 months after *Y. enterocolitica* enteritis.[141] Another patient had intermittent enteritis and chronic arthritis 20 years after an attack of appendicitis. The factors that contribute to the development of such chronic infections in certain persons are not apparent.[142]

Miscellaneous Infections and Reputed Disease Associations

Y. enterocolitica has been rarely reported to cause such extraintestinal infections as meningitis,[143] peritonitis,[144] osteomyelitis,[143] panophthalmitis, and glomerulonephritis.[96] It has been linked but not proven to cause uveitis[145,146] and thyroid disease.[2]

LABORATORY IDENTIFICATION

Isolation

The debate concerning routine culture using specific media for *Yersinia,* in light of current health maintenance organization restrictions, is as vigorous today as at any time. In 1988, Kachoris and associates[73] found that such testing was not cost-effective for their institution in Boston, Massachusetts. Over a 1-year period they found only 7 (0.2%) of 3622 stool specimens to be positive for *Yersinia.* Of the five patients from whom *Y. enterocolitica* was isolated, three were available for clinical assessment. Two of the three patients required treatment because they were still symptomatic at the time of reporting, indicating that testing was relevant for these individuals. The third patient's infection had resolved spontaneously by the time of the report. Laboratories in other geographic locations, such as Vancouver, British Columbia,[147] and Atlanta, Georgia,[148] have found *Yersinia* culture to be relevant and cost-effective. Metchock and colleagues[148] isolated 76 strains of *Y. enterocolitica* from 7290 of rectal swab specimens over a 1-year period. Fifty-nine (78%) of the positive specimens were from children younger than 1 year of age, and all were pathogenic serotypes. For this age group, *Yersinia* was second only to *Salmonella* in frequency of isolation. Given the wide geographic distribution of *Yersinia,* these authors recommended that all institutions, especially those with a large pediatric constituency, assess their own need for routine *Yersinia* culture. Other factors argue in favor of the use of specific media for *Yersinia* isolation. They include increased sensitivity and speed of recovery of this organism, which facilitates the detection of outbreaks and the ability to detect other enteric pathogens, such as *Aeromonas* spp.[73] and *Plesiomonas shigelloides* (personal observation), on the same medium (see later discussion).

As with any specimen for the isolation of enteric pathogens, if culture is not performed within 2 to 4 hours of collection, the specimen should be refrigerated or placed in transport

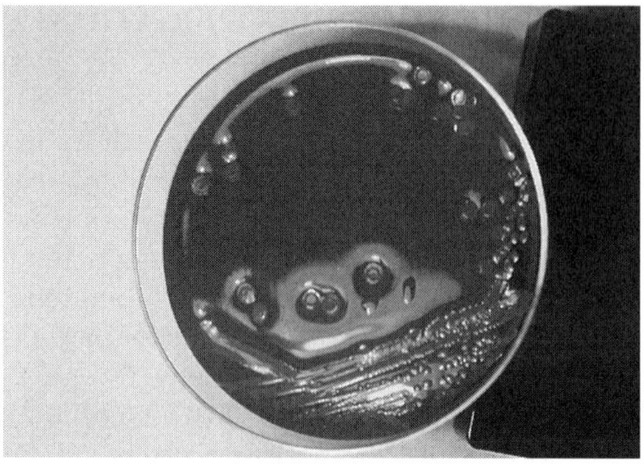

FIG. 4. Pectinase production by colonies of *Y. enterocolitica* causing pitting (dissolving) on pectin agar. See also Color Plate 17.

medium. Cary-Blair, Amies, and Stuart's, buffered glycerol in saline, or any of the commercially available transport media is suitable for holding specimens awaiting inoculation.

A variety of isolation media specific for *Yersinia* have been developed over the years, including pectin (Fig. 4; Color Plate 17), Y medium, and cellobiose-arginine-lysine (CAL) and cefsoludin-irgasan-novobiocin (CIN) agars.[149–152] Because of the complexity of their formulation or the need for autoclave and pH meters for preparation, DYS medium was developed.[153] Because this medium does not require any of the aforementioned materials or equipment, it is suitable for use in laboratories in developing countries, and it has been found to be superior to MAC or Salmonella-Shigella (SS) agar. Ratnam[154] and Head[150] and their associates studied isolation rates on a variety of media designed for *Yersinia* and on MAC and SS agars, for which variable isolation rates have been reported. In their study, Head and colleagues[150] compared recovery rates for a variety of *Y. enterocolitica* serotypes on MAC, SS, pectin, CAL, Y medium, and CIN agar, first in pure culture and then diluted to 10^2 and 10^1 cfu/g of stool (Table 7). In every phase of their testing, CIN was superior to the other media. They noted, as have others, that the ability to grow on different media varied from serotype to serotype; however, they found variability in growth within serotypes as well, indicating that this characteristic is not antigen dependent. Culture conditions were 29°C for Y medium and room temperature for all other agars; incubation was 20 to 40 hours regardless of medium. Ratnam and colleagues[154] also found the best recovery on CIN, compared with CAL or MAC. In contrast to these results, Fukushima and Gomyoda[155] found that 48% of 58 strains of *Y. pseudotuberculosis* from stool specimens were inhibited on CIN. Recovery of this species was excellent on MAC, with SS slightly less acceptable (107% and 72%, respectively, compared with trypticase soy agar). Also, half of serotype O:3 strains that belonged to an unusual biotype (3B) were inhibited on CIN. These results prompted Fukushima and Gomyoda[155] to recommend concomitant use of MAC for isolation of *Yersinia.* Enteric media unsatisfactory for the isolation of *Y. enterocolitica* include eosin–methylene blue, xylose-lysine-desoxycholate, and Hektoen enteric agars. Fermentation of the sucrose contained in these media by *Y. enterocolitica* renders colonies indistinguishable from those of normal flora. Colonies on CIN are described as having a "bull's-eye" morphology, with a well-defined, dark red or pink center and a

TABLE 7. *Comparison of isolation rates for* Yersinia enterocolitica

Plating medium	Recovery Rate (%), All Serotypes (n=35): Pure Culture	Recovery Rate (%), All Serotypes (n=35): 10^2 in Stool	Recovery Rate (%), Serotype 3 (n=11): Pure Culture	Recovery Rate (%), Serotype 3 (n=11): 10^2 in Stool
MacConkey	75	0	50	0
Salmonella-Shigella	48	11	40	10
Cefsoludin-irgasan-novobiocin	85	100	87	100

Data from Head et al.[150]

transparent apron. Other enteric flora can grow on this medium, the most common of which are *Citrobacter* spp., *Proteus/Morganella,* and *Pseudomonas. Citrobacter* colonies are extremely similar in appearance to those of *Yersinia,* although they are generally larger; we have noted that the *Citrobacter* colony has an entire edge, whereas *Yersinia* colonies have an irregular edge. *Aeromonas* colonies are indistinguishable from those of *Yersinia. Proteus/Morganella* and *Pseudomonas* colonies are colorless because they fail to ferment mannitol. However, since *P. shigelloides,* a pathogenic member of the *Vibrionaceae,* also produces colorless colonies on CIN, it is necessary to pick these colonies to screening media as well.

Ratnam and coworkers[154] noted that the inability to repress competing normal flora in plating or enrichment media resulted in decreased isolation of *Y. enterocolitica.* In about 20% of stool specimens in their study, CIN agar failed to adequately suppress other flora, and consequently *Yersinia* was not isolated. In a report studying the antagonism of gram-negative flora in relation to the growth of *Y. enterocolitica,* Schiemann and Olsen[156] concluded that metabolic crowding was responsible for the inability of *Y. enterocolitica* to compete in mixed culture. Their findings indicated that *Yersinia,* which grew more slowly than the other gram-negative organisms, stopped growing when the other bacteria reached the stationary phase. Use of colder temperatures (15°C) allowed *Y. enterocolitica* to grow at a pace equivalent to that of other bacteria; that is, the other bacterial colonies grew more slowly at that temperature while *Yersinia* grew faster, resulting in equivalent growth rates. These data support the use of colder temperatures (22°–29°C) for isolation. Even if low-temperature incubators are not available, plating media incubated at 35°C should be examined for *Yersinia*-like colonies. Colonies are usually smaller at this temperature and colorless on MAC or SS. On these agars after overnight incubation at 35°C, colonies may be very small (1 mm) or pinpoint (Fig. 5; Color Plate 18) and may be disregarded as enterococci. If such colonies are seen on these media, the plate should be reincubated overnight and reexamined. Colonies that have increased in size should be picked to screening media.

Other methods to improve the isolation of *Y. enterocolitica* and suppress normal fecal flora include alkaline treatment of the stool,[157] cold enrichment,[158] and the use of standard enrichment broths (eg, selenite broth, gram-negative [GN] broth). However, alkaline treatment of stools before inoculation onto plating media (a 1:2 ratio of stool to 0.5% potassium hydroxide incubated for 2 minutes) was not found to be effective in increasing isolation rates.[154] Cold enrichment (the incubation of stool in phosphate-buffered saline for 21 days at 4°C) does increase the isolation of *Yersinia,* but most of the strains recovered are considered nonpathogenic biotypes of *Y. enterocolitica* or other avirulent *Yersinia* species. Even the isolation of pathogenic biotypes may not be clinically relevant given the long time required for their recovery. Van Noyen and colleagues[158] recommended limiting cold enrichment to detection of *Yersinia* in convalescent patients and in those patients experiencing later manifestations of yersiniosis such as arthritis or erythema nodosum. Of the conventional enrichment broths used most commonly in clinical laboratories, selenite and

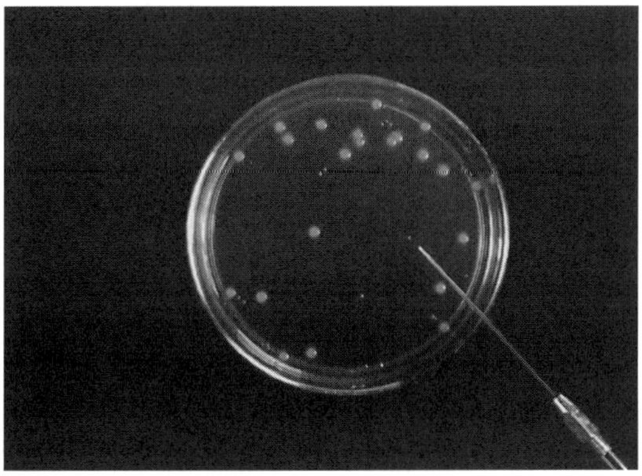

FIG. 5. Pinpoint colonies of *Y. enterocolitica* on MacConkey agar. See also Color Plate 18.

modified Rappaport were the most successful, although there are conflicting reports as to which of them is better. Van Noyen and colleagues[158] found modified Rappaport broth to be superior to selenite, whereas Ratnam and associates[154] found that subculture from selenite yielded 11% to 34% recovery of *Yersinia,* compared with 5% to 25% for tetrathionate, trypticase soy, GN, or modified Rappaport broth.

A number of media can be used to effectively screen colonies for *Y. enterocolitica* and *Y. pseudotuberculosis* (Table 8). In triple sugar iron (TSI) agar, *Y. enterocolitica* are acid (yellow) throughout the tube and do not produce gas or H_2S. *Y. enterocolitica* produces gas in TSI only very rarely (we have seen 3 such cases in a collection of 600 strains), and then only as a tiny bubble, whereas they readily produce gas in Durham vials in conventional glucose or other carbohydrate broths. Neither *Y. enterocolitica* nor any other *Yersinia* species decarboxylate lysine, so the butt of lysine-iron agar (LIA) is yellow with these organisms, indicating acid from glucose fermentation. In motility-indol-ornithine (MIO) medium, *Y. enterocolitica* is nonmotile or extremely weakly motile if incubated at 35°C (motile if incubated at 25°), indole-variable, and ODC-positive (purple color in butt of tube). *Y. pseudotuberculosis* does not ferment sucrose or produce gas or H_2S and appears *Shigella*-like in TSI (K\A$^-$). In MIO, *Y. pseudotuberculosis* motility is similar to that of *Y. enterocolitica,* but it is otherwise indol- and ODC-negative. *Y. pestis* is not found in stool but would appear identical to *Y. pseudotuberculosis* in TSI, LIA, or MIO.

Stolk-Engelaar and Hoogkamp-Korstanje[159] found that stool culture was satisfactory for the isolation of *Y. enterocolitica* from patients of uncomplicated enteritis (162/164 cases) or pseudoappendicitis (16/20 cases). However, for complicated enteritis (accompanied by septicemia, arthritis, erythema nodosum, generalized or localized lymphadenitis, or liver dysfunction), recovery from stool cultures fell to 66% (21/32 cases), and for colitis or ileitis isolations, rates were even lower (7/14 and 1/8 cases, respectively). Because fecal specimens become negative after migration of the organism from the intestine and circulating antibodies disappear after acute disease, these authors found that an anti-Yop (*Yersinia* outer membrane protein) rabbit antiserum was necessary to detect the organism in later manifestations of yersiniosis. With this test, 91% to 100% of complicated cases of enteritis, ileitis, and colitis were diagnosed. This assay, however, is not commercially available.

Biochemical Methods

Commercial Systems

Within the *Enterobacteriaceae,* identification problems involving commercial kit systems are most alarming in regard to members of the genus *Yersinia.* These problems are further magnified when such limitations are compared to a conventional testing format. Although yersiniae are infrequently encoun-

TABLE 8. *Differential reactions in screening media for* Yersinia *and other* Enterobacteriaceae *picked from CIN*

Organism	TSI[a]	LIA	MIO	Citrate	Urea (4 h at 25°C)	PPA
Y. enterocolitica	A/A$^-$	K/A$^-$	–/V/+	–	+	–
Y. pseudotuberculosis	K/A$^-$	K/A$^-$	–/–/–	–	+	–
Citrobacter	A/AG[b]	K/A$^-$	+/–/V	+	–	–
Proteus	K/AG$^+$, A/AG[b]	K/A$^+$	+/V/V	+	+	+
Morganella	K/AG$^-$	K/A$^-$	+/+/+	–	+	+

[a]All reactions at 35°C unless otherwise noted.
[b]H_2S production may be variable.
TSI, triple sugar iron agar; LIA, lysine-iron agar; MIO, motility-indol-ornithine agar; PPA, phenylpyruvate agar; A, acid, yellow; K, alkaline, purple; G, gas; V, variable; +, ≥90%; –, ≤10%.

tered in the clinical setting, their medical importance (plague, pseudoappendicitis, gastroenteritis outbreaks) mandates their rapid and accurate identification.

Virtually all commercial systems include both *Y. pseudotuberculosis* and *Y. enterocolitica* in some format in their databases (Table 9). *Y. pseudotuberculosis* has been reported to be correctly identified almost 100% of the time. The only exception to this rule was a study by Holmes and colleagues[160] using the Biolog GN microplate system (Biolog, Hayward, CA) and an automated reader. In that investigation, 10 of 15 *Y. pseudotuberculosis* strains were correctly identified in 24 hours; 5 isolates produced a "no identification" code based on similarity indices of 0.500 or less. When all 15 strains were manually read, all were correctly identified. Even though *Y. pseudotuberculosis* appears to be identified correctly by most kits, their relative rarity has prevented rigorous challenge of systems with large numbers of strains. A more disturbing problem concerns identification of *Y. enterocolitica.* Although all of these products include *Y. enterocolitica* in their databases, several major systems (MicroScan W/A, Vitek, Crystal E/NF, R/E) do so only as part of a group designation that typically includes several nonpathogenic but phenotypically similar species such as *Y. intermedia* and *Y. frederiksenii.* Even when kits identify yersiniae as *Y. enterocolitica* sensu stricto (eg, API 20E, RapID onE), such designations provide the laboratory with little usable information in regard to clinical significance when organisms are recovered from stools in association with gastrointestinal disease. Within *Y. enterocolitica* sensu stricto, strains can be subdivided into pathogenic and nonpathogenic groups based on biotype (see Biotyping). No current system distinguishes between these two groups in their databases. This means that when yersiniae are recovered from a fecal specimen and identified as *Y. enterocolitica* by a commercial kit, further biochemical testing (biotyping) is necessary to establish whether the isolate in question is clinically relevant. Although accuracies of more than 80% are usually reported for identification of *Y. enterocolitica,* most surveys have failed to challenge these systems with a large number of phenotypically distinct strains.

A third problem concerns the etiologic agent of bubonic plague, *Y. pestis.* At this time, only three systems approved by the U.S. Food and Drug Administration (FDA) include *Y. pestis* in their databases (API 20E, MicroScan W/A, Vitek); a fourth system (Biolog) is not licensed for clinical use. Although plague is rare, its clinical and public health significance (buboes, blood, respiratory specimens) warrants its rapid identification. It is presently unclear how such isolates would key out on systems that do not include it in their matrix. A 1996 study by Wilmouth and colleagues[161] found that 25 *Y. pestis* strains produced a unique code on the Crystal E/NF system, suggesting that this organism could be identified accurately by this product. A preliminary study of this group found that Crystal E/NF identified 11 of 12 isolates correctly.[161] Re-

TABLE 9. Yersinia *databases in common commercial identification systems*

Commercial System	Incubation (h)	*Yersinia* Database
API 20E	21–24	*Y. enterocolitica, Y. frederiksenii, Y. kristensenii, Y. intermedia, Y. pseudotuberculosis, Y. pestis* (presumptive), *Y. ruckeri*
Crystal E/NF, RS/E	18–20; 3	*Y. enterocolitica* group,[a] *Y. pseudotuberculosis*
RapID onE	4	*Y. enterocolitica, Y. frederiksenii, Y. kristensenii, Y. intermedia, Y. pseudotuberculosis*
Biolog	4–24	*Y. enterocolitica, Y. frederiksenii, Y. kristensenii, Y. intermedia, Y. pseudotuberculosis, Y. pestis, Y. ruckeri, Y. aldovae, Y. rohdei*
MicroScan W/A	2	*Y. enterocolitica* group, *Y. pseudotuberculosis, Y. pestis, Y. ruckeri*
Vitek	4–13	*Y. enterocolitica* group, *Y. pestis*

[a]Includes *Y. enterocolitica, Y. intermedia, Y. frederiksenii.*

gardless of the system used, a result that indicates *Y. pestis* as a first choice on any kit should be considered only a presumptive identification. Such tentatively identified isolates should immediately be forwarded to a state public health reference laboratory for biochemical and serologic confirmation.

A final point regarding *Yersinia* identification concerns the number of species presently not included. With the exception of the Biolog GN panel (not FDA-approved), no other system identifies *Y. aldovae* or *Y. rohdei;* some identify *Y. ruckeri.* Furthermore, at present no system includes *Y. mollaretti* or *Y. bercovieri,* even though the former species is relatively common clinically. When included in matrices, *Y. intermedia* is often identified accurately, but other species such as *Y. frederiksenii* and *Y. kristensenii* require 48-hour incubation for accurate identification results.

Conventional Biochemicals

Speciation of members of the genus *Yersinia* is usually quite straightforward when conventional media are used (Table 10). *Y. pestis* and *Y. pseudotuberculosis* can easily be differentiated from *Y. enterocolitica* by their negative reactions for ODC and sucrose fermentation. Results for fermentation of rhamnose, raffinose, and melibiose separate *Y. frederiksenii* and *Y. intermedia* from *Y. enterocolitica,* whereas negative sucrose and Voges-Proskauer (VP) reactions differentiate *Y. kristensenii. Y. bercovieri* and *Y. mollaretii,* previously biotypes 3A and 3B of *Y. enterocolitica,* are more difficult to separate from *Y. enterocolitica* and involve tests not readily available in most clinical laboratories. They are, however, VP-negative at both 25°C and 35°C, whereas *Y. enterocolitica* are positive at 25°C and variable at 35°C. Negative indole and positive lipase, pyrazinamidase, proline peptidase tests are useful in characterizing these species. Two other species only rarely encountered in humans, *Y. rohdei* and *Y. aldovae,* are also difficult to separate from *Y. enterocolitica* but are in general ODC-, maltose-, and cellobiose-negative. Table 10 lists the biochemical reactions most useful in separating members of the genus *Yersinia.*

Biotyping

Biotyping of strains of *Y. enterocolitica* is extremely important because the biotype serves as an indicator of pathogenicity. The first biotyping schema was introduced by Niléhn and was later modified by Wauters and then Bercovier.[162] In 1987, Wauters and colleagues[162] introduced a revised schema that is commonly used to biotype or biogroup strains

TABLE 10. *Biochemical separation of the members of the genus* Yersinia

Organism	ODC	SUC	RHAM	MEL	VP (25°C)	IND	α-MG
Y. enterocolitica	+	+	–	–	+	V	–
Y. pseudotuberculosis	–	–	V	V	–	–	–
Y. pestis	–	–	–	V	–	–	–
Y. kristensenii	+	–	–	–	–	V	–
Y. frederiksenii	+	+	+	–	+	+	–
Y. intermedia	+	+	+	V	+	+	V
Y. rohdei[a]	V	+	–	V	NA	–	–
Y. aldovae[a]	V	–	–	–	NA	–	–
Y. bercovieri[b]	V	+	–	–	–	–	–
Y. mollaretii[b]	V	+	–	–	–	–	–
Y. ruckeri	+	–	–	–	NA	–	–

[a]*Y. rohdei* gives a variable reaction for ONPG and raffinose fermentation; *Y. aldovae* is negative for both tests.

[b]These two organisms require positive reactions in lipase, pyrazinamidase, and proline peptidase to separate them from *Y. enterocolitica.*

ODC, ornithine decarboxylase; SUC, sucrose; RHAM, rhamnose; MEL, melibiose; VP, Voges-Proskauer; IND, indol; α-MG, α-methyl-D-glucoside; ONPG, *o*-nitrophenyl-D-pyranogalactoside; NA, not available; +, ≥90%; V, 10%–90%; –, ≤10%.

TABLE 11. *Abbreviated version of a revised scheme[a] for biogrouping of* Yersinia enterocolitica *strains*

Test	1A	1B	2	3	4	5
Lipase	+	+	–	–	–	–
Esculin/salicin (24 h)	+	–	–	–	–	–
Indol	+	+	+	–	–	–
D-Xylose	+	+	+	+	–	–
Trehalose/nitrate	+	+	+	+	+	–
Pyrazinamidase	+	–	–	–	–	–
β-D-Glucosidase	+	–	–	–	–	–

[a]Wauters et al.[162]

(Table 11). Isolates belonging to biotype 1A are considered nonpathogenic, and those of biotypes 1B and 2 through 5 are thought to be virulent. This scheme included a sixth biogroup, which contained strains that later were placed into the species *Y. bercoveri* and *Y. mollaretii.*[21] Biotype 1A can essentially be separated from the other biotypes by a positive reaction in esculin or salicin at 24 hours and by the pyrazinamidase test. This fact prompted Farmer and colleagues[163] to propose a scheme that relied on these two tests plus D-xylose and Congo red–magnesium oxalate (CR-MOX) agar. Positive reactions in pyrazinamidase and esculin/salicin identify nonpathogenic strains in this scheme. Pathogenic biotypes yield negative reactions in these two tests along with small red colonies on CR-MOX agar (indicating Congo red uptake and calcium-dependent growth at 36°C, characteristics which depend on the presence of the virulence plasmid). Lack of fermentation of D-xylose indicates that the strain belongs to serogroup O:3, the most common serotype now isolated in the United States. Strains that are esculin/salicin-negative but positive for pyrazindamidase are difficult to interpret and may require the more extensive testing found in Wauters's scheme or plasmid analysis. Farmer and colleagues,[163] however, found serotype 1,2a,3, a pathogenic serotype, to be 100% pyrazinamidase-positive and esculin/salicin-negative. Likewise, Cimolai and associates[164] isolated such a strain in pure culture from a mesenteric node, indicating it was pathogenic. The slightly abbreviated version shown in Table 11 varies from Wauters's schema only in that VP (all biotypes except 6, which no longer exists, are positive) and proline peptidase (only biotype 1A is positive, and the test is not readily available) were deleted. Fukushima and coworkers[165] reported that a subset of biotype 3 (serotype 3) strains are sucrose-negative; like their sucrose-positive counterparts, these strains are pathogenic.

Although determination of the biotype or biogroup to which an isolate belongs is generally recognized as a simple means of identifying virulent versus nonvirulent strains, not all studies support this concept. Results of a study on young children in Santiago, Chile, by Morris and colleagues[166] did not find a correlation between illness and proposed virulence factors. Nonpathogenic (biogroup 1A) *Y. enterocolitica* were found in stool cultures from seven children with diarrhea but in none of the control cultures (P=0.02). None of the seven isolates were invasive in HEp-2 assays, reacted with probes to the virulence plasmid, or were positive in any other assays used to determine pathogenicity in *Y. enterocolitica.* Two strains did cause fluid accumulation in infant mice, indicating a possible alternative means of ascertaining virulence in these strains. A study done in 1996 by Burnens and colleagues[167] looked at clinical data and isolate characteristics in 66 patients. The authors concluded that neither severity of diarrhea nor other symptoms correlated with the presence of the virulence plasmids (ie, biogroup). They further found that biogroup 1A strains could be divided into two subgroups based on ribotyping. One group was associated with human isolates, and the other contained predominantly environmental isolates. Although organisms of serogroup O:3 biotype 4 (pathogenic strains) now predominate in California, in the past we found 40% of *Y. enterocolitica* in patients with gastroenteritis and other infections belonging to biogroup 1A.[54]

Immunodiagnostics

Winblad devised the first serogrouping scheme for *Yersinia* based on cell wall lipopolysaccharide (O somatic antigens) in 1967.[168] Before 1990 there were 56 recognized O groups;[169] however, Kawaoka and associates[170] reported that the O:15 antigen is actually a rough antigen, and O:34 was found to be identical to O:10.[171] In 1991, Wauters and coworkers[172] reported on an additional 19 new O and 6 new H (flagellar) antigens in *Yersinia.* Eleven somatic groups correlate with virulence in strains of *Y. enterocolitica.* Of these O:3; O:5,27; O:8; and O:9 are the most common. The remaining groups are O:1,2a,3; O:2a,2b,3; O:4,32; O:13a,13b; O:18; O:20; and O:21. Occasionally O:3 and O:9 or other pathogenic serotypes may be encountered in a species other than *Y. enterocolitica*[172]; therefore, serogrouping should be performed only on isolates that have been fully characterized biochemically. Because *Yersinia* strains become rough on repeated cultivation, especially at 37°C, isolates should be maintained at 22°C to 25°C if serogrouping is to be performed. Only a limited number of antisera are commercially available (Accurate Chemical and Scientific Corp., Westbury, NY). A more complete serogrouping analysis must be performed at a reference center. Flagellar or H typing is performed only at a limited number of reference laboratories and usually is helpful only as an epidemiologic marker in outbreak investigations.[173] In 1984, Aleksić and Bockemühl[171] proposed a revised scheme restricted to typing of *Y. enterocolitica* strains only, as opposed to the broader schema that is applicable to *Y. frederiksenii, Y. intermedia,* and *Y. kristensenii.* In this system, there are 18 O serogroups containing 20 O factors and 12 H factors. As is typical of members of the *Enterobacteriaceae,* strains from other genera may cross-react in *Yersinia* antisera. Thomas and associates[174] found no cross-reaction between *Shigella* and *Yersinia,* but *Salmonella* may cross-react weakly, and *Escherichia coli* O97 and O98 antigens were found to be identical to *Y. enterocolitica* O:5,27 and O:11 antigens, respectively.

Y. pseudotuberculosis has its own serogrouping scheme based on six groups labelled I through VI. Groups I, II, IV, and V have a and b subgroups.[174] Brewer and Corbel[175] reported O group I to be the most common in Europe, followed by II and III. We have seen only types I and III in humans in California. Groups IV and V are rare isolates in Europe and Japan, and O group VI has been found only in Japan. Serotypes VII and VIII have been isolated from animals in Japan.[176]

Typing

Kaneko and Maruyama[176] developed an enzyme immunoassay (EIA) to detect pathogenic strains of *Y. enterocolitica* and *Y. pseudotuberculosis.* They prepared antiserum to *Y. enterocolitica* O:3 plasmid-encoded proteins and performed EIAs against temperature-inducible proteins released from cultures of these two species. This system reacted with virulent *Y. enterocolitica* types O:3; O:5,27; O:8; and O:9 and all serogroups of *Y. pseudotuberculosis.* DNA hybridization using probes from the virulence plasmid of *Y. enterocolitica* and the *ail* and *inv* chromosomal loci has also been used to determine virulence in *Yersinia* species.[177]

Bacteriophage typing has been used with *Yersinia.* Two schemes were developed in Europe.[178] However, these systems do not lyse serotype O:8 strains found in the United States, and one phage pattern accounts for 87% and 57% of strains in the two schemes, respectively.[174,178] Baker and Farmer[178] developed a typing system employing 24 phages, which typed more than 90% of all *Y. enterocolitica, Y. frederiksenii, Y. intermedia,* and *Y. kristensenii.* With this system only 22% of strains fell into the most common type, with most types containing fewer than 5% of isolates.

Molecular techniques may be used to subtype *Y. enterocolitica* strains.[179–181] Najdenski and colleagues[179] used pulsed-field gel electrophoresis successfully to differentiate strains of the same bioserotype. They also found that pathogenic *Y. enterocolitica* strains showed homogeneity specific for biotypes. Ribotyping has produced a limited number of ribotypes but did link a food source to human cases in an outbreak investigation.[180] Restriction of the virulence plasmid has also been used to subtype *Yersinia*.[181]

PATHOGENESIS

Probably no group of enteric bacteria, excluding *E. coli* and possibly the salmonellae, has been scrutinized more intensely at the molecular level regarding pathogenicity than have the yersiniae. Such activity has produced a voluminous number of general reviews on this topic[30,48,96,182–184] and many minireviews on selected aspects of gene organization, structure, and function as they relate to microbial pathogenicity.[185–192] Most of these molecular investigations have appeared in the past 10 to 12 years, although the foundation for this research clearly dates back to experimental data predating this molecular revolution. Despite significant advances in the understanding of *Yersinia* pathogenesis, many questions and issues remain unresolved.

Invasive disease is the flagship of *Yersinia* infections, and the predilection of yersiniae for lymphoid tissues is manifested by their ability to cause serious systemic disease, including bacteremia and chronic, long-term sequelae such as arthritis.[182,185] Paradoxically, although *Y. pestis* and *Y. pseudotuberculosis* are more closely related to each other on a genetic basis than to *Y. enterocolitica,* it is the latter two species that share more similarities in regard to pathogenesis, including routes of infection (oral) and primary clinical presentation (enterocolitis). These clinical observations lead to the rather obvious conclusion that although pathogenic yersiniae share some common pathogenetic mechanisms, differential expression of virulence factors among these organisms does occur, since their environmental niches, modes of acquisition, routes of infection, and disease presentations differ appreciably from species to species.

The overall goal of any pathogenic mechanism is to enable the bacterium to establish an infection in the host, multiply, and subsequently reinfect secondary hosts for continued dissemination and perpetuation of the microbial species. In vivo, this process requires damage to host cell functions and avoidance of immunologic defense mechanisms. For yersiniae, these tasks are formidable; an insect vector (flea) is involved in the life cycle of one species *(Y. pestis),* and others must circumvent normal physiologic and anatomic barriers in the stomach and intestines in order to initiate a localized infection. Currently, the best evidence for the role of most virulence factors in *Yersinia* pathogenesis is circumstantial—that is, the presence of a gene or gene product in a recognized pathogenic species (*Y. pestis, Y. pseudotuberculosis,* or *Y. enterocolitica*) and its absence in environmental members. An extension of this rule involves *Y. enterocolitica,* in which pathogenic (O:8, O:3) and nonpathogenic serogroups are recognized.[186] Even within the pathogenic serovars, groups can be further subdivided based on whether they represent American (O:8, O:5,27) or non-American (O:3, O:9) serotypes.[187] Such a geographic distinction has a pathogenic basis because American serotypes traditionally cause more serious human disease and lethal infections in murine models than their less virulent counterparts. However, even when reputed virulence factors are identified, it is often difficult to firmly establish their role in pathogenesis. For one thing, multiple functions have been ascribed to individual virulence factors in vitro, making it difficult to determine which attributes are critical in the infective process. In some instances, several different gene products serve the same function (eg, resistance to complement-mediated lysis of fresh serum). This further complicates matters because gene inactivation of one product does not necessarily inactivate the phenotype itself. In many instances, it is not known whether in vitro functions correlate with in vivo activity. A prime example involves the low-calcium response in pathogenic *Yersinia.* At low Ca^{++} levels, certain outer membrane proteins (OMPs) are synthesized at high rates, but it is unclear whether yersiniae ever encounter physiologic calcium concentrations in the host that would induce such a response.[188] Many of these virulence determinants are under complex positive and negative feedback control mechanisms, complicating their genetic manipulation and analysis. Furthermore, in vitro genome instability has been detected in *Y. pestis,* and to a lesser extent *Y. pseudotuberculosis,* creating additional obstacles.[193] Finally, some virulence determinants, particularly those associated with *Y. pestis,* produce their effect only in the insect vector and as a consequence have only an indirect effect on in vivo human pathogenicity. Some of the major proposed virulence determinants for the pathogenic yersiniae are listed in Table 12.

Iron Acquisition

Iron is an essential micronutrient for microbial growth and virulence gene regulation, and bacteria have evolved specific mechanisms to facilitate iron acquisition under adverse conditions.[194] Virulent strains of *Y. pestis* appear pigmented (Pgm^+) on media containing 100 μg/mL of hemin. Pigmentation is a thermoregulated trait, with the Pgm^+ phenotype expressed at 26°C but not at 37°C.[30] Despite macroscopic appearances, *Y. pestis* colonies on hemin-containing agar actually are not pigmented. Rather, coloration (greenish-brown) results from the intense intracellular uptake of hemin into the bacterial outer membrane. Pgm^+ strains also produce red colonies on Congo red agar because the dye sterically resembles the binding site for porphyrins, including protoporphyrin IX and hemin.[30,193] Mutants defective in this phenotype (Pgm^-) lose the ability to adsorb hemin, grow on iron-chelated media, be susceptible to pesticin, and express iron-repressible proteins. Pgm^- colonies (white) are additionally avirulent in mouse models unless inocula are fortified with an exogenous iron source.

The *pgm* locus of *Y. pestis* is located on a pathogenicity island (102 kb) that contains genes for other virulence traits such as hemin storage, the yersiniabactin receptor, and iron-regulated proteins.[195] One of these loci, located on a 7-kb stretch of chromosomal DNA, codes for hemin storage *(hms)* and is flanked by insertion sequences.[190] In vivo, this *hms* locus plays a critical role in infectivity by allowing for extensive adsorption of iron by plague bacilli in the hemin-rich gut of the flea. This adsorption results in the blockage of the flea foregut by a massive bolus of plague bacilli, a prelude to their introduction into human hosts during the flea's subsequent blood meal. Hms-negative mutants fail to colonize the flea's proventriculus, where this blockage normally occurs, thereby changing the insect's blood feeding behavior and disease transmission.[196] In humans, the role for Hms is less clear, since mutants do not appreciably alter mouse pathogenicity. Proposed roles for Hms in humans include increased uptake by eucaryotic cells, escape from bactericidal phagocytic mechanisms, and as a storage source for iron reserves.[30] However, there is no compelling evidence to support any of these hypotheses at present.

Pathogenic yersiniae can also acquire the essential micronutrient by at least two other mechanisms. One of these is siderophore independent and involves the direct utilization of iron-containing compounds such as hemin, hemopexin, hemoglobin, and ferritin.[30,183] This process is energy dependent and involves cell-bound activities.[183] A second pathway for iron acquisition involves siderophore synthesis—that is, small, low-molecular-mass compounds (<1000 Da) are produced that actively transport iron from the environment into the bacterial cytosol. Although early studies were inconclusive regarding the ability of pathogenic yersiniae to synthesize such

TABLE 12. *Characteristics of selected virulence factors in pathogenic* Yersinia *species*

System	Factor	Genetic Location	Genetic Organization	Phenotype/Genotype	Present in:		
					Yent	Ypstb	Ypest
Iron assimilation		Chromosome	Pathogenicity island/loci		+[a]	+	+
	Pigmentation/hemin storage	Chromosome	Pathogenicity island	Pgm	–	–	+
	Yersiniabactin	Chromosome	Locus	Ybt	+	+	+
	Regulation	Chromosome	Locus	Fur	+	+	+
Cellular invasion	Invasin	Chromosome	Locus	*inv*	+	+	–
	Invasin	Chromosome	Locus	*ail*	+	–	–
LCR	Low-calcium response	Plasmid	Stimulon		+	+	+
	OMPs/secreted proteins	Plasmid	Regulon/operons/loci	Yops	+	+	+
	V antigen	Plasmid	Operon	LcrV	+	+	+
Adherence	Adhesin	Plasmid	Locus	YadA	+	+	–
Enzymes	Plasminogen activator	Plasmid	Locus	Pla	–	–	+
	Urease	Chromosome	Operon	*ure*	+	+	–

[a]Present in some strains but not necessarily in all isolates.

Yent, *Y. enterocolitica;* Ypstb, *Y. pseudotuberculosis;* Ypest, *Y. pestis;* LCR, low-calcium response; OMPs, outer membrane proteins.

compounds, several studies employing the universal siderophore-detecting medium Chrome Azurol S agar (CAS) found selected strains of *Y. enterocolitica* and *Y. pseudotuberculosis* to be CAS^{+}.[197] In 1993, Haag and coinvestigators[198] purified a siderophore termed yersiniabactin from strain WA-C of *Y. enterocolitica.* Yersiniabactin was isolated by ethyl acetate extraction and has a molecular mass of 482 Da. It is apparently a catechol-like compound. Yersiniabactin appears to be expressed by *Y. enterocolitica* serogroups O:8, O:13, and O:40 but not by O:3, O:9, or O:5,27. Chambers and Sokol[199] isolated similar compounds from *Y. enterocolitica* O:8, O:4,32, and O:21 by ethyl acetate extraction. Although they could not confirm that this siderophore was a catechol-like compound, it appears likely it is similar if not identical to that described by Haag and colleagues.[198] Additional unrelated water-soluble siderophores may also be produced by other environmental strains of *Y. enterocolitica.*[199] *Y. enterocolitica* can also utilize siderophores produced by other microbial species, such as enterobactin, ferrioxamine B and E, and ferrichrome.[190] Although hydroxamate-type siderophores have not been found in pathogenic *Yersinia* species, aerobactin-positive strains of *Y. frederiksenii, Y. intermedia,* and *Y. kristensenii* have been isolated.[200]

Iron-regulated gene expression in yersiniae is under the control of a Fur-like (*f*erric *u*ptake *r*egulation) protein analogous to that found in *E. coli.*[194] At least two major high-molecular-weight molecules (~190 and 240 kDa) are induced in *Y. enterocolitica* under iron-limited conditions.[190,194,201] These proteins are found only in pathogenic serogroups such as O:8. Their exact functions are unknown but may involve uptake of inorganic iron.[190]

Cellular Invasion

A cardinal feature of pathogenic yersiniae, particularly *Y. enterocolitica,* is the ability to translocate from the gastrointestinal lumen into lymphoid tissues (Peyer's patches), which results in secondary sequelae such as mesenteric lymphadenitis. Transcytosis is thought to be an active process involving key bacterial determinants responsible for host cell invasion. At least two major chromosomal loci responsible for epithelial cell penetration have been identified.[202] One such chromosomal determinant, originally detected by Ralph Isberg and colleagues in *Y. pseudotuberculosis,* was subsequently found in *Y. enterocolitica.* This determinant, named invasin, is located on a single transferable gene designated *inv.*[187,192] Phenotypically, this gene can be detected in strains of *Y. enterocolitica* that efficiently invade and replicate in nonprofessional phagocytic cell lines such as HEp-2, HeLa, Chinese hamster ovary (CHO), Caco-2, and Henle 407.[187] It is maximally expressed at lower rather than higher (37°C) temperatures. Invasin is a 92-kDa OMP that preferentially binds to the β_1-family of integrins on the surface of target cells before triggering internalization.[192] Crossover appears to preferentially involve attachment to and invasion of phagocytic M cells overlying the Peyer's patches of the gastrointestinal lining.[184] Genetic evidence supports the concept that invasin is the key molecule facilitating entry of yersiniae into epithelial cells. On the other hand, *inv* mutants are 80 times less invasive in HEp-2 cell monolayers and colonize Peyer's patches less efficiently (10^7 fewer organisms) than their wild-type counterparts.[203] Although *inv* sequences are found in all *Y. enterocolitica* strains, this gene is apparently nonfunctional in nonpathogenic serogroups, thereby explaining this apparent paradox.[204]

A second locus, designated *ail* (*a*ttachment-*i*nvasion *l*ocus), promotes attachment and invasion of a number of eucaryotic cell lines but to a much lesser extent than *inv.*[192,202] The *ail* locus codes for a 17-kDa OMP that is associated exclusively with pathogenic serogroups of *Y. enterocolitica.*[205] This association makes Ail an attractive candidate as a virulence determinant in *Y. enterocolitica* pathogenesis. In addition to an invasion function, Ail also confers resistance to the bactericidal activity of normal human serum.[206,207]

TABLE 13. *Characteristics of invasion determinants in* Yersinia enterocolitica

Characteristic	Inv	Ail
Molecular mass (kDa)	92	17
Location		
Gene	Chromosome	Chromosome
Protein	Outer membrane	Outer membrane
Maximal transcription	23°C	37°C
Tissue tropism[a]		
HEp-2 invasion	+++	+
CHO cell invasion	+++	+++
Gene distribution[b]		
Pathogenic *Y. enterocolitica*	+	+
Nonpathogenic *Y. enterocolitica*	+	–

[a] +++, strongly invasive; +, weakly invasive
[b] +, gene present; –, gene absent

Resistance to complement-mediated lysis may be a critical virulence factor in the invasion and dissemination of pathogenic yersiniae from local infections. Ail also shows functional and structural homology to the Rck protein of *Salmonella* Typhimurium, which is involved in serum resistance in that species.[206] Because of the many described functions for Ail, its exact role in *Yersinia* pathogenesis is presently unclear.[192] Key differences between Inv and Ail are listed in Table 13. There appears to be at least a third minor pathway by which yersiniae conceivably could enter eucaryotic cells. This involves expression of a plasmid-derived protein, YadA (see later discussion). Although such activity accounts for only 1% of cellular invasion when compared to wild-type strains, gene introduction and complementation studies performed in noninvasive *E. coli* strains indicate that this invasive capability is a result of YadA expression.[192] The biologic relevance of this rather inconsequential uptake of bacteria due to YadA remains unknown.

LCR

The LCR (*Yersinia* low-calcium response) regulon is a 20-kb region of DNA whose expression is triggered by fluctuations in calcium ion concentration.[184–186,188,191] LCR resides on a 70- to 75-kb plasmid designated pYV (plasmid involved in *Yersinia* virulence). At 37°C and at Ca^{++} concentrations lower than 10^{-4} mol/L, upregulation of LCR ensues, resulting in growth restriction and the synthesis of massive amounts of selected proteins (Yops). When the [Ca^{++}] increases beyond this threshold, downregulation occurs.[188] In addition to [Ca^{++}], LCR responds to changes in nucleotide levels (adenosine triphosphate) and thermal gradients.[186] How [Ca^{++}] affects bacterial growth in vitro is unknown, as is the relevancy of this stimulon to in vivo situations. Perhaps LCR mimics some other, as yet unknown in vivo trigger.

Yops

Under the control of the LCR regulon are a series of 10 to 11 proteins called Yops that reside in monocistronic and polycistronic operons throughout the virulence plasmid.[189] Most but not all of the operons within the Yop regulon/virulon are under LCR control.[188] The designation Yop is an abbreviation for *Yersinia* outer membrane protein, based on the original belief that these were surface-exposed structural proteins.[185] However, more recent evidence indicates that Yops are in all likelihood secreted proteins.[184] Yops are present in all three pathogenic species and, with minor exceptions, are highly related on an immunologic and structural basis.[184,185] They are released through a type III secretory pathway[195] and under low [Ca^{++}] can accumulate to levels where they represent 10% of the total bacterial protein.[185]

The exact role Yops play in disease manifestation is unknown, although loss of specific Yops correlates with increased 50% lethal dose (LD_{50}) values in mice.[185] Putative functions for several Yops have been defined. On contact with host cells, yersiniae release effector molecules through a type III secretion process.[195] One of these effector molecules is YopE, which appears to inactivate immune host cell function. YopE is a 23-kDa cytotoxin that is released only after bacterial attachment to the surface of the target cell.[189] YopE acts on cell microfilaments and is antiphagocytic. YopH is a 51-kDa protein with tyrosine phosphatase activity.[208] It is also antiphagocytic, presumably through disruption of host cell cellular functions.[184,189] YopM is a 41- to 42-kDa protein that binds thrombin and inhibits platelet aggregation, which may produce an antiinflammatory effect.[184,189] YopO (YpkA) is a protein kinase that is essential for virulence in mice.[209] Functions for several other known Yops have not yet been described.

V Antigen

LcrV (formerly V antigen) is a 37- to 41-kDa protein with antihost function.[189] It has traditionally been considered distinct from Yops, but in reality they are the same. Although LcrV is under the control of LCR, it also has a positive regulatory effect on virulence operons, including an autoregulatory role.[186] LcrV has classically been associated with resistance to phagocytosis. However, new data indicate that it may have an immunosuppressive function through inhibition of cytokine production during the normal inflammatory response.[30]

Adherence (YadA)

YadA is an OMP to which a large variety of biologic functions have been attributed. In the older medical and scientific literature it was referred to by a number subjective synonyms, including protein 1, P1, YopA, Yop 1, and OMP 1.[183,184] In vivo, YadA forms a fibrillar matrix on the cell surface. This matrix consists of a 200- to 400-kDa polymer composed of 50-kDa subunits.[184] The gene encoding YadA is located on the 70-kb virulence plasmid. It is expressed at 37°C, but, like the lipoprotein YlpA, it is not directly involved in the LCR, although it plays a role in cell–cell adhesion triggering this stimulon.[188] YadA is highly hydrophobic and is responsible for bacterial clumping. In the diagnostic laboratory, key tests used to separate virulent from nonpathogenic strains of *Y. enterocolitica* (autoagglutination, Congo red binding) are linked to expression of YadA.

YadA binds to such biologic molecules as collagen (types I–V and XI), laminin, fibronectin, and intestinal mucin.[184] It has also been demonstrated to play a major role in adherence to rabbit intestinal tissue and brush border membranes.[210] Binding to mammalian tissues appears to occur through β_1-integrin receptors.[192] YadA also reduces deposition of complement component C3b on the bacterial cell surface, inducing resistance to complement-mediated lysis and decreased opsonization and phagocytosis.[184,191,192,211] Inhibition of the antiinvasive effect of γ-interferon has also been ascribed to YadA.[184] Which of these functions in addition to adhesion are important is presently not known.

Other Enzymatic Activities

Several other enzymes may play important roles in *Yersinia* pathogenesis. The plasminogen activator Pla is found only in *Y. pestis.* It is a 36-kDa OMP with fibrinolytic and coagulase activities.[30,191] The gene encoding this function is located on a 9.5-kb plasmid.[191] Possible functions for this protein include cleavage of fibrin that traps yersiniae, degradation of mechanical support structures allowing bacterial dissemination, and reduction in chemoattractants.[30] Koning-Ward and Robins-Browne[212] proposed that the urease of *Y. enterocolitica* may promote intestinal colonization and infection by allowing yersiniae to bypass the bactericidal effect of gastric acidity. Acid tolerance was related to urease activity: urease-negative mutants were 1000-fold less viable than wild-type strains at pH 2.5 to pH 3.5 in the presence of 3.4 mmol/L urea.

Miscellaneous Factors

Several other bacterial structures or enzymes may play roles in *Yersinia* pathogenesis. The lipopolysaccharide of pathogenic yersiniae may promote serum resistance, block access to bile salts, support adhesion and invasion of host tissues, and play a role in reactive arthritis.[191,213] Rough mutants have higher LD_{50} values in mice than do wild-type strains with smooth surfaces. *Y. enterocolitica* produces an enterotoxin that is maximally expressed at 26°C but cannot be detected when bacterial cultures are grown at 30°C or higher.[191,214] The fact that the enterotoxin gene, *yst,* is found only in pathogenic serogroups of *Y. enterocolitica* and induces diarrhea in young rabbits supports its role in gastroenteritis despite the temperature regulation paradox.[184,214] Other factors thought potentially to play a role in *Yersinia* pathogenesis include the pH6 antigen and Myf fibrillae.[183,184]

ANTIMICROBIAL SUSCEPTIBILITY

Y. enterocolitica generates a fairly consistent antimicrobial susceptibility profile, with only a few minor exceptions noted.[215–219] Most strains are inherently resistant to penicillins and aminopenicillins (including amoxicillin), carbenicillin (also ticarcillin), and first-generation cephalosporins (Table 14). Resistance to penicillin- and cephalosporin-like compounds is mediated predominantly through the production of two β-lactamases, designated A and B.[96,215] Virtually all *Y. enterocolitica* strains produce at least one β-lactamase, as determined by the nitrocefin test.[215,218] The most common pattern is coexpression of β-lactamases A and B, although a few strains originating from Canada, Australia, and New Zealand produce A only.[218] β-Lactamase A is a noninducible, broad-spectrum compound with predominantly penicillinase activity; β-lactamase B is a cephalosporinase-type compound whose activity can be enhanced by prior induction with imipenem.[215,218] Some studies have noted significant differences in the in vitro susceptibility of *Y. enterocolitica* to selected drugs based on biotype and/or serogroup.[215,217,218] These differences commonly involve aminopenicillins, some extended-spectrum

TABLE 14. *Susceptibility profile for pathogenic* Yersinia enterocolitica *strains*

Susceptible (~90%–99% of Strains)			Resistant (>50% of Strains)		
Agent	MIC_{50}	MIC_{90}	Agent	MIC_{50}	MIC_{90}
Cefuroxime (≤0.5–≥16)	2	4	Ampicillin (4–128)	32	64
Ceftazidime (0.3–4)	≤0.5	≥1,0	Carbenicillin (2–1024)	256	256
Cefizoxime (≤0.016–1)	0.06	0.25	Cefoxitin (1–64)	16	64
Ceftriaxone (0.03–0.25)	0.06	0.125	Cephalothin (8–512)	256	256
Cefotaxime (0.030–0.5)	0.06	0.25	Cefalexin (4–256)	64	128
Cefepime (≤0.008–0.128)	0.03	0.03			
Cefpirome (≤0.5–2)	≤0.5	≤0.5			
Piperacillin (0.25–16)	2	≤8			
Mezlocillin (0.5–8)	1	4			
Imipenem (0.06–0.5)	0.25	≤0.5			
Aztreonam (0.06–2)	0.5	2			
Doxycycline (≤0.125–8)	1	2			
Gentamicin (≤0.125–2)	0.5	1			
Tobramycin (≤0.125–2)	0.5	1			
Amikacin (0.25–4)	2	2			
Ciprofloxacin (≤0.008–0.32)	0.016	0.016			
Chloramphenicol (≤0.5–8)	2	4			

Data from references 215, 217–220.
Numbers in parentheses indicate range of MICs reported for strains.

penicillins, first-generation cephalosporins, and antibiotic–β-lactamase inhibitor combinations.[215,217,218] Although these differences are interesting from a scientific standpoint, they are not of practical importance because these drugs would not be used routinely to treat *Y. enterocolitica* infections under most circumstances. It should also be mentioned that a 1995 study of 335 *Y. enterocolitica* strains found no such phenotypic correlation and could not demonstrate any major changes in resistance within the species over the previous 10 years.[219]

Good in vitro activity has been demonstrated against *Y. enterocolitica* with most groups of antimicrobial agents, including third- and fourth-generation cephalosporins, carbapenems, monobactams, aminoglycosides, and quinolones (see Table 14). The most active in vitro agents are ceftizoxime, ceftriaxone, cefotaxime, cefepime, and ciprofloxacin, based on 50% and 90% minimal inhibitory concentrations. In one study, only imipenem was found to have excellent in vitro bactericidal activity.[215] However, a caveat regarding this cumulative data involves noted discrepancies between in vitro and in vivo activity. Hoogkamp-Korstanje[216] assessed the clinical performance of a variety of antimicrobial agents in treating 88 patients with gastrointestinal *Y. enterocolitica* infection. A high failure rate was noted when treatment involved a cephalosporin only (75%) or gentamicin with or without a β-lactam.[67] Use of ciprofloxacin to treat four patients with primary treatment failure produced two cures and two findings of clinical improvement. Similar results were observed in a retrospective study of 43 cases of *Y. enterocolitica* septicemia.[84] In that investigation, high failure rates were found when therapy involved amoxicillin only (100%), amoxicillin/clavulanate (100%), or a benzylpenicillin (100%). The highest success rates were found with the use of a quinolone or doxycycline only (100%), a quinolone plus a third-generation cephalosporin (100%), or a third-generation cephalosporin in combination with an aminoglycoside (83%). Part of the discrepancy between in vitro and in vivo results may result from the fact that yersiniae are intracellular pathogens. Certain agents, such as aminoglycosides, cannot pass through the eucaryotic cell membrane, and intracellular bacteria are therefore protected from their deleterious effects.[216]

Persons with or suspected to harbor *Y. pestis* (plague) should be placed in isolation and treated with appropriate therapeutic regimens based on age, dosing schedule, and route of administration.[30] Appropriate chemotherapeutic agents include streptomycin, gentamicin, the tetracyclines, and chloramphenicol. Large susceptibility studies on *Y. pseudotuberculosis* have not been performed recently. Using an invasive mouse model, the best results for treatment of *Y. pseudotuberculosis* infections have been obtained with quinolones, doxycycline, and gentamicin.[220]

CONCLUSIONS

As pointed out previously, *Yersinia* is one of only a select few genera within the *Enterobacteriaceae* that have been studied extensively, particularly with regard to virulence mechanisms. Despite this, a question still remains as to whether certain species—*Y. frederiksenii, Y. kristensenii,* and *Y. rohdei*—are truly pathogenic for humans, opportunists, or simply transiently present in the gastrointestinal tract. The same can be asked of the nonpathogenic serogroups/biotypes within the genus *Y. enterocolitica.* Identification of alternative virulence mechanisms in these strains, coupled with case-control studies, would aid in our understanding.

REFERENCES

1. Van Loghem JJ. The classification of the plague bacillus. *Ant Van Leeuw J Serol Microbiol* 1944;10: 15–16.
2. Bottone EJ. The genus *Yersinia* (Excluding *Y. pestis*). In: Balows A, Trüper HG, Dworkin M, Harder W, Schleifer K-H, eds. *The prokaryotes: A handbook on the biology of bacterial ecophysiology, isolation, identification, applications.* 2nd ed. Berlin: Springer-Verlag, 1991;2863–2887.
3. Bercovier H, Mollaret HH. Genus XIV: *Yersinia* Van Loghem 1944, 15[AL]. In: Krieg NR, Holt JG, eds. *Bergey's manual of systematic bacteriology.* Baltimore: Williams & Wilkins, 1984;498–505.

4. Schleifstein JI, Coleman MB. An unidentified microorganism resembling *B. ligneri* and *Past. pseudotuberculosis,* and pathogenic for man. *N Y State J Med* 1939;39:1749–1753.
5. Schleifstein J, Coleman MB. *Bacterium enterocoliticum. N Y State Dept Health* 1943;56.
6. Frederiksen W. A study of some *Yersinia pseudotuberculosis*-like bacteria (*"Bacterium enterocoliticum"* and *"Pasteurella X"*). *Scand Congr Pathol Microbiol Proc* 1964;14:103–104.
7. Niléhn B. Studies on *Yersinia enterocolitica* with special reference to bacterial diagnosis and occurrence in human acute enteric disease. *Acta Pathol Microbiol Scand Suppl* 1969;206:1–48.
8. Knapp W, Thal E. Differentiation of *Yersinia enterocolitica* by biochemical reactions. *Contrib Microbiol Immunol* 1973;2:10–16.
9. Bercovier H, Alonzo JM, Bentaiba ZN, Brault J, Mollaret HH. Contribution to the definition and the taxonomy of *Yersinia enterocolitica. Contrib Microbiol Immunol* 1979;5:12–22.
10. Stevens M, Mair NS. A numerical taxonomic study of *Yersinia enterocolitica* strains. *Contrib Microbiol Immunol* 1973;2:17–22.
11. Brenner DJ, Steigerwalt AG, Falcao DP, Weaver RE, Fanning GR. Characterization of *Yersinia enterocolitica* and *Yersinia pseudotuberculosis* by deoxyribonucleic acid hybridization and by biochemical reactions. *Int J Syst Bacteriol* 1976;26:180–194.
12. Brenner DJ, Ursing J, Bercovier H, et al. Deoxyribonucleic acid relatedness in *Yersinia enterocolitica* and *Yersinia enterocolitica*-like organisms. *Curr Microbiol* 1980;4:195–200.
13. Bercovier H, Brenner DJ, Ursing J, et al. Characterization of *Yersinia enterocolitica* sensu stricto. *Curr Microbiol* 1980;4:201–206.
14. Brenner DJ, Bercovier H, Ursing J, et al. *Yersinia intermedia:* A new species of *Enterobacteriaceae* composed of rhamnose-positive, melibiose-positive, raffinose-positive strains (formerly called *Yersinia enterocolitica* or *Yersinia enterocolitica*-like). *Curr Microbiol* 1980;4:207–212.
15. Ursing J, Brenner DJ, Bercovier H, et al. *Yersinia frederiksenii:* A new species of *Enterobacteriaceae* composed rhamnose-positive strains (formerly called atypical *Yersinia enterocolitica* or *Yersinia enterocolitica*-like). *Curr Microbiol* 1980;4:213–217.
16. Bercovier H, Brault J, Barré N, Treignier M, Alonso JM, Mollaret HH. Biochemical, serological, and phage typing characteristics of 459 *Yersinia* strains isolated from a terrrestial ecosystem. *Curr Microbiol* 1978;1: 353–357.
17. Bercovier H, Ursing J, Brenner DJ, et al. *Yersinia kristensenii:* A new species of *Enterobacteriaceae* composed of sucrose-negative strains (formerly called atypical *Yersinia enterocolitica* or *Yersinia enterocolitica*-like). *Curr Microbiol* 1980;4:219–224.
18. Bercovier H, Steigerwalt AG, Guiyoule A, Huntley-Carter G, Brenner DJ. *Yersinia aldovae* (formerly *Yersinia enterocolitica*-like group X2): A new species of *Enterobacteriaceae* isolated from aquatic ecosystems. *Int J Syst Bacteriol* 1984;34:166–172.
19. Ewing WH, Ross AJ, Brenner DJ, Fanning GR. *Yersinia ruckeri* sp. nov., the Redmouth (RM) bacterium. *Int J Syst Bacteriol* 1978;28:37–44.
20. Aleksić S, Steigerwalt AG, Bockemühl J, Huntley-Carter GP, Brenner DJ. *Yersinia rohdei* sp. nov. isolated from human and dog feces and surface water. *Int J Syst Bacteriol* 1987;37:327–332.
21. Wauters G, Janssens M, Steigerwalt AG, Brenner DJ. *Yersinia mollaretii* sp. nov. and *Yersinia bercovieri* sp. nov., formerly called *Yersinia enterocolitica* biogroups 3A and 3B. *Int J Syst Bacteriol* 1988;38:424–429.
22. Ibrahim A, Goebel BM, Liesack W, Griffiths M, Stackebrandt E. The phylogeny of the genus *Yersinia* based on 16S rDNA sequences. *FEMS Microbiol Lett* 1993;114:173–178.
23. Bercovier H, Mollaret HH, Alonso JM, et al. Intra- and interspecies relatedness of *Yersinia pestis* by DNA hybridization and its relationship to *Yersinia pseudotuberculosis. Curr Microbiol* 1980;4:225–229.
24. Dolina M, Peduzzi R. Population genetics of human, animal, and environmental *Yersinia* strains. *Appl Environ Microbiol* 1993;59:442–450.
25. Jensen WI, Owen CR, Jellison WL. *Yersinia philomiragia* sp. n., a new member of the *Pasteurella* group of bacteria, naturally pathogenic for the muskrat *(Ondatr azibethica)*. *J Bacteriol* 1969;100:1237–1241.
26. Ursing J, Steigerwalt AG, Brenner DJ. Lack of genetic relatedness between *Yersinia philomiragia* (the "Philomiragi" bacterium) and *Yersinia* species. *Curr Microbiol* 1980;4:231–233.
27. Hollis DG, Weaver RE, Steigerwalt AG, Wenger JD, Moss CW, Brenner DJ. *Francisella philomiragia* comb. nov. (formerly *Yersinia philomiragia*) and *Francisella tularensis* biogroup novicida (formerly *Francisella novicida*) associated with human disease. *J Clin Microbiol* 1989;27:1601–1608.
28. Brubaker RR. Molecular biology of the dread black death. *ASM News* 1984;50:240–245.
29. Quan TJ. *Yersinia pestis.* In: Balows A, Trüper HG, Dworkin M, Harder W, Schleifer K-H, eds. *The prokaryotes: A handbook on the biology of bacteria ecophysiology, isolation, identification, applications.* 2nd ed. Berlin: Springer-Verlag, 1991;2888–2989.
30. Perry RD, Fetherston JD. *Yersinia pestis*: Etiologic agent of plague. *Clin Microbiol Rev* 1997;10:35–66.
31. Weniger BG, Warren J, Forseth V, et al. Human bubonic plague transmitted by a domestic cat scratch. *JAMA* 1984;251:927–928.
32. Werner SB, Weidmer CE, Nelson BE, Nygaard GS, Goerhal RM, Poland JD. Primary plague pneumonia contracted from domestic cat at South Lake Tahoe, Calif. *JAMA* 1984;251:929–931.
33. Doll JM, Zeitz PS, Ettestad P, Bucholtz AL, Davis T, Gage K. Cat-transmitted fatal pneumonic plague in a person who traveled from Colorado to Arizona. *Am J Trop Med Hyg* 1994;51:109–114.
34. Centers for Disease Control. Human plague—India, 1994. *JAMA* 1994;272:1162.
35. Kumar S. Plague in India. *Lancet* 1994;344:941–942.
36. Nandan G. Troops battle to contain India's outbreak of plague. *Br Med J* 1994;309:827.
37. Tsubokura M, Otsuki K, Tanaka M, et al. Special features of distribution of *Yersinia pseudotuberculosis* in Japan. *J Clin Microbiol* 1989;27:790–791.
38. Greenwood MH. Human carriage of *Yersinia* species and incidence in foods. *Contrib Microbiol Immunol* 1995;13:74–76.
39. Aleksić S, Bockemühl J, Wuthe H-H. Epidemiology of

Y. pseudotuberculosis in Germany. *Contrib Microbiol Immunol* 1995;13:55–58.
40. Wuthe H-H, Aleksić S, Kwapil S. *Yersinia* in the European brown hare of northern Germany. *Contrib Microbiol Immunol* 1995;13:51–54.
41. Hurvell B. Chapter 14. Zoonotic *Yersinia enterocolitica* infection: Host range, clinical manifestations, and transmission between animals and man. In: Bottone EJ, ed. *Yersinia enterocolitica.* Boca Raton: CRC Press, 1981;145–159.
42. Fukushima H, Tsubokura M, Otsuki K, Kawaoka Y. Biochemical heterogeneity of serotype O3 strains of 700 *Yersinia* strains isolated from humans, other mammals, flies, animal feed, and river water. *Curr Microbiol* 1984;11:149–154.
43. Shayegani M, DeForge I, McGlynn DM, Root T. Characteristics of *Yersinia enterocolitica* and related species isolated from human, animal, and environmental sources. *J Clin Microbiol* 1981;14:304–312.
44. Botzler RG, Wetzler TF, Cowan AB. *Yersinia enterocolitica* and yersinia-like organisms isolated from frogs and snails. *Bull Wild Dis Assoc* 1968;4:110–115.
45. Lee WH, Vanderzant C, Stern N. The occurrence of *Yersinia enterocolitica* in foods. In: Bottone EJ, ed. *Yersinia enterocolitica.* Boca Raton: CRC Press, 1981;161–171.
46. Escudero ME, Velázquez L, de Guzmán AMS. *Yersinia enterocolitica* and related species isolated from animals slaughtered for human consumption. *Food Microbiol* 1996;13:201–204.
47. Logue CM, Sheridan JJ, Wauters G, McDowell DA, Blair LS. *Yersinia* spp. and numbers, with particular reference to *Y. enterocolitica* bio/serotypes, occurring on Irish meat and meat products, and the influence of alkali treatment on their isolation. *Int J Food Microbiol* 1996;33:257–274.
48. Cover TL, Aber RC. *Yersinia enterocolitica. N Engl J Med* 1989;321:16–24.
49. Swaminathan B, Harmon MC, Mehlman IJ. A review: *Yersinia enterocolitica. J Appl Bacteriol* 1982;52: 151–183.
50. Kapperud G. *Yersinia enterocolitica* in food hygiene. *Int J Food Microbiol* 1991;12:53–66.
51. Tsai S, Chen L. Occurrence of *Yersinia enterocolitica* in pork products from Northern Taiwan. *Contrib Microbiol Immunol* 1991;12:56–62.
52. Tauxe RV, Wauters G, Goossens V, et al. *Yersinia enterocolitica* infections and pork: The missing link. *Lancet* 1987;1:1129–1132.
53. Bottone EJ, Gullans CR, Sierra MF. Disease spectrum of *Yersinia enterocolitica* serogroup O:3, the predominant cause of human infection in New York City. *Contrib Microbiol Immunol* 1987;8:56–60.
54. Bissett ML, Powers C, Abbott SL, Janda JM. Epidemiologic investigations of *Yersinia enterocolitica* and related species: Sources, frequency, and serogroup distribution. *J Clin Microbiol* 1990;28:910–912.
55. Lee LA, Taylor J, Carter GP, Quinn B, Farmer JJ III, Tauxe RV, and the *Yersinia enterocolitica* Collaborative Study Group. *Yersinia enterocolitica* O:3: An emerging cause of pediatric gastroenteritis in the United States. *J Infect Dis* 1991;163:660–663.
56. Lee LA, Gerber AR, Lonsway DR, et al. *Yersinia enterocolitica* O:3 infections in infants and children, associated with household preparation of chitterlings. *N Engl J Med* 1990;322:984–987.
57. Ichinohe H, Yoshioka M, Fukushima H, Kaneko S, Maruyama T. First isolation of *Yersinia enterocolitica* serotype O:8 in Japan. *J Clin Microbiol* 1991;29: 846–847.
58. Hayashidani H, Ohtomo Y, Toyokawa Y, et al. Potential sources of sporadic human infection with *Yersinia enterocolitica* serovar O:8 in Aomori prefecture, Japan. *J Clin Microbiol* 1995;33:1253–1257.
59. Sulakvelidz A, Dalakishvili K, Barry E, et al. Analysis of clinical and environmental *Yersinia* isolates in the republic of Georgia. *J Clin Microbiol* 1996;34: 2325–2327.
60. Bottone EJ. Atypical *Yersinia enterocolitica:* Clinical and epidemiological parameters. *J Clin Microbiol* 1978;7:562–567.
61. Tertti R, Granfors K, Lehtonen O-P, et al. An outbreak of *Yersinia pseudotuberculosis* infection. *J Infect Dis* 1984;149:245–250.
62. Black RE, Jackson RJ, Tsai T, et al. Epidemic *Yersinia enterocolitica* infection due to contaminated chocolate milk. *N Engl J Med* 1978;298:76–79.
63. Shayegani M, Morse D, DeForge I, et al. Microbiology of a major foodborne outbreak of gastroenteritis caused by *Yersinia enterocolitica* serogroup O:8. *J Clin Microbiol* 1983;17:35–40.
64. Bottone EJ. *Yersinia enterocolitica.* In: Ellner PD, ed. *Infectious diarrheal diseases: Current concepts and laboratory procedures.* New York: Marcel Dekker, 1984;13–48.
65. Morse DL, Shayegani M, Gallo RJ. Epidemiologic investigation of a *Yersinia* camp outbreak linked to a food handler. *Am J Pub Health* 1984;74:589–592.
66. Tacket CO, Narain JP, Sattin R, et al. A multistate outbreak of infections caused by *Yersinia enterocolitica* transmitted by pasteurized milk. *JAMA* 1984;251: 483–486.
67. Rose FB, Camp CJ, Antes EJ. Family outbreak of fatal *Yersinia enterocolitica* pharyngitis. *Am J Med* 1987; 82:636–637.
68. Naqvi SH, Swierkosz EM, Gerard J, Mills JR. Presentation of *Yersinia enterocolitica* enteritis in children. *Pediatr Infect Dis J* 1993;12:386–389.
69. Marks MI, Pai CH, Lafleur L, Lackman L, Hammerberg O. *Yersinia enterocolitica* gastroenteritis: A prospective study of clinical, bacteriologic, and epidemiologic features. *J Pediatr* 1980;96:26–31.
70. Hoogkamp-Korstanje JAA, Stolk-Engelaar VMM. *Yersinia enterocolitica* infection in children. *Pediatr Infect Dis J* 1995;14:771–775.
71. Saeb A, Lassen J. Acute and chronic gastrointestinal manifestations associated with *Yersinia enterocolitica* infection. A Norwegian 10-year follow-up study on 458 hospitalized patients. *Ann Surg* 1992;215: 250–255.
72. Mingrone MG, Fantasia M, Figura N, Guglielmetti P. Characteristics of *Yersinia enterocolitica* isolated from children with diarrhea in Italy. *J Clin Microbiol* 1987; 25:1301–1304.
73. Kachoris M, Ruoff KL, Welch K, Kallas W, Ferraro MJ. Routine culture of stool specimens for *Yersinia enterocolitica* is not a cost-effective procedure. *J Clin Microbiol* 1988;26:582–583.
74. Ogunsanya TI, Rotimi VO, Adenuga A. A study of aetiological agents of childhood diarrhoea in Lagos, Nigeria. *J Med Microbiol* 1994;40:10–14.

75. Sethi SK, Khuffash FA, Al-Nakib W. Microbial etiology of acute gastroenteritis in hospitalized children in Kuwait. *Pediatr Infect Dis J* 1989;8:593–597.
76. Utsunomiya A, Elio-Calvo D, Reyes AAB, et al. Major enteropathogenic bacteria isolated from diarrheal patients in Bolivia: A hospital-based study. *Microbiol Immunol* 1995;39:845–851.
77. Pai CH, Gillis F, Marks MI. Infection due to *Yersinia enterocolitica* in children with abdominal pain. *J Infect Dis* 1982;146:705.
78. Monno R, Valenza MA, Quarto M, et al. *Yersinia enterocolitica* infection in a boy with beta thalessemia major. *Pediatr Infect Dis J* 1994;13:233–234.
79. Krober MS, Bass JW, Barcia PJ. Scarlatiniform rash and pleural effusion in a patient with *Yersinia pseudotuberculosis* infection. *J Pediatr* 1983;102:879–881.
80. El-Maraghi NRH, Mair NS. The histopathology of enteric infection with *Yersinia pseudotuberculosis. Am J Clin Pathol* 1979;71:631–639.
81. Sato K, Ouchi K, Taki M. *Yersinia pseudotuberculosis* infection in children, resembling Izumi fever and Kawasaki syndrome. *Pediatr Infect Dis* 1983;2: 123–126.
82. Sato K. *Yersinia pseudotuberculosis* infection in children: Clinical manifestations and epidemiology. *Contrib Microbiol Immunol* 1987;9:111–116.
83. Tertti R, Vuento R, Mikkola P, Granfors K, Mäkelä A-L, Toivanen A. Clinical manifestations of *Yersinia pseudotuberculosis* infection in children. *Eur J Clin Microbiol Infect Dis* 1989;8:587–591.
84. Gayraud M, Scavizzi MR, Mollaret HH, Guillevin L, Hornstein MJ. Antibiotic treatment of *Yersinia enterocolitica* septicemia: A retrospective review of 43 cases. *Clin Infect Dis* 1993;17:405–410.
85. Jensen KJ, Arpi M, Frederiksen W. *Yersinia enterocolitica* septicemia in Denmark 1972–1991: A report of 100 cases. *Contrib Microbiol Immunol* 1995;13:11–15.
86. Foberg U, Fryden A, Kihlström E, Persson K, Weiland O. *Yersinia enterocolitica* septicemia: Clinical and microbiological aspects. *Scand J Infect Dis* 1986;18: 269–279.
87. Cohen JI, Rodday P. *Yersinia enterocolitica* bacteremia in a patient with the acquired immunodeficiency syndrome. *Am J Med* 1989;86:254–255.
88. Vadillo M, Corbella X, Pac V, Fernandez-Viladrich P, Pujol R. Multiple liver abscesses due to *Yersinia enterocolitica* discloses primary hemochromatosis: Three case reports and review. *Clin Infect Dis* 1994;18: 938–941.
89. Boelaert JR, van Landuyt HW, Valcke WF, et al. The role of iron overload in *Yersinia enterocolitica* and *Yersinia pseudotuberculosis* bacteremia in hemodialysis patients. *J Infect Dis* 1987;156:384–387.
90. Chiu HY, Flynn DM, Hoffbrand AV, Politis D. Infection with *Yersinia enterocolitica* in patients with iron overload. *Br Med J* 1986;292:97.
91. Melby K, Slørdahl S, Gutteberg TJ, Nordbø SA. Septicaemia due to *Yersinia enterocolitica* after oral overdoses of iron. *Br Med J* 1982;285:467–468.
92. Milteer RM, Sarpong S, Poydras U. *Yersinia enterocolitica* septicemia after accidental oral iron overdose. *Pediatr Infect Dis J* 1989;8:537–538.
93. Beresford AM. Transfusion reaction due to *Yersinia enterocolitica* and review of other reported cases. *Pathology* 1995;27:133–135.
94. Wagner SJ, Friedman LI, Dodd RY. Transfusion-associated bacterial sepsis. *Clin Microbiol Rev* 1994;7: 290–302.
95. Jacobs J, Jamaer D, Vandeven J, Wouters M, Vermylen C, Vandepitte J. *Yersinia enterocolitica* in donor blood: A case report and review. *J Clin Microbiol* 1989;27: 1119–1121.
96. Bottone EJ. *Yersinia enterocolitica*: The charisma continues. *Clin Microbiol Rev* 1997;10:257–276.
97. Sire JM, Michelet C, Mesnard R, et al. Septic shock due to *Yersinia enterocolitica* after autologous transfusion. *Clin Infect Dis* 1993;17:954–955.
98. Jones BL, Saw MH, Hanson MF, Mackie MJ, Murphy WG. *Yersinia enterocolitica* septicaemia from transfusion of red blood cell concentrate stored for 16 days. *J Clin Pathol* 1993;46:477–478.
99. Woernle CH, Hoffman RE, Smith JD, et al. *Yersinia enterocolitica* bacteremia and endotoxin shock associated with red blood cell transfusions—United States, 1991. *JAMA* 1991;265:2174–2175.
100. Haverly RM, Harrison CR, Doughterty TH. *Yersinia enterocolitica* bacteremia associated with red blood cell transfusion. *Arch Pathol Lab Med* 1996;120:499–500.
101. Sazama K. Bacteria in blood for transfusion: A review. *Arch Pathol Lab Med* 1994;118:350–365.
102. Shapiro ED. *Yersinia enterocolitica* septicemia in normal infants. *Am J Dis Child* 1981;135:477–478.
103. Antonio-Santiago MT, Kaul A, Lue Y, Nichols L, Gromisch DS. *Yersinia enterocolitica* septicemia in an infant presenting as fever of unknown origin. *Clin Pediatr* 1986;25:213.
104. Cherchi GB, Pacifico L, Cossellu S, et al. Prospective study of *Yersinia enterocolitica* infection in thalassemic patients. *Pediatr Infect Dis J* 1995;14:579–584.
105. Challapalli M, Cunningham DG. *Yersinia enterocolitica* septicemia in infants younger than three months of age. *Pediatr Infect Dis J* 1993;12:168–169.
106. Pallant A, Jenkins M. A 13-day-old with fever and diarrhea. *Pediatr Infect Dis J* 1994;13:79–80,82–83.
107. McCabe EB. *Yersinia enterocolitica* septicemia in a normal child [letter]. *Am J Dis Child* 1983;137:305.
108. AlMohsen I, Luedtke G, English BK. Invasive infections caused by *Yersinia enterocolitica* in infants. *Pediatr Infect Dis J* 1997;16:253–255.
109. Rabson AR, Hallett AF, Koornhof HJ. Generalized *Yersinia enterocolitica* infection. *J Infect Dis* 1975; 131:447–451.
110. Arpi M, Jensen KT, Frederiksen W. *Yersinia enterocolitica* bacteremia: An analysis of thirty-six patients in Denmark and a review of the literature. *Contrib Microbiol Immunol* 1991;12:260–264.
111. Clarridge J, Roberts C, Peters J, Musher D. Sepsis and empyema caused by *Yersinia enterocolitica. J Clin Microbiol* 1983;17:936–938.
112. Coppens L, Sztern B, Korman D, Lustman F. *Yersinia enterocolitica* bacteremia with intracranial extension. *Scand J Infect Dis* 1995;27:409–410.
113. Kellogg CM, Tarakji EA, Smith M, Brown PD. Bacteremia and suppurative lymphadenitis due to *Yersinia enterocolitica* in a neutropenic patient who prepared chitterlings. *Clin Infect Dis* 1995;21:236–237.
114. Ljungberg P, Valtonen M, Harjola VP, Kaukoranta-Tolvanen SS, Vaara M. Report of four cases of *Yersinia pseudotuberculosis* septicemia and a literature review. *Eur J Clin Microbiol Infect Dis* 1995;14:804–809.

115. Farrer W, Kloser P, Ketyer S. Case report: *Yersinia pseudotuberculosis* sepsis presenting as multiple liver abscesses. *Am J Med Sci* 1988;295:129–132.
116. Davenport A, O'Connor B, Finn R. Acute renal failure following *Yersinia pseudotuberculosis* septicaemia. *Postgrad Med J* 1987;63:815–816.
117. Giamarellou H, Antoniadou A, Kanavos K, Papaioannou C, Kanatakis S, Papadaki K. *Yersinia enterocolitica* endocarditis: Case report and literature review. *Eur J Clin Microbiol Infect Dis* 1995;14:126–130.
118. Lecomte F, Eustache M, Lemeland J-F, Oksenhendler G, Humbert G. Purulent pericarditis due to *Yersinia enterocolitica. J Infect Dis* 1989;159:363.
119. Plotkin GR, O'Rourke JN. Case report: Mycotic aneurysm due to *Yersinia enterocolitica. Am J Med Sci* 1981;281:35–42.
120. Hagensee ME. Mycotic aneurysm due to *Yersinia enterocolitica* [correspondence]. *Clin Infect Dis* 1994; 19:801–802.
121. Tacket CO, Davis BR, Carter GP, Randolph JF, Cohen ML. *Yersinia enterocolitica* pharyngitis. *Ann Intern Med* 1983;99:40–42.
122. Blinkhorn RJ, Marino JA. Lateral pharyngeal abscess due to *Yersinia enterocolitica. Am J Med* 1988;85: 851–852.
123. Aho K, Ahvonen P, Laitinen O, Leirisalo M. Arthritis associated with *Yersinia enterocolitica* infection. In: Bottone EJ, ed. *Yersinia enterocolitica.* Boca Raton: CRC Press, 1981;113–124.
124. Borg AA, Gray J, Dawes PT. *Yersinia*-related arthritis in the United Kingdom: A report of 12 cases and review of the literature. *Q J Med* 1992;304:575–582.
125. Crowe M, Ashford K, Ispahani P. Clinical features and antibiotic treatment of septic arthritis and osteomyelitis due to *Yersinia enterocolitica. J Med Microbiol* 1996;45:302–309.
126. Krogstad P, Mendelman PM, Miller VM, et al. Clinical and microbiologic characteristics of cutaneous infection with *Yersinia enterocolitica. J Infect Dis* 1992; 165:740–743.
127. Kahn FW, Glasser JE, Agger WA. Psoas muscle abscess due to *Yersinia enterocolitica. Am J Med* 1984; 76:947–949.
128. Brennessel DJ, Robbins N, Hindman S. Pyomyositis caused by *Yersinia enterocolitica. J Clin Microbiol* 1984;20:293–294.
129. Gautier E, Carniel E, Génin J, Tancrède C. Tumor-like pyomyositis of the thigh caused by *Yersinia enterocolitica. Clin Infect Dis* 1996;23:658–659.
130. Cropp AJ, Gaylord SF, Watanakunakorn C. Case report: Cavitary pneumonia due to *Yersinia enterocolitica* in a healthy man. *Am J Med Sci* 1984;288:130–132.
131. Sodervik H, Syrjala N, Raisanen S. Interstitial pneumonia and sepsis caused by *Yersinia enterocolitica* serotype O:3. *Scand J Infect Dis* 1986;241–243.
132. Kane DR, Reuman PD. *Yersinia enterocolitica* causing pneumonia and empyema in a child and a review of the literature. *Pediatr Infect Dis J* 1992;11:591–593.
133. Ettensohn DB, Roberts NJ Jr. *Yersinia enterocolitica* pneumonia. *NY State J Med* 1981;81:791–794.
134. Sinnott JT IV , Multhopp H, Leo J, Rechtine G. *Yersinia enterocolitica* causing spinal osteomyelitis and empyema in a nonimmunocompromised host. *South Med J* 1989;82:399–400.
135. Clarridge J, Roberts C, Peters J, Musher D. Sepsis and empyema caused by *Yersinia enterocolitica. J Clin Microbiol* 1983;17:936–938.
136. Baert F, Knockaert D, Bobbaers H. Bilateral hilar lymphadenopathy associated with *Yersinia enterocolitica. Clin Infect Dis* 1994;19:197–198.
137. Nemoto H, Murabayashi K, Kawamura Y, et al. Multiple liver abscesses secondary to *Yersinia enterocolitica. Intern Med* 1992;31:1125–1127.
138. Tak PP, Visser LG, Hoogkamp-Korstanje JAA, et al. Unusual manifestations of *Yersinia enterocolitica* infections diagnosed using novel methods. *Clin Infect Dis* 1992;15:645–649.
139. Winblad S. Chapter 12. Erythema nodosum associated with infection with *Yersinia enterocolitica.* In: Bottone EJ, ed. *Yersinia enterocolitica.* Boca Raton: CRC Press, 1981;125–133.
140. Luqmani RA, Dawes PT. *Yersinia* arthritis with erythema nodosum. *Postgrad Med J* 1986;62:405.
141. Hoogkamp-Korstanje JAA, de Koning J, Heesemann J. Persistence of *Yersinia enterocolitica* in man. *Infection* 1988;16:81–85.
142. Read RC, Barry RE. Relapsing yersinia infection. *Br Med J* 1990;300:1694.
143. Casey MF, Gilligan PH, Smiley ML. *Yersinia enterocolitica* meningitis and osteomyelitis: A case report. *Diagn Microbiol Infect Dis* 1987;8:47–50.
144. Flament-Saillour M, de Truchis P, Risbourg M, Nordmann P. *Yersinia enterocolitica* peritonitis in a patient infected with the human immunodeficiency virus. *Clin Infect Dis* 1994;18:655–656.
145. Wakefield D, Stahlberg TH, Toivanen A, Granfors K, Tennant C. Serologic evidence of *Yersinia* infection in patients with anterior uveitis. *Arch Ophthalmol* 1990; 108:219–221.
146. Tanaka T, Yamakawa N, Yamaguchi H, et al. Common antigenicity between *Yersinia enterocolitica*-derived heat-shock protein and the retina, and its role in uveitis. *Ophthalmic Res* 1996;28:284–288.
147. Barteluk RL, Noble MA. Routine culturing of stool specimens for *Yersinia enterocolitica. J Clin Microbiol* 1988;26:1616.
148. Metchock B, Lonsway DR, Carter GP, Lee LA, McGowan JE Jr. *Yersinia enterocolitica:* A frequent seasonal stool isolate from children at an urban hospital in the southeast United States. *J Clin Microbiol* 1991; 29:2868–2869.
149. Bowen JH, Kominos SD. Evaluation of a pectin agar mdium for isolation of *Yersinia enterocolitica* within 48 hours. *Am J Clin Path* 1979;72:586–590.
150. Head CB, Whitty DA, Ratnam S. Comparative study of selective media for recovery of *Yersinia enterocolitica. J Clin Microbiol* 1982;16:615–621.
151. Dudley MV, Shotts EB Jr. Medium for isolation of *Yersinia enterocolitica. J Clin Microbiol* 1979;10: 180–183.
152. Schiemann DA. Synthesis of a selective agar medium for *Yersinia enterocolitica. Can J Microbiol* 1979;25: 1298–1304.
153. Agbonlahor DE, Odugbemi T, Dosunmu-Ogunbi O. Differential and selective medium for isolation of *Yersinia enterocolitica* from stools. *J Clin Microbiol* 1982;15:599–602.
154. Ratnam S, Looi CL, Patel TR. Lack of efficacy of alkali treatment for isolation *Yersinia enterocolitica* from feces. *J Clin Microbiol* 1983;18:1092–1097.

155. Fukushima H, Gomyoda M. Growth of *Yersinia pseudotuberculosis* and *Yersinia enterocolitica* biotype 3B serotype O3 inhibited on cefsulodin-irgasan-novobiocin agar. *J Clin Microbiol* 1986;24:116–120.
156. Schiemann DA, Olson SA. Antagonism by gram-negative bacteria to growth of *Yersinia enterocolitica* in mixed culture. *J Clin Microbiol* 1984;48:539–544.
157. Weissfeld AS, Sonnenwirth AC. Rapid isolation of *Yersinia* spp. from feces. *J Clin Microbiol* 1982;15:508–510.
158. Van Noyen R, Vandepitte J, Wauters G, Selderslahgs R. *Yersinia enterocolitica:* Its isolation by cold enrichment from patients and healthy subjects. *J Clin Pathol* 1981;34:1052–1056.
159. Stolk-Engelaar VMM, Hoogkamp-Korstanje JAA. Clinical presentation and diagnosis of gastrointestinal infections by *Yersinia enterocolitica* in 261 Dutch patients. *Scand J Infect Dis* 1996;28:571–575.
160. Holmes B, Costas M, Ganner M, On SLW, Stevens M. Evaluation of Biolog system for identification of some gram-negative bacteria of clinical importance. *J Clin Microbiol* 1994;32:1970–1975.
161. Wilmouth BA, Chu MC, Quan TJ. Identification of *Yersinia pestis* by BBL Crystal enteric/nonfermenter identification system. *J Clin Microbiol* 1996;34:2829–2830.
162. Wauters G, Kandolo K, Janssens. Revised biogrouping scheme of *Yersinia enterocolitica. Contrib Microbiol Immunol* 1987;9:14–21.
163. Farmer JJ III, Carter GP, Miller VL, Falkow S, Wachsmuth IK. Pyrazinamidase, CR-MOX agar, salicin fermentation-esculin hydrolysis, and D-xylose fermentation for identifying pathogenic serotypes of *Yersinia enterocolitica. J Clin Microbiol* 1992;30:2589–2594.
164. Cimolai N, Trombley C, Blair GK. Implications of *Yersinia enterocolitica* biotyping. *Arch Dis Child* 1994;70:19–21.
165. Fukushima H, Maruyama K, Omori I, Ito K, Kaneko S. Isolation of sucrose-negative *Yersinia enterocolitica* biotype 3 serotype O3 strains and their pathogenicity. *Curr Microbiol* 1988;17:199–202.
166. Morris JG Jr, Prado V, Ferreccio C, et al. *Yersinia enterocolitica* isolated from two cohorts of young children in Santiago, Chile: Incidence of and lack of correlation between illness and proposed virulence factors. *J Clin Microbiol* 1991;29:2784–2788.
167. Burnens AP, Frey A, Nicolet J. Association between clinical presentation, biogroups and virulence attributes of *Yersinia enterocolitica* strains in human diarrhoeal disease. *Epidemiol Infect* 1996;116:27–34.
168. Kapperud G, Bergan T. Biochemical and serological characterization of *Yersinia enterocolitica. Methods Microbiol* 1984;15:296–343.
169. Aleksić S, Bockemühl J. Diagnostic importance of H-antigens in *Yersinia enterocolitica* and other *Yersinia* species. *Contrib Microbiol Immunol* 1987;9:279–284.
170. Kawaoka Y, Wauters G, Otsuki, Tsubokura M. Identification of *Yersinia enterocolitica* O15 lipopolysaccharide as a rough antigen. *J Clin Microbiol* 1986;24:272–274.
171. Aleksić S, Bockemühl J. Proposed revision of the Wauters et al. antigenic scheme for serotyping of *Yersinia enterocolitica. J Clin Microbiol* 1984;20:99–102.
172. Wauters G, Aleksić S, Charlier J, Schulze. Somatic and flagellar antigens of *Yersinia enterocolitica* and related species. *Contrib Microbiol Immunol* 1991;12:239–243.
173. Aleksić S, Bockemühl J, Lange F. Studies on the serology of flagellar antigens of *Yersinia enterocolitica* and related *Yersinia* species. *Zentralbl Bakteriol* 1986;261:299–310.
174. Thomas LV, Gross RJ, Cheasty T, Shipp CR, Rowe B. Antigenic relationships among type strains of *Yersinia enterocolitica* and those of *Escherichia coli, Salmonella* spp., and *Shigella* spp. *J Clin Microbiol* 1983;17:109–111.
175. Brewer RA, Corbel MJ. *Yersinia enterocolitica* and related species. *Soc Appl Microbiol Technical Series* 1985;83–104.
176. Kaneko S, Maruyama T. Evaluation of enzyme imunoassay for the detection of pathogenic *Yersinia enterocloitica* and *Yersinia pseudotuberculosis* strains. *J Clin Microbiol* 1989;27:748–751.
177. Robins-Browne RM, Miliotis MD, Cianciosi S, Miller VL, Falkow S, Morris JG Jr. Evaluation of DNA colony hybridization and other techniques for detection of virulence in *Yersinia* species. *J Clin Microbiol* 1989;27:644–650.
178. Baker PM, Farmer JJ III. New bacteriophage typing system for *Yersinia enterocolitica, Yersinia kristensenii, Yersinia frederiksenii,* and *Yersinia intermedia:* Correlation with serotyping, biotyping and antibiotic susceptibility. *J Clin Microbiol* 1982;15:491–502.
179. Najdenski H, Iteman I, Carniel. Efficient subtyping of pathogenic *Yersinia enterocolitica* strains by pulsed-field gel electrophoresis. *J Clin Microbiol* 1994;32:2913–2920.
180. Blumber HM, Kielbauch JA, Wachsmuth IK. Molecular epidemiologiy of *Yersinia enterocolitica* O:3 infections: Use of chromosomal DNA restriction fragment polymorphism of rRNA genes. *J Clin Microbiol* 1991;29:2368–2374.
181. Fukushima H, Gomyoda M, Aleksi S, Tsubokura M. Differentiation of *Yersinia enterocolitica* serotype O:5,27 strains by phenotypic and molecular techniques. *J Clin Microbiol* 1993;31:1672–1674.
182. Cornelis G, Laroche Y, Balligand G, Sory M-P, Wauters G. *Yersinia enterocolitica,* a primary model for bacterial invasiveness. *Rev Infect Dis* 1987;9:64–87.
183. Brubaker RR. Factors promoting acute and chronic diseases caused by yersiniae. *Clin Microbiol Rev* 1991;4:309–324.
184. Cornelis GR. *Yersinia* pathogenicity factors. *Curr Top Microbiol Immunol* 1994;192:243–263.
185. Cornelis GR, Biot T, Lambert de Rouvroit C, et al. The *Yersinia yop* regulon. *Mol Microbiol* 1989;3:1455–1459.
186. Straley SC. The low-Ca^{2+} response virulence regulon of human-pathogenic yersiniae. *Microb Pathog* 1991;10:87–91.
187. Miller VM. *Yersinia* invasion genes and their products. *ASM News* 1992;58:26–33.
188. Straley SC, Plano GV, Skrzypek E, Haddix PL, Fields KA. Regulation by Ca^{2+} in the *Yersinia* low-Ca^{2+} response. *Mol Microbiol* 1993;8:1005–1010.
189. Straley SC, Skrzypek E, Plano GV, Bliska JB. Yops of *Yersinia* spp. pathogenic for humans. *Infect Immun* 1993;61:3105–3110.

190. Perry RD. Acquisition and storage of inorganic iron and hemin by the yersiniae. *Trends Microbiol* 1993;1:142–147.
191. Straley SC, Perry RD. Environmental modulation of gene expression and pathogenesis in *Yersinia. Trends Microbiol* 1995;3:310–317.
192. Isberg RR. Uptake of enteropathogenic *Yersinia* by mammalian cells. *Curr Top Microbiol Immunol* 1996;209:1–24.
193. Najdenski H, Iteman I, Carniel E. The genome of *Yersinia enterocolitica* is the most stable of the three pathogenic species. *Contrib Microbiol Immunol* 1995;13:281–284.
194. Litwin CM, Calderwood SB. Role of iron in regulation of virulence genes. *Clin Microbiol Rev* 1993;6:137–149.
195. Mecsas J, Strauss EJ. Molecular mechanisms of bacterial virulence: Type III secretion and pathogenicity islands. *Emerg Infect Dis* 1996;2:271–288.
196. Hinnebusch BJ, Perry RD, Schwan TG. Role of the *Yersinia pestis* hemin storage *(hms)* locus in the transmission of plague by fleas. *Science* 1996;273:367–370.
197. Heesemann J. Chromosomal-encoded siderophores are required for mouse virulence of enteropathogenic *Yersinia enterocolitica. FEMS Microbiol Lett* 1987;48:229–233.
198. Haag H, Hantke K, Drechsel H, Stojilkovic I, Jung G, Zähner H. Purification of yersiniabactin: A siderophore and possible virulence factor of *Yersinia enterocolitica. J Gen Microbiol* 1993;139:2159–2165.
199. Chambers CE, Sokol PA. Comparison of siderophore production and utilization in pathogenic and environmental isolates of *Yersinia enterocolitica. J Clin Microbiol* 1994;32:32–39.
200. Stuart SJ, Prpic JK, Robins-Browne RM. Production of aerobactin by some species of the genus *Yersinia. J Bacteriol* 1986;166:1131–1133.
201. Carniel E, Mercereau-Puijalon O, Bonnefoy S. The gene coding for the 190,000-dalton iron-regulated protein of *Yersinia* species is present only in highly pathogenic strains. *Infect Immun* 1989;57:1211–1217.
202. Miller VL, Falkow S. Evidence for two genetic loci in *Yersinia enterocolitica* that can promote invasion of epithelial cells. *Infect Immun* 1988;56:1242–1248.
203. Pepe JC, Miller VL. *Yersinia enterocolitica* invasin: A primary role in the initiation of infection. *Proc Natl Acad Sci U S A* 1993;90:6473–6477.
204. Pierson DE, Falkow S. Nonpathogenic isolates of *Yersinia enterocolitica* do not contain functional *inv*-homologous sequences. *Infect Immun* 1990;58:1059–1064.
205. Miller VL, Farmer JJ III, Hill WE, Falkow S. The *ail* locus is found uniquely in *Yersinia enterocolitica* serotypes commonly associated with disease. *Infect Immun* 1989;57:121–131.
206. Bliska JB, Falkow S. Bacterial resistance to complement killing mediated by the Ail protein of *Yersinia enterocolitica. Proc Natl Acad Sci U S A* 1992;89:3561–3565.
207. Pierson DE, Falkow S. The *ail* gene of *Yersinia enterocolitica* has a role in the ability of the organism to survive serum killing. *Infect Immun* 1993;61:1846–1852.
208. Guan K, Dixon JE. Protein tyrosine phosphatase activity of an essential virulence determinant in *Yersinia. Science* 1990;249:553–556.
209. Galyov EE, Håkansson S, Forsberg Å, Wolf-Watz H. A secreted protein kinase of *Yersinia pseudotuberculosis* showing homology with eukaryotic Ser/Thr protein kinases is an indispensable virulence determinant. *Nature* 1993;361:730–732.
210. Pærregaard A, Espersen F, Skurnik M. Role of *Yersinia* outer membrane protein YadA in adhesion to rabbit intestinal tissue and rabbit brush border membrane vesicles. *Acta Pathol Microbiol Scand* 1991;99:226–232.
211. Visser LG, Hiemstra PS, van den Barselaar MT, Ballieux PA, van Furth R. Role of YadA in resistance to killing of *Yersinia enterocolitica* by antimicrobial polypeptides of human granulocytes. *Infect Immun* 1996;64:1653–1658.
212. Koning-Ward TF, Robins-Browne RM. Contribution of urease to acid tolerance in *Yersinia enterocolitica. Infect Immun* 1995;63:3790–3795.
213. Skurnik M, Zhang L. Molecular genetics and biochemistry of *Yersinia* lipopolysaccharide. *Acta Pathol Microbiol Scand* 1996;104:849–872.
214. Delor I, Kaeckenbeeck A, Wauters G, Cornelis GR. Nucleotide sequence of *yst,* the *Yersinia enterocolitica* gene encoding the heat-stable enterotoxin, and prevalence of the gene among pathogenic and nonpathogenic yersiniae. *Infect Immun* 1990;58:2983–2988.
215. Hornstein MJ, Jupeau AM, Scavizzi MR, Philippon AM, Grimont PAD. In vitro susceptibility of 126 clinical isolates of *Yersinia enterocolitica* to 21 β-lactam antibiotics. *Antimicrob Agents Chemother* 1985;27:806–811.
216. Hoogkamp-Korstanje JAA. The possible role of quinolones in yersiniosis. *Drugs* 1987;34(Suppl):134–138.
217. Pham JN, Bell SM, Lanzarone JYM. Biotype and antibiotic sensitivity of 100 clinical isolates of *Yersinia enterocolitica. J Antimicrob Chemother* 1991;28:13–18.
218. Pham JN, Bell SM, Hardy MJ, Martin L, Guiyoule A, Carniel E. Susceptibility to β-lactam agents of *Yersinia enterocolitica* biotype 4, serotype O3 isolated in various parts of the world. *J Med Microbiol* 1995;43:9–13.
219. Stolk-Engelaar VMM, Meis JFGM, Mulder JA, Loeffen FLA, Hoogkamp-Korstanje JAA. In-vitro antimicrobial susceptibility of *Yersinia enterocolitica* isolates from stools of patients in the Netherlands from 1982–1991. *J Antimicrob Chemother* 1995;36:839–843.
220. Lemaitre BC, Mazigh DA, Scavizzi MR. Failure of β-lactam antibiotics and marked efficacy of fluoroquinolones in treatment of murine *Yersinia pseudotuberculosis* infection. *Antimicrob Agents Chemother* 1991;35:1785–1790.

14
The Genus *Proteus*

TAXONOMY

The genus *Proteus* is one of the three oldest genera in the family *Enterobacteriaceae*.[1] Originally described by Hauser in 1885, the generic name originates from the sea god Proteus, an attendant to Poseidon, who could change his shape at will according to Greek mythology. This latter attribute reminded Hauser of the way protei exhibited morphologic variability on subculture, often swarming over the entire agar surface. By 1919, two *Proteus* species were recognized, namely, *Proteus mirabilis* and *Proteus vulgaris*.[2]

Over the intervening decades, a number of other species were included in the genus *Proteus* for various periods based on urease production, reputed swarming activity, or other phenotypic features associated with *P. mirabilis* and *P. vulgaris* (see Chapters 15 and 16). Rauss[3] summarized a number of these taxonomic proposals dating between 1945 and 1960. One consistent feature of almost all such schemes was the inclusion of both *P. mirabilis* and *P. vulgaris* within the genus *Proteus*. In 1966, a potential new species was isolated from living and dead gypsy moth larvae collected from New York and Connecticut.[4] Bacteria isolated from larvae were biochemically distinct from both *P. mirabilis* and *P. vulgaris* and produced slime when grown in trypticase soy broth. The name *Proteus myxofaciens* ("slime-producing") was proposed for this organism by Cosenza and Podgwaite.[4] Twelve years later, Brenner and coworkers[5] confirmed by DNA–DNA reassociation kinetics that *P. myxofaciens* was indeed a legitimate member of the genus *Proteus*.

During the early to middle 1970s, questions regarding the correct taxonomic position for such groups as *Morganella morganii ("Proteus morganii")* and *Providencia rettgeri ("Proteus rettgeri")* continued to surface. These controversies were resolved by Brenner and associates[5] in a systematic investigation of more than 120 *Proteus* and *Providencia* strains. Several major conclusions regarding the genus *Proteus* emanated from this study. *P. mirabilis* was found to be an extremely homogeneous group at the DNA level, with most strains more than 90% related to the reference strain PR14 (1% divergence). In contrast, at least two distinct DNA groups were found within *P. vulgaris,* represented by reference strains PR1 and CDC 931-72. Interrelatedness between *P. mirabilis* and *P. vulgaris* approached 55%. *P. myxofaciens,* originally described in 1966, was also confirmed to be a valid *Proteus* species since it was 44% to 46% related to *P. mirabilis* and *P. vulgaris.* Other taxa previously included in the genus were reassigned to other genera *(P. rettgeri)* or placed in a genus of their own *(Morganella).*

Although Brenner's investigation indicated that *P. vulgaris* was genetically heterogeneous, all strains were retained within this phenospecies pending further study and the identification of biochemical tests useful in separating these genomospecies. In 1982, Hickman and associates[6] identified three biogroups within *P. vulgaris* based on indole production, esculin hydrolysis, and fermenta-

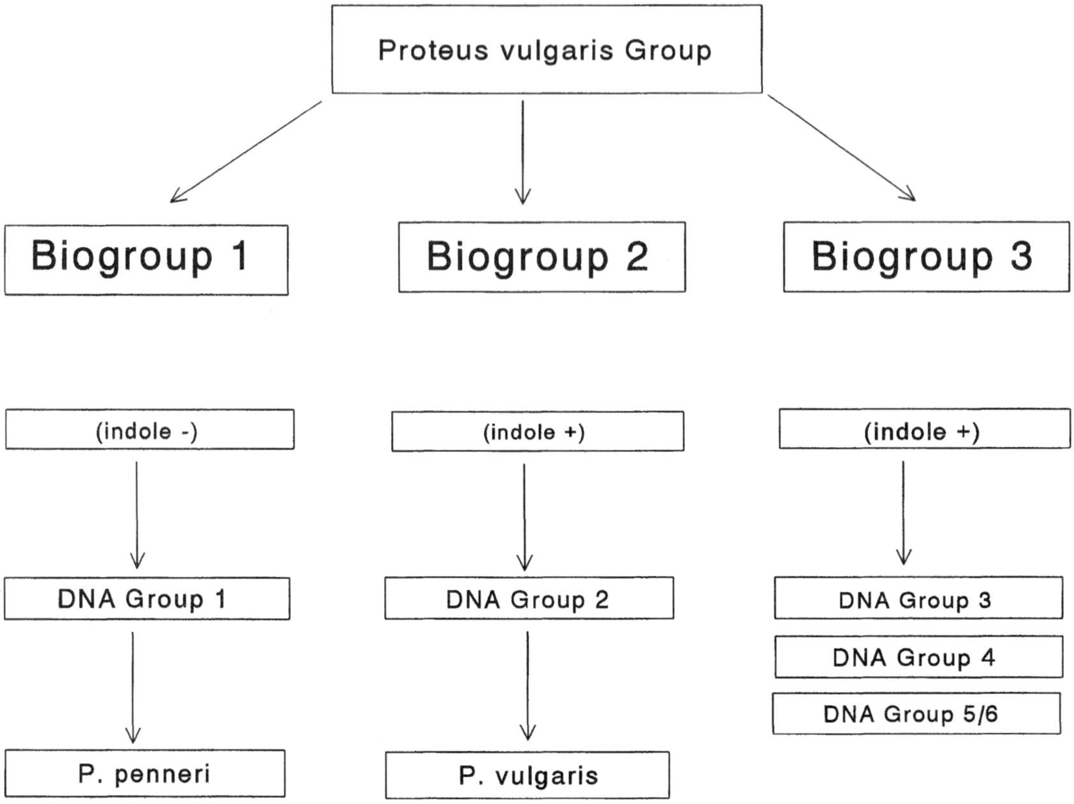

FIG. 1. Correlation of *P. vulgaris* biogroups with nomenspecies and genomospecies.

tion of salicin (Fig. 1). Biogroup 1, which consisted of 20 indole-negative isolates, were 88% to 99% related at 60°C. This group included CDC 931-72, which corresponded to Brenner's second hybridization group. The name *Proteus penneri* was proposed for these microorganisms.[6] Biogroup 2 strains were indole-, esculin-, and salicin-positive, whereas biogroup 3 strains were indole-positive and esculin- and salicin-negative.

The *P. vulgaris* biogroups 2 and 3 were subsequently studied by electrophoretic analysis of their protein patterns for relative homogeneity.[7] When the principal protein bands were omitted, all *P. vulgaris* biogroup 2 strains clustered together and the biogroup 3 strains formed two distinct groups comprising 4 and 12 strains. This study was followed by an international collaborative effort in which biogroup 2 and 3 strains were subjected to DNA–DNA hybridization.[8] As found by protein profile analysis, *P. vulgaris* biogroup 2 was shown to be homogeneous and was determined to represent a single genetic species (DNA group 2). Within biogroup 3, however, four DNA groups were detected (DNA groups 3–6). One problem arising from the revelation that *P. vulgaris* contained multiple genomospecies concerned the status of the type strain ATCC 13315 (NCTC 4175). The type strain was found to reside in genomospecies 3, which consisted of only the type strain and one other isolate, CDC 1732-80. Therefore, if the type strain for *P. vulgaris* were to remain in DNA group 3, the majority of strains that previously would have been identified as *P. vulgaris* would now belong to a new species. Such a change would cause major confusion in regard to the taxonomic literature of *P. vulgaris* dating back more than 50 years. To

TABLE 1. *Current composition of the genus* Proteus

Species	Type Strain	Isolated from Humans	Ref.
P. mirabilis	ATCC 29906	Yes	1
P. vulgaris	ATCC 29905[a]	Yes	9
P. myxofaciens	ATCC 19692	No	1
P. penneri	ATCC 33519	Yes	6

[a]Proposed replacement type strain.

avoid this situation, Brenner and colleagues[8] proposed that ATCC 13315 (NCTC 4175) be replaced with ATCC 29905 (CDC PR1) as the type strain for *P. vulgaris.* Since ATCC 29905 belongs to DNA group 2 (biogroup 2), a commonly isolated phenotype within the clinical laboratory, the integrity of the species name could be maintained. Current composition of the genus *Proteus* is listed in Table 1.

EPIDEMIOLOGY

As with many *Enterobacteriaceae* genera, much of our knowledge concerning the environmental distribution of protei comes from older literature, which predates recent taxonomic changes within the genus. This fact somewhat clouds the significance of many earlier studies, which often included the providence group and morganellae as members of the genus *Proteus.* Despite this fact, a general pattern of the ecologic distribution of the major *Proteus* species can be determined by a careful scrutiny of the biochemical and phenotypic criteria used to identify protei in these surveys.

Phillips[9] reported on the distribution of 200 swarming *Proteus* strains recovered from a number of homeothermic and poikilothermic species. Although the sources of these strains were not mentioned, the gastrointestinal tract is the presumed site of isolation for most isolates. *P. mirabilis* was the most frequently isolated species (>75%), commonly recovered from dogs, cows, and birds. Cows and birds were also common reservoirs for *P. vulgaris.* However, in contrast to *P. mirabilis, P. vulgaris* was more frequently isolated from pigs and less often from dogs. Other species reported to harbor protei included cats and reptiles.[9,10] A number of pathologic conditions in animals have been attributed to *Proteus* species; these include urinary tract infections (UTIs) in dogs, equine abortions and hoof canker, and abscesses in rabbits.[9,11]

Proteus species are often associated with fish, including both freshwater and saltwater varieties.[12–14] Because of their ability to secrete the enzyme L-histidine decarboxylase, they have been implicated as histamine producers in the spoilage of certain fish such as tuna and sardines.[13,14] Consumption of such contaminated fish containing high levels of histamine leads to the clinical condition known as scombroid fish poisoning. Although microbiology studies aimed at enumerating the relative concentrations of protei in various consumable products have not been undertaken recently, their isolation from both bovine and porcine species implies their presence in meat products. Von Holy and colleagues[15] found that *Proteus* species were commonly associated with the manufacture of vacuum-packed Vienna sausages. Among enteric bacteria, *Proteus* species were one of the three most common genera isolated (along with *Serratia* and *Enterobacter*), accounting for between 4.0% and 4.3% of the total microbial burden.

P. mirabilis is an important hospital-associated pathogen. In the most recent National Nosocomial Infections Surveillance System (NNIS) survey, *P. mirabilis* was responsible for 4% of all reported infections.[16] Most often this species is implicated as a cause of UTIs, where it accounts for 5% of all nosocomially transmitted disease. Despite its overall importance, recent figures indicate that its prevalence as a nosocomial pathogen is decreasing. This is thought to be in part a result of the susceptibility of this organism to many commonly prescribed antibiotics.[16]

CLINICAL SIGNIFICANCE

Protei are important agents of human disease despite an NNIS study indicating that their overall incidence as nosocomial pathogens is decreasing.[16] At least two thirds of all *Proteus* infections are hospital acquired, with most of these involving *P. mirabilis.*[17] The single most common anatomic site from which protei are recovered is the urinary tract, although they are also isolated from sites involved in extraintestinal diseases such as septicemia and meningitis. On occasion, *Proteus* species have been implicated as a causative agent in a number of miscellaneous illnesses including wound, bone, and joint infections and diseases of the respiratory tract.

Urinary Tract Infections

P. mirabilis and related species are major uropathogens. Although the NNIS survey estimates the overall prevalence of *P. mirabilis* nosocomial UTIs at 5%, rates may vary significantly from hospital to hospital depending on the type of medical facility, the patient population serviced, and the geographic locale. A 1-year (1991–1992) study of 580 inpatients at a Nigerian teaching hospital found an overall urine positivity rate of 38%.[18] Of 220 persons with UTIs, 30 involved *Proteus* species for an incidence of 13.6%. In most cases of UTI, protei are present in large numbers ($>10^5$ cfu/mL), either as the sole etiologic agent or as part of a polymicrobic infection often involving *Klebsiella* or *Serratia.*[19] In approximately 20% to 30% of cultures positive for *Proteus,* concentrations of protei in urine do not meet acceptable threshold criteria for bacteriuria ($>10^3$–10^5 cfu/mL). In these instances, the role of protei in UTIs must be determined in conjunction with other ancillary findings including patient symptomatology.[18,20]

P. mirabilis is the second most common bacterial species (next to *Escherichia coli*) recovered from urine cultures.[16,21,22] Bacteriuria due to *P. mirabilis* develops primarily in two distinct clinical settings. The first of these involves uncomplicated UTIs in healthy adult women and, less frequently, young children.[23,24] Dysuric females typically present with acute signs of cystitis including pyuria, increased frequency and urgency to void, suprapubic pain, and, occasionally, hematuria.[25] Such infections typically respond to an appropriate short course (3 days) of antimicrobial chemotherapy. In some instances mild, uncomplicated UTIs may progress to more moderate disease (eg, pyelonephritis). Such a progression is often the result of resistance by *Proteus* strains to multiple drugs. Treatment of such uncomplicated cases of pyelonephritis requires prolonged antimicrobial intervention.[25] In approximately 3% of women with acute UTI symptomatology, midstream urine specimens fail to yield an etiologic agent.[26] Bladder aspirates, however, contain *P. mirabilis* in concentrations ranging from less than 10^2 to more than 10^5 cfu/mL. In children, one study found that *Proteus* UTIs were more commonly associated with boys.[23,27] This gender difference may be related to the fact that protei are more often recovered from the preputial sac than from the vulva.

The second major arena in which protei play a role as uropathogens involves complicated cases of UTI in which the patient has underlying disease or risk factors predisposing to possible infection with *P. mirabilis.* Like UTIs caused by other members of the tribe *Proteeae* (ie, *Morganella, Providencia*), these conditions commonly include structural, physiologic, or neurologic abnormalities involving the urinary tract. Often-cited underlying conditions leading to *P. mirabilis* infection include diabetes mellitus, prostatic hypertrophy, sickle cell anemia, neurogenic bladder, polycystic renal disease, urologic malignancy, and surgery.[18,23] The common thread connecting most of these diseases to *P. mirabilis*–associated UTIs is catheterization. An early, classic study by Adler and coworkers[17] found that 68% of persons with nosocomial UTIs caused by *Proteus* had indwelling Foley catheters or had recently undergone urogenital manipulations before their bacteriuric episode. Most patients acquired their nosocomial *Proteus* infection within the first 28 days of hospitalization. Subsequent studies have chronicled this association between catheterization and *Proteus* bacteriuria.[23,25,28,29] Nicolle and associates,[30] in a review of the microbiology of bacteriuria in nursing home residents, estimated the rate of *P. mirabilis* infection in these populations at 20% to 38%. Such high prevalence rates are undoubtedly a result of the large number of elderly nursing home residents who are transiently or permanently catheterized (15%–53%). In catheterized women, infection occurs predominantly via the periurethral route; in men, most illnesses result from intraluminal contamination.[29]

One of the most serious complications associated with *P. mirabilis* UTI is long-term catheterization (>30 days). In addition to treatment problems, long-term catheterization produces an increased risk for development of bacteriuria with urease-positive organisms including *P. mirabilis*.[31] The high level of urease activity produced by *P. mirabilis* in the bladder causes alkalization of the urine. This rise in pH leads to precipitation of organic (Tamm-Horsfall protein, uric acid) and inorganic (cations, anions) compounds. These components then coalesce in a crystallization process to produce struvite ($MgNH_4PO_4 \bullet 6H_2O$, magnesium ammonium phosphate hexahydrate) and carbonate-apatite [$Ca_{10}(PO_4CO_3OH)_6(OH_2)$] crystals.[23,29,31,32] Formation of such deposits can lead to biofilm development and encrustation of catheters, resulting in occlusion and blockage of lines. Kidney stones may also develop. A recent investigation of 75 catheters removed from patients with long-term bladder management found bacterial biofilms to be present in 92% of these lines.[33] Of the 69 biofilms, 18 (26%) contained *P. mirabilis*, 10 of which (56%) were in pure culture. Stone formation can lead to additional medical complications. One report described a case of *P. mirabilis* septicemia in a 61-year-old woman with polycystic kidney disease secondary to a staghorn calculus, which developed after repeated bacteriuric episodes with *Proteus* over an 11-year period.[34]

In addition to *P. mirabilis*, both *P. vulgaris* and *P. penneri* have been linked to UTIs. Several studies have estimated the incidence of *P. vulgaris*–associated UTI at 0.5% to 1.8%.[18,20,21] Because it is such an infrequent cause of bacteriuria, specific studies addressing epidemiologic issues related to infection have not been attempted. The newest member of the genus *Proteus*, *P. penneri*, has also been associated with infections of the urogenital system. In the original publication describing this species, more than 50% of all *P. penneri* strains were recovered from urine, although clinical information regarding the significance of these isolates was unavailable.[6] In 1984, Krajden and colleagues[35] described the repeated recovery of *P. penneri* from the urine of a 92-year-old man with hematuria. Because repeated doses of chemotherapeutic agents failed to eradicate the focus of infection, further tests were performed. A radiographic study of the bladder revealed three stones, one of which yielded pure growth of *P. penneri*. A reservoir for this unusual infection could not be identified. Three years later, a follow-up epidemiologic study of *P. penneri* infections in Canada reported 12 more urine isolates.[36] A majority of these strains were recovered from elderly patients (mean age 70.2 years), 83% of whom had indwelling catheters. In 7 of these 12 cases, *P. penneri* was isolated in pure culture.

Bacteremia

Among members of the *Enterobacteriaceae*, *Proteus* species are one of the most frequent causes of bacteremia. General surveys of positive blood cultures conducted between 1980 and 1983 found *Proteus* to be the third most common enteric genus involved in sepsis after *E. coli* and *Klebsiella*.[37,38] Later studies (1984–1987) from Vancouver, British Columbia, found *Proteus* septicemia to rank fifth among enteric pathogens, following *E. coli, Klebsiella, Enterobacter,* and *Serratia.*[39] Regardless of the findings reported in these individual investigations, *Proteus* bacteremia continues to be an important and frequent infectious complication arising in hospitalized persons.[16]

Positive *Proteus* blood cultures are almost invariably (93%–100%) clinically significant,[38,39] with only rare cases of transient bacteremia noted.[39] Based on total positive blood cultures, the overall frequency of *Proteus* bacteremia ranges from 1.2% to 4.7%, according to several clinical series.[37–40] Similar frequencies (2%) have been reported in pediatric populations.[41] A higher prevalence of *Proteus* bacteremia, however, can be found in older persons in long-term care facilities. Setia and colleagues[42] retrospectively studied 100 episodes of bacteremia occurring during a 2-year period in older patients (mean age 79 years) in a long-term care facility in New Jersey. In this study 14% of all bacteremias were caused by *Proteus* species. Rudman and coworkers[43] found a similarly high rate of *Proteus* sepsis (15%) in older, chronically institutionalized patients in a Veterans Administration extended-care facility. The increased rates of *Proteus* sepsis cited in these surveys are most likely a consequence of higher colonization rates in geriatric patients, which in turn is in part attributable to catheterization.[43]

Few studies have looked specifically at *Proteus* bacteremia. Adler and colleagues,[17] in 1971, studied 68 cases of *Proteus* sepsis over an 8-month period. Most infections were hospital acquired (66%) and occurred in older persons (median age 71 years), and 70% developed within 24 days of hospitalization (median 14 days). Underlying diseases often associated with protei bacteremia included diabetes, congestive heart failure, acute abdominal crisis, malignancy, and severe chronic pulmonary disease. At least 83% of patients had additional risk factors for infection such as Foley catheterization, urogenital instrumentation, central venous catheters, and tracheostomies. Other significant conditions noted by Adler to predispose patients to *Proteus* septicemia included recent treatment with antimicrobials or with penicillin alone. A 1994 publication by Watanakunakorn and Perni[44] described 176 cases of *Proteus* bacteremia at a community hospital over a 10-year period. In that study, *Proteus* bacteremia was found to be a disease of older persons (median age 75 years), 64% of whom were catheterized, and who most often had been transferred to the hospital from a nursing home. As in the earlier study of Adler and associates,[17] more than 70% of all septicemias developed within the first month of hospitalization (28 days). In virtually all studies involving *Proteus* sepsis, the single most common site from which septicemia disseminates is the urinary tract. The percentage of blood isolates originating from this source was estimated to be between 33% and 100% in selected investigations.[17,37–39,42–44] Other common sources from which septicemia develops, in decreasing order of frequency, are the respiratory tract, wounds (including decubitus), the skin, abscesses, and the hepatobiliary tree.[17,38,39,42,44] Although one early study[17] cited an exceptionally high fatality rate associated with *Proteus* bacteremia (72%), more recent investigations have generally reported much lower rates (20%–29%), perhaps because of better medical care and more active antimicrobial agents.[39,42,44] Higher mortality rates among patients with *Proteus* sepsis have been linked to the development of septic shock, inappropriate chemotherapy, and the failure to immediately recognize septic individuals.[17] Watanakunakorn and Perni[44] found that a number of additional factors were associated with an elevated mortality rate, including nosocomially derived infection, the presence of an ultimately fatal underlying disease, hypotension, increased levels of serum creatinine and bilirubin, and a source of sepsis other than the urinary tract.

Microbiologic analyses of *Proteus* blood cultures indicate that 70% to 83% represent monomicrobic infections.[37,39,40,44] Mixed infections most often involve *E. coli* (>50%). Other microorganisms commonly associated with polymicrobic *Proteus* bacteremias include *Klebsiella pneumoniae,* coagulase-negative staphylococci, enterococci, and *Providencia stuartii.*[40,44] *P. mirabilis* is by far the most common species recovered from patients with *Proteus* bacteremia, although 2% to 5% of septicemic episodes are caused by *P. vulgaris.*[17,40] *P. penneri* has been isolated from blood on three separate occasions. The first report provided no clinical information regarding the significance of the isolate.[6] Engler and coinvestigators[45] provided the first description of *P. penneri* sepsis in a 59-year-old man with acute lymphocytic leukemia. After being hospitalized for 16 days and treated with antineoplastic agents, he developed a subcutaneous abscess of the lateral left thigh during treatment with cefoperazone and metronidazole. *P. penneri* was isolated from his blood and skin lesion, and the infection resolved with a 14-day course of amikacin. A second episode of *P. penneri* bacteremia occurred in a 30-year-old woman who sustained massive injuries as a result of a motorcycle accident.[46] The infection commenced 30 days after her surgery, and *P. penneri* was isolated from her blood (with *K. pneumoniae*), urine, and a vascular catheter. She was treated with ceftazidime and responded favorably.

Central Nervous System

Another serious complication of *P. mirabilis* infection is pediatric meningitis. Most cases of pediatric meningitis occur during the neonatal period. Early-onset disease (<14 days), which carries with it a graver prognosis, typically manifests itself within the first 4 to 5 days of life. Classic signs of meningitis include irritability, failure to thrive, lethargy, convulsions/seizures, and apnea. It has been estimated that *P. mirabilis* accounts for between 0.2% and 4% of all cases of neonatal meningitis.[47]

A number of cases of *P. mirabilis* meningitis since 1980 are listed in Table 2. Most of these case reports share a number of features in common. Infants with *Proteus* meningitis usually do not exhibit significant risk factors associated with their delivery. The apparent portal of infection in most situations is the umbilicus, usually documented by a positive culture,[48] the diagnosis of omphalitis,[47] or a notation of a foul odor emanating from the umbilical cord.[52] A second consistent attribute that *P. mirabilis* shares with *Citrobacter koseri* is a propensity for producing cerebral abscesses of the frontal area. In virtually all cases of neonatal meningitis due to *P. mirabilis,* there is medical evidence of abscess formation or porencephalic[50] or epidermoid[51] cysts as determined by cranial ultrasonography or computed tomography (CT). In some instances these secondary lesions are not initially suspected or detected, leading to recurrent central nervous system (CNS) disease.[48,51] No specific therapeutic regimen for treating *P.*

TABLE 2. *Recent cases of* Proteus mirabilis *meningitis*

Sex/Age	Risk Factors	Apgar Score	Cerebral Abscess	Recurrent Disease	Positive Cultures[a]	Treatment[b]	Outcome	Ref.
F/1 d	None	—	+	+	Umbilicus, blood, pus	Pen/Gm/ Chl–Gm/Chl/Cot	Died	48
M/1 mo	None	—	+	–	Pus	Gm/Amp/Cot	Survived	48
M/4 d	None	8	?	–	None	Gm/Chl	Died	48
M/4 d	None	—	+	–	None	Amp/Ctx	Survived	49
F/4 d	Prolonged membrane rupture	9/9	–	–	Blood	Amp/Gm-Amp/ Ctx-Ctx/Chl-Amp/ Sxt	Survived	50
M/3 d	None	4/7	+	–	None	Amp/Gm-Ctx/Gm	Survived	47
F/25 mo	None	—	+	+	Pus	Cftr/Ctz	Survived	51
F/4 d	Preterm labor	9/10	+	–	Blood	Clox/Amp/Gm	Died	52

[a]Sites other than cerebrospinal fluid.
[b]Hyphens indicate a change in chemotherapy during the course of infection.
Pen, penicillin; Gm, gentamicin; Chl, chloramphenicol; Cot, cotrimoxazole; Amp, ampicillin; Ctx, cefotaxime; Sxt, trimethoprim–sulfamethoxazole; Cftr, ceftriaxone; Ctz, ceftazidime; Clox, cloxacillin.

mirabilis CNS disease is universally accepted. Most protocols include both antimicrobial therapy and neurosurgical drainage, although one report describes the resolution of a patient's meningitis and cerebral abscesses with nonsurgical management.[47] Velvis and colleagues[50] described a case of *P. mirabilis* meningitis in a 4-day-old infant that was refractory to treatment with cefotaxime.[50] Substitution of trimethoprim-sulfamethoxazole and ampicillin for cefotaxime and chloramphenicol brought about resolution of symptomatology and sterilization of the patient's cerebrospinal fluid. Overall, the mortality rate in early-onset cases recently reported in the literature can approach 50%. Neonates who survive infection are often left with severe learning deficits, hydrocephalus, and impaired visual and auditory responses.

In addition to pediatric meningitis, *P. penneri* was reported to be associated with an epidural abscess in a 17-year-old man.[53] The young man had an 8-year history of chronic suppurative otitis media and presented to the hospital with a 2-week history of headaches, fever, and intermittent vomiting. The diagnosis was suppurative otitis media with epidural abscess formation and otogenic meningitis. Treatment included a radical mastoidectomy and multiple chemotherapeutic agents (cotrimoxazole, gentamicin), to which he responded favorably.

Wound Infections

P. mirabilis has been implicated in 4% of nosocomial wound infections. In hospital-acquired disease, protei are most frequently recovered from infected wounds resulting from surgical procedures, from decubitus ulcers, or from infections involving the skin or subcutaneous tissues.[17] The clinical significance of many of these isolates is difficult to determine because the infections are often polymicrobic.

Protei are occasionally involved in abscesses of the skin and soft tissues. One survey estimated the frequency of *P. mirabilis* abscesses in non–drug users at 4%.[54] There have been two case reports of primary psoas abscesses due to *P. mirabilis*. The first involved a 29-year-old woman who had previously undergone a right nephrectomy.[55] She presented with a 2-day history of fever, weak right upper quadrant abdominal pain, anemia, and an elevated erythrocyte sedimentation rate (ESR). An abdominal CT scan indicated an enlarged psoas muscle with a pus-filled abscess (500 mL) containing *P. mirabilis*. Drainage of the abscess along with a multidrug regimen (azlocillin, gentamicin, synthomycin) brought about a satisfactory resolution. The second case concerned an 87-year-old man with a 3-month history of diffuse abdominal pain, fever, and weight loss.[56] Significant laboratory findings included anemia and an elevated ESR. A CT scan revealed a hypodense region in the psoas muscle. Pus from the abscess again grew *P. mirabilis*. Similar curative procedures (percutaneous drainage, antibiotics) allowed the patient to be discharged within 3 days. In both instances, infection of the psoas muscle was thought to result via hematogenous spread of protei from occult sources. Balkan and associates[57] described a lumbar abscess caused by *P. vulgaris* (indole-positive) in a 19-year-old man secondary to cholesteatoma of the right mastoid process. His chief complaints included headaches, irritability, high fever, nausea, vomiting, and right-sided facial paralysis. A CT scan indicated a cavitary process running from the base of the skull downward to the right side of the neck. In another case, *Proteus* and *Staphylococcus aureus* were isolated from a recurrent right-side neck abscess in a elderly woman with adult-onset Langerhans cell histiocytosis.[58] Cure required repeated drainage and extensive systemic antibiotic therapy.

Members of the genus *Proteus* are also responsible for wound infections associated with invasive disease. These include synergistic infections of the subcutaneous soft tissues and fasciae (necrotizing fasciitis) and of skin and muscle (necrotizing cellulitis), where the disease commonly involves one or more anaerobes and an enteric organism.[59] Musher[60] described a fatal case of *P. mirabilis* peripheral gangrene in a middle-aged quadriplegic man with pyelonephritis and kidney stones. Cyanosis and necrosis of peripheral tissues (nose, fingers, toes) occurred in conjunction with disseminated intravascular coagulation. *P. mirabilis* was isolated from the patient's urine and blood.

Gastroenteritis

There is only superficial evidence implicating protei as possible human enteropathogens. Two studies by Müller found *P. mirabilis* to be more frequently recovered from diarrheic stool specimens than from controls.[61,62] In the second of these studies, the association was found to be independent of the recovery of most traditional enteropathogens except *Yersinia entercolitica*.[62] However, other species such as *M. morganii* and *P. vulgaris*, though not approaching statistical significance, were also more frequently isolated from symptomatic as opposed to asymptomatic persons. A second line of in vitro evidence suggesting that *P. mirabilis* may be an enteropathogen is the fact that some strains are invasive in tissue culture cell assays (see Pathogenesis). Since the isolation of *P. mirabilis* has not yet been linked to bloody diarrhea, the relevance of this in vitro finding remains questionable. Various studies have reported a wide range in frequencies (0.45%–6%) of the isolation of *P. vulgaris* from stools.[61–63] Explanations for such differences are not readily apparent. *P. penneri* has also been isolated from feces, although at a reduced rate (0.4%–1.2%).

Arthritis

A controversial topic is the possible relation between *P. mirabilis* infection and rheumatic disease. A 1985 publication by Ebringer and colleagues[64] found higher serum antibody titers to *P. mirabilis* in patients with rheumatoid arthritis than in persons with active or inactive ankylosing spondylitis or controls. This study was supported by the work of later investigators, who sometimes also detected elevated immunoglobulin A (IgA) titers to *P. mirabilis* in patients with the axial type of ankylosing spondylitis.[65,66] Gaston[67] summarized the current literature concerning the relation between *Proteus* and polyarthritis in an editorial. In addition to the fact that DR4-specific sera bind *Proteus* cells, it has been demonstrated that the *Proteus* hemolysin and human leukocyte antigen (HLA) DR4 subtypes associated with rheumatoid arthritis share a small amino acid peptide sequence in common. The sera of patients with rheumatoid arthritis also possess antibodies that react with the DR4/*Proteus* hemolysin peptide sequence.[67] However, in opposition to these findings is the fact that *Proteus* antibody levels are not dramatically different in those with disease and in controls. Furthermore, cross-absorption studies of reactive sera using purified hemolysin of HLA-DR have not as yet been carried out. Gaston summarized the data by stating that the present evidence only indicates that some immune process in the course of rheumatic arthritis causes the enhancement of patient serum binding of intact *Proteus* bacteria.[67]

Miscellaneous Infections

Proteus has rarely been associated with pulmonary disease and chronic osteomyelitis.[17] Clinical studies have failed to detect such infections. A severe case of *P. mirabilis* endophthalmitis in an elderly monocular patient resulted in complete loss of sight.[68] Infection of the patient's ocular prosthesis was thought to have resulted from fecally contaminated fingers.

LABORATORY IDENTIFICATION

Isolation

Proteus is one of the easiest genera in the *Enterobacteriaceae* to identify in the laboratory. They are immediately recognizable by microbiologists on most enteric media because of their unique ability to swarm on agar surfaces (Fig. 2). Swarming on a cellular level results from the bacterial transformation of

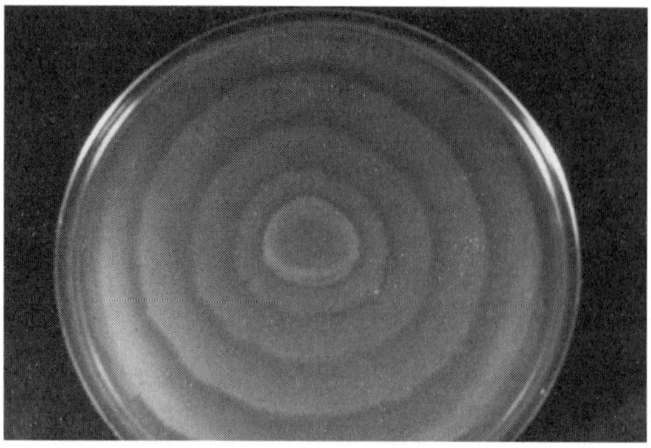

FIG. 2. Swarming of *P. mirabilis* on MacConkey agar as evidenced by concentric rings of growth. (Courtesy of E.J. Bottone).

swimmer cells (in broth) to swarmer cells (in agar) in a process involving cellular elongation, polyploidy, and increased flagellin synthesis.[69] Macroscopically, swarming appears as concentric rings of growth emanating from a single colony or inoculum. These outgrowths with irregular edges consist of halos of swarmer cells. The literature indicates that between 85% and 100% of *P. mirabilis* and *P. vulgaris* strains swarm, whereas only 65% of *P. penneri* have been reported to do so.[6]

Most protei are lactose-negative (98%–99%) and, with the exception of *P. mirabilis,* sucrose-negative. Although they produce hydrogen sulfide (except most *P. penneri*), its formation is not apparent on most enteric media because of sucrose fermentation (in the case of *P. mirabilis*) and swarming motility. Most common media (eg, MacConkey, Salmonella-Shigella [SS], Hektoen enteric, xylose-lysine-desoxycholate, bismuth sulfite) support the growth of protei.[70] Since protei typically swarm, they can usually be purified from contaminated plates by picking growth from the advancing edge of a swarm. They can also be isolated on differential media specifically designed for the recovery of *Proteeae,* as described by Hawkey and coworkers.[71] This medium, *Proteeae* isolation medium (PIM), relies on the ability of *Proteeae* to degrade tyrosine crystals and produce a brown pigment in the presence of DL-tryptophan. PIM has been used most often to recover members of the *Proteus-Providencia-Morganella* group from contaminated specimens such as feces. For enrichment of *P. mirabilis* and *P. vulgaris,* both tetrathionate and selenite broths have been found satisfactory.[70]

When protei are recovered from contaminated specimens, they present a major headache to microbiologists. Their swarming behavior often causes them to overgrow colonies of other gram-negative bacilli, making isolation and purification of non-*Proteus* organisms an extremely difficult and tedious task. Over the years, a number of compounds have been found to inhibit *Proteus* swarming, including high concentrations of agar, bile salts, Teepol, neomycin, *p*-nitrophenyl glycerol, and sodium azide.[70,72] Sodium salicylate at concentrations approximating 15 to 20 mmol/L in tryptone plates inhibits swarming of both *P. mirabilis* and *P. vulgaris* strains.[73] In the clinical laboratory, the simplest method for purification probably involves restreaking cultures or growth to SS agar. SS agar tends to inhibit the swarming of most *Proteus* isolates long enough that individual colonies of other organisms can be picked, restreaked, and tested.

Identification

The protei share a common taxonomic ancestry with other members of the tribe *Proteeae (Providencia, Morganella).* For various periods, members of one genus were intermingled with species of another based on morphologic, biochemical, or immunologic characteristics. Such assignments, which predate modern molecular taxonomic methods, were often the result of weighted phenotypic tests. As an example, *P. rettgeri* and *M. morganii* resided in the genus *Proteus* because of positive urea reactions. Later, systematic methods employing genetic techniques resolved these issues and members of the *Proteeae* were realigned into three distinct genera.

Despite taxonomic changes, the *Proteus-Providencia-Morganella* group concept still holds value because these genera share some phenotypic properties rarely encountered in other *Enterobacteriaceae* (Table 3). The most intriguing of these properties is the ability to oxidatively deaminate certain amino acids to keto acids.[74] Amino acids deaminated by *Proteeae* include leucine, norleucine, isoleucine, methionine, and tryptophan.[75] Diagnostically, phenylalanine has traditionally been used as the amino acid substrate of choice. Deamination of phenylalanine to its corresponding keto acid (phenylpyruvic acid) in the presence of ferric chloride produces a green color. Addition of ferric chloride to keto acids resulting from the deamination of other amino acids produces colors ranging from gray-violet to cherry-red.[75] No other genera in the family *Enterobacteriaceae* has this capability. A second exclusive feature associated with the *Proteeae* is the formation of a reddish-brown pigment on media (nutrient agar) containing 0.1% to 1% DL-tryptophan.[76] Using electron spin resonance analysis, Agodi and colleagues[77] concluded that the brown-black pigment elaborated by *P. mirabilis* is chemically and ultrastructurally similar to melanin. A tyrosinase enzyme is thought to play a role in pigment production based on inhibitor studies. Finally, all *Proteeae,* most *Citrobacter koseri,* and some *Citrobacter freundii* degrade L-tyrosine crystals, presumably by a tyrosine phenol-lyase.[78] Use of these three tests can clearly separate members of the *Proteeae* from all other genera in the family.

Once it has been determined that a strain belongs to the *Proteeae,* identification to genus can be accomplished with the use of one or more differential tests (see Table 3). *P. mirabilis* and *P. vulgaris* typically swarm and produce alkaline over acid reactions with H_2S formation (K/A^+) or an acid over acid pattern with H_2S production on triple sugar iron (TSI) slants. Most *P. penneri* are H_2S^-. Another diagnostic feature that differentiates *Proteus* and *Providencia* from *Morganella* is the formation of a red slant on lysine-iron agar (LIA). This reaction involves the deamination of lysine, yielding keto acids that produce an orange color in the presence of ferric chloride.[74] On an LIA slant, this orange color apparently blends with the brom cresol indicator to produce a red color. Other useful tests separating genera within the *Proteeae* include gelatinase activity and acid production from mannose fermentation.

Commercial Systems

Because of the unique biochemical characteristics of the *Proteeae* (eg, tryptophan deaminase), it is not surprising that most commercial systems do an excellent job of identifying *Proteus* species.[79–91] Table 4 lists the cumulative results of some studies employing rapid, automated, and manual systems in identification of protei to species. With minor exceptions, most systems are highly accurate (>90%) in the identification of all three clinically significant species, including *P. penneri.* When sample size was adequate

TABLE 3. *General characteristics of the* Proteeae

Characteristic	*Proteus*	*Providencia*	*Morganella*
Common features			
Phenylalanine deaminase	+	+	+
Reddish-brown pigment[a]	+	+	+
Degradation of L-tyrosine crystals	+	+	+
Differential features			
Swarming	+	–	–
Red slant (LIA)	+	+	–
H_2S production (TSI)	+	–	–
Gelatinase	+	–	–
Mannose fermentation	–	+	+

[a]On media containing DL-tryptophan.
LIA, lysine-iron agar; TSI, triple sugar iron agar.

TABLE 4. *Identification of* Proteus *species on commercial systems*

System	Incubation	*P. mirabilis*		*P. vulgaris*		*P. penneri*	
		No.	CI (%)	No.	CI (%)	No.	CI (%)
API 20E	21–24 h	90	97	41	98	24	100
Crystal E/NF	18–24 h	67	94	32	97	8	100
Crystal RS/E	3 h	5	100	5	100	5	100
Biolog GN	24 h[a]	31	87	4	75	2	100
MicroScan W/A	2 h	136	93	29	79	20	80
Vitek	4–18 h	89	98	32	97	9	89
RapID onE	4 h	17	94	5	80	—	—
Cobas	21 h	8	100	6	83	5	0

Data from references 79–91.
[a]Manual reading only.
CI, correct identification (does not include correct with low probability); h, hours of incubation.

(n≥10), only one significant problem was evident, that concerning the MicroScan W/A system and *P. vulgaris* identification. When misidentifications do occur, protei are most commonly identified as other *Proteus* species, *M. morganii,* or *P. rettgeri.*[80,85]

Conventional Media

The genus *Proteus* consists of motile, gram-negative bacilli that swarm on many common laboratory media and oxidatively deaminate phenylalanine and tryptophan.[1] All species hydrolyze urea; this fact, coupled with the ability to produce H_2S, makes their generic identification simple with a TSI slant and Christensen's urea agar. Three tests help separate *Proteus* species of clinical importance (Table 5). Of particular interest is the indole test. Although both *P. mirabilis* and *P. penneri* are indole-negative, Müller found that all *P. penneri* and 25% of *P. mirabilis* strains grown in tryptone broth produce a green color when tested by Kovac's method.[93] This reaction, which requires a 3-day incubation to detect, is thought to involve an intermediate product in the metabolism of tryptophan to indole. The reaction is more pronounced in media containing large amounts of tryptophan; only *P. penneri* strains are positive after 7 days of incubation. Farmer and collaborators[92] suggested that susceptibility to chloramphenicol may also be useful in protei species identification. *P. penneri* is chloramphenicol resistant, whereas *P. mirabilis* and *P. vulgaris* biogroup 3 strains are not. *P. vulgaris* biogroup 2 isolates are variable in their response. Pompei and coworkers[94] investigated the use of the phosphatase reaction as a method to separate *Proteus* species. The test involves the incorporation of phenolphthalein monophosphate or diphosphate into agar containing methyl green. With either phenolphthalein compound, *P. vulgaris* strains are uniformly phosphatase-negative; on the other hand, *P. mirabilis* isolates are phosphatase-positive only with the monophosphate substrate. The difficulty lies in the fact that about 33% of *P. penneri* are positive (green colonies and halo) in either media. Vasyurenko and Chernyav-

TABLE 5. *Biochemical identification of* Proteus *species of clinical importance*

Test	*P. mirabilis*	*P. penneri*	*P. vulgaris* Complex
Indole	–	–	+
ODC	+	–	–
Acid from maltose	–	+	+

Data from Farmer et al.[92]
ODC, ornithine decarboxylase.

skaya[95] looked at the fatty acid composition of 25 *P. mirabilis* and *P. vulgaris* strains. Although the lipopolysaccharide fatty acid composition of the two species was very similar, overall cellular fatty acid analysis indicated that *P. vulgaris* contained a higher concentration of hexadecanoic acid ($C_{16:1}$) and lower methylenehexadecanoic acid ($C_{17\nabla}$) than *P. mirabilis*. Cyclopropane fatty acids ($C_{19\nabla}$) were not detected in *P. vulgaris*, but they amounted to 1.9% of *P. mirabilis*. *P. penneri* was not studied.

Studies indicate that *P. vulgaris* as it presently exists is actually composed of a number of genomospecies that cannot be separated biochemically.[5,8] A 1994 study by O'Hara and associates[96] identified several tests that could aid in the separation of these DNA groups. In addition to those listed in Table 6, tartrate and citrate utilization and lipase production were found helpful in separating members of the *P. vulgaris* complex. However, two of the largest groups (DNA groups V and VI) could not be unambiguously separated from one another even using all of the tests in the proposed schema.

Immunodiagnostics

Several serotyping schemes have been devised for *Proteus* species. The most recent of these, by Penner and Hennessy,[97] recognizes 27 distinct serogroups of *P. vulgaris* and 37 serogroups within *P. mirabilis* based on somatic (O) antigens. No single serogroup in either system accounted for more than 15% of about 200 strains of each species typed from two different institutions. Chemotaxonomic studies of the somatic antigen components of *Proteus* species indicate several unusual features, including the presence of galacturonic acid and glucuronic acid in the lipopolysaccharide; L- and D-glycero-D-mannoheptose have also been determined to be characteristic components of the core region of protei.[98] Because of the unique cell wall chemistry, cross-reactions between protei and antisera derived from other bacteria are rare events.[97] However, there is one well-known serologic cross-reaction involving protei and patients with rickettsial infections. The Weil-Felix reaction is a serologic test to detect epidemic typhus based on the reactivity of a patient's serum with certain *Proteus* strains causing bacterial agglutination.[98] This reaction has been expanded to detect other rickettsial infections such as spotted fever, scrub typhus, and murine typhus using specific *Proteus* strains (O:X19, O:X2, O:XK).

Most studies looking at a humoral response to *Proteus* infection have involved persons with rheumatoid arthritis. The technique most commonly employed is an enzyme-linked immunosorbent assay (ELISA) that uses whole bacterial antigens.[65,66] Senior and colleagues[65] found that more than 50% of persons with rheumatoid arthritis had IgM antibodies that reacted with a 90-kDa *P. mirabilis* protein on immunoblotting. This antibody was not detected in rheumatoid factor–positive individuals with systemic lupus erythematosus or Sjögren's disease.

TABLE 6. *Separation of DNA groups within the* Proteus vulgaris *complex*

	P. vulgaris DNA Groups			
Test	DNA Group II	DNA Group III	DNA Group IV	DNA Groups V/VI
Esculin	+	–	–	–
Acid from				
Salicin	+	–	–	–
Rhamnose	–	–	+	–
DNase	+	–	+	V

Data from Brenner et al.[8] and O'Hara et al.[96]
DNase, deoxyribonuclease; V, variable.

Typing

Typing schemes for both *P. mirabilis* and *P. vulgaris* are fraught with difficulties. Many of the methods more readily available to laboratories, such as biotyping, predate recent taxonomic changes within the *Proteeae* and have not been used with any regularity in recent years. Other methods that are more discriminatory in nature and can provide important information regarding strain relatedness are not easily accessible to most medical institutions. Molecular techniques, used for typing of many enteric bacteria, have not found their place in *Proteus* epidemiology because of a lack of reported outbreaks associated with these bacteria over the past two decades. In most instances, traditional techniques are still the major methods by which protei are typed in epidemiologic situations.

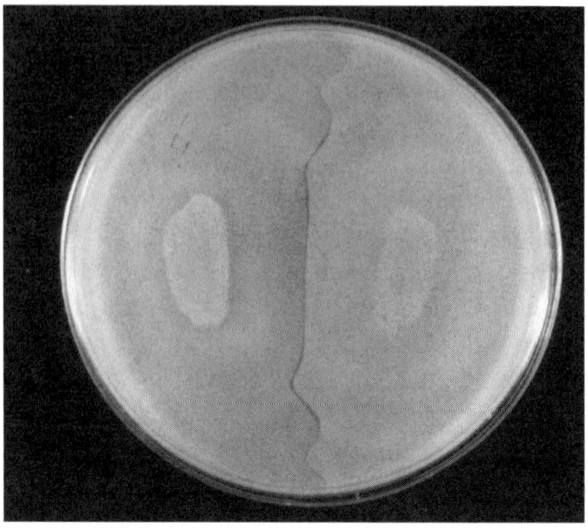

FIG. 3. A positive Dienes test on trypticase soy agar. The line of demarcation *(center)* indicates that the two strains are not related. (Courtesy of E.J. Bottone.)

Dienes Test

In 1946, Dienes[99] made an empiric observation that motile, spreading cultures of distinct *Proteus* strains inoculated onto blood agar were separated by a line of demarcation where the two swarms met. Subsequently referred to as the Dienes test, this reaction still has clinical relevance a half-century later in determining strain relatedness. When two nonidentical *Proteus* strains are inoculated onto a suitable medium (eg, trypticase soy, blood agar, MacConkey), a clear line of demarcation is evident after 24 hours of incubation (Fig. 3). Such a reaction is referred to as being Dienes$^+$ (incompatible), indicating that the two strains are not genetically related.[100] If no line is observed (Dienes$^-$, compatible), it implies that the isolates are identical. Although the exact basis for the Dienes reaction is not known, it is believed to be related to production of and susceptibility to bacteriocins by protei.[101]

Although the test is simple and straightforward, there are a number of caveats. First, some *P. mirabilis* and *P. vulgaris* strains (7%) fail to swarm or lose this property rapidly on initial subculture, thus precluding the use of this test in determining strain relatedness.[100] Second, distinct strains (as determined by other assays) sometimes do not produce a demarcation line for some unexplained reason. Therefore, it has been stated that a positive Dienes test (nonidentity) is more reliable than a negative one.[70] Although this test is still an excellent first approximation of strain relatedness, Dienes$^-$ isolates must be further typed by different methods.

Other Systems

Hickman and Farmer[100] devised a *P. mirabilis* bacteriophage typing scheme using 23 lytic bacteriophages. Of 200 strains studied, more than 90% were typable. An additional advantage of this system was its highly discriminatory capability; these 200 strains fell into more than 100 different lytic patterns. Senior and Larsson[101] used a combined typing scheme to type both *P. mirabilis* and *P. vulgaris* based on serogroup and bacteriocin (proticine) activity. O serogrouping was found to be independent of proticine production–proticine sensitivity typing; 133 urinary and fecal isolates of protei produced 81 different types with this combined typing scheme. The Dienes reaction was used as a supplementary adjunct in some cases in which identical serogroups and proticine patterns produced by separate strains were not thought to indicate strain identity. Engstrand and colleagues[102] used a molecular approach in an investigation of a case of *P. mirabilis* septicemia thought possibly to be caused by transfusion sepsis. The bacteremia developed in a 50-year-old patient with hepatorenal syndrome who had received a platelet concentrate transfusion. Using arbitrarily primed polymerase chain reaction technology, agarose gel electrophoresis revealed the amplified *P. mirabilis* profiles of both the blood and platelet concentrate isolates to be identical.

PATHOGENESIS

Potential virulence factors associated with *Proteus* infections have been reviewed.[103] As with most human pathogens, the pathogenicity of *Proteus* species is multifaceted, involving the expression of multiple virulence-associated characteristics.[104] Further, many protei simultaneously produce distinct gene products that phenotypically mask each other (eg, multiple hemolysins, hemagglutinins). This makes the creation of genetic mutants to study virulence-associated factors difficult and hinders association of specific biologic functions with a given gene product. To date, almost all pathogenicity studies have centered on *P. mirabilis*. Many factors potentially linked to the virulence of *P. mirabilis* show significant homology with similar products produced by *Serratia*. As for other protei, results obtained with *P. mirabilis* can be extrapolated to a limited extent to other proticine species because they often possess similar phenotypic markers (eg, urease, swarming).

Besides swarming, the most distinctive feature of *Proteus* isolates is their ability to hydrolyze urea, a characteristic shared by all species. Urease activity is thought to be a major virulence factor in the establishment and progression of urinary tract disease caused by *P. mirabilis*. Potential ways that this enzyme could be involved in disease include generation of energy sources (nitrogen) from the breakdown of urea and release of toxic end products of metabolism (ammonia). Release of ammonia through urea hydrolysis can cause alkalinization of the urine (increased pH), leading to crystal deposition (struvite), stone formation, and cellular damage to the renal epithelium.[105,106] The *P. mirabilis* urease is composed of three distinct subunits, one large and two small.[107] The native urease molecule has been estimated to have a molecular mass of about 215,000 Da; it shows significant amino acid sequence homology (58%–73%) with the jack bean urease.[107] A mutant of strain HI4320 with a defect in the synthesis of the large subunit of the *P. mirabilis* urease (*ureC*) failed to infect the urine, bladder, and kidneys of transurethrally challenged CBA mice as well as the parental strain.[105] This urease-defective isolate was also cleared from urine quicker than its parental counterpart. A follow-up investigation using a murine model of ascending UTI found that the *P. mirabilis* HI4320 *ureC* mutant required higher infective doses (>1000-fold), produced milder histopathologic lesions in the kidney, and colonized and persisted in both the bladder and kidney less well than the progenitor strain.

Another cardinal feature of most protei is their ability to swarm on agar surfaces. Flagellin synthesis in *P. mirabilis* is controlled by a global regulon containing the *flaA* locus.[108] This locus contains in tandem two almost identical genes *(flaA, flaB)*, which codc for the flagellin protein. In wild-type cells, only *flaA* is expressed; *flaB* is silent.[108] Swarm-defective mutants have been found that are unable to colonize and infect the bladder and kidney of intravesically inoculated mice.[109] Since swarmer bacteria (elongated filamentous forms) are the principal cell type associated with the kidney parenchyma of infected mice and vegetative (swimmer) cells predominate in pus, experimental evidence suggests that swarming migration is required in order for *P. mirabilis* to gain access to the kidney in a murine model.[109] A second attribute of the swimmer-to-swarmer cell differentiation is hyperproduction of other virulence-associated factors such as hemolysin, protease, and urease.[110] Such overproduction of virulence factors may help facilitate initial colonization and infection of the bladder and kidneys by swarmer cells. The gene *cmfA,* which stands for capsular migration factor, has been implicated in the rapid migration of swarmer cells.[111] This gene encodes for a putative carbohydrate transferase that is required for assembly of an acidic surface capsular polysaccharide.[111] A Tn*phoA cmfA* mutant is swarm-defective and unable to infect the kidneys and colonize the bladder in as high numbers as wild-type isolates.

A number of other products have been implicated in *P. mirabilis* pathogenesis. Early studies demonstrated that most *P. mirabilis, P. vulgaris,* and *P. penneri* produce hemolytic substances.[112–115] This activity was not readily apparent on blood agar plates after overnight culture but rather required prolonged incubation periods (>48 hours) to be detected.[112,114] Hemolytic activity can be more easily demonstrated (15 minutes to 3 hours) by incubating *Proteus* broth cultures with sheep, rabbit, horse, or human erythrocyte suspensions.[112–115] The difference in the rapidity of detection of hemolytic activity is caused by the fact that the hemolysin of *P. mirabilis* is primarily cell associated; the product is not released in appreciable quantities into the extracellular milieu under routine conditions.[113–116] Several different hemolytic activities have been described for *Proteus* species based on growth kinetics and calcium dependency.[113,114,117] The more common hemolysin is produced by most *P. mirabilis* (94%–100%) and *P. vulgaris* (84%–96%) strains during early to middle log phase.[118] This calcium-independent hemolysin, called the hemolysin of *P. mirabilis* (HpmA), can be detected in cell-free supernatants.[117,118] It shares significant homology with nucleotide sequences specifying the *Serratia marcesens* hemolysin.[119] HpmA is cytotoxic to T24 bladder epithelial cells, lymphomas, human monocytes, and Vero cells.[120] An *hpmA* mutant, however, produced colonization and histopathologic damage to mouse kidneys equivalent to that caused by its parental counterpart.[120] The exact role of this product in pathogenicity is presently unknown. A second calcium-dependent hemolysin is found chiefly in *P. penneri* (98%) and some *P. vulgaris* (8%) isolates.[114,116,118] The hemolysin displays both genetic and antigenic similarities to the α-hemolysin of *E. coli* (HlyA) based on size, immunoblotting studies, and DNA–DNA hybridization assays using *hly* probes.[116–118,121,122] The *P. vulgaris* HlyA forms pores in lipid-bilayer membranes and channels in a fashion analogous to the α-hemolysin of *E. coli.*[123] In *P. penneri,* expression of HlyA is a chromosomally determined trait[121] that is maximally produced in early exponential growth phase and lyses erythrocytes, neutrophils, and human lung fibroblasts.[122] However, in at least one instance, HlyA was linked to a 17-kb plasmid in *P. vulgaris.*[117] Because of the limited distribution of HlyA in protei, it is likely to play a very minor role (if any) in pathogenesis.

P. mirabilis produces a number of fimbriae, some of which cause the agglutination of red blood cells (hemagglutinins) in vitro. Adegbola and collaborators[124] recognized three different hemagglutination patterns in a collection of 48 *P. mirabilis* strains. Two of these adhesins agglutinated erythrocytes in the presence of D-mannose. One of these adhesins agglutinated tanned but not fresh ox erythrocytes in a manner similar to that observed for *Klebsiella* and was designated MR/K (mannose-resistant, *Klebsiella*-like). A second hemagglutinin, named MR/P (mannose-resistant, *Proteus*-like), agglutinated both fresh and tanned cells of a number of animal species. A third, mannose-sensitive (MS) hemagglutinin was infrequently encountered. All three hemagglutinins were associated with the presence of bacterial fimbriae, as determined by transmission electron microscopy of negatively stained preparations. Strains producing only the MR/P hemagglutinin have been more frequently recovered in cases of pyelonephritis than from catheter-associated bacteriurias or from feces.[125] The gene that codes for these fimbriae has been cloned *(mrpA)* and shows 57% amino acid sequence homology with the major mannose-resistant fimbrial subunit of *S. marcescens.*[126] The MR/K hemagglutinin also exhibits higher amino acid sequence homology to a fimbrial subunit of *S. marcescens* (36%) than it does to that of *K. pneumoniae* (15%). The gene has been designated *pmfA.*[127] The *pmfA* mutants are poor colonizers of bladder-infected mice compared with wild-type strains, although kidney colonization numbers were similar in both groups.[128] In addition to these fimbriae, two others in *P. mirabilis* have recently been identified. Massad and associates[129] described an ambient-temperature fimbriae (ATF) that, as its name implies, is optimally expressed at 23°C in Luria broth. It does not exhibit significant homology with any other fimbrial protein; all eight *P. mirabilis* strains studied produced ATF. In 1995, a nonagglutinating fimbriae (NAF) was isolated that showed partial homology (50%–55%) to an S-fimbrial protein of *E. coli* and fimbriae of *Hemophilus influenzae.*[130] Sequence analysis of NAF indicated slight variation (2 of 20 residues) in the N-terminus of one of five *P. mirabilis* strains. *P. penneri* strains produce all three of the major hemagglutinin phenotypes, namely MR/K, MR/P, and MS.[131] The majority of *P. penneri* strains (67%) express MR/K only; at least two distinct antigenic forms of MR/K have been identified.

Protei also have the capacity to invade nonprofessional phagocytic cell lines. Peerbooms and colleagues[132] were the first to report invasive *P. mirabilis* strains on Vero cell monolayers. Subsequently a number of other cell lines, including human bladder (T24,EJ/28,5637), embryonic intestinal epithelial (INT 407), and ileocecal epithelial (HCT-8) lines, were found to support *P. mirabilis* invasion.[133,134] In most studies, vegetative cells appear to be more efficient invaders than are swarmers[134,135] and invasive capabilities decrease on continued passage of strains in vitro. It is postulated that this invasive capability into urothelial cells may manifest itself in vivo by causing serious damage to the renal epithelium.[133,135] Rozalski and Kotelko[115] also reported that two *P. penneri* strains are invasive in Vero cell assays.

Proteus species produce a number of proteases, one of which degrades IgA. In one study IgA protease activity was detected in all *P. mirabilis* and *P. penneri* strains and in most *P. vulgaris.*[136] The IgA protease of *P. mirabilis* degrades IgA1, IgA2, and IgG isotypes.[137] Both IgA and IgG are known to occur in urine, and urine samples from *P. mirabilis*-infected persons have demonstrated *P. mirabilis* protease activity.[138] Two other activities produced by protei, which may play a role in virulence, include a vascular permeability factor[139] and amino acid deaminases that act as siderophores.[140] The general properties of most of these virulence-associated factors are listed in Table 7.

A number of models, including bacterial encrustation investigations, have been developed to study the uropathogenicity of protei.[141,142] Mobley and Belas[104] proposed a model for virulence of *P. mirabilis* as a urinary tract pathogen (Fig. 4). In this model, *P. mirabilis* colonizes the bladder by means of fimbriae. Swimmer cells then differentiate into swarmer cells and ascend the ureters to the kidneys, where MR/P adhesions help bind bacteria to the renal epithelium.[104] Once attachment occurs, swarmer cells hypersecrete a number of enzymes (urease, hemolysin), which can lead to pyelonephritis, urolithiasis, and renal damage. *P. myxofaciens* has been recovered from living and dead gypsy moth larvae, although its pathologic significance in this setting is unknown.[4]

ANTIMICROBIAL SUSCEPTIBILITY

Few recent studies have comprehensively looked at the in vitro susceptibility of *Proteus* species to newer antimicrobial agents.[143,144] Table 8 contains cumulative data on the in vitro susceptibility of protei to a variety of chemotherapeutic agents obtained from broad-based surveys of drug resistance among members of the family *Enterobacteriaceae.*[143–150] As with most *Proteeae,* indole-

TABLE 7. *Virulence factors of* Proteus *species*

Factor	Phenotype	Genotype[a]	Protein Size[b]	Expressed in:[c]			Proposed Functions
				P. mirabilis	*P. penneri*	*P. vulgaris*	
Urease	Ure	*ureA, ureB, ureC*	11, 12.2, 61	+	+	+	Stone formation, kidney damage
Swarming	Mot	*flaA*	39.4	+	+	+	Bladder colonization, kidney infection
	CP	*cmfA*	40.6	+	–	–	Bladder colonization, kidney infection
Hemolysin	HpmA	*hpmA*	166	+	–	+	?
	HlyA	*hlyA*	110	–	+	+	?
Fimbriae	MR/P	*mrpA*	16.4	+	+	–	Bladder colonization, kidney colonization
	MR/K	*pmfA*	19.5	+	+	–	Bladder colonization
	ATF	*atfA*	24	+	–	–	?
	NAF	—	27	+	–	–	?
Protease	—	*zapA*	50	+	+	+	Degrades IgA, IgG
Deaminase	—	*aad*	—	+	+	+	Iron acquisition

Data from references 104, 107–109, 111, 113–119, 126–134, 136, and 140.

[a]Protein structural genes.

[b]For subunits or native molecules.

[c]+, species possessing phenotype/genotype; –, factor absent or not tested for.

Ure, urease; Mot, motility; CP, capsular polysaccharide; HpmA, hemolysin of *P. mirabilis;* HlyA, α-hemolysin of *E. coli;* MR/P, mannose-resistant, *Proteus*-like fimbriae; MR/K, mannose-resistant, *Klebsiella*-like fimbriae; ATF, ambient temperature fimbriae; NAF, nonagglutinating fimbriae.

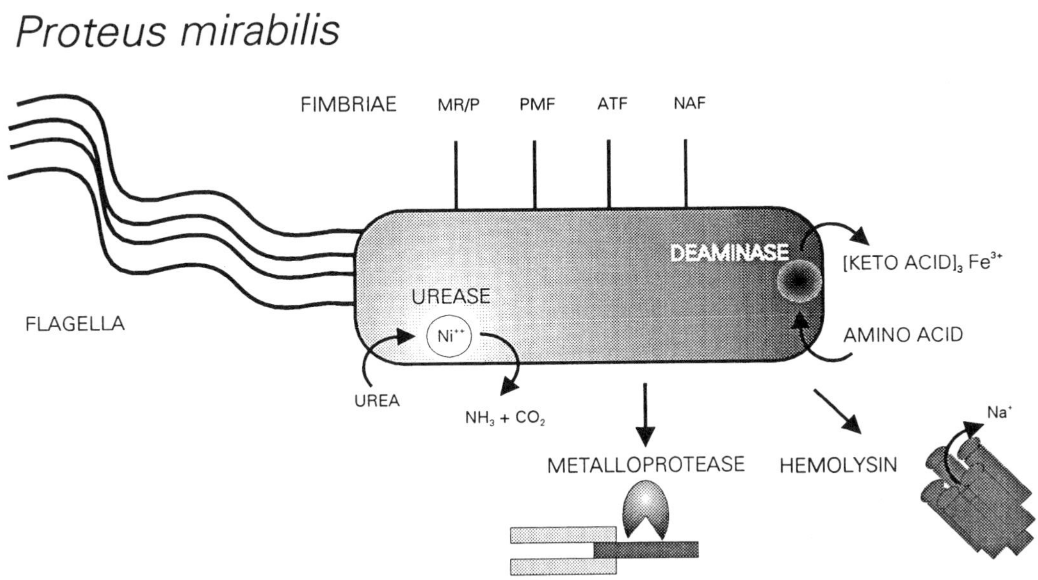

FIG. 4. Current model of *P. mirabilis* uropathogenicity involving expression of multiple virulence-associated factors. From Mobley and Belas[104] with permission of Elsevier Science Ltd.

TABLE 8. *In vitro susceptibility of* Proteus *species to selected antimicrobial agents*[a]

	P. mirabilis		*P. penneri*		*P. vulgaris*	
Agent	MIC_{50}	MIC_{90}	MIC_{50}	MIC_{90}	MIC_{50}	MIC_{90}
Ampicillin	1	>256	ND	ND	>256	>256
Carbenicillin	0.4	3.1	6.25	100	1.6	12.5
Piperacillin	0.4	3.1	2–50	16–200	0.8	1.6
Cephalothin	4	16	ND	ND	>256	>256
Cefoxitin	3.1	6.25	1.6–2	4–12.5	6.25	12.5
Cefuroxime	1.6	3.1	6.25	100	3.1	12.5
Cefoperazone	0.5	1–2	2	32	2	16
Cefotaxime	0.015–≤0.25	0.03–≤0.25	1	>16	0.06–0.5	2–8
Ceftazidime	0.06–0.2	0.12–1.6	0.1–0.12	0.5–0.8	0.1–≤0.12	0.2–1
Ceftizoxime	≤0.25	≤0.25	0.03	0.25	≤0.25	≤0.25
Ceftriaxone	0.015–0.025	0.06–0.2	0.2–1	1.6–64	0.05–0.5	0.2–>32
Cefepime	0.06	0.12	ND	ND	ND	ND
Gentamicin	1	4	0.5	1	1	2
Tobramycin	0.4	0.8	0.8–1	2–3.1	0.2	0.4
Amikacin	0.8	3.1	1.6–4	8–25	0.2	0.4
Ciprofloxacin	≤0.03–0.5	0.12–1	≤0.06	0.12	≤0.06–0.13	0.13–0.25
Ofloxacin	0.12–0.13	0.25–0.5	0.12	0.12	≤0.06–0.13	0.13–0.25
Lomefloxacin	0.25	0.5	0.25	0.25	0.12	0.25
Imipenem	0.2–0.5	0.8–2	0.4	3.1	0.4	1.6
Aztreonam	0.1	0.4	0.8	6.25	0.4	1.6

Data from references 143–150.
[a]All concentrations are given in μg/mL.
MIC, minimal inhibitory concentration (50%, 90%); ND, not determined.

negative groups *(P. mirabilis)* are generally more susceptible to many antimicrobial agents than are indole-positive species *(P. vulgaris)*. The greater susceptibility to therapeutic drugs by *P. mirabilis* is thought to be partially responsible for the declining incidence of this species as a major cause of nosocomial infections.[16] The exception to this rule is the indole negative species *P. penneri* (formerly biogroup 1 *P. vulgaris*), which acts more like an indole-positive species in terms of its greater resistance to many antimicrobial agents.

As reported by Piccolomini and coinvestigators,[143] *P. mirabilis* is more susceptible to the penicillins, extended-spectrum penicillins, and cephalosporins (including cefazolin) than either *P. penneri* or *P. vulgaris,* despite the fact that β-lactamase–producing strains of *P. mirabilis* are fairly common. Overall, most protei are susceptible to the aminoglycosides, quinolones, imipenem, and aztreonam.[40,44,143–145,148,151] They are also susceptible to ticarcillin/clavulanate[151] and trimethoprim-sulfamethoxazole[44] and resistant to tetracyclines.[40,44] One study of bacteremic *P. mirabilis* isolates[40] and another involving a selection of multiresistant *P. mirabilis* strains[145] noted higher frequencies (30%–60%) of gentamicin resistance. A multicenter survey of increasing drug resistance in microbial pathogens reported a 17.9% prevalence (range 3%–32%) of imipenem resistance in *P. mirabilis* strains.[150] However, on retesting, these strains were found to be susceptible to this carbapenem. False resistance in many instances was thought to be caused by the use of automated susceptibility systems (Vitek). Extended-spectrum β-lactamase resistance in both *P. mirabilis* and *P. vulgaris* has not been reported.[152]

Not much is known about resistance mechanisms in proticine species. Imipenem resistance in *P. mirabilis* was linked to the loss of a 24-kDa outer membrane protein in one report[153] and to altered penicillin-binding proteins in another.[154] A gene designated *pqrA* (*Proteus* quinolone resistance) has been cloned from a quinolone-resistant strain of *P. vulgaris.*[155] It may function as a transcriptional activator involved in an efflux mechanism. Ten isolates of *P. mirabilis* were found to express an inhibitor-resistant TEM β-lactamase (IRT) in association with a TEM-1 enzyme.[156] The IRT enzyme was found to be derived from TEM-2 by a single point mutation.

CONCLUSIONS

Much work remains to be done with the genus *Proteus.* Although considerably more is known regarding the factors that regulate *P. mirabilis* as a uropathogen today than a decade ago, there is very little information concerning which determinants are critical in the causation of extraintestinal disease such as bacteremia. For other species, virtually no information is presently available on pathogenicity. When the *P. vulgaris* group is redefined on a genetic basis, studies will need to be undertaken to see whether the disease spectrum, epidemiology, or pathogenicity of this complex is significantly changed based on newly established genomospecies.

REFERENCES

1. Penner JL. Genus XI: *Proteus* Hauser 1885, 12[AL]. In: Krieg NR, Holt JG, eds. *Bergey's manual of systematic bacteriology.* Baltimore: Williams & Wilkins, 1984; 491–494.
2. Penner JL. The Genera *Proteus, Providencia,* and *Morganella.* In: Balows A, Trüper HG, Dworkin M, Harder W, Schleifer K-H, eds. *The prokaryotes: A handbook on the biology of bacteria ecophysiology, isolation, identification, applications.* 2nd ed. Berlin: Springer-Verlag, 1991;2849–2862.
3. Rauss K. A proposal for the nomenclature and classification of the proteus and providencia groups. *Int Bull Bacteriol Nomencl Taxon* 1962;12:53–64.
4. Cosenza BJ, Podgwaite JD. A new species of *Proteus* isolated from larvae of the gypsy moth *Porthetria dispar* (L). *J Microbiol Serol* 1966;32:187–191.
5. Brenner DJ, Farmer JJ III, Fanning GR, et al. Deoxyribonucleic acid relatedness of *Proteus* and *Providencia* species. *Int J Syst Bacteriol* 1978;28:269–282.
6. Hickman FW, Steigerwalt AG, Farmer JJ III, Brenner DJ. Identification of *Proteus penneri* sp. nov., formerly known as *Proteus vulgaris* indole negative or as *Proteus vulgaris* biogroup 1. *J Clin Microbiol* 1982;15: 1097–1102.

7. Costas M, Holmes B, Frith KA, Riddle C, Hawkey PM. Identification and typing of *Proteus penneri* and *Proteus vulgaris* biogroups 2 and 3, from clinical sources, by computerized analysis of electrophoretic protein patterns. *J Appl Bacteriol* 1993;75:489–498.
8. Brenner DJ, Hickman-Brenner FW, Holmes B, et al. Replacement of NCTC 4175, the current type strain of *Proteus vulgaris,* with ATCC 29905. Request for an opinion. *Int J Syst Bacteriol* 1995;45:870–871.
9. Phillips JE. In vitro studies on *Proteus* organisms of animal origin. *J Hyg Camb* 1955;53:26–31.
10. Goldstein EJC, Agyare EO, Vagvolgyi AE, Halpern M. Aerobic bacterial oral flora of garter snakes: Development of normal flora and pathogenic potential for snakes and humans. *J Clin Microbiol* 1981;13: 954–956.
11. Garcia E, Hamers AM, Bergmans HEN, van der Zeijst BAM, Gaastra W. Adhesion of canine and human uropathogenic *Escherichia coli* and *Proteus mirabilis* strains to canine and human epithelial cells. *Curr Microbiol* 1988;17:333–337.
12. Youssef H, El-Timawy AK, Ahmed S. Role of aerobic intestinal pathogens of fresh water fish in transmission of human diseases. *J Food Protect* 1992;55:739–740.
13. López-Sabater EI, Rodríguez-Jerez JJ, Roig-Sagués AX, Mora-Ventura MAT. Bacteriological quality of tuna fish *(Thunnus thynnus)* destined for canning: Effect of tuna handling on presence of histidine decarboxylase bacteria and histamine level. *J Food Protect* 1994;57:318–323.
14. Ababouch L, Afilal ME, Rhafiri S, Busta FF. Identification of histamine-producing bacteria isolated from sardine *(Sardina pilchardus)* stored in ice and at ambient temperature (25°C). *Food Microbiol* 1991;8: 127–136.
15. von Holy A, Holzapfel WH, Dykes GA. Bacterial populations associated with Vienna sausage packaging. *Food Microbiol* 1992;9:45–53.
16. Schaberg DR, Culver DH, Gaynes RP. Major trends in the microbial etiology of nosocomial infection. *Am J Med* 1991;91(Suppl):72S–75S.
17. Adler JL, Burke JP, Martin DF, Finland M. *Proteus* infections in a general hospital. II: Some clinical and epidemiologic characteristics with an analysis of 71 cases of *Proteus* bacteremia. *Ann Intern Med* 1971;75: 531–536.
18. Ozumba UC, Dosunmi-Ogunbi O, Onile B. Urinary tract infection with *Proteus* species in a teaching hospital. *East Afr Med J* 1995;72:72–74.
19. Sekaninová G, Kolářová M, Semrádová S, Táborská D, Zajícová V. Nosocomial infections caused by selected gram-negative bacteria at the anaesthesiology and intensive care unit of the teaching hospital in Brno. *Centr Eur J Public Health* 1995;3:80–83.
20. Vila J, Gene A, Rullan J, Jimenez de Anta MT. Rapid detection of urinary tract infection caused by *Escherichia coli* or *Proteeae* species. *Eur J Clin Microbiol Infect Dis* 1991;10:922–926.
21. Markowitz SM. Urinary tract specimens. In: Dalton HP, Nottebart HC Jr, eds. *Interpretive Medical Microbiology.* New York: Churchill Livingstone, 1986;523–592.
22. Flournoy DJ. Incidence of bacteria from clean-catch and catheterized urine of nursing home residents. *J Okla State Med Assoc* 1993;86:550–552.
23. Kunin CM. Urinary tract infections in females. *Clin Infect Dis* 1994;18:1–12.
24. Shapiro ED. Infections of the urinary tract. *Pediatr Infect Dis J* 1992;11:165–168.
25. Stamm WE, Hooton TM. Management of urinary tract infections in adults. *N Engl J Med* 1993;329: 1328–1334.
26. Fairley KF, Birch DF. Detection of bladder bacteriuria in patients with acute urinary symptoms. *J Infect Dis* 1989;159:226–231.
27. Naylor GRE. A 16-month analysis of urinary tract infection in children. *J Med Microbiol* 1984;17:31–36.
28. Stark RP, Maki DG. Bacteriuria in the catheterized patient. What quantitative level of bacteriuria is relevant? *N Engl J Med* 1984;311:560–564.
29. Stamm WE. Catheter-associated urinary tract infections: Epidemiology, pathogenesis, and prevention. *Am J Med* 1991;91(Suppl 3B):65S–71S.
30. Nicolle LE, Strausbaugh LJ, Garibaldi RA. Infections and antibiotic resistance in nursing homes. *Clin Microbiol Rev* 1996;9:1–17.
31. Mobley HLT, Warren JW. Urease-positive bacteriuria and obstruction of long-term urinary catheters. *J Clin Microbiol* 1987;25:2216–2217.
32. Prien EL Jr. Kidney stones and urease-producing bacteria. *Clin Microbiol Newslett* 1990;12:129–132.
33. Stickler D, Ganderton L, King J, Nettleton J, Winters C. *Proteus mirabilis* biofilms and the encrustation of urethral catheters. *Urol Res* 1993;21:407–411.
34. Hess B, Metzger RM, Ackermann D, Montandon A, Jaeger P. Infection-induced stone formation in a renal allograft. *Am J Kidney Dis* 1994;24:868–872.
35. Krajden S, Fuksa M, Lizewski W, Barton L, Lee A. *Proteus penneri* and urinary calculi formation. *J Clin Microbiol* 1984;19:541–542.
36. Krajden S, Fuksa M, Petrea C, Crisp LJ, Penner JL. Expanded clinical spectrum of infections caused by *Proteus penneri. J Clin Microbiol* 1987;25:578–579.
37. Roberts FJ. A review of positive blood cultures: Identification and source of microorganisms and patterns of sensitivity to antibiotics. *Rev Infect Dis* 1980;2: 329–339.
38. Weinstein MP, Reller LB, Murphy JR, Lichenstein KA. The clinical significance of positive blood cultures: A comprehensive analysis of 500 episodes of bacteremia and fungemia in adults. I: Laboratory and epidemiologic observations. *Rev Infect Dis* 1983;5:35–53.
39. Roberts FJ, Geerre IW, Coldman A. A three-year study of positive blood cultures, with emphasis on prognosis. *Rev Infect Dis* 1991;13:34–46.
40. Berger SA. *Proteus* bacteraemia in a general hospital 1972–1982. *J Hosp Infect* 1985;6:293–298.
41. Levy I, Leibovici L, Drucker M, Samra Z, Konisberger H, Ashkenazi S. A prospective study of gram-negative bacteremia in children. *Pediatr Infect Dis J* 1996;15: 117–122.
42. Setia U, Serventi I, Lorenz P. Bacteremia in a long-term care facility: Spectrum and mortality. *Arch Intern Med* 1984;144:1633–1635.
43. Rudman D, Hontanosas A, Cohen Z, Mattson DE. Clinical correlates of bacteremia in a Veterans Administration extended care facility. *J Am Geriatr Soc* 1988; 36:726–732.
44. Watanakunakorn C, Perni SC. *Proteus mirabilis* bac-

teremia: A review of 176 cases during 1980–1992. *Scand J Infect Dis* 1994;26:361–367.
45. Engler HD, Troy K, Bottone EJ. Bacteremia and subcutaneous abscess caused by *Proteus penneri* in a neutropenic host. *J Clin Microbiol* 1990;28:1645–1646.
46. Schwalbe RS, Verma PR. Disseminated infection caused by *Proteus penneri. Clin Microbiol Newslett* 1991;13:93–95.
47. Hervás JA, Ciria L, Henales V, López P, de la Fuente A, Del Valle JM. Nonsurgical management of neonatal multiple brain abscesses due to *Proteus mirabilis. Helv Paediatr Acta* 1987;42.451–456.
48. Smith ML, Mellor D. *Proteus mirabilis* meningitis and cerebral abscess in the newborn period. *Arch Dis Child* 1980;55:308–310.
49. Gray PH, O'Reilly C. Neonatal *Proteus mirabilis* meningitis and cerebral abscess: Diagnosis by real-time ultrasound. *J Clin Ultrasound* 1984;12:441–443.
50. Velvis H, Carrasco N, Hetherington S. Trimethoprim-sulfamethoxazole therapy of neonatal *Proteus mirabilis* meningitis unresponsive to cefotaxime. *Pediatr Infect Dis J* 1986;5:591–593.
51. Ceccarelli M, Balestri M, Fontani C, Lupetti L, Ughi C. Recurrent meningitis: A case report. *Eur J Pediatr* 1989;148:646–647.
52. Grahnquist L, Lundberg B, Tullus K. Neonatal meningoencephalitis. *Acta Pathol Microbiol Scand* 1992; 100:734–736.
53. Li Z, Wang X, Bian Z, et al. *Proteus penneri* isolated from the pus of a patient with epidural abscess. *J Jpn Assoc Infect Dis* 1992;2:144–147.
54. Sumanen PH, Talan DA, Strong C, et al. Bacteriology of skin and soft-tissue infections: Comparison of infections in intravenous drug users and individuals with no history of intravenous drug use. *Clin Infect Dis* 1995;20(Suppl):S279–S282.
55. Knobel B, Sommer I, Schwartz G. Primary psoas abscess three years after ipsilateral nephrectomy. *Infection* 1985;13:27–28.
56. Gruenwald I, Abrahamson J, Cohen O. Psoas abscess: Case report and review of the literature. *J Urol* 1992; 147:1624–1626.
57. Balkan E, Dinc O, Fisenk F, Bircan O. Imaging quiz case. *Arch Otolaryngol Head Neck Surg* 1994;120: 668–671.
58. Watson LC, Mendelsohn SS, Nash J. Adult-onset Langerhans cell histiocytosis associated with severe abscesses. *Clin Exp Dermatol* 1995;20:415–418.
59. Kahn RM, Goldstein EJC. Common bacterial skin infections: Diagnostic clues and therapeutic options. *Postgrad Med* 1993;93:175–182.
60. Musher DM. Cutaneous and soft-tissue manifestations of sepsis due to gram-negative enteric bacilli. *Rev Infect Dis* 1980;2:854–866.
61. Müller HE. Occurrence and pathogenic role of *Morganella-Proteus-Providencia* group bacteria in human feces. *J Clin Microbiol* 1986;23:404–405.
62. Müller HE. The role of *Proteeae* in diarrhea. *Zentralbl Bakteriol* 1989;272:30–35.
63. Senior BW, Leslie DL. Rare occurrence of *Proteus vulgaris* in faeces: A reason for its rare association with urinary tract infections. *J Med Microbiol* 1986;21:139–144.
64. Ebringer A, Corbett M, Macafee Y, et al. Antibodies to *Proteus* in rheumatoid arthritis. *Lancet* 1985;2: 305–307.
65. Senior BW, McBride PDP, Morley KD, Kerr MA. The detection of raised levels of IgM to *Proteus mirabilis* in sera from patients with rheumatoid arthritis. *J Med Microbiol* 1995;43:176–184.
66. Mäki-Ikola O, Nissilä M, Lehtinen K, Leirisalo-Repo M, Toivanen P, Granfors K. Antibodies to *Klebsiella pneumoniae, Escherichia coli* and *Proteus mirabilis* in the sera of patients with axial and peripheral form of ankylosing spondylitis. *Br J Rheumatol* 1995;34: 413–417.
67. Gaston H. *Proteus:* Is it a likely aetiological factor in chronic polyarthritis? *Ann Rheum Dis* 1995;54: 157–158.
68. Morris R, Camesasca FI, Byrne J, John G. Postoperative endophthalmitis resulting from prothesis contamination in a monocular patient. *Am J Ophthalmol* 1993; 116:346–349.
69. Belas R. The swarming phenomenon of *Proteus mirabilis. ASM News* 1992;58:15–22.
70. Penner JL. The tribe *Proteeae.* In: Starr MP, Stolp H, Trüper HG, Balows A, Schlegel HG (ed). *The prokaryotes: A handbook on habitats, isolation, and identification of bacteria.* Berlin: Springer-Verlag, 1981; 1204–1224.
71. Hawkey PM, McCormick A, Simpson RA. Selective and differential medium for the primary isolation of members of the *Proteeae. J Clin Microbiol* 1986;23: 600–603.
72. Williams FD, Schwarzhoff RH. Nature of the swarming phenomenon in *Proteus. Annu Rev Microbiol* 1978;32:101–122.
73. Kunin CM, Hua TH, Bakaletz LO. Effect of salicylate on expression of flagella by *Escherichia coli* and *Proteus, Providencia* and *Pseudomonas* spp. *Infect Immun* 1995;63:1796–1799.
74. Ewing WH. *Edwards and Ewing's identification of* Enterobacteriaceae. 4th ed. New York, Elsevier, 1986.
75. Singer J, Volcani BE. An improved ferric chloride test for differentiating *Proteus*-providence group from other *Enterobacteriaceae. J Bacteriol* 1955;69:303–306.
76. Polster M, Svobodová M. Production of reddish-brown pigment from DL-tryptophan by enterobacteria of the *Proteus*-providence group. *Experientia* 1964;20: 637–638.
77. Agodi A, Stefani S, Corsaro C, Campanile F, Gribaldo S, Sichel G. Study of a melanic pigment of *Proteus mirabilis. Res Microbiol* 1996;147:167–174.
78. Sheth NK, Kurup VP. Evaluation of tyrosine medium for the identification of *Enterobacteriaceae. J Clin Microbiol* 1:483–485.
79. Miller JM, Rhoden DL. Preliminary evaluation of Biolog, a carbon source utilization method for bacterial identification. *J Clin Microbiol* 1991;29:1143–1147.
80. Pfaller MA, Sahm D, O'Hara C, et al. Comparison of the autoSCAN-W/A rapid bacterial identification system and the Vitek automicrobic system for identification of gram-negative bacilli. *J Clin Microbiol* 1991; 29:1422–1428.
81. O'Hara CM, Rhoden DL, Miller JM. Reevaluation of the API 20E identification system versus conventional biochemicals for identification of members of the family *Enterobacteriaceae*: A new look at an old product. *J Clin Microbiol* 1992;30:123–125.
82. Kelly MT, Leicester C. Evaluation of the autoscan walkaway system for rapid identification and suscepti-

bility testing of gram-negative bacilli. *J Clin Microbiol* 1992;30:1568–1571.

83. Visser MR, Bogaards L, Rozenberg-Arska M, Verhoef J. Comparison of the autoSCAN-W/A and Vitek automicrobic systems for identification and susceptibility testing of bacteria. *Eur J Clin Microbiol Infect Dis* 1992;11:979–984.
84. York MK, Brooks GF, Fiss EH. Evaluation of the autoSCAN-W/A rapid system for identification and susceptibility testing of gram-negative fermentative bacilli. *J Clin Microbiol* 1992;30:2903–2910.
85. O'Hara CM, Tenover FC, Miller JM. Parallel comparison of accuracy of the API 20E, Vitek GNI, MicroScan walk/away rapid ID, and Becton Dickinson Cobas micro ID-E/NF for identification of members of the family *Enterobacteriaceae* and common gram-negative, non–glucose-fermenting bacilli. *J Clin Microbiol* 1993;31:3165–3169.
86. Kitch TT, Jacobs MR, Appelbaum PC. Evaluation of RapID onE system for identification of 379 strains in the family *Enterobacteriaceae* and oxidase-negative, gram-negative nonfermenters. *J Clin Microbiol* 1994; 32:931–934.
87. Holmes B, Costas M, Ganner M, On SLW, Stevens M. Evaluation of Biolog system for identification of some gram-negative bacteria of clinical importance. *J Clin Microbiol* 1994;32:1970–1975.
88. Holmes B, Costas M, Thaker T, Stevens M. Evaluation of two BBL Crystal systems for identification of some clinically important gram-negative bacteria. *J Clin Microbiol* 1994;32:2221–2224.
89. Robinson A, McCarter YS, Tetreault J. Comparison of Crystal enteric/nonfermenter system, API 20E system, and Vitek automicrobic system for identification of gram-negative bacilli. *J Clin Microbiol* 1995;33: 364–370.
90. Wauters G, Boel A, Voorn GP, et al. Evaluation of a new identification system, Crystal enteric/non-fermenter, for gram-negative bacilli. *J Clin Microbiol* 1995;33:845–849.
91. Rhoads S, Marinelli L, Imperatrice CA, Nachamkin I. Comparison of MicroScan walkaway system for identification of gram-negative bacteria. *J Clin Microbiol* 1995;33:3044–3046.
92. Farmer JJ III, Davis BR, Hickman-Brenner FW, et al. Biochemical identification of new species and biogroups of *Enterobacteriaceae* isolated from clinical specimens. *J Clin Microbiol* 1985;21:46–76.
93. Müller HE. *Proteus* penneri showing a green colour reaction with Kovacs' indole reagent. *Zentralbl Bakteriol Hyg A* 1986;261:198–203.
94. Pompei R, Cornaglia G, Ingianni A, Satta G. Use of a novel phosphatase test for simplified identification of species of the tribe *Proteeae. J Clin Microbiol* 1990; 28:1214–1218.
95. Vasyurenko ZP, Chernyavskaya YN. Confirmation of *Morganella* distinction from *Proteus* and *Providencia* among *Enterobacteriaceae* on the basis of cellular and lipopolysaccharide fatty acid composition. *J Hyg Epidemiol Microbiol Immunol* 1990;34:81–90.
96. O'Hara C, Hickman-Brenner F, Steigerwalt A, et al. *Proteus vulgaris* DNA groups 3, 4, 5, and 6. Abstract C-253. *Abstracts of the 94th General Meeting of the American Society for Microbiology.* Las Vegas: 1994;535.
97. Penner JL, Hennessy JN. Separate O-serogrouping schemes for serotyping clinical isolates of *Proteus vulgaris* and *Proteus mirabilis. J Clin Microbiol* 1980;12: 304–309.
98. Larsson P. Serology of *Proteus mirabilis* and *Proteus vulgaris. Methods Microbiol* 1984;14:187–214.
99. Dienes L. Reproductive processes in *Proteus* cultures. *Proc Soc Exp Biol Med* 1946;62:265–270.
100. Hickman FW, Farmer JJ III. Differentiation of *Proteus mirabilis* by bacteriophage typing and the Dienes reaction. *J Clin Microbiol* 1976;3:350–358.
101. Senior BW, Larsson P. A highly discriminatory multi-typing scheme for *Proteus mirabilis* and *Proteus vulgaris. J Med Microbiol* 1983;16:193–202.
102. Engstrand M, Engstrand L, Högman, Hambraeus A, Branth S. Retrograde transmission of *Proteus mirabilis* during platelet transfusion and the use of arbitrarily primed polymerase chain reaction for bacteria typing in suspected cases of transfusion transmission of infection. *Transfusion* 1995;35:871–873.
103. Różalski A, Sidorczyk Z, Kotelko K. Potential virulence factors of *Proteus* bacilli. *Microbiol Mol Biol Rev* 1997;61:65–89.
104. Mobley HLT, Belas R. Swarming and pathogenicity of *Proteus mirabilis* in the urinary tract. *Trends Microbiol* 1995;3:280–284.
105. Jones BD, Lockatell CV, Johnson DE, Warren JW, Mobley HLT. Construction of a urease-negative mutant of *Proteus mirabilis:* Analysis of virulence in a mouse model of ascending urinary tract infection. *Infect Immun* 1990;58:1120–1123.
106. Johnson DE, Russell RG, Lockatell CV, Zulty JC, Warren JW, Mobley HLT. Contribution of *Proteus mirabilis* urease to persistence, urolithiasis, and acute pyelonephritis in a mouse model of ascending urinary tract infection. *Infect Immun* 1993;61:2748–2754.
107. Jones BD, Mobley HLT. *Proteus mirabilis* urease: Nucleotide sequence determination and comparison with jack bean urease. *J Bacteriol* 1989;171:6414–6422.
108. Belas R. Expression of multiple flagellin-encoding genes in *Proteus* mirabilis. *J Bacteriol* 1994;176: 7169–7181.
109. Allison C, Emödy L, Coleman N, Hughes C. The role of swarm cell differentiation and multicellular migration in the uropathogenicity of *Proteus mirabilis. J Infect Dis* 1994;169:1155–1158.
110. Allison C, Lai HC, Hughes C. Coordinate expression of virulence genes during swarm cell differentiation and population migration of *P. mirabilis. Mol Microbiol* 1992;6:1583–1591.
111. Gygi D, Rahman MM, Lai H-C, Carlson R, Guard-Petter J, Hughes C. A cell-surface polysaccharide that facilitates rapid population migration by differentiated swarm cells of *P. mirabilis. Mol Microbiol* 1995;17: 1167–1175.
112. Peerbooms PGH, Verweij AMJJ, MacLaren DM. Investigation of the haemolytic activity of *Proteus mirabilis* strains. *Antonie van Leeuwenhoek* 1983;49:1–11.
113. Koronakis V, Cross M, Senior B, Koronakis E, Hughes C. The secreted hemolysins of *Proteus mirabilis, Proteus vulgaris,* and *Morganella morganii* are genetically related to each other and to the alpha-hemolysin of *Escherichia coli. J Bacteriol* 1987;169: 1509–1515.
114. Senior BW, Hughes C. Production and properties of

haemolysins from clinical isolates of the *Proteeae. J Med Microbiol* 1987;24:17–25.
115. Rozalski A, Kotelko K. Hemolytic activity and invasiveness in strains of *Proteus penneri. J Clin Microbiol* 1987;25:1094–1096.
116. Lukomski S, Rózalski LSA, Dziadek J, Staczek P, Jaworski A. Cell-free and cell-bound hemolytic activities of *Proteus penneri* determined by different Hly determinants. *Can J Microbiol* 1991;37:419–424.
117. Welch RA. Identification of two different hemolysin determinants in uropathogenic *Proteus* isolates. *Infect Immun* 1987;55:2183–2190.
118. Swihart KG, Welch RA. The HpmA hemolysin is more common than HlyA among *Proteus* isolates. *Infect Immun* 1990;58:1853–1860.
119. Uphoff TS, Welch RA. Nucleotide sequencing of the *Proteus mirabilis* calcium-independent hemolysin genes (hpmA and hpmB) reveals sequence similarity with the *Serratia marcesens* hemolysin genes (*shlA* and *shlB*). *J Bacteriol* 1990;172:1206–1216.
120. Swihart KG, Welch RA. Cytotoxic activity of the *Proteus* hemolysin HpmA. *Infect Immun* 1990;58:1861–1869.
121. Lukomski S, Müller B, Schmidt G. Extracellular haemolysin of *Proteus penneri* coded by chromosomal *hly* genes is similar to the α-haemolysin of *Escherichia coli. Zentralbl Bakteriol* 1990;273:150–155.
122. Senior BW. The production of HlyA toxin by *Proteus penneri* strains. *J Med Microbiol* 1993;39:282–289.
123. Benz R, Hardie KR, Hughes C. Pore formation in artificial membranes by the secreted hemolysins of *Proteus vulgaris* and *Morganella morganii. Eur J Biochem* 1994;220:339–347.
124. Adegbola RA, Old DC, Senior BW. The adhesins and fimbriae of *Proteus mirabilis* strains associated with high and low affinity for the urinary tract. *J Med Microbiol* 1983;16:427–431.
125. Mobley HLT, Chippendale GR. Hemagglutinin, urease, and hemolysin production by *Proteus mirabilis* from clinical sources. *J Infect Dis* 1990;161:525–530.
126. Bahrani FK, Mobley HLT. *Proteus mirabilis* MR/P fimbriae: Molecular cloning, expression, and nucleotide sequence of the major fimbrial subunit gene. *J Bacteriol* 1993;175:457–464.
127. Bahrani FK, Cook S, Hull RA, Massad G, Mobley HLT. *Proteus mirabilis* fimbriae: N-terminal amino acid sequence of a major fimbrial subunit and nucleotide sequences of the genes from two strains. *Infect Immun* 1993;61:884–891.
128. Massad G, Lokatell CV, Johnson DE, Mobley HLT. *Proteus* mirabilis fimbriae: Construction of an isogenic pmfA mutant and analysis of virulence in a CBA mouse model of ascending urinary tract infection. *Infect Immun* 1994;62:536–542.
129. Massad G, Bahrani FK, Mobley HLT. *Proteus mirabilis* fimbriae: Identification, isolation, and characterization of a new ambient-temperature fimbria. *Infect Immun* 1994;62:1989–1994.
130. Tolson DL, Barrigar DL, McLean RJC, Altman E. Expression of a nonagglutinating fimbria by *Proteus mirabilis. Infect Immun* 1995;63:1127–1129.
131. Yakubu DE, Old DC, Senior BW. The haemagglutinins and fimbriae of *Proteus penneri. J Med Microbiol* 1989;30:279–284.
132. Peerbooms PGH, Verweij AMJJ, MacLaren DM. Vero cell invasiveness of *Proteus mirabilis. Infect Immun* 1984;43:1068–1071.
133. Allison C, Coleman N, Jones PL, Hughes C. Ability of *Proteus mirabilis* to invade human urothelial cells is coupled to motility and swarming differentiation. *Infect Immun* 1992;60:4740–4746.
134. Oelschlaeger TA, Tall BD. Uptake pathways of clinical isolates of *Proteus mirabilis* into human epithelial cell lines. *Microb Pathog* 1996;21:1–16.
135. Chippendale GR, Warren JW, Trifillis AL, Mobley HLT. Internalization of *Proteus mirabilis* by human renal epithelial cells. *Infect Immun* 1994;62:3115–3121.
136. Senior BW, Albrechtsen M, Kerr MA. A survey of IgA protease production among clinical isolates of *Proteeae. J Med Microbiol* 1988;25:27–31.
137. Loomes LM, Senior BW, Kerr MA. A proteolytic enzyme secreted by *Proteus mirabilis* degrades immunoglobulins of the immunoglobulin A1 (IgA1), IgA2, and IgG isotypes. *Infect Immun* 1990;58:1979–1985.
138. Loomes LM, Senior BW, Kerr MA. Proteinases of *Proteus* spp.: Purification, properties, and detection in urine of infected patients. *Infect Immun* 1992;60:2267–2273.
139. Hoštacká A, Čižnár I, Markovicé J, Karolček J. Toxinogenic potential of *Proteus mirabilis* strains. *Zentralbl Bakteriol* 1984;257:83–92.
140. Drechsel H, Thieken A, Reissbrodt R, Jung G, Winkelmann G. α-Keto acids are novel siderophores in the genera *Proteus, Providencia,* and *Morganella* and are produced by amino acid deaminases. *J Bacteriol* 1993;175:2727–2733.
141. Peerbooms PG, Verweij AMJJ, MacLaren DM. Urinary virulence of *Proteus mirabilis* in two experimental mouse models. *Infect Immun* 1982;36:1246–1248.
142. Sarangapani S, Cavedon K, Gage D. An improved model for bacterial encrustation studies. *J Biomed Materials Res* 1995;29:1185–1191.
143. Piccolomini R, Cellini L, Allocati N, Di Girolama A, Ravagnan G. Comparative in vitro activities of 13 antimicrobial agents against *Morganella-Proteus-Providencia* group bacteria from urinary tract infections. *Antimicrob Agents Chemother* 1987;31:1644–1647.
144. Fuksa M, Krajden S, Lee A. Susceptibilities of 45 clinical isolates to *Proteus penneri. Antimicrob Agents Chemother* 1984;26:419–420.
145. Fernandes CJ, Ackerman VP. In vitro studies of ciprofloxacin and survey of resistance patterns in current isolates. *Diagn Microbiol Infect Dis* 1990;13:79–91.
146. Jones RN, Pfaller MA, Allen SD, Gerlach EH, Fuchs PC, Aldridge KE. Antimicrobial activity of cefpirome: An update compared to five third-generation cephalosporins against nearly 6000 recent clinical isolates from five medical centers. *Diagn Microbiol Infect Dis* 1991;14:361–364.
147. Thornsberry C. Susceptibility of clinical bacterial isolates to ciprofloxacin in the United States. *Infection* 1994;22(Suppl 2):S80–S89.
148. Prosser BLT, Beskid G. Multicenter in vitro comparative study of fluoroquinolones against 25,129 gram-positive and gram-negative clinical isolates. *Diagn Microbiol Infect Dis* 1995;21:33–45.
149. Doern GV. The in vitro activity of cefotaxime versus bacteria involved in selected infections of hospitalized

patients outside of the intensive care unit. *Diagn Microbiol Infect Dis* 1995;22:13–17.
150. Kessler RE, Fung-Tomc J. Susceptibility of bacterial isolates to β-lactam antibiotics from U.S. clinical trials over a 5-year-period. *Am J Med* 1996;100(Suppl 6A): 13S–19S.
151. Jones RN, Kehrberg EN, Erwin ME, Anderson SC, and the fluoroquinolone resistance surveillance group. Prevalence of important pathogens and antimicrobial activity of parenteral drugs at numerous medical centers in the United States: I. Study on the threat of emerging resistances: Real or perceived? *Diagn Microbiol Infect Dis* 1994;19:203–215.
152. Sirot DL, Goldstein FW, Soussy CJ, et al. Resistance to cefotaxime and seven other β-lactams in members of the family *Enterobacteriaceae:* A 3-year survey in France. *Antimicrob Agents Chemother* 1992;36:1677–1681.
153. Mehtar S, Tsakris A, Pitt TL. Imipenem resistance in *Proteus mirabilis. J Antimicrob Chemother* 1991;28: 612–615.
154. Neuwirth C, Siébor E, Duez J-M, Péchinot A, Kazmierczak A. Imipenem resistance in clinical isolates of *Proteus mirabilis* associated with alterations in penicillin-binding proteins. *J Antimicrob Chemother* 1995;36:335–342.
155. Ishida H, Fuziwara H, Kaibori Y, Horiuchi T, Sato K, Osada Y. Cloning and multidrug resistance gene pqrA from *Proteus vulgaris. Antimicrob Agents Chemother* 1995;39:453–457.
156. Bret L, Chanal C, Sirot D, Labia R, Sirot. Characterization of an inhibitor-resistant IRT-2 derived from TEM-2 β-lactamase produced by *Proteus mirabilis* strains. *J Antimicrob Chemother* 1996;38:183–191.

15

The Genus *Morganella*

TAXONOMY

In 1906, Morgan described a group of bacteria associated with cases of infantile gastroenteritis that we now recognize as belonging to the genus *Morganella.*[1] These organisms were originally referred to by the vernacular names of "Morgan's No. 1 bacillus" or the "Morgan bacillus" and were characterized as gram-negative, indole-positive, nonproteolytic bacteria that produced acid and gas from glucose metabolism but failed to ferment a number of common sugars including sucrose, lactose, mannitol, and maltose.[2] Subsequently, the "Morgan bacillus" was placed in a number of different genera in the family *Enterobacteriaceae,* including *Salmonella.* In 1936, Rauss[3] proposed that this group be included in the genus *Proteus* as *Proteus morganii* based on antigenic characteristics and their reputed ability to swarm on solid media.

The "Morgan bacillus" remained in the genus *Proteus* for the next 40 years despite considerable controversy regarding its correct taxonomic position. Fulton, as early as 1943, proposed that there were sufficient phenotypic differences between these bacteria and other members of the genus *Proteus* to warrant their inclusion in a genus of their own.[4] He suggested the name *Morganella morganii.* Later investigations, most notably by Kauffman in 1956 and Rauss and Vörös in 1959, supported this proposal primarily on the basis of significant phenotypic differences between *P. morganii* and other protei.[2] In 1976, Brenner and coinvestigators provided the first detailed taxonomic study of the genetic relatedness of species residing in the genera *Proteus* and *Providencia.*[5] These systematic investigations concluded that *P. morganii* strains were no more closely related (13%–29% relatedness) to protei (eg, *Proteus mirabilis, Proteus rettgeri*) than they were to other members of the *Enterobacteriaceae* (18%–24% relatedness) such as *Salmonella* Typhimurium and *Citrobacter freundii.* Brenner and colleagues[5] then proposed the transfer of *P. morganii* to a new genus. The genus name *Morganella* was resurrected, based on Fulton's[4] original proposal that these organisms deserved distinct generic status. All *M. morganii* strains studied in this seminal investigation were 89%–100% related at 60°C, indicating a single, homogeneous DNA group.[5]

Despite evidence indicating a single genomospecies (DNA group) within *M. morganii,* a number of reports continued to surface describing strains with unusual biochemical characteristics. Rauss and Vörös[2] identified rare *M. morganii* strains that either fermented sucrose, raffinose, sorbitol, or sorbose or failed to attack galactose. Later Siboni[6] described 21 unusual strains of *M. morganii* that fermented trehalose. In 1980, the Centers for Disease Control and Prevention reported on two groups of *M. morganii* characterized by lysine decarboxylase activity (LDC^+) in one instance and the failure to decarboxylate ornithine (ODC^-) in another.[7] The LDC^+ group was associated with two other phenotypic markers (nonmotility, rapid glycerol fermentation), suggesting

TABLE 1. *Current taxonomic status of the genus* Morganella

Species	Subspecies	DNA Relatedness Group	Trehalose Fermentation	No. Biogroups	Type Strain
morganii	*morganii*	1	–	4	ATCC 25830
	sibonii	2	+	3	ATCC 49948
	sibonii	3	+	1	

the possibility that new, previously unrecognized species existed within *M. morganii.* However, DNA–DNA reassociation kinetics reaffirmed the genetic homogeneity of *M. morganii* and indicated that these variants represented new biogroups within the species.[7]

The genetic homogeneity of *M. morganii* was reinvestigated in an international collaborative effort using a more extensive collection of strains of different biogroups and diverse phenotypic characteristics. Based on DNA–DNA hybridization results, three DNA relatedness groups were detected, all of which were interrelated just below the species level.[8] One DNA group, which included all trehalose-negative *M. morganii,* was proposed as the subspecies *M. morganii* ssp. *morganii.* A second taxon, containing two DNA relatedness groups, was proposed as a second subspecies, *M. morganii* ssp. *sibonii;* all of these strains were trehalose-positive.[8] Within each subspecies, multiple biogroups exist. The current taxonomic makeup of the genus *Morganella* is listed in Table 1.

EPIDEMIOLOGY

Most information on the ecologic distribution of *M. morganii* from nonhuman sources comes from studies published before 1960. As such, the validity of accurate species identification in some of these reports has been questioned. Cumulative data from these investigations indicate that dogs, other mammals, and reptiles (snakes) can occasionally harbor morganellae in their intestinal tract, although definitive data regarding relative concentrations and incidence in such species are lacking.[1,9] Morganellae are also strong histamine producers in fish such as tuna and salted anchovies.[10,11] They have been incriminated indirectly in scrombroid food poisoning and the consumption of fish with high histamine levels.

In humans, *M. morganii* can be recovered from a number of anatomic specimens, including the urinary and respiratory tracts, wounds, skin and subcutaneous tissues, and sterile body fluids.[12] In most instances, *M. morganii* is nosocomially transmitted (secondary pathogen), although community-acquired infections occasionally occur. Morganellae have been isolated from several clusters of hospital infections, but in only one instance has *Morganella* been proven to be responsible for a documented outbreak.[13,14] In that instance, a series of microbiologically typed strains were epidemiologically linked to cases of sepsis and wound infection in a cardiac surgery unit.[14] Surveillance cultures from clusters or outbreaks failed to uncover an environmental hospital reservoir for such infections. One possible source for nosocomially transmitted infections may be inpatients who excrete *Morganella* (from draining wounds) for prolonged periods despite aggressive antimicrobial intervention.[13,14] Such patients may act as reservoirs of infection by seeding persons who are in close physical proximity with morganellae. Another likely cause of many hospital-acquired illnesses is breaches in sterility, including cross-infections of patients with contaminated medical devices or failure of hospital staff to observe good aseptic (hand-washing) practices.[13] At one Long Island medical center, development of nosocomial *M. morganii* infections was linked to prolonged hospitalization, which increased patient recovery times and hospital costs.[13]

CLINICAL SIGNIFICANCE

Urinary Tract Infections

Urine is the most common specimen from which *M. morganii* is isolated, although morganellae are relatively infrequent pathogens of the urinary tract. In a series of more than 14,000 bacterial isolates recovered from unselected urines at the Medical College of Virginia in 1981, *M. morganii* accounted for only 0.48% of all gram-negative aerobic isolates.[15] The frequency is somewhat higher in selected patient populations. One Danish survey looked at urine specimens of men admitted to a urologic ward for prostate enlargement and urolithiasis. Among gram-negative bacilli, the annual incidence of *M. morganii* varied from 0% to 5% over a 9-year period.[16]

The involvement of *M. morganii* in urinary tract infections (UTIs) is most often associated with catheterization. Among noncatheterized persons, occasional episodes of low-level bacteriuria (10^2–10^4 cfu/mL) with *M. morganii* have been described in ambulatory individuals precolonized with members of the tribe *Proteeae (Proteus, Providencia, Morganella)* on epithelial surfaces of the groin area.[17] UTI-associated conditions caused by *Morganella* can range from uncomplicated cases of bacteriuria to catheter obstruction as a result of development of encrustations or biofilms on indwelling medical devices.[18]

Bacteriuria, an asymptomatic condition in which the urine is colonized by one or more bacterial species, is most commonly seen in elderly nursing home patients receiving long-term urethral catheterization (≥30 days) for medical conditions such as urinary incontinence or bladder outlet obstruction.[19] Colonization arises in such persons predominantly by the periurethral route in females or the intraluminal route in males.[18] Persons undergoing long-term catheterization invariably develop bacteriuria, often of polymicrobic etiology, and colony counts of the urine often exceed 10^5 cfu/mL.[20] Such conditions can lead to more serious infectious and medical complications, including pyelonephritis, abscess formation, kidney stones, and renal dysfunction.[20] In these clinical settings, *M. morganii* is the fifth leading cause of bacteriuria in persons with indwelling or replacement catheters, accounting for approximately 7% of all microbial isolates.[19] Mobley and Warren[21] found that more than 50% of all urine specimens collected from 32 patients with long-term catheterization contained *M. morganii;* more than 90% of all patients studied during a 1-year interval were colonized with morganellae on one or more occasions. One prospective investigation found that *M. morganii* colonized the urine for an average of 3.3 weeks per episode.[20]

Bacteremia

M. morganii bacteremia is a relatively uncommon event, and this may partially explain the surprisingly low number of clinical reports documenting morganellae as causative agents of systemic disease. In one study from a New York Veterans Administration hospital,[22] *M. morganii* was responsible for 1.6% of all bacteremias recorded over a 5-year period (3.2% of all gram-negative aerobic bacteremias). A study of more than 280,000 blood culture sets drawn at the Mayo Clinic between 1984 and 1992 found the overall incidence of *M. morganii* bacteremia to be 0.5%.[23] Little variation in the frequency of *Morganella* bacteremia was noted during the 8-year study period.

Morganella bacteremia shares a number of clinical features with bacteremias caused by other members of the *Enterobacteriaceae. Morganella* septicemia most often occurs in older persons (mean age 56.4 years) with one of several underlying disorders. *Morganella* septicemia is predominantly a male disease, with a male/female ratio of approximately 8:1. This figure may be somewhat misleading because the single largest collection of morganellae-associated bacteremias (n=19) originated from a Veterans Administration hospital, a fact that may have biased the overall data.[22] The underlying conditions most commonly associated with *M. morganii* sepsis, in decreasing order of frequency, are diabetes mellitus, liver cirrhosis (including ethanol abuse), and malignancy[13,14,22]; more than 70% of bacteremias are nosocomially transmitted.[14,22]

Despite the similarities between cases of *Morganella* bacteremia and sepsis caused by other enteric microorganisms, there are several differences. Although the urinary tract is the most common site from which morganellae are recovered, most septicemias develop from infected postsurgical (open) wounds.[22] In the study of McDermott and Mylotte,[22] 11 of 13 nosocomial bacteremias occurred in surgical patients; in 8 of the 11 cases, sepsis occurred after surgery (mean 11.4 days) or concurrent with debridement of infected tissues. Another difference concerns nosocomial versus community-acquired infections. Nosocomial bacteremias caused by *M. morganii* are more often seen in older persons (58 versus 52 years) and more commonly involve patients with polymicrobic sepsis (56% versus 14%) and recent antimicrobial therapy (62% versus 16%) than do septicemias that are community associated. Prior antibiotic exposure most commonly involves β-lactam antibiotics. The overall fatality rate for *Morganella* septicemia in one study was 22%.[22]

Cases of unusual sequelae associated with *Morganella* sepsis have occasionally been reported. Two of these involved infections of the central nervous system (CNS). Mastroianni and colleagues[24] described a fatal case of *M. morganii* meningitis in a 45-year-old man with the acquired immunodeficiency syndrome (AIDS). *M. morganii* was isolated once from his urine and twice from the cerebrospinal fluid during the course of his disease. Despite antimicrobial intervention (netilmicin, ceftriaxone), he died 5 days after admission. A second case of morganellae-associated meningitis involved brain abscess formation in a 23-day-old infant.[25] The child, who was delivered vaginally without complications, presented to the medical center with irritability, opisthotonos, and a bulging anterior fontanel; *Morganella* was recovered from the cerebrospinal fluid but not from pus aspirated from the abscess. The infant improved on gentamicin and cefotaxime therapy, although he had delayed psychomotor development at a 1-year follow-up. A source for his infection could not be identified, but vertical transmission (eg, urogenital) was suspected, despite the fact that the mother had been asymptomatic. A fatal case of *M. morganii* septicemia accompanied by hemorrhagic bullae formation occurred in a 64-year-old woman with lymphoma.[26] Two weeks after a course of chemotherapy for malignancy, she developed signs of sepsis (shaking chills, fever) accompanied by cellulitis with bullae development on the lower extremities of both legs. Blood and bullae cultures yielded pure growth of *M. morganii.* Despite radical debridement and combination antimicrobial chemotherapy (tobramycin, ticarcillin, cefoxitin), she again became septic and died 10 days later. Kurtz and associates[27] described a case of urosepsis in a 67-year-old woman with babesiosis. She presented to the emergency room with fever, chills, nausea, and abdominal and back discomfort. During a subsequent fever spike (40°C), *M. morganii* was recovered from her urine (>100,000 cfu) and blood. Laboratory analysis of blood samples revealed hemolytic anemia and babesiosis. She was treated with an aminoglycoside and cephalosporin with a positive outcome. The authors speculated that the patient's parasitemia decreased her humoral and cell-mediated immune responses, making her more susceptible to sepsis caused by *M. morganii.* Further support for this concept of *M. morganii* as an opportunistic, bloodborne pathogen in persons with transitory immunodeficiency conditions comes from the case report of Carmona and colleagues.[28] A 32-year-old pregnant woman developed *M. morganii* chorioamnionitis and septicemia after delivery of a full-term infant. Two days before admission she had undergone amnioscopy, which may have triggered the initial infection. Both blood and amniotic fluid cultures grew morganellae. After her septic crisis, she developed adult respiratory distress syndrome, but it is unclear whether *M. morganii* precipitated this postinfectious sequela. After prolonged hospitalization, which included antimicrobial chemotherapy, intubation, and mechanical ventilation, she recovered.

Gastroenteritis

Studies conducted before 1935 suggested a possible association between the recovery of morganellae from feces and diarrheal disease.[1] Over the next 50 years, however, few data were accumulated to support this largely anecdotal information. In 1986, Müller[29] analyzed more than 2500 fecal specimens for the presence of *Morganella.* Although the overall carriage rate was low (0.8%), isolation of morganellae was significantly associated with symptomatic patients ($P<0.01$). A subsequent prospective investigation of traveler's diarrhea by Åhrén and coworkers[30] identified two enterotoxigenic strains of *M. morganii.* One of the two patients from whom toxigenic morganellae were recovered presented with loose stools only, and the second had clinically defined gastroenteritis. The latter case was complicated by the simultaneous isolation of *Campylobacter* and enterotoxigenic *M. morganii* from the stool. Such limited evidence does not presently support a major role for morganellae in diarrheal disease.

Miscellaneous Infections

Several cases of *Morganella*-associated bone or joint disease have been published. Schonwetter and Orson[31] reported a case of septic arthritis in a 95-year-old man with rectal adenocarcinoma and an indwelling catheter for incontinence. Although he had a relatively benign presentation (no fever, no significant radiographic changes), a single colony of *M. morganii* was repeatedly isolated from multiple arthrocentesis specimens of his right shoulder. He had had a positive urine culture for *M. morganii* 3.5 months before his arthritic episode, and this may have been the initial focal point of infection. Two cases of reactive arthropathies attributed to *Morganella* infection have been observed, in a 21-year-old woman and in an 85-year-old woman with bilateral knee effusions.[32] Both patients had chronic diarrhea before seeking medical attention for their arthritic conditions; the elderly woman had a heavy growth of *M. morganii* cultured from her feces. Evidence suggesting morganellae as the etiologic agent of both arthropathies included elevated serologic titers to *M. morganii.* In 1992, Murphey and colleagues[33] described a 51-year-old man who developed septic arthritis and osteomyelitis as complications of a wound trauma sustained to the third finger by a catfish spine. Despite several medical procedures (multiple debridements, irrigation, antibiotics), he failed to respond, necessitating amputation of the finger; *M. morganii* and *Edwardsiella tarda* were recovered from surgical material.

M. morganii is also associated with wound infections, causing disease in hospitalized patients primarily through superinfection of preexisting open surgical wounds.[22] One review lists *M. morganii* as an organism associated with freshwater and saltwater injuries and infection.[33] However, most of these wound isolates involved multiple bacterial species, and the exact role of each organism in the pathogenesis of the disease is difficult to determine.[34] A case of *M. morganii* pyomyositis and septicemia was reported in a 31-year-old man with human immunodeficiency virus (HIV) infection, hepatic cirrhosis, and a history of alcoholism.[35] He died during hospitalization from severe gastrointestinal bleeding as a result of esophageal varices. There is one report of empyema due to *M. morganii* in a 55-year-old man with underlying liver disease.[36] This infection developed 70 days after the patient was initially admitted to the hospital for abdominal distention. A rapid diagnosis with appropriate medical treatment resulted in a favorable outcome. *M. morganii* pericarditis was observed in a 26-year-old man with a history of acute lymphoblastic leukemia and immune pancytopenia. The infection developed 1 year after he had received an allogeneic bone marrow transplantation.[37]

LABORATORY IDENTIFICATION

Isolation

Morganellae grow extremely well on most selective or differential agars designed for the primary isolation of gram-negative aerobic bacilli. Unlike *Proteus,* they fail to swarm on solid media. Almost all strains are both sucrose- and lactose-negative, and they typically present as colorless colonies on MacConkey's, xylose-lysine-deoxycholate, eosin–methylene blue, and Salmonella-Shigella agars. Some strains are highly pleomorphic, producing colonial variants such as smooth and rough varieties on many media (Fig. 1). These morphotypes may be mistaken for culture contaminants during the purification of individual colonies from primary plates. On nonselective media (eg, blood agar), most morganellae appear as nonhemolytic, buff-colored, convex colonies, 2 to 3 mm in diameter, after overnight incubation at 35°C to 37°C; hemolytic activity on blood agar can be detected by prolonged incubation (48–72 hours).

Because *M. morganii* grow on most selective media, their recovery from sterile body sites or specimens containing a limited number of bacterial species (eg, wounds, sputum) does not typically present a problem. However, recovery of morganellae from heavily contaminated specimens (ie, feces), particularly if in low numbers, may be difficult. A number of differential media have been devised for the isolation of morganellae based on their phenotypic response to the presence of aromatic amino acids in common basal media. These include the production of a reddish-brown pigment in DL-tryptophan agar and the degradation of L-tyrosine crystals in nutrient agar.[38,39] One study evaluated several of these media for the isolation of members of the tribe *Proteeae* from fecal specimens.[29] Blood agar was found to be superior to tryptophan agar in this investigation, although the results may simply reflect the suitability of blood agar for detection of swarming *Proteus* species. A modified MacConkey-based agar containing methyl blue and phenolphthalein diphosphate (phosphatase substrate) was used successfully to separate *Providencia stuartii* and *Morganella* from other gram-negative bacilli.[40] Both genera produce red colonies on this medium because they are lactose-negative and phosphatase-positive; all other enteric bacteria produce violet or clear colonies because they are either lactose-positive or lactose- and phosphatase-negative.

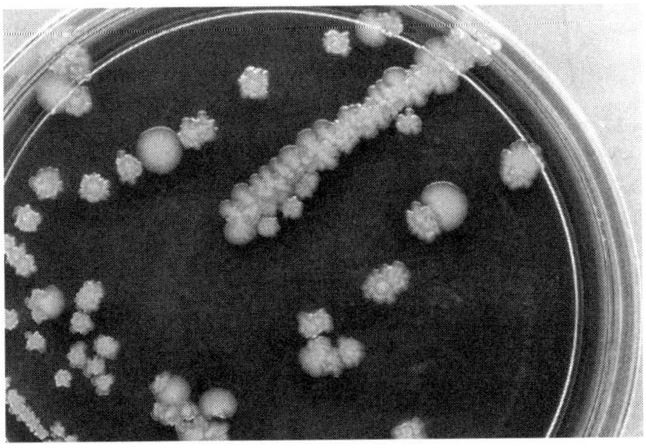

FIG. 1. Colonial dissociants of *M. morganii* on Salmonella-Shigella agar.

On Kligler's or triple sugar iron (TSI) slants, *M. morganii* gives a K/AG$^-$ result at 18 to 24 hours. Although hydrogen sulfide production typically is absent, morganellae are H_2S^+ on other media such as cysteine-based soft agars containing sodium thiosulfate and ferric ammonium citrate. If TSI slants are incubated for an additional 24- to 48-hour period, an H_2S-like reaction (browning) appears at the junction between butt and slant (Fig. 2; Color Plate 19), a phenomenon that may be related to production of small amounts of a reddish-brown pigment. This reaction is reminiscent of the "thumbnail" H_2S production on TSI that is classically exhibited by strains of *Salmonella* Typhi. On lysine-iron agar, another common screening medium, morganellae do not produce a deep reddening of the slant as *Proteus* and *Providencia* species do; this failure reflects the inability of most *M. morganii* to oxidatively deaminate lysine.[41]

Identification

Commercial Systems

M. morganii is a phenotypically tight species exhibiting little strain-to-strain varia-

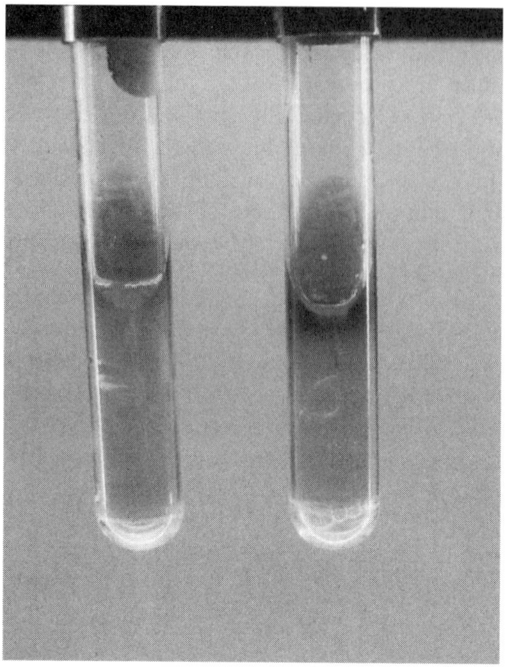

FIG. 2. Darkening of TSI slant by *Morganella*. See also Color Plate 19.

tion in biochemical properties. Because of this uniformity in biochemical characteristics, commercial systems rarely have difficulties in providing a correct identification of these organisms. To date, only a few minor problems have been noted. Kelly and Leicester[42] found that only 7 of 11 *M. morganii* isolates were correctly identified by the Autoscan Walkaway (W/A) system (Dade MicroScan, West Sacramento, CA) at 2 hours. A similar observation was recorded by Pfaller and colleagues,[43] who noted that only four of six *Morganella* strains were correctly identified by W/A over a similar time frame; an additional fifth isolate was correctly identified but with a low probability index. O'Hara and coinvestigators[44] reported that the Cobas system (Becton Dickinson Diagnostic Instrument Systems, Sparks, MD) misidentified two of seven *M. morganii* isolates after overnight incubation (21 hours). The Biolog GN microplate system (Biolog, Hayward, CA) had difficulties identifying 15 *Morganella* isolates by the automated method.[45] Although all 15 strains were correctly identified at 24 hours with manual readings, more than 40% of these strains generated a "no identification" profile when scored with an automated reader at both 4 and 24 hours.

Conventional Media

Morganellae are defined biochemically by their inclusion in the tribe *Proteeae* of the family *Enterobacteriaceae,* along with companion genera *Proteus* and *Providencia.*[41] Defining reactions for this tribe include a positive methyl red test, a negative Voges-Proskauer test, growth in potassium cyanide broth, and production of phenylalanine deaminase.

Several biochemical properties are uniquely associated with the genus *Morganella* and related tribe members or are rarely encountered in other genera within the *Enterobacteriaceae.* Deamination of phenylalanine is one of the properties restricted to this group, a fact that has been used successfully to rapidly detect *Proteeae* from urine specimens, with a reported sensitivity of 0.92 and specificity of 0.99.[46] A second unique property of the *Proteus-Providencia-Morganella* group is the production of a reddish-brown pigment on media containing DL-tryptophan.[38] This melanin-like compound is water soluble and has an estimated molecular weight of less than 12,000.[29] It is most easily observed on slants, and color intensification occurs on prolonged incubation (Fig. 3; Color Plate 20). An almond-like odor is also emitted by morganellae when the bacteria are grown on media containing aromatic amino acids such as L-phenylalanine.[29] Finally, only the *Proteus-Providencia-Morganella* group and some species in the genus *Citrobacter* are capable of degrading L-tyrosine.[39] When inoculated onto media containing L-tyrosine, *Morganella* isolates degrade (clear) insoluble tyrosine crystals. This reaction presumably is mediated through a bacterial tyrosine phenol lyase.

M. morganii is a relatively inert species that attacks a limited number of carbohydrate compounds. Usually, acid is produced from

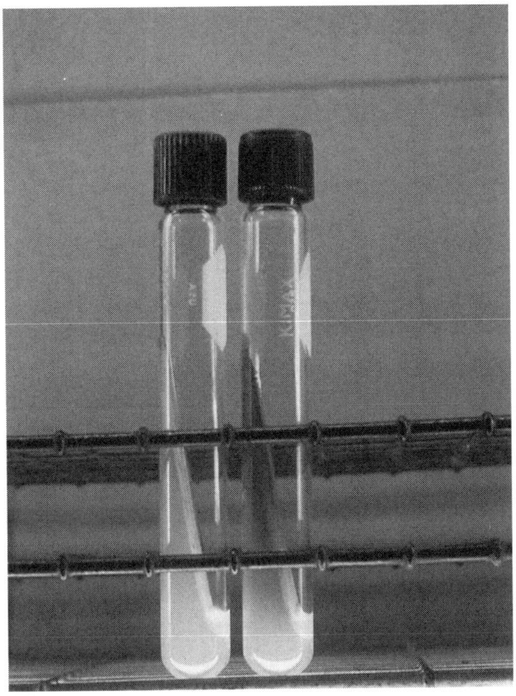

FIG. 3. Production of reddish-brown pigment on DL-tryptophan slants by *Morganella* after prolonged incubation. See also Color Plate 20.

the fermentation of glucose, mannose, fructose (levulose), and galactose only. Trehalose and glycerol fermentation, both infrequently observed, are related to subspecies and biogroup designations. Rare strains have been reported to ferment sucrose (in a delayed fashion), but this activity is apparently unrelated to subspecies or biogroup designation. Production of acid from lactose and LDC$^+$ activity are both unusual phenotypic properties within the genus *Morganella* and have been shown to be plasmid mediated[47,48]; plasmids carrying the LDC$^+$ trait appear to be relatively stable over prolonged periods. Lactose-positive *M. ("Proteus") morganii* strains have been isolated from urine on rare occasions.[49] Biochemical reactions useful in the separation of *Morganella* from related genera are listed in Table 2.

Biotypes and Subspecies

Rauss and Vörös[2] were the first investigators to systematically investigate phenotypic variation in *M. morganii.* On the basis of 6 reactions, 12 biogroups were recognized among a collection of 222 strains. Biogroup 1 strains, which exhibited only typical reactions, accounted for 83% of the entire collection; among the remaining 11 biotypes, certain phenotypes, such as a positive Voges-Proskauer reaction, appeared to be inconsistent with known biochemical traits of the genus. Of the remaining tests, the most commonly observed variants included isolates that were galactose-negative (5%), delayed sucrose-positive (4%), and delayed citrate-positive (3%). This biogrouping scheme, however, did not receive widespread use. After a number of additional reports on phenotypic variation within *M. morganii* in the late 1970s and early 1980s, Jensen and coworkers[8] systematically looked at a large collection of biochemically aberrant

TABLE 2. *Distinguishing phenotypic properties of the* Proteus-Providencia-Morganella *group*[a]

Test	*Proteus*	*Providencia*	*Morganella*
Swarming (BAP)	+	–	–
Indole	–[b]	+	+
Gelatinase	+	–	–
Acid from			
D-Mannose	–	–	+
Xylose	+	–	–
Urea hydrolysis	+	+[c]	+

[a]For clinically relevant species.
[b]Except *Proteus vulgaris.*
[c]Except *Providencia stuartii.*
BAP, blood agar plate.

strains. This study distinguished two subspecies and seven biogroups within *M. morganii.* Subspecies designations were based on the ability of individual isolates to ferment trehalose. This unusual trait had been previously recognized by Siboni[6] to be linked to tetracycline resistance and a longer flagellar wavelength. A biogrouping system was established that was based primarily on ODC and LDC reactions and susceptibility to a 30-μg disk of tetracycline (Table 3). By far the greatest number of *M. morganii* strains isolated in clinical laboratories belong to biogroup A. These strains are commonly referred to as typical *M. morganii* (ODC^+, LDC^-). Biogroup A strains are responsible for most of the serious human infections caused by morganellae. Biogroup B strains, which are ODC^+ and LDC^+, are linked to two other unusual properties for morganellae, nonmotility and rapid (24-hour) glycerol fermentation. These strains are also designated as *M. morganii* biogroup 1 by Farmer and colleagues.[50] *M. morganii* ssp. *sibonii,* which is trehalose-positive, has been estimated to occur in clinical specimens at a frequency of 10% to 14%[8]; biogroup G, which is ODC^+ and LDC^-, is the predominant biogroup within the trehalose-positive subspecies.[50,51] A clinical study of 73 *M. morganii* strains found 90% of all isolates to be *M. morganii* ssp. *morganii,*[51] with biogroup A (ODC^+, LDC^-) predominating (78%). Rapid glycerol fermentation was linked to biogroup B isolates (ODC^+, LDC^+) of *M. morganii* ssp. *morganii* in this study.[51] Two DNA relatedness groups exist within biogroup G strains[8]; these groups cannot be separated phenotypically. Other rare *M. morganii* ssp. *sibonii* strains, which are LDC^- and ODC^- and biochemically fall in between biogroups F and G, are known to exist.

Immunodiagnostics

Various serologic schemes for *M. morganii* have been proposed based on the presence of somatic (O), flagellar (H), and capsular (K) antigens.[2,52] Rauss and Vörös[2] recognized 29 somatic groups and 19 different flagellar antigens among 222 strains of *M. morganii* studied; various combinations of O and H antigens resulted in 57 different serotypes. This system has been expanded to include 55 O antigens, 38 H, and 4 K antigens; more than 77 different serotypes are now recognized.[53] Somatic antigen determinations can be performed in slide agglutination assays[52,53]; some groups, including *Shigella flexneri,* are known to cross-react with common *M. morganii* serogroups.[54] The fatty acid composition of *Morganella* lipopolysaccharide is similar to that of *Escherichia* and *Shigella* and includes major amounts of 3-hydroxytetradecanoic (3-OH-$C_{12:0}$), tetradecanoic ($C_{14:0}$), hexadecanoic ($C_{16:0}$), and dodecanoic ($C_{12:0}$) acids[55]; high levels of ($C_{12:0}$) in *M. morganii* separate these organisms from *Proteus* and *Providencia* species.

Only one study to date has used serology as an aid in the diagnosis of *Morganella* infections.[32] Two patients with suspected arthro-

TABLE 3. *Biogroups of* Morganella morganii

		Phenotypic Characteristics			
Biogroup	Subspecies	LDC	ODC	Tet	Synonyms
A	*morganii*	–	+	V	Typical
B	*morganii*	+	+	S	Farmer's biogroup 1
C	*morganii*	–	–	R	Hickman's ODC^- group
D	*morganii*	+	–	S	
E	*sibonii*	+	+	R	
F	*sibonii*	+	–	R	
G	*sibonii*	–	+	S	

From Jensen et al.,[8] reprinted with permission from the American Society for Microbiology.

LDC, lysine decarboxylase; ODC, ornithine decarboxylase; Tet, tetracycline susceptibility; V, variable; S, susceptible; R, resistant.

pathies due to *Yersinia enterocolitica* were found by a saline tube agglutination assay to possess rising serologic titers to *M. morganii* during the more severe stage of their illness. Although cross-reacting antibodies to *Y. enterocolitica* were present, up to eightfold higher titers to *Morganella* antigens were detected during their protracted course of illness (10 months to 2 years).

Typing

A number of typing systems in addition to serologic properties have been developed to determine strain relatedness in *M. morganii.* These include older systems using the lytic activity of bacteriophages and serologic characteristics.[1] Senior[56] devised a typing system based on production of and sensitivity to bacteriocins (termed morganocins) elaborated by *M. morganii.* The original system recognized 49 morganocin types, of which type 1 (35%) and type 0 (17.5%, nonproducers) predominated. Because of the large preponderance of type 1 morganocin producers found in a random collection of 160 strains, isolates were further subdivided on the basis of their susceptibility to morganocins produced by 7 indicator strains; this allowed for 90 different morganocin production/sensitivity types to be detected among the 160 strains. In 1989, this system was expanded to include 15 new morganocins (64 total) and 1 new morganocin sensitivity type (32 total).[57]

A more recent approach to the epidemiologic monitoring of *M. morganii* strains is cellular protein profile analysis. Costas and associates[58] found 21 different electrophoretic whole-cell protein patterns among 65 strains analyzed. Another study of 150 strains identified a similar number of distinct outer membrane protein (OMP) profiles, although published results suggest that these were probably whole-cell (cellular) rather than OMP patterns.[59] One common finding in these two studies was that unusual *M. morganii* subspecies or biotypes tend to produce characteristic protein patterns (phenoms) on sodium dodecyl sulfate–polyacrylamide gel electrophoresis (SDS-PAGE) analysis.

One of the major problems regarding typing of *M. morganii* strains is the relative predominance of a single phenom regardless of which typing system is employed. Biotyping, serotyping, phage typing, and protein profiles all generate specific phenotypic clusters that predominate, thereby limiting their potential applicability in many situations. In some systems (morganocins), further subtyping is available but time-consuming and laborious. Serotyping, a potentially useful technique, is performed only in a limited number of laboratories throughout the world. The single technique that lends itself most readily to laboratory use (biotyping) is handicapped by the infrequent occurrence of phenotypic variants among *M. morganii* isolates. Molecular approaches, commonly used for other enteric organisms, have not been applied to *Morganella* because of the limited number of outbreak-related situations reported in the literature. Because morganellae exhibit similar if not identical susceptibility patterns to various antibiotics (excluding tetracycline), antibiograms are of limited value.[13]

With the use of a single typing system (serogrouping), Williams and coinvestigators[14] were able to trace the emergence of a single *M. morganii* strain in a cardiac surgery unit. In a limited study of nine cases, Vörös and Senior[53] found that use of two typing systems (serogroup, morganocin) in combination had superior epidemiologic discriminatory power than either test applied alone.

PATHOGENESIS

M. morganii strains produce very few extracellular enzymatic activities. Of these, β-hemolysin is probably the most attractive candidate as a virulence factor. Hemolytic activity is expressed by more than 50% of *Morganella* strains,[60] and a high incidence of hemolysin-positive isolates from extraintestinal sources has been noted by Emödy and coworkers.[61] The *M. morganii* hemolysin is genetically, immunologically, and functionally related to the α-hemolysin of *Escherichia coli,* although extensive divergence in hemolysin gene sequences between these two species is known to exist.[61,62] Hemolytic strains are more pathogenic for mice (median lethal dose, mean survival time, hematuria) than are nonhemolytic isolates or hemolysin-negative mutants derived from wild-type strains after exposure to actinomycin D[61]; similar results have been obtained using chick embryos. This cytolytic toxin may function in extraintestinal disease by promoting invasion and dissemination or, alternatively, it may cause cell death and aid in the infection process of local wounds by increasing bacterial replication through the release of metabolites from eucaryotic cells.

A second factor likely to be involved in pathogenesis of UTI is the constituitively produced *Morganella* urease.[63] Ureases promote the alkalinization of urine, leading to crystal formation and encrustation[18] and ultimately to catheter-associated obstructions. *P. mirabilis* strains multiply quicker and produce alkaline conditions faster in urine than does *M. morganii.*[53] This may partially explain the more frequent association of *P. mirabilis* with UTI rather than morganellae. The urease of *M. morganii* may also promote gastrointestinal tract colonization by facilitating acid tolerance under conditions of extreme acidity such as those encountered in the stomach.[64]

Morganellae do not produce catecholate- or hydroxamate-type siderophores. However, α-keto acids, byproducts of deaminase activity, may function in iron transport.[65] Elevated levels of keto or amino acids with siderophore-like activity may be present in UTIs associated with underlying diseases such as diabetes and may help to promote infection.

ANTIMICROBIAL SUSCEPTIBILITY

Composite susceptibility data for *M. morganii* are listed in Table 4. Almost all *M. morganii* strains are ampicillin resistant, and many show widespread resistance to extended-spectrum penicillins and first-generation cephalosporins. Conversely, more than 90% of all *M. morganii* are susceptible to aztreonam, the quinolones, and aminoglycosides, although one Italian study noted that some strains were resistant to amikacin but susceptible to tobramycin.[66] Although some surveys have found morganellae susceptible to imipenem, Jones and colleagues[69] found a 20.8% resistance rate to imipenem among strains from four hospitals participating in a national surveillance study. This increasing trend in imipenem resistance is an alarming note if true. An additional problem concerns the susceptibility testing of *M. morganii* by disk diffu-

TABLE 4. *Susceptibility of* Morganella morganii *to selected antimicrobial agents (in μ/mL)*

Antibiotic	Range	MIC_{50}	MIC_{90}
Ampicillin	12.5–>200	>200	>200
Piperacillin	1.56–≥200	12.5	100
Carbenicillin	1.56–100	3.12	50
Cefoxitin	3.13–≥200	>200	>200
Ceftazidime	0.05–3.12	0.1	1.56
Ceftriaxone	0.025–0.78	0.0254	0.2
Cefuroxime	0.2–12.5	1.56	12.5
Gentamicin	0.2–≥200	0.78	1.6
Tobramycin	0.2–0.78	0.39	0.78
Amikacin	1.56–50	3.12	25
Aztreonam	0.39–12.5	0.78	6.25
Imipenem	0.1–12.5	3.13	6.25
Ciprofloxacin	≤0.06–>8	≤0.06	≤0.06
Ofloxacin	0.06–>8	0.12	0.25

Data compiled from references 66–69.
MIC, minimal inhibitory concentration.

sion against some of the newer cephems. For at least four cephems (cefuroxime, cefixime, cefpodoxime, and cefetamet), unacceptable rates of minor errors (resistant organism reported as intermediate) or very major errors (resistant organism reported as susceptible) were found when disk susceptibility results were compared with minimal inhibitory concentration values generated by agar dilution.[70]

CONCLUSIONS

Even though *M. morganii* is a commonly isolated bacterium in the clinical laboratory, very little is known concerning its ecologic distribution and role in causing human infections other than UTIs. Few studies have looked at the role of these microorganisms in causation of wound infections and disseminated diseases, particularly in reference to susceptible populations, modes and routes of infection, and disease pathogenesis. It is highly likely that many more nosocomially transmitted *M. morganii* infections of a serious nature occur than are suspected or reported.[22] The recent medical literature provides few reports on the relative effectiveness of newer antimicrobial agents; this point is particularly distressing in light of emerging resistance to imipenem and the failure of disk diffusion assays to accurately predict susceptibility to various cephems.

REFERENCES

1. Penner JL. The genera *Proteus, Providencia,* and *Morganella.* In: Balows A, Trüper HG, Dworkin M, Harder W, Schleifer K-H, eds. *The prokaryotes: A handbook on the biology of bacteria ecophysiology, isolation, identification, applications.* 2nd ed. Berlin: Springer-Verlag, 1991;2849–2862.
2. Rauss K, Vörös S. The biochemical and serological properties of *Proteus morganii. Acta Microbiol Hung* 1959;6:233–248.
3. Rauss KF. The systematic position of Morgan's bacillus. *J Pathol Bacteriol* 1936;42:183–192.
4. Fulton M. The identity of *Bacterium columbensis* Castellani. *J Bacteriol* 1943;46:79–82.
5. Brenner DJ, Farmer III JJ, Fanning GR, et al. Deoxyribonucleic acid relatedness of *Proteus* and *Providencia* species. *Int J Syst Bacteriol* 1976;28:269–282.
6. Siboni K. Correlation of the characters fermentation of trehalose, non-transmissable resistance to tetracycline, and relatively long flagellar wavelength in *Proteus morganii. Acta Pathol Microbiol Scand B* 1976;84: 421–427.
7. Hickman FW, Farmer JJ III, Steigerwalt AG, Brenner DJ. Unusual groups of *Morganella ("Proteus") morganii* isolated from clinical specimens: lysine-positive and ornithine-negative biogroups. *J Clin Microbiol* 1980;12:88–94.
8. Jensen KT, Frederiksen WM, Hickman-Brenner FW, Steigerwalt AG, Riddle CF, Brenner DJ. Recognition of *Morganella* subspecies, with proposal of *Morganella morganii* subsp. *morganii* subsp. nov. and *Morganella morganii* subsp. *sibonii* subsp. nov. *Int J Syst Bacteriol* 1992;42:613–620.
9. Goldstein EJC, Agyare EO, Vagvolgyi AE, Halpern M. Aerobic bacterial oral flora of garter snakes: Development of normal flora and pathogenic potential for snakes and humans. *J Clin Microbiol* 1981;13:954–956.
10. López-Sabater EI, Rodriguez-Jerez JJ, Roig-Sagués AX, Mora-Ventura MAT. Bacteriological quality of tuna fish *(Thunnus thynnus)* destined for canning: Effect of tuna handling on presence of histidine decarboxylase bacteria and histamine level. *J Food Prot* 1994;57:318–323.
11. Rodriguez-Jerez JJ, Mora-Ventura MT, Lopez-Sabater EI, Hernandez-Herrero M. Histidine, lysine and ornithine decarboxylase bacteria in Spanish salted semi-preserved anchovies. *J Food Protect* 1994;57:784–791.
12. Adler JL, Burke JP, Martin DF, Findland M. *Proteus* infections in a general hospital. I: Biochemical characteristics and antibiotic susceptibility of the organisms, with special reference to proticine typing and the Dienes phenomenon. *Ann Intern Med* 1971;75:517–530.
13. Tucci V, Isenberg HD. Hospital cluster epidemic with *Morganella morganii. J Clin Microbiol* 1981;14: 563–566.
14. Williams EW, Hawkey PM, Penner JL, Senior BW, Barton LJ. Serious nosocomial infection caused by *Morganella morganii* and *Proteus mirabilis* in a cardiac surgery unit. *J Clin Microbiol* 1983;18:5–9.
15. Markowitz SM. Urinary tract specimens. In: Dalton HP, Nottebart HC Jr, eds. *Interpretive medical microbiology.* New York: Churchill Livingstone, 1986;523–592.
16. Søgaard H, Zimmermann-Nielsen C, Siboni K. Antibiotic-resistant gram-negative bacilli in a urological ward for male patients during a nine-year period: Relationship to antibiotic consumption. *J Infect Dis* 1974;130: 646–650.
17. Ehrenkranz NJ, Alfonso BC, Eckert DG, Moskowitz LB. *Proteeae* species bacteriuria accompanying *Proteeae* species groin skin carriage in geriatric outpatients. *J Clin Microbiol* 1989;27:1988–1991.
18. Stamm WE. Catheter-associated urinary tract infections: Epidemiology, pathogenesis, and prevention. *Am J Med* 1991;91(Suppl 3B):65S–71S.
19. Warren JW. Catheter-associated urinary tract infections. *Infect Dis Clin North Am* 1987;1:823–854.
20. Warren JW, Tenney JH, Hoopes JM, Muncie HL, Anthony WC. A prospective microbiologic study of bacteriuria in patients with chronic indwelling urethral catheters. *J Infect Dis* 1982;146:719–723.
21. Mobley HLT, Warren JW. Urease-positive bacteriuria and obstruction of long-term urinary catheters. *J Clin Microbiol* 1987;25:2216–2217.
22. McDermott C, Mylotte JM. *Morganella morganii:* Epidemiology of bacteremic disease. *Infect Control* 1984; 5:131–137.

23. Cockerill FR III, Hughes JG, Vetter EA, et al. Analysis of 281,797 consecutive blood cultures performed over an eight-year period: Trends in microorganisms isolated and the value of anaerobic culture of blood. *Clin Infect Dis* 1997;24:403–418.
24. Mastroianni A, Coronado O, Chiodo F. *Morganella morganii* meningitis in a patient with AIDS [letter]. *J Infect* 1994;29:356–357.
25. Verboon-Maciolek M, Vandertop WP, Peters ACB, Roord JJ, Geelen SPM. Neonatal brain abscess caused by *Morganella morganii. Clin Infect Dis* 1995;20:471.
26. Bagel J, Grossman ME. Hemorrhagic bullae associated with *Morganella morganii* septicemia. *J Am Acad Dermatol* 1985;12:575–576.
27. Kurtz B, Michaels I, Kim D, Mattana J. Urosepsis from an atypical pathogen in a patient with babesiosis. *Ann Emerg Med* 1994;24:1199.
28. Carmona F, Fábregues F, Alvarez R, Vila J, Cararach V. A rare case of chorioamnionitis by *Morganella morganii* complicated by septicemia and adult respiratory distress syndrome. *Eur J Obstet Gynecol* 1992;45:67–70.
29. Müller HE. Occurrence and pathogenic role of *Morganella-Proteus-Providencia* group bacteria in human feces. *J Clin Microbiol* 1986;23:404–405.
30. Åhrén CM, Jertborn M, Herclik L, Kaijser B, Svennerholm AM. Infection with bacterial enteropathogens in Swedish travellers to South-East Asia: A prospective study. *Epidemiol Infect* 1990;105:325–333.
31. Schonwetter RS, Orson FM. Chronic *Morganella morganii* arthritis in an elderly patient. *J Clin Microbiol* 1988;26:1414–1415.
32. Cafferkey MT, Buckley TF, O'Morain CA. Association of reactive arthropathy and *Morganella morganii* cross reacting with *Yersinia enterocolitica. Scand J Infect Dis* 1988;20:439–442.
33. Murphey DK, Septimus EJ, Waagner DC. Catfish-related injury and infection: Report of two cases and review of the literature. *Clin Infect Dis* 1992;14:689–693.
34. von Graevenitz A, Spector H. Observations on indol-positive *Proteus. Yale J Biol Med* 1969;14:434–445.
35. Arranz-Caso JA, Cuadrado-Gomez LM, Romanik-Cabrera J, Garcia-Tena J. Pyomyositis caused by *Morganella morganii* in a patient with AIDS. *Clin Infect Dis* 1996;22:372–373.
36. Isobe H, Motomura K, Kotou K, Sakai H, Satoh M, Nawata H. Spontaneous bacterial empyema and peritonitis caused by *Morganella morganii. J Clin Gastroenterol* 1994;18:87–88.
37. Sica S, Di Mario A, Salutari P, et al. *Morganella morganii* pericarditis after resolvent splenectomy for immune pancytopenia following allogeneic bone marrow transplantation for acute lymphoblastic leukemia. *Clin Infect Dis* 1995;21:1052–1053.
38. Polster M, Svobodová M. Production of reddish-brown pigment from DL-tryptophan by enterobacteria of the *Proteus-Providencia* group. *Experientia* 1964;20: 637–638.
39. Sheth NK, Kurup VP. Evaluation of tyrosine medium for the identification of *Enterobacteriaceae. J Clin Microbiol* 1975;1:483–485.
40. Thaller MC, Berlutti F, Pantanella F, Pompei R, Satta G. Modified MacConkey medium which allows simple identification of *Providencia stuartii. J Clin Microbiol* 1992;30:2054–2057.
41. Ewing EH. *Edwards and Ewing's identification of* Enterobacteriaceae. 4th ed. New York: Elsevier, 1986.
42. Kelly MT, Leicester C. Evaluation of the Autoscan Walkaway system for rapid identification and susceptibility testing of gram-negative bacilli. *J Clin Microbiol* 1992;30:1568–1571.
43. Pfaller MA, Sahm D, O'Hara C, Ciaglia C, Yu M, Yamane N, Scharnweber G, Rhoden D. Comparison of the AutoSCAN-W/A rapid bacterial identification system and the Vitek AutoMicrobic system for identification of gram-negative bacilli. *J Clin Microbiol* 1991;29:1422–1428.
44. O'Hara CM, Tenover FC, Miller JM. Parallel comparison of accuracy of API 20E, Vitek GNI, MicroScan Walk/Away Rapid ID, and Becton Dickinson Cobas Micro ID-E/NF for identification of members of the family *Enterobacteriaceae* and common gram-negative, non–glucose-fermenting bacilli. *J Clin Microbiol* 1993; 31:3165–3169.
45. Holmes B, Costas M, Ganner M. On SLW, Stevens M. Evaluation of Biolog system for identification of some gram-negative bacteria of clinical importance. *J Clin Microbiol* 1994;32:1970–1975.
46. Vila J, Rullan J, Jiminez de Anta MT. Rapid detection of urinary tract infection caused by *Escherichia coli* and *Proteeae* species. *Eur J Clin Microbiol Infect Dis* 1991; 10:922–926.
47. Le Minor L, Coynault C. Determinise plasmidique du caractere atypique lactose positif de *Enterobacter hafniae* et de *Proteus morganii. Ann Microbiol Inst Past* 1976;127A:213–221.
48. Cornelis G, van Bouchaute M, Wauters G. Plasmid-encoded lysine decarboxylation in *Proteus morganii. J Clin Microbiol* 1981;14:365–369.
49. Tierno PM, Steinberg P. Isolation and characterization of a lactose-positive strain of *Proteus morganii. J Clin Microbiol* 1975;1:108–109.
50. Farmer JJ III, Davis BR, Hickman-Brenner FW, et al. Biochemical identification of new species and biogroups of *Enterobacteriaceae* isolated from clinical specimens. *J Clin Microbiol* 1985;21:46–76.
51. Janda JM, Abbott SL, Khashe S, Robin T. Biochemical investigations of biogroups and subspecies of *Morganella morganii. J Clin Microbiol* 1996;34:108–113.
52. Penner JL, Hennessy JN. O antigen grouping of *Morganella morganii (Proteus morganii)* by slide agglutination. *J Clin Microbiol* 1979;10:8–13.
53. Vörös S, Senior BW. New O antigens of *Morganella morganii* and the relationships between haemolysin production, O antigens and morganocin types of strains. *Acta Microbiol Hung* 1990;37:341–349.
54. Penner JL. The tribe *Proteeae.* In: Starr MP, Stolp H, Trüper HG, Balows A, Schlegel HG, eds. *The prokaryotes: A handbook on habitats, isolation, and identification of bacteria.* Berlin: Springer-Verlag, 1991;1204–1224.
55. Vasyurenko ZP, Chernyavskaya YN. Confirmation of *Morganella* distinction from *Proteus* and *Providencia* among *Enterobacteriaceae* on the basis of cellular and lipopolysaccharide fatty acid composition. *J Hyg Epidemiol Microbiol Immunol* 1990;34:81–90.
56. Senior BW. The typing of *Morganella morganii* by bacterocin production and sensitivity. *J Med Microbiol* 1987;23:33–39.
57. Senior BW, Vörös S. Discovery of new morganocin types of *Morganella morganii* strains of diverse serotype and

the apparent independence of bacteriocin type from serotype of strains. *J Med Microbiol* 1989;29:89–93.
58. Costas M, Holmes B, Wood AC. Numerical analysis of electrophoretic protein patterns of *Morganella morganii* strains from faeces, wound, urine and other clinical sources. *J Appl Bacteriol* 1990;69:426–438.
59. Senior BW, Vörös S. Protein profile typing: A new method of typing *Morganella morganii* strains. *J Med Microbiol* 1990;33:259–264.
60. Koronakis V, Cross M, Senior B, Koronakis E, Hughes C. The secreted hemolysins of *Proteus mirabilis, Proteus vulgaris,* and *Morganella morganii* are genetically related to each other and to the alpha-hemolysin of *Escherichia coli. J Bacteriol* 1987;169:1509–1515.
61. Emödy L, Vörös S, Pál T. Alpha-haemolysin, a possible virulence factor in *Proteus morganii. FEMS Microbiol Lett* 1982;13:329–331.
62. Eberspächer B, Hugo F, Pohl M, Bhakdi S. Functional similarity between the haemolysins of *Escherichia coli* and *Morganella morganii. J Med Microbiol* 1990;33:165–170.
63. Senior BW. *Proteus morganii* is less frequently associated with urinary tract infections than *Proteus mirabilis:* An explanation. *J Med Microbiol* 1983;16:317–322.
64. Young GM, Amid D, Miller VL. A bifunctional urease enhances survival of pathogenic *Yersinia enterocolitica* and *Morganella morganii* at low pH. *J Bacteriol* 1996;178:6487–6495.
65. Drechsel H, Thieken A, Reissbrodt R, Jung G, Winkelmann G. α-Keto acids are novel siderophores in the genera *Proteus, Providencia,* and *Morganella* and are produced by amino acid deaminases. *J Bacteriol* 1993;175:2727–2733.
66. Piccolomini R, Cellini L, Allocati N, di Girolamo A, Ravagnan G. Comparative in vitro activities of 13 antimicrobial agents against *Morganella-Proteus-Providencia* group bacteria from urinary tract infections. *Antimicrob Agents Chemother* 1987;31:1644–1647.
67. Igari J, Hayashi Y, Shitara S, et al. Antimicrobial susceptibility patterns of the *Proteeae* in Japan, 1989. *Jpn J Antibiot* 1993;46:285–294.
68. Prosser BLT, Beskid G. Multicenter in vitro comparative study of fluoroquinolones against 25,129 gram-positive and gram-negative clinical isolates. *Diagn Microbiol Infect Dis* 1995;21:33–45.
69. Jones RN, Kehrberg EN, Erwin ME, Anderson SC, and the Fluoroquinolone Resistance Surveillance Group. Prevalence of important pathogens and antimicrobial activity of parenteral drugs at numerous medical centers in the United States: I. Study on the threat of emerging resistances: Real or perceived? *Diagn Microbiol Infect Dis* 1994;19:203–215.
70. Biedenbach DJ, Jones RN, Erwin ME. Interpretive accuracy of the disk diffusion method for testing newer orally administered cephalosporins against *Morganella morganii. J Clin Microbiol* 1993;31:2828–2830.

16

The Genus *Providencia*

TAXONOMY

The origin of the genus *Providencia* dates back to a colorful period of bacterial taxonomy when many present-day *Enterobacteriaceae* genera were initially recognized as phenotypically distinct entities. These groups, often poorly characterized, had intermediate biochemical properties which most often did not resemble those of established genera. Such transitional forms were commonly referred to as "paracolon bacteria." In 1943, a number of paracolon bacteria were biochemically and antigenically characterized, and several new groups were established as a result of this analysis.[1] One such cluster of 23 anaerogenic paracolon bacteria, typified by type culture 29911, were indole-positive, Voges-Proskauer- and urea-negative and unable to ferment D-mannitol, sucrose, or lactose. Three years later the paracolon 29911 group was further characterized biochemically and found to have properties similar to those of *Proteus* species.[2] In 1951, Kauffmann coined the vernacular term "Providence group" for these bacteria.[3] This name originates from the fact that Stuart and others originally isolated many of the paracolon 29911 bacteria from Providence and the local vicinity including Rhode Island Hospital. Over the next decade various proposals were entertained in an attempt to classify the Providence group. These proposals included elevation of the group to genus status as *Providencia providenciae* or to species rank in the genus *Proteus* as *Proteus inconstans*.[4,5] In 1955, Proom[6] concluded from a nutritional study that the only alternative was to establish for these paracolon bacteria a new genus *(Providencia)*, which would include the species previously named *Proteus rettgeri*. However, this proposal did not receive widespread acceptance, and the Providence group continued to vacillate between group, species, and genus rank under different schema.

In 1962, Ewing[7] formally proposed establishing the genus *Providencia* to include various strains and biogroups previously classified in the Providence group. Two species were recognized, namely *Providencia alcalifaciens* (formerly providence subgroup A) and *Providencia stuartii* (providence subgroup B). Despite this proposal, many investigators continued to refer to these bacteria as Providence bacilli, Providence group or, in some instances, *P. inconstans*.[8] The name *P. inconstans* had been previously rejected by Ewing on the grounds that the original culture was no longer available and the neotype strain was found not to be a member of this group.[7] To add to this confusion was the question of the correct taxonomic position of *P. rettgeri*. This species was isolated originally by Rettger in 1904 and had been placed in several genera, including *Shigella*.[3] In 1943, Rustigian and Stuart[9] proposed that *Shigella rettgeri* be transferred to the genus *Proteus* as *P. rettgeri*, primarily on the basis of its positive urease reaction. Over the next several decades various taxonomic proposals were made for these organisms, including placing them in a genus of their own, *Rettgerella*.[4] Preliminary studies of the DNA-based composition of these groups indicated that *P.*

TABLE 1. *Current composition of the genus* Providencia

Species	Type Strain	Isolated from Humans	Synonyms	Ref.
P. alcalifaciens	ATCC 9886	Yes		7
P. stuartii	ATCC 29914	Yes		7
P. rettgeri	ATCC 29944	Yes	*Proteus rettgeri*	10
P. rustigianii	ATCC 33673	Yes	*P. friedericiana, P. alcalifaciens* biogroup 3	11, 13
P. heimbachae	ATCC 35613	No	Enteric group 78	14

rettgeri was not closely related to other protei such as *Proteus mirabilis* and *Proteus vulgaris.* Finally, the urease reaction, which traditionally had been used to separate *P. rettgeri* from the Providence group, was found to be an unreliable marker by Ursing.[8] These questions left considerable controversy regarding the appropriate taxonomic position of members residing in the *Proteus-Providencia* groups.

In 1978, Brenner and associates[10] published a detailed taxonomic assessment of members of the genera *Proteus* and *Providencia* using DNA reassociation kinetics. The major conclusion of this study was that three species should be recognized in the genus *Providencia*—namely, *P. alcalifaciens, P. stuartii,* and *P. rettgeri.* This study also found that a previous biogroup of *P. alcalifaciens* represented a new, as yet unnamed species within the genus. In 1983, a formal proposal to designate *P. alcalifaciens* biogroup 3 as *Providencia rustigianii* was made by Hickman-Brenner and colleagues.[11] This was followed 4 months later by a publication by Müller[12] describing a new *Providencia* species, *Providencia friedericiana,* which was isolated from five different species of penguins residing in German zoos. Subsequent collaborative efforts by both laboratories indicated that *P. rustigianii* and *P. friedericiana* were highly related and were subjective synonyms for the same organisms. Based on its earlier date of publication and validation, *P. rustigianii* is the senior synonym and has priority over *P. friedericiana.*[13] In 1986, a fifth species in the genus was described. *Providencia heimbachae* was the name proposed for a group of indole-negative strains recovered from penguins and a cow.[14] In addition to these established species, others may exist. In the most recent study of *P. alcalifaciens,* Picard and collaborators[15] suggested that enough heterogeneity exists within this group to warrant the establishment of another *Providencia* species. Studying 27 previously hybridized strains, Picard found two distinct groups within *P. alcalifaciens* based on zymotype (esterase-βa, acid phosphatase), rDNA polymorphism (ribotype), and ΔT_m values (thermal instability of DNA–DNA hybrid molecules). These subdivisions closely match the two remaining biogroups within *P. alcalifaciens.* The authors indicated that these collective facts argue for the creation of a new species within *P. alcalifaciens.* The current taxonomic composition of the genus *Providencia* is listed in Table 1.

EPIDEMIOLOGY

Few studies have reported on the isolation of the Providence group from animals or environmental samples. An early Centers for Disease Control and Prevention (CDC) investigation[16] found that only 3 of more than 500 *Providencia* isolates originated from nonhuman sources (cat tissue, dog, irrigation water). A much later Egyptian study detected providenciae in the fecal contents of many freshwater fish caught in a branch of the Nile River.[17] The highest recovery rate recorded in this survey was for *P. rettgeri* (17.8%). *P. rettgeri* has additionally been isolated from nematodes,[18] the pupa of carrot flies,[19] and from sheep liver.[19] *P. alcalifaciens* has been recovered from penguins (like *P. heimbachae*), from the eye of a guinea pig, and from duck feces at a zoo.[12,14,20] *P. stuartii* has been implicated as the etiologic agent of neonatal diarrhea in dairy and beef calves; it was isolated from 14 separate herds in Idaho.[21] Stickler and coworkers[22] attempted to determine the natural reservoir of *P. stuartii* using a selective medium. *P. stuartii* was found on the groin area, in urine, and/or in feces of 12 patients with long-term indwelling catheters. However, inanimate hospital objects sampled were negative for providenciae unless they had been in intimate contact with colonized persons (eg, urine jugs, groin bowls). Nonhospital samples such as sewage and sewage-contaminated sea and river water failed to yield *Providencia.* These facts suggest that providenciae do not survive well in the environment in a free-living form but rather favor a more intimate association with both vertebrate and invertebrate species in protected sites such as the gastrointestinal or alimentary tract.

CLINICAL SIGNIFICANCE

The Providence group is infrequently encountered in clinical material, and there are few comprehensive clinical investigations on this genus. *Providencia* have a very limited anatomic distribution in humans. The sites from which providenciae are most frequently recovered are the urinary tract and feces. Rarely, other extraintestinal specimens such as blood and respiratory tract secretions yield providenciae. Regardless of the source of isolation, *Providencia* species are often recovered as part of mixed infections, commonly involving either *Proteus* species or *Pseudomonas aeruginosa.* In a 2-year study (1968–1969) of the Providence group at the University of Minnesota Medical Center, 158 isolates were recovered from hospitalized patients.[23] Overall, the Providence group accounted for 0.8% to 1.4% of all *Enterobacteriaceae* isolated from several different body sites. Seventy-five percent of strains originated from urine. Based on clinical criteria, only 61 (39%) of the 158 strains were judged to be clinically significant. A decade later, McHale and associates[24] collected similar data in an 8-year survey of *P. stuartii* infections at several local hospitals in Dublin. Of 241 isolates, 60% were from urine and an additional 39% were associated with skin infections. Again, as in the Minnesota study, fewer than 50% of all isolates were clearly involved in infections. Such data indicate that many isolates of *Providencia* are of dubious or questionable clinical significance.

Urinary Tract Infections

Urine is the most common specimen yielding *Providencia* in the clinical laboratory. Despite this fact, *Providencia* are uncommon causes of urinary tract infections (UTIs) in the general population (Table 2). In persons with long-term or permanent renal dysfunction, providenciae can be important uropathogens. Early studies found a strong association (96%) between the recovery of Providence bacilli from urine and the presence of underlying urologic disorders.[23] This association was corroborated by subsequent investigators who demonstrated a striking correlation between neurologic disease and/or urologic abnormalities and *Providencia* UTI.[24–26] Persons especially prone to development of such illnesses include those in chronic care facilities (nursing homes, geriatric wards, rehabilitation centers) and those in neurology wards.

Bacteriuria is the most common urologic complication associated with *Providencia* infection in debilitated hosts. Although the condition is most often asymptomatic, fever may be present.[25] The predominant risk factor tied to a high incidence of *Providencia* bacteriuria is the use of long-term indwelling urethral catheters. Solberg and Matsen[23] found that 50 (93%) of 54 patients with *Providencia* UTIs had indwelling catheters. In 1986, Warren[25] reviewed eight well controlled international studies involving 215 adults with long-term catheterization from whom more than 1000 urine cultures were collected. *P. stuartii* was one of the five most common microbial species isolated from catheterized adults. In six of the studies, *P. stuartii* was recovered from urine at a frequency of 0% to 61% (overall rate approximately 40%). Another survey found *P. stuartii* to be the most common species (59% of all

TABLE 2. *Clinical infections associated with the genus* Providencia

Infection	Susceptible Populations	Predisposing Conditions	Risk Factors	Incidence: General Population	Incidence: High-Risk Groups	Agent
Urinary tract infections, bacteriuria	Elderly, nursing home patients	Neurogenic bladder, congenital urologic abnormalities, prostatic hypertrophy	Long-term catheterization	0.3%–0.6%	>30%	*P. stuartii*
Bacteremia	Elderly	Cancer, heart disease, stroke, prostatic hypertrophy	Urinary tract obstruction, paraplegia	0.1%–0.5%	6%–24%	*P. stuartii*
Gastroenteritis	Children <5 y, adults	None	Foreign travel	0.5%–2.5%	0.5%–10%	*P. alcalifaciens*

Data from references 23–27, 31–33, 40–43.

urines) isolated from 19 persons with indwelling urinary or condom catheters.[26] Once providenciae gain access to the bladder via introduction through a urinary catheter, they can persist for extremely long intervals; mean durations of infection were 9 weeks and 6.4 months in two studies.[26,27] Persistent bacteriurias, commonly of polymicrobic etiology, are a major concern because they can lead to a number of serious sequelae, including obstructed catheters, acute pyelonephritis, renal abscesses, bladder and kidney stones, vesiculoureteral reflux, chronic renal failure, bacteremia, and death.[26,27] However, some studies of catheterized persons have not reported such high prevalence rates of *Providencia* species. Flournoy[28] studied urine isolates from persons in more than 50 different Oklahoma nursing homes during a 3-year period. The frequency of *P. stuartii* in clean-catch (4%–5%) versus catheterized (4%–9%) urines was not dramatically different. Furthermore, urea-hydrolyzing strains of *P. stuartii* were less common in the catheterized group (10%–12%) than in the general population (30%).

Pockets of UTI caused by *P. stuartii* have occasionally been reported.[26,29] In most instances, these clusters were demonstrated to be caused by multiple strains. Environmental sampling of various inanimate surfaces within the hospital failed to detect potential reservoirs for most of these illnesses, fueling speculation that the colonized patient rather than medical staff is the source for most infections. One small outbreak of *P. stuartii* UTI in patients in a Veterans Administration neurology ward was described by Fierer and Ekstrom in 1981.[30] Fourteen patients with neurologic deficits had positive urines for *P. stuartii* over an 8-month period. Infections developed in 3 of these patients, and the remaining 11 were colonized. Only those who subsequently were catheterized (Foley catheter) became infected. Surveillance cultures demonstrated *P. stuartii* in urinary bags (38%) and urinals (29%). Although these fomites were the implicated sources for the ongoing outbreak, person-to-person transmission by hospital personnel could not be excluded entirely. Before infection control procedures were completely implemented, 6 (12%) of 52 patients at risk for development of infection (ie, with condom catheters) became colonized with *P. stuartii* in their urine, including 4 colonized on the perineal skin area. The outbreak was eventually abolished by institution of multiple controls, including the assignment of urinals to only a single patient and the physical separation of infected, colonized, and susceptible persons from each other.

Bacteremia

Most of the present knowledge regarding *Providencia* as bloodborne pathogens stems from a handful of case reports[34,35] and several small clinical series[23,24,36–38] However, Woods and Watanakunakorn[44] reported on a large series of patients with *P. stuartii* bacteremia seen at a community teaching hospital over a 12-year period. As in the case of UTI, there is an extremely low incidence of *Providencia* sepsis in the general population; one study estimates the incidence at 0.17 cases per 1000 admissions.[44] However, in certain high-risk groups (eg, nursing home residents), the relative frequency of septicemia may increase 10- to 50-fold.[33,39,44]

Bacteremia caused by providenciae typically occurs in older men (mean age 62.5 years, male/female ratio 2:1), although septicemia in children as young as 7 years of age has been described (Table 3). Virtually 100% of patients presenting with *Providencia* bacteremia have serious underlying disease. The most common conditions noted to accompany providenciae sepsis are cardiovascular and cerebrovascular problems, malignancy, diabetes, and multiple sclerosis. Bryan[31] identified two major risk factors associated with *Providencia* septicemia, urologic obstruction and paraplegia. A Canadian study found that 50% of patients with providenciae sepsis developed the infection shortly after some urologic manipulation;[38] this observation was confirmed by others.[34–36] Signs of sepsis (fever, chills) often appeared within 2 to 3 hours after termina-

TABLE 3. *Clinical studies of* Providencia *bacteremia*

Study	Date	No. Patients	Mean Age in Years (Range)	Portal of Entry: Urine	Wounds	Other	Mortality
Solberg & Matsen[23]	1971	5	45.8 (7–75)	3	1	1	60%
Milstoc & Steinberg[37]	1973	8	67.6 (36–83)	8	0	0	75%
Prentice & Robinson[38]	1979	6	76.2 (50–85)	6	0	0	33%
McHale et al.[24]	1981	5	65.2 (32–85)	2	3	0	80%
Woods & Watankunakorn[44]	1996	49	75 (29–96)	40	0	9	25%

tion of the urologic manipulation. Procedures linked to the rapid development of *Providencia* septicemia include catheterization, urethral dilation, transurethral resection, and passage of stones.

Providenciae typically are isolated from blood in pure culture, although polymicrobic infections involving gram-positive cocci, gram-negative rods, and anaerobes have been reported,[38,39] and one 1996 study found that 51% of all *P. stuartii* bacteremias involved multiple pathogens.[44] Most of these blood infections arose from primary UTI (see Table 3). In a long-term study of bacteremia in a Veterans Administration medical center, Muder and colleagues[39] found that *P. stuartii* was the most common cause of urosepsis, accounting for 19% of all episodes. Less frequently, the gastrointestinal tract and pulmonary tree have been cited as sites from which dissemination occurred.[36] Regardless of the portal of entry, persons who develop *Providencia* septicemia present with signs of sepsis that are commonly observed with gram-negative bacillary infections. Fever (39.2°C–40.6°C) is the most prominent finding (100%), and leukocytosis (12,200–19,500 cells/mm^3) is often present. Other symptoms frequently include nausea, vomiting, chills, tachycardia, diaphoresis, lethargy, and pallor. Clinical signs of septic shock (blood pressure drop, rapid pulse, pallor, vomiting) are seen in approximately 50% of people with providenciae bacteremia, and the mortality rate in this group approaches 80%. One fatal case of sepsis with septic shock caused by a second episode of *Providencia* septicemia (possibly a relapse) in an 85-year-old man with benign prostatic hypertrophy was described by Prentice and Robinson.[38] The overall mortality rate for all cases of *Providencia* bacteremia is about 67%. This value is probably a reflection of several factors, including the dissemination of multiple antibiotic-resistant strains from urine and the inherent pathogenicity of the genus.

Gastroenteritis

The role of providenciae as etiologic agents of acute gastroenteritis in humans is highly speculative (see Table 2). The early studies of members of the Providence group contain many anecdotal reports implicating these bacteria as potential causes of diarrhea. Most of these reports involve nothing more than the isolation of *Providencia* species from diarrheal stools in conjunction with other genera or paracolon bacteria. In the late 1950s and 1960s, a number of published reports implicated providenciae in acute gastroenteritis based on case-control studies. Many of these investigations originated from the Indian subcontinent.[40]

Renewed interest in *Providencia*-associated gastroenteritis surfaced in the 1980s. Müller analyzed the frequency of *Providencia* species in both healthy subjects and patients with diarrhea and found no significant differences in isolation rates between these two groups.[41,42] In 1989, Haynes and Hawkey[43] found a significantly higher isolation rate for *P. alcalifaciens* in the diarrheal stools of people who had traveled to the Mediterranean area than in those of patients with gastroenteritis who did not travel abroad. In the 1990s, Albert and associates at the International

Centre for Diarrheal Disease Research (Dhaka, Bangladesh) investigated a number of *P. alcalifaciens* strains recovered in pure culture or as predominant growths from children and adults with secretory diarrhea.[45–47] Characterization of these strains led to the identification of potential enteropathogenic mechanisms (see Pathogenesis) which could potentially play a role in vivo.

At present many unanswered questions regarding *P. alcalifaciens* and diarrheal disease remain. Although indirect evidence suggests that *P. alcalifaciens* is associated with a secretory enteritis, there is very little recent information documenting symptoms accompanying bouts of diarrheal disease associated with this species. Furthermore, on many occasions *P. alcalifaciens* has been isolated from diarrheal specimens in conjunction with other recognized or reputed enteric pathogens such as *Campylobacter, Salmonella, Shigella,* and *Aeromonas.*[46] Because most published studies have originated from underdeveloped countries, information on the occurrence of *P. alcalifaciens*-associated gastroenteritis in industrialized nations is lacking. These issues need to be addressed by well controlled epidemiologic investigations specifically targeting this species and its association with acute gastroenteritis.

Miscellaneous Infections

Providencia species occasionally are involved in wound infections that result from surgical procedures or severe burns. Solberg and Matsen[23] described four cases of postoperative wound infection; three patients treated with kanamycin survived, while the fourth developed bronchopneumonia and septicemia and died. In a retrospective study of patients with serious burns at the U.S. Army Institute of Surgical Research in Houston, Texas, *P. stuartii* was identified as the pathogen most frequently associated with fatal episodes of sepsis.[48] Of those patients with *Providencia* sepsis, between 75% and 80% died as a result of their infections. Multiple anatomic sites such as sputum and urine were often found to contain Providence bacilli. Regardless of where the patient was initially treated before transfer to the Institute, the recovery of *P. stuartii* from burn wounds began to increase dramatically 6 days after the initial thermal injury. Local environmental sources at the Institute (eg, water supply, air conditioning systems, hydrotherapy tanks, medical staff) were negative for providenciae, suggesting a general increase in the frequency of these bacteria throughout the environment of many hospitals rather than at one specific medical center. McHale and colleagues[24] found that most of 37 *P. stuartii* wound infections occurred in men and involved multiple pathogens including *Proteus* species and *P. aeruginosa.* Of 37 such infections, 59% were surgery-related and the remainder were associated with thermal injuries.

Respiratory infections are infrequently caused by *Providencia* bacilli. In an 18-month study at one Belgian hospital, tracheal secretions or sputum was the primary site of isolation for 15 of 19 *Providencia* strains.[36] Most of these isolates were recovered from neurosurgical patients who had undergone a tracheotomy and who were receiving aminoglycoside therapy. Two of these patients died from fulminant bronchopulmonary disease. The study also noted that those who developed nosocomial *Providencia* infections had been hospitalized for prolonged periods (mean 54.2 days). Other studies have infrequently reported providenciae as causes of lobar pneumonia or bronchopneumonia.[23] Miscellaneous sites from which *Providencia* species were recovered included ulcerated tumors of the skin, rectal abscesses, prostatic fluid, a suprapubic abscess, and pericardium at autopsy.[16,36] Details on these isolates are not available.

LABORATORY IDENTIFICATION

Isolation

Growth of *Providencia* species is supported on most enteric media, including MacConkey, Salmonella-Shigella, and eosin–methylene blue agars. On most media, *Providencia* phenotypically resemble either *Salmonella* or *Shigella,* since they are lactose-negative. Providenciae also exhibit good growth on bismuth sulfite agar, appearing as colorless to olive green colonies with or without a metallic sheen.[16] Growth on desoxycholate agar is variable, with only 40% to 50% of *P. stuartii* and *P. rettgeri* strains able to grow.[49] Brilliant green agar is highly inhibitory.[16] On common protein-based nonselective agars (eg, nutrient, trypticase soy), *Providencia* appear as smooth, white colonies with an entire edge.[49] Often on nonselective and selective media *Providencia* strains produce distinct colonial morphologies (Fig. 1). The most common variations involve colonial transformations from opaque to translucent or smooth to rough after multiple laboratory passages.[16] On media containing L-phenylalanine, an almond-like odor may be present.[41] In liquid media, providenciae often form pellicles (60%–90%) without evidence of granularity.[49]

Providencia species have been reported to swarm on a variety of media. McKell and Jones[49] found that 14% to 22% of *P. stuartii* and 22% to 36% of *P. rettgeri* strains swarmed on nutrient and MacConkey agars. Rauss[4] noted that 60% of providenciae could be induced to swarm when strains were incubated at lower temperatures (30°C) or in media with reduced agar concentrations (1.3%). However, under these conditions the swarming phenomenon appears to involve pseudopodial outgrowths (emanations, arborizations) and not the classic swarming with concentric ring formation exhibited by *Proteus* species.[50,51] Swarming by *Providencia* on tryptone plates containing 0.25% agar can be inhibited by addition of 15 to 20 mmol/L sodium salicylate.[52]

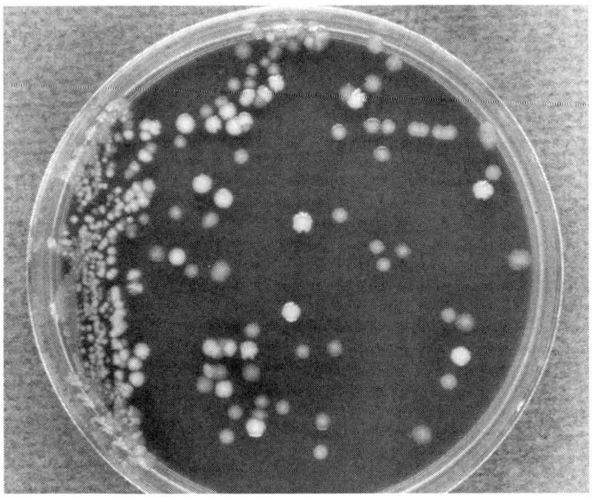

FIG. 1. *Providencia* isolate on heart infusion agar displaying opaque and translucent colony types.

Two types of selective media which isolate providenciae have been described. The first involves media which are not specifically designed for *Providencia* but rather take advantage of unusual features or characteristics of the genus that are present in only a few enteric bacteria. Because members of the tribe *Proteeae* (*Proteus, Morganella,* and *Providencia*) produce a brownish pigment in the presence of aromatic amino acids and also degrade L-tyrosine crystals, media incorporating these compounds have been developed to aid in the isolation of providenciae from contaminated specimens. Müller[41] examined isolations of the *Morganella-Proteus-Providencia* group from more than 2600 fecal specimens on a variety of differential media. Overall, 361 strains were recovered, including 13 *Providencia* isolates. Media containing sheep blood agar and/or tryptophan agar were superior to proteose peptone-based agars with either L-tyrosine or L-phenylalanine for the isolation of all proteeae, including *Providencia.*[41] A modification of this medium, called *Proteeae* isolation agar, contains both L-tyrosine and DL-tryptophan.[53] Other media which have been proposed for the recovery of aminoglycoside-resistant strains include MacConkey agar with 5 μg/mL of gentamicin.[54]

The second type of selective medium has been formulated primarily for the recovery of *P. stuartii,* the species most frequently isolated from clinical material. Stickler and colleagues[22] used CI agar (100 μg/mL of colistin and 1% inositol in nutrient agar with bromthymol blue) to recover *P. stuartii* from feces, urine, and the skin of persons with long-term indwelling catheters and from environmental hospital samples. On CI agar, *P. stuartii* appears as large, yellow (inositol-positive) colonies. In 1992, Thaller and associates[55] developed a modified MacConkey phosphatase medium (MCP), which detects phosphatase activity by the incorporation of the substrate phenolphthalein diphosphate and methyl blue into MacConkey agar with crystal violet. *P. stuartii,* one of only a handful of enteric bacteria (except *Morganella morganii*) with high cell-associated and extracellular phosphatase activity,[56] produces red colonies (phosphatase-positive, lactose-negative) on MCP. In contrast, lactose-positive and phosphate-negative organisms appear violet, and lactose- and phosphatase-negative genera are colorless, allowing for the easy differentiation of *P. stuartii–M. morganii* from all other common enteric groups. The newest selective agar for providenciae was designed by Senior[57] to recover *P. alcalifaciens* from feces based on its reputed enteropathogenicity. Designated PAM (*P. alcalifaciens* medium), the medium relies on the inability of *P. alcalifaciens* to ferment any of its three sugars (xylose, galactose, and mannitol). Fecal isolates of *P. alcalifaciens* subcultured from tetrathionate broth to PAM appear as red colonies after overnight incubation; virtually all other oxidase-negative enteric bacteria appear lemon-yellow.

Little work has been performed on enrichment procedures. Most *Providencia* grow in selenite (90%–100%) and in tetrathionate broths.[49] Hawkey and coinvestigators[54] discovered that the recovery rate of *P. stuartii* from gastrointestinal samples using a semiselective medium could be enhanced by preincubation of feces in heart infusion broth. Such an enrichment caused initial *P. stuartii* concentrations as low as 4 cfu/mL to reach levels in excess of 10^7 cfu/mL after an 18-hour incubation at 37°C.

Identification

Commercial Systems

Table 4 lists the cumulative results for testing of *Providencia* members on manual and automated microidentification panels. A majority of isolates of most *Providencia* species are correctly identified by commercial systems.[58–69] When misidentifications occur, they most commonly result in correct species identification with low probability, correct genus identification only, or a final report of no identification. When *Providencia* are completely misidentified (wrong genus and species), they are most often reported as other members of the tribe *Proteeae* (ie, *Proteus* or *Morganella*). The lower values for correct identification of *P. rettgeri* strains by the MicroScan W/A system in Table 4 are based largely on a single CDC study in which only two of six strains were correctly identified in 24 hours[63]; further information on misidentifications was not provided. For *P. stuartii,* most panels report slightly lower correct iden-

TABLE 4. *Identification of* Providencia *species by commercial systems*

		P. alcalifaciens		*P. rettgeri*		*P. stuartii*	
System	Read Time (h)	No.	CI (%)	No.	CI (%)	No.	CI (%)
Biolog[a]	24	17	13 (76)	20	19 (95)	19	16 (84)
MicroScan W/A	2	10	10 (100)	19	14 (74)	48	45 (94)
Vitek	4–18	7	6 (86)	21	21 (100)	23	21 (91)
API 20E	24	16	14 (88)	22	22 (100)	40	33 (83)
Cobas	21	6	5 (83)	6	6 (100)	5	4 (80)
Crystal E/NF	18–20	6	6 (100)	13	13 (100)	27	26 (96)
Crystal RS/E	3	6	6 (100)	5	3 (60)	5	5 (100)
RapID onE	4	4	4 (100)	10	10 (100)	15	15 (100)

Data from references 58–69.
[a]Automated reading only.
CI, correct identification.

tification values. This decrease in accuracy may be a reflection of the difficulty most systems have in correctly identifying urea-positive isolates. O'Hara and colleagues[60] found that only one of three urea-positive *P. stuartii* strains was correctly identified to species by the API 20E system in 24 hours; the remaining two strains required additional tests and further incubation. Other systems such as Vitek and Cobas have also misidentified urea-positive *P. stuartii* as *P. rettgeri* or reported them as unidentified bacteria.[63] Only rarely have *P. rustigianii* strains been tested on commercial panels.[58,63] In most instances, these isolates are identified as *P. alcalifaciens.*

One study investigated how well several systems commonly used in Europe identified *Providencia* species. As in the U.S. studies, all five systems evaluated (API 20E, Enterotube II, Cobas, Sceptor, Titertek-Enterobac-Rapid Automated System) were extremely accurate in the identification of *P. alcalifaciens* and *P. rettgeri*[70]; only the Enterotube II required the use of additional biochemical tests to accurately identify about 25% of *P. alcalifaciens* to species. However, when 80 *P. stuartii* isolates were tested on these same five systems, incorrect identification rates ranged from a low of 1.3% (Titertek) to a high of 16.3% (API 20E). Furthermore, most panels required supplementary tests to correctly identify *P. stuartii* to species (19%–51% of all isolates).

Conventional Tests

Members of the genus *Providencia* are defined as indole-positive bacilli that produce phenylalanine deaminase (PDA), are positive on Simmons' citrate agar, and ferment one or more of several alcoholic sugars including *m*-inositol.[71] Although the common *Providencia* species (*P. alcalifaciens, P. rettgeri,* and *P. stuartii*) are generally considered to be motile, strains recovered on primary isolation often appear nonmotile or sluggishly motile on semisolid media. All *Providencia* species are lysine-, ornithine decarboxylase (ODC)-, and arginine dihydrolase-negative. On triple sugar iron (TSI) slants, *Providencia* are alkaline over acid (K/A) with or without the evolution of gas. H_2S is not produced. Gelatinase, lipase, and hemolysin activities are not expressed.[71,72]

In addition to PDA activity, a number of other unique properties exclusively associated with the tribe *Proteeae* are found in the genus *Providencia.* On lysine-iron agar (LIA), only *Providencia* and *Proteus* species produce a red slant and yellow butt. The red slant, indicative of lysine deamination, is a trait found only in these two genera among members of the *Enterobacteriaceae.* The red slant results from oxidative deamination of lysine, which yields keto acids.[73] Keto acids then blend with ferric chloride, producing an orange color which mixes with the brom cresol purple indicator in LIA to produce a red slant (Fig. 2; Color Plate 21). Most *Providencia* produce a red slant on LIA, but only about 64% of *P. rustigianii* isolates appear to do so.[11] All other *Enterobacteriaceae,* including *Morganella,* produce a purple slant on LIA. *Providencia* also excrete a reddish-brown pigment on basal media into which 0.5% to 1.0% DL-tryptophan has been added.[74] Virtually all providenciae tested produce this pigment.[53] Some studies indicate that L-phenylalanine can be substituted for DL-tryptophan, but this amino

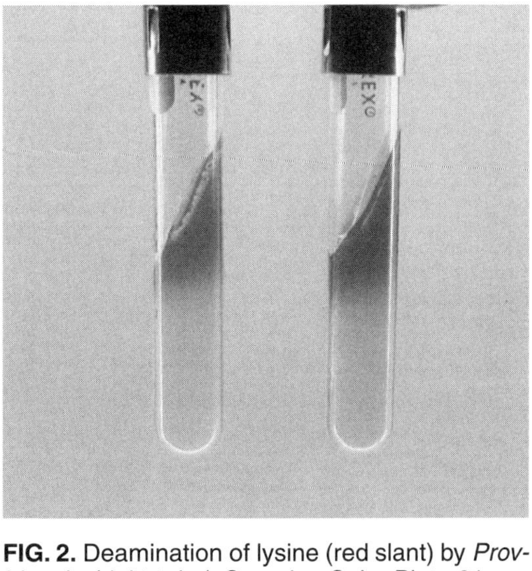

FIG. 2. Deamination of lysine (red slant) by *Providencia (right tube).* See also Color Plate 21.

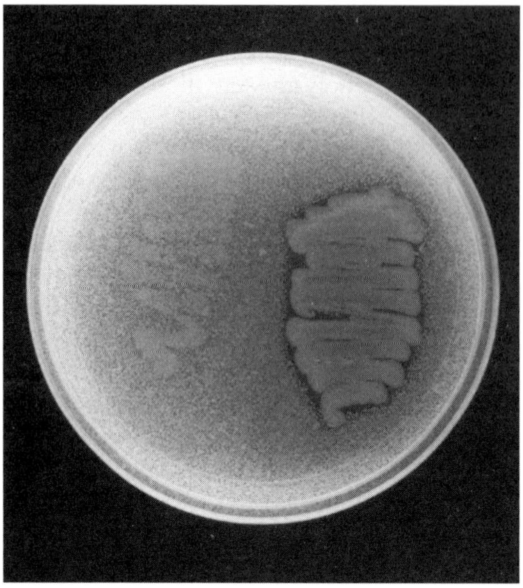

FIG. 3. Degradation of L-tyrosine crystals by *Providencia.*

acid appears to work less well.[41] A final unusual attribute of the proteeae and some citrobacters is the ability to degrade L-tyrosine crystals, presumably through a tyrosine lyase.[75] On nutrient or trypticase soy agars containing 0.4% to 0.5% L-tyrosine (insoluble), providenciae (including *P. rustigianii*) elaborate a tyrosinase that dissolves these crystals (Fig. 3). A positive reaction is indicated by the appearance of a clear zone surrounding growth after a 24- to 48-hour incubation.[53] These tests are extremely useful when aberrant *Providencia* strains (eg, PDA$^-$) are encountered and aid in their correct genus identification.

Providencia can be identified to species based on fermentation of selected carbohydrates (hexoses, disaccharides, alcoholic sugars), urea hydrolysis, and utilization of sodium citrate (Table 5). Fatty acid methyl ester (FAME) analysis of the cellular fatty acid composition of *P. rettgeri* and *P. stuartii* as determined by gas chromatography, while distinct from *Morganella* strains, is very similar in overall composition to *P. mirabilis.*[76] However, *P. alcalifaciens* can be differentiated from *P. stuartii* and *P. rettgeri* based on an elevated hexadecenoic acid ($C_{16:1}$) peak and the lack of detectable methylenehexadecanoic acid ($C_{17\nabla}$). The lipopolysaccharide fatty acid composition of each *Providencia* species is again similar, although an elevated hexadecanoic acid peak separates *P. alcalifaciens* from the other two species.

On occasion, problems can arise in the identification of *Providencia* isolates to species because of unusual biochemical reactions. Fischer and coworkers[77] found 10 inositol-positive *Providencia* strains that had phenotypic properties intermediate between those of *P. stuartii* and *P. rettgeri.* Although these strains fermented trehalose, suggesting *P. stuartii,* they also attacked D-arabitol and *m*-erythritol, which resembles *P. rettgeri.* Serogrouping of 9 of these 10 isolates indicated that they possessed O antigens charac-

TABLE 5. *Identification of indole-positive* Providencia *species*[a]

Character	*P. alcalifaciens*	*P. rettgeri*	*P. rustigianii*	*P. stuartii*
Simmons' citrate	+	+	V	+
Urea hydrolysis	–	+	–	V
Decarboxylation of L-glutamic acid	–	+	ND	–
Fermentation of				
D-Galactose	–	+	+	+
Trehalose	–	–	–	+
D-Mannitol	–	+	–	–
Adonitol	+	+	–	–
D-Arabitol	–	+	–	–
m-Inositol	–	+	–	+

Data from Hickman-Brenner et al.,[11] Fischer et al.,[77] and Farmer et al.[78]
[a]*P. heimbachae* is indole-, citrate-, and urea-negative.
V, variable; ND, no data.

teristic of *P. stuartii.* DNA–DNA homology studies on 5 of these 10 strains confirmed serogrouping results. The authors suggested that both trehalose fermentation and glutamic acid decarboxylation (by thin-layer chromatography) were useful adjunct tests for correct species assignment of biochemically aberrant strains of *P. stuartii.* A second biochemically unusual group involves urea-positive strains of *P. stuartii.* Approximately 30% of all *P. stuartii* are urease-positive,[78] and some anecdotal reports indicate that this phenotype is unstable in this species. A 1981 Canadian study found that urease activity in two strains of *P. stuartii* could be transferred to *Escherichia coli* with the aid of helper plasmids.[79] A third *P. stuartii* strain transferred urease activity and ampicillin resistance directly to *E. coli,* suggesting that the two markers were closely linked on a plasmid of about 92 MDa. Later studies by Mobley and associates[80] confirmed these observations and extended the number of transferable plasmid-encoded traits to include fermentation of sucrose (plasmid of ~54 MDa) and lactose (plasmid of ~100 MDa). The plasmid-encoded urease gene of *P. stuartii* has a high degree of DNA homology and similar restriction endonuclease cleavage sites with urease genes found in *E. coli* and *Salmonella* Cubana[81]; all are closely related to the urease genes found in *P. mirabilis.* Currently it is unknown whether all such phenotypes are carried on conjugative extrachromosomal elements or whether some may be chromosomally determined. Other unusual *Providencia* phenotypes described in the literature include PDA$^-$ *P. stuartii* and adonitol-negative *P. rettgeri.*[77]

Molecular Identification

Because of the difficulty of accurate species identification of some *P. stuartii* isolates.[70,77,79,80] Thaller and colleagues[82] developed a species-specific molecular probe for this group. The probe, an 0.4-kb DNA fragment, is derived from a *Hind*III digest of the *phoN* gene which codes for a nonspecific acid phosphatase. In slot blot hybridizations performed under high stringency conditions, only *P. stuartii* strains react. Under more relaxed conditions, both *M. morganii* and *Rahnella aquatilis* produce weak signals. The probe was also tested against 30 *P. stuartii* strains in filter hybridization assays, and the results were comparable to those obtained in slot blot experiments. These findings suggest a potentially useful probe that may aid in the accurate species identification of aberrant *P. stuartii.*

Biotypes

Biotypes (biogroups) exist within each of the three major *Providencia* species. The importance of most of these groups is of minimal clinical significance. *P. alcalifaciens* was originally composed of four biogroups,[73] two of which were later reclassified as *P. rustigianii* (biogroup 3) and *P. stuartii* (biogroup 4). The two remaining biogroups are distinguished only by their ability to produce gas from glucose, with biogroup 1 being aerogenic. At present, the only potential relevance these biogroups may have is their possible association with new species within *P. alcalifaciens.* In the study of Picard and associates,[15] all six biogroup 2 strains were clustered within one genetic taxon based on DNA–DNA reassociation kinetics, ribotypes, and zymotypes; four of seven biogroup 1 strains were linked to the other genomovar. In 1975, Penner and collaborators[83] proposed the recognition of five *P. rettgeri* biogroups based on the analysis of 674 strains. These biogroups (excluding biogroup 5) could be separated from one another by fermentation of L-rhamnose and salicin. Groups could be further subdivided based on whether *m*-erythritol was fermented. Biogroup 5 (urea-positive) has since been determined to reside within *P. stuartii* by Brenner and colleagues.[10] For *P. stuartii,* at least three biogroups exist based on adonitol fermentation and urease activity.[10,11,73]

Immunodiagnostics

Providencia species produce somatic antigens similar to those found in other genera within the *Enterobacteriaceae.* Penner[84,85] has reclassified the O-serogrouping scheme for *Providencia* to include more than 100 distinct serogroups consisting of *P. alcalifaciens* (46 O antigens), *P. stuartii* (17 O antigens), and *P. rettgeri* (84 O antigens). Studies on the human immune response to *Providencia* infection have not been published to date.

Typing

Outbreaks of disease caused by *Providencia* species, almost invariably by *P. stuartii,* have been reported rarely.[29,30] Because of this fact, extensive investigations aimed at developing suitable typing systems for these bacteria have not been undertaken. Traditional typing systems previously used for *P. stuartii* include biotyping, antibiograms, and serogrouping.[29,82] Biotyping, although useful with some enteric bacteria, has a number of drawbacks with *P. stuartii.* Few markers are available that express sufficient variability to make them helpful. Several traits, such as urease activity and sucrose fermentation, are sometimes unstable because they can be plasmid mediated, thus making their use unreliable (see Biotypes). Antibiograms can be a first step in determining strain relatedness.[26,29] However, susceptibility patterns of the same strain can vary over time, particularly in patients with chronic UTIs who receive multidrug regimens. An additional concern in regard to antibiograms is the fact that some resistance markers are plasmid-mediated. Serogrouping is an excellent technique with high resolution in determining strain identity. However, its limited availability (reference centers) precludes its general use by most laboratories.

Several molecular strategies have been investigated for their potential use in typing of *P. stuartii* strains. Hawkey and colleagues[86] looked at plasmid carriage in 123 *P. stuartii* isolates recovered from male patients on a geriatric ward. Of the 123 isolates screened, 40 (33%) contained a single cryptic plasmid of about 24 MDa. When this plasmid was cut by restriction endonucleases (*Eco*R1, *Hin*CII), two separate fragment patterns were observed. Contour length studies of these molecules indicated two distinct plasmids of 24 MDa and 22 MDa which were not distinguishable by agarose gel electrophoresis. This suggests that plasmid analysis as a typing tool for *P. stuartii* may be hampered by the large number of plasmid-free strains and by the inability of agarose gel electrophoresis to adequately distinguish among several extrachromosomal elements of similar mass. A 1994 investigation employed restriction fragment length polymorphism (RFLP) as a molecular technique to study successive *P. stuartii* urinary isolates recovered from persons with indwelling catheters.[26] Using a 16S/23S ribosomal RNA (rRNA) probe, RFLP patterns of successive isolates from the same patient over an 18-month period remained constant even though susceptibility profiles changed. Therefore, whereas antibiograms incorrectly suggested that each patient acquired two or three different strains during the survey period, RFLP proved that each person was persistently colonized by a single strain. Other RFLP studies in which a DNA probe was used to detect rRNA cistrons in *P. stuartii* have been described.[87] Using a biotinylated cDNA probe to 16S and 23S rRNA, the authors were able to identify 11 separate *Eco*R1 patterns among 23 *P. stuartii* strains by RFLP analysis.[88] *Hind*III, *Bam*H1, and *Pvu*II digests were less useful. These collective results indicate that RFLP analysis is a promising technique for typing of *P. stuartii* strains, although a larger number of isolates must be evaluated.

PATHOGENESIS

The roles played by providenciae in human infections have been investigated along several lines, including the relative virulence of strains in animal models and the relation between specific species and organotrophic disease. Several early studies looked at the pathogenic potential of Providence bacilli in Swiss-Webster mice. Solberg and Matsen[23] found that the median lethal dose (LD_{50}) for seven *Providencia* strains injected with 5% hog gastric mucin into Swiss-Webster mice ranged from 2.9×10^2 to

7.4×10^7 cfu. This wide variation in mouse pathogenicity was confirmed by Janis and coinvestigators,[34] who found that seven other *Providencia* strains yielded LD_{50} values from 10 to 6.1×10^6 cfu under comparable conditions. In the latter study, two segregating colonial morphotypes from a patient with fatal septicemia had dramatically different LD_{50} values (>100-fold difference) in mice. In both studies, isolates recovered from invasive infections (bacteremia, pneumonia) were more likely to have low LD_{50} values than isolates recovered from UTIs.

The strong association of *P. stuartii* with bacteriuria in persons with long-standing catheterization has been addressed in several articles. Strains recovered from nursing home residents attached strongly to uroepithelial cells, but less so if repeatedly washed, suggesting that cell surface structures are involved in the adherence process.[89] Follow-up investigations led to the discovery that strains that express the mannose-resistant/*Klebsiella*-like hemagglutinin (MR/K, type 3) are more likely to be associated with long-term bacteriuric episodes than short-term ones.[90] In samples from 27 persons with long-term bacteriuric disease (>12 weeks), more than 50% of all *P. stuartii* isolates recovered were MR/K-positive, compared with only 21% of isolates from short-term (1-week) episodes. A model to study the relation between catheterization and polymicrobic bacteriuria *(P. stuartii, E. coli)* has been developed using artificial urine medium.[91] In such a system, *P. stuartii* apparently enhances the growth in urine of *E. coli* while inhibiting its ability to adhere to catheters. Reasons for such paradoxical effects are not well understood. *Providencia* spp. also produce fimbriae.[92] At least six different types have been described, but their role in pathogenesis is unknown.

Providenciae have been implicated as potential pathogens in diarrheal disease for several decades.[84] Although the species implicated in many of these gastroenteritis cases is not always clear, *P. alcalifaciens* is the agent most commonly reported to be involved in such infections. Until recently, most of the evidence supporting such a role for *Providencia* was highly speculative, relying on individual case reports and limited case controlled studies. In 1992, however, Albert and colleagues, in a collaborative effort,[45] provided the first direct evidence associating *P. alcalifaciens* with infectious gastroenteritis. Three strains of *P. alcalifaciens,* obtained from a Bangladeshi child with fulminant diarrhea and two Australian adults with diarrhea, were found to be invasive in HEp-2 cell assays. Although none of these strains produced enterotoxins or cytotoxins, whole-cell cultures of each strain inoculated into the Removable Intestinal Tie Adult Rabbit Diarrhea (RITARD) model produced diarrhea in 16 (76%) of 21 rabbits. Microbiologic analysis of the gastrointestinal tract of diarrheagenic rabbits showed that the lower ileum was the major site of bacterial multiplication, and histologic studies found that the ileum and cecum were the primary locations of bacterial attachment and invasion. Electron microscopic studies of ileal tissue from infected animals demonstrated two modes of entry, one by endocytosis and the other by disruption of tight junctions with subsequent multiplication in intracellular spaces.[47] An analysis of 14 additional *P. alcalifaciens* strains from diarrheal stools indicated that all isolates were invasive in HEp-2 cell monolayers, although the degree of invasiveness varied significantly from strain to strain.[46] These studies were confirmed by Guth and Perella.[93] Although *P. alcalifaciens* strains are invasive, they do not appear to survive long in HEp-2 cells, since viable numbers decrease within 4 to 6 hours after infection. One strain of *P. stuartii* fed in high concentrations (~10^{10} cfu) to newborn calves caused diarrhea in 12 (67%) of 18 animals.[21] Mice fed *Providencia* failed to develop diarrhea.[40]

Several other virulence-associated properties in *Providencia* have been described. Invasive *P. alcalifaciens* strains are positive in the fluorescent actin test,[45,46] although the type of fluorescence observed is different from that seen with enteropathogenic *E. coli* (EPEC). The significance of this observation is unknown. Nawrot and associates[94] evaluated the serum susceptibilities of four *P. rettgeri* strains. All four isolates were susceptible to 50% normal serum. *P. rettgeri* strains were killed by several different serum factors, including both the classic and alternative complement pathways and lysozyme. Other *Providencia* species were not investigated. Most strains of *Providencia* also secrete keto acids. These α-keto acids produced by amino acid deaminases possess significant siderophore activity.[95] Their biologic role is unknown.

ANTIMICROBIAL SUSCEPTIBILITY

Providencia as a group are extremely resistant microorganisms. They are often refractory to treatment with commonly administered antibiotics, disinfectants (chlorhexidine), and topical heavy metal–containing medications such as silver sulfadiazine.[96] Antibiotics to which providenciae often exhibit high-level resistance include penicillin G,[97] ampicillin,[24,96,97] tetracycline,[23,96,97] chloramphenicol,[96] nalidixic acid,[24,37] colistin,[23,24] polymyxin B,[23,96] and nitrofurantoin.[23,96]

Resistance mechanisms in providenciae are varied. Because most clinical isolates originate from urine, they are often exposed to elevated concentrations of chemotherapeutic agents. In persons with long-term catheterization, this can lead to the in vivo development of resistance which can then be spread to other susceptible individuals. In vivo resistance presumably is acquired through mutations that lead to alterations in membrane permeability to drugs or induction of antibiotic-modifying enzymes. *P. stuartii,* a particularly troublesome member of the genus, produces several β-lactamases which make many cephalosporins ineffective. Aminoglycoside resistance in *P. stuartii* is most often the result of antibiotic-inactivating enzymes such as aminoglycoside-*N*-acetyltransferase[96] and aminoglycoside-3′-*O*-phosphotransferase type VI.[98] Plasmids carrying transferable resistance markers to ampicillin, carbenicillin, kanamycin, and amikacin have been reported on a number of occasions.[80,97,98]

The literature regarding the antimicrobial susceptibility patterns of providenciae is somewhat confusing. Many surveys have chosen to report all strains under the heading of "*Providencia* spp." rather than as individual species. This is unfortunate because most indole-positive members of the tribe *Proteeae* (including *Providencia*) are inherently more resistant to antibiotics than are indole-negative species (Table 6). Even when *Providencia* isolates are identified to species, *P. alcalifaciens* is often omitted, and there are surprisingly few susceptibility studies in which data on all the common species are reported.[100,104]

Of the three major *Providencia* species, *P. stuartii* is the most resistant to antimicrobial agents and *P. alcalifaciens* is the most susceptible. Penner and associates[107] used three antibiotics, cephalothin (16 μg/mL), cefamandole (8 μg/mL), and gentamicin (1 μg/mL), as screening drugs for species assignment of *Providencia* isolates. Most *P. alcalifaciens* (SSS), *P. rettgeri* (RSS), and *P. stuartii* (RSR) could easily be separated by this method from one another and from phenotypically similar genera *(Proteus, Morganella)*. Table 6 summarizes the cumulative data on the susceptibility of *Providencia* species to agents active against this group. Based on 90% minimal inhibitory concentration (MIC_{90}) data, the results confirm previous observations regarding the relative resistance of *P. stuartii* to many antibiotics. Although second-generation cephalosporins and extended-spectrum penicillins are active against *P. alcalifaciens* and most *P. rettgeri,* many *P. stuartii* are typically resistant. Biedenbach and Jones[108] found an unacceptably high number of very major errors (false reports of susceptibility) when *Providencia* were tested against certain antimicrobial agents by disk diffusion. These discrepancies included second-generation cephalosporins (cefprozil) and carbacephems (loracarbef). The authors recommended that (as with loracarbef) warnings be placed in the footnote section of the National Committee for Clinical Laboratory Standards table regarding this problem. For third-generation cephalosporins good activity against all three species has

TABLE 6. *Susceptibility of* Providencia *species to selected antimicrobial agents (in µm/mL)*

	P. alcalifaciens		*P. rettgeri*		*P. stuartii*	
Antibiotic	MIC_{50}	MIC_{90}	MIC_{50}	MIC_{90}	MIC_{50}	MIC_{90}
Cefuroxime	0.1	0.4	0.4	12.5	1.6–32	12.5–>128
Cefoxitin	1.6	3.1	3.1	50	3.1	50
Cefotaxime	ND	ND	0.06	≤0.25–0.5	≤0.25–0.25	1–4
Ceftazidime	0.05	0.1	0.05	0.2–0.5	0.05	0.1–4
Ceftriaxone	<0.012	0.05	<0.012	0.2	0.1	0.4
Carbenicillin	≤0.5–1.6	12.5–16	≤0.5–0.8	1–3.1	2–3.1	50–≥256
Piperacillin	≤0.5–0.8	6.25–8	0.4	≤0.5–3.1	6.3–8	12.5–≥256
Ticarcillin	≤0.5	64	≤0.5	1	1	≥256
Gentamicin	ND	ND	1	4	32	128
Tobramycin	0.2	1.6	0.4	3.1	6.3	25
Amikacin	0.8	1.6	0.4	1.6	0.8	3.1
Ciprofloxacin	0.25	2	≤0.06	0.06–1	0.25–4	2–64
Ofloxacin	0.5	2	0.25	0.25–2	0.5–1	2–>8
Aztreonam	0.4	3.1	0.4	≤0.12–6.3	0.8	≤0.12–3.1
Imipenem	0.8	3.1	1–1.6	2–12.5	0.4–1	2–3.1

Data from references 99–106.
ND, no data.

been found. Cefpirome, the first fourth-generation cephalosporin, also exhibits excellent activity against *Providencia,* with low MIC_{50} (≤0.12 µg/mL) and MIC_{90} (1 µg/mL) values.[109] Other agents active against all major *Providencia* species include amikacin, aztreonam, and imipenem. A disturbing observation concerns the wide variation in *P. stuartii* susceptibility to ciprofloxacin reported in the literature (Table 7). Although some of the differences in susceptibility may be related to the source of strains tested (eg, institutionalized elderly patients), in other investigations the reasons for an unusually low incidence of susceptible strains is not clear.[104,105] Further studies with *P. stuartii* recovered from different sources and tested by various methods for ciprofloxacin susceptibility should be undertaken to clarify this issue.

CONCLUSIONS

The genus *Providencia* is an important human pathogen, typified by the associations of *P. stuartii* with bacteriuria in nursing home residents, with long-term catheterization, and with drug resistance. Reports that *P. alcalifaciens* possesses an enteropathogenic mechanism (invasion) that may be operative in diarrheal disease have yet to be independently confirmed. If data are provided to support these findings, studies on the frequency, clinical symptoms, and duration of diarrheal disease associated with *P. alcalifaciens* infection need to be started, as do case-control investigations. Despite the fact that it is one of the oldest legitimate bacterial species in the scientific literature, very little is known regarding the role of *P. rettgeri* in human infections.

TABLE 7. *Susceptibility of* Providencia stuartii *to ciprofloxacin*

				P. stuartii		
Study	Date	Assay Method	No. Isolates	MIC_{50}	MIC_{90}	Susceptible (%)
Fernandes & Ackerman[103]	1990	Agar dilution	14	0.25	2	93
Chin et al.[110]	1991	Agar dilution	15	0.06	0.25	100
Thornsberry[111]	1994	Varied	536	1	1	91
Prosser & Beskid[105]	1995	MicroScan	169	0.25	>8	64
Cornaglia et al.[104]	1995	Agar dilution	202	4	64	33

Other laboratory issues that seem to warrant further study include the development of better commercial and conventional tests to identify *P. stuartii* and more comprehensive investigations of the susceptibility profiles of the major *Providencia* species to newer antimicrobial agents including the quinolones.

REFERENCES

1. Stuart CA, Wheeler KM, Rustigian R, Zimmerman A. Biochemical and antigenic relationships of the paracolon bacteria. *J Bacteriol* 1943;45:101–119.
2. Stuart CA, Wheeler KM, McGann V. Further studies on one anaerogenic paracolon organism, type 29911. *J Bacteriol* 1946;52:431–438.
3. Coetzee JN. Genetics of the proteus group. *Ann Rev Microbiol* 1972;28:23–54.
4. Rauss K. A proposal for the nomenclature and classification of the proteus and providencia groups. *Int Bull Bacteriol Nomen Taxon* 1962;12:53–64.
5. Kauffmann F, Edwards PR. Classification and nomenclature of *Enterobacteriaceae. Int Bull Bacteriol Nomen Taxon* 1952;2:2–8.
6. Proom H. Amine production and nutrition in the Providence group. *J Gen Microbiol* 1955;13:170–175.
7. Ewing WH. The tribe *Proteeae:* Its nomenclature and taxonomy. *Int Bull Bacteriol Nomen Taxon* 1962;12: 93–102.
8. Ursing J. Biochemical study of *Proteus inconstans (Providencia). Acta Path Microbiol Scand B* 1974;82: 527–532.
9. Rustigian R, Stuart CA. Taxonomic relationships in the genus *Proteus. Proc Soc Exper Biol Med* 1943;47: 108–112.
10. Brenner DJ, Farmer JJ III, Fanning GR, et al. Deoxyribonucleic acid relatedness of *Proteus* and *Providencia* species. *Int J Syst Bacteriol* 1978;28:269–282.
11. Hickman-Brenner FW, Farmer JJ III, Steigerwalt AG, Brenner DJ. *Providencia rustigianii:* A new species in the family *Enterobacteriaceae* formerly known as *Providencia alcalifaciens* biogroup 3. *J Clin Microbiol* 1983;17:1057–1060.
12. Müller HE. *Providencia friedericiana:* A new species isolated from penguins. *Int J Syst Bacteriol* 1983;33: 709–715.
13. Hickman-Brenner FW, Fanning GR, Müller HE, Brenner DJ. Priority of *Providencia rustigianii* Hickman-Brenner, Farmer, Steigerwalt, and Brenner 1983 over *Providencia friedericiana* Müller 1983. *Int J Syst Bacteriol* 1986;36:565.
14. Müller HE, O'Hara CM, Fanning GR, Hickman-Brenner FW, Swenson JM, Brenner DJ. *Providencia heimbachae,* a new species of *Enterobacteriaceae* isolated from animals. *Int J Syst Bacteriol* 1986;36:252–256.
15. Picard B, Picard-Pasquier N, Krishnamoorthy R, Goullet P. Correlation between DNA polymorphism argues in favour of the delineation of two species within *Providencia alcalifaciens. Res Microbiol* 1991;142: 965–969.
16. Ewing WH, Tanner KE, Dennard DA. The Providence group: An intermediate group of enteric bacteria. *J Infect Dis* 1954;94:134–140.
17. Youssef H, El-Timmawy AK, Ahmed S. Role of aerobic intestinal pathogens of fresh water fish in transmission of human diseases. *J Food Protect* 1992;55:739–740.
18. Jackson TJ, Wang H, Nugent MJ, Griffin CT, Burnell AM, Dowds BCA. Isolation of insect pathogenic bacteria, *Providencia rettgeri,* from *Heterorhabditis* spp. *J Appl Bacteriol* 1995;78:237–244.
19. Costas M, Holmes B, Wood AC, On SLW. Numerical analysis of electrophoretic protein patterns of *Providencia rettgeri* strains from human faeces, urine and other specimens. *J Appl Bacteriol* 1989;67:441–452.
20. Holmes B, Costas M, Sloss LL. Numerical analysis of electrophoretic protein patterns of *Providencia alcalifaciens* strains from human faeces and veterinary specimens. *J Appl Bacteriol* 1988;64:27–35.
21. Waldhalm DG, Meinershagen WA, Frank FW. *Providencia stuartii* as an etiologic agent in neonatal diarrhea in calves. *Am J Vet Res* 1969;30:1573–1575.
22. Stickler DJ, Fawcett C, Chawla JC. *Providencia stuartii:* A search for its natural habitat. *J Hosp Infect* 1985; 6:221–223.
23. Solberg CO, Matsen JM. Infections with Providence bacilli: A clinical and bacteriologic study. *Am J Med* 1971;50:241–246.
24. McHale PJ, Walker F, Scully B, English L, Keane CT. *Providencia stuartii* infections: A review of 117 cases over an eight year period. *J Hosp Infect* 1981;2: 155–165.
25. Warren JW. *Providencia stuartii:* A common cause of antibiotic-resistant bacteriuria in patients with long-term indwelling catheters. *Rev Infect Dis* 1986;8: 61–67.
26. Rahav G, Pinco E, Silbaq F, Bercovier H. Molecular epidemiology of catheter-associated bacteriuria in nursing home patients. *J Clin Microbiol* 1994;32: 1031–1034.
27. Warren JW, Tenney JH, Hoopes JM, Muncie HL, Anthony WC. A prospective microbiologic study of bacteriuria in patients with chronic indwelling urethral catheters. *J Infect Dis* 1982;146:719–723.
28. Flournoy DJ. Incidence of bacteria from clean-catch and catheterized urine of nursing home residents. *J Okla State Med Assoc* 1993;86:550–552.
29. Kocka FE, Srinivasan S, Mowjood M, Kantor HS. Nosocomial multiply resistant *Providencia stuartii:* A long-term outbreak with multiple biotypes and serotypes at one hospital. *J Clin Microbiol* 1980;11: 167–169.
30. Fierer J, Ekstrom M. An outbreak of *Providencia stuartii* urinary tract infections: Patients with condom catheters are a reservoir of the bacteria. *JAMA* 1981; 245:1553–1555.
31. Bryan CS. Clinical implications of positive blood cultures. *Clin Microbiol Rev* 1989;2:329–353.
32. Roberts FJ, Geere IW, Coldman A. A three-year study of positive blood cultures, with emphasis on prognosis. *Rev Infect Dis* 1991;13:34–46.
33. Nicolle LE, Strausbaugh LJ, Garibaldi RA. Infections and antibiotic resistance in nursing homes. *Clin Microbiol Rev* 1996;9:1–17.
34. Janis B, Evans RG, Hoeprich PD. Providence bacillus

bacteremia and septicopyemia. *Am J Med* 1968;45: 943–947.
35. Keren G, Tyrrel DLJ. Gram-negative septicemia caused by *Providencia stuartii. Int J Pediatr Nephrol* 1987;8:91–94.
36. Klastersky J, Bogaerts A-M, Noterman J, Van Laer E, Daneau D, Mouawad E. Infections caused by Providence bacilli. *Scand J Infect Dis* 1974;6:153–160.
37. Milstoc M, Steinberg P. Fatal septicemia due to Providence group bacilli. *J Am Geriatr Soc* 1973;21:159–163.
38. Prentice B, Robinson BL. A review of *Providencia* bacteremia in a general hospital, with a comment on patterns of antimicrobial sensitivity and use. *Can Med Assoc J* 1979;121:745–749.
39. Muder RR, Brennen C, Wagener MM, Goetz AM. Bacteremia in a long-term-care facility: A five-year prospective study of 163 consecutive episodes. *Clin Infect Dis* 1992;14:647–654.
40. Prakash OM, Dayal S, Kalra SL. Bacterial aetiology of infantile diarrhoea in a village population with observations on some providence strains isolated from diarrhoea and non-diarrhoea cases. *Indian J Med Res* 1966;54:705–713.
41. Müller HE. Occurrence and pathogenic role of *Morganella-Proteus-Providencia* group bacteria in human feces. *J Clin Microbiol* 1986;23:404–405.
42. Müller HE. The role of proteae in diarrhea. *Zbl Bakt* 1989;272:30–35.
43. Haynes J, Hawkey PM. *Providencia alcalifaciens* and travellers' diarrhoea. *Br Med J* 1989;299:94–95.
44. Woods TD, Watanakunakorn C. Bacteremia due to *Providencia stuartii:* Review of 49 episodes. *South Med J* 1996;89:221–224.
45. Albert MJ, Alam K, Ansaruzzaman M, et al. Pathogenesis of *Providencia alcalifaciens*-induced diarrhea. *Infect Immun* 1992;60:5017–5024.
46. Albert MJ, Ansaruzzaman M, Bhuiyan NA, Neogi PKB, Faruque ASG. Characteristics of invasion of HEp-2 cells by *Providencia alcalifaciens. J Med Microbiol* 1995;42:186–190.
47. Mathan MM, Mathan VI, Albert MJ. Electron microscopic study of the attachment and penetration of rabbit intestinal epithelium by *Providencia alcalifaciens. J Pathol* 1993;171:67–71.
48. Curreri PW, Bruck HM, Lindberg RB, Mason AD Jr, Pruitt BA. *Providencia stuartii* sepsis: A new challenge in the treatment of thermal injuries. *Ann Surg* 1973;177:133–138.
49. McKell J, Jones D. A numerical taxonomic study of proteus-providence bacteria. *J Appl Bacteriol* 1976; 41:143–161.
50. Coetzee JN, de Klerk HC. Effect of temperature on flagellation, motility and swarming of *Proteus. Nature* 1964;202:211–212.
51. Penner JL. The tribe *Proteeae.* In: Starr MP, Stolp H, Trüper HG, Balows A, Schlegel HG (eds.) *The prokaryotes: A handbook on habitats, isolation, and identification of bacteria.* Berlin: Springer-Verlag, 1981;1204–1224.
52. Kunin CM, Hua TH, Bakaletz LO. Effect of salicylate on expression of flagella by *Escherichia coli* and *Proteus, Providencia,* and *Pseudomonas* spp. *Infect Immun* 1995;63:1796–1799.
53. Hawkey PM, McCormick A, Simpson RA. Selective and differential medium for the primary isolation of members of the *Proteeae. J Clin Microbiol* 1986;23: 600–603.
54. Hawkey PM, Potten MR, Stephens M. The use of pre-enrichment for the isolation of small numbers of gentamicin-resistant *Providencia stuartii* from feces. *J Hosp Infect* 1982;3:369–374.
55. Thaller MC, Berlutti F, Pantanella F, Pompei R, Satta G. Modified MacConkey medium which allows simple and reliable identification of *Providencia stuartii. J Clin Microbiol* 1992;30:2054–2057.
56. Pompei R, Cornaglia G, Ingianni A, Satta G. Use of a novel phosphatase test for simplified identification of species of the tribe *Proteeae. J Clin Microbiol* 1990; 28:1214–1218.
57. Senior BW. Media for the detection and recognition of the enteropathogen *Providencia alcalifaciens* in faeces. *J Med Microbiol* 1997;46:524–527.
58. Miller JM, Rhoden DL. Preliminary evaluation of Biolog, a carbon source utilization method for bacterial identification. *J Clin Microbiol* 1991;29:1143–1147.
59. Pfaller MA, Sahm D, O'Hara CM, et al. Comparison of the AutoSCAN-W/A rapid bacterial identification system and the Vitek AutoMicrobic system for identification of gram-negative bacilli. *J Clin Microbiol* 1991;29:1422–1428.
60. O'Hara CM, Rhoden DL, Miller JM. Reevaluation of the API 20E identification system versus conventional biochemicals for identification of members of the family *Enterobacteriaceae:* A new look at an old product. *J Clin Microbiol* 1992;30:123–125.
61. Kelly MT, Leicester C. Evaluation of the Autoscan Walkaway system for rapid identification and susceptibility testing of gram-negative bacilli. *J Clin Microbiol* 1992;30:1568–1571.
62. York MK, Brooks GF, Fiss EH. Evaluation of the autoSCAN-W/A rapid system for identification and susceptibility testing of gram-negative fermentative bacilli. *J Clin Microbiol* 1992;30:2903–2910.
63. O'Hara CM, Tenover FC, Miller JM. Parallel comparison of accuracy of the API 20E, Vitek GNI, MicroScan Walk/Away rapid ID, and Becton Dickinson Cobas ID-E/NF for identification of members of the family *Enterobacteriaceae* and common gram-negative, non–glucose-fermenting bacilli. *J Clin Microbiol* 1993;31:3165–3169.
64. Kitch TT, Jacobs MR, Appelbaum PC. Evaluation of RapID onE system for identification of 379 strains in the family *Enterobacteriaceae* and oxidase-negative, gram-negative nonfermenters. *J Clin Microbiol* 1994; 32:931–934.
65. Holmes B, Costas M, Ganner M, On SLW, Stevens M. Evaluation of Biolog system for identification of some gram-negative bacteria of clinical importance. *J Clin Microbiol* 1994;32:1970–1975.
66. Holmes B, Costas M, Thaker T, Stevens M. Evaluation of two BBL Crystal systems for identification of some clinically important gram-negative bacteria. *J Clin Microbiol* 1994;32:2221–2224.
67. Robinson A, McCarter YS, Tetreault J. Comparison of Crystal Enteric/Nonfermenter system, API 20E system, and Vitek Automicrobic system for identification of gram-negative bacilli. *J Clin Microbiol* 1995;33: 364–370.

68. Wauters G, Boel A, Voorn GP, et al. Evaluation of a new identification system, Crystal Enteric/Non-Fermenter, for gram-negative bacilli. *J Clin Microbiol* 1995;33:845–849.
69. Rhoads S, Marinelli L, Imperatrice CA, Nachamkin I. Comparison of MicroScan WalkAway system for identification of gram-negative bacilli. *J Clin Microbiol* 1995;33:3044–3046.
70. Cornaglia G, Dainelli B, Berlutti F, Thaller MC. Commercial identification systems often fail to identify *Providencia stuartii. J Clin Microbiol* 1988;26: 323–327.
71. Penner JL. Genus XII: *Providencia* Ewing 1962, 96[AL]. In: Krieg NR, Holt JG, eds. *Bergey's manual of systematic bacteriology.* Baltimore: Williams & Wilkins, 1984;494–496.
72. Senior BW, Hughes C. Production and properties of haemolysins from clinical isolates of *Proteeae. J Med Microbiol* 1987;24:17025.
73. Ewing WH, Davis BR, Sikes JV. Biochemical characterization of Providencia. *Public Health Lab* 1972;30: 25–38.
74. Polster M, Svobodová M. Production of reddish-brown pigment from DL-tryptophan by enterobacteria of the *Proteus-Providencia* group. *Experientia* 1964;20: 637–638.
75. Sheth NK, Kurup VP. Evaluation of tyrosine medium for the identification of *Enterobacteriaceae. J Clin Microbiol* 1975;1:483–485.
76. Vasyurenko ZP, Chernyavskaya YN. Confirmation of *Morganella* distinction from *Proteus* and *Providencia* among *Enterobacteriaceae* on the basis of cellular and lipopolysaccharide fatty acid composition. *J Hyg Epidemiol Microbiol Immunol* 1990;34:81–90.
77. Fischer R, Penner JL, Zurinaga G, Riddle C, Sämisch W, Brenner DJ. Usefulness of trehalose fermentation and L-glutamic acid decarboxylation for identification of biochemically abberant *Providencia stuartii* strains. *J Clin Microbiol* 1989;27:1969–1972.
78. Farmer JJ III, Davis BR, Hickman-Brenner FW, et al. Biochemical identification of new species and biogroups of *Enterobacteriaceae* isolated from clinical specimens. *J Clin Microbiol* 1985;21:46–76.
79. Grant RB, Penner JL, Hennessy JN, Jackowski BJ. Transferable urease activity in *Providencia stuartii. J Clin Microbiol* 1981;13:561–565.
80. Mobley HLT, Chippendale GR, Fraiman MH, Tenney JH, Warren JW. Variable phenotypes of *Providencia stuartii* due to plasmid-encoded traits. *J Clin Microbiol* 1985;22:851–853.
81. D'Orazio SEF, Collins CM. Characterization of a plasmid-encoded urease gene cluster found in members of the family *Enterobacteriaceae. J Bacteriol* 1993;175: 1860–1864.
82. Thaller MC, Berlutti F, Riccio ML, Rossolini M. A species-specific DNA probe for *Providencia stuartii* identification. *Mol Cell Probes* 1992;6:417–422.
83. Penner JL, Hinton NA, Hennessy J. Biotypes of *Proteus rettgeri. J Clin Microbiol* 1975;1:136–142.
84. Penner JL. Chapter 151. The Genera *Proteus, Providencia,* and *Morganella.* In: Balows A, Trüper HG, Dworkin M, Harder W, Schleifer K-H, eds. *The prokaryotes: a handbook on the biology of bacteria ecophysiology, isolation, identification, applications.* 2nd ed. Berlin: Springer-Verlag, 1991;2849–2862.
85. Penner JL, Fleming PC, Whiteley GR, Hennessy JN. O-serogrouping of *Providencia alcalifaciens. J Clin Microbiol* 1979;10:761–765.
86. Hawkey PM, Bennett PM, Hawkey CA. Cryptic plasmids in hospital isolates of *Providencia stuartii. J Med Microbiol* 1984;18:277–284.
87. Pitcher DG, Owen RJ, Dayal P, Beck A. Synthesis of a biotinylated DNA probe to detect ribosomal RNA cistrons in *Providencia stuartii. FEMS Microbiol Lett* 1987;48:283–287.
88. Owen RJ, Beck A, Dayal PA, Dawson C. Detection of genomic variation in *Providencia stuartii* clinical isolates by analysis of DNA restriction fragment length polymorphisms containing rRNA cistrons. *J Clin Microbiol* 1988;26:2161–2166.
89. Mobley HLT, Chippendale GR, Tenney JH, Warren JW. Adherence to uroepithelial cells of *Providencia stuartii* isolated from catheterized urinary tract. *J Gen Microbiol* 1986;132:2863–2872.
90. Mobley HLT, Chippendale GR, Tenney JH, et al. MR/K hemagglutination of *Providencia stuartii* correlates with adherence to catheters and with persistence in catheter-associated bacteriuria. *J Infect Dis* 1988; 157:264–271.
91. Fletcher M, Oppenheimer SR, Warren JW. Colonization of urinary catheters by *Escherichia coli* and *Providencia stuartii* in a laboratory model system. *J Urol* 1994;152:232–236.
92. Old DC, Scott SS. Hemagglutinins and fimbriae of *Providencia* spp. *J Bacteriol* 1981;146:404–408.
93. Guth BEC, Perrella E. Prevalence of invasive ability and other virulence-associated characteristics in *Providencia alcalifaciens* strains isolated from São Paulo, Brazil. *J Med Microbiol* 1996;45:459–462.
94. Nawrot U, Mokracka-Latajka G, Grzybek-Hryncewicz J, Krzyzanowska B, Jankowski S. Bactericidal activity of normal human serum against *Morganella, Proteus,* and *Providencia* strains. *Acta Microbiol Pol* 1995;44: 55–61.
95. Drechsel H, Thieken A, Reissbrodt R, Jung G, Winkelmann G. α-Keto acids are novel siderophores in the genera *Proteus, Providencia,* and *Morganella* and are produced by amino acid deaminases. *J Bacteriol* 1993; 175:2727–2733.
96. Hawkey PM. *Providencia stuartii:* A review of a multiply antibiotic-resistant bacterium. *J Antimicrob Chemother* 1984;13:209–226.
97. McHale PJ, Keane CT, Dougan G, Antibiotic resistance in *Providencia stuartii* isolated in hospitals. *J Clin Microbiol* 1981;13:1099–1104.
98. Lambert T, Gerbaud G, Courvalin P. Characterization of transposon Tn1528, which confers amikacin resistance by synthesis of aminoglycoside 3′-*O*-phosphotransferase type VI. *Antimicrob Agents Chemother* 1994;38:702–706.
99. Hawkey PM, Pedler SJ, Turner A. Comparative in vitro activity of semisynthetic penicillins against *Proteeae. Antimicrob Agents Chemother* 1983;23:619–621.
100. Piccolomini R, Cellini L, Allocati N, Di Girolamo A, Ravagnan G. Comparative in vitro activities of 13 antimicrobial agents against *Morganella-Proteus-Providencia* group bacteria from urinary tract infections. *Antimicrob Agents Chemother* 1987;31:1644–1647.
101. Neu HC. Aztreonam: The first monbactam. *Med Clin North Am* 1988;72:555–565.

102. Lipman B, Neu HC. Imipenem: A new carbapenem antibiotic. *Med Clin North Am* 1988;72:567–579.
103. Fernandes CJ, Ackerman VP. In vitro studies of ciprofloxacin and survey of resistance patterns in current isolates. *Diagn Microbiol Infect Dis* 1990;13: 79–91.
104. Cornaglia G, Frugoni S, Mazzariol A, Piacentini E, Berlusconi A, Fontana R. Activities of oral antibiotics on *Providencia* strains isolated from institutionalized elderly patients with urinary tract infections. *Antimicrob Agents Chemother* 1995;39:2819–2821.
105. Prosser BLT, Beskid G. Multicenter in vitro comparative study of fluoroquinolones against 25,129 gram-positive and gram-negative clinical isolates. *Diagn Microbiol Infect Dis* 1995;21:33–45.
106. Doern GV. The in vitro activity of cefotaxime versus bacteria involved in selected infections of hospitalized patients outside of the intensive care unit. *Diagn Microbiol Infect Dis* 1995;22:13–17.
107. Penner JL, Preston MA, Hennessy JN, Goodbody MM. Species differences in susceptibilities of *Proteeae* spp. to six cephalosporins and three aminoglycosides. *Antimicrob Agents Chemother* 1982;22:218–221.
108. Biedenbach DJ, Jones RN. Predictive accuracy of disk diffusion test for *Proteus vulgaris* and *Providencia* species against five newer orally administered cephalosporins, cefdinir, cefetamet, cefprozil, cefuroxime, loracarbef. *J Clin Microbiol* 1994;32:559–562.
109. Jones RN, Pfaller MA, Allen SD, Gerlach EH, Fuchs PC, Aldridege KE. Antimicrobial activity of cefpirome: An update compared to five third-generation cephalosporins against nearly 6000 recent clinical isolates from five medical centers. *Diagn Microbiol Infect Dis* 1991;14:361–364.
110. Chin N-X, Gu J-W, Yu K-W, Zhang Y-X, Neu HC. In vitro activity of sparfloxacin. *Antimicrob Agents Chemother* 1991;35:567–571.
111. Thornsberry C. Susceptibility of clinical bacterial isolates to ciprofloxacin in the United States. *Infection* 1994;22(Suppl):S80–S89.

17

The Genus *Edwardsiella*

TAXONOMY

In the late 1950s, several different groups of investigators began independently to study clusters of phenotypically distinct gram-negative, fermentative bacilli that appeared to represent a new taxon within the family *Enterobacteriaceae.* In Japan, Sakazaki and Murata[1] reported on a collection more than 150 such isolates recovered almost exclusively from snakes. All of these strains shared a number of features in common, including hydrogen sulfide formation and failure to ferment D-mannitol. The vernacular name "Asakusa group" was chosen, a reference to a ward of Tokyo where these strains were originally isolated.[2] In 1964, the first report of an apparent association between these bacteria and human infection (gastroenteritis) was published by King and Adler.[3] They referred to this unidentified member of the *Enterobacteriaceae* as belonging to the "Bartholemew group," the name originating from the county in Indiana where the strain was isolated. Quickly thereafter, Radovsky and Aldová[4] identified three fecal isolates of the "Bartholemew group" from Cuba. One of these strains was associated with diarrhea in a 1-month-old infant.

The Centers for Disease Control and Prevention (CDC) had also been studying strains with similar biochemical characteristics since 1959. The CDC strains had been categorized simply as biotype 1483-59. In contrast to the Japanese collection, most CDC isolates were of human origin. In 1965, Ewing and colleagues[5] at the CDC proposed a genus and species name, *Edwardsiella tarda,* to replace several vernacular names that had appeared in the literature. The epithet *tarda* was a reference to the relative inactivity of this species in attacking most fermentable carbohydrates. The taxonomic position of *E. tarda* was confirmed by Brenner and associates[6] at the Walter Reed Army Institute of Research in 1974 using DNA–DNA reassociation kinetics. Nineteen *E. tarda* strains were 82% to 95% related to the reference strain CDC 3592-64 at 60°C but only 20% to 29% related to other species in the family. At 75°C, the relative binding percentages were 81% to 93% and less than 8%, respectively. These values were found to be consistent with the establishment of a separate genus and species.

In 1975 the correct species designation for this group was challenged. Sakazaki and Tamura[7] indicated that Hoshina in 1962[8] had described a new species, *Paracolobactrum anguillimortiferum,* that was associated with diseased eels. Although the original strain was no longer extant, the biochemical properties described for these strains, as reported by Sakazaki and Tamura,[7] identically matched those of *E. tarda.* Because "anguillimortiferum" predates "tarda" by 3 years in the literature, a proposal was made to change the name to *Edwardsiella anguillimortiferum* comb. nov. Farmer and collaborators[9] argued, however, against this change on the basis of the obscurity of the citation, the general acceptance of the name *E. tarda* in the literature, and the need for stability and avoidance of confusion. Later, Farmer and McWhorter[10] showed that there were at least several poten-

TABLE 1. *Current species composition within the genus* Edwardsiella

Species	Type Strain	Isolated from Clinical Material	Human Pathogen	Synonyms	Ref.
E. hoshinae	ATCC 33379	Yes[a]	No		14
E. ictaluri	ATCC 33202	No	No		15
E. tarda	ATCC 15947	Yes	Yes	Asakusa group, Bartholemew group, *Edwardsiella anguillimortifera*	5, 7

[a]Only two isolates from humans (fecal samples) have been reported.[10]

tial differences between these two groups, including the methyl red test, pathogenicity for trout, and growth requirements. Although this controversy continued for a number of years,[11] in 1992 Sakazaki and Tamura[12] recommended that *E. tarda* be retained as the legitimate name in light of the general use of this epithet in all recent papers characterizing these organisms.

Although the legitimacy of the *E. tarda* group had been proven, some still questioned the appropriateness of having a genus with a single species residing within it. In a large international numerical taxonomic investigation of the family *Enterobacteriaceae,* a second cluster of edwardsiellae were identified that were 95% similar to *E. tarda* but which had not been recognized previously.[13] These strains were isolated by Sakazaki and were referred to in this study as "*Edwardsiella* spp." They differed from typical *E. tarda* in their ability to produce acid from the fermentation of sucrose and salicin. In 1980, Grimont and others characterized these atypical *Edwardsiella* isolates and additionally found a new group of malonate-positive edwardsiellae that produced acid from the fermentation of sucrose, D-mannitol, and trehalose.[14] By DNA hybridization, the former group ("*Edwardsiella* spp.") was found to represent a new biogroup (biogroup 1) of *E. tarda,* and the latter malonate-positive cluster was found to merit species status. The name *Edwardsiella hoshinae* was proposed for this latter group.[14] A third species of *Edwardsiella, Edwardsiella ictaluri,* was proposed 1 year later by Hawke and colleagues.[15] *E. ictaluri* is the causative agent of enteric septicemia of channel catfish. The current members of the genus *Edwardsiella* are listed in Table 1.

EPIDEMIOLOGY

Edwardsiellae have a limited environmental distribution and host range. Most human infections are a result of direct contact with or exposure to edwardsiellae in their natural habitat. For *E. tarda* and related species, freshwater and marine environments serve as reservoirs for potential infections, as do animals that are intimately associated with such ecological niches, including fish, reptiles, amphibians, and aquatic birds.

Edwardsiella species are occasionally isolated from the marine environment and from freshwater sources such as rivers, well water, and lake shores.[16,17] In one 15-month survey of streams and lakes in north central Florida, *E. tarda* was isolated from 12% of more than 100 water samples tested.[17] On rare occasions these organisms have also been isolated from sewage.[16,18] Despite such isolations, they apparently do not survive long in the free-living form in such environments. King and Shotts[19] demonstrated a more than 100-fold reduction in *E. tarda* numbers within 72 hours of in vitro cultivation in saline solutions. However, if freshwater ciliated protozoa were cocultivated with *E. tarda,* a 100-fold increase in edwardsiellae colony-forming units was seen over a similar time frame. This fact suggests that *Edwardsiella* species require a commensal association with other living freshwater inhabitants in order to persist and survive for prolonged periods.

The most common species found to endogenously harbor edwardsiellae are poikilothermic (cold-blooded) animals such as reptiles and fish. Early Japanese investigations isolated *E. tarda* primarily from snake feces.[1,2,20] Later, Kourany and associates[18] cultured the intestinal tracts of 27 different genera of snakes indigenous to Panama for *E. tarda* and found an overall incidence of 5%. However, when Goldstein and coworkers[21] cultured the oropharnygeal flora of 82 garter snakes, only one isolate of *E. tarda* was recovered, from an adult snake. Fish also appear to be common sources for *E. tarda.* One Vietnamese investigation found 44% of normal healthy fish to be endogenously colonized with edwardsiellae in their intestinal tract.[22] Similar results were obtained by Van Damme and Vandepitte,[23] who isolated *E. tarda* from 57% of Zairese freshwater fish; larger fish yielded higher isolation rates of *E. tarda.* Other poikilotherms associated with high *E. tarda* isolation rates include water tortoises (100%), turtles (19%), and toads (6%), based on selected surveys.[16,18,24]

E. tarda has been isolated from a number of other vertebrates, including mammals (Table 2). Most of these isolations have been sporadic and are not indicative of each vertebrate type as a common host for edwardsiellae. In a survey of more than 2000 mammals, Kourany and collaborators[18] isolated only six strains of *E. tarda;* five of these were recovered from opossums (relative incidence 1.6%). Rats, bats, sloths, and all but one monkey were negative for *E. tarda* in this study, and all isolates were obtained from apparently healthy wild animals. Swine have occasionally been found to harbor *E. tarda,* but its role in disease is unclear.[31,32]

Meyer and Bullock[33] described a foul-smelling infection of channel catfish caused by *E. tarda* that they termed emphysematous putrefactive disease. Pathologic manifestations of the disease in catfish ranged from small cutaneous lesions to large-cavity gas-filled abscesses. *E. tarda* also was associated with an outbreak of chronic enteritis in Rockhopper penguins at the Cheyenne Mountain Zoological Park.[26] Of eight penguins that died from an enteritis-type infection (including failure to feed), five were submitted for bacteriologic analysis on autopsy. *E. tarda* was isolated from the intestinal contents of four of these five birds. Other pathologic conditions sporadically associated with *E. tarda* infections include hemorrhagic enteritis in pelicans and loons, enteritis in ostriches and pigs, nephritis in alligators, and septicemia in a pigeon.[17,28,29,32,34]

CLINICAL SIGNIFICANCE

E. tarda is the only currently recognized species within the genus that is pathogenic for humans. Clinical syndromes associated with

TABLE 2. *Animal groups found to harbor* Edwardsiella tarda

Marine/Freshwater	Reptiles/Amphibians	Mammals	Birds
Clams	Alligators	Cow	Sandbill crane
Eels	Lizards	Dog	Bald eagle
Fish:	Skink	Monkey	Ring-billed gull
Bass	Snakes:	Opossums	Blue heron
Red sea bream	Cobra	Panther	Loon
Channel catfish	Garter snake	Pig	Ostrich
Flounder	Grass snake	Skunk	Brown pelican
Red tail	Viper		Penguin
Salmon	Water tortoises		King vulture
Sardines	Turtles		
Trout			
Mussels			
Sea lions			
Seals			

Data from references 16, 18, 20–23, 25–30.

E. tarda infection have been reviewed.[35] The single most common site from which edwardsiellae are isolated is the gastrointestinal tract, which accounts for more than 80% of all clinical strains.[35] Other sites from which *E. tarda* can occasionally be recovered are wounds and blood. *E. tarda* has been implicated as an infrequent cause of hepatobiliary disease, bacteriuria, osteomyelitis, and peritonitis. Table 3 lists the sources from which *E. tarda* was isolated in several clinical series.[30,36–41] Although the earliest two reports are from the United States, the most recent studies have all originated from Southeast Asia, particularly Thailand.[30,38–41] Although the gastrointestinal tract remains the most common site of isolation of edwardsiellae in these later investigations, extraintestinal disease (53%) appears to be a more common manifestation of *E. tarda* infection in reports from Southeast Asia. The increased incidence of extraintestinal disease described in these publications may be a reflection of the populations studied,[41] geographic factors (environmental frequency), or ethnic differences related to occupational exposure and dietary habits.

Gastroenteritis

Almost from its inception the genus *Edwardsiella* has been associated with diarrheal disease. From a collection of more than 250 *E. tarda* strains recovered almost exclusively from nonhuman sources, Sakazaki[2] described five fecal isolates associated with human cases of gastroenteritis. This observation was strengthened by data from the CDC indicating that 74% of 54 *E. tarda* fecal isolates received between 1959 and 1967 were isolated from persons with diarrhea.[36] Thereafter, other clinical studies and several case histories were published linking edwardsiellae with the diarrheal disease state.[35] Most reports on *E. tarda*–associated diarrheal disease have come from Southeast Asia and Central America, suggesting a possibly higher incidence of the disease in tropical and subtropical regions of the world. A seasonal pattern has not been demonstrated.

A number of independent lines of evidence implicate *E. tarda* as a bona fide enteropathogen. Asymptomatic persons rarely harbor edwardsiellae in their feces, and a mean carrier rate of 0.16% has been postulated based on the results of several surveys.[35] It has also been demonstrated in a number of case-control investigations from Southeast Asia and Central America that *E. tarda* is consistently recovered in higher numbers from symptomatic individuals than from control subjects (Table 4). Furthermore, at least one study has shown that persons infected with *E. tarda* and *Entamoeba histolytica* have significantly higher circulating anti–*E. tarda* agglutinating antibodies than normal healthy people. An additional piece of evidence

TABLE 3. *Anatomic sites of isolation of* Edwardsiella tarda *from selected studies*

				Anatomic Sites				
Study	Period	Site	No. Isolates	Feces	Wounds[a]	Blood	Urine	Others[b]
Ewing et al.[36]	1959–1967	USA	148	120	0	5	2	21
Jordan & Hadley[37]	1964–1967	USA	9	4	4	1	0	0
Bockemühl et al.[30]	Not specified	Thailand	25	23	0	1	0	1
Koshi & Lalitha[38]	1972–1975	India	12	3	5	2	0	2
Tan et al.[39]	Not specified	Singapore	11	5	3	0	2	1
Ovartlarnporn et al.[40]	1980–1981	Thailand	23	21	0	2	0	0
Jaruratanasirikul & Kalnauwakul[41]	1982–1989	Thailand	44[c]	2	21	3	13	5
TOTALS			272	178	33	14	17	30

[a]Includes abscesses and tissue.
[b]Includes bile (n=7), peritoneal fluid (n=1), sputum (n=1), and unspecified (n=21).
[c]This study involved adult patients only.

TABLE 4. *Studies on the role of edwardsiellae as a cause of gastroenteritis*

Study	Period	Site	Population	Setting	*E. tarda* Isolation Rate (No. Screened)	
					Symptomatic	Controls
Bhat et al.[42]	1963–1965	India	Children	Rural, urban	0.8% (513)[a]	0% (319)
Kourany et al.[18]	1965–1972	Panama	All ages	Rural, deep forest, urban	0.6% (5891)	0% (535)
Kourany et al.[43]	1967	Panama	Children	Urban, rural, semirural	0.4% (1178)	—
Sakazaki et al.[44]	1969–1970	India	All ages	Urban	0.2% (3433)	—
Nguyen-Van-Ai et al.[22]	1969–1973	Vietnam	All ages	Urban	1.3% (13947)	—
Gilman et al.[45]	1971	Malaysia	Aborigines	Jungle	13.9% (208)	0.8% (120)
Vandepitte et al.[46]	1972	Mali	Not given	Not given	8.7% (323)	5.4% (56)
Iverson[47]	1973	Australia	All ages	Urban, rural, semitropical	0.3% (12807)	—

[a]Controls for urban areas only.

implicating *E. tarda* as a gastrointestinal pathogen is the demonstration of an enteropathogenic mechanism (invasion) in most isolates of this species (see Pathogenesis). Finally, a number of severe gastrointestinal infections linked to *E. tarda* have required antimicrobial therapy.[39,48–50] In each of these cases, antimicrobial therapy specifically directed against *E. tarda* resulted in resolution of symptomatology and the disappearance of edwardsiellae from feces. Although this collective evidence is compelling, there are still a number of unresolved issues regarding the role of *Edwardsiella* in bacterial gastroenteritis. To date, there has been no reported outbreak of gastroenteritis attributed to this bacterium. However, this fact may be more related to its natural habitat and low clinical incidence than to an inherent inability to produce outbreak-related disease. Finally, although few studies have looked at the pathogenicity of this group, there is at present no animal model for *E. tarda*–associated gastroenteritis and therefore Koch's postulates remain unfulfilled.

In 1969, Jordan and Hadley[37] drew a comparison between enteric infections caused by *E. tarda* and salmonellosis. Three main disease states associated with the gastrointestinal tract were described, namely (1) gastroenteritis, (2) typhoid-like illnesses, and (3) the asymptomatic/colonization state. Patients with *E. tarda*–associated gastroenteritis, both children and adults, typically present with a secretory (watery) enteritis. Most reports of *E. tarda*–associated diarrhea describe a mild to moderate self-limiting illness of 3 to 5 days' duration. On occasion, longer episodes of diarrheal disease (≥21 days) have been reported.[42] Milder disease commonly involves 4 to 5 diarrheal episodes per day, whereas in more severe cases of *E. tarda* gastroenteritis the frequency of evacuation may range as high as 10 to 20 bowel movements per day.[48,49] Fever is usually absent or, if present, is of low grade (37.8°–38.3°C [100°–101°F]).[18] Abdominal pain (cramps), vomiting, and nausea are infrequent findings. Kourany and coworkers[18] studied 10 persons with diarrhea in Panama, 7 with the clinical diagnosis of acute gastroenteritis. Of these 10 patients, 70% had vomiting, and most had a low-grade fever and five or more bowel movements per day. Most gastrointestinal illnesses lasted only 2 to 3 days. The high rate of vomiting observed in this study has not been reported by other groups. Dehydration is occasionally associated with *E. tarda* diarrhea.[18,30,38,39] It has primarily been described in children and adults living in Southeast Asia or in tropical environments; values as high as 10% have been cited in the literature.[38]

Less frequently *E. tarda* can cause more severe gastrointestinal illnesses. Bockemühl and colleagues[30] noted two age groups (infants, adults ≥50 years) who were prone to development of protracted courses of severe diarrheal disease caused by *E. tarda.* All of the older adults in this study had bloody diarrhea with or without mucus; most (70%) had fever, and three (42%) were dehydrated. In contrast, younger adults (20–49 years) with *E. tarda* gastroenteritis had milder infections and were less likely to be febrile and dehydrated (33%). In a Malaysian study involving 30 people with *E. tarda* isolated from their feces, 29 (97%) were symptomatic with bloody diarrhea. Twenty-five (86%) of these 29 ill persons concurrently harbored the protozoan pathogen *E. histolytica* in their stools. The reason for this association is poorly understood. In addition to these two investigations, a number of other reports have described *E. tarda* as the cause of enterocolitis, colitis, dysentery, or chronic diarrhea.[18,35,37,48] On rare occasions, severe *E. tarda* infections of the gut have mimicked other clinical syndromes. Nagel and associates[49] described a case of watery diarrhea of 11 days' duration in a 37-year-old woman with fever. Proctoscopy indicated pathologic changes consistent with pseudomembranous colitis. When fecal cultures yielded *E. tarda,* the patient was queried regarding possible environmental exposure to this organism and revealed that she had accidentally drunk from a glass containing turtle water left over from her son's cleaning of their turtle tank. She was subsequently treated with amoxicillin and recovered. A second case concerned a 31-year-old physician who complained of gastroenteritis of 9 days' duration which progressed to grossly bloody diarrhea.[48] Sigmoidoscopy detected sigmoid and rectal ulcerations and colonic thickening consistent with Crohn's disease. When *E. tarda* was detected in his feces, a history of handling pet turtles on Cape Cod was elicited. He was treated with cotrimoxazole and recovered uneventfully. *E. tarda* was isolated repeatedly from the stool of a premature infant with intermittent (2 months) gastroenteritis which subsequently progressed to bloody diarrhea.[51] The infant's infection was traced to an ornamental fish housed in a aquarium, from which *E. tarda* was also isolated. In five cases, all from Asia, severe diarrhea caused by *E. tarda* was associated with a fatal outcome. Bockemühl and colleagues[30] described three of these episodes, which involved residents of Thailand. In one case a 10-month-old girl died of malarial infection 1 month after admission to the hospital. In the other two cases, a 63-year-old woman and a 55-year-old man presented with bloody diarrhea and dehydration; both died within 1 week of admission to the hospital. A pure culture of *E. tarda* was isolated from the woman's feces. Koshi and Lalitha[38] described a second case of fatal *E. tarda* infection in an infant. The 1-year-old child had gastroenteritis with 10% dehydration and acidosis. Despite fluid correction and antimicrobial therapy, the child died within 24 hours of admission. No pathogens other than *E. tarda* were isolated. A final fatal adult case of *E. tarda*–associated gastroenteritis was described by Ovartlarnporn and colleagues.[40] This illness involved a 60-year-old man with hematemesis resulting from a gastric ulcer. During treatment for this condition he developed fever and diarrhea, and rectal biopsy indicated colitis. *E. tarda* was recovered from his stool. Despite therapy with multiple antibiotics, he continued to have diarrhea, and died from massive gastrointestinal bleeding 3 days after surgery.

Edwardsiellae have also been described as causing "enteric fcver" or "typhoid fever-like" conditions. King and Adler[3] reported the isolation of an *E. tarda* strain from the feces of a man hospitalized with acute gastroenteritis and enteric fever. A second case was published by Jordan and Hadley[37] and involved a 14-year-old girl from Nicaragua. Her initial symptoms were similar to those described by King and Adler,[3] but her condition deteriorated to include septicemia and hepatic abscess caused by edwardsiellae. *E. tarda* was recovered from her blood and from the liver abscess but not from feces. In 1980, Clarridge and collaborators[52] described a third case of typhoid fever–like disease in a 46-year-old man from Latin America. His wife had had similar symptoms 6 weeks earlier after both had returned from a trip to Mexico. Blood and fecal cultures from the man grew *E. tarda*. No additional cases have since been reported.

Although originally thought to be uncommon, numerous reports have described a carrier or asymptomatic colonization state associated with the recovery of edwardsiellae from feces.[18,30,38,39,42,45] One review of previous reports on the isolation of *E. tarda* from feces cited an infection/colonization ratio of approximately 3:1.[35] The high percentage (25%) of asymptomatic persons with *E. tarda* in their stools may be somewhat misleading, because most reports documenting such a carrier state have originated from rural to semirural areas of Southeast Asia. In these geographic settings, a higher colonization rate could be a result of increased cultural, occupational, or recreational exposure to these bacteria, resulting in the development of local immunity in some persons. In 1990, Desenclos and coinvestigators[53] described a cluster of eight *E. tarda* isolates recovered from children and a teacher at a Florida day care center. Four of these isolates were submitted for serologic analysis, and all belonged to the same serotype, O4:H1,11. None of these persons had diarrhea in the 2 weeks preceding or the 1 week after isolation of *E. tarda* from their feces. Sometimes the significance of the recovery of *E. tarda* from feces has been clouded by the simultaneous isolation of other enteropathogens or the presence of underlying medical conditions. Common microbial pathogens isolated with *E. tarda* include *E. histolytica*,[38] *Shigella flexneri*,[18,30,42] *Salmonella* spp.,[37,44,45] *Plesiomonas shigelloides*,[41] and *Vibrio cholerae*.[44] Isolation of *E. tarda* from the gut has also been associated with colonic carcinoma[37] and maladsorption syndrome.[38]

Bacteremia

Septicemia is a life-threatening complication of extraintestinal infection with *E. tarda*. At least 21 cases have been described in the literature in some detail.[18,30,37–41,54–62] *Edwardsiella* bacteremia spans all age groups (range 3 days to 83 years; mean 40.4 years), although it is an infection primarily associated with men (male/female ratio 2.5:1). At least 50% of all cases of *E. tarda* sepsis occur in older adults (>50 years) or younger children (<1 year), a pattern strikingly similar to that described by Bockemühl and colleagues[30] for severe gastrointestinal illnesses caused by edwardsiellae.

The major clinical features associated with published cases of *E. tarda* bacteremia are listed in Table 5. Persons with positive *E. tarda* blood cultures present with typical signs of gram-negative sepsis (eg, fever, chills, hypotension). In four instances, sepsis was accompanied by meningeal involvement.[54,55,57,60] Two fatal cases of neonatal meningitis have been described. Okubadejo and Alausa[55] reported on an 11-day-old Nigerian girl with *E. tarda* meningitis. Despite prompt medical attention and aggressive chemotherapy, she died 1 month later from a variety of meningitis-associated complications including hypothermia and oliguria. In 1988, a second report of neonatal meningitis, in an Hispanic female infant in New York City, was described.[60] The prematurely born infant developed jaundice within 1 day of birth, and her clinical condition rapidly deteriorated (lethargy, hypotonia, cyanosis, bradycardia), leading to death 24 hours later. Sonnenwirth and Kallus[54] reported a fatal case of *E. tarda* meningitis in a 31-year-old woman with lupus erythematosus. She died less than 24 hours after admission. The only person reported to have survived *E. tarda* meningitis was a 17-year-old boy with sickle hemoglobinopathy who was treated with ampicillin.[57] Why a favorable outcome occurred in this case is unknown.

Hepatobiliary disease is the chief predisposing condition linked to *E. tarda* sepsis (see Table 5). In 50% of all cases there is some history of liver disease, such as alcoholic cirrhosis, ethanol abuse, hepatoma, hepatosplenomegaly, or icterus[35]; therefore, a predisposition for development of *E. tarda* septicemia in such

TABLE 5. *Selected characteristics of patients with* Edwardsiella tarda *septicemia*

Characteristic	% of Patients[a]
Presenting diagnosis (n=20)	
Septicemia only	55
Septicemia with meningitis	20
Gastroenteritis	10
Typhoid fever	5
Pneumonia	5
Pancreatitis	5
Major underlying condition	
Hepatobiliary disease	29
Malignancy	14
Sickle cell disease	10
Neonatality	10
Lupus erythematosus	5
Pancytopenia	5
Rheumatic heart disease	5
Gastrectomy	5
None or unknown	19
Presenting symptoms (n=14)	
Fever	100
Concurrent or antecedent diarrhea	50
Chills	43
Hypotension	43
Epigastric pain	43
Nausea	36
Vomiting	36
Headache	7
Laboratory findings	
Monomicrobic bacteremia	90
Polymicrobic bacteremia	10
Additional sites positive for *E. tarda*	43
Therapy (n=14)	
Multiple antibiotics	71
Multiple-agent chemotherapy and surgery	7
Single-agent chemotherapy	14
None	7
Fatality rate (n=20)	40

Adapted from Janda & Abbott.[35]

[a]Based on 21 cases, 16 cases reported in Janda & Abbott[35] and 5 additional cases from Jaruratanasirikul & Kalnauwakul,[41] Martínez,[59] and Wu et al.[62]

persons has been proposed.[62] Interestingly, many persons with *Edwardsiella* septicemia have underlying conditions associated with iron overload states, such as red cell sickling, neonatality, leukemia, or cirrhosis.[35] Disseminated intravascular coagulation is an infrequent complication of *E. tarda* septicemia.[18,61] The overall fatality rate associated with *E. tarda* sepsis ranges from 40% to 47%.[35,61]

Virtually all cases of *E. tarda* bacteremia are monomicrobic. Although the portal of entry for most of these illnesses is unknown, colonization and infection of the gastrointestinal tract before systemic invasion appears to be the main route of infection. Evidence pointing to this as the main mechanism includes the high percentage of patients with a history of diarrhea immediately preceding or concurrent with their septic episode (see Table 5) and the simultaneous isolation of *E. tarda* from feces during bacteremic episodes.[52,58,62] Factors thought to have led to exposure to edwardsiellae in confirmed cases of bacteremia include various occupational activities (farmers,[18] gardeners,[56] and zoo workers[59]) and the consumption of raw seafood[62] or snake flesh.[55]

Wounds

There are a large number of reports in the literature describing the recovery of *E. tarda* from wound specimens.[37–39] However, the significance of these isolates is lost in most instances because of a lack of information or the simultaneous recovery of other gram-negative and gram-positive microbial species. Two types of wound infections have been principally associated with *E. tarda.* The first involves lesions that result from direct trauma to mucosal surfaces (eg, penetrating injuries, major trauma, surgery). Most penetrating injuries have had a common underlying theme of exposure to freshwater environments or to inhabitants of such ecologic niches as aquatic flora, fish, or reptiles. Most documented cases of penetrating injuries associated with *E. tarda* infection involved local trauma to the feet or legs from punctures by the dorsal fin or spines of catfish (Table 6). These infections are commonly polymicrobic, often involving other recognized waterborne pathogens such as *Aeromonas hydrophila.*[65,66] Frequently these lesions appear indolent or innocuous on initial examination, requiring only irrigation and sometimes debridement, foreign body removal, or antimicrobial therapy.[52,65,67,68] However, such benign-appearing wounds often progress to more severe and life-threatening infections within 12 to 72 hours after initial insult,[65,67,68] sometimes involving infection of

TABLE 6. *Puncture wounds or penetrating injuries associated with* Edwardsiella tarda *infection*

Case	Age/Sex	Clinical Diagnosis	Affected Area	Precipitating Event	Pure Culture	Clinically Significant	Treatment	Ref.
1	52/M	Cellulitis	Left foot	Catfish injury	No	Yes	Cph	52
2	15/M	Gas gangrene	Right thigh	Log puncture (lake)	Yes	No	Pen, Chl, antitoxin, debridement	63
3	67/M	Osteomyelitis, septic arthritis	Right/left hip	Snake remains puncture	Yes	Yes	Cph, Ox, Gm, Cb, arthroplasty	64
4	24/M	Cellulitis	Right arm/thigh	Catfish puncture	Yes	Unknown	Cef, Gm	65
5	15/M	Cellulitis	Left foot	Branch puncture (lake)	Yes	Yes	Cph, drainage	66
6	25/F	Abscess	Right foot	Catfish puncture	No	Yes	Cip, Pip, Amp, drainage, debridement	67
7	51/M	Septic arthritis, osteomyelitis	Third finger	Catfish puncture	Yes	Yes	Cph, drainage, debridement, surgery	68

Cph, cephalothin; Pen, penicillin; Chl, chloramphenicol; Ox, oxacillin; Gm, gentamicin; Cb, carbenicillin; Cef, ceftriaxone; Cip, ciprofloxacin; Pip, piperacillin; Amp, ampicillin.

the bone[64,68] or requiring additional surgical procedures[65] including amputation.[68] Penetrating injuries can also be caused by a variety of other insults, such as swimming and diving accidents in lakes,[37,63] fishing,[66] and reptile puncture wounds.[64] On occasion major traumas have resulted in *E. tarda* infections. Jordan and Hadley[37] described an *E. tarda* infection in an 18-year-old boy who sustained a laceration to his right leg in an automobile accident. A purulent infection developed, and both *E. tarda* and coagulase-negative staphylococci were isolated from pus. Drainage of the wound and chemotherapy (chloramphenicol, tetracycline) produced a favorable outcome. Surgical or incisional wounds sometimes contain *E. tarda*.[37,38] Such infections appear to result from spillover of edwardsiellae from the gastrointestinal tract after small bowel perforations or abdominal surgery. The first case of a vascular prosthesis infection caused by *E. tarda* was reported by Coutlée and colleagues[69] in a 69-year-old man. Computed tomography detected the formation of a soft mass surrounding an abdominal aortic Dacron prosthesis which apparently resulted from an aortoduodenal fistula. Granulation tissue from this mass yielded *E. tarda,* anaerobes, and viridans streptococci. Debridement, prosthesis replacement, and antimicrobial therapy brought about resolution of symptomatology.

The second type of wound infection associated with edwardsiellae involves pus-filled cavitary lesions (abscess formation). In a study of 44 *E. tarda* infections published by Jaruratanasirikul and Kalnauwakul,[41] 19 isolates originated from pus, and many of these were associated with abscess formation or appendicitis. The abscess site from which edwardsiellae have most frequently been recovered is the liver. At least eight cases of hepatic abscesses caused by *E. tarda* have been described (Table 7). Most of these illnesses occurred in healthy people with no predisposing factors. However, in two instances prior liver disease (hepatoma, hepatitis) was noted.[38,71] Most persons with hepatic abscesses present with fever and severe abdominal or right quadrant pain.[35] The reported fatality rate is about 38%. Other wound infections associated with *E. tarda* include pelvic,[16] subphrenic,[16] and subgaleal[72] abscesses. Pien and Jackson[73] described an *E. tarda* infection in a 42-year-old Hawaiian native. The woman was admitted to the hospital with flu-like symptoms including fever and lower abdominal pain after having consumed a meal of raw fish called "poki." After admission, a possible diagnosis of acute appendicitis was entertained. However, during surgery the appendix was found to be normal, and a right-sided mass involving the right tube, ovary, peritoneal wall, and bowel was observed. The tuboovarian ab-

TABLE 7. Edwardsiella tarda *and abscesses of the liver*

Case	Age/Sex	Location	Clinical Diagnosis	Positive Blood Culture	Pure Culture	Treatment	Outcome	Ref.
1	71/F	Panama	Liver abscess	–	Yes	Not given	Died	18
2	35/M	Panama	Sepsis, cholangitis, choledocholithiasis	+	Yes	Gm, Pen, Amp	Died	18
3	14/F	USA	Sepsis, liver abscess	+	Yes	Amp, Chl, Km	Survived	37
4	45/M	India	Liver abscess	–	Yes	Not given	Died	38
5	52/M	Thailand	Liver abscess, intrahepatic stones	+	Yes	Not given	Survived	41
6	55/F	Thailand	Liver abscess	–	Yes	Not given	Survived	41
7	58/F	Italy	Sepsis, liver abscess	–	No	Pip, Ctx, Met	Survived	70
8	40/M	USA	Pneumonia	–	Yes	Imp, Ery	Survived	71

Gm, gentamicin; Pen, penicillin; Amp, ampicillin; Chl, chloramphenicol; Km, kanamycin; Pip, piperacillin; Ctx, cefotaxime; Met, metronidazole; Imp, imipenem; Ery, erythromycin.

scess complex was removed and cultured, revealing a pure culture (heavy growth) of *E. tarda*. The patient recovered after chemotherapy and additional corrective surgery.

Miscellaneous Infections

E. tarda has been isolated from urine and bile, but little clinical information is available regarding such isolates.[30,39,41] Of 13 urine isolates reported in one investigation,[41] 9 were recovered in pure culture and all were associated with urinary tract infections. Biliary isolates have often been recovered from persons with bile duct or gallstones who are undergoing cholecystectomy or have an underlying malignancy.[30,41] Most often they are recovered from bile as part of a mixed microbial flora. One case of salpingitis was listed without any details,[16] and several cases of peritonitis have been described. Clarridge and associates[52] reported a fatal case of peritonitis due to *E. tarda* in a middle-aged man with alcoholic liver disease and pancreatitis. A second fatal case of peritonitis, in a 20-year-old man with strangulated volvulus caused by polymicrobic infection (including *E. tarda*), was described by Koshi and Lalitha.[38] A case of osteomyelitis precipitated by lacerations from flying glass in an automobile accident was published in 1977.[74] Despite appropriate treatment, multiple medical problems developed while the patient was hospitalized, and she died from septicemia (unspecified) 6 weeks later.

LABORATORY IDENTIFICATION

Isolation

Systematic investigations on the plating efficiency of edwardsiellae on various selective media have not been attempted. No doubt part of the reluctance to undertake such a project is the low incidence of these organisms in clinical specimens. Another contributing factor may be their unique biochemical properties (lactose- and sucrose-negative, H_2S-positive), which enable them to be readily distinguished from normal enteric flora on selective and differential media primarily designed for the recovery of *Salmonella* and *Shigella*. However, both *E. hoshinae* and *E. ictaluri* are H_2S-negative, and their isolation would depend on the use of lactose-containing media. Most *E. tarda* strains grow well on MacConkey, xylose-lysine-desoxycholate, desoxycholate, Hektoen enteric, and Salmonella-Shigella agars.[10,12] However, at least two media, bismuth sulfite and brilliant green agars, appear unsatisfactory for their primary isolation from feces.[12,75] Two studies by Iveson[47,75] found the recovery of edwardsiellae from reptile cloacal samples and from human feces to be superior with use of an enrichment broth rather than direct plating onto either desoxycholate or Salmonella-Shigella agars. In these studies, enrichment of feces in strontium chloride B medium generated a 3.2- to 3.5-fold higher isolation rate than direct plating. The efficiency of strontium chloride B broth ranged from 84% to 95% in these surveys. Other enrichment media, such as selenite F and gram-negative (GN) broth, yielded no isolates of *E. tarda*. More than 80% of all *E. tarda* isolates are resistant to 100 μg/mL of colistin.[76] Although this observation has not been exploited, it has been suggested that this attribute in conjunction with peptone iron agar (to detect H_2S production) might make a useful selective medium for the recovery of *E. tarda* from contaminated material.[10]

Biochemical Identification

Commercial Systems

Despite the limited number of strains tested on commercial systems (~60), cumulative results indicate that most panels identify *E. tarda* accurately and with a high discriminatory value. A couple of minor exceptions have been noted. O'Hara and colleagues[77] found that the API 20E strip (Vitek bioMérieux, Hazelwood, MO) failed to identify two of five *E. tarda* strains when read at either 24 or 48 hours, both generating unacceptable profile numbers. One discrepancy was caused by a false-negative

TABLE 8. *Differential properties among* Edwardsiella *species*

Character	*E. tarda*	*E. hoshinae*	*E. ictaluri*
Isolated from humans	+	+	−
Pathogenic for humans	+	−	−
Prefers temperatures below 35°C for growth	−	−	+
TSI (H_2S)	+	−	−
Indole	+	V	−
Motility	+	+	−
Malonate	−	+	−
Trehalose	−	+	−

From Janda and Abbott.[79]
V, variable; TSI, triple sugar iron agar.

H_2S reaction, and the other involved the inability to identify an indole-negative isolate. In a second CDC publication, 9 of 10 *E. tarda* strains were correctly identified in 2 hours using the AutoSCAN-W/A system (Dade MicroScan, West Sacramento, CA).[78] The single exception was a biogroup 1 strain (see Biotypes) which was identified as *Salmonella* Typhi with a high probability value (93.9%).

Conventional Tests

Although more than 99% of all edwardsiellae isolated in the microbiology laboratory are *E. tarda,* at least one report indicates recovery of *E. hoshinae* from the feces of two asymptomatic persons.[10] All three *Edwardsiella* species can grow at 35° to 37°C (albeit not equally), suggesting that on rare occasions either *E. hoshinae* or *E. ictaluri* may be isolated. Differential characteristics separating these three species are listed in Table 8.

E. tarda, like *Hafnia alvei* (see Chapter 18), is a phenotypically tight species. Most biochemical reactions are either positive or negative for all isolates within the species. Typically, *E. tarda* strains are motile, produce indole, are lysine– and ornithine decarboxylase–positive, and are Voges-Proskauer– and Simmons' citrate–negative. They ferment few sugars other than D-glucose (with gas), maltose, and glycerol. Few extracellular enzymes are expressed besides chondroitinase and hemolysin.[80] On primary isolation from feces, they should invariably be picked for further workup because they are H_2S-positive. However, they can be misidentified as other H_2S-positive bacteria (including *Salmonella*) if an indole reaction is not included as part of the screen. Differential tests useful in separating *E. tarda* from other common H_2S-producing bacteria are listed in Table 9.

Edwardsiellae exhibit several unusual properties. Most strains are more susceptible to the vibriostatic agent O/129 (2,4-diamino-6,7-diisopropyl pteridine) than are other members of the *Enterobacteriaceae.*[10] Although most studies report *E. tarda* as nonhemolytic on sheep blood agar after 24 to 48 hours' incubation, some reports have de-

TABLE 9. *Differential tests useful in separating H_2S-positive bacteria*

Character	*E. tarda*	*Salmonella*	*Citrobacter*	*Proteus*[a]
Indole	+	−	V	V
LDC	+	+	−	−
ODC	+	+	V	V
Simmons' citrate	−	+	+	V
Arabinose	−	+	+	−

[a]For the most common *Proteus* species.
LDC, lysine decarboxylase; ODC, ornithine decarboxylase; V, variable.

TABLE 10. *Aberrant biochemical properties reported for* Edwardsiella tarda *strains*

Characteristic	Aberrant Reaction	No. Strains	Sources	Ref.
Indole	–	3	Blood, feces	58, 77, 79
Arabinose	+	6	Blood, feces, abscess, lizard	30, 84
Mannitol	+	2	Fish, feces	23, 76
Sucrose	+	3	Fish, bile, bat	23, 83, 84
H_2S	–	1	Bile	83

scribed this species as hemolytic. This conflicting information is explained by the fact that *E. tarda* produces a cell-associated hemolysin which requires intimate contact with red blood cells in order for hemolytic activity to be detected.[81] This can be demonstrated in the laboratory by mixing viable bacteria with red blood cells in an appropriate buffer; within a 1- to 2-hour period, sparkling hemolysis results. With the use of a number of different techniques (colony agar overlay, contact-dependent hemolysis), more than 90% of all *E. tarda* have been shown to elaborate a β-hemolysin.[82] Finally, a number of edwardsiellae spontaneously autoagglutinate in common media such as brain heart infusion broth in a manner analogous to that of pathogenic *Yersinia enterocolitica* strains. The significance of this observation is unknown.

Although *E. tarda* is a phenotypically tight species, a number of aberrant reactions for individual biochemical properties have been described (Table 10). These reactions usually involve the fermentation of sugars not generally attacked by the group. Some of these reports may actually reflect the isolation of atypical biogroups (see later discussion) not recognized when the studies were published. Because of these atypical reactions, some strains have been misidentified, most commonly as salmonellae.[77,83]

Biotypes

In 1980, Grimont and coinvestigators[14] discovered a new group of *E. tarda* strains with biochemical properties distinct from those previously described for classic or wild-type edwardsiellae. Seven strains belonging to this group were identified, and most differed from classic *E. tarda* in their ability to ferment several sugars and their inability to reduce tetrathionate.[14] In 1985, Farmer and colleagues[85] gave this group the trivial name *E. tarda* biogroup 1. Biogroup 1 strains are exceedingly rare. Major differences between biogroup 1 and classic isolates are listed in Table 11.

Immunodiagnostics

An international typing scheme for *E. tarda* has been established.[86] This scheme is based on the recognition of 61 distinct somatic (O) and 45 unique flagellar (H) antigens. Significant cross-reactions between *E. tarda* O and H antigens and antisera raised against similar

TABLE 11. *Biotypes of* Edwardsiella tarda

Character	*E. tarda* biogroups: Classic	Biogroup 1
Sources	Humans, reptiles, fish, water	Snakes, water
Acid from		
L-Arabinose	–	+
D-Mannitol	–	+
Sucrose	–	+

From Janda and Abbott.[79]

components of *Salmonella, Escherichia coli, Citrobacter freundii, Proteus vulgaris, Proteus mirabilis, Enterobacter cloacae,* and *H. alvei* have not been observed.[87]

Immune responses to *E. tarda* infection were detected in one survey and two case studies. Gilman and collaborators[45] found agglutinating antibodies to *E. tarda* in the sera of Malaysian aborigines suffering from bouts of bloody diarrhea. However, interpretation of these results is clouded by the fact that no difference in antibody levels was observed between persons harboring both *E. tarda* and *E. histolytica* in their stools and those containing *E. histolytica* alone. Chatty and Gavan[63] found agglutinating antibodies to *E. tarda* in the serum of a 53-year-old man with chronic intermittent diarrhea. Five months after his bout with severe diarrhea, a 1:160 tube agglutination titer for both O and H antigens was detected; sera from normal individuals were nonreactive. In 1983, a similar case of chronic intermittent bloody diarrhea was reported in a 2-month-old infant.[51] Two months after the disappearance of *E. tarda* in the baby's stool, his serum yielded a 1:80 tube agglutination titer against the infecting strain.

Typing

No outbreaks of disease due to *E. tarda* have been reported. Desenclos and coworkers[53] described a cluster of *E. tarda* isolates recovered from asymptomatic children and one adult at a day care center. Four of these isolates were available for further study, and all were identified as belonging to a single serotype, O4:H1,11. No further characterization was attempted.

PATHOGENESIS

Considerable progress has been made during the past decade in identifying potential virulence factors associated with human infections caused by *E. tarda.*[35] Two prominent candidates for major roles in *E. tarda* pathogenesis are invasiveness and expression of a β-hemolysin. Although several early studies failed to find any invasive activities in edwardsiellae, Marques and associates[88] found that all 12 *E. tarda* strains studied were invasive in HeLa cells. This study was subsequently confirmed and expanded by others using HEp-2 cells.[89] Virtually all *E. tarda* strains appear to be invasive in in vitro assays (Fig. 1), and the failure to detect common plasmids in these

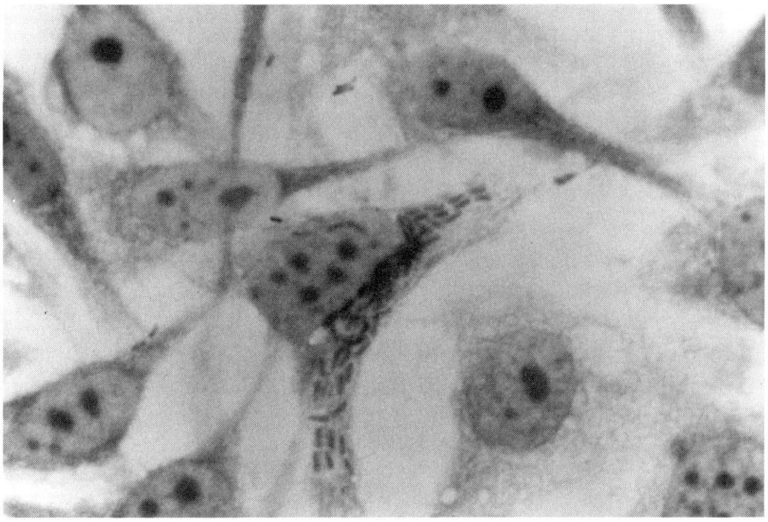

FIG. 1. Invasion of HEp-2 cells by edwardsiellae. From Janda et al.[89]

strains suggests that this is a chromosomally encoded trait.[90] Invasion appears to be a microfilament-dependent process, since cytochalasins B and D both inhibit invasion.[89] This invasive characteristic could conceivably be responsible for sporadic cases of bloody diarrhea, enterocolitis, colitis, and dysentery caused by *E. tarda.* However, such activity would not explain the majority of *E. tarda* illnesses associated with secretory diarrhea.

Although *E. tarda* strains are most commonly described as being nonhemolytic on blood agar plates, early studies by Watson and White[81] and Ullah and Arai[91] using modified assays detected β-hemolytic activity in most *E. tarda.* In 1996, the β-hemolysin of one *E. tarda* strain was cloned, characterized, and sequenced.[92] The hemolysin gene of *E. tarda* is part of an operon encoding three open reading frames (ORFs), one of which (ORF III) encodes for a 34-kDa protein with hemolytic activity. This protein contains no cysteine moieties, and sequence analysis revealed no major homology between this gene product and the hemolysins of *E. coli* and *A. hydrophila.* A second iron-regulated hemolysin similar or identical to that described by Janda and Abbott[82] has been cloned and sequenced by Hirono and colleagues.[93] A 4782-bp ORF, *ethA,* encodes for the 165.3-kDa secreted protein. The deduced amino acid sequence of this gene shows homology to the hemolysins of *P. mirabilis* and *Serratia marcescens.* This iron-regulated hemolysin appears to be the predominant hemolysin expressed by most *E. tarda* strains. The *E. tarda* β-hemolysin could conceivably function as a virulence factor in human cases of dysentery or hemorrhagic septicemia in cold-blooded animals, or, alternatively, it may play a role in the replication and release of progeny from infected eucaryotic cells.

A heat-stable enterotoxin was found in 3 of 25 *E. tarda* strains.[84] The enterotoxin, which required concentration by ultrafiltration, was detectable in suckling mouse and rabbit ileal loop assays. Other reports have not detected such activity in cell-free (unconcentrated) filtrates.[88,90,91] A heat-stable enterotoxin could be responsible for the common form of diarrhea (enteritis) associated with *E. tarda* infection. In addition to enterotoxigenic activity, a number of other properties have been postulated to possibly play a role in the pathogenicity of edwardsiellae. Some strains produce slime,[91] although one study which included ultrastructural analysis found no significant glycocalyx associated with ruthenium tetroxide-stained cells.[94] Fimbriae in *E. tarda* have not been described,[91,94] but afimbrial hemagglutinins have been detected.[94] Ullah and Arai[95] partially purified two dermatonecrotic exotoxins from *E. tarda* that produced erythema or edema and necrosis when injected into the depilated backs of rabbits. Further characterization of these substances has not occurred. Median lethal doses (LD_{50}) for edwardsiellae in outbred Swiss-Webster mice range from 3.1×10^6 to 1.3×10^9 cfu.[90]

ANTIMICROBIAL SUSCEPTIBILITY

There is good agreement in the literature that all *E. tarda* strains, regardless of source or site of isolation, are susceptible to all groups of antimicrobial agents commonly administered in the treatment of gram-negative bacillary infections.[96–98] This includes the penicillins (ampicillin), cephalosporins, extended-spectrum penicillins (ticarcillin, piperacillin), cephamycins (cefoxitin), β-lactamase inhibitor combinations, the aminoglycosides, tetracycline, chloramphenicol, the quinolones, monobactams (aztreonam), carbapenems (imipenem), and sulfonamides (trimethoprim-sulfamethoxazole). Reinhardt and associates[96] found that ciprofloxacin, norfloxacin, and cefotaxime had the highest activities against *E. tarda* on a weight basis. Clark and colleagues[97] found that third-generation cephalosporins (cefoperazone, ceftriaxone), aztreonam, and ciprofloxacin had the lowest 90% minimal inhibitory concentrations (MIC_{90}) for 22 *E. tarda* strains tested. All *E. tarda* strains have been found to produce β-lactamases by one of several methods.[96–98] In light of their uniform susceptibility to ampicillin, however, it is unclear what the significance of these observations is.

CONCLUSION

The significance of the genus *Edwardsiella* in clinical microbiology rests on final proof that this species is a human enteropathogen. This will require development of an animal model in which to faithfully reproduce gastrointestinal disease. Persons with severe underlying illnesses, including liver disease and iron overload conditions, should avoid activities through which they may come in direct contact with these bacteria, because severe extraintestinal infections may ensue.

REFERENCES

1. Sakazaki R, Murata Y. The new group of the *Enterobacteriaceae,* the Asakusa group. *Jpn J Bacteriol* 1962;17:616–617.
2. Sakazaki R. A proposed group of the family *Enterobacteriaceae,* the Asakusa group. *Int Bull Bacteriol Nomen Taxon* 1965;15:45–47.
3. King BM, Adler DL. A previously undescribed group of *Enterobacteriaceae*. *Am J Clin Pathol* 1964;41: 230–232.
4. Radovsky J, Aldová E. Isolation of strains of the new *Enterobacteriaceae* group "Bartolomew" in Cuba. *J Hyg Epidemiol Microbiol Immunol* 1965;9:112–114.
5. Ewing WH, McWhorter AC, Escobar MR, Lubin AH. *Edwardsiella,* a new genus of *Enterobacteriaceae* based on a new species, *E. tarda. Int Bull Bacteriol Nomen Taxon* 1965;15:33–38.
6. Brenner DJ, Fanning GR, Steigerwalt AG. Polynucleotide sequence relatedness in *Edwardsiella tarda. Int J Syst Bacteriol* 1974;24:186–190.
7. Sakazaki R, Tamura K. Priority of the specific epithet *anguillimortiferum* over the specific epithet *tarda* in the name of the organism presently known as *Edwardsiella tarda. Int J Syst Bacteriol* 1975;25:219–220.
8. Hoshina T. On a new bacterium, *Paracolobactrum anguillimortiferum* n. sp. *Bull Jpn Soc Sci Fish* 1962;28: 162–164.
9. Farmer JJ III, Brenner DJ, Clark WA. Proposal to conserve the specific epithet *tarda* over the epithet *anguillimortiferum* in the name of the organism presently known as *Edwardsiella tarda. Int J Syst Bacteriol* 1976; 26:293–294.
10. Farmer JJ III, McWhorter AC. Genus X: *Edwardsiella* Ewing and McWhorter 1965, 37^{AL} In: Kreig NR, Holt JG, eds. *Bergey's manual of systematic bacteriology.* Baltimore: Williams & Wilkins, 1984;486–491.
11. Sakazaki R, Tamura K. Comment on a proposal of Farmer et al. to conserve the specific epithet *tarda* over the specific epithet *anguillimortiferum* in the name of the organism known as *Edwardsiella tarda. Int J Syst Bacteriol* 1978;28:130–131.
12. Sakazaki R, Tamura K. Chapter 143. The genus *Edwardsiella.* In: Balows A, Trüper HG, Dworkin M, Harder W, Schleifer K-H, eds. *The prokaryotes: A handbook on the biology of bacteria ecophysiology, isolation, identification, applications.* 2nd ed. Berlin: Springer-Verlag, 1991;2737–2743.
13. Johnson R, Colwell RR, Sakazaki R, Tamura K. Numerical taxonomy study of the *Enterobacteriaceae. Int J Syst Bacteriol* 1975;25:12–37.
14. Grimont PAD, Grimont F, Richard C, Sakazaki R. *Edwardsiella hoshinae,* a new species of *Enterobacteriaceae. Curr Microbiol* 1980;4:347–351.
15. Hawke JP, McWhorter AC, Steigerwalt AG, Brenner DJ. *Edwardsiella ictaluri* sp. nov., the causative agent of enteric septicemia of catfish. *Int J Syst Bacteriol* 1981; 31:396–400.
16. Sechter I, Shmilovitz M, Altmann G, et al. *Edwardsiella tarda* isolated in Israel between 1961 and 1980. *J Clin Microbiol* 1983;17:669–671.
17. White FH, Simpson CF, Williams LE Jr. Isolation of *Edwardsiella tarda* from aquatic animal species and surface waters in Florida. *J Wildl Dis* 1973;9:204–208.
18. Kourany M, Vasquez MA, Saenz R. Edwardsiellosis in man and animals in Panama: Clinical and epidemiological characteristics. *Am J Trop Med Hyg* 1977;26: 1183–1190.
19. King CH, Shotts EB. Enhancement of *Edwardsiella tarda* and *Aeromonas salmonicida* through ingestion by the ciliated protozoan *Tetrahymena pyriformis. FEMS Microbiol Lett* 1988;51:95–100.
20. Sakazaki R. Studies on the Asakusa group of *Enterobacteriaceae (Edwardsiella tarda). Jpn J Med Sci Biol* 1967;20:205–212.
21. Goldstein EJC, Agyare EO, Vagvolgyi AE, Halpern M. Aerobic bacterial oral flora of garter snakes: Development of normal flora and pathogenic potential for snakes and humans. *J Clin Microbiol* 1981;13:954–956.
22. Nguyen-Van-Ai, Nguyen-Duc-Hank, Le-Tien-Van, Nguyen-Van-Le, Nguyen-Thi Lan-Huong. Contribution à l'étude des *"Edwardsiella tarda"* isolés au Viet-Nam. *Bull Soc Pathol Exot* 1975;68:355–359.
23. Van Damme LR, Vandepitte J. Frequent isolation of *Edwardsiella tarda* and *Plesiomonas shigelloides* from healthy Zairese freshwater fish: A possible source of sporadic diarrhea in the tropics. *Appl Environ Microbiol* 1980;39:475–479.
24. Otis VS, Behler JL. The occurrence of salmonellae and *Edwardsiella* in the turtles of the New York zoological park. *J Wildl Dis* 1973;9:4–6.
25. Brenner DJ. Chapter 88. Introduction to the family *Enterobacteriaceae.* In: Starr MP, Stolp H, Trüper HG, Balows A, Schlegel HG, eds. *The prokaryotes: A handbook on habitats, isolation, and identification of bacteria.* Berlin: Springer-Verlag, 1981;1105–1127.
26. Cook RA, Tappe JP. Chronic enteritis associated with *Edwardsiella tarda* infection in Rockhopper penguins. *J Am Vet Med Assoc* 1985;187:1219–1220.
27. Nomura J, Aoki T. Morphological analysis of lipopolysaccharide from gram-negative fish pathogenic bacteria. *Fish Pathol* 1985;20:193–197.
28. Wallace LJ, White FH, Gore HL. Isolation of *Edwardsiella tarda* from a sea lion and two alligators. *J Am Vet Med Assoc* 1966;149:881–883.
29. White FH, Neal FC, Simpson CF, Walsh AF. Isolation of *Edwardsiella tarda* from an ostrich and an Australian skink. *J Am Vet Med Assoc* 1969;155:1057–1058.
30. Bockemühl J, Pan-Urai R, Burkhardt F. *Edwardsiella tarda* associated with human disease. *Pathol Microbiol* 1971;37:393–401.

31. Arambulo PV, Westerlund NC, Sarmiento RV, Abaga AS. Isolation of *Edwardsiella tarda:* A new genus of *Enterobacteriaceae* from pig bile in the Philippines. *Far East Med J* 1967;3:385–386.
32. Owens DR, Nelson SL, Addison JB. Isolation of *Edwardsiella tarda* from swine. *Appl Microbiol* 1974;27: 703–705.
33. Meyer FP, Bullock GL. *Edwardsiella tarda,* a new pathogen of channel catfish *(Ictalurus punctatus)*. *Appl Microbiol* 1973;25:155–156.
34. Chamoiseau G. Note sur le pouvior pathogene *d'Edwardsiella tarda.* Un cas de septicemie mortelle du pigeon. *Rev Élev Med Vét Pays Trop (Paris)* 1967;20: 493–495.
35. Janda JM, Abbott SL. Infections associated with the genus *Edwardsiella:* The role of *Edwardsiella tarda* in human disease. *Clin Infect Dis* 1993;17:742–748.
36. Ewing WH, McWhorter AC, Ball MM, Bartes SF. The biochemical reactions of *Edwardsiella tarda,* a new genus of *Enterobacteriaceae.* Atlanta: United States Public Health Service, Bureau of Disease Prevention and Environmental Control, National Communicable Disease Center, 1967:1–8.
37. Jordan GW, Hadley WK. Human infection with *Edwardsiella tarda. Ann Intern Med* 1969;70:283–288.
38. Koshi G, Lalitha MK. *Edwardsiella tarda* in a variety of human infections. *Indian J Med Res* 1976;64: 1753–1759.
39. Tan RJS, Lim E-W, Yeo M. Occurrence and medical importance of *Edwardsiella tarda* from clinical specimens. *Ann Acad Med Singapore* 1977;6:337–341.
40. Ovartlarnporn B, Chayakul P, Suma S. *Edwardsiella tarda* infection in Hat Yai hospital. *J Med Assoc Thai* 1986;69:599–603.
41. Jaruratanasirikul S, Kalnauwakul S. *Edwardsiella tarda:* A causative agent in human infections. *Southeast Asian J Trop Med Public Health* 1991;22:30–34.
42. Bhat P, Myers RM, Carpenter KP. *Edwardsiella tarda* in a study of juvenile diarrhoea. *J Hyg Camb* 1967;65: 293–298.
43. Kournay M, Vasquez MA, Mata LJ. Prevalence of pathogenic enteric bacteria in children of 31 Panamanian communities. *Am J Trop Med Hyg* 1971;20:608–615.
44. Sakazaki R, Tamura K, Prescott LM, Bencic Z. Bacteriological examination of diarrheal stools in Calcutta. *Indian J Med Res* 1971;59:1025–1034.
45. Gilman RH, Madasamy M, Gan E, Mariappan M, Davis CE, Kyser KA. *Edwardsiella tarda* in jungle diarrhoea and a possible association with *Entamoeba histolytica. Southeast Asian J Trop Med Pubic Health* 1971;2: 186–189.
46. Vandepitte J, Van Damme L, Fofana Y, Desmyter J. *Edwardsiella tarda* et *Plesiomonas shigelloides.* Leur rôle comme agents de diarrhées et leur épidémiologie. *Bull Soc Pathol Exot* 1980;73:139–149.
47. Iveson JB. Enrichment procedures for the isolation of *Salmonella, Arizona, Edwardsiella,* and *Shigella* from faeces. *J Hyg Camb* 1973;71:349–361.
48. Marsh PK, Gorbach SL. Invasive enterocolitis caused by *Edwardsiella tarda. Gastroenterology* 1982;82: 336–338.
49. Nagel P, Serritella A, Layden TJ. *Edwardsiella tarda* gastroenteritis associated with a pet turtle. *Gastroenterology* 1982;82:1436–1437.
50. Reina J, Ros MJ. Gastroenteritis associated with *Edwardsiella tarda* in a patient infected by human immunodeficiency virus. *Clin Microbiol Newslett* 1996;18:29–30.
51. Vandepitte J, Lemmens P, De Swert L. Human edwardsiellosis traced to ornamental fish. *J Clin Microbiol* 1983;17:165–167.
52. Clarridge JE, Musher DM, Fainstein V, Wallace RJ Jr. Extraintestinal human infection caused by *Edwardsiella tarda. J Clin Microbiol* 1980;11:511–514.
53. Desenclos J-CA, Conti L, Junejo S, Klontz KC. A cluster of *Edwardsiella tarda* infection in a day-care center in Florida [letter]. *J Infect Dis* 1990;162:782–783.
54. Sonnenwirth AC, Kallus BA. Meningitis due to *Edwardsiella tarda:* First report of meningitis caused by *E. tarda. Am J Clin Pathol* 1968;49:92–95.
55. Okubadejo OA, Alausa KO. Neonatal meningitis caused by *Edwardsiella tarda. Br Med J* 1968;3:357–358.
56. Pankey GA, Seshul MB. Septicemia caused by *Edwardsiella tarda. J La State Med Soc* 1969;121:41–43.
57. Sachs JM, Counts GW. Sickle hemoglobinopathy and *Edwardsiella tarda* meningitis. *Am J Dis Child* 1974;128:387–388.
58. Le Frock JL, Klainer AS, Zuckerman K. *Edwardsiella tarda* bacteremia. *South Med J* 1976;69:188–190.
59. Martínez LM. *Edwardsiella* bacteremia [letter]. *Eur J Clin Microbiol* 1987;6:599–600.
60. Vohra K, Torrijos E, Jhaveri R, Gordon H. Neonatal sepsis and meningitis caused by *Edwardsiella tarda. Pediatr Infect Dis J* 1988;7:814–815.
61. Wilson JP, Waterer RR, Wofford JD Jr, Chapman SW. Serious infections with *Edwardsiella tarda:* A case report and review of the literature. *Arch Intern Med* 1989;149:208–210.
62. Wu M-S, Shyu R-S, Lai M-Y, Huang G-T, Chen D-S, Wang T-H. A predisposition toward *Edwardsiella tarda* bacteremia in individuals with preeexisting liver disease [letter]. *Clin Infect Dis* 1995;21:705–706.
63. Chatty HB, Gavan TL. *Edwardsiella tarda:* Identification and clinical significance. *Cleve Clin Q* 1968;35:223–228.
64. Rao KRP, Shah J, Rajashekaraiah KR, Patel AR, Miskew DB, Fennewald PS. *Edwardsiella tarda* osteomyelitis in a patient with SC hemoglobinopathy. *South Med J* 1981;74:288–292.
65. Hargreaves JE, Lucey DR. Life-threatening *Edwardsiella tarda* soft-tissue infection associated with catfish puncture wound [letter]. *J Infect Dis* 1990;162:1416–1417.
66. Vartian CV, Septimus EJ. Soft-tissue infection caused by *Edwardsiella tarda* and *Aeromonas hydrophila* [letter]. *J Infect Dis* 1990;161:816.
67. Banks AS. A puncture wound complicated by infection with *Edwardsiella tarda. J Am Podiatr Med Assoc* 1992;82:529–531.
68. Murphey DK, Septimus EJ, Waagner DC. Catfish-related injury and infection: Report of two cases and review of the literature. *Clin Infect Dis* 1992;14:689–693.
69. Coutlée F, Saint-Jean LA, Plante R. Infection with *Edwardsiella tarda* related to a vascular prosthesis [letter]. *Clin Infect Dis* 1992;14:621–622.
70. Maserati R, Farina C, Valenti L, Filice C. Liver abscess caused by *Edwardsiella tarda. Boll 1st Sieroter Milan* 1985;64:419–421.
71. Zighelboim J, Williams TW Jr, Bradshaw MW, Harris RL. Successful medical management of a patient with multiple hepatic abscesses due to *Edwardsiella tarda. Clin Infect Dis* 1992;14:117–120.
72. Gonzalez AB, Ruffolo EH. *Edwardsiella tarda:* Etio-

logic agent in post-traumatic subgaleal abscess. *South Med J* 1966;59:340,346.
73. Pien FD, Jackson MT. Tuboovarian abscess caused by *Edwardsiella tarda. Am J Obstet Gynecol* 1995;173: 964–965.
74. Ruff WL, Scott-Cora J, Prout JD. *Edwardsiella tarda* associated with osteomyelitis. *J Natl Med Assoc* 1977;69: 871–872.
75. Iveson JB. Strontium chloride B and E.E. enrichment broth media for the isolation of *Edwardsiella, Salmonella* and *Arizona* species from tiger snakes. *J Hyg Camb* 1971;69:323–330.
76. Muyembe T, Vandepitte J, Desmyter J. Natural colistin resistance in *Edwardsiella tarda. Antimicrob Agents Chemother* 1973;4:521–524.
77. O'Hara CM, Rhoden DL, Miller JM. Reevaluation of the API 20E identification system versus conventional biochemicals for identification of members of the family *Enterobacteriaceae:* A new look at an old product. *J Clin Microbiol* 1992;123–125.
78. O'Hara CM, Miller JM. Evaluation of the autoSCAN-W/A system for rapid (2-hour) identification of members of the family *Enterobacteriaceae. J Clin Microbiol* 1992;30:1541–1543.
79. Janda JM, Abbott SL. *Edwardsiella* gastroenteritis. *ASCP Tech Sample* 1995;MB–6:1–7.
80. Waltman WD, Shotts EB, Hsu TC. Biochemical and enzymatic characterization of *Edwardsiella tarda* from the United States and Taiwan. *Fish Pathol* 1986;21:1–8.
81. Watson JJ, White FH. Hemolysins of *Edwardsiella tarda. Can J Comp Med* 1979;43:78–83.
82. Janda JM, Abbott SL. Expression of an iron-regulated hemolysin by *Edwardsiella tarda. FEMS Microbiol Lett* 1993;111:275–280.
83. Walton DT, Abbott SL, Janda JM. Sucrose-positive *Edwardsiella tarda* mimicking a biogroup 1 strain isolated from a patient with cholelithiasis. *J Clin Microbiol* 1993;31:155–156.
84. Bockemühl J, Aleksić V, Wokatsch R, Aleksić S. Pathogenicity tests with strains of *Edwardsiella tarda:* Detection of a heat-stable enterotoxin. *Zentralbl Bakteriolt Hyg I Abt Orig A* 1983;255:464–471.
85. Farmer JJ III, Davis BR, Hickman-Brenner FW, et al. Biochemical identification of new species and biogroups of *Enterobacteriaceae* isolated from clinical specimens. *J Clin Microbiol* 1985;21:46–76.
86. Tamura K, Sakazaki R, McWhorter AC, Kosako Y. *Edwardsiella tarda* serotyping scheme for international use. *J Clin Microbiol* 1988;26:2343–2346.
87. Sakazaki R. Serological typing of *Edwardsiella tarda. Methods Microbiol* 1984;15:213–225.
88. Marques LRM, Toledo MRF, Silva NP, Magalhães M, Trabulsi LR. Invasion of HeLa cells by *Edwardsiella tarda. Curr Microbiol* 1984;10:129–132.
89. Janda JM, Abbott SL, Oshiro LS. Penetration and replication of *Edwardsiella* spp. in HEp-2 cells. *Infect Immun* 1991;59:154–161.
90. Janda JM, Abbott SL, Kroske-Bystrom S, et al. Pathogenic properties of *Edwardsiella* species. *J Clin Microbiol* 1991;29:1997–2001.
91. Ullah A, Arai T. Pathological activities of the naturally occurring strains of *Edwardsiella tarda. Fish Pathol* 1983;18:65–70.
92. Chen J-D, Lai S-Y, Huang S-L. Molecular cloning, characterization, and sequencing of the hemolysin gene from *Edwardsiella tarda. Arch Microbiol* 1996;165: 9–17.
93. Hirono I, Tange N, Aoki T. Iron-regulated haemolysin gene from *Edwardsiella tarda. Mol Microbiol* 1997;24: 851–856.
94. Wong JD, Miller MA, Janda JM. Surface properties and ultrastructure of *Edwardsiella* species. *J Clin Microbiol* 1989;27:1797–1801.
95. Ullah A, Arai T. Exotoxic substances produced by *Edwardsiella tarda. Fish Pathol* 1983;18:71–75.
96. Reinhardt JF, Fowlston S, Jones J, George WL. Comparative in vitro activities of selected antimicrobial agents against *Edwardsiella tarda. Antimicrob Agents Chemother* 1985;27:966–967.
97. Clark RB, Lister PD, Janda JM. In vitro susceptibilities of *Edwardsiella tarda* to 22 antibiotics and antibiotic–β-lactamase inhibitor agents. *Diagn Microbiol Infect Dis* 1991;14:173–175.
98. Reger PJ, Mockler DF, Miller MA. Comparison of antimicrobial susceptibility, β-lactamase production, plasmid analysis and serum bactericidal activity in *Edwardsiella tarda, E. ictaluri* and *E. hoshinae. J Med Microbiol* 1993;39:273–281.

18
The Genus *Hafnia*

TAXONOMY

As with many enteric bacteria, the genus *Hafnia* has a complex past in regard to taxonomy and nomenclature. Early studies conducted between the years 1905 and 1919 reportedly identified these bacteria under the epithets *"Bacillus asciaticus (var. mobilis)"* or *"Bacillus paratyphi alvei."*[1] However, these identifications have been questioned because of the failure to maintain extant isolates from these studies and because of reported biochemical reactions which are inconsistent for inclusion in the genus *Hafnia.*[2] Probably the first legitimate description of this group comes from the studies of Stuart and coworkers[3] on the so-called "paracolon" bacteria. Stuart and colleagues studied lactose-negative or late lactose-fermenting bacteria and found an atypical biotype within the genus *Aerobacter,* which they designated 32011.[3] Biotype 32011 produced gas from glucose, was Voges-Proskauer–positive (VP^+), failed to ferment sucrose and lactose, and was citrate-negative. All of these properties are consistent with the genus *Hafnia.* In 1954, Møller[4] published one of a series of seminal investigations on decarboxylase activity of members within the family *Enterobacteriaceae.* One group of 30 VP^+ strains which were lysine decarboxylase (LDC)– and ornithine decarboxylase (ODC)–positive and arginine dihydrolase (ADH)–negative were found to represent a new group. Møller coined the name "Hafnia group" to refer to these phenotypically distinct bacteria. Included within the Hafnia group were several strains previously identified as *"Bacillus cadaveris"* and *"Paratyphus alvei."* Møller proposed that since *"Paratyphus alvei"* had been originally isolated from the intestinal tract of bees in 1919, the appropriate type species for the genus should be *Hafnia alvei.* In 1961, Sakazaki performed biochemical and serologic studies on 294 cultures of the Hafnia group and concluded that they deserved species rank; he proposed that they be included in the recently redesignated genus *Enterobacter ("Aerobacter")* as *Enterobacter alvei.*[5] In 1968, Ewing and Fife[2] proposed the name *Enterobacter hafniae* after a systematic review of the relevant taxonomic literature.

Over the next few years the correct taxonomic position of the Hafnia group was in question, and its assignment vacillated between the genera *Enterobacter* and *Hafnia.* However, this controversy was eventually decided by an increasing volume of experimental data indicating that this group was genetically distinct. Johnson and collaborators[6] studied 10 *H. alvei* strains using numerical taxonomy and found that this was a biochemically homogenous group which was phenotypically distinct from the *Enterobacter-Klebsiella-Serratia* complex. This finding seemed to warrant its residing in its own genus, *Hafnia,* rather than within the genus *Enterobacter.* Later numerical taxonomy studies supported these general conclusions by grouping all hafniae within a single taxon at a similarity value of 87.5%, using a simple matching coefficient and an unweighted average-linkage clustering algorithm.[7] A second supporting piece of evidence came from observed differences in G+C

TABLE 1. *Current taxonomic makeup within* Hafnia alvei

DNA Group	Reference Strains	Biochemical Properties	Source	Other Designations
1	ATCC 13337	Typical	Feces	CDC 434-68; Stuart's type 32011
	ATCC 29926	Typical	Human[a]	CDC 5632-72
	CDC1946-74	Typical	—	
2	ATCC 29927	Typical	Human[a]	CDC 4510-73
	CDC 4798-72	Typical	—	
3	None	Atypical	—	

Data from Brenner,[12] Brenner,[13] and Gelev et al.[14]
[a]Anatomic site not specified.
—, not specified.

content. Although *Enterobacter* species have G+C contents ranging from 52 to 58 mol%, several independent studies measured the G+C content of *H. alvei* strains at 45 to 49 mol%.[6–8] This indicated that *H. alvei* strains were not closely related to members of the genus *Enterobacter.* Further support for this argument came from DNA relatedness studies conducted on members of the *Enterobacteriaceae.* Brenner and coinvestigators[9,10] at the Centers for Disease Control and Prevention (CDC) found that the labeled DNAs of reference strains of *H. alvei* were no more closely related to *Enterobacter* species (18%–23%) than they were to other members of this family (17%–24%). Therefore, the collective facts from three independent lines of investigation (numerical taxonomy, G+C content, DNA relatedness) indicated that the Hafnia group was sufficiently distinct to warrant its own genus, *Hafnia.*

The genus *Hafnia* presently consists of a single species, *H. alvei.* Although this species has been thought to be relatively homogenous, one report found two distinct taxa within *H. alvei* based on the rapidity of fermentation of lactose, raffinose, and adonitol.[11] In 1976, Steigerwalt and associates[9] conducted DNA–DNA hybridization experiments on four strains of *H. alvei.* Using labeled DNA from strain CDC 4360-67, two distinct relatedness groups within *H. alvei* could be discerned. One group, typified by 4360-67, was found to be 85% to 100% related to labeled DNA at 60°C, with a 5.5% divergence (mismatch). A second group was found to be only 51% to 55% related to 4360-67 under identical conditions, with more than 13% divergence. The reference strain for this second group was ATCC 13337. These results suggested the possible existence of significant genetic diversity within *H. alvei.* Subsequent investigations by Brenner and colleagues[10,12] confirmed these initial results and indicated that at least three DNA relatedness groups exist within *H. alvei* (Table 1). Two of these relatedness groups conform to the general definition of *H. alvei* (Table 2), while the third appears to be more closely related to members of *Salmonella, Escherichia, Citrobacter,* and *Enterobacter* than to *Hafnia.*[12] *H. alvei* DNA groups 1 and 2 appear to occur with approximately equal fre-

TABLE 2. *Genetic relatedness of DNA groups within* Hafnia alvei

	DNA Relatedness (%)			
H. alvei Labeled DNA	*H. alvei* DNA 1	*H. alvei* DNA 2	*H. alvei* DNA 3	Other Enterics
DNA group 1	78–85	57	27	17–24
DNA group 2	51–59	86	27	—
DNA group 3	27	27	90	—

Adapted from Steigerwalt et al.,[9] Brenner,[12] and Gelev et al.[14]

TABLE 3. *Phenotypic properties separating DNA groups within* Hafnia alvei

Characteristic	DNA Group 1	DNA Group 2	DNA Group 3
Susceptibility to *Hafnia* phage 1672	+	+	−
Voges-Proskauer	+	+	−
Catalase	+	+	−
Acid from D-fructose	+	+	−

Data from Brenner.[12]

quency in clinical specimens.[13] In addition to these observations, it has also been demonstrated that the previously recognized *Obesumbacterium proteus* biogroup 1[8] is genetically indistinguishable from *H. alvei* DNA group 1[12]; therefore the type strain of this group, ATCC 12841, is actually an anaerogenic biogroup of *H. alvei.*[13] In summary then, at least three distinct DNA groups presently exist within the *"H. alvei"* complex; two of these belong in the genus *Hafnia,* and the third (*"Hafnia"* species 3) eventually may be placed in *Escherichia* or in a new genus.[12] Since the type strain of *H. alvei* (ATCC 13337) resides within DNA group 1, this group would retain the name *H. alvei* (sensu stricto); DNA group 2 would in all likelihood constitute a second species within *Hafnia.* Biochemical properties useful in separating true *Hafnia* organisms from "Species 3" (DNA group 3) are listed in Table 3.

EPIDEMIOLOGY

H. alvei has been recovered from many different environmental and clinical sources. Aquatic habitats found to contain hafniae include drinking water, fish farm water, and sewage.[15,16] Among animals, *H. alvei* has been isolated from birds (including chickens), fish, snakes, swine, horses, rabbits, and rodents.[4,15–18] On at least two separate occasions *H. alvei* has been reported to cause disease in animals. Kume[17] reported a case of equine abortion associated with *H. alvei;* hafniae were recovered in almost pure culture from both the aborted fetus and lochia (vaginal discharge) of the dam. In 1990, an epizootic outbreak of hemorrhagic septicemia in rainbow trout was linked to a gram-negative bacillus referred to as FTB (fish pathogenic bacterium). FTB-like bacteria were recovered in pure culture from both parenchymal organs (acute phase) and subcutaneous necrotic tissues (subacute phase). Genetic analysis of these strains by DNA–DNA hybridization assays revealed extensive homology (84% at 60°C) with *H. alvei* DNA group 1, and biochemical test results were consistent with that final identification.[14] In addition to these sources, *H. alvei* has been recovered from a variety of food samples, including dairy products[16] and refrigerated meats.[19]

H. alvei can be isolated as part of the normal resident microflora of humans and from a number of infected anatomic sites including blood, wounds, and urine. They are commonly isolated from the gastrointestinal tract as normal flora. Matsumoto[20] screened more than 1900 stools of healthy Japanese subjects for the presence of hafniae and found an overall prevalence rate of 13.0%. Anecdotal information from other surveys indicates *Hafnia* fecal isolation rates as high as 42%.[21] Nakaya and colleagues[22] found *H. alvei* to be one of the most frequently isolated *Enterobacteriaceae* among the gram-negative vaginal flora of women with preterm labor. Dibb[23] sampled the outer ear canals of 77 Norwegian men and women for microbial flora. *H. alvei* was the only gram-negative bacillus found (in approximately 4% of subjects).

CLINICAL SIGNIFICANCE

Hafniae are relatively uncommon clinical isolates (Table 4). They are most frequently isolated from fecal specimens and only occasionally recovered from other body sites.[24]

TABLE 4. *Clinical distribution of* Hafnia alvei

Anatomic Site	Occurrence[a]	Clinical Frequency (%)[b]	Clinical Significance[c]
Gastrointestinal tract	++	0–4	4
Respiratory tract	+	<1	3
Genitourinary tract	+	<0.1–0.6	2
Wounds/abscesses	+	<0.1	2
Blood	R	<0.5	1
Miscellaneous	R	<0.01	1

[a]Relative frequency of *H. alvei* isolation at different body sites: ++, common site; +, occasionally isolated; R, rarely recovered.
[b]Frequency of *H. alvei* isolation relative to other gram-negative, facultative bacilli.
[c]Clinical significance: 1, usually significant; 2, often significant; 3, occasionally significant but most often not; 4, significance undetermined.

Selected surveys show extreme variation in the fecal isolation rate of *H. alvei,* ranging from 0% to 42%. Such differences may be partially explained by geographic factors, patient populations (symptomatic, asymptomatic), dietary habits, and detection methods. Günthard and Pennekamp[25] reviewed the association of *H. alvei* with human disease with particular emphasis on extraintestinal infections. They found that most hafniae were recovered in mixed culture from persons with underlying illnesses (>90%). Few strains were judged to be clinically significant (~5%).

Gastroenteritis

The earliest descriptions of *H. alvei* as a possible agent of diarrheal disease come from the studies of Stuart and colleagues[3] on the paracolon biogroup 30211. They reported that one laboratory worker became infected with a biogroup 30211 strain during a serologic analysis of paracolon bacteria and subsequently developed gastroenteritis. Over the next several decades, sporadic reports appeared associating *H. alvei* with diarrheal disease.[5,26] In the late 1970s, Ratnam and associates[27] described two outbreaks of *H. alvei*–associated gastroenteritis in a Newfoundland hospital which occurred within 1 month of each other (Table 5). Overall, 40 adults were affected, and although the diarrhea was self-limiting in nature, symptoms included diarrhea, abdominal cramps, and nausea. Follow-up cultures on symptomatic patients were negative for *H. alvei* with only one exception. Further evidence supporting the association of hafniae with bacterial gastroenteritis comes from a reported case of diarrhea in a 9-month-old girl whose fecal culture yielded pure growth of *H. alvei.*[29] The *H. alvei* strain isolated from this infant produced diarrhea in 75% of adult rabbits tested (n=12) in an intestinal tie model. Histologic changes

TABLE 5. *Outbreaks of* Hafnia alvei*–associated gastroenteritis*

Character	Outbreak 1	Outbreak 2
No. patients	15	25
No. with fecal culture	8	15
No. positive for *H. alvei*	5 (63%)	8 (53%)
No. with pure/heavy growth of *H. alvei*	5 (63%)	8 (53%)
Duration of symptoms (days)	1–2	1–2
H. alvei serotype	O3:H$^-$	O3:H$^-$
No. controls screened (asymptomatic)	75	73
No. positive for *H. alvei*	3 (4%)	2 (3%)
No. with pure/heavy growth of *H. alvei*	0 (0%)	0 (0%)

Data from Ratnam et al.[27] and Ratnam.[28]

noticed in infected rabbits included loss of the intact microvillus border in both ileum and colon; ultrathin sections of the ileum demonstrated hafniae adhering to the intestinal epithelium in a pedestal-like formation similar to that previously reported for enteropathogenic *Escherichia coli* (EPEC).[29]

Since these reports, additional lines of evidence have been published that lend credence to the belief that hafniae are indeed involved in diarrheal illness. Case studies of persons presenting with *H. alvei*–associated gastroenteritis have demonstrated fairly consistent symptomatology, including diarrhea and cramping abdominal pain.[27,30] On rare occasions, more severe symptoms have been described, including fever, chills, myalgia, and subsequent arthropathies.[31] In children, symptoms of *H. alvei* infection often include nausea and vomiting, with six to eight bowel movements per day noted[32]; mild to moderate dehydration may also be present.[29,32] Symptoms have been reported to persist for between 3 to 7 days or longer. In each of these reports, hafniae were isolated from symptomatic persons either in pure culture or as heavy predominant growth in the absence of other recognized bacterial enteropathogens. Resolution of symptoms in most instances was accompanied by the disappearance of hafniae from feces. Ridell and coinvestigators studied Finnish tourists returning from travel to Morocco.[33] Sixteen percent of these travellers with diarrhea had *H. alvei* in their stools, 5% as the only potential enteropathogen. Asymptomatic persons (n=321) were uniformly negative for hafniae. However, a Canadian study of 11 *H. alvei* recovered from children with diarrhea failed to find any strain with in vitro virulence-associated characteristics similar to those described by Albert and collaborators[29] for his prototype diarrheagenic strain 19982.[34] These 11 strains also differed from 19982 in their outer membrane protein profiles, plasmid composition, and pulsed-field gel electrophoresis patterns,[34] raising the possibility that the Bangladesh strains may be extremely biochemically aberrant *H. alvei.*

Respiratory Tract Infections

Although the respiratory tract is one of the more common anatomic sites from which hafniae can be isolated, few pulmonary infections have been documented to be caused by this bacterium. Most *H. alvei* isolates from the respiratory tract arise from patients with one of several underlying medical complications such as chronic respiratory illness, malignancy, or heart disease; intubation is a frequent predisposing factor.[25] Most true infections appear to be nosocomially transmitted. In one retrospective investigation, only two of eight respiratory cultures (mouth, throat, sputum, trachea) containing hafniae were judged to be clinically significant.[24] Both infections involved *H. alvei* as secondary pathogens in fatal cases of bronchopneumonia; one of these two patients additionally had *H. alvei* recovered from postmortem specimens of the lung. Klapholz and coworkers[35] examined seven respiratory isolates of *H. alvei* recovered from their hospital over a 3-year period and could find compelling evidence for clinical significance in only one case. This culture was recovered from a 68-year-old woman with a left lower lung pneumonia. She was treated with aztreonam and showed marked improvement, although she later died from unspecified causes related to other serious underlying conditions (chronic obstructive pulmonary disease, heart disease, renal failure). A case of *H. alvei* pneumonia was reported in 1990 by Frick and associates[36]; in this instance, hafniae were isolated in pure culture from a bronchoscopic specimen from a person who had required mechanical ventilation for longer than a week. Most recently, a case of *H. alvei* community-acquired pneumonia, empyema, and bacteremia was reported in a 54-year-old woman who was infected with the human immunodeficiency virus (HIV).[37] Sputum and blood cultures were positive for *H. alvei.* She was treated intravenously with ticarcillin/clavulanate, to which she responded favorably.

Bacteremia

Invasive disease has been an infrequently reported medical complication of *H. alvei* infection. Among the 11 cases of *Hafnia* septicemia reported in the last 20 years, 8 occurred in children (cases 3, 4, 6, 9, and 11) or in adults older than 55 years of age (Table 6). One adult case involved *H. alvei* sepsis in a 20-year-old soldier who received a gunshot wound during the Middle East war.[42] In two of the three children with *Hafnia* bacteremia (cases 4 and 6), sepsis was concomitantly accompanied by pneumatosis intestinalis, a condition of submucosal or subserosal gas-filled cystic lesions of the gastrointestinal tract.[43] All 11 patients survived their septic crisis, although the infant with meningitis (case 3) died from *Pseudomonas aeruginosa* bacteremia less than 1 week after her *Hafnia* infection.

Most persons with *Hafnia* septicemia have presented with a variety of clinical symptoms and not with classic signs of gram-negative sepsis (ie, fever, chills, shock). In at least two instances (cases 1 and 2), catheterization of the urogenital tract was thought to be the mechanism by which hafniae entered the bloodstream; in two other cases, the portal of entry was judged to be the gastrointestinal tract (cases 4 and 6). All hafniae bacteremias (except case 11) were monomicrobic, and most persisted for only 2 to 3 days. However, Englund[38] described a case of *H. alvei* septicemia in an older adult that lasted for 30 days despite aggressive intervention which included aminoglycoside therapy. This strain of *H. alvei* was susceptible in vitro to all antibiotics administered (including ampicillin), and the reasons for its persistence in blood despite appropriate chemotherapeutic intervention are unknown.

Miscellaneous Infections

H. alvei is sometimes recovered from wounds or abscess material as a secondary pathogen.[24] In most instances, the significance of the isolate is clouded by the fact that it is recovered as part of a mixed microbial flora. Predisposing factors associated with hafniae wound infections include surgery, cancer, and trauma.[24,41] Agustin and Cunha[46] described a wound infection in a 44-year-old man who sustained multiple puncture wounds to his buttocks from carpet nails. He

TABLE 6. *Cases of* Hafnia alvei *bacteremia*

Case	Age/Sex	Symptoms	Underlying Disease	Positive Sites	Catheterized	Diarrhea	Ref.
1	58/M	Chest pain	None	Blood	Yes	No	38
2	69/M	Dehydration; failure to eat	Diabetes, chronic pyelonephritis	Blood	Yes	No	39
3	1/F	FUO, diarrhea	None	CSF	No	Yes	40
4	20 d/M	Necrotizing enterocolitis	Preterm delivery	Blood, feces	Yes	Yes	41
5	20/M	Fever	Trauma	Blood	No	No	42
6	6/M	Fever, abdominal pain	Leukemia	Blood	No	No	43
7	37/M	Sepsis, perforated colon	Acute pancreatitis	Blood, cyst fluid	No	No	25
8	49/F	Sepsis, perforated colon	Sigmoid diverticulitis	Blood, CVC	Yes	No	25
9	8/M	Fever, vomiting	HIV-positive	Blood	No	Yes	44
10	61/F	Liver rejection, cholestasis	Liver transplantation	Blood, liver	No	No	45
11	2/M	Acute liver rejection, fever	Liver transplantation, autoimmune hepatitis	Blood, liver	No	No	45

FUO, fever of unknown origin; CSF, cerebrospinal fluid; CVC, central venous catheter; HIV, human immunodeficiency virus.

subsequently developed a localized abscess which, when drained, yielded a pure culture of *H. alvei.* Although he remained afebrile, an elevated leukocyte count (15,000 cells/mm^3) was recorded. He was treated with ceftizoxime and recovered without any complications. An ophthalmic infection with *Salmonella* subgroup 3 and *H. alvei* was reported in a 55-year-old Hispanic woman who complained of blurred vision lasting longer than 4 months.[47] A vitreous tap grew *Salmonella* and *H. alvei,* whereas a previous blood culture was positive for *Salmonella* only. Despite intravenous chemotherapy, the condition of her left eye deteriorated and eventually required enucleation; only *Salmonella* was recovered from a second vitreous culture. These cumulative results in conjunction with the patient's previous medical history (snake powder use) suggest that *H. alvei* probably played an insignificant role in the development of her endophthalmitis. Günthard and Pennekamp[25] reported a case of *H. alvei* peritonitis in a 50-year-old man who had undergone endoscopic retrograde cholangiopancreatography for stenosis of the choledochus duct. The gastrointestinal tract was the suspected portal of entry. Hafniae are rarely isolated from the urogenital system. Most persons found to have *H. alvei* in their urine have underlying illnesses such as nephrosis, chronic glomerulonephritis, pyelonephritis, and cirrhosis.[24] Washington and colleagues[24] reported on three urine cultures containing *H. alvei,* two of which contained hafniae as the sole microorganism. All three cultures were judged to represent commensal flora and were thought to be clinically insignificant.

LABORATORY IDENTIFICATION

Isolation

Hafniae grow well on many less inhibitory media used for the isolation of gram-negative bacilli. On MacConkey, Hektoen enteric, desoxycholate, and xylose-lysine-desoxycholate agars they appear as transparent (colorless) colonies because of their inability to ferment disaccharides such as sucrose and lactose. *H. alvei* is also D-sorbitol–negative and is commonly recovered as a contaminant on sorbitol-MacConkey agar, a medium designed for the recovery of *E. coli* O157:H7 from feces.[48] On most media after overnight incubation, hafniae typically appear as smooth, convex colonies 2 to 3 mm in diameter with an entire edge; some isolates may exhibit an irregular border. On rare occasions lactose-positive variants of *H. alvei* have been isolated from clinical and environmental sources, including urine, pus, and bile.[49] The lactose-positive phenotype in almost half of these strains was transmissible to *Salmonella* Typhi, indicating that fermentation of this disaccharide was a plasmid-encoded trait.

H. alvei strains often fail to grow on more inhibitory media such as Salmonella-Shigella (25%–60% inhibition) or brilliant green (75%–100% inhibition) agar, and in some instances on bismuth sulfite agar also.[1,6,7] Selective media specifically designed for the isolation of hafniae from feces have not been developed. Although Salmonella-Shigella agar is inhibitory, it was used to determine the fecal carrier rate for *H. alvei* in normal persons in one survey.[20] Sakazaki[21] achieved a 42% isolation rate for hafniae from approximately 600 stools screened with the use of desoxycholate-lactose-sucrose-sorbitol agar. By comparison, only a 2% isolation rate for *H. alvei* was obtained on MacConkey agar. Suspect *H. alvei* colonies picked and screened to Kligler's or triple sugar iron (TSI) slants invariably exhibit an alkaline over acid reaction with gas and without H_2S formation (K/AG$^-$) because of their lactose/sucrose-negative phenotype.

Identification

Commercial Systems

Most commercial systems identify *H. alvei* fairly accurately (87%–100%), although an extensive number of strains have not been tested on most systems. Of the systems listed

TABLE 7. *Identification of* Hafnia alvei *using commercial systems*

System	Read Time (h)	No. Strains	CI	Inc	Correct ID (%)
Walk/Away	2	29	29	0	100
RapID onE	4	3	3	0	100
Vitek	4–18	14	13	1	93
Crystal E/NF	18–20	10	9	1	90
Micro ID-E/NF	21	9	8	1	89
API 20E	21–24	31	27	4	87
Crystal RS/E	5	5	4	1	80
Biolog GN	24	21	12	9	57

Data from references 50–60.
CI, correct identification; Inc, incorrect identification (also includes no identification category); ID, identification.

in Table 7, only the Biolog GN microplate had serious difficulties identifying hafniae. Regardless of whether the GN microplate was read manually or with an automated reader, about 50% of all *H. alvei* tested were not identified or were misidentified after a 24-hour incubation. The percentage of correct identifications was even lower when the reading was taken at less than 24 hours.[49] A small percentage of strains on some of the other systems listed in Table 7 require additional biochemical tests or extended incubation periods (48 hours) for correct identification.[52,54]

Conventional Tests

H. alvei isolates exhibit fairly consistent biochemical features which include motility, production of β-galactosidase, growth in KCN broth, and lack of gelatinase activity. They are also ODC- and LDC-positive and ADH-negative. Most strains produce acetylmethylcarbinol (VP^+), although this activity may be somewhat temperature dependent.[61] Citrate (Simmons') is not utilized as a carbon source for energy. Gas is produced from D-glucose, and mannitol, rhamnose, and xylose are additionally fermented; more than 50% of all hafniae also ferment cellobiose. Most of the other carbohydrates that are commonly used in the diagnostic laboratory are not attacked by *H. alvei.*

In 1968, Guinée and Valkenburg[62] isolated a bacteriophage, designated phage 1672, that was specific for the genus *Hafnia*. Phage 1672 lysed all 100 cultures of *H. alvei* but no isolates of other *Enterobacteriaceae* genera, including *Salmonella, Enterobacter, Citrobacter,* and *Klebsiella.* Another unusual property present in members of the genus *Hafnia* is L-proline aminopeptidase activity.[63] Only strains of *H. alvei* and certain *Serratia* species express this activity (yellow-green color) within 30 to 60 minutes of exposure to the chromogenic substrate, L-proline-4-nitroanilide.

Genomospecies and Biotypes

Studies indicate that the nomenspecies *H. alvei* is genetically heterogenous at the DNA level.[9,12] Although the two DNA groups cannot be unequivocally separated from each other by biochemical methods, most isolates can be assigned to one or the other based on a small number of phenotypic properties. As determined by Brenner,[12] motility at 24 hours is the single best differential test which separates the two DNA groups within *H. alvei* (Table 8). More than 90% of all DNA group 1 strains (*H. alvei* sensu stricto) are nonmotile

TABLE 8. *Separation of DNA groups within* Hafnia alvei

Character	DNA Group 1	DNA Group 2
Motility at 24 hours	–	+
β-Glucosidase activity	V	–
Fermentation of inulin	–	V
Sodium acetate utilization	–	V
Urease activity	V	–

+, >90% positive; –, <10% positive; V, variable (10%–90% positive).

at 24 hours, whereas all DNA group 2 strains are positive.[12] Another useful test is β-glucosidase activity. Approximately half of all DNA group 1 strains can hydrolyze glycosidic compounds such as salicin, esculin, and arbutin. DNA group 2 strains fail to attack these aglycone compounds. Conversely, more than 70% of all DNA group 2 strains produce acid from inulin, whereas 10% or fewer of the DNA group 1 strains do so.

The fact that certain biochemical properties (eg, β-glucosidase activity) are variable in frequency within *H. alvei* suggests that multiple biotypes may exist within each genomospecies. Farmer and colleagues[64] proposed the recognition of *H. alvei* biogroup 1. This group, previously designated *O. proteus* biogroup 1, is genetically related to the genus *Hafnia.*[12] Biogroup 1 strains, also referred to as the "brewery biogroup," are slow-growing, anaerogenic, nonmotile rods that fail to grow in KCN broth and do not produce acid from sugars commonly fermented by typical hafniae (eg, L-arabinose, L-rhamnose, D-xylose). They have not been recovered from clinical material to date. Studies to establish a biotyping system for clinical isolates also have not been published. Because of the likely possibility that certain strains of *H. alvei* are enteropathogenic, Ridell and colleagues[15] searched for biochemical markers that might identify this group. In a study of 16 *H. alvei* isolates bearing an EPEC-associated virulence gene *(eaeA),* they found a number of properties that separate these strains from the more predominant *eaeA*-negative isolates. These tests included assimilation of histidine, 2-keto-gluconate, and 3-hydroxy-benzoate; utilization of indoxyl phosphate and malonate; and fermentation of rhamnose.

Immunodiagnostics

A number of different antigenic schema have been developed for *H. alvei.*[61] Most of these date back to the early 1950s to mid-1960s. Such schemes have recognized between 7 and 39 separate somatic groups, 2 to 36 different flagellar antigens, and up to 130 distinct serotypes. Cross-reactions are known to occur between certain O groups of *H. alvei* and selected strains of *Salmonella, Enterobacter cloacae, E. coli,* and *Citrobacter freundii.* None of these serologic typing systems has found widespread use in either clinical or reference laboratories. Some strains of *H. alvei* have been reported to produce a K (slime) antigen, but this is not a widely accepted fact.[1]

Studies of the core oligosaccharide region (heptose residues, 3-deoxy-D-manno-octulosonic acid) of *H. alvei* by chemical analysis, gas chromatography–mass spectrometry, and nuclear magnetic resonance spectroscopy indicate that many strains share a common hexasaccharide skeleton, although the phosphorylation pattern is different in individual strains.[65] The O-specific region of *H. alvei* strains consists of either hexasaccharide or octasaccharide units[66]; some of these polysaccharide structures contain sialic acid at the nonreducing terminus.

Immune responses to *Hafnia* infection in humans have not been demonstrated to date. However, Kume[17] demonstrated a rise in an agglutinin titer to *H. alvei* in the acute and convalescent sera of an infected female horse.

Specialized Assays

Currently, specialized assays are required to identify *H. alvei* strains implicated as causative agents of infectious gastroenteritis. These strains can be recognized by using in vitro assays specifically developed to detect EPEC. The characteristic adherence mechanism (attaching and effacing lesions) produced by EPEC strains have also been demonstrated in some *H. alvei* isolates. Localized adherence, the in vitro adherence marker for EPEC, can be sought by screening of hafniae in HEp-2 cells. However, the localized adherence pattern produced by hafniae is significantly weaker than that exhibited by classic EPEC (or, in some cases, negative).[67] The fluorescent actin stain (FAS) test, which employs fluorescein isothiocyanate (FITC)–conjugated phalloidin to stain actin filaments, is positive both for EPEC and diarrhea-associated *H. alvei*. The immunofluorescent signal in FAS$^+$ *H. alvei* is weak to poor when compared with that of EPEC.[67]

The *eae* gene is responsible for the characteristic attaching and effacing lesions produced by EPEC and other selected enteric bacteria. Diarrheagenic *H. alvei* hybridize to the *eae* gene or probes of its associated locus with one of several molecular techniques, including Southern hybridization and polymerase chain reaction amplification.[15,33,67,68]

Typing

Although rarely implicated as a cause of outbreak-related disease,[27,28] *H. alvei* isolates can be typed for epidemiologic purposes by one of several methods. Biotyping not only provides a good approximation of genetic relatedness (DNA group) but also helps determine strain identity within a genomospecies. Phenotypic tests useful for such purposes include esculin and arbutin hydrolysis, salicin fermentation, and malonate utilization. Strains may also be typed serologically by one of several different antigenic schema. However, serotyping is a technique with limited utility at present because it is restricted to a few research or reference laboratories. One other reasonable approach to typing of *H. alvei* is plasmid analysis. Although this technique has not been investigated as an epidemiologic tool, some strains are known to carry one to four plasmids ranging in molecular mass from 1.5 to 130 MDa.[67]

PATHOGENESIS

Most of the limited research on the pathogenicity of hafniae has centered on the role of these bacteria as agents of diarrheal disease. In a removable intestinal tie model, an isolate of *H. alvei* recovered from a 9-month-old girl with gastroenteritis induced diarrhea in 8 of 12 adult rabbits injected with approximately 10^{10} cfu of this strain.[29] Diarrhea in these animals developed 1 to 4 days after inoculation and was characterized by soft stools containing mucus without blood. Histologic sectioning of selected animals demonstrated only *H. alvei* in the midjejunum and midileum sections. Electron micrographs of thin sections of the infected ileum showed destruction of the microvilli border and cupping or pedestal formation.

A number of additional *H. alvei* strains implicated in diarrhea were also studied by Albert and his colleagues.[29] All of these strains are noninvasive in both cell culture (HeLa) and animal (Serény) models and fail to produce heat-stable enterotoxins (suckling mice), heat-labile enterotoxins (Y1 adrenal cells), or Shiga-like cytotoxins.[67] However, they consistently produce attaching and effacing lesions in rabbit ileal loops and are FAS$^+$.[67] Although they do not hybridize to the EAF probe (plasmid-mediated) of EPEC, they show homologous sequences to the *eae* gene of EPEC.[67,68] The C-terminal portions of the *eae*-like genes of *H. alvei, Citrobacter rodentium* (formerly *C. freundii* biotype 4280), enterohemorrhagic (Shiga-toxin–producing) *E. coli* (STEC), and EPEC show extensive sequence similarities of about 80%.[69] These *eae*-encoded proteins, termed intimins, are homologous to the invasion gene products of *Yersinia enterocolitica* and *Yersinia pseudotuberculosis.*

Outside of research on the role of hafniae as a cause of diarrhea, very little additional information is known regarding the pathogenic potential of *H. alvei. H. alvei* has been shown to produce an aerobactin-like siderophore by bioassay, but only one strain was studied.[70] Other studies, by Rutz and associates,[71] showed buried (outer membrane bilayer) and exposed (cell-surface) ferric enterobactin epitopes in *H. alvei* using immunoblot and cytofluorimetric analyses with monoclonal antibodies. Capsules have not been detected. Fewer than 20% of hafniae tested (n=26) were found to be resistant to the normal bactericidal activity of fresh bovine serum.[72] The differential susceptibility of certain hafniae to complement-mediated lysis appears to be related to their lipopolysaccharide architecture.

TABLE 9. *Susceptibility of* Hafnia alvei *to newer antimicrobial agents*

Agent	MIC_{50} (µg/mL)	MIC_{90} (µm/mL)	Susceptible (%)	Ref.
Aztreonam	NA	0.12	>90	74
Cefotaxime	2.9	21	88	75,76
Ceftizoxime	4	>32	88	75
Ceftazidime	16	>16	59	75
Cefoperazone	16	>32	76	75
Ciprofloxacin	1	1	97	77
Ofloxacin	0.06	0.025	100	78
Imipenem	0.12	1	>90	79
Amoxicillin-clavulanic acid	128	>128	0	80

NA, not available.

ANTIMICROBIAL SUSCEPTIBILITY

Because of their low incidence in human disease, large-scale investigations of the antimicrobial susceptibility profiles of *H. alvei* have not been performed to date. Early studies indicate that most hafniae are susceptible to the aminoglycosides, tetracycline, chloramphenicol, and carbenicillin but often are resistant to ampicillin and first-generation cephalosporins such as cephalothin.[1] Kapholz and coworkers[35] reviewed susceptibility results on seven respiratory isolates and found that the majority of strains were resistant to ampicillin and first- and second-generation cephalosporins. Some strains were also resistant to third-generation cephalosporins, with resistance noted in single isolates to piperacillin and in one instance to the monobactam aztreonam. Pancoast[73] reviewed the literature regarding the susceptibility of *H. alvei* to aminoglycosides and found that 80% to 90% of all strains surveyed were susceptible to gentamicin, tobramycin, amikacin, and netilmicin. Older aminoglycosides, such as streptomycin and kanamycin, were ineffective. Other studies have generally supported these observations.[25] Reports addressing the susceptibility of hafniae to the newer antimicrobial agents are listed in Table 9. In most of these studies, the number of *H. alvei* strains tested was relatively small (<20 isolates), with the exception of a ciprofloxacin survey.[77] Resistance to third-generation cephalosporins, and in particular ceftazidime, appears to be increasing. In some instances, ceftazidime resistance in *H. alvei* has been directly attributed to the production of an inducible Bush group 1 β-lactamase.[81]

CONCLUSIONS

No doubt the lack of current knowledge regarding *H. alvei* comes from its infrequent association with human disease. This may change if reports suggesting a possible role in gastrointestinal disease are confirmed. However, strains implicated in gastroenteritis to date possess very unusual biochemical properties for the genus *Hafnia.*[15,29] This suggests either that these strains represent a very rare biotype within the genus *Hafnia* or, alternatively, that they may not fall within *Hafnia* at all. Although serious disease with this group is uncommon, there are an increasing number of reports associating hafniae with extraintestinal infections.[25] Such reports suggest that more definitive studies should be undertaken to establish guidelines with respect to the proper selection of antimicrobial agents for treatment. Because significant resistance to third-generation cephalosporins has already been reported for this species, increased surveillance of emerging resistance within *H. alvei* seems warranted.

REFERENCES

1. Sakazaki R, Tamura K. The genus *Hafnia.* In: Balows A, Trüper HG, Dworkin M, Harder W, Schleifer K-H, eds. *The prokaryotes: A handbook on the biology of bacteria ecophysiology, isolation, identification, applications.* 2nd ed. Berlin: Springer-Verlag, 1991;2816–2821.

2. Ewing WH, Fife MA. *Enterobacter hafniae* (the "Hafnia group"). *Int J System Bacteriol* 1968;18:263–271.
3. Stuart CA, Wheeler KM, Rustigian R, Zimmerman A. Biochemical and antigenic relationships of the paracolon bacteria. *J Bacteriol* 1943;45:101–119.
4. Møller V. Distribution of amino acid decarboxylases in *Enterobacteriaceae. Acta Pathol Microbiol Scand* 1954; 35:259–277.
5. Sakazaki R. Studies on the hafnia group of *Enterobacteriaceae. Jpn J Med Sci Biol* 1961;14:223–241.
6. Johnson R, Colwell RR, Sakazaki R, Tamura K. Numerical taxonomy study of the *Enterobacteriaceae. Int J Syst Bacteriol* 1975;25:12–37.
7. Greipsson S, Priest FG. Numerical taxonomy of *Hafnia alvei. Int J Syst Bacteriol* 1983;33:470–475.
8. Priest FG, Somerville HJ, Cole JA, Hough JS. The taxonomic position of *Obesumbacterium proteus,* a common brewery contaminant. *J Gen Microbiol* 1973;75: 295–307.
9. Steigerwalt AG, Fanning GR, Fife-Asbury MA, Brenner DJ. DNA relatedness among species of *Enterobacter* and *Serratia. Can J Microbiol* 1976;22:121–137.
10. Brenner DJ. Characterization and clinical identification of *Enterobacteriaceae* by DNA hybridization. *Prog Clin Pathol* 1977;7:71–117.
11. Gavini F, Ferragut C, Lefebvre B, Leclerc H. Étude taxonomique d'entérobactéries appartenant ou apparentées au genre *Enterobacter. Ann Microbiol (Inst Pasteur)* 1976;127B:317–335.
12. Brenner DJ. Introduction to the family *Enterobacteriaceae.* In: Starr MP, Stolp H, Trüper HG, Schlegel HG, eds. *The prokaryotes: A handbook on habitats, isolation, and identification of bacteria.* Berlin: Springer-Verlag, 1981;1105–1127.
13. Brenner DJ. Chapter 155. Additional genera of *Enterobacteriaceae.* In: Balows A, Trüper HG, Dworkin M, Harder W, Schleifer K-H, eds. *The prokaryotes: A handbook on the biology of bacteria ecophysiology, isolation, identification, applications.* 2nd ed. Berlin: Springer-Verlag, 1991;2922–2937.
14. Gelev I, Gelev E, Steigerwalt AG, Carter GP, Brenner DJ. Identification of the bacterium associated with haemorrhagic septicaemia in rainbow trout as *Hafnia alvei. Res Microbiol* 1990;141:573–576.
15. Ridell J, Siitonen A, Paulin L, Lindroos O, Korkeala H, Albert MJ. Characterization of *Hafnia alvei* by biochemical tests, random amplified polymorphic DNA PCR, and partial sequencing of 16S rRNA gene. *J Clin Microbiol* 1995;33:2372–2376.
16. Sakazaki R. Genus IX: *Hafnia* Møller 1954, 272[AL]. In: Krieg NR, Holt JG, eds. *Bergey's manual of systematic bacteriology*. Baltimore: Williams & Wilkins, 1984; 484–486.
17. Kume T. A case of abortion possibly due to *Hafnia* organism. *J Hokkaido Vet Assoc* 1962;6:1–4.
18. Goldstein EJC, Agyare EO, Vagvolgyi AE, Halpern M. Aerobic bacterial oral flora of garter snakes: Development of normal flora and pathogenic potential for snakes and humans. *J Clin Microbiol* 1981;13:954–956.
19. Ridell J, Korkeala H. Minimum growth temperatures of *Hafnia alvei* and other Enterobacteriaceae isolated from refrigerated meat determined with a temperature gradient incubator. *Int J Food Microbiol* 1997;35:287–292.
20. Matsumoto H. Studies on the *Hafnia* isolated from normal human. *Jpn J Microbiol* 1963;7:105–114.
21. Sakazaki R. The genus *Hafnia.* In: Starr MP, Stolp H, Trüper HG, Schlegel HG, eds. *The prokaryotes: A handbook on habitats, isolation, and identification of bacteria.* Springer-Verlag, 1981;1181–1186.
22. Nakaya T, Hanada S, Yagami Y. Implications of maternal blood endotoxins and preterm labor. *Int J Feto-Matern Med* 1990;3:196–204.
23. Dibb WL. The normal microbial flora of the outer ear canal in healthy Norwegian individuals. *Niph Annals* 1990;13:11–16.
24. Washington JA II, Birk RJ, Ritts RE Jr. Bacteriologic and epidemiologic characteristics of *Enterobacter hafniae* and *Enterobacter liquefaciens. J Infect Dis* 1971; 124:379–386.
25. Günthard H, Pennekamp A. Clinical significance of extraintestinal *Hafnia alvei* isolates from 61 patients and review of the literature. *Clin Infect Dis* 1996;22: 1040–1045.
26. Harada K, Shimizu K, Matsuyama T. *Hafnia* isolated from men. *Gunma J Med Sci* 1957;6:109–112.
27. Ratnam S, Butler RW, March S, Parsons RT. *Enterobacter hafniae*–associated gastroenteritis—Newfoundland. *Can Dis Wkly Rep* 1979;5:231–232.
28. Ratnam S. Etiologic role of *Hafnia alvei* in human diarrheal illness. *Infect Immun* 1991;59:4744–4745.
29. Albert MJ, Alam K, Islam M, et al. *Hafnia alvei,* a probable cause of diarrhea in humans. *Infect Immun* 1991;59:1507–1513.
30. Westblom TU, Milligan TW. Acute bacterial gastroenteritis caused by *Hafnia alvei. Clin Infect Dis* 1992;14: 1271–1272.
31. Newmark JJ, Hobbs WN, Wilson BE. Reactive arthritis associated with *Hafnia alvei* enteritis. *Arthritis Rheum* 1994;6:960–962.
32. Reina J, Hervas J, Borrell N. Acute gastroenteritis caused by *Hafnia alvei* in children. *Clin Infect Dis* 1993;16:443.
33. Ridell J, Siitonen A, Paulin L, Mattila L, Korkeala H, Albert HJ. *Hafnia alvei* in stool specimens from patients with diarrhea and healthy controls. *J Clin Microbiol* 1994;32:2335–2337.
34. Ismaili A, Bourke B, de Azavedo JCS, Ratnam S, Karmali M, Sherman PM. Heterogeneity in phenotypic and genotypic characteristics among strains of *Hafnia alvei. J Clin Microbiol* 1996;34:2973–2979.
35. Klapholz A, Lessnau K-D, Huang B, Talavera W, Boyle JF. *Hafnia alvei* respiratory tract isolates in a community hospital over a three-year period and a literature review. *Chest* 1994;104:1098–1100.
36. Frick T, Kunz M, Vogt M, Turina M. Typische nosokomiale infektion mit seltenem erregeri *Hafnia alvei. Schweiz Rundshau Med Prax* 1990;79:1092–1094.
37. Fazal BA, Justman JE, Turett GS, Telzak EE. Community-acquired *Hafnia alvei* infection. *Clin Infect Dis* 1997;24:527–528.
38. Englund GW. Persistent septicemia due to *Hafnia alvei. Am J Clin Pathol* 1969;51:717–719.
39. Mobley DF. *Hafnia* septicemia. *South Med J* 1971;64: 505–506.
40. Mojtabee A, Siadati A. *Enterobacter hafnia* meningitis. *J Pediatr* 1978;93:1062–1063.
41. Ginsberg HG, Goldsmith JP. *Hafnia alvei* septicemia in an infant with necrotizing enterocolitis. *J Perinatol* 1988;8:122–123.
42. Berger SA, Edberg SC, Klein RS. *Enterobacter hafniae*

infection: Report of two cases and review of the literature. *Am J Med Sci* 1977;277:101–104.
43. Yeager AM, Kanoff ME, Kramer SS, Jones B, Saral R, Lake AM, Santos GW. Pneumatosis intestinalis in children after allogeneic bone marrow transplantation. *Pediatr Radiol* 1987;17:18–22.
44. Conte M, Castagnola E, Venzano P, Tasso L, Giacchino R. Bacteremia caused by *Hafnia alvei* in a human immunodeficiency virus–infected child [letter]. *Pediatr Infect Dis J* 1996;15:182–183.
45. Barry JW, Dominguez EA, Boken DJ, Preheim LC. *Hafnia alvei* infection after liver transplantation. *Clin Infect Dis* 1997;24:1263–1264.
46. Agustin ET, Cunha BA. Buttock abscess due to *Hafnia alvei. Clin Infect Dis* 1995;20:1426.
47. Caravalho J Jr, McMillan VM, Ellis RB, Betancourt A. Endogenous endophthalmitis due to *Salmonella arizonae* and *Hafnia alvei. South Med J* 1990;83:325–327.
48. Feng P. *Escherichia coli* serotype O157:H7: Novel vehicles of infection and emergence of phenotypic variants. *Emerg Infect Dis* 1995;1:47–52.
49. Le Minor L, Coynault C. Déterminisme plasmidique du caractère atypique lactose positif de *Enterobacter hafniae* et de *Proteus morganii. Ann Microbiol (Inst Past)* 1976;127A:213–221.
50. Pfaller MA, Sahm D, O'Hara C et al. Comparison of the AutoSCAN-W/A rapid bacterial identification system and the Vitek AutoMicrobic system for identification of gram-negative bacilli. *J Clin Microbiol* 1991;29:1422–1428.
51. O'Hara CM, Tenover FC, Miller JM. Parallel comparison of accuracy of API 20E, Vitek GNI, MicroScan Walk/Away Rapid ID, and Becton Dickinson Cobas Micro ID-E/NF for identification of members of the family *Enterobacteriaceae* and common gram-negative, non-glucose-fermenting bacilli. *J Clin Microbiol* 1993;31:3165–3169.
52. Miller JM, Rhoden DL. Preliminary evaluation of Biolog, a carbon source utilization method for bacterial identification. *J Clin Microbiol* 1991;29:1143–1147.
53. O'Hara CM, Rhoden DL, Miller JM. Reevaluation of the API 20E identification system versus conventional biochemicals for identification of members of the family *Enterobacteriaceae:* A new look at an old product. *J Clin Microbiol* 1992;30:123–125.
54. O'Hara CM, Miller JM. Evaluation of the autoSCAN-W/A system for rapid (2-hour) identification of members of the family *Enterobacteriaceae. J Clin Microbiol* 1992;30:1541–1543.
55. York MK, Brooks GF, Fiss EH. Evaluation of the autoSCAN-W/A rapid system for identification and susceptibility testing of gram-negative fermentative bacilli. *J Clin Microbiol* 1992;30:2903–2910.
56. Kitch TT, Jacobs MR, Appelbaum PC. Evaluation of RapID onE system for identification of 379 strains in the family *Enterobacteriaceae* and oxidase-negative, gram-negative nonfermenters. *J Clin Microbiol* 1994;32:931–934.
57. Holmes B, Costas M, Ganner M, On SLW, Stevens M. Evaluation of Biolog system for identification of some gram-negative bacteria of clinical importance. *J Clin Microbiol* 1994;32:1970–1975.
58. Holmes B, Costas M, Thaker T, Stevens M. Evaluation of two BBL Crystal systems for identification of some clinically important gram-negative bacteria. *J Clin Microbiol* 1994;32:2221–2224.
59. Robinson A, McCarter YS, Tetreault J. Comparison of Crystal Enteric/Nonfermenter system, and Vitek Automicrobic system for identification of gram-negative bacilli. *J Clin Microbiol* 1995;33:364–370.
60. Wauters G, Boel A, Voorn GP, et al. Evaluation of a new identification system, Crystal Enteric/Non-Fermenter, for gram-negative bacilli. *J Clin Microbiol* 1995;33:845–849.
61. Ewing WH. *Identification of Enterobacteriaceae.* New York: Elsevier, 1986.
62. Guinée PAM, Valkenburg JJ. Diagnostic value of a *Hafnia*-specific bacteriophage. *J Bacteriol* 1968;96:564.
63. Fanghänel S, Reissbrodt R, Giesecke H. L-Prolineaminopepidase activity as a tool for identification and differentiation of *Serratia marcescens, Serratia liquefaciens* and *Hafnia alvei* strains. *Zentralbl Bakteriol* 1991;275:11–15.
64. Farmer JJ III, Davis BR, Hickman-Brenner FW, et al. Biochemical identification of new species and biogroups of *Enterobacteriaceae* isolated from clinical specimens. *J Clin Microbiol* 1985;21:46–76.
65. Gamian A, Katzenellenbogen E, Romanowska E, Dabrowski U, Dabrowski J. Lipopolysaccharide core region of *Hafnia alvei:* Structure elucidation using chemical methods, gas chromatography-mass spectrometry, and NMR spectroscopy. *Carbohydr Res* 1995;266:221–228.
66. Ravenscroft N, Gamian A, Romanowska E. 3-Deoxyoctulosonic-acid-containing hexasaccharide fragment of unusual core type isolated from *Hafnia alvei* 2 lipopolysaccharide. *Eur J Biochem* 1995;227:889–896.
67. Albert MJ, Faruque SM, Ansaruzzaman M, et al. Sharing of virulence-associated properties at the phenotypic and genetic levels between enteropathogenic *Escherichia coli* and *Hafnia alvei. J Med Microbiol* 1992;37:310–314.
68. McDaniel TK, Jarvis KG, Donnenberg MS, Kaper JB. A genetic locus of enterocyte effacement conserved among diverse enterobacterial pathogens. *Proc Nat Acad Sci U S A* 1995;92:1664–1668.
69. Frankel G, Candy DCA, Everest P, Dougan G. Characterization of the C-terminal domains of intimin-like proteins of enteropathogenic and enterohemorrhagic *Escherichia coli, Citrobacter freundii,* and *Hafnia alvei. Infect Immun* 1994;62:1835–1842.
70. Martinez JL, Cercenado E, Baquero F, Pérez-Díaz, Delgado-Iribarren A. Incidence of aerobactin production in gram-negative hospital isolates. *FEMS Microbiol Lett* 1987;43:351–353.
71. Rutz JM, Abdullah T, Singh SP, Kalve VI, Klebba PE. Evolution of the ferric enterobactin receptor in gram-negative bacteria. *J Bacteriol* 1991;173:5964–5974.
72. Jankowski S, Rowi ski S, Cisowska A, Gamian A. The sensitivity of *Hafnia alvei* strains to the bactericidal effect of serum. *FEMS Immunol Med Microbiol* 1996;13:59-64.
73. Pancoast SJ. Aminoglycoside antibiotics in clinical use. *Med Clin North Am* 1988;72:581–612.
74. Neu HC. Aztreonam: The first monobactam. *Med Clin North Am* 1988;72:555–566.
75. Jones RN, Pfaller MA, Allen SD, Gerlach EH, Fuchs PC, Aldridge KE. Antimicrobial activity of cefpirome: An update compared to five third-generation cephalosporins against nearly 6000 recent clinical isolates from

five medical centers. *Diagn Microbiol Infect Dis* 1991; 14:361–364.
76. Cormican MG, Jones RN. Antimicrobial activity of cefotaxime tested against infrequently isolated pathogenic species (unusual pathogens). *Diagn Microbiol Infect Dis* 1995;22:43–48.
77. Thornsberry C. Susceptibility of clinical bacterial isolates to ciprofloxacin in the United States. *Infection* 1994;22:S80–S89.
78. Chin N-X, Gu J-W, Yu K-W, Zhang Y-X, Neu HC. In vitro activity of sparfloxacin. *Antimicrob Agents Chemother* 1991;35:567–571.
79. Lipman B, Neu HC. Imipenem: A new carbapenem. *Med Clin North Am* 1988;72:567–579.
80. Hohl P, Lüthy-Hottenstein J, Zollinger-Iten J, Altwegg M. In vitro activities of fleroxacin, cefetamet, ciprofloxacin, ceftriaxone, trimethoprim-sulfamethoxazole, and amoxicillin-clavulanic acid against rare members of the family *Enterobacteriaceae* primarily of human (clinical) origin. *Antimicrob Agents Chemother* 1990;34:1605–1608.
81. Thomson KS, Sanders CC, Washington JA II. Ceftazidime resistance in *Hafnia alvei. Antimicrob Agents Chemother* 1993;37:1375–1376.

19

The Genera *Cedecea* and *Kluyvera*

TAXONOMY

Cedecea

Cedecea is a novel epithet derived from the initials of the Centers for Disease Control and Prevention (CDC), Atlanta, Georgia. The name honors its work in recognizing this group of organisms which, in 1977, they designated enteric group 15 (Table 1). Originally, the genus was comprised of two named species, *Cedecea davisae* and *Cedecea lapagei,* and three unnamed species, *Cedecea* species 001, *Cedecea* species 002, and *Cedecea* species 012.[1] The three unnamed species were also designated as species 3, 4 and 5, respectively.[2] In 1982, *Cedecea* species 4 was named *Cedecea neteri*[3]; species 3 and 5 remain unnamed for a lack of strains. The first described isolate of *Cedecea* species 3 was from sputum. In 1982, an additional isolation of species 3 was reported from the postmortem heart blood of an elderly patient with pancreatitis and cirrhosis of the liver.[4] The significance of this *Cedecea* strain was unclear because *Candida albicans* was found in blood cultures before the patient's death. The first isolate of *Cedecea* species 5 was from a toe, and a literature search revealed no further isolates.

Cedecea biochemically most closely resemble *Serratia,* but DNA hybridization studies show them to be less than 23% related to other *Enterobacteriaceae* and 32% to 100% related to each other at 60°C, with ΔT_m values ranging between 3.3% and 13.3%.[1] Guanine plus cytosine (G+C) content for *C. davisae* is 49 to 50 mol%, and for *C. lapagei* it is 48% to 52%. The type species for the genus is *C. davisae.*[1]

Kluyvera

Kluyvera was first proposed in 1956 by Asai and coworkers[5,6] as a genus designation for a group of polar flagellated, α-ketoglutaric acid–producing isolates from soil and sewage. Two species, *Kluyvera citrophila* and *Kluyvera noncitrophila,* were named.[5] However, in 1962 these strains were found to be peritrichously flagellated, and subsequently were moved to the genus *Escherichia.* The genus name, *Kluyvera,* was retracted by the authors[7] and had no standing in the literature as of January, 1980. In 1981, the CDC redefined the genus *Kluyvera*[8] and proposed it for a group of organisms previously termed by them as enteric group 8 (see Table 1). The species names assigned by Asai and coworkers,[5] based on the utilization of citric acid as a sole carbon source, were not employed because they did not fit the CDC's species descriptions. The designations chosen, *Kluyvera ascorbata* and *Kluyvera cryocrescens,* were based on the use of ascorbate and the ability to ferment glucose at 5°C, characteristics which separate the two species. A third group of strains, *Kluyvera* species 3, was unnamed. However, in 1996 the species designation *Kluyvera georgiana* was proposed for this group, and a fourth species isolated from mollusks was described and named *Kluyvera cochleae.*[9]

All five of the original strains described in 1956 by Asai and colleagues[5] were included in DNA hybridization studies; one of the

TABLE 1. *Current taxonomic composition of the genera* Cedecea *and* Kluyvera

Previous Epithet	Current Nomenclature	Type Strain	Clinical Specimen/Clinically Significant
Enteric group 15	*C. davisae*	ATCC 33431	Yes/probable opportunist
	C. lapagei	ATCC 33432	Yes/probable opportunist
	C. neteri	ATCC 33856	Yes/yes
	Cedecea species 3	—	Yes/unknown
	Cedecea species 5	—	Yes/unknown
Enteric group 8,	*K. ascorbata*	ATCC 33433	Possibly/sites other than sputum
K. citrophila,	*K. cryocrescens*	ATCC 33435	Yes/unknown
K. noncitrophila	*K. georgiana*	ATCC 51603	Yes/unknown
	K. cochleae	ATCC 51609	No

strains hybridized with *K. ascorbata* and the other four with *K. cryocrescens.* Strains of *K. ascorbata* showed 65% to 100% relatedness at 75°C, with ΔT_m values ranging from 1% to 6%; *K. cryocrescens* were 63% to 100% related, with ΔT_m values of 1% to 8%. The G+C content for *Kluyvera* is 55 to 57 mol%. The type strain for the genus is *K. ascorbata.*[8]

EPIDEMIOLOGY

Cedecea

Since there have been relatively few numbers of isolates or reports of this organism, little is known of its epidemiology. All 17 strains comprising the original report were human isolates; however, *Cedecea* was subsequently isolated from *Dacus dorsalis,* the oriental fruit fly.[10] *C. neteri* was reported to have caused a nosocomial bacteremia in a patient with systemic lupus erythematosus.[11] The possible sources of the patient's strain were not discussed.

Kluyvera

Kluyvera strains seem to be more ubiquitous than *Cedecea,* with isolations from sources as varied as soil, water, sewage, milk, a cow, kitchen food, a hospital sink,[8] oriental fruit fly,[10] and two genera of captive raptors.[12] Human isolations are most frequently made from sputum, urine, stool, and throat cultures but have also been obtained from blood, peritoneal fluid, a Broviac catheter, and wound specimens. *K. cochleae* is the only species not yet isolated from human clinical specimens; the three described isolates were made from snails and a slug.[9] Müller and associates[9] contend that mollusks are the normal ecologic niche for all *Kluyvera* and that soil, water, and humans are all incidental sources. Although *Kluyvera* infections have been acquired during hospitalization, there is no evidence of nosocomial transmission. From information in at least two reports,[13,14] patients appeared to be colonized, even if only temporarily, suggesting that this may be a source of infection. In one case, the organism was recovered from a cut wound made by a tin can when the patient retrieved the item from her garbage.[15]

CLINICAL SIGNIFICANCE

Cedecea

The six human strains of *C. lapagei* described in the literature were isolated exclusively from the throat, sputum, or lung. For five of these, the isolates included in the paper proposing *Cedecea* as a genus, no clinical information is available.[1] The sixth isolate, reported in 1987 by Coudron and Markowitz,[16] was from the lung of a 60-year-old man with a history of tuberculosis and left upper lung lobectomy. In this case, bronchial washings were positive for *Torulopsis glabrata* and *C. albicans,* so significance could not be established for the *C. lapagei* isolate.

The sources of the *C. davisae* isolates are more diverse; five were obtained from sputum, two from wounds, and one each from blood, scrotum, urine, eye, stool, and gallbladder. All of the patients involved were compromised, and the organism has never been isolated in pure culture from any source. Four histories, taken from three reports in the literature, indicated preexisting medical conditions were involved in each case and all of the patients were older adults (50–76 years). One patient with an isolate from a scrotal abscess also had cardiac and alcoholic liver disease[17]; two patients with sputum isolates had various heart conditions, one complicated by diabetes[18]; and the last patient, whose isolate was from the blood, had chronic obstructive pulmonary disease.[19] The scrotal abscess, yielding only rare *Staphylococcus epidermidis* in addition to *C. davisae,* was attributed to the *Cedecea* isolate. *Cedecea* was also the predominant isolate in both sputum specimens (scant *Klebsiella oxytoca* was present in one specimen, few *Neisseria* spp. and α-hemolytic streptococci in the other), but in neither instance could its importance be determined. In the bacteremic patient, *S. epidermidis* was isolated simultaneously with *C. davisae.* The patient recovered after removal of a contaminated central venous pressure catheter and Swan-Ganz lines and treatment with antibiotics to which *C. davisae* was susceptible in vitro. She was, however, colonized with her *C. davisae* strains for weeks after her recovery.

Although accounts of *C. neteri* infections are scant, two cases of bacteremia have been reported. One was a community-acquired case of possible endocarditis.[3] *C. neteri* was isolated in pure culture from three of four blood cultures taken before the institution of antibiotic therapy; the patient recovered uneventfully from the infection. In the second case,[11] which was fatal, *Cedecea* was contracted during hospitalization for an exacerbation of systemic lupus erythematosus and was believed to be directly involved in the patient's death.

An unusual strain of *Cedecea* spp. was isolated from extensive heel and ankle ulcers[20] in a patient with a history of diabetic arteritis. This strain, biochemically aberrant from all described strains of *Cedecea,* died before it could be identified by a reference laboratory. It was isolated repeatedly from the ulcers, but it was not stated whether any other agent was also present.

Kluyvera

The most common source of *Kluyvera* is sputum. However, Farmer and colleagues[8] concluded that *Kluyvera* was probably not significant at this site when they described the genus. This was subsequently corroborated by Schwach,[21] who reported that throat, bronchial washing, and sputum isolates from three respective patients did not have any role in disease.

Kluyvera is also not infrequently isolated from stool, where its role is perhaps more difficult to determine. Fainstein and collaborators[13] described a series of isolates from neutropenic cancer patients. Five patients were symptomatic with mild to severe diarrhea, and five others were colonized and asymptomatic. *Clostridium difficile* and *Salmonella* Enteritidis were isolated from two of the cancer patients who had severe diarrhea. However, one 71-year-old man with neither an underlying disease nor other etiologic agents had four or more episodes of diarrhea per day. *K. ascorbata* was found in pure culture in an otherwise healthy 3-week-old baby with profuse diarrhea.[22] Our laboratory received an isolate that we identified as *K. ascorbata* from a patient with diarrhea of 2 weeks' duration. Since the stool yielded a pure culture of *Kluyvera,* we obtained a convalescent serum from the patient to test against her isolate. There was no evidence of antibody formation to her *Kluyvera* strain by either a tube agglutination test or Western blot (unpublished data). However, because this patient had a secretory diarrhea, in which antibodies are not always produced, no conclusion could be reached regarding the role of *Kluyvera* in her illness.

Two isolations have been made from gallbladder specimens. In one case, gallbladder drainage from a patient with pancreatitis yielded *K. ascorbata,*[23] and in another *K. cryocrescens* was isolated in pure culture from a patient with cholecystitis.[24] Both isolates were considered significant.

Of the 144 strains of *Kluyvera* reported by Farmer and colleagues,[8] five were from blood, three were *K. ascorbata,* and one each were *K. cryocrescens* and *Kluyvera* species group 3 *(K. georgiana)*. Insufficient data were received on these strains to assess their significance. Both a 1987 report of *K. cryocrescens* from the blood in a Broviac catheter infection[25] and the isolation of *Kluyvera* (not speciated) from the urine and blood of a child with an urinary tract infection and subsequent sepsis[26] suggest that this organism is an opportunist, if not a true pathogen. A pathogenic role was attributed to *K. ascorbata* in three apparently immunocompetent patients. One of these patients developed peritonitis and subsequently died from her infection[27]; the other two patients had acute pyelonephritis[28] and a soft tissue infection of the forearm,[15] respectively.

LABORATORY IDENTIFICATION

Isolation

Cedecea

Cedecea are nonfastidious facultative anaerobes that grow well on most media. Isolation from blood has been achieved with both vented and unvented blood culture bottles incubated at 35°C in O_2 using either Columbia or tryptic soy broths, and also with Bactec bottles (Johnston Laboratories, Towson, MD). Cultures grew within 24 hours after subculturing to a variety of media, including chocolate, colistin-nalidixic acid, sheep blood, MacConkey, brain-heart infusion, and phenylethyl alcohol agars. Most strains of *Cedecea* are lactose-negative and appear as nonfermenting, colorless colonies on lactose-containing differential media.

Kluyvera

Members of this genus grow readily on a variety of plating media: chocolate, sheep blood, MacConkey, eosin–methylene blue, xylose-lysine-desoxycholate, and cefsoludin-irgasan-novobiocin (CIN) agars. They have been isolated from blood with the use of Isolator tubes (DuPont, Wilmington, DE) or tryptic soy broth culture bottles. In stool cultures, *Kluyvera* colonies are usually picked because they are growing in pure culture rather than because of a recognizable differential characteristic. Since they are lactose- and sucrose-positive, they appear as fermenting colonies on most selective and differential enteric isolation media.

Biochemical Methods

Cedecea

Cedecea species are fairly distinctive biochemically and therefore should be readily identifiable by commercial systems or conventional biochemicals. In a report by O'Hara and coinvestigators,[29] two isolates each of *C. lapagei* and *C. davisae* were correctly identified by Vitek (Vitek bioMérieux, Hazelwood, MO), MicroScan (Dade MicroScan, West Sacramento, CA) and Cobas (Becton Dickinson Instrument Systems, Sparks, MD) systems. In that same study, API (Vitek bioMérieux) misidentified one of the two *C. lapagei* isolates. In three other reports evaluating commercial systems,[30–32] 11 isolates of *Cedecea* were tested with only 1 strain of *Cedecea* species 3 misidentified, and it was called *C. lapagei.* These systems were PASCO (PASCO Laboratories, Wheat Ridge, CO), which missed the *Cedecea* species 3 identification but which also tested the greatest number of strains (6 of the 11); RapID onE (Innovative Diagnostic Systems, Norcross, GA); and Crystal Enteric/Non-Fermenter (Becton Dickinson, Cockeysville, MD). In one case report, Vitek failed to identify a patient isolate of *C. lapagei* that both API and MicroScan correctly identified.

Cedecea can be separated from other members of the *Enterobacteriaceae* by negative tests for lysine decarboxylase (LDC), gelatinase, and deoxyribonuclease; their inability to

TABLE 2. *Biochemical separation of the species within the genus* Cedecea

Test	*C. davisae*	*C. lapagei*	*C. neteri*	*Cedecea* Species 3	*Cedecea* Species 5
ODC	V	–	–	–	+
KCN	V	+	–	+	+
NaAc	V	+	+	+	–
Acid from					
Sucrose	+	–	+	+	+
Sorbitol	–	–	+	–	+
Xylose	+	–	+	+	+
Melibiose	–	–	–	+	+
α-Methylglucoside	V	–	+	+	–

Compiled from Grimont et al.[1] and Farmer et al.[2]
+, ≥90% positive; V, 10%–90% positive; –, ≤10% positive; ODC, ornithine decarboxylase; KCN, potassium cyanide; NaAc, sodium acetate.

ferment L-arabinose and L-rhamnose; and a positive test for arginine dihydrolase. The five species within *Cedecea* are separable by their ornithine, malonate, sucrose, sorbitol, xylose, and melibiose reactions (Table 2). In addition, some strains of *Cedecea* produce a distinctive, "potato-like" odor (actually a rather pungent, unpleasant, musty smell), a trait they share with some strains of *Serratia*. Although easily detected and remembered by the nose, the complex mixture of pyrazines responsible for this odor[33] requires the use of more sophisticated equipment (gas liquid chromatography, GLC) to make an identification.

Kluyvera

Most laboratories using either commercial systems or conventional biochemicals have no difficulty identifying *Kluyvera* to genus. Speciation, however, usually requires a reference laboratory at which ascorbate utilization tests and GLC are available. *Kluyvera* must be separated from other indole-positive, Voges-Proskauer (VP)–negative members of the *Enterobacteriaceae* (Table 3). Notably, these include *Escherichia coli,* which can be separated from *Kluyvera* by negative tests for citrate, malonate, and KCN, and two *Citrobacter* species (*Citrobacter koseri* and *Citrobacter amalonaticus*), which give negative reactions on lysine, lactose, and salicin tests (for which *Kluyvera* are positive). Although indole-negative, *Buttiauxella agrestis* can also be confused with *Kluyvera.* A negative LDC reaction and the inability of *Buttiauxella* to ferment α-methylglucoside are useful in distinguishing between the two genera. The newest species of *Kluyvera, K. cochleae,* is indole-negative but a positive VP reaction differentiates it from *Buttiauxella.*

Speciation of *Kluyvera*[8] is based on reactions in ascorbate medium and fermentation of glucose at 5°C (Table 4). Ascorbate, because of its rapid oxidation, is a very difficult medium to make and maintain, and results of growth or fermentation tests can vary greatly depending on the ability to maintain a stable temperature. For these reasons, Altwegg and colleagues[34] recommended susceptibility to

TABLE 3. *Separation of* Kluyvera *from other common indole-positive, Voges-Proskauer–negative* Enterobacteriaceae

Test	*Kluyvera*	*E. coli*	*C. koseri*	*Morganella*	*Providencia*
Citrate	–	–	+	–	V
Urea	–	–	+	+	+
LDC	+	+	–	–	–
KCN	+	–	–	+	+

+, ≥90% positive; V, 10%–90% positive; –, ≤10% positive; LDC, lysine decarboxylase; KCN, potassium cyanide.

TABLE 4. *Biochemical separation of* Kluyvera *species*

Test	*K. ascorbata*	*K. cryocrescens*	*K. georgiana*	*K. cochleae*
Ascorbate	+	–	NA	NA
Acid glucose at 5°C	–	+	NA	NA
Irgasan susceptibility	R	S	NA	NA
Indole	+	+	+	NA
VP	–	–	–	+
Malonate	+	–	+	–

Data from Farmer et al.,[8] Müller et al.,[9] and Gallios & Grimont.[33]
+, ≥90% positive; –, ≤10% positive; R, resistant; S, susceptible; VP, Voges-Proskauer; NA, not available.

irgasan as an adjunct test for speciation. In testing 18 strains, all *K. ascorbata* were resistant to more than 125 µg/L, whereas all *K. cryocrescens* were susceptible to 0.25 µg/L of irgasan. These authors further recommended plating of strains on CIN medium as a simple method to test for irgasan resistance. In our laboratory, we have also found GLC to be helpful in the separation of *Kluyvera* species; *K. ascorbata* produces 2OH-12.0, and *K. cryocrescens* does not (unpublished data). *K. cochleae* is separated from other *Kluyvera* by positive VP and negative indole tests; however, *K. georgiana* cannot be distinguished from either *K. ascorbata* or *K. cryocrescens* by any readily available test.

ANTIMICROBIAL SUSCEPTIBILITY

Few data are available on antimicrobial susceptibility testing for *Cedecea* and *Kluyvera;* when given, the information is extremely variable from report to report regarding antimicrobials tested. Often information such as the method of testing, concentrations tested, zone sizes produced, and minimal inhibitory concentration (MIC) end points used is not included.

Cedecea

In those case reports that included susceptibility testing for ampicillin and cephalothin, all *Cedecea* strains were resistant except one strain of *C. lapagei,*[16] which was susceptible to both drugs. Generally, *Cedecea* species were reported to be susceptible to trimethoprim-sulfamethoxazole, carbenicillin, tetracycline, amikacin, gentamicin, and tobramycin. However, one isolate of *C. neteri* was resistant to aminoglycosides,[11] and one *C. davisae* strain was resistant to chloramphenicol.[17] Strains were usually resistant to cefoxitin and cefamandole, but susceptibility to other second- and third-generation cephalosporins was variable. Berkowitz and Metchock[35] reported resistance to aztreonam, ampicillin/sulbactam, ceftriaxone, and ceftazidime in two of three isolates of *Cedecea,* and all three were resistant to piperacillin, cefazolin, cefotetan, ceftizoxime, and cefotaxime. In a national surveillance study of bacterial resistance to ciprofloxacin, Thornsberry[36] reported two strains of *C. davisae* to be susceptible, with MICs of 0.5 and 1 mg/L.

Kluyvera

Susceptibility testing data for isolates of *Kluyvera* species from case reports[14,15,22,26,27] are varied. When ampicillin was tested, all strains were resistant. Susceptibility varied from laboratory to laboratory for carbenicillin, cephalosporins, piperacillin, ticarcillin, and aztreonam. Resistance was not reported to trimethoprim-sulfamethoxazole, tetracycline, chloramphenicol, or any aminoglycoside.

In a study of antimicrobial susceptibilities of new species in the family *Enterobacteriaceae,* Freney and associates[37] tested five isolates of *K. ascorbata.* All were susceptible to ampicillin, amoxicllin-clavulanic acid, ticarcillin, cefoxitin, and cefotaxime. Four of the five strains were susceptible to cephalothin. In another study,[36] five strains of *Kluyvera* species tested against ciprofloxacin were found to be susceptible, with MICs ranging from less than 0.5 to 1 mg/L.

CONCLUSIONS

The roles that *Cedecea* or *Kluyvera* may play in human disease are little understood at this time. It seems clear that these organisms can cause opportunistic infections in compromised hosts, as do a wide variety of bacteria. Undoubtedly because of their relatively infrequent isolation from human specimens, nothing has been done to determine what (if any) virulence characteristic they may possess. Although this information would be valuable, it is probably more important to expand our knowledge of the ecology of these bacteria if we are to have any success in preventing future infections.

REFERENCES

1. Grimont PAD, Grimont F, Farmer JJ III, Asbury MA. *Cedecea davisae* gen. nov., sp. nov. and *Cedecea lapagei* sp. nov., new *Enterobacteriaceae* from clinical specimens. *Int J Syst Bacteriol* 1981;31:317–326.
2. Farmer JJ III, Davis BR, Hickman-Brenner FW, et al. Biochemical identification of new species and biogroups of *Enterobacteriaceae* isolated from clinical specimens. *J Clin Microbiol* 1985;21:46–76.
3. Farmer JJ III, Sheth NK, Hudzinski JA, Rose HD, Asbury MA. Bacteremia due to *Cedecea neteri* sp. nov. *J Clin Microbiol* 1982;16:775–778.
4. Mangum ME, Radisch D. *Cedecea* species: Unusual clinical isolate. *Clin Microbiol Newslett* 1982;4: 117–119.
5. Asai T, Okumura S, Tsunoda T. On a new genus, *Kluyvera. Proc Jpn Acad* 1956;32:488–493.
6. Asai T, Okumura S, Tsunoda T. On the classification of the α-ketoglutaric acid accumulating bacteria in aerobic fermentation. *J Gen Appl Microbiol* 1957;3:13–33.
7. Asai T, Iizuka H, Komagata K. The flagellation of genus *Kluyvera. J Gen Appl Microbiol* 1962;8:187–191.
8. Farmer JJ III, Fanning GR, Huntley-Carter GP, et al. *Kluyvera,* a new (redefined) genus in the family *Enterobacteriaceae:* Identification of *Kluyvera ascorbata* sp. nov. and *Kluyvera cryocrescens* sp. nov. in clinical specimens. *J Clin Microbiol* 1981;13:919–933.
9. Müller HE, Brenner DJ, Fanning GR, Grimont PAD, Kämpfer P. Emended description of *Buttiauxella agrestis* with recognition of six new species of *Buttiauxella* and two new species of *Kluyvera: Buttiauxella ferragutiae* sp. nov., *Buttiauxella gaviniae* sp. nov., *Buttiauxella brennerae* sp. nov., *Buttiauxella izardii* sp. nov., *Buttiauxella noackiae* sp. nov., *Buttiauxella warmboldiae* sp. nov., *Kluyvera cochleae* sp. nov., and *Kluyvera georgiana* sp. nov. *Int J Syst Bacteriol* 1996; 46:50–63.
10. Jang EB, Nishijima KA. Identification and attractancy of bacteria associated with *Dacus dorsalis* (Diptera: Tephritidae). *Environ Entomol* 1990;19:1726–1731.
11. Aguilera A, Pascual J, Loza E, et al. Bacteraemia with *Cedecea neteri* in a patient with systemic lupus erythematosus. *Postgrad Med J* 1995;71:179–180.
12. Bangert RL, Ward AC, Stauber EH, Cho BR, Widder PR. A survey of the aerobic bacteria in the feces of captive raptors. *Avian Dis* 1988;32:53–62.
13. Fainstein V, Hopfer RL, Mills K, Bodey GP. Colonization by or diarrhea due to *Kluyvera* species. *J Infect Dis* 1982;145:127.
14. Sierra-Madero J, Pratt K, Hall GS, Stewart RW, Scerbo JJ, Longworth DL. *Kluyvera* mediastinitis following open-heart surgery: A case report. *J Clin Microbiol* 1990;28:2848–2849.
15. Luttrell RE, Rannick GA, Soto-Hernandez JL, Verghese A. *Kluyvera* species soft tissue infection: Case report and review. *J Clin Microbiol* 1988;26:2650–2651.
16. Coudron PE, Markowitz SM. *Cedecea lapagei* isolated from lung tissue. *Clin Microbiol Newslett* 1987;9: 171–172.
17. Bae BHC, Sureka SB. *Cedecea davisae* isolated from scrotal abscess. *J Urology* 1983;1:148–149.
18. Bae BHC, Sureka SB, Ajamy JA. Enteric group 15 *(Enterobacteriaceae)* associated with pneumonia. *J Clin Microbiol* 1981;14:596–597.
19. Perkins SR, Beckett TA, Bump CM. *Cedecea davisae* bacteremia. *J Clin Microbiol* 1986;24:675–676.
20. Hansen MW, Glupczynski GY. Isolation of an unusual *Cedecea* species from a cutaneous ulcer. *Eur J Clin Microbiol* 1984;3:152–153.
21. Schwach T. Case report. *Clin Microbiol Newslett* 1979; 1:4–5.
22. Aevaliotis A, Belle AM, Chanione JP, Serruys E. *Kluyvera ascorbata* isolated from a baby with diarrhea. *Clin Microbiol Newslett* 1985;7:51.
23. Braunstein H, Tomasulo M, Scott S, Chadwick MP. A biotype of *Enterobacteriaceae* intermediate between *Citrobacter* and *Enterobacter. Am J Clin Pathol* 1980;73:114–116.
24. Thaller R, Berletta F, Thaller MC. A *Kluyvera cryocrescens* strain from a gall-bladder infection. *Eur J Epidemiol* 1988;1:124–126.
25. Wong VK. Broviac catheter infection with *Kluyvera cryocrescens:* A case report. *J Clin Microbiol* 1987;25: 1115–1116.
26. Tristram DA, Forbes BA. *Kluyvera:* A case report of urinary tract infection and sepsis. *Pediatr Infect Dis J* 1988;7:297–298.
27. Yogev R, Kozlowski S. Peritonitis due to *Kluyvera ascorbata:* Case report and review. *Rev Infect Dis* 1990; 12:399–402.
28. Dollberg S, Gandacu A, Klar A. Acute pyelonephritis due to a *Kluyvera* species in a child. *Eur J Clin Microbiol* 1990;9:281–282.
29. O'Hara CM, Tenover FC, Miller JM. Parallel comparison of accuracy of API 20E, Vitek GNI, MicroScan Walk/Away Rapid ID, and Becton Dickinson Cobas Micro ID-E/NF for identification of members of the family *Enterobacteriaceae* and common gram-negative, non–glucose-fermenting bacilli. *J Clin Microbiol* 1993; 31:3165–3169.

30. Rhoden DL, Schable B, Smith PB. Evaluation of PASCO MIC-ID system for identifying gram-negative bacilli. *J Clin Microbiol* 1987;25:2363–2366.
31. Kitch TT, Jacobs MR, Appelbaum PC. Evaluation of RapID onE system for identification of 379 strains in the family *Enterobacteriaceae* and oxidase-negative, gram-negative nonfermenters. *J Clin Microbiol* 1994; 32:931–934.
32. Wauters G, Boel A, Voorn GP, et al. Evaluation of a new identification system, Crystal Enteric/Non-fermenter, for gram-negative bacilli. *J Clin Microbiol* 1995;33: 845–849.
33. Gallios A, Grimont PAD. Pyrazine responsible for the potatolike odor produced by some *Serratia* and *Cedecea* strains. *Appl Environ Microbiol* 1985;50:1048–1051.
34. Altwegg M, Zollinger-Iten J, von Graevenitz A. Differentiation of *Kluyvera cryocrescens* from *Kluyvera ascorbata* by irgasan susceptibility testing. *Ann Microbiol (Inst Pasteur)* 1986;137A:159–168.
35. Berkowitz FE, Metchock B. Third generation cephalosporin-resistant gram-negative bacilli in the feces of hospitalized children. *Pediatr Infect Dis* 1995;14: 97–100.
36. Thornsberry C. Susceptibility of clinical bacterial isolates to ciprofloxacin in the United States. *Infection* 1994;22(Suppl 2):S80–S89.
37. Freney J, Husson MO, Gavini F, et al. Susceptibilities to antibiotics and antiseptics of new species of the family *Enterobacteriaceae. Antimicrob Agents Chemother* 1988;32:873–876.

20

Miscellaneous Genera Associated with Clinical Specimens

EWINGELLA

Ten organisms comprising enteric group 40 formed the basis for one of a series of new genera and species that have been placed within the family *Enterobacteriaceae* since 1981. All 10 strains, now known as *Ewingella americana,* were 86%±10% related (at 75°C) to each other but only 1% to 21% related to other members of the family. The G+C content of *Ewingella* is 54.4±0.8 mol%, and the type strain is American Type Culture Collection (ATCC) 33852 (Table 1).

The genus epithet was selected to honor American bacteriologist William Ewing; the epithet *americana* denoted that all known strains were from America.[1] However, *E. americana* has subsequently been found worldwide, with isolates reported from both the European and African continents.[2,3] All 10 of the isolates described in 1983[1] were from human specimens (6 sputa/throat, 2 wounds, 1 blood, 1 unknown). However, 3 of 44 isolates listed by Farmer and colleagues[4] as *E. americana* were from nonhuman sources. These included one from food and two unspecified environmental isolates. Human sources of the organism included the respiratory tract (n=18), blood (n=11), wounds (n=7), urine (n=1), stool (n=2); unknown (n=2). In other case reports, *Ewingella* was recovered from bacteremic patients,[3,5,6] a leg wound,[2] and the conjunctiva of both eyes of a young woman.[7]

In 1987, *E. americana* was found in pure culture in unused coagulation tubes after epidemiologic studies implicated these organisms in an outbreak of pseudobacteremia involving 20 patients.[8] *Ewingella* was not isolated from tubes without anticoagulant, from the ingredients used in the citrate anticoagulant solution, or from any water supply. Water was thought to be the source of the contamination despite the fact that no isolations were made. This assumption may have been influenced by an earlier report of a nosocomial outbreak that linked *Ewingella* to ice baths used to cool syringes for cardiac output determinations.[6] *Ewingella* was present on all of the metal wells of the ice bath and in cardiac output injectate solution taken from a syringe cooled in the bath. All other environmental surfaces tested were negative for *Ewingella;* however, water was not tested.

It was not until 1995 that the first isolation of *Ewingella* from animals was reported.[9] Twenty-three strains were cultured from snails and slugs. Because fewer than 10% of the mollusks examined harbored the organism, the reporting authors believed *Ewingella* was not normal flora in these animals. This report and the two previous reports indicate a probable soil or aquatic source for *Ewingella.*

In their 1983 publication, Grimont and associates[1] characterized *E. americana* as a rare pathogen or colonizer isolated from immunocompetent persons. Time seems to have proved this observation to be the case, because *Ewingella* is seldom seen in clinical laboratories. It also appears to be of limited

TABLE 1. *Taxonomy and source of unusual* Enterobacteriaceae *associated with clinical specimens*[a]

Previous Designations	Current Taxonomy	ATCC Number	Specimen Sources: Human	Specimen Sources: Animal/ Environment
Enteric group 40	*Ewingella americana*	33852	*Respiratory tract, *blood, *wounds, feces, urine	Food, ice baths, snails/slugs, coagulation tubes
Escherichia adecarboxylata, Enterobacter agglomerans, enteric group 41	*Leclercia adecarboxylata*	23216	*Blood, *sputum, *urine, feces, wounds	*Water, *foods
Enteric group 57	*Leminorella grimontii*	33999	*Feces, urine	Mouse feces
	Leminorella richardii	33998		
	Leminorella species 3	—		
Enteric group 46	*Moellerella wisconsensis*	35017	*Feces, gallbladder	Water
Group H2	*Rahnella aquatilis*	33071	*Blood, wounds, bronchial washing, urine	Water
EF-9	*Tatumella ptyseos*	33301	*Respiratory tract, blood, feces, urine	Gastric feeding tube
Enteric group 90	*Trabulsiella guamensis*	49490	*Feces	*Vacuum cleaner dust, wheat
Koserella trabulsii	*Yokenella regensburgei*	35313	*Wounds, sputum, knee fluid, feces, urine	*Insects, water
Achromobacter nematophilus	*Xenorhabdus nematophilus*	19061		*Nematodes
Xenorhabdus nematophilus ssp.	*Xenorhabdus bovienii*	UQM 2211		*Nematodes
	Xenorhabdus poinarii	UQM 2216		
	Xenorhabdus beddingii	UQM 2872		
Xenorhabdus luminescens	*Photorhabdus luminescens*	29999	Wounds	*Nematodes

[a]More common isolations are marked with asterisks.
UQM, University of Queensland Medical Collection.

pathogenic potential, as evidenced by the recovery of *Ewingella* from the leg wound of a healthy person 4 months after an automobile accident.[2] Thought to be a colonizer, the organism cleared from the wound without treatment. Cephradine therapy for an earlier staphylococcal infection probably provided a favorable environment for the *Ewingella* strain, which was resistant to that drug. No source could be found for the isolate from either the patient's own flora or that of other patients on the same ward.

There are reports supporting *Ewingella* as a pathogen or opportunist as well as a colonizer. The most convincing is the isolation of *Ewingella* from the conjunctiva of both eyes of a patient with a 6-month history of conjunctivitis[7]; swab specimens yielded a pure culture of *E. americana.* This otherwise healthy 30-year-old woman experienced itching with edematous upper and lower eyelids and swelling in the areas above the cheekbones. The infection was successfully treated with neomycin/polymyxin drops and oral cotrimoxazole, but symptoms recurred 2 weeks later. A further 2-week course of amoxicillin-clavulanate was required for complete recovery. There have also been a series of reports of bacteremias caused by *Ewingella.* One patient was a previously healthy 75-year-old man with cholecystitis in whom sepsis developed after a series of unrelated febrile episodes.[3] Blood culture revealed *Ewingella,* and the infection was ascribed to an undocumentable abdominal source after gallbladder surgery. The patient responded to temocillin therapy and apparently recovered uneventfully. In another case of *Ewingella* bacteremia, both *Pseudomonas* spp. and *Ewingella* were isolated from the blood of a diabetic patient who had undergone heart surgery.[5] Neither organism could be isolated from sputum, urine, or the mediastinal wound. Although they were not isolated from the catheter, a contaminated Swan-Ganz line was the suspected source for both organisms.

Since the infection was polymicrobic, it was not possible to ascribe a role for the *Ewingella* isolate. Lastly, four patients with high fever and leukocytosis had *E. americana* isolated from blood; all patients were in the same intensive care unit for cardiac or vascular surgery.[6] This nosocomial outbreak, discussed previously, was caused by contaminated syringes used to determine cardiac output. Although all patients recovered after therapy with multiple antibiotics, one required a prolonged hospital stay.

In relation to most other members of the *Enterobacteriaceae, Ewingella* can be a poor grower even on chocolate and blood agars.[7] In addition to these two agars, it has also been isolated from trypticase soy, heart infusion, eosin–methylene blue, and MacConkey agars (personal observation). Blood cultures of *E. americana* have been detected by Bactec media 6A and 7B (Bactec Systems 660, Becton Dickinson Diagnostic Instrument Systems, Towson, MD)[3] and in vented and unvented brain-heart infusion broth with 0.25% sodium polyanetholesulfonate.[5]

Ewingella is a biochemically inactive organism and notably is negative for lysine decarboxylase (LDC), ornithine decarboxylase (ODC), and arginine dihyrolase (ADH). Therefore it must be differentiated from *Enterobacter agglomerans (Pantoea)* and a number of the other more unusual members of the *Enterobacteriaceae* that also show these traits (Table 2). This lack of activity in most biochemical substrates can also cause these organisms to generate a numerical profile (1005101) that codes for *Yersinia pestis* in API (Vitek bioMérieux, Hazelwood, MO).[2,4] Several cultures of *E. americana* in our collection were submitted with an API identification of *Y. pestis,* including isolates from throat, sputum, and a cat bite wound. The addition of strains of *Ewingella* to the API system database appears to have rectified this problem.[3,7,10] Numerical code 1205101 should yield an identification of *E. americana,* but the code depends on citrate utilization, a test which is frequently variable in commercial systems. With conventional biochemicals, motility readily separates *E. americana* (+) from *Y. pestis* (−); furthermore, *Ewingella* ferments arabitol while *Y. pestis* does not. In studies evaluating clinical identification systems, the autoSCAN-W/A (Dade MicroScan, West Sacramento, CA)[11] and Crystal Enteric/Non-Fermenter (Crystal E/NF, Becton Dickinson, Cockeysville, MD)[10] systems correctly identified 10 of 10 and 1 of 1 strains, respectively.

Ewingella are generally susceptible to most antimicrobials. However, strains from two reports by Pien and collaborators[5,6] were resistant to cephalothin and cefoxitin. The strain of Bear and associates[2] was resistant to ampicillin and cephradine. The only other resistance reported was to cefaclor.[7] Other antimicrobials to

TABLE 2. *Comparison of unusual H_2S-negative, LDC-negative, ODC-negative, and ADH-negative* Enterobacteriaceae *with* Enterobacter agglomerans

Organism	MOT	Gas/ Glucose	KCN	VP	Fermentation: Sucrose	Fermentation: Arabinose	Fermentation: Trehalose	Yellow Pigment
Ewingella	V	−	−	+	−	−	+	−
Leclercia	+	+	+	−	+	+	+	V
Moellerella	+	+	V	−	−	+	+	−
Rahnella	−	+	−	+	+	+	+	−
Tatumella	−	−	−	−	+	−	+	−
Xenorhabdus/Photorhabdus	+	−	−	−	−	−	−	V
Enterobacter agglomerans	V	V	V	V	V	+	+	V

H_2S, hydrogen sulfide; LDC, lysine decarboxylase; ODC, ornithine decarboxylase; ADH, arginine dihydrolase; MOT, motility; KCN, potassium cyanide; VP, Voges-Proskauer; +, ≥90% positive; V, 10%–90% positive; –, ≤10% positive.

which the organisms were susceptible included all aminoglycosides, tetracycline, carbenicillin, trimethoprim-sulfamethoxazole, chloramphenicol, ofloxacin, piperacillin, mezlocillin, imipenem, aztreonam, moxalactam, and amoxicillin-clavulanate. Second- and third-generation cephalosporins to which *Ewingella* were susceptible were cefoperazone, cefamandole, cefotaxime, cefuroxime, and ceftazidime.

LECLERCIA

Strains of *Leclercia adecarboxylata* bounced around within the *Enterobacteriaceae* before settling in a new genus with a single species. Previous epithets included *Escherichia adecarboxylata,*[12] *E. agglomerans,* and enteric group 41.[4] Leclerc placed this organism within *Escherichia* because it resembles *Escherichia coli* in the IMViC (indole, methyl red, Voges-Proskauer, and citrate) test.[13] The Centers for Disease Control and Prevention (CDC) originally considered them to be *E. agglomerans-Erwinia* because they produce yellow pigment and do not utilize lysine, ornithine, or arginine; later, the CDC referred to these strains as enteric group 41 because of the taxonomic uncertainty surrounding them.[4] The nomenclatural problem was eventually solved when in 1986 Tamura and colleagues[13] proposed a new genus and species, *Leclercia adecarboxylata,* for this organism (see Table 1). Their proposal followed an extensive study of 86 strains accumulated over a 13-year period. All isolates either had previously been identified by conventional methods or commercial systems as *E. agglomerans* or did not fit any known *Enterobacteriaceae.* A panel of conventional biochemical tests and a computer-assisted probability matrix clearly refuted an association with *Enterobacter* or *Escherichia.* DNA–DNA hybridization showed *Leclercia* strains to be 71% to 100% related to each other at 75°C but only 3.2% to 32% related to other *Enterobacteriaceae.*[13] Their G+C content is 52 to 54 mol%, and the type strain is ATCC 23126.[14]

This organism has been isolated from a variety of foods[12] and water.[4] However most of the 86 strains from Tamura's study[13] were from human clinical specimens. These included 18 isolates from blood, 27 from sputum, 5 from urine, and 2 each from stools and wounds. No clinical information was available for these strains. Nevertheless, isolation from sterile sites (blood and urine) certainly suggests that *Leclercia* is an opportunistic pathogen. This assumption is supported by a 1991 case report[14] of an isolate from an infant. Having had a jejunojejunostomy performed after birth to correct congenital abnormalities of the abdominal wall and intestines, the baby received total parenteral nutrition via a central venous catheter. Shaking chills and fever appeared 1 hour after feeding, and the child was admitted to the hospital the next morning. Blood drawn from the baby was culture-negative, but blood from the central venous line yielded *L. adecarboxylata.* The patient's physicians considered the *Leclercia* to be significant because no other etiologic agent was recovered.

Because *Leclercia* is lactose- and sucrose-positive, colonies appear either pink or yellow depending on the enteric isolation medium used. On blood or infusion agar plates, colonies may be yellow-pigmented. Because this organism is indole-positive, ferments lactose, and is negative for citrate utilization, it resembles decarboxylase-negative *E. coli.* However, positive malonate and potassium cyanide (KCN) reactions and fermentation of cellobiose and arabitol easily separate *Leclercia* from all *Escherichia* species. *E. agglomerans (Pantoea)* and *Xenorhabdus* spp. are decarboxylase-negative and, if yellow-pigmented, will at first glance resemble *Leclercia. Xenorhabdus* are distinguishable by their failure to grow in the presence of KCN or to ferment carbohydrates. For separation from *Pantoea,* which are biochemically extremely variable from strain to strain, citrate, VP, KCN, and dulcitol are the most useful tests (see Table 2). *Leclercia* is included in a number of commercial identification systems. In separate studies challenging their systems, RapID onE

(Innovative Diagnostic Systems, Norcross GA)[15] and PASCO MIC-ID (PASCO Laboratories, Wheat Ridge, CO)[16] correctly identified one and three strains of *Leclercia,* respectively. Otani and Bruckner[14] reported that their isolate was correctly identified by API and MicroScan (Dade MicroScan) systems but not by Vitek (Vitek bioMérieux).

Tamura and colleagues[13] reported their strains to be sensitive to ampicillin, carbenicillin, cephaloridine, gentamicin, and nalidixic acid. Variable susceptibilities were reported for tetracycline and chloramphenicol. In the only case report[14] of an *L. adecarboxylata* infection, the isolate was susceptible to amikacin, ampicillin, cefazolin, cefoperazone, ceftazidime, gentamicin, mezlocillin, piperacillin, tobramycin, and trimethoprim-sulfamethoxazole.

LEMINORELLA

Formerly enteric group 57, *Leminorella* found a place in the *Enterobacteriaceae* as a new genus in 1985.[17] Two species, *Leminorella richardii* and *Leminorella grimontii,* were recognized. *Leminorella* species 3, a third species defined by DNA hybridization and of which there was only a single strain, was biochemically indistinguishable from *L. grimontii.* Because the relative binding ratio of species 3 to other *Leminorella* is only 21% to 51% at 75°C, this strain has remained unnamed and awaits additional similar strains for clarification. All other *Leminorella* are 78% to 100% related to each other by DNA–DNA hybridization. *L. grimontii* is the type species for the genus. ATCC 33999 is the type strain for *L. grimontii,* and ATCC 33998 the type strain for *L. richardii* (see Table 1).

Eleven *Leminorella* strains were included in the original report,[17] and a literature search revealed no further clinical cases. All isolates were from human stool (eight) or urine (two), except for one strain that was from mouse feces. Six of the strains were from females, four from males. Patient age ranged from 2 months to 80 years, but most (6/8) of those for whom an age was given were younger than 10 years old. No clinical information was available on any of the patients, but most strains were isolated during a search for *Salmonella* in diarrheal stools. Our laboratory's experience is similar with regard to the 11 strains of *Leminorella* we have received for identification. Only one strain was identified as *L. richardii;* the remainder were *L. grimontii.* All isolates were from feces, although one was additionally isolated from urine, and all were found while looking for enteric pathogens in diarrheal stools.

Leminorella should grow relatively well on enteric isolation media, producing colorless colonies with or without a black center because it does not ferment lactose or sucrose and is H_2S-positive.

As with the other H_2S-positive organisms covered in this chapter, *Leminorella* must be distinguished from *Salmonella, Proteus, Citrobacter,* and *Edwardsiella,* as well as *Budvicia, Trabulsiella,* and *Pragia* (Table 3). Neither species of *Leminorella* is very active biochemically. Besides producing H_2S in triple sugar iron agar (TSI), both species utilize mucate and Jordan's tartrate and ferment glucose, arabinose, and xylose. Their most notable trait is the delayed fermentation of glucose, which invariably requires 48 to 72 hours for acid production.

Citrate utilization and dulcitol fermentation separate strains of *L. grimontii* (+) from *L. richardii* (−). Some strains of *Leminorella* agglutinate weakly in *Salmonella* antisera.[17] This should create few problems, because the two genera are readily separable by biochemical tests (see Table 3).

One strain of *L. richardii* was included in an evaluation of the RapID onE system and was correctly identified.[15] Other commercial systems may also include this organism in their databases, but their ability to identify *Leminorella* has not been documented in the literature.

All 11 strains described by Hickman-Brenner were resistant to penicillin, ampicillin, carbenicillin, and cephalothin but susceptible to gentamicin and chloramphenicol.[17] *L. grimontii* was variable in its susceptibility to

TABLE 3. *Unusual H_2S-positive* Enterobacteriaceae *from clinical specimens*

Organism	LDC	ODC	Urea	Arabinose	Citrate	Dulcitol	KCN	ONPG
Leminorella grimontii	–	–	–	+	+	V	–	–
Leminorella richardii	–	–	–	+	–	–	–	–
Trabulsiella guamensis	+	+	–	+	V	–	+	+
Salmonella subgroup I	+	+	–	+	+	+	–	–
Edwardsiella tarda	+	+	–	–	–	–	–	–
Citrobacter	–	V	+	+	+	V	+	+
Proteus	–	V	+	–	V	–	+	–

LDC, lysine decarboxylase; ODC, ornithine decarboxylase; KCN, potassium cyanide; ONPG, *o*-nitrophenol-β-D-galactopyranoside; +, ≥90% positive; V, 10%–90% positive; –, ≤10% positive.

tetracycline, while *L. richardii* was uniformly susceptible.

MOELLERELLA

In 1994, the CDC formed yet another new genus for nine strains known as enteric group 46.[18] These strains, which came to be designated as *Moellerella wisconsensis,* were 80% to 93% DNA related (0% to 1.5% divergent) at 60°C. They were also 23% to 32% related to *Providencia* at that temperature but were not placed in that genus because of their phenotypic dissimilarity and the fact that DNA relatedness dropped to 7% or less at 75°C. The type strain is ATCC 35017 (see Table 1).

The species name, *wisconsensis,* was coined because six of the nine known isolates were from Wisconsin. Eight strains were from stool, and five of the patients from whom they were isolated reported diarrheal symptoms. In the initial case, the patient suffered severe bloody diarrhea, fever, abdominal pain, and cramping. The ninth strain was obtained from nonchlorinated drinking water from a city water supply, suggesting a possible source for the stool isolates. In 1985, an additional isolate was reported from a 71-year-old patient with symptoms of acute cholecystitis.[19] At surgery the gallbladder was found to be inflamed but otherwise healthy. *M. wisconsensis* was isolated in equal numbers along with enterococci. After treatment with tetracycline, to which both organisms were resistant in vitro, the patient's symptoms subsided. Because it was a mixed infection and the patient recovered despite inappropriate therapy, it was not clear what role, if any, was played by *Moellerella.* Serum collected several weeks after cholecystectomy did not have antibodies to *M. wisconsensis.* In 1986, Marshall and colleagues[20] tested 400 diarrheal stools using a medium specifically designed for *Moellerella.* In three patients, fecal samples were positive for *M. wisconsensis* and for no other bacterial pathogens. The first of these three patients was a 3-year-old boy receiving chemotherapy for acute lymphatic leukemia. Although his symptoms, abdominal pain and distention, dissipated after treatment for constipation, a stool taken 4 days later was positive for *Moellerella.* Of the other two patients, also both males, one binged on alcohol and the other had traveled to India. Both had diarrhea that was relatively mild and short-lived. An isolate has also been reported from a bronchial aspirate obtained from a 20-year-old woman who had undergone cardiac arrest and was diagnosed with aspiration pneumonia.[21] No diarrhea was noted, blood cultures were negative, and ciliated bronchial cells in the aspirate indicated that the organism was probably introduced from digestive secretions into the lung.

Moellerella are lactose and sucrose fermenters and grow as pink or yellow colonies on enteric differential media, where they may be mistaken for *E. coli* and not investigated. Because of this, Marshall and coworkers[20] made a peptone-based medium to which they added 30 g/L of polymyxin B and 500,000 U of bacitracin, drugs to which *Moellerella* strains are universally resistant. Their medium also contained lactose and an indicator dye. Lactose-fermenting (pink) colonies that grew in the presence of the antimicrobials were picked for identification as potential *Moellerella* strains. Wittke and coworkers[19] recommended placing colistin or polymyxin B discs on (nonenteric) plating media for specimens from the biliary system to select for this organism.

Moellerella must be separated from other bacteria that are negative for LDC, ODC, and ADH (see Table 2). *M. wisconsensis* strains are nonmotile, negative for H_2S production, and do not produce gas from glucose or acetoin from acetylmethylcarbinol. Other than lactose and sucrose, sugars fermented by most members of the *Enterobacteriaceae* are not utilized. Both Crystal E/NF[10] and RapID onE[15] have correctly identified *Moellerella.* In the study using Crystal E/NF, API 20E (bioMérieux, La Balme les Grottes, France) was used for comparison and did not identify the strain used. However, API-ID 32E (bioMérieux, France) was used in the identification of the isolate from the bronchial washing.[21]

Moellerella is notably resistant to colistin and polymyxin B, information probably more useful for isolation and identification than therapy. Hickman-Brenner and coworkers[18] reported that 33% of their strains were resistant to tetracycline and 78% to both carbenicillin and ampicillin. Only minor differences in susceptibility were reported between the gallbladder[19] and bronchial washing[21] isolates; one was resistant to tetracycline and fosfomycin while the other was not. Both isolates were susceptible to amoxicillin, ticarcillin, piperacillin, all aminoglycosides, and cotrimoxazole. Susceptibility also was reported for aztreonam, ciprofloxacin, imipenem, and chloramphenicol, as well as cefotaxime, ceftazidime, cefazolin, and cefoperazone.

RAHNELLA

Before 1979, the strains known collectively as group H2 were clustered together based on numerical taxonomy and phenotypic characteristics. Izard and colleagues[22] found that group H2 strains constituted a new genus based on DNA hybridization studies which showed them to be 69% to 100% related. *Enterobacter,* which group H2 most closely resembled, was only 37% related. Based on this information, a new genus was established in 1979, but *Rahnella,* with its single species *aquatilis,* was not validated in the *International Journal of Systematic Bacteriology* until 1981.[4] *Rahnella* has a G+C ratio of 51 to 56 mol%; the type strain is ATCC 33071 (see Table 1). As could be predicted from its name, all of the original and most of the later environmental isolates of *Rahnella* are from its natural habitat, water, although one isolate has been reported from pine tree dust.[23]

Although the organism is rarely encountered in a clinical setting, the role of *Rahnella* as an opportunistic pathogen is well substantiated by case reports in the literature. Five cases of *Rahnella* bacteremia have been reported; all occurred in immunocompromised patients. Underlying diseases included acute lymphoblastic leukemia,[24] neuroblastoma requiring bone marrow transplantation,[25] human immunodeficiency virus (HIV) infection with drug abuse,[26] diabetes,[27] and laryngeal carcinoma.[27] One case was associated with a contaminated Hickman catheter.[25] *Rahnella* was isolated in pure culture from each of these infections except one, in which *Staphylococcus epidermidis* was also present. In all but the patients with diabetes,[27] *Rahnella* was considered significant. In that instance, the underlying infection, an ileocecal abscess, yielded *E. coli* and two species of *Acinetobacter* but not *Rahnella,* making the significance of the *Rahnella* isolate from the blood unclear.

Miscellaneous isolation sites include a burn wound,[4] bronchial washings,[23] urine,[28] a knee wound,[29] and stool.[30] No information was available regarding the burn case. The bronchial washings were from a patient with acquired immunodeficiency syndrome (AIDS) whose X-ray films showed bilateral infiltrates. Because both *Klebsiella pneumoniae* (a known pulmonary pathogen) and *Rahnella* were isolated, it was difficult to assess the significance of the *Rahnella* in this patient. Both organisms disappeared after treatment with gentamicin. In 1992, a case report was published in which

Rahnella was isolated in pure culture from the urine of a renal transplant recipient. The patient, who had received a kidney several years earlier, had reported for a routine check-up and was asymptomatic at the time the isolation was made. The dilemma for the patient's physicians was not a question of significant numbers (>100,000 organisms per milliliter of urine), but rather whether the presence of *Rahnella* was significant. A repeat culture on urine obtained by subrapubic aspiration was positive for *Rahnella* at more than 10^5 cfu/mL and was convincing enough to initiate treatment with amoxicillin. The patient eventually required a 2-week course of ciprofloxacin to resolve his infection. A *Rahnella* isolate also was obtained from a patient with a fractured knee which required surgical intervention.[29] A necrotic wound over the surgical site failed to heal even with routine cleansing with hydrogen peroxide and an iodophor compound. Wound cultures were initially negative, but at 8 months a purulent discharge appeared and pus from the wound yielded a pure culture of *Rahnella.* The patient was treated first with intravenous trimethoprim-sulfamethoxazole followed by oral administration of the same drug for a total of 45 days. At no time were urine or blood cultures positive for any bacterial agent, and osteomyelitis did not develop. One month after therapy, a new skin graft was performed and the wound healed. The two stool isolates were from children with self-limited diarrhea from which no other enteric pathogen could be isolated. Both patients had a secretory diarrhea of 3 to 4 days' duration which resolved spontaneously without treatment. One of the children had AIDS. It is impossible to determine the significance of these isolates in relation to the children's disease; only small numbers of organisms were recovered, and since *Rahnella* is a water isolate its presence could represent merely transient gut flora.

This organism grows relatively well, albeit somewhat slowly, on sheep blood agar prepared from a variety of basal media or on chocolate agar. Colonies are lactose-positive on MacConkey agar and usually require 48 hours for growth.

There are no unusual or distinguishing characteristics that can readily be associated with *Rahnella.* In common with other seldom-seen genera described in this chapter, it is LDC-, ODC-, and ADH-negative (see Table 2) and must be separated from *E. agglomerans (Pantoea).* Lack of yellow pigment, motility at 25°C but not at 36°C, and fermentation of dulcitol differentiate it from *E. agglomerans.*

API 20E was reported to have correctly identified *Rahnella* in five of six studies.[24–26,28,29] The numerical code, given in four of those reports, was 1005573. In the one instance in which API 20E failed to identify the isolate as *Rahnella,*[23] the numerical code 1105523 was generated, which indicated the organism now known as *Enterobacter asburiae.* In that same report, a MicroScan Combo Plus panel (Dade MicroScan) gave no identification for this isolate. Wauters and associates[10] reported correct identification of two strains of *R. aquatilis* by Crystal E/NF. However, because of the paucity of strains available, some commercial systems may not contain *Rahnella* in their databases.

Strains were universally resistant to ampicillin and cefazolin whenever susceptibilities were determined for those agents. Only one of six strains was resistant to cotrimoxazole.[28] Resistance to cefuroxime and cephalothin was reported two and three times, respectively. All strains were susceptible to all aminoglycosides, quinolones, imipenem, tetracycline, piperacillin, and aztreonam. Variable susceptibilities were noted with amoxicillin-clavulanate, chloramphenicol, and azlocillin. The isolate of Maraki and coworkers[29] was similar to *Enterobacter* in that β-lactamase production was increased in the presence of cefoxitin; for example, the minimal inhibitory concentration (MIC) of cefotaxime increased from 0.5 to 32 mg/L.

TATUMELLA

The genus *Tatumella* was established in 1981 to contain a number of the strains previously included in group EF-9 (see Table 1). The type strain, ATCC 33301, has a G+C ratio of 54 mol%. Although there is only a single species, two groups of strains emerged from DNA hybridization testing. At 75°C, 19 isolates were 89% to 100% related (0% to 0.5% divergence) and 6 others were 68% to 78% related (5.4% to 7.8% divergence).[31] Despite this grouping, two factors supported the existence of a only single species. First, the percent relatedness of all the strains fell within values that define a species. Second, all strains were phenotypically similar. The species name, *ptyseos,* was derived from the Greek word for "spitting." Since all but 6 of 44 strains studied were from the respiratory tract, mostly from sputum, it would be hard to find a more fitting epithet. Other isolation sites included three from blood and one each from urine, stool, and a gastric feeding tube.[31] As with other miscellaneous genera of the *Enterobacteriaceae* isolated from clinical specimens, *Tatumella* appears to be an opportunistic pathogen or a simple colonizer. Two of the three blood isolates were from patients with leukemia or an abdominal tumor. The person from whom the stool isolate was obtained had gastroenteritis, and the gastric tube isolate was from a patient with neuromuscular disease. Those patients with respiratory isolates were also immunocompromised (heart disease, Wegener's granulomatosis, chronic lung disease). Age was known for 20 of these persons, and approximately a third of them (six cultures) were children younger than 15 years old. There were no nonhuman sources other than the gastric tube, and this was the only such report found in the literature. By 1985, the CDC had 12 additional strains of *Tatumella.*[4] In addition to more respiratory isolates, there was one additional urine isolate and two human strains listed as source unknown. Two nonhuman isolates were also listed but not further specified. Our laboratory has received only three *Tatumella* isolates, two from the same patient. *Tatumella* was isolated from the blood of a 41-year-old man hospitalized for pneumonia and sepsis, twice in pure culture and a third time with *Proprionibacterium acne.* The second patient was a 55-year-old man who reported dizziness and was dehydrated. Although *T. ptyseos* was isolated in pure culture from his blood, he showed no clinical signs of sepsis. No underlying condition was given for the first patient, but the second patient reportedly consumed large amounts of alcohol.

Uncharacteristically for a member of the *Enterobacteriaceae,* this organism grows poorly on laboratory media. Frequently, 2 days are required for growth.[31] By 48 hours all reported strains grew on MacConkey and eosin–methylene blue agars, two grew on xylose-lysine-desoxycholate agar, and none grew on Hektoen enteric medium. In addition to its slow growth, *Tatumella* survives poorly on laboratory media held at room temperature. The CDC recommends that strains be stored at –40° to –70°C. However, two of our three strains failed to survive more than 2 years even stored in blood at –70°C.

The flagellar arrangement of this organism is another unusual characteristic. Unlike most *Enterobacteriaceae,* which are peritrichous (ie, with multiple flagella around the entire cell), *Tatumella* generally possess only a single flagellum per cell. However, because the flagella are difficult to demonstrate, flagellar stains are not especially helpful for identification purposes.

Again, as with other species that are negative for LDC, ODC, and ADH, *Tatumella* must be separated from *E. agglomerans (Pantoea)* (see Table 2). Helpfully, *Tatumella* is exceptionally inactive biochemically. It fails to produce gas from glucose and may be either nonmotile or weakly motile only at 25°C. It does not grow in the presence of KCN and is malonate-negative. Of the commonly used carbohydrates, only sucrose and trehalose are fermented at least 90% of the time within 48 hours.

Because *Tatumella* is a rare clinical isolate, minimal numbers of strains are included in commercial system databases. Those that do include this organism are infrequently challenged with this genus, so it is difficult to assess their capabilities. Kitch and associates[15] reported that RapID onE correctly identified two isolates used in their study.

Tatumella grows poorly on Mueller-Hinton agar, so large zones of inhibition are seen with the disk diffusion test. This includes penicillin, for which most other *Enterobacteriaceae* show resistance. One isolate was resistant to trimethoprim-sulfamethoxazole by disk diffusion. MICs were 0.12 to 1 µg/mL for gentamicin, tobramycin, and amikacin; 0.25 to 0.5 µg/mL for ampicillin; 0.25 to 2 µg/mL for tetracycline.[31]

TRABULSIELLA

The epithet *Trabulsiella* was proposed for a small group of strains first recognized in 1985 as enteric group 90.[32] These organisms had been referred to the CDC as *Salmonella* but failed to agglutinate in *Salmonella* typing sera and varied biochemically from existing *Salmonella* subgroups. In 1991, a new genus with a single species, *Trabulsiella guamensis*, was created after DNA–DNA hybridization showed the eight strains to be 85% to 100% related (ΔT_m values 0.7%–2.6%) at 75°C.[32] Relatedness to other *Enterobacteriaceae* varied from 6% to 41% at 60°C. The type strain of *T. guamensis* is ATCC 49490. The G+C content was not given.

Six of the eight *T. guamensis* strains were isolated from vacuum cleaner dust in a study undertaken in Guam to assess levels of *Salmonella* in households and commercial sites. The remaining two strains were from human diarrheal stools. No other reports were found in a literature search. Strains reported in the addendum of this original study were from wheat, human feces, and unspecified environmental material. Although three of the additional isolates were from human stool, no clinical significance was determined for them. Two of these patients had diarrhea, and one of them, a 45-year-old woman, had experienced recurrent bouts of diarrhea. The third patient, a 44-year-old man, had a fever of unknown origin, and the organism was isolated from his stool.

Trabulsiella should grow well on routine laboratory media such as blood and infusion agars. When growth occurs on differential enteric isolation media, the colonies resemble those of *Salmonella,* because the organism is lactose- and sucrose-negative and H_2S-positive.

Biochemically, *Trabulsiella* must be separated from other H_2S-positive members of the *Enterobacteriaceae* (see Table 3). *Proteus, Citrobacter,* and *Budvicia* can be distinguished from *Trabulsiella* by their inability to decarboxylate lysine and a positive urea test. *Leminorella* and *Pragia* are also LDC-negative, and they fail to ferment rhamnose, maltose, or trehalose. *Trabulsiella* most closely resemble *Edwardsiella tarda* and *Salmonella.* However, fermentation of arabinose and rhamnose separates this genus from *E. tarda,* while a positive KCN reaction and a positive *o*-nitrophenyl-β-D-galactopyranoside (ONPG) test separate it from *Salmonella.* Additionally, these bacteria fail to react in *Salmonella* antisera and are not lysed by *Salmonella*-specific bacteriophage 01.

Most commercial identification systems have not included *Trabulsiella* in their databases, and because of their similarity to *Salmonella* these strains may easily be misidentified by these methods.

McWhorter and colleagues[32] performed susceptibility studies using a taxonomic battery of antimicrobials. They found *Trabulsiella* to be sensitive to trimethoprim-sulfamethoxazole, gentamicin, and chloramphenicol and resistant to penicillin, ampicillin, and cephalothin.

YOKENELLA

This genus has the distinction of having been given two different names from two distant parts of the world almost simultaneously. In 1982 and 1983, Dr. Friedrich Haas of Regensburg, Germany, sent a number of *Hafnia*-like strains isolated from insects to the National Institute of Health (NIH) in Japan for study. In June 1984, six of the German strains and five similar strains from the NIH collection identified as NIH biogroup 9 were published as a new genus and species, *Yokenella regensburgei.*[33] Unaware of the Japanese report, the CDC proposed a new genus and species, *Koserella trabulsii,* for a group of strains previously identified as atypical *Haf-*

nia alvei.[34] Their proposal, published in January of 1985, was submitted for publication before the Japanese report was published. The CDC's 12 *Koserella* strains, previously called enteric group 45, were phenotypically similar to *H. alvei* but were only 15% to 16% related to *Hafnia* by DNA hybridization. The Japanese and U.S. strains were biochemically similar, both differing from *Hafnia* by a negative VP reaction and resistance to the *Hafnia*-specific bacteriophage of Guinee and Valkenburg.[34] In April of 1985, both *Y. regensburgei* and *K. trabulsii* were recognized in the nomenclature following publication on Validation List No. 17 in the *International Journal of Systematic Bacteriology.*[35] The similarity of the organisms was recognized, and an exchange of strains occurred between the two centers in an attempt to clarify the nomenclature. In 1987, *Y. regensburgei* and *K. trabulsii* were found after extensive molecular and phenotypic testing to be subjective synonyms.[36] At 65°C, the type strain of *Yokenella* was 80% related to the type strain of *Koserella,* with the remainder of the strains ranging from 79% to 96% related. Conversely, *Koserella* was 99% to 100% related to the type strain of *Yokenella.* The Japanese type strain of *Yokenella* was deposited in the Japanese NIH as NIH 725-83. The G+C content of the DNA was 59 mol%. The type strain of *Koserella* was deposited in the ATCC as number 35313. In 1991, Kosako and Sakazaki[37] proposed priority of *Y. regensburgei* over *K. trabulsii,* citing Rule 24b of the *International Code of Nomenclature of Bacteria.* Rule 24b establishes the priority of an epithet based on the date of receipt for submission when two epithets are published on the same Validation List. *Yokenella,* submitted for validation 3 months before *Koserella* was first published, had priority.

Of the 11 strains of Kosako and collaborators,[33] 6 were from insect intestines and 5 were human isolates. Three of the human strains were from urine, one from a wound, and one from an abscess. The isolates from the CDC were more varied and all but one were from human sources.[34] Five were from wounds on extremities, two from sputa, and one each from throat, knee fluid, and stool. No source was available for one of the human strains, and one isolate was from well water. The only other published report found on *Yokenella* described two cases.[38] In one of these, *Y. regensburgei* was isolated from the knee fluid of a 75-year-old man with a history of alcohol abuse who had previously undergone surgery for a prosthetic knee. At the time of isolation, the knee was swollen and septic. No other etiologic agent was present. The patient was treated with amikacin and presumably recovered. The second patient was a 35-year-old immunocompromised woman hospitalized for gastrointestinal bleeding. Her medical history was significant for pleural effusion of 6 months' duration, pancreatitis, and liver disease from alcohol abuse. Although she did not experience any symptoms of sepsis, *Yokenella* was isolated from her blood. Although her bacteremia was apparently transient, she was nonetheless treated with ciprofloxacin and released. No source was ascribed for either case.

Like most *Enterobacteriaceae, Yokenella* grows well on routine laboratory media. On differential enteric media, colonies appear colorless because they ferment neither sucrose nor lactose and do not produce H_2S.

Biochemically, *Yokenella* most closely resembles *H. alvei.* However, *Yokenella* varies from *Hafnia* by negative reactions for VP and glycerol, a positive citrate test, and fermentation of cellobiose and melibiose (Table 4).

TABLE 4. *Key characteristics separating* Yokenella regensburgei *and* Hafnia alvei

Organism	VP	Citrate	Cellobiose	Melibiose	Glycerol	*Hafnia* Phage
Yokenella	–	+	+	+	–	–
Hafnia	V	–	V	–	+	+

VP, Voges-Proskauer; +, ≥90% positive; V, 10%–90%; –, ≤10% positive.

The scarcity of strains limits their inclusion in commercial system databases. Abbott and Janda[38] indicated that one of the strains included in their paper was misidentified as *Salmonella* by Vitek analysis at the submitting laboratory.

The only susceptibility data available is from the taxonomic battery of antimicrobials used by the CDC.[34] Their strains were susceptible to nalidixic acid, gentamicin, and chloramphenicol. Only 25% of strains were susceptible to ampicillin or carbenicillin, 8% to cephalothin, and 83% to tetracycline. *Yokenella* are resistant to colistin, which may be helpful in identifying strains.

XENORHABDUS AND *PHOTORHABDUS*

Strains of *Xenorhabdus* and *Photorhabdus* share both an unusual and a chaotic taxonomic history that remains uncertain. The genus *Xenorhabdus* was created in 1979[39] after the genus *Achromobacter,* to which these organisms were first assigned as *Achromobacter nematophilus,* was rejected. A second species, *Xenorhabdus luminescens,* was also introduced at that time. *Xenorhabdus* was then placed in the *Enterobacteriaceae* despite the inability of this organism to reduce nitrates or produce catalase. Some strains are bioluminescent, and all strains contain either crystal-like structures or amorphous structures associated with mesosomes, properties no other members of the *Enterobacteriaceae* possess. Additionally, *X. nematophilus* and *X. luminescens* inhabit the intestines of nematodes and act with them as symbiont pairs to devour insects. Even though all of these characteristics, as well as an unusual fatty acid profile,[40] make the appropriateness of their placement tenuous, *Xenorhabdus* remains within the *Enterobacteriaceae.*

In 1980, Akhurst[41] complicated taxonomic analysis of this genus by describing two colony types (dimorphism), which he termed primary and secondary forms (opaque and translucent colonies, respectively). The primary form is the only form found in infective juvenile nematodes, and it is unstable in vitro. Production of extracellular, destructive enzymes such as protease and lipase also varies between colony types.[41,42]

In 1983, after the discovery of a new organism fitting the description of *Xenorhabdus,* three subspecies were established: *X. nematophilus* ssp. *poinarii, X. nematophilus* ssp. *nematophilus,* and *X. nematophilus* ssp. *bovienii.*[43]

Next, Grimont and coinvestigators[44] undertook the study of 30 clones from 20 strains of *Xenorhabdus,* using both phenotypic and DNA hybridization techniques. Strains of *X. nematophilus* formed two DNA hybridization groups. Group 1 was 78% to 100% related (by the hydroxyapatite method) at 75°C, with ΔT_m values of 0.0% to 2.5%, and corresponded to *X. nematophilus* ssp. *nematophilus* phenotypically. Group 2 strains were also 78% to 100% related to each other at 75°C (hydroxyapatite method, ΔT_m values <3.0%), but they were only 21% to 42% related to the strains in group 1. Although the authors believed that these organisms formed a new genomospecies, the number of strains was too few to assign a species designation, and they were left unnamed as *Xenorhabdus* spp. The bioluminescent strains of *X. luminescens* fell into four DNA relatedness groups. Again, insufficient numbers of strains prevented species assignment for any of these groups. In addition, luminescent strains were only 3% to 27% related to *X. nematophilus* strains at 75°C, raising the question of their inclusion within the genus *Xenorhabdus*.

Two years later, in 1986, a fourth subspecies of *X. nematophilus,* discovered at an earlier date but unnamed because no primary form had been found, was named *X. nematophilus* ssp. *beddingii* after isolation of a primary form was finally accomplished.[45]

In 1988, supported by numerical taxonomy findings, Akhurst and Boemare[46] elevated the four subspecies of *X. nematophilus* to species status as *X. nematophilus, X. bovienii, X. poinarii,* and *X. beddingii*. The year 1989 heralded not only the first report of human clinical specimens containing *X. luminescens* but also the news of a fifth DNA hybridization group within this species. Farmer and associates[47] described five isolates from wounds and blood that were clearly *X. luminescens* phenotypically but were only 34% to 40% related (ΔT_m values >16%) to known strains of *X. luminescens.*

Finally, in 1993, a culmination to these taxonomic struggles was realized that was almost if not quite satisfactory. The elevation of what were previously subspecies of *X. nematophilus* to species status was upheld by DNA relatedness studies, and a new genus, *Photorhabdus,* was proposed for *X. luminescens* strains,[48] with the species epithet being retained. One or two points, however, are still left to be resolved. First, more species of *Xenorhabdus* are conceivable, because six strains in this study could not be assigned to any known species (<38% related to known species and only 27% related to each other). Second, although the authors agreed with previous data[44,47] indicating that there are multiple species within the genus *Photorhabdus,* they examined too few strains to clearly define subgroups. Therefore more additions, and perhaps changes, are in store for the future.

The G+C ratio for both genera is 43 to 44 mol%. The type strains for *X. nematophilus* and *P. luminescens* are ATCC 19061 and ATCC 29999, respectively (see Table 1). The type strains for *X. poinarii, X. beddingii,* and *X. bovienii* are deposited in the University of Queensland Microbiology Collection as strains UQM 2216, UQM 2872, and UQM 2211, respectively.

The ecology of *Xenorhabdus* and *Photorhabdus* is no less amazing than their taxonomy.[42,49] Both genera share a symbiotic relation with nematodes that is lethal to insects and has been used to advantage as a natural insecticide. *Xenorhabdus* are found in nematodes of the family *Steinernematidae* and *Photorhabdus* in those of the *Heterorhabditidae.* Each nematode species associates with only one bacterial species, but a given bacterial species may be found within a number of different nematodes. For instance, the nematode *Steinernema bibionis* associates only with *X. bovienii,* but *X. bovienii* can be found in six species of *Steinernema.* The symbiotic cycle begins in infective juvenile nematodes, in which *Xenorhabdus* exist within specialized vesicles or *Photorhabdus* throughout the gut. The nematodes, which are present in the soil, invade insects and release their bacteria into the insect hemocoel. Septicemia develops as a result of nematode/bacterial activity, and the insect dies within 48 hours. The bacteria secrete lipases, phospholipases, and proteases to break down insect tissue, providing nutrients as well as antimicrobial compounds that inhibit invading competitive bacteria. Both actions provide optimal growth conditions for the nematode. As the nutrient material in the insect cadaver is used up, the nematode/bacterium association reforms and a new infective juvenile, once again in the soil, is ready to start the cycle again. Neither *Xenorhabdus* nor *Photorhabdus* has been found free-living in the soil, but both can be grown easily in vitro in the laboratory.

Until the findings of Farmer and associates,[47] neither genus had been reported in clinical material (or from any source other than nematodes), and *Xenorhabdus* still has not. The first isolate from a human specimen was from the leg wound of a 78-year-old man with a history of an ulcer below the left knee that was accompanied by pain in the thigh/groin area. His medical history also was remarkable for insulin-dependent diabetes, heart disease, and prostate/bladder cancer, and he was a heavy smoker. Blood cultures were negative, but swabs of the ulcer taken in the emergency room yielded light growth of both *Staphylococcus aureus* and a gram-negative rod. Aspirates from the thigh taken 2 days later yielded the same gram-negative rod, and both of these isolates ultimately were identified as *X. luminescens* (hereafter referred to as *P. luminescens*). The patient was treated with intravenous cefoxitin and oxacillin, and the infection resolved. A search of the CDC culture collections found three more strains of *P. luminescens,* all from adult women. The first was from blood and skin cultures obtained from an 80-year-old woman with heart disease, the second was from blood, and the last was from the abdomen and submandible of an insulin-dependent woman with "disseminated infection." Approximately 1 year after the first case occurred, a second case was identified by the same laboratory.[47] An apparently previously healthy 45-year-old man had been bitten 6 weeks earlier on the foot by a spider. Although the bite healed within 2 weeks, an ulcer formed at that site and shortly thereafter at a second site close by. The wounds were debrided, and the patient was treated with intravenous oxacillin and gentamicin. Cultures taken yielded light but pure growth of *P. luminescens.* A sixth wound infection was reported in an addendum to this paper. Specific information was unavailable, except that it was a groin isolate. Both it and the other two reported cases originated from one area (San Antonio, Texas).

In addition to these cases and isolates, there are also historical accounts of luminescent wounds in battlefield hospitals in the 1800s.[47] Since such wounds usually led to a positive outcome for the patient, it can be speculated that perhaps *P. luminescens,* with its propensity for antimicrobial production, was present. In any case, although *P. luminescens* is rarely found, its presence in blood cultures and in pure culture in the wound of an immunocompetent patient indicates a clinical significance for this organism.

Both genera grow in the laboratory, but they are somewhat fragile and are susceptible to superoxides present in growth media.[42] Addition of pyruvate and storage of media in the dark help to reduce this problem. *Xenorhabdus* and *Photorhabdus* strains isolated from nematodes grow 1.5- to 3 mm colonies on nutrient agar within 48 hours at 30°C.[41,44] Two colony types occur in both genera. Colonies of the primary form are white, opaque, convex and have irregular margins on nutrient agar; secondary-form colonies are translucent and flat. On media containing bromthymol blue and triphenyltetrazolium chloride (TTC), primary colonies are red overlaid by blue and eventually are surrounded by a clear zone; the secondary colonies are red with no blue color or clear zones.[41] Differences in colony appearance on this medium occur because the primary forms absorb bromthymol blue and reduce TTC, whereas secondary forms reduce TTC but do not absorb bromthymol blue. Both forms from the same culture are 100% DNA related,[48,49] but they differ in other characteristics besides colony morphology and dye absorption. Akhurst[41] reported that primary forms were phenylalanine deaminase- and lipase-negative and secondary forms were positive for both. Antimicrobial production, bioluminescence, and development of protein crystal inclusions also vary and are found in primary forms but low or lacking in secondary forms.[42,49] Only primary forms are found in infective juvenile nematodes, but in vitro these convert to more stable secondary forms. Reversion from secondary to primary forms has also been noted. To date no explanations for this phase variation have been found,[42] but loss of either phages or plasmids has been ruled out. The presence of transitional forms seems to rule out a "master switch" theory, leaving the possibility of a complex mechanism that may be difficult to work out.

Photorhabdus isolated from human specimens grew within 24 hours at 35°C on sheep blood trypticase soy, chocolate, and MacConkey agars and in thioglycolate broth. Colonies on MacConkey agar were lactose-negative. Although growth occurred between 19° and 37°C on nutrient agar, it was best at 30°C; however, on blood agar better growth was obtained at 36°C. An unusual form of hemolysis on sheep blood agar was noted for all strains at 25°C[47]; this consisted of a line of β-hemolysis which encircled the colony but did not touch it. This hemolysis was not noted at 36°C, at which temperature strains produced α-hemolysis by 48 hours. Yellow pigment production occurred in four strains within 48 hours at 36°C, and the fifth organism, although delayed, produced a soluble tan pigment. Blood agar plate cultures were visibly luminescent within 6 minutes in a darkroom (compared with luminescent *Vibrio* spp.

which glow within seconds) and ranged from 90,000 to 1,900,000 counts per minute in a scintillation counter.[47]

Biochemically, both *Xenorhabdus* and *Photorhabdus* are inactive and neither reduces nitrates. *P. luminescens* strains are all catalase-positive and luminescent,[48] while *Xenorhabdus* spp. are negative for both characteristics. Citrate utilization and production of urease can also be used to separate the two genera although these characteristics vary even within the genus *Photorhabdus* (see Table 2). At 7 days (36°C), all five strains reported from humans[47] were motile, indol-negative, mannose- and gelatin-positive, and bioluminescent. Only one strain was citrate-positive, but four were urea-positive. Unusual characteristics included growth in the presence KCN for three strains and a positive ONPG test in one strain. Neither genus is included in the databases of commonly used clinical commercial systems.

All strains are reported to be resistant to penicillin and β-lactam antibiotics.[42] Susceptibility results reported for one of the human isolates[47] indicated the strain was sensitive to chloramphenicol, tetracycline, gentamicin, and cefoxitin but resistant to cephalothin and ampicillin. Strains of *Xenorhabdus* and *Photorhabdus* are powerful producers of antimicrobial compounds themselves.[42,49,50] It is their antimicrobial activity that prevents putrefaction of the insect cadaver, allowing for nematode maturation. Akhurst[50] reported inhibition of a variety of bacteria and yeast, including *S. aureus,* several *Bacillus* spp., *E. coli,* other miscellaneous *Enterobacteriaceae, Candida,* and *Saccharomyces* spp. Compounds identified as xenorhabdins (a member of the pyrrothine family of antibiotics) are active against gram-positive organisms and to a lesser extent against gram-negative organisms. Two other groups of antimicrobials have been identified from *Xenorhabdus* spp., one of water-soluble antibiotics called xenocoumacins, the other of indole derivatives. *Photorhabdus* spp. produce two classes of antimicrobials known as hydroxystilbenes and anthraquinones. Many more such antimicrobials may yet be identified, because only a very small number of strains have been studied thus far.

CONCLUSIONS

The series of heterologous genera within the *Enterobacteriaceae* described in this chapter at first glance appear to be strikingly different, but they all share a common bond: a singular lack of knowledge concerning most aspects of these organisms, especially with regard to their involvement in clinical disease. Some groups are still in a state of taxonomic flux, and some have so few isolates that little is known of their ecology or their specific role within their ecological niche. Which organisms are significant in human and animal infections, and from which sites or clinical specimens? Do they cause plant disease? What virulence characteristics might they possess? For these questions to be answered, a concerted effort is required on the part of researchers, clinicians, and laboratory personnel alike. Laboratories must isolate, correctly identify, and forward clinical strains (along with adequate histories) to reference centers such as the CDC so that sufficient strains are available to answer these questions.

REFERENCES

1. Grimont PAD, Farmer JJ III, Grimont F, Asbury MA, Brenner DJ, Deval C. *Ewingella americana* gen. nov., sp. nov., a new *Enterobacteriaceae* isolated from clinical specimens. *Ann Microbiol (Inst Pasteur)* 1983; 134A:39–52.
2. Bear N, Klugman KP, Tobiansky L, Koornhof HJ. Wound colonization by *Ewingella americana. J Clin Microbiol* 1986;23:650–651.
3. Devreese K, Claeys G, Verschraegen G. Septicemia with *Ewingella americana. J Clin Microbiol* 1992;30: 2746–2747.
4. Farmer JJ III, Davis BR, Hickman-Brenner FW, et al. Biochemical identification of new species and biogroups of *Enterobacteriaceae* isolated from clinical specimens. *J Clin Microbiol* 1985;21:46–76.
5. Pien FD, Farmer JJ III, Weaver RE. Polymicrobial bacteremia caused by *Ewingella americana* (family *Enterobacteriaceae*) and an unusual *Pseudomonas* species. *J Clin Microbiol* 1983;18:727–729.
6. Pien FD, Bruce AE. Nosocomial *Ewingella americana* bacteremia in an intensive care unit. *Arch Intern Med* 1986;146:111–112.
7. Heizmann WR, Michel R. Isolation of *Ewingella amer-*

icana from a patient with conjunctivitis. *Eur J Clin Microbiol Infect Dis* 1991;10:957–959.
8. McNei MM, Davis BJ, Soloman SL, et al. *Ewingella americana:* Recurrent pseudobacteremia from a persistent environmental reservoir. *J Clin Microbiol* 1987;25: 498–500.
9. Müller HE, Fanning GR, Brenner DJ. Isolation of *Ewingella americana* from mollusks. *Curr Microbiol* 1995;31:287–290.
10. Wauters G, Boel A, Voorn GP, et al. Evaluation of a new identification system, Crystal Enteric/Non-Fermenter, for gram-negative bacilli. *J Clin Microbiol* 1995;33: 845–849.
11. O'Hara CM, Miller JM. Evaluation of the autoScan-W/A system for rapid (2-hour) identification of members of the family *Enterobacteriaceae. J Clin Microbiol* 1992;30:1541–1543.
12. Leclerc H. Etude biochimique d'*Enterobacteriaceae* pigmentees. *Ann Inst Pasteur* 1962;102:726–741.
13. Tamura K, Sakazaki R, Kosako Y, Yoshizaki E. *Leclercia adecarboxylata* gen., comb. nov., formerly known as *Escherichia adecarboxylata. Curr Microbiol* 1986;13: 179–184.
14. Otani E, Bruckner DA. *Leclercia adecarboxylata* isolated from a blood culture. *Clin Microbiol Newslett* 1991;13:20.
15. Kitch TT, Jacobs MR, Appelbaum PC. Evaluation of RapID onE system for identification of 379 strains in the family *Enterobacteriaceae* and oxidase-negative, gram-negative nonfermenters. *J Clin Microbiol* 1994; 32:931–934.
16. Rhoden DL, Schable B, Smith PB. Evaluation of PASCO MIC-ID system for identifying gram-negative bacilli. *J Clin Microbiol* 1987;25:2363–2366.
17. Hickman-Brenner FW, Vohra MP, Huntley-Carter GP, et al. *Leminorella,* a new genus of *Enterobacteriaceae:* Identification of *Leminorella grimontii* sp. nov. and *Leminorella richardii* sp. nov. found in clinical specimens. *J Clin Microbiol* 1985;21:234–239.
18. Hickman-Brenner FW, Huntley-Carter GP, Saitoh Y, Steigerwalt AG, Farmer JJ III, Brenner DJ. *Moellerella wisconsensis,* a new genus and species of *Enterobacteriaceae* found in human stool specimens. *J Clin Microbiol* 1984;19:460–463.
19. Wittke J-W, Aleksić S, Wuthe H-H. Isolation of *Moellerella wisconsensis* from an infected human gallbladder. *Eur J Clin Microbiol* 1985;4:351–352.
20. Marshall AR, Al-Jumaili IJ, Bint AJ. The isolation of *Moellerella wisconsensis* from stool samples in the U.K. *J Infect* 1986;12:31–33.
21. Wallet F, Fruchart A, Bouvet PJM, Courcol RJ. Isolation of *Moellerella wisconsensis* from bronchial aspirate. *Eur J Clin Microbiol Infect Dis* 1994;13:182–183.
22. Izard D, Gavini F, Trinel PA, Leclerc H. *Rahnella aquatilis,* nouveau membre de la famille des *Enterobacteriaceae. Ann Microbiol (Inst Pasteur)* 1979; 130A:163–177.
23. Harrell LJ, Cameron ML, O Hara CM. *Rahnella aquatilis,* an unusual gram-negative rod isolated from the bronchial washing of a patient with acquired immunodeficiency syndrome. *J Clin Microbiol* 1989;27:1671–1672.
24. Goubau P, Van Aelst F, Verhaegen J, Boogaerts M. Septicaemia caused by *Rahnella aquatilis* in an immunocompromised patient. *Eur J Clin Microbiol Infect Dis* 1988;7:697–699.
25. Hoppe JE, Herter M, Aleksić S, Klingbiel T, Niethammer D. Catheter-related *Rahnella aquatilis* bacteremia in a pediatric bone marrow transplant recipient. *J Clin Microbiol* 1993;31:1911–1912.
26. Funke G, Rosner H. *Rahnella aquatilis* bacteremia in an HIV-infected intravenous drug abuser. *Diagn Microbiol Infect Dis* 1995;22:293–296.
27. Oh HML, Tay L. Bacteraemia caused by *Rahnella aquatilis:* Report of two cases and review. *Scand J Infect Dis* 1995;27:79–80.
28. Alballaa SR, Hussein Quadri SM, Al-Furayh O, Al-Qatary K. Urinary tract infection due to *Rahnella aquatilis* in a renal transplant patient. *J Clin Microbiol* 1992;30:2948–2950.
29. Maraki S, Samonis G, Marnelakis E, Tselentis Y. Surgical wound infection caused by *Rahnella aquatilis. J Clin Microbiol* 1994;32:2706–2708.
30. Reina J, Lopez A. Clinical and microbiological characteristics of *Rahnella aquatilis* strains isolated from children. *J Infect* 1996;33:135–137.
31. Hollis DG, Hickman FW, Fanning GR, Farmer JJ III, Weaver RE, Brenner DJ. *Tatumella ptyseos* gen. nov., sp. nov., a member of the family *Enterobacteriaceae* found in clinical specimens. *J Clin Microbiol* 1981; 14:79–88.
32. McWhorter AC, Haddock Rl, Nocon FA, et al. *Trabulsiella guamensis,* a new genus and species of the family *Enterobacteriaceae* that resembles *Salmonella* subgroups 4 and 5. *J Clin Microbiol* 1991;29:1480–1485.
33. Kosako Y, Sakazaki R, Yoshizaki E. *Yokenella regensburgei* gen. nov., sp. nov.: A new genus and species in the family *Enterobacteriaceae. Jpn J Med Sci Biol* 1984;37:117–124.
34. Hickman-Brenner FW, Huntley-Carter GP, Fanning GR, Brenner DJ, Farmer JJ III. *Koserella trabulsii,* a new genus and species of *Enterobacteriaceae* formerly known as enteric group 45. *J Clin Microbiol* 1985;21: 39–42.
35. Anonymous. Validation of the publication of new names outside the IJSB: List No. 17. *Int J Syst Bacteriol* 1985;35:223–225.
36. Kosako Y, Sakazaki R, Huntley-Carter GP, Farmer JJ III. *Yokenella regensburgei* and *Koserella trabulsii* are subjective synonyms. *Int J Syst Bacteriol* 1987;37: 127–129.
37. Kosako Y, Sakazaki R. Priority of *Yokenella regensburgei* Kosako, Sakazaki, and Yoshizaki 1985 over *Koserella trabulsii* Hickman-Brenner, Huntley-Carter, Brenner, and Farmer 1985. *Int J Syst Bacteriol* 1991;41: 171.
38. Abbott SL, Janda JM. Isolation of *Yokenella regensburgei ("Koserella trabulsii")* from a patient with transient bacteremia and from a patient with a septic knee. *J Clin Microbiol* 1994;32:2854–2855.
39. Thomas GM, Poinar GO Jr. *Xenorhabdus* gen. nov., a genus of entomopathogenic, nematophilic bacteria of the family *Enterobacteriaceae. Int J Syst Bacteriol* 1979;29:352–360.
40. Janse JD, Smits PH. Whole cell fatty acid patterns of *Xenorhabdus* species. *Lett Appl Microbiol* 1990;10: 131–135.
41. Akhurst RJ. Morphological and functional dimorphism in *Xenorhabdus* spp., bacteria symbiotically associated with the insect pathogenic nematodes *Neoplectana* and *Heterorhabditis. J Gen Microbiol* 1980;121:303–309.

42. Forst S, Nealson K. Molecular biology of the symbiotic-pathogenic bacteria *Xenorhabdus* spp. and *Photorhabdus* spp. *Microbiol Rev* 1996;60:21–43.
43. Thomas GM, Poinar GO Jr. Amended description of the genus *Xenorhabdus* Thomas and Poinar. *Int J Syst Bacteriol* 1983;33:878–879.
44. Grimont PAD, Steigerwalt AG, Boemare N, et al. Deoxyribonucleic acid relatedness and phenotypic study of the genus *Xenorhabdus*. *Int J Syst Bacteriol* 1984; 34:378–388.
45. Akhurst RJ. *Xenorhabdus nematophilus* subsp. *beddingii* (*Enterobacteriaceae*): A new subspecies of bacteria mutualistically associated with entomopathogenic nematodes. *Int J Syst Bacteriol* 1986;36:454–457.
46. Akhurst RJ, Boemare NE. A numerical taxonomic study of the genus *Xenorhabdus (Enterobacteriaceae)* and proposed elevation of the subspecies of *X. nematophilus* to species. *J Gen Microbiol* 1988;134:751–761.
47. Farmer JJ III, Jorgensen JH, Grimont PAD, et al. *Xenorhabdus luminescens* (DNA hybridization group 5) from human clinical specimens. *J Clin Microbiol* 1989; 27:1594–1600.
48. Boemare NE, Akhurst RJ, Mourant RG. DNA relatedness between *Xenorhabdus* spp. *(Enterobacteriaceae)*, symbiotic bacteria of entomopathogenic nematodes, and a proposal to transfer *Xenorhabdus luminescens* to a new genus, *Photorhabdus* gen. nov. *Int J Syst Bacteriol* 1993;43:249–255.
49. Hurlbert RE. Investigations into the pathogenic mechanisms of the bacterium-nematode complex. *ASM News* 1994;60:473–478.
50. Akhurst RJ. Antibiotic activity of *Xenorhabdus* spp., bacteria symbiotically associated with insect pathogenic nematodes of the families *Heterorhabditidae* and *Steinernematidae*. *J Gen Microbiol* 1982;128: 3061–3065.

21

Enterobacteriaceae Not Associated with Human Disease

BUDVICIA

This group of organisms was first described and named in 1983[1] but was not formally proposed as a new genus with a single species until 1985 (Table 1). The epithet *Budvicia* was derived from the Latin name of the Czechoslovakian city of České Budějovice, the location of the first isolation. The species name, *aquatica,* refers to water, the primary source for most of these organisms. Strains are 87.2%±7.6% related at 60°C by DNA–DNA hybridization. *Xenorhabdus* spp. are the most closely related members within the *Enterobacteriaceae,* with 8% homology by the S1 nuclease method of determination. The G+C content is 46±1 mol%. The type strain is ATCC 25567.[1]

Primarily isolated from water and less often from small rodents, *Budvicia* has not yet been isolated from humans.[1] There have been no reports of isolations from ill or diseased animals or plants.[1]

Budvicia grow at temperatures ranging from 4° to 37°C but not at 42°C. Rather small colonies (0.1 mm) are produced on nutrient agar at 37°C, with slightly larger colonies produced if 30°C incubation is used. Forty-eight hours may be required for good growth. *Budvicia* is variable for lactose fermentation and negative for sucrose fermentation. Therefore, on enteric isolation agars, it may appear as pink or yellow fermenting colonies or it may be colorless. The organism does grow on MacConkey agar; if growth occurs on other differential or selective agars that detect H_2S

TABLE 1. *Taxonomy and ecological niche of non–human-associated* Enterobacteriaceae

Current Taxonomy	Previous Epithet	ATCC Number	Environmental Association
Budvicia aquatica	None	25567	Wells, pipes; streams; rivers; brooks; small wild mammals
Buttiauxella	Group F		
B. agrestis		33320	Water; unpolluted soil; slugs; snails
B. ferragutiae		51602	Water; soil
B. gaviniae		51604	Soil; slugs; snails
B. brennerae		51605	Slugs; snails
B. izardii		51606	Slugs; snails; earthworms
B. noackiae	Enteric group 59	51607	Slugs; snails; human specimens; food (ham)
B. warmboldiae		51608	Snails
Obesumbacterium proteus[a]	*Flavobacterium proteus, O. proteus* biogroup 2	12841	Breweries
Pragia fontium	*Budvicia*-like	None[b]	Well water, pipes; human stool

[a]Biogroup 1 strains are now designated *Hafnia alvei* biogroup 1.
[b]The type strain is Czechoslovak National Culture Type Collection no. CNCTC Eb 11/82, Institut Pasteur no. IP 20125, or Centers for Disease Control and Prevention no. CDC 963-84.

TABLE 2. *Separation of H_2S-positive members of the* Enterobacteriaceae

Test	*Budvicia*	*Pragia*	*Leminorella*	*Edwardsiella*	*Salmonella*	*Citrobacter*	*Proteus*
LDC	–	–	–	+	+	–	–
ODC	–	–	–	+	+	V	V
Urea	+	–	–	–	–	+	+
ONPG	+	–	–	–	–[a]	V	–
Indol	–	–	–	+	–	–	V
Citrate	–	(+)	V	–	+	V	V
Motile	V	+	–	+	+	V	+
KCN	–	–	–	–	–[b]	+	+
Ara	V	–	+	V	+	V	–

Compiled from Bouvet et al.,[1] Aldová et al.,[10] and Farmer.[11]
[a]Except subgroup 3a, 3b, and 6 strains, which are + or V.
[b]Except subgroup 4 and 5 strains, which are +.
+, ≥90% positive; V, 10%–90% positive; (+), ≥90% positive in 3–7 d; –, ≤10% positive; LDC, lysine decarboxylase; ODC, ornithine decarboxylase; ONPG, *o*-nitrophenyl-β-D-galactopyranoside; KCN, potassium cyanide; Ara, arabinose.

production, *Budvicia* may appear *Salmonella*-like because it is H_2S-positive.

Biochemically, this organism can be separated from *Salmonella* and *Edwardsiella* by negative lysine decarboxylase (LDC) and ornithine decarboxylase (ODC) reactions and a positive test for urease. *Budvicia* must also be separated from other H_2S-positive members of the *Enterobacteriaceae,* especially *Pragia fontium,* which is also a H_2S- and decarboxylase-negative water isolate (Table 2). *B. aquatica* is unlikely to be included in the database of any of the commonly employed commercial identification systems used in clinical laboratories.

As would be expected, antimicrobial susceptibility data are not available for these environmental strains.

BUTTIAUXELLA

In 1981, 28 isolates previously lumped together and designated simply as group F were subjected to DNA hybridization by Ferragut and colleagues.[2] The resulting data indicated that group F was actually a hetereogeneous cluster of strains. Homology ranged from 82% to 96% for 17 isolates, 62% to 66% for 4 other isolates, and 51% or less for the remaining 7 isolates. Strains with homology of 80% or greater were selected to form a new genus and species, *Buttiauxella agrestis.*[2] The second genetic group was believed to represent an additional species; however, although they were placed in the genus *Buttiauxella* by Ferragut and colleagues,[2] these strains were not given a species name. The Centers for Disease Control and Prevention (CDC), in studying these strains, arbitrarily placed the four isolates into enteric groups 63 (three strains) and 64 (one strain) until more strains could be studied.[3] Isolates with 50% or less relatedness (the third genetic group) were not included in the genus.

In 1996, six species were added to the genus (see Table 1). Most of these species were discovered as a result of a study of organisms isolated from snails and slugs collected worldwide.[4] *B. agrestis,* previously isolated only from water and unpolluted soil, was found in 18 snails and slugs. The type strain for *B. agrestis* is ATCC 33320. Enteric group 63 (LDC-positive) *Buttiauxella* were not found in mollusks because during prescreening testing all LDC-positive strains were eliminated. However, previously isolated strains were subjected to hybridization in this study. These strains, all soil and water isolates, were, at last, given a name, *Buttiauxella ferragutiae.* The type strain is ATCC 51602. The single strain of enteric group 64 from unpolluted soil was joined by 35 isolates from mollusks. The species name for this group is *Buttiauxella gaviniae* (type

TABLE 3. *Separation of* Buttiauxella *and* Kluyvera[a]

Test	Ba	Bf	Bg	Bb	Bi	Bn	Bw	Kco	Kg	Ka	Kcr
*ADH	–	–	+	V	–	+	+	–	V	–	–
CIT	V	–	V	V	+	+	–	+	V	+	V
DUL	–	–	V	–	–	–	–	+	V	V	–
*IND	–	–	V	–	–	V	–	–	+	+	+
LAC	+	–	–	+	+	–	–	V	V	+	+
MAL	+	–	+	+	+	+	+	+	V	+	V
MEL	V	V	–	+	V	–	–	+	+	+	+
*SUC	–	–	–	–	–	–	–	V	+	+	–
*LDC	–	+	–	–	–	–	–	–	+	+	+
ODC	+	+	–	V	+	–	–	+	+	+	+
*NAG	+	+	+	+	+	+	+	–	–	–	–

Adapted from Müller et al.[4]
[a]Tests most useful in separating *Buttiauxella* and *Kluyvera* are marked with asterisks.
Ba, *B. agrestis;* Bf, *B. ferragutiae;* Bg, *B. gaviniae;* Bb, *B. brennerae;* Bi, *B. izardii;* Bn, *B. noackiae;* Bw, *B. warmboldiae;* Kco, *K. cochleae;* Kg, *K. georgiana;* Ka, *K. ascorbata;* Kcr, *K. cryocrescens;* ADH, arginine dihydrolyase; CIT, citrate; DUL, dulcitol; IND, indol; LAC, lactose; MAL, malonate, MEL, melibiose; SUC, sucrose; LDC, lysine decarboxylase; ODC, ornithine decarboxylase; NAG, *N*-acetyl-glutamate utilization; +, ≥90% positive; V, 10%–90% positive; –, ≤10% positive.

strain ATCC 51604). Three other hybridization groups represented new species and contained a total of 21 strains from snails or slugs and an earthworm. These newly designated species are *Buttiauxella brennerae* (ATCC 51605), *Buttiauxella izardii* (ATCC 51606) and *Buttiauxella warmboldiae* (ATCC 51608). The last species, *Buttiauxella noackiae,* includes strains previously recognized as enteric group 59. Fifteen snail/slug isolates fell into this group. Enteric group 59 isolates from the CDC collection included six strains; sources listed were sputa, a foot wound, and ham.[3] None of the human isolates was from a sterile site, and no information other than source was available; further human isolations have not been reported. Because the preponderance of *Buttiauxella* isolates are from water, soil, or mollusks and because no significance could be ascribed to any human isolates, *Buttiauxella* was included in this chapter. The type strain for this last species is ATCC 51607.

Shortly after the genus *Buttiauxella* was formed in 1981, Gavini and colleagues[5] feared confusion might arise between this genus and another relatively newly described genus, *Kluyvera.* Therefore, further hybridization and G+C content studies were performed on both genera. *Kluyvera* spp. were found to be only 30% to 36% DNA-related to *Buttiauxella* by the nitrocellulose filter method at 52.8°C. Furthermore, its G+C content of 55% to 57% was much higher than that of *Buttiauxella* (48%–50%). Both results soundly supported the separation of these organisms as distinct genera.

This organism grows well on ordinary plating media at a range of 4° to 37°C but not at 41°C. *Buttiauxella* is lactose-positive and may resemble *Escherichia coli* on MacConkey agar. It is easily separable from *E. coli* by negative indol and LDC reactions and a positive malonate test.

Buttiauxella has been described as most closely resembling *Kluyvera* phenotypically.[4,5] However, positive reactions for indol production, LDC, and fermentation of sucrose and a negative arginine dihydrolase (ADH) reaction readily identify strains as *Kluyvera* (Table 3). Equally problematic is the differentiation of other H_2S-negative genera whose natural habitat, like that of *Buttiauxella,* is water and soil (Table 4). Like most of the unusual genera in the *Enterobacteriaceae,* this genus is not included in the databases of clinical commercial systems.

No antimicrobial susceptibility data have been generated for these organisms because of their lack of clinical relevance.

TABLE 4. *Biochemical characteristics of H_2S-negative water and soil organisms*

Organism	LDC	ODC	Urea	VP	Sorbitol	Sucrose	Dulcitol
Buttiauxella	–	–	–	–	–	V	–
Enterobacter amnigenus biogroup 1	–	V	–	+	–	+	–
Enterobacter amnigenus biogroup 2	–	+	–	+	+	–	–
Enterobacter intermedium	–	V	–	+	+	V	+
Rahnella aquatilis	–	–	–	+	+	+	V
Serratia fonticola	+	+	–	–	+	V	+
Budvicia aquatica	–	–	V	–	–	–	–
Klebsiella terrigena	+	–	–	+	+	+	+
Klebsiella planticola	+	–	+	+	+	+	+

LDC, lysine decarboxylase; ODC, ornithine decarboxylase; VP, Voges-Proskauer; +, ≥90% positive; V, 10%–90% positive; –, ≤10% positive.

OBESUMBACTERIUM

Perhaps having a brewery as its sole ecologic niche makes *Obesumbacterium's* "tipsy" past history less inappropriate than it would otherwise seem (see Table 1). This docile bacterium was given the decidedly undistinguished name of "Bacterium Y" in 1936.[6] Later the same year it was placed in the genus *Flavobacterium* as *Flavobacterium proteus* simply because this genus was slightly less unsuitable than any other; this placement was not completely without merit, however, since growth on potato slices was noted as yellowish. In 1964 it was realized that the inclusion of this organism in *Flavobacterium* or in any other existing genera was inappropriate, so a new genus and species was proposed.[6] The new genus, *Obesumbacterium,* relates to the rather fat nature of this small, rod-shaped bacterium. The species epithet, *proteus,* was retained because of the pleomorphic nature of this organism; it refers to the god Proteus, who was known for the ability to change form at will.

In 1973, Priest and associates,[7] in attempting to clarify the taxonomic position of *Obesumbacterium,* added further to the confusion. Although they found two distinct biogroups by DNA/DNA competition studies, they proposed a single species on the basis that all strains demonstrated similar activity in the brewery (ie, growth during wort fermentation). These authors also recommended dropping the name *Obesumbacterium* in favor of the epithet *Hafnia protea,* since strains of *Obesumbacterium* biogroup 1 were highly related to *Hafnia alvei.*[7] Somewhat later, *Obesumbacterium proteus* biogroup 1 was again shown to share homology (75% relatedness) with one of three known groups of *Hafnia alvei,* and it was then placed in the genus *Hafnia* as *H. alvei* biogroup 1.[8] Further evidence of relatedness was demonstrated by the fact that *Obesumbacterium* biogroup 1 strains were lysed by *Hafnia*-specific bacteriophage and were biochemically similar (anaerogenic and xylose-negative variants). *O. proteus* biogroup 2 isolates, however, were only 20% to 24% related to *Hafnia* and 25% to 30% related to *O. proteus* biogroup 1.[8] For this reason, they have remained in their own genus as a single species, *Obesumbacterium proteus.* Nonetheless, these strains are 55% related to *Escherichia blattae,* and so their taxonomic status may be changed at some future time.

O. proteus are plump, short, nonmotile rods that are capable of competing with yeast during brewery wort fermentation.[6] They are found exclusively, and rather profusely, in breweries. They are slow-growing and biochemically inactive.[3] They decarboxylate lysine and ornithine and ferment only glucose, trehalose, and mannose with any consistency. Table 5 lists those biochemicals most useful in separating metabolically inactive, H_2S-negative members of the *Enterobacteriaceae.* *Obesumbacterium* is not in the database of commercial systems commonly used in clinical laboratories.

TABLE 5. *Reactions of metabolically inactive, H_2S-negative* Enterobacteriaceae

Test	*O. proteus*	*H. alvei* Biogroup 1	*T. ptyseos*	*Xenorhabdus*	*E. americana*
Lysine decarboxylase	+	+	–	–	–
Ornithine decarboxylase	+	V	–	–	–
Voges-Proskauer	–	V	–	–	+
Phenylpyruvic acid	–	–	+	–	–
Citrate	–	–	–	V	+
Sucrose	–	–	+	–	–
Mannitol	–	V	–	–	+
Hafnia bacteriophage susceptibility	–	+	NA	NA	NA
Found in human specimens	No	No	Yes	Yes	Yes

+, ≥90% positive; V, 10%–90% positive; –, ≤10% positive; NA, not applicable.

Since it is not involved in human disease, no antimicrobial susceptibility data are available.

PRAGIA

Yet another water organism, comprising strains now called *Pragia fontium,* were discovered in Czechoslovakia at the same time as *B. aquatica.*[9] Biochemically dissimilar, they were called *Budvicia*-like until Bouvet and coworkers[1] showed them to be only 9% to 22% DNA-related to *Budvicia.* By 1986, 17 *Budvicia*-like strains had been isolated, and 16 of these were recognized to form a single genomospecies by DNA hybridization (85%–93% related at 75°C) and numerical taxonomy (similarity coefficient 0.9).[10] One of the 17 strains was only 37% related to the type strain of *Pragia* and shared biochemical characteristics with both *Pragia* and *Budvicia;* ultimately, DNA relatedness showed this strain to belong to the genus *Budvicia.* Aldová and colleagues[10] believed that the remaining strains deserved placement in a separate genus because they were biochemically distinct, only 13% to 17% related to *Budvicia* and 2% to 16% related to other *Enterobacteriaceae.* The G+C content of *Pragia* is 46 to 47 mol%. The type strain is CNCTC Eb 11/82 (Czechoslovak National Culture Type Collection), IP 20125 (Institut Pasteur), or CDC 963-84.[10]

To date, strains have been isolated almost exclusively from well water or water pipes. A strain isolated from the stool of a healthy woman was thought to represent transient gut flora.[10]

Pragia produce small colonies (≤0.5 mm) on nutrient and MacConkey agars. This non–lactose-fermenting, non–sucrose-fermenting organism forms colorless colonies on enteric isolation media. On media detecting H_2S production, colonies are clear with a black center. Growth occurs at 4°, 22°, or 37°C but not at 42°C.[10]

Relatively metabolically inactive, *Pragia* are separable from other H_2S-positive organisms by negative results on LDC, ODC, urea, *o*-nitrophenyl-β-D-galactopyranoside (ONPG), and arabinose fermentation tests and failure to grow in KCN broth (see Table 2). Although readily identified by commonly employed biochemicals, this organism is not found in commercial identification system databases.

Antimicrobial susceptibility data are irrelevant (even if they were available) because there is no known association with human clinical disease.

CONCLUSIONS

It would appear that the essentially environmental group of bacteria discussed in this chapter have no apparent consequence for humans. Nevertheless, it is necessary for microbiologists to be aware of their presence, if only to recognize them as possible transient flora or asymptomatic colonizers in humans or animals should they be encountered. Exactly what roles these organisms play in their respective ecologic niches has yet to be determined.

REFERENCES

1. Bouvet OM, Grimont PAD, Richard C, Aldová E, Hausner O, Gabrhelová M. *Budvicia aquatica* gen. nov., sp. nov.: A hydrogen sulfide-producing member of the *Enterobacteriaceae. Int J Syst Bacteriol* 1985;35:60–64.
2. Ferragut C, Izard D, Gavini F, Lefebvre B, Leclerc H. *Buttiauxella,* a new genus of the family *Enterobacteriaceae. Zentralbl Bakteriol* 1981;C2:33–44.
3. Farmer JJ III, Davis BR, Hickman-Brenner FW, et al. Biochemical identification of new species and biogroups of *Enterobacteriaceae* isolated from clinical specimens. *J Clin Microbiol* 1985;21:46–76.
4. Müller HE, Brenner DJ, Fanning GR, Grimont PAD, Kämpfer P. Emended description of *Buttiauxella agrestis* with recognition of six new species of *Buttiauxella* and two new species of *Kluyvera: Buttiauxella ferragutiae* sp. nov., *Buttiauxella gaviniae* sp. nov., *Buttiauxella brennerae* sp. nov., *Buttiauxella izardii* sp. nov., *Buttiauxella noackiae* sp. nov., *Buttiauxella warmboldiae* sp. nov., *Kluyvera cochleae* sp. nov., and *Kluyvera georgiana* sp. nov. *Int J Syst Bacteriol* 1996;46:50–63.
5. Gavini F, Izard D, Ferragut C, Farmer JJ III, Leclerc H. Separation of *Kluyvera* and *Buttiauxella* by biochemical and nucleic acid methods. *Int J Syst Bacteriol* 1983;33:880–882.
6. Shimwell JL. *Obesumbacterium,* a new genus for the inclusion of *Flavobacterium proteus. J Inst Brew* 1964;70:247–248.
7. Priest FG, Somerville HJ, Cole JA, Hough JS. The taxonomic position of *Obesumbacterium proteus,* a common brewery contaminant. *J Gen Microbiol* 1973;75:295–307.
8. Brenner DJ. Introduction to the family *Enterobacteriaceae.* In: Starr MP, Stolp H, Trüper HG, Balows A, Schlegel HG, eds. *The prokaryotes: A handbook on habitats, isolation, and identification of bacteria.* Berlin: Springer-Verlag, 1981;1173–1180.
9. Aldová E, Hausner O, Gabrhelová M, Schindler J, Petrá P, Braná H. A hydrogen sulphide producing gram-negative rod from water. *Zentralbl Bakteriol* 1983;A254:95–108.
10. Aldová E, Hausner O, Brenner DJ, et al. *Pragia fontium* gen. nov., sp. nov. of the family *Enterobacteriaceae,* isolated from water. *Int J Syst Bacteriol* 1988;38:183–189.
11. Farmer JJ III. *Enterobacteriaceae:* Introduction and identification. In: Murray PR, Barron EJ, Pfaller MA, Tenover FC, Yolken RH, eds. *Manual of clinical microbiology.* 6th ed. Washington, DC: ASM Press, 1995;438–449.

Subject Index

Page numbers followed by *t* indicate tables; page numbers followed by *f* indicate figures.

A

Abdominal aneurysm, 218
"Aberrant coliforms," 2
Abscess
  Cedecea, 338
  Citrobacter, 186*t,* 187
  E. coli, 34
  Edwardsiella, 309, 312*t,* 313
  Enterobacter, 160
  Hafnia, 325*t,* 327, 328
  Klebsiella, 114, 116, 118–119
  Proteus, 251, 252
  typhoid fever, 104
  Yersinia, 222
Acid tolerance
  Morganella, 280
  Salmonella, 93
  Salmonella Typhi, 106
  Shigella, 75
  Yersinia, 236
Actin, 46
  in *Shigella* pathogenicity, 75
Adenitis, mesenteric, 207
Adherence. *See also* Attaching and effacing adherence
  E. coli, 28, 45–46, 47
  Hafnia, 330–331
  Klebsiella, 123–124
  Providencia, 297
  Yersinia, 236
Adherence assay, 45–46
AE adherence. *See* Attaching and effacing adherence
Aerobacter, 5, 110, 148
Aerobactin, 49–50
  Citrobacter, 196
  Enterobacter, 169
  Klebsiella, 123
  Serratia, 142
  Yersinia, 234
  Aeromonas, 225, 311
Agars. *See also specific types*
  in *Salmonella* isolation, 85–88, 89*t*
Aged. *See* Elderly
Aggregative adherence
  enteroadherent *E. coli,* 28, 45, 46
  Klebsiella, 123–124
AIDS
  Salmonella infections, 84, 92, 102
  Yersinia septicemia, 214
Alcoholism
  Klebsiella infection, 115
  Yersinia septicemia, 218
Alkyl-methoxypyrazine, 141
Alveolar macrophage, in
    Klebsiella infections, 123
Amber disease, 133
Ambient-temperature fimbriae, 262
Amikacin
  Cedecea, 341
  Citrobacter, 197
  Hafnia, 332
  Morganella, 280
  Providencia, 299
  Serratia, 143
Aminoglycosides
  Cedecea, 341
  Citrobacter, 197
  Edwardsiella, 318
  for *Enterobacter* bacteremia, 156
  Hafnia, 332
  Klebsiella, 125, 126
  Kluyvera, 341
  Morganella, 280
  Proteus, 264
  Providencia, 298
  Serratia, 143, 144
  Yersinia, 238
Aminopenicillins, 237
Amoxicillin, 237, 238
Amoxicillin/clavulanate, 50, 51
  Klebsiella, 125, 126
  Kluyvera, 341
  Salmonella, 96
  Serratia, 143
  Yersinia, 238
Amp-C-type enzymes, 144
Amphibians, *Edwardsiella* and, 305
Ampicillin
  Cedecea, 341
  Citrobacter, 196, 197*t*
  E. coli, 50, 51
  Hafnia, 332
  Kluyvera, 341
  Morganella, 280
  Providencia, 298
  Salmonella, 96
  Salmonella Typhi, 106, 107
  Shigella, 77
Ampicillin/sulbactam, 341
Aneurysms, 218

Animals
Citrobacter, 184, 195
Edwardsiella, 304, 305–306
Enterobacter, 151–152
Ewingella, 344
Hafnia, 324
Morganella, 271
Proteus, 247
Providencia, 285
Serratia, 133
Yersinia, 206–207
Ankylosing spondylitis, 253
Anorexia, shigellosis, 68, 69
Anthraquinones, 358
Antibiograms
Providencia typing, 296
Shigella typing, 74
Antibiotics. *See* Antimicrobials
Antigens
Citrobacter, 194
E. coli, 42–43
Edwardsiella, 316–317
Klebsiella, 121–122
Morganella, 278
Proteus, 257
Providencia, 295–296
Salmonella, 90–91
Serratia, 141
Yersinia, 230
Antimicrobials
Cedecea, 341
Citrobacter, 196–198
Edwardsiella, 318
Enterobacter, 170–172
Escherichia, 50–51, 64–65
Ewingella, 346–347
Hafnia, 332
Kluyvera, 341–342
Leclercia, 348
Leminorella, 348–349
Moellerella, 350
Morganella, 280–281
Photorhabdus, 358
Proteus, 263–264
Providencia, 298–299
Rahnella, 351
Salmonella, 92, 95–96, 97
Salmonella Typhi, 106–107
Serratia, 134–135, 143–144
Shigella, 76–77
Tatumella, 353
therapy for *Enterobacter* bacteremia, 156
therapy for plague, 238
Trabulsiella, 353
Xenorhabdus, 358
Yersinia, 237–238
Yokenella, 355
Aorta, *Salmonella* mycotic aneurysm, 85
API identification systems. *See* Microbial identification systems
Aquatic habitats
Citrobacter, 184
E. coli, 14, 25–26
Edwardsiella, 305
Hafnia, 324
Morganella, 271
Proteus, 247
Providencia, 285
Serratia, 132, 133
Arthritis
Edwardsiella, 312*t*
Klebsiella, 119
Morganella, 274
Proteus, 253, 257
Serratia, 138
Yersinia, 213, 219–220, 223
Arthropathy, *Morganella,* 274
Ascites, 162
Ascorbate utilization test, 340
Assays. *See also* Biochemical characteristics
for *E. coli,* 43–47
Asymptomatic bacteriuria, 118
Atrophic rhinitis, 116, 117
Attaching and effacing adherence (AE adherence), 46, 47
Citrobacter, 196
E. coli, 45–46
Hafnia, 330–331
Attachment-invasion locus, 234–235
AutoScan WalkAway system. *See* Microbial identification systems
Aztreonam
Cedecea, 341
Edwardsiella, 318
Enterobacter, 171
Kluyvera, 341
Morganella, 280
Proteus, 264
Providencia, 299
Serratia, 143

B

Bacillus, 1
Bacteremia, 9–10, 11*f*
Cedecea, 338
Citrobacter, 187–188
E. coli, 15, 31–33, 43*t,* 49
Edwardsiella, 310–311
Enterobacter, 154–158
Ewingella, 345–346
Hafnia, 325*t,* 326–327
iron excess, 214–215
Klebsiella, 114–116
Morganella, 272–274
Proteus, 249–251
Providencia, 287*t,* 288–289
Rahnella, 350
recurrent, 33
Salmonella, 84
Serratia, 136–137
Shigella, 68–69
Yersinia, 214–215, 216–217, 218
Yokenella, 354
Bacteriocins
Enterobacter, 166
Klebsiella, 122
Morganella, 279
Proteus, 258
Bacteriophage typing
Enterobacter, 166, 168
Hafnia, 329

Klebsiella, 122
Proteus, 258
Salmonella, 91
Salmonella Typhi, 105–106
Yersinia, 230
Bacterium, 1, 13
Bacteriuria. *See also* Urinary tract infections
asymptomatic, 31, 118
E. coli, 15, 30–31
Edwardsiella, 307
Enterobacter, 161
Klebsiella, 118, 161
Morganella, 272
Proteus, 248, 249
Providencia, 286, 287*t,* 288, 297, 299
Salmonella, 84
Bacteroides, 160, 188
Beef products. *See* Food; Meat products
Benzylpenicillin, 238
Biochemical characteristics
Budvicia, 362
Buttiauxella, 363, 364*t*
Cedecea, 339–340
Citrobacter, 191–192
E. coli, 38–39
Edwardsiella, 315–316
Enterobacter, 163–164
Escherichia spp., 64
Ewingella, 346
Hafnia, 329, 330
Kluyvera, 340–341
Leclercia, 346*t,* 347
Leminorella, 348
Moellerella, 350
Morganella, 276–287
Obesumbacterium, 364
Pantoea, 180–181
Photorhabdus, 358
Pragia, 365
Proteus, 254–257
Providencia, 293–295
Rahnella, 351
Salmonella, 89–90, 105
Serratia, 139–141
Shigella, 71
Tatumella, 346*t,* 352
Trabulsiella, 353
Xenorhabdus, 358
Yersinia, 226–230
Yokenella, 354–355
Biofilms, 249
Biogroups. *See also* Biotyping
Edwardsiella, 316
Enterobacter, 164–165
Hafnia, 329–330
Morganella, 271
Proteus vulgaris, 245–247
Yersinia enterocolitica, 202–204
Biolog identification system. *See* Microbial identification systems
Biotinylated probes, 36–37, 296
Biotyping. *See also* Biogroups
Citrobacter, 192–193
Edwardsiella, 316
enterovirulent *E. coli,* 40–42
Hafnia, 331
Klebsiella, 121
Morganella, 277–278
Pantoea, 181
Providencia, 295, 296
Serratia, 141
Shigella, 73
Yersinia, 228–230
Birds
E. coli, 13
Edwardsiella, 305
Proteus, 247
Yersinia, 206
Bismuth salts, 126
Bismuth sulfite agar, 85, 86*t,* 88
Citrobacter, 189
Hafnia, 328
Providencia, 291
Salmonella Typhi, 104
Black Death, 206
Bloodborne infections. *See* Bacteremia; Septicemia
Blood culture
Morganella isolation, 275
Salmonella Typhi isolation, 104
Serratia isolation, 138
Bone infections, 161
Bone marrow culture, 104
Brain abscess
Citrobacter, 186*t,* 187
Klebsiella, 114
Proteus, 251
Brain lesions, 158
Breweries, *Obesumbacterium,* 364
Brilliant green agar, 85, 86*t,* 291, 328
Bronchopneumonia, 290, 326
Broths. *See also specific types*
Edwardsiella, 314
Providencia, 292
Salmonella, 88
Shigella, 70
Yersinia, 225–226
Broviac catheter infections, 339
Bubonic plague, epidemiology, 206
Budvicia, 89*t,* 361–362
Bullae, hemorrhagic, 160–161, 273
Burn wounds
Enterobacter, 154, 160
Klebsiella, 119
Providencia, 290
Bursitis, 138, 161–62
Buttiauxella, 340, 361*t,* 362–363

C
Calcifications, 222
Calves. *See* Cattle
Campylobacter, 23, 97
Cancer
Citrobacter bacteremia, 188
Klebsiella infection, 115
Yersinia septicemia, 214
Caprylate (MUCAP), 87
Caprylate-thallous agar, 138
Capsular migration factor, 260
Capsular polysaccharides, in *Klebsiella* pathogenesis, 122–123

Capsular typing
  Klebsiella, 121–122
  Serratia, 141
Carbacephem, 298
Carbapenem, 238, 318
Carbenicillin
  Cedecea, 341
  Citrobacter, 196, 197*t*
  Hafnia, 332
  Kluyvera, 341
  Moellerella, 350
  Yersinia, 237
Carbon source utilization assay
  Citrobacter, 192
  Klebsiella, 121
  Serratia, 140, 141
Cardiovascular infections. *See* Bacteremia; Septicemia
Catheterization
  antimicrobial resistance, 298
  enteric infections, 8
  Hafnia, 327
  Kluyvera, 339
  Morganella UTIs, 272
  Proteus septicemia, 250
  Proteus UTIs, 248–249
  Providencia UTIs, 286, 287*t,* 288, 299
Cathra Repliscan replicator system. *See* Microbial identification systems
Cats
  Proteus, 247
  Yersinia, 206
Cattle
  colisepticemia in, 14
  Klebsiella mastitis in, 113
  Proteus, 247
  Providencia, 285
  STEC gastroenteritis, 25
  Yersinia, 207
Cedecea, 342
  antimicrobials, 341
  clinical sources, 337–338
  epidemiology, 337
  identification, 339–340
  isolation, 339
  taxonomy, 336, 337*t*
Cedecea davisae, 336, 337–338, 339
Cedecea lapagei, 336, 337*t,* 339
Cedecea neteri, 336, 337*t,* 338
Cefamandole, 298, 341
Cefoperazone, 108
Cefotaxime
  Edwardsiella, 318
  Enterobacter, 170, 171
  Kluyvera, 341
  Salmonella Typhi, 108
  Yersinia, 238
Cefoxitin, 341
Cefpirome, 299
Cefprozil, 298
Cefsoludin-irgasan-novobiocin agar, 224–225
Ceftazidime, 50, 51
  Cedecea, 341
  Citrobacter, 197, 198
  Enterobacter, 161, 170
  Hafnia, 332
Ceftizoxime, 238
Ceftriaxone
  Cedecea, 341
  Citrobacter, 198
  Salmonella Typhi, 108
  Yersinia, 238
Cellobiose-arginine-lysine agar, 224
Cellular invasion
  E. coli, 47–50
  Edwardsiella, 317–318
  Proteus, 262–263
  Providencia, 297
  Shigella, 75–76
  Yersinia, 234–235
Cellulitis
  Citrobacter, 186*t*
  Edwardsiella, 312*t*
  Enterobacter, 160, 161
  Proteus, 252
  Serratia, 138
  Yersinia, 220
Central nervous system infections
  Citrobacter, 185, 186*t,* 187
  Enterobacter, 158–159
  Morganella, 273
  Proteus, 251–252
  Serratia, 137
Cephalosporinase, 170
Cephalosporins
  Citrobacter, 197
  E. coli, 50
  Edwardsiella, 318
  Enterobacter, 170
  Hafnia, 332
  Kluyvera, 341
  Morganella, 280
  Proteus, 264
  Salmonella, 96
  Salmonella Typhi, 107–108
  Serratia, 144
  Shigella, 77
  Yersinia, 237, 238
Cephalothin
  Cedecea, 341
  Citrobacter, 196, 197*t*
  Kluyvera, 341
  Providencia, 298
  Salmonella, 96
  Serratia, 143
Cephems, 281
Cerebral abscess. *See* Brain abscess
Chickens. *See* Poultry
Children. *See also* Infants
  bloodborne infections, 10, 11*f*
  E. coli gastroenteritis, 17–18, 19, 20, 27
  E. coli infections, 15, 30, 33
  Enterobacter infections, 157
  Hafnia infections, 326, 327
  Proteus infections, 248, 251–252
  Providencia infections, 287*t*
  Salmonella infections, 82
  Shigella infections, 69
  Yersinia infections, 209, 210–213, 216–217, 220
Chloramphenicol
  Cedecea, 341

Edwardsiella, 318
Hafnia, 332
Kluyvera, 341
Proteus, 256
Providencia, 298
Salmonella Typhi, 106, 107
Serratia, 143
Yersinia, 238
Cholangitis, 162, 186*t*
Cholelithiasis, 27
Chondroitinase, 315
Choriomanionitis, 273
Christensen's urea agar, 256
Chrome Azurol S agar, 234
Chronic care facilities. *See* Long-term care; Nursing homes
CI agar, 291–292
CIN agar, 190, 341
Ciprofloxacin
Citrobacter, 197
E. coli, 50, 51
Edwardsiella, 318
Hafnia, 332
Klebsiella, 126
Kluyvera, 342
Salmonella, 95, 96
Salmonella Typhi, 107
Serratia, 143
Yersinia, 238
Citrate utilization
E. coli, 39
Enterobacter, 164*t*
Klebsiella, 120*t*
Leminorella, 348
Salmonella, 89*t*
Salmonella Typhi, 105*t*
Citrobacter
antimicrobials, 196–198
bacteremia, 187–188
biochemical identification, 190–193
biotyping, 192–193
central nervous system infections, 185, 186*t,* 187
clinical infections, 185–189
distinguishing from *Salmonella,* 88, 89, 89*t,* 90
distinguishing from *Yersinia,* 225, 226*t*
epidemiology, 184–185
genomospecies, 183–184
identification, 349*t,* 362*t*
immunodiagnostics, 193–194
nosocomial infections, 184–185, 188
pathogenesis, 195–196
taxonomy, 183–184
typing, 194–195
urinary tract infections, 185, 188–189
Citrobacter amalonaticus, 183, 191, 192, 193, 196, 197
Citrobacter braakii, 192
Citrobacter diversus, 183
Citrobacter farmeri, 184, 191, 193
Citrobacter freundii
antimicrobials, 196–197, 198
bacteremia, 188
central nervous system infections, 187
emphysematous pyelonephritis, 189
enteric fever, 188
epidemiology, 185
identification, 191–192
invasive mechanism, 196
serotyping, 194
taxonomy, 5, 6, 183
toxigenic activity, 195–196
wound infections, 189
Citrobacter koseri, 12
aerobactin, 196
antimicrobials, 196, 197, 198
bacteremia, 188
biotyping, 193
brain abscesses, 187
epidemiology, 185
identification, 191
neonatal meningitis, 185, 187
outer membrane proteins in, 194–195
pathogenesis, 195
plasmid analysis in, 194
serotyping, 194
taxonomy, 183, 184
ventriculitis, 187
wound infections, 189
Citrobacter rodentium, 183–184, 189–190, 196, 331
Citrobacter sedlakii, 192
Citrobacter werkmanii, 192
Citrobacter youngae, 192
CLARE. *See* Contact lens-induced acute red eye
CLDT. *See* Cytolethal distending toxin
CNF. *See* Cytotoxic necrotizing factor
Cobas identification systems. *See* Microbial identification systems
Cold-blooded animals, *Edwardsiella* and, 304, 305, 306
Colibacillary toxemia, 14
Colibacillosis, 13–14
Coliform bacteria, 2
Colisepticemia, 14
Colistin, 298, 314, 350
Colitis
E. coli, 11, 17*t,* 22–24
Klebsiella, 118, 125
ulcerative, clinical features, 23*t*
Colon, *Shigella* infection, 68
Colonic carcinoma, 310
ColV plasmid, 49
Commensalism, *Edwardsiella* and, 305
Commercial identification systems. *See* Microbial identification systems
Community-acquired infections
E. coli, 9, 15, 30, 31–32, 52
Enterobacter, 159–160, 172
Klebsiella, 116
Morganella, 271
Yersinia, 209
Congo red-magnesium oxalate agar, 229, 232
Conjunctivitis, 69
Contact lens-induced acute red eye (CLARE), 135–136, 138
Corneal ulceration, 69, 162
Corn oil lipase agar, 162
Corticosteroids, *Yersinia* septicemia, 214
Cows. *See* Cattle
Crepitant myonecrosis, 119
Crohn's disease, 23*t*
Cruise ships, shigellosis, 67
Crystal identification systems. *See* Microbial identification systems

CT-SMAC, 35–36
Cystitis, 30, 31, 118
Cytolethal distending toxin (CLDT), 29–30, 44*t*
Cytotoxic necrotizing factor (CNF), 30, 44*t*, 50
Cytotoxins
  Citrobacter, 196
  Klebsiella, 124–125
  Serratia, 142
  Shiga toxin-producing *E. coli*, 21
  Yersinia, 236

D
DAEC. *See* Diffuse adhering *E. coli*
Dairy products, *Yersinia*, 210, 219
Day care centers
  Edwardsiella, 317
  shigellosis, 67
Deaminase, 255, 261*t*, 262*f*
Deferoxamine, 215
Dehydration
  Edwardsiella enteritis, 308, 309
  Shigella gastroenteritis, 68
  Yersinia gastroenteritis, 211
Desoxycholate agar, 291, 314
Desoxycholate-lactose-sucrose-sorbitol agar, 328
Desoxyribonuclease-toluidine blue-cephalothin agar, 138
Developing countries. *See* Third World
2,4-Diamino-6,7-diisopropyl pteridine, 315
Diarrhea, 8, 11–12
  Citrobacter, 189
  E. coli, 15 16
    enteroadherent, 28–29
    enteroinvasive, 19
    enteropathogenic, 16–17
    enterotoxigenic, 19–21
    foodborne, 14–15
    pathogenesis, 47
    serotypes in, 43*t*
  E. fergusonii, 63, 65
  Edwardsiella, 307
  Hafnia, 325, 331
  Kluyvera, 338
  Moellerella, 349
  Morganella, 274
  Proteus, 253
  Providencia, 289–290, 297, 299
  Salmonella, 11, 83
  traveler's, 11, 20–21, 28
  in typhoid fever, 102–103
  Yersinia, 211, 212
Diarrhea, bloody. *See also* Hemorrhagic colitis
  E. coli, 16, 20, 22, 24, 28
  Edwardsiella, 309
  Yersinia, 211, 218
Dienes test, 258
Diffuse adherence
  E. coli, 45–46
  Klebsiella, 123
Diffuse adhering *E. coli* (DAEC), 28, 29, 45–46, 48*t*
Disinfectants, *Providencia*, 298
Disk diffusion test
  Morganella, 280–281
  Providencia, 298
  Tatumella, 353
DL-Tryptophan, 276
DNA groups
  Enterobacter cloacae, 150–151
  Hafnia, 323–324, 329
  Morganella, 271
  Proteus vulgaris, 245–247
DNA hybridization, 5, 148–149, 178, 230
Dogs
  Morganella, 271
  Proteus, 247
Doxycycline, 238
Drinking water. *See* Water
Ducks, *Yersinia* and, 206
Dulcitol fermentation, 90*t*, 348
Duodenal string capsules, 104
Dysentery
  enteroinvasive *E. coli*, 17*t*, 18
  Shigella, 23*t*, 68
DYS media, 224
Dysuria, 31

E
EAEC. *See* Enteroadherent *E. coli*
EAF plasmid, 47
EAggEC. *See* Enteroaggregative *E. coli*
Eberthella, 5
Ecthyma gangrenosum, 162
Edema disease, in pigs, 14
Edwardsiella
  antimicrobials, 318
  bacteremia, 310–311
  biotyping, 316
  commensalism, 305
  distinguishing from *Salmonella*, 89, 89*t*, 90
  epidemiology, 305–306
  gastroenteritis, 307–310
  identification, 314–316, 349*t*, 362*t*
  immunodiagnostics, 316–317
  isolation, 314
  pathogenesis, 317–318
  peritonitis, 314
  sources, 307
  taxonomy, 304–305
  typing, 317
  urinary tract infections, 314
  wound infections, 311–314
Edwardsiella hoshinae, 305, 314, 315
Edwardsiella ictaluri, 305, 314, 315
Edwardsiella tarda. *See Edwardsiella*
EHEC. *See* Enterohemorrhagic *E. coli*
EIA. *See* Enzyme immunoassay
EIEC. *See* Enteroinvasive *E. coli*
Elderly
  Enterobacter bacteremia, 154
  Hafnia septicemia, 327
  Morganella infections, 272–273
  Proteus infections, 249, 250
  Providencia bacteremia, 287*t*, 288
  risk of enterobacterial infections, 8, 12
  Yersinia bacteremia, 214
Electrophoresis. *See* Pulsed-field gel electrophoresis
ELISA. *See* Enzyme-linked immunosorbent assay
Empyema, 221, 274
Encephalopathy, 19, 69
Endocarditis
  Citrobacter, 186*t*

Enterobacter, 157–158
Serratia, 135, 136
Yersinia, 218
Endocytosis, in *Providencia* pathogenicity, 297
Endophthalmitis
endogenous, 34
Enterobacter, 162
Hafnia, 328
Klebsiella, 118
Proteus, 253
Endotoxins, in colibacillary toxemia, 14
Enrichment broths. *See also specific types*
Edwardsiella, 314
Providencia, 292
Salmonella, 88
Shigella isolation, 70
Yersinia, 225–226
Entamoeba histolytica, 307, 309, 310, 317
Enteric colibacillosis, 13–14
Enteric fever, 84, 309. *See also* Typhoid fever
Enteritis. *See* Gastroenteritis
Enteroadherent *E. coli* (EAEC)
adherence assays, 45
biotyping, 40*t*
gastroenteritis, 16, 17*t,* 18*t,* 28–29
pathogenic factors, 48*t,* 49
Enteroaggregative *E. coli* (EAggEC), 28, 29
adherence assays, 45
biotyping, 40*t,* 41
pathogenic factors, 48*t*
Enterobacter
antimicrobials, 170–172
bacteremia, 10, 154–158
biogroups, 164–165
central nervous system infections, 158–159
community-acquired infections, 172
epidemiology, 151–153
gastroenteritis, 172
genomospecies, 165–166
identification, 162–166, 364*t*
immunodiagnostics, 166
isolation, 162
nosocomial infections, 153–154, 157, 159, 161
pathogenesis, 168–169
respiratory infections, 159–160
taxonomy, 148–151
typing, 166–168
urinary tract infections, 161
wound infections, 160–161
Enterobacteraceae, 3
Enterobacter aerogenes
antimicrobial resistance, 170, 171, 172
bacteremia, 156
biogroups, 165
crepitant cellulitis, 160
food products, 152
meningitis, 158–159
nosocomial infections, 154
pneumonia, 159
taxonomy, 148, 149, 150
typing, 166
Enterobacter agglomerans, 148
antimicrobials, 171*t,* 172
distinguishing from *Ewingella,* 346
food products, 152
taxonomy, 151, 178–180
Enterobacter agglomerans-Erwinia herbicola complex, 178–180
Enterobacter amnigenus, 157, 165
Enterobacter asburiae, 157
Enterobacter cancerogenus
antimicrobials, 171*t,* 172
bacteremia, 157
osteomyelitis, 161
taxonomy, 149, 150
urinary tract infections, 161
wound infections, 161
Enterobacter cloacae
antimicrobial resistance, 170, 171, 172
bacteremia, 156–157
biochemical features, 164
biogroups, 165
DNA groups, 150–151
food products, 152
genomospecies, 165–166
hemorrhagic bursitis, 161–162
meningitis, 158–159
nosocomial infections, 154
outer membrane proteins, 169
respiratory infections, 159
serotyping, 166
taxonomy, 1, 148, 149
typing, 166, 168
vertebral osteomyelitis, 161
wound infections, 160
Enterobacter dissolvens, 164
Enterobacter gergoviae, 157, 170
Enterobacter hormaechei, 157
Enterobacteriaceae
bacteremia, 10
discovery, 1
gastrointestinal disease, 8, 11–12
geriatric population, 8, 12
major phenotypic features, 4*t*
microbial identification systems, 9
nosocomial infections, 9
septicemia, 9–10
taxonomy, 1–6
Enterobacterial repetitive intergenic consensus sequences (ERIC), 168
Enterobacter intermedium, 157, 164
Enterobacter kobei, 149, 165
Enterobacter nimipressuralis, 149, 164
Enterobacter pyrinus, 149
Enterobacter sakazakii, 153, 157, 158, 180
Enterobacter taylorae, 150
Enterobactin, 123, 142, 234
Enterocolitis, 210–213, 219
Enterohemolysin, 36, 44*t*
Enterohemorrhagic *E. coli* (EHEC), 11, 21, 331. *See also* Shiga toxin-producing *E. coli*
Enteroinvasive *E. coli* (EIEC)
biotyping, 40*t,* 41
diarrheagenic characteristics, 17*t*
distinguishing from *Shigella,* 39
gastroenteritis, 16, 18–19
pathogenic factors, 48*t*
serotypes, 43*t*
Enteropathogenic *E. coli* (EPEC)
adherence assays, 45–46

Enteropathogenic *E. coli* (EPEC) *(contd.)*
  biotyping, 40*t*
  cytolethal distending toxin, 30
  eae-like gene, 331
  fluorescent actin stain test, 46–47
  gastroenteritis, 16–18
  pathogenesis, 47, 48*t*
  serotypes, 43*t*
Enterotoxigenic *E. coli* (ETEC), 11, 67
  biotyping, 40*t,* 41–42
  in colibacillary toxemia, 14
  gastroenteritis, 16, 17*t,* 18*t,* 19–21
  pathogenesis, 47, 48*t*
  serotypes, 43*t*
  toxin assays, 43–44
Enterotoxins, 20, 47
  Citrobacter, 196
  E. coli, 47
  Edwardsiella, 318
  Klebsiella, 124
  Shigella, 76
  Yersinia, 237
Enzyme immunoassay (EIA), 141, 230
Enzyme-linked immunosorbent assay (ELISA), 122, 257
EPEC. *See* Enteropathogenic *E. coli*
Epidemiology
  Cedecea, 337
  Citrobacter, 184–185
  Edwardsiella, 305–306
  Enterobacter, 151–153
  Escherichia coli, 13–15
  Ewingella, 344
  Hafnia, 324
  Klebsiella, 111–113
  Kluyvera, 337
  Leclercia, 347
  Moellerella, 349
  Morganella, 271
  Photorhabdus, 356
  Proteus, 247–248
  Providencia, 285–286
  Rahnella, 350
  of rhinoscleroma, 113
  Salmonella, 81–83, 91–92
  Serratia, 132–136
  of shigellosis, 67–68
  Trabulsiella, 353
  of typhoid fever, 101–102
  Xenorhabdus, 356
  Yersinia, 205–210
  Yokenella, 354
Epidural abscess, 252
Epithelial cells
  Salmonella, 94
  Shigella, 76
ERIC. *See* Enterobacterial repetitive intergenic consensus sequences
Erwinia, 151, 178–180
Erythema nodosum, 213, 222–223
ESBLs. *See* Extended-spectrum β-lactamase
Escherichia blattae, 61–62
Escherichia cloacae, 6
Escherichia coli. See *also* Enteroadherent *E. coli;* Enteroaggregative *E. coli;* Enteroinvasive *E. coli;* Enteropathogenic *E. coli;* Enterotoxigenic *E. coli;* Shiga toxin-producing *E. coli*
  adherence assay, 45–46
  antimicrobials, 50–51
  antiphagocytic factors, 49
  assays, 43–47
  bacteremia, 10, 31–33
  cholecystitis, 119
  cysteine-dependent strains, 35, 38, 51
  cytolethal distending toxin, 29–30
  cytotoxic necrotizing factor, 30
  distinguishing from *Salmonella,* 89*t,* 90
  distinguishing from *Shigella,* 5, 39, 66, 72, 73
  enterohemorrhagic, 11, 21
  enterovirulent biotypes, 40–42
  epidemiology, 13–15
  fluorescent actin stain test, 46–47
  food adulteration, 14–15
  gastroenteritis, 15–31
  identification, 37–42
  immunodiagnostics, 42–43
  invasion assay, 44–45
  isolation, 34–37
  K1 capsules, 33, 50
  Kluyvera identification, 340
  meningitis, 33
  non-O157:H7 serotypes, 27–28
  nosocomial infections, 9, 15, 30, 31–32, 52
  ocular infections, 34
  O157:H7, 11
    biotyping, 40–41
    gastroenteritis, 21–27
    identification, 38, 39
  pathogenesis, 47–50
  peritonitis, 34
  polymicrobic infections, 157, 188, 250
  respiratory infections, 34
  septicemia, 32
  serotypes, 42–43
  taxonomy, 1, 13
  thymidine-dependent strains, 35
  toxin assay, 43–44
  typing, 43
  urinary tract infections, 30–31
Escherichia coli ADO, 66
Escherichia fergusonii, 61–65
Escherichia freundii, 5
Escherichia hermanii, 61–65
Escherichia spp.
  antimicrobials, 64–65
  disease, 61–63
  isolation/identification, 63–64
  taxonomy, 61
Escherichia vulneris, 61–65
Esculin, 140*t,* 229
ETEC. *See* Enterotoxigenic *E. coli*
Ewingella, 344–347, 365*t*
Exotoxins, 318
Extended-spectrum β-lactamase
  E. coli, 50, 51
  Enterobacter, 170
  Klebsiella, 125–126
  Proteus, 264
  Serratia, 144
Extended-spectrum penicillins, 264, 318
Eye infections
  E. coli, 34
  Enterobacter, 162

Ewingella, 345
Hafnia, 328
Serratia, 138
Shigella, 69

F
FAME analysis. *See* Fatty acid methyl ester analysis
Fasciitis, 160, 252
Fast food, STEC gastroenteritis, 25
Fatty acid analysis
Morganella, 278
Proteus, 257
Providencia, 294
Fatty acid methyl ester (FAME) analysis, 294
Feces. *See* Stool
Fermentation
Citrobacter, 189–190, 191, 192
Edwardsiella, 315, 316
Enterobacter, 164*t*
Hafnia, 330
Leminorella, 348
Moellerella, 349–350
Morganella, 277, 278
Proteus, 254
Providencia, 294, 295
Yersinia, 228
Fever
associated with typhoid fever, 103
in *Yersinia* gastroenteritis, 212
Fimbriae, 49
Klebsiella, 124
Proteus, 260, 261*t,* 262, 263
Salmonella, 93–94
Serratia, 142
Fish
Citrobacter, 184
E. coli, 14
Edwardsiella, 305
Enterobacter, 152
Hafnia, 324
Morganella, 271
Proteus, 247
Flagellar typing
Edwardsiella, 316–317
Salmonella, 90–91
Serratia, 141
Yersinia, 230
Flagellin, *Proteus,* 259
Fleas, *Yersinia pestis,* 206, 232
Fluorescent actin stain, 46–47, 297, 330–331
Fluoroquinolones, 50–51, 107–108
Food
Citrobacter, 184
E. coli, 14–15, 21, 25
Enterobacter, 152
Morganella, 271
Salmonella, 83, 97
Yersinia, 207, 209
Francisella philomiragia, 204

G
Gangrene, 116, 253, 312*t*
Gas gangrene, 312*t*
Gastric acid. *See also* Acid tolerance
Salmonella, 93
Salmonella Typhi, 106
Shigella, 75
Gastroenteritis
Citrobacter, 186*t,* 189
Edwardsiella, 307–310
Enterobacter, 172
Enterobacteriaceae, 8, 11–12
Hafnia, 325–326, 332
Klebsiella, 118
Morganella, 274
Proteus, 253
Providencia, 287*t,* 289–290, 297
Salmonella, 11, 83–84, 92
Shigella, 68
Yersinia, 210–213
Gastroenteritis, *E. coli*-associated, 15–31
in children, 17–18, 19, 20, 27
cytolethal distending toxin, 29–30
cytotoxic necrotizing factor, 30
enteroadherent *E. coli,* 16, 17*t,* 18*t,* 28–29
enteroinvasive *E. coli,* 16, 17*t,* 18–19
enteropathogenic *E. coli,* 16–18
enterotoxigenic *E. coli,* 16, 17*t,* 18*t,* 19–21
pathogenesis, 47
shiga toxin-producing *E. coli,* 16, 17*t,* 18*t,* 21–28
Gel electrophoresis. *See* Pulsed-field gel electrophoresis
Genes
in antimicrobial resistance, 76–77
expression, in *Yersinia,* 234
in pathogenicity, 75–76, 92–93
Genomospecies, 6
Citrobacter, 183–184
Enterobacter, 165–166
Hafnia, 329–330
Proteus vulgaris, 246
Gentamicin
Cedecea, 341
Hafnia, 332
Proteus, 264
Providencia, 298
Serratia, 143
Yersinia, 238
Geriatric patients. *See* Elderly
Glomerulonephritis, 223
β-Glucosidase, 229*t,* 330
β-D-Glucuronidase, 36, 37, 40
Glutamic acid decarboxylation, 295
Glycerol fermentation, 278
Gram-negative broth, 314
Ground beef. *See also* Food; Meat products
Enterobacter, 152
STEC gastroenteritis, 25

H
Hafnia
antimicrobials, 332
bacteremia, 326–327
clinical sources, 324–325
distinguishing from *Salmonella,* 89*t*
distinguishing from *Yokenella,* 354
DNA groups, 323–324, 329
epidemiology, 324
gastroenteritis, 325–326, 332
identification, 328–330, 365*t*
immunodiagnostics, 330
isolation, 328

Hafnia (contd.)
nosocomial infections, 326
pathogenesis, 331
respiratory tract infections, 326
specialized assays for, 330–331
taxonomy, 322–324
typing, 331
Hafnia alvei. See Hafnia
Hamburger. *See* Ground beef; Meat products
Hansens's disease, 115–116
Hares, *Yersinia,* 206
Hektoen agar, 70, 85, 87, 88, 314
Hemagglutinin
Edwardsiella, 318
Enterobacter, 168–169
Proteus, 260, 262
Providencia, 297
Serratia, 142
Hematomas, *Enterobacter* infections, 160, 161
Hemin, 232
Hemochromatosis, 214, 215, 217, 218
Hemolysin, 36, 44*t,* 49–50
Edwardsiella, 315, 317, 318
Morganella, 279–280
Proteus, 253, 260, 261*t,* 262*f,* 263
Serratia, 142
Hemolytic-uremic syndrome
Citrobacter, 189
Enterohemorrhagic *E. coli,* 11
Shigella, 69
STEC gastroenteritis, 24–25, 26–27, 28
Hemorrhagic colitis, 11, 17*t,* 22–24, 125
Hemosiderin, 215
Hemosiderosis, 215, 218
Hepatic abscess. *See* Liver abscess
Hepatic disease. *See* Liver disease
Hepatobiliary disease, 307, 310
Hepatomegaly, 27
HEp-2 cells, 297
Histamines, 247, 271
"Honeymoon cystitis," 30
Horses, *Klebsiella* metritis in, 113
Hospitals. *See also* Nosocomial infections
transmission of *Serratia marcescens,* 134–135
Yersinia testing, 223
Hybridization groups, in *Pantoea* classification, 178–179
Hydrogen sulfide
Citrobacter, 191, 192*t*
E. coli, 39
Edwardsiella, 314, 315
Morganella, 275
Proteus, 254, 256
Salmonella, 85, 89–90
Salmonella Typhi, 104
Hydroxystilbenes, 358
Hypoglycemia, shigellosis and, 69
Hyponatremia, 69
Hypoproteinemia, 68

I

Identification. *See* Biochemical characteristics; Isolation; Microbial identification systems
Ileal perforation, typhoid fever, 103
Ileitis, 210, 212–213
Ileum, *Salmonella* infection, 93
Imipenem
Enterobacter, 171
Morganella, 280
Proteus, 264
Providencia, 299
Serratia, 143
Yersinia, 238
Immunity, to *E. coli* O157:H7, 27
Immunodiagnostics
Citrobacter, 193–194
E. coli, 42–43
Edwardsiella, 316–317
Enterobacter, 166
Hafnia, 330
Klebsiella, 121–122
Morganella, 278–279
Proteus, 257
Providencia, 295–296
Salmonella, 90–91, 105
Serratia, 141
Shigella, 73–74
Yersinia, 230
IMP-1 enzyme, 143
Incubation period, for *Salmonella,* 82, 102
Indole assay
Edwardsiella, 315
Enterobacter, 163
Klebsiella, 120*t,* 121
Salmonella, 89*t*
Yersinia, 229*t*
Infants. *See also* Children; Meningitis, neonatal; Neonates
Citrobacter koseri, 185, 187
E. coli gastroenteritis, 15, 17–18, 19, 20
Klebsiella infections, 112, 113
shigellosis, 67
Yersinia infections, 209, 216–217
Infections, chronic, *Yersinia,* 223
Infections, polymicrobic
due to *Citrobacter,* 188
due to *Enterobacter,* 157
due to *Proteus,* 248, 250–251
Providencia, 286, 289, 297
Infectious doses
Salmonella, 82
Salmonella Typhi, 102
Shigella, 67
Insects, *Serratia* and, 132, 133
Intestinal adhesions. *See* Adherence
Intestinal perforation
Shigella, 68
typhoid fever, 103
Intimins, 331
Intubation, *Hafnia* infections, 326
Inulin, 330
Invasin, 234, 235*t*
Invasion assays, *E. coli,* 44–45
Invasiveness. *See* Cellular invasion
Irgasan, 341
Iritis, 69, 213
Iron acquisition
Morganella, 280
Yersinia, 232, 234
Iron chelators. *See* Siderophores

Iron excess
  Edwardsiella bacteremia, 311, 319
  Yersinia bacteremia, 214–215
Iron-regulated gene expression, 234
Irritable bowel syndrome, 83
Isolation techniques
  Budvicia, 361–362
  Buttiauxella, 363
  Cedecea, 339
  Edwardsiella, 314
  Enterobacter, 162
  Escherichia coli, 34–37
  Escherichia spp., 63–64
  Ewingella, 346
  Hafnia, 328
  Klebsiella, 119–120
  Kluyvera, 339
  Leminorella, 348
  Moellerella, 349–350
  Morganella, 275
  Photorhabdus, 357–358
  Pragia, 365
  Proteeae, 291
  Proteus, 253–254
  Providencia, 290–292
  Rahnella, 351
  Salmonella, 85–90
  Salmonella Typhi, 85, 104–105
  Serratia, 138
  Tatumella, 352
  Trabulsiella, 353
  Xenorhabdus, 357
  Yersinia, 223–226
  Yokenella, 354
Isolator tube, 34, 138
IV drug use
  Enterobacter infections, 160
  Serratia infection, 135, 136
IV solutions
  Enterobacter infections, 152–153
  Klebsiella infections, 115
Izumi fever, 213

J

Japan, STEC gastroenteritis in, 25

K

Kanamycin, 332
Kawasaki syndrome, 213
K1-capsule, 33, 50
Keratitis, 69
Keratoconjuctivitis test, 41
Keto acids, 280, 298
Kidney, *Proteus* pathogenicity, 259–263
Klebsiella
  antimicrobial resistance, 124, 125–126
  bacteremia, 10, 32, 114–116
  bacteriuria, 161
  biochemical characteristics, 120–121
  biotyping, 121
  central nervous system infections, 113–114
  clinical infections, 113–119
  cysteine-dependent strains, 119
  discovery, 1
  epidemiology, 111–113
  gastroenteritis, 118
  identification, 364*t*
  immunodiagnostics, 121–122
  isolation, 119–120
  pathogenesis, 122–125
  in polymicrobic infections, 157, 248
  pyogenic liver abscess, 118–119
  relatedness to *Enterobacter,* 150
  respiratory infections, 116–117
  taxonomy, 110–111
  typing, 122
  urinary tract infections, 117–118
Klebsiella aerogenes, 110
Klebsiella edwardsii, 110
Klebsiella ornithinolytica, 111, 121, 126
Klebsiella oxytoca, 110, 111, 187
  antimicrobials, 126
  identification, 120, 121
  nonpseudomembranous colitis, 118
  toxins, 124–125
Klebsiella ozaenae, 110, 111*t*
  antimicrobials, 126
  bacteremia, 115, 116
  bacteriuria, 118
  central nervous system infections, 114
  identification, 120
  isolation, 119
  pyogenic liver abscess, 119
  respiratory infections, 117
Klebsiella planticola, 111, 120–121, 126
Klebsiella pneumoniae
  antimicrobial resistance, 125, 126
  bacteremia, 115
  cholecystitis, 119
  gastroenteritis, 118
  identification, 120–121
  meningitis, 113–114
  pathogenesis, 123–124
  in polymicrobic infections, 188, 250
  pyogenic liver abscess, 118–119
  retropharyngeal abscess, 119
  taxonomy, 1, 2*t,* 110–111, 148
  toxins, 124
  urinary tract infections, 117
Klebsiella rhinoscleromatis, 12
  antimicrobials, 126
  identification, 120
  isolation, 119
  respiratory infections, 116–117
  rhinoscleroma, 113
  taxonomy, 110, 111*t*
Klebsiella terrigena, 111, 118, 120–121, 126
Kligler's agar, 88, 275, 328
Kluyvera
  antimicrobials, 341–342
  clinical sources, 337, 338
  distinguishing from *Buttiauxella,* 363
  epidemiology, 337
  identification, 340–341
  isolation, 339
  taxonomy, 336–337
Kluyvera ascarbata, 337*t,* 338, 339, 341
Kluyvera cochleae, 337*t,* 340
Kluyvera cryocrescens, 337*t,* 339, 341

L
β-Lactam antibiotics
  Enterobacter infections, 156, 159
  Klebsiella resistance to, 125–126
β-Lactamase
  Citrobacter, 197
  E. coli, 50
  Edwardsiella, 318
  Enterobacter, 170, 171
  Hafnia, 332
  Klebsiella, 124, 125–126
  Providencia, 298
  Salmonella, 96
  Serratia, 143–144
  Yersinia, 237
Lactose fermentation, 2, 189–190
Lambs, colisepticemia in, 14
Leclercia, 178, 345*t,* 346*t,* 347–348
Leminorella, 89*t,* 345*t,* 348–349, 362*t*
Lesions. *See also* Wound infections
  Enterobacter, 158
  Shigella, 75
  Yersinia, 222–223
Leukocytes, fecal, in STEC gastroenteritis, 23
Levinea, 183
Life expectancy, enterobacterial infections and, 8
Lipase test, 164*t,* 229*t*
Lipopolysaccharides
  Hafnia, 331
  Klebsiella, 122–123
  Providencia, 294
  Yersinia, 237
Liver abscess, 215, 309, 313
Liver disease, 214, 222, 310, 319
Lobar pneumonia, 290
Localized adherence
  enteroadherent *E. coli,* 28, 45
  Hafnia, 330–331
Locus of enterocyte effacement, 47
Long-term care. *See also* Nursing homes
  E. coli infections, 15
  Enterobacter infections, 154
  Morganella UTIs, 272
  Proteus bacteremia, 250
Loracarbef, 298
Low-calcium response, in *Yersinia,* 235
LT enterotoxin, 47
Lung, *Cedecea* isolation, 337
Lung abscess, 116
Lupus erythematosus, 338
Lymphadenitis, mesenteric, 210, 212–213, 234
Lymphadenopathy, 222
Lymphoid tissue, *Yersinia* invasion, 234
Lysine decarboxylase test
  Citrobacter, 191
  E. coli, 41
  Enterobacter, 164*t*
  Morganella, 277, 278
  Serratia, 139, 140
Lysine-iron agar
  E. coli, 37
  Morganella, 275
  Proteus, 255
  Providencia, 293
  Salmonella, 88, 89*t*
  Salmonella Typhi, 105
  Shigella, 70, 71*t*
  Yersinia, 226

M
MacConkey agar
  E. coli, 35
  Edwardsiella, 314
  Hafnia, 328
  Morganella, 275
  Providencia, 290, 292
  Salmonella, 85, 88
  Shigella, 70
  Yersinia, 219, 224, 225
Macrophage
  Klebsiella infection, 123
  Salmonella infection, 94, 95
Maladsorption syndrome, 310
Malnutrition, in shigellosis, 68
Mannitol fermentation, 72
Mannose-resistant hemagglutinins, 169, 297
Mannose-sensitive hemagglutinins, 142, 169
Marine habitats. *See* Aquatic habitats
Mastitis, 113
M cells. *See* Microfold cells
Meat products
  E. coli, 14
  Enterobacter, 152
  Proteus, 247
  Yersinia, 207, 217
Megacolon, toxic, 68
Melanin, 255
Meningitis, 10–11, 69
  Citrobacter, 186*t,* 187, 195
  E. coli, 15, 33, 49*t,* 50
  Enterobacter, 158 159
  Klebsiella pneumoniae, 113–114
  Morganella, 273
  Salmonella, 84–85
  Serratia, 137
  Yersinia, 223
Meningitis, neonatal, 12
  Citrobacter, 185, 186*t,* 187
  E. coli, 15, 33, 43*t,* 50
  Edwardsiella, 310
  Enterobacter, 158, 159*t*
  Morganella, 273
  Proteus, 251–252
  Salmonella, 84–85
  Serratia, 137
Metallo-β-lactamases, 143–144
Methyl red test
  Citrobacter, 191
  Enterobacter, 164
  Morganella, 276
4-Methyl-umbelliferyl-β-D-glucuronide (MUG), 36, 37, 40–41
Metritis, 113
Microbial identification systems
  Cedecea, 339
  Citrobacter, 190–191
  E. coli, 37–38
  Edwardsiella, 314–315
  Enterobacter, 162–163
  Enterobacteriaceae, 9

Escherichia spp., 63–64, 71–73
Ewingella, 346
Hafnia, 328–329
Klebsiella, 120
Leclercia, 347–348
Leminorella, 348
Moellerella, 350
Morganella, 275–276
Pantoea, 180, 181
Proteus, 255–256
Providencia, 292–293
Rahnella, 351
Salmonella, 88–89
Serratia, 138–139
Shigella, 71–73
Trabulsiella, 353
Yersinia, 226–228
Yokenella, 355
Microfold cells, 106
MicroScan WalkAway identification system. *See* Microbial identification systems
Microvilli, *E. coli* AE lesions, 46, 47
Modified Rappaport broth, 226
Modified semisolid Rappaport-Vassailidis agar, 86*t,* 87, 88
Moellerella, 345*t,* 346*t,* 349–350
Mollusks
Ewingella, 344
Kluyvera, 337
Monobactams
Edwardsiella, 318
Hafnia, 332
Yersinia, 238
"Montezuma's revenge." *See* Traveler's diarrhea
Morganella, 245
antimicrobials, 280–281
bacteremia, 272–274
distinguishing from *Yersinia,* 225, 226*t*
epidemiology, 271
gastroenteritis, 274
identification, 275–278, 295
immunodiagnostics, 278–279
isolation, 275
nosocomial infections, 271, 273, 281
pathogenesis, 279–280
taxonomy, 270–271
typing, 279
urinary tract infections, 272
Morganella morganii. See Morganella
Morganella *morganii* ssp. *morganii,* 271, 278
Morganella morganii ssp. *sibonii,* 271, 278
Morganella-Proteus-Providencia group, 291
Morganocins, 279
Mortality rates
Citrobacter bacteremia, 188
Edwardsiella bacteremia, 311
Enterobacter bacteremia, 154, 156
Proteus bacteremia, 250
Providencia bacteremia, 289
Serratia bacteremia, 136
Yersinia septicemia, 218
Motility
enteroinvasive *E. coli,* 41
Hafnia, 329–330
Proteus, 253–254, 259, 260, 261*t,* 263
Providencia, 291
Motility-indole-ornithine medium
E. coli, 37
Salmonella, 88, 89*t*
Salmonella Typhi, 104–105
Shigella, 70, 71*t*
Yersinia, 226
Mouse. *See* Rodents
MUCAP. *See* Caprylate
MUG. *See* 4-Methyl-umbelliferyl-β-D-glucuronide
Multi-drug resistance. *See also* Resistance, antimicrobial
in *Salmonella* Typhi, 106–107
Multilocus enzyme electrophoresis, 167*t,* 168, 195
Muscle
Proteus infection, 252
Yersinia infection, 220–221
Mycotic aneurysm, 85, 218
Myonecrosis, crepitant, 119

N

Nalidixic acid, 77, 298
Neonatal meningitis. *See* Meningitis, neonatal
Neonates. *See also* Meningitis, neonatal
Citrobacter infections, 187, 188, 195
Enterobacter infections, 153, 157
Klebsiella infections, 112, 113
Nephritis, 213
Netilmicin, 332
Neurology wards, *Providencia* infections, 286, 288
Neuropsychiatric symptoms, in typhoid fever, 103
Nitrofurantoin, 298
Nonagglutinating fimbriae, 262
Nonpseudomembranous colitis, 118
Norfloxacin, 318
Nose
atrophic rhinitis, 117
carriage of *Klebsiella,* 112
Nosocomial infections
Citrobacter, 184–185, 188
E. coli, 9, 15, 30, 31–32, 52
Enterobacter, 152–153, 157, 159, 161
Enterobacteriaceae, 9
Ewingella, 344, 346
Hafnia, 326
Klebsiella, 111–113, 114–115, 126
Morganella, 271, 273, 281
Proteus, 247–249, 250
Providencia, 286, 288–289, 290
Serratia, 134–135, 136–138
Novobiocin-brilliant green-glycerol-lactose agar, 86*t,* 87, 88
Nucleotidyltransferase, 197
Nursing homes. *See also* Long-term care
E. coli infections, 15
Morganella UTIs, 272
Proteus UTIs, 249
Providencia UTIs, 286, 287*t,* 299

O

O/129, 315
Obesumbacterium, 324, 361*t,* 364–365
Ocular infections. *See* Eye infections
Odor
Cedecea, 340
Serratia, 141

Ofloxacin, 108
OMPs. *See* Outer membrane proteins
ONPG (*o*-Nitrophenyl-β-D-galactopyranoside), 71, 90*t*, 163
Ophthalmic infection. *See* Eye infections
Oral vaccines, for *Salmonella* Typhi, 108
Ornithine decarboxylase test, 72
Enterobacter, 164*t*
Klebsiella, 120*t*, 121
Morganella, 278
Salmonella Typhi, 105*t*
Serratia, 139, 140
Yersinia, 228
Osteomyelitis
Citrobacter, 186*t*
Edwardsiella, 307, 312*t*, 314
Enterobacter, 161
Morganella, 274
Proteus, 253
Salmonella, 84
Serratia, 138
typhoid fever, 104
vertebral, 69
Yersinia, 220, 223
Outer membrane proteins (OMPs)
Citrobacter, 194–195
E. coli, 49
Morganella, 279
Yersinia, 234, 235–236
Oxgall medium, 104
Ozena, 117

P
PAM medium, 292
Pancreatic abscess, 34, 160
Pancreatic thrombotic microangiopathy, 27
Pancytopenia, 160
Panophthalmitis, 223
Pantoea, 151, 178–181
Paracolobactrum, 2
Paracolon bacteria, 2
Paratyphoid fever. *See* Typhoid fever
PASCO systems. *See* Microbial identification systems
Pasteurella, 202, 203
Pathogenesis
Citrobacter, 195–196
E. coli, 47–50
Edwardsiella, 317–318
Enterobacter, 168–169
Hafnia, 331
Morganella, 279–280
Providencia, 296–298
Salmonella, 92–95
Serratia, 142
Shigella, 75–76
Yersinia, 231–237
Pectin, 224
Pedestals, in EPEC pathogenesis, 46, 47
Penicillins
Edwardsiella, 318
Klebsiella, 124, 125
Morganella, 280
Proteus, 264
Providencia, 298
Serratia, 143
Yersinia, 237
Peptone iron agar, 314
Peptostreptococcus, 160
Pericarditis, 218, 274
Peritonitis
Citrobacter, 186*t*
E. coli, 34
Edwardsiella, 307, 314
Hafnia, 328
Kluyvera, 339
Yersinia, 223
Periungual desquamation, 213
Peyer's patches, 106, 210, 211*t*, 234
pH, urinary, 249, 259, 280
Phage typing. *See* Bacteriophage typing
Phallotoxin, 46–47
Pharyngitis, 210, 219
Phenospecies, 6
Phenylalanine, 255
Phenylalanine deaminase, 276, 293
Phenylpyruvic acid test, 39
Phosphatase reaction, 256
Photorhabdus, 5, 345*t*, 346*t*, 355–358
Phytopathogens, *Pantoea* as, 180
Pigmentation
Enterobacter, 149, 162
Escherichia, 63, 64
Pantoea, 180, 181
Salmonella, 85
Serratia, 140–141
Yersinia, 232
Pigs
colibacillary toxemia, 14
Proteus, 247
Yersinia, 206, 207, 208
Piliation, in *Citrobacter,* 195
Pimples, *Yersinia* infections, 220
Pineapple, pink disease, 181
Pink disease, 181
Piperacillin
Citrobacter, 197, 198
Hafnia, 332
Kluyvera, 341
Plague, 222
antimicrobial therapy, 238
epidemiology, 205–206, 209
Plants
Enterobacter, 149, 151, 152
Klebsiella, 113
Pantoea, 180
Serratia, 132, 133
Plasmid analysis
Citrobacter, 194
Enterobacter, 166, 168
Hafnia, 331
Providencia, 296
Salmonella, 92
Serratia, 142
Shigella, 74, 106
Plasmids
Salmonella Typhi multidrug resistance, 107
Salmonella virulence, 94*t*, *95*
Shigella infection, 75–76
Plesiomonas shigelloides, 6, 73, 310

Pneumatosis intestinalis, 327
Pneumonia, 9. *See also* Nosocomial infections; Respiratory infections
- *Citrobacter,* 186*t*
- *E. coli,* 15, 34
- *Enterobacter,* 152, 154, 159–160
- *Hafnia,* 326
- *Klebsiella,* 116
- *Providencia,* 290
- *Serratia,* 137–138
- *Yersinia,* 221

Pneumonic plague, 206, 222
Poikilotherms, *Edwardsiella* infections, 304, 305, 306
Polymicrobic infections. *See* Infections, polymicrobic
Polymorphonuclear leukocytes, *Klebsiella* infections, 123
Polymyxin B, 298, 350
Pork. *See also* Meat products
- *Enterobacter,* 152
- *Yersinia,* 207, 208

Potassium cyanide
- *E. coli,* 39, 64
- *Morganella,* 276
- *Salmonella,* 89*t,* 90*t*

Poultry
- colisepticemia, 14
- *Enterobacter,* 152

Pragia, 89*t,* 361*t,* 362*t,* 365
Pregnancy, neonatal meningitis, 33
Preputial sac, 248
Primates, shigellosis, 67
Prodigiosin, 140–141
L-Proline aminopeptidase, 329
Prosthesis infection, 313
Protease
- *Proteus,* 261*t,* 263
- *Serratia,* 142

Proteeae
- enteric infections, 8
- identification, 254–255, 276
- isolation, 291

Proteeae isolation agar, 254, 291
Proteus
- antimicrobials, 263–264
- arthritis, 253, 257
- bacteremia, 10*f,* 249–251
- central nervous system infections, 251–252
- discovery, 1
- distinguishing from *Salmonella,* 89, 89*t,* 90
- distinguishing from *Yersinia,* 225, 226*t*
- epidemiology, 247–248
- gastroenteritis, 253
- identification, 254–257, 349*t,* 362*t*
- immunodiagnostics, 257
- isolation, 253–254
- nosocomial infections, 247–248, 248–249, 250
- pathogenesis, 259–263
- in polymicrobic infections, 286
- swarming, 253–254
- taxonomy, 245–247, 284–285
- typing, 258–259
- urinary tract infections, 248–249
- wound infections, 252–253, 290

Proteus mirabilis
- antimicrobials, 263–264
- arthritis, 253, 257
- bacteremia, 251
- epidemiology, 247–248
- gastroenteritis, 253
- hemolysin, 260
- identification, 255, 256–257
- IgA protease, 263
- immunodiagnostics, 257
- pathogenesis, 259–263
- pediatric meningitis, 251–252
- swarming, 254
- taxonomy, 245, 247*t*
- typing, 258
- urinary tract infections, 248–249
- wound infections, 252–253

Proteus myxofaciens, 245, 247*t,* 263
Proteus penneri, 246, 247*t*
- antimicrobials, 264
- bacteremia, 251
- gastroenteritis, 253
- hemagglutinin, 262
- hemolysin, 260
- identification, 256–257
- IgA protease, 263
- urinary tract infections, 249

Proteus-Providencia-Morganella group, 254–255, 276
Proteus vulgaris
- antimicrobials, 264
- epidemiology, 247
- gastroenteritis, 253
- genomospecies, 257
- hemolysin, 260
- identification, 255, 256–257
- IgA protease, 263
- immunodiagnostics, 257
- swarming, 254
- taxonomy, 245–247
- typing, 258
- urinary tract infections, 249

Proticine, 258
Providencia, 245
- antimicrobials, 298–299
- bacteremia, 287*t,* 288–289
- biotyping, 295
- epidemiology, 285–286
- gastroenteritis, 287*t,* 289–290
- identification, 292–295
- immunodiagnostics, 295–296
- isolation, 290–292
- nocosomial infections, 286, 288–289, 290
- pathogenesis, 296–298
- respiratory infections, 290
- taxonomy, 284–285
- typing, 296
- urinary tract infections, 286, 287*t,* 288
- wound infections, 290

Providencia alcalifaciens
- antimicrobials, 298, 299*t*
- biogroups, 295
- diarrhea, 289–290, 299
- epidemiology, 285
- gastroenteritis, 297
- identification, 293, 294
- isolation, 292
- pathogenicity, 297

Providencia alcalifaciens (contd.)
somatic antigens, 296
taxonomy, 284, 285
Providencia alcalifaciens medium, 292
Providencia friedericiana, 285
Providencia heimbachae, 285
Providencia mirabilis, 294
Providencia rettgeri
antimicrobials, 298, 299*t*
epidemiology, 285
identification, 292, 293, 294
pathogenicity, 298
somatic antigens, 296
swarming, 291
taxonomy, 284–285
Providencia rustigianii, 285, 293, 295
Providencia stuartii
antimicrobials, 298, 299
bacteremia, 287*t,* 288
bacteriuria, 297, 299
biogroups, 295
epidemiology, 285–286
identification, 292, 293, 294, 295
isolation, 291–292
in polymicrobic infections, 251
septicemia, 289
somatic antigens, 296
swarming, 291
taxonomy, 284, 285
typing, 296
urinary tract infections, 286, 287*t,* 288
wound infections, 290
Pseudoappendicitis, 209, 210, 212–213
Pseudobacteremia, 344
Psoas muscle
Proteus infection, 252
Yersinia infection, 220–221
Public health, in *Salmonella* outbreaks, 82
Pulmonary gangrene, 116
Pulsed-field gel electrophoresis
Enterobacter typing, 167*t,* 168
Klebsiella typing, 122
Salmonella typing, 92
Shigella typing, 74–75
Yersinia typing, 231
Purpura fulminans, 162
Pyelonephritis, 30–31
Citrobacter, 188–189
Klebsiella, 118, 124
Kluyvera, 339
Proteus, 248, 263
Pyogenic liver abscess, 118–119
Pyomyositis
Citrobacter, 186*t,* 189
Enterobacter, 160
Morganella, 274
Yersinia, 221
Pyrazinamidase, 229
Pyrazines, 340
Pyrimine, 140
L-Pyrrolidonyl peptidase, 88, 90, 191

Q
Quellung reaction, 122, 141
Quinolones
Citrobacter, 197
Edwardsiella, 318
Morganella, 280
Proteus, 264
Shigella, 77
Yersinia, 238

R
Rabbits, *Yersinia,* 206
Rahnella, 178, 295, 345*t,* 346*t,* 350–351, 364*t*
Rambach agar, 86*t,* 87, 88
Random amplified polymorphic DNA (RAPD), 122
RapID onE identification system. *See* Microbial identification systems
Rash, enteric *Yersinia* infections, 213
Rat model, for *Citrobacter* pathogenicity, 195
Reactive arthritis, 219–220
Rectal swab, *Shigella* isolation, 69–70
Red blood cell transfusion, *Yersinia* bacteremia, 215–216
Red mouth bacterium, 204
Renal dysfunction, *Providencia,* 286
Renal failure, STEC gastroenteritis, 27
Repetitive extragenic palindromic sequences (REP), 168, 195
Reptiles
Edwardsiella, 304, 305
Morganella, 271
Proteus, 247
transmission of *Salmonella,* 83
Resistance, antimicrobial
Cedecea, 341
Citrobacter, 196–197
Enterobacter, 170–172
Ewingella, 346–347
Hafnia, 332
Kluyvera, 341–342
Leminorella, 348–349
Moellerella, 350
Morganella, 280–281
Photorhabdus, 358
Proteus, 264
Providencia, 298–299
Rahnella, 351
Salmonella, 92, 96, 106–107
Serratia, 134–135, 143–144
Shigella, 76–77
Xenorhabdus, 358
Yersinia, 237
Respiratory infections. *See also* Pneumonia
Citrobacter, 189
E. coli, 34
Enterobacter, 159–160
Hafnia, 325*t,* 326
Klebsiella, 116–117
Proteus, 253
Providencia, 290
Serratia, 137–138
Yersinia, 221–222
Rhamnose, 35–36, 139
Rheumatic arthritis, 253, 257
Rhinoscleroma, 12, 113, 116–117
Ribotyping
Enterobacter, 167*t,* 168

Salmonella, 92
Shigella, 75
Yersinia, 231
Rickettsial infections, 257
Rodents
Budvicia, 361
Yersinia, 206, 207, 208–209

S
Salads, *Serratia* adulteration, 133
Salicin, 192, 229
Salmonella, 23, 310
antimicrobials, 92, 95–96
biochemical identification, 88–90
clinical disease, 83–85
cross-reactivity with *Citrobacter freundii,* 194
distinguishing from *Citrobacter,* 88, 89, 90
distinguishing from *E. coli,* 90
economic costs of, 11, 82
epidemiology, 81–83, 91–92
gastroenteritis, 11, 83–84
identification, 349*t,* 362*t*
isolation, 85–90
pathogenesis, 92–95
serotyping, 90–91
taxonomy, 80–81
transmission, 95
typing, 91–92
virulence plasmids, 94*t,* 95
Salmonella choleraesuis, 101
Salmonella Enteritidis, 91, 93
Salmonella isolation and identification agar, 86*t,* 87, 88
Salmonella Java, 101
Salmonella Paratyphi A, 101, 105
Salmonella Paratyphi B, 101, 105
Salmonella Paratyphi C, 101, 105
Salmonella pathogenicity island, 94
Salmonella-Shigella agar, 70
Citrobacter, 189
Edwardsiella, 314
Hafnia, 328
Salmonella, 85, 87, 88
Shigella, 69–71
Yersinia, 224, 225
Salmonella-Shigella-novobiocin agar, 86*t,* 87
Salmonella Typhi
epidemiology, 101–102
immunodiagnostics, 105
isolation, 85, 104–105
multi-drug resistance in, 106–107
pathogenesis, 106
taxonomy, 101
typhoid fever, 102–104
typing, 105–106
vaccines for, 108
Salmonella Typhimurium, 91–92, 93–94, 96
Salmonellosis. *See Salmonella*
Salpingitis, 314
Scombroid food poisoning, 247, 271
Screening tests. *See* Isolation techniques
Scrotal abscess, 338
SDS-PAGE. *See* Sodium dodecyl sulfate-polyacrylamide gel electrophoresis
Secretory enteritis, 290, 308
Seizures, shigellosis, 69
Selenite broth
Edwardsiella, 314
Proteus, 254
Providencia, 292
Yersinia, 225
Septic arthritis
Edwardsiella, 312*t*
Klebsiella, 119
Morganella, 274
Serratia, 138
Yersinia, 220
Septicemia
Citrobacter, 186*t,* 187–188
E. coli, 15, 19, 32
Edwardsiella, 309, 310–311
Enterobacter, 154
Enterobacteriaceae, 9–10
Hafnia, 327
Morganella, 272–274, 274
Proteus, 249–251
Providencia, 288–289
Salmonella, 84
Yersinia, 207, 213–218, 222
Septi-Chek, 34, 138
Septic shock, 289
Sereny test, 41, 76
Serotyping. *See also* Immunodiagnostics
E. coli, 42–43
Hafnia, 331
Klebsiella, 121–122
Providencia, 295–296
Salmonella, 90–91
Shigella, 73–74
Serratia, 5
antimicrobial resistance, 134–135, 143–144
bacteremia, 10*f,* 136–137
biotyping, 141
central nervous system infections, 137
epidemiology, 132–136
eye infections, 135–136, 138
hygiene, 144
identification, 138–141, 364*t*
immunodiagnostics, 141
isolation, 138
muscoskeletal infections, 138
nosocomial infections, 136–138
pathogenesis, 142
in polymicrobic infections, 248
respiratory infections, 137–138
taxonomy, 131–132
typing, 141–142
wound infections, 138
Serratia entomophila, 132, 138, 139
Serratia ficaria, 132, 137, 138, 141
Serratia fonticola, 132, 133, 138, 139
Serratia grimesii, 132
Serratia liquefaciens, 131, 133, 136, 138, 139–140
Serratia marcescens
antimicrobial resistance in, 134–135
bacteremia, 136
central nervous system infections, 137
contact lens-induced acute red eye, 135–136, 138
discovery, 1
epidemiology, 133

Serratia marcescens (contd.)
hospital transmission, 134–135
identification, 138–139
isolation, 138
IV drug use, 135, 136
pathogenesis, 142
pigmentation, 140
respiratory infections, 137
taxonomy, 131
use in transmission studies, 133
Serratia odorifera, 132, 137, 141
Serratia plymuthica
bacteremia, 136–137
epidemiology, 133
identification, 139
isolation, 138
pigmentation, 140
pneumonia, 137–138
taxonomy, 132
Serratia proteamaculans, 131, 132, 133
Serratia proteamaculans ssp. *quinovora,* 137–138
Serratia rubidaea
bacteremia, 136, 137
epidemiology, 133
isolation, 138
odor, 141
pigmentation, 140
serotyping, 141
taxonomy, 132
Serum resistance/sensitivity
Providencia, 298
Salmonella, 95
Serratia, 142
Sewage
Citrobacter, 184
Hafnia, 324
Sexual transmission
E. coli UTIs, 30
shigellosis, 67
ShET toxins, 76
Shiga-like toxin, 169
Shiga's bacillus. *See Shigella dysenteriae*
Shiga toxin, 44*t,* 76
Shiga toxin-producing *E. coli* (STEC)
biotyping, 40–41
diarrheagenic qualities, 17*t*
fluorescent actin stain test, 46–47
gastroenteritis, 16, 18*t,* 21–28
identification, 39
isolation, 35–37
non-O157:H7 serotypes, 27–28
O157:H7 serotype, 21–27
pathogenesis, 47–49
serotyping, 42–43
toxin assays, 42–43
Shigella, 284
antimicrobials, 76–77
bacteremia, 68–69
biotyping, 73
distinguishing from *E. coli,* 39, 72, 73
epidemiology, 67–68
eye infections, 69
gastrointestinal illness, 68
hemolytic-uremic syndrome, 69
identification, 71–73
isolation, 69–71
pathogenesis, 75–76
relatedness to *E. coli,* 5, 66
serotyping, 73–74
taxonomy, 66–67
typing, 74–75
vaccines for, 77
Shigella alkalescens, 66
Shigella boydii, 66, 67, 69, 73
Shigella dispar, 66
Shigella dysenteriae, 66
antimicrobials, 77
bacteremia, 68
dysentery, 23*t,* 68
E. coli toxins, 16, 21
epidemiology, 67–68
hemolytic-uremic syndrome, 69
identification, 72
serotypes, 73
Shiga toxin, 76
Shigella flexneri, 66, 310
antimicrobials, 77
bacteremia, 68
biotyping, 72
dysentery, 68
identification, 72–73
isolation, 70
serotypes, 73
Shiga toxin, 76
urinary tract infections, 69
vaccines, 77
Shigella sonnei, 66, 68
antimicrobials, 77
in contaminated water, 67
identification, 71–72, 73
isolation, 70
serotypes, 73
urinary tract infections, 69
vaccines, 77
Shigellosis. *See Shigella*
Sickle cell disease, *Salmonella* osteomyelitis, 84
Siderophores
Citrobacter, 196
Enterobacter, 169
Hafnia, 331
Klebsiella, 123
Providencia, 298
Serratia, 142
Yersinia, 214–215, 232, 234
Silver sulfadiazine, 298
Skin
carriage of *Klebsiella,* 112–113
Yersinia infections, 222–223
Slime, in *Edwardsiella,* 318
slt-II gene, 196
SMAC. *See* Sorbitol-MacConkey agar
Snakes. *See* Reptiles
Sodium dodecyl sulfate-polyacrylamide gel electrophoresis (SDS-PAGE), 279
Sodium salicylate, 126, 254
Soil
Enterobacter, 151
Yersinia, 209
Somatic antigens
Edwardsiella, 316–317

Morganella, 278
Proteus, 257
Providencia, 295–296
Serratia serotyping, 141
Yersinia, 230
Sorbitol, 40, 90*t,* 139*t*
Sorbitol-MacConkey agar (SMAC), 35–36, 40, 328
Sorbose, 41–42
Spondylitis, 189
Sputum
Cedecea isolation, 337–338
Kluyvera isolation, 338
STEC. *See* Shiga toxin-producing *E. coli*
ST enterotoxin, 47
Stool
Citrobacter isolation, 189–190
E. coli isolation, 35–37
Klebsiella in, 112
Morganella isolation, 275
Salmonella in, 82, 95
Salmonella Typhi isolation, 104
Shigella isolation, 69–70
Yersinia isolation, 226
Streptomycin
Hafnia, 332
Salmonella Typhi, 106
Yersinia, 238
Strontium chloride B broth, 314
Struvite, 249, 259
Stx toxins, 21, 27
Sucrose fermentation, 228
Sulfonamide, 106, 318
Summer diarrhea, 15
Swarming
Proteus, 253–254, 259, 260, 261*t,* 263
Providencia, 291
Swimmer cells, *Proteus,* 259
Swine. *See* Pigs

T

Tatumella, 345*t,* 346*t,* 351–353, 365*t*
Taxonomy
Aerobacter, 148
Budvicia, 361
Buttiauxella, 361*t,* 362–363
Cedecea, 336, 337*t*
Citrobacter, 183–184
Edwardsiella, 304–305
Enterobacter, 148–151
Enterobacteriaceae, 1–6
Escherichia, 13, 61
Ewingella, 344
Hafnia, 322–324
Klebsiella, 110–111
Kluyvera, 336–337
Leclercia, 345*t,* 347
Leminorella, 345*t,* 348
Moellerella, 349
Morganella, 270–271
Obesumbacterium, 361*t,* 364
Pantoea, 178–180
Photorhabdus, 345*t,* 356
Pragia, 361*t,* 365
Proteus, 245–247
Providencia, 284–285
Rahnella, 345*t,* 350
Salmonella, 80–81
Salmonella Typhi, 101
Serratia, 131–132
Shigella, 66–67
Tatumella, 345*t,* 351–352
Trabulsiella, 345*t,* 353
Xenorhabdus, 345*t,* 355–356
Yersinia, 202–205
Yokenella, 345*t,* 353–354
Temperature, in *Yersinia* isolation, 225
Tenesmus, 68
Tetracycline
Cedecea, 341
Edwardsiella, 318
Hafnia, 332
Kluyvera, 341
Proteus, 264
Providencia, 298
Salmonella Typhi, 106, 107
Serratia, 143
Shigella, 77
Yersinia, 238
Tetrathionate broth
Edwardsiella, 316
Proteus, 254
Providencia, 292
Thalassemia major, 215
Third World
enterotoxigenic *E. coli* in, 20
Shigella, 67, 68
Throat
Cedecea isolation, 337
Klebsiella isolation, 112
Thrombin, 236
Thyroid disease, 223
Ticarcillin, 264, 341
Tobramycin
Cedecea, 341
Enterobacter, 161
Hafnia, 332
Morganella, 280
Toxic epidermal necrosis, 116
Toxic megacolon, 68
Toxin assays, *E. coli,* 43–44
Toxins. *See also* Endotoxins; Enterotoxins
cholera, 20
Citrobacter, 195–196
cytolethal distending, 29–30, 44*t*
in *Klebsiella* pathogenesis, 124–125
released by enterotoxigenic *E. coli,* 20
Shiga, 44*t,* 76
STEC, 16, 21
Stx, 21, 27
Trabulsiella, 89*t,* 345*t,* 349*t,* 353
Transcytosis, 234
Transfusions, *Yersinia* bacteremia, 215–216
Transmission. *See also* Epidemiology
Citrobacter, 184, 195
E. coli UTIs, 30
Salmonella, 83, 95
Serratia in hospitals, 134–135
Shigella, 67
STEC, 25–26

Transmission *(contd.)*
studies using *Serratia marcescens,* 133
yersinosis, 208
Traveller's diarrhea, 11, 20–21, 28, 274
Trehalose fermentation, 277, 278, 294–295
Trimethoprim, 106, 107
Trimethoprim-sulfamethoxazole
Cedecea, 341
Enterobacter meningitis, 159
Klebsiella, 126
Kluyvera, 341
Proteus, 264
Salmonella, 96
Serratia, 143
Shigella, 77
Triple sugar iron agar
E. coli, 37
Hafnia, 328
Morganella, 275
Proteus, 255
Salmonella, 88, 89*t*
Salmonella Typhi, 104, 105
Shigella, 70, 71*t*
Yersinia, 226
Trypticase soy broth, 104
Tryptone broth, 256
Tryptophan, 89*t,* 291
"Turista." *See* Traveller's diarrhea
Turkey. *See* Poultry
Turtles, *Edwardsiella,* 305
Two-disk test, 51
Typhoid fever, 101–104, 106, 188. *See also* Enteric fever
Typhus, Weil-Felix reaction, 257
Typing
Citrobacter, 194–195
E. coli, 43
Edwardsiella, 317
Enterobacter, 166–168
Hafnia, 331
Klebsiella, 122
Morganella, 279
Proteus, 258–259
Providencia, 296
Salmonella, 91–92
Salmonella Typhi, 105–106
Serratia, 141–142
Shigella, 74–75
Yersinia, 230–231
Tyrosine lyase, 190, 276, 291, 294
Tyrosine phosphate, 236

U
Umbilicus, *Proteus* meningitis, 251
Underdeveloped nations. *See* Third World
United States
enterotoxigenic *E. coli,* 21
epidemiology of plague, 206
epidemiology of typhoid fever, 101–102
incidence of STEC gastroenteritis, 24, 26
Urea
Klebsiella, 120*t*
Proteus, 256
Salmonella, 89*t*
Urease
E. coli, 39
Morganella, 280
Proteus, 259, 261*t,* 262*f,* 263
Providencia, 295
Yersinia, 236
Urinary tract infections
Citrobacter, 185, 186*t,* 188–189
E. coli, 15, 30–31, 32, 43*t,* 49
Edwardsiella, 314
Enterobacter, 161
Enterobacteriaceae, 9
Hafnia, 325*t*
Klebsiella, 111, 117–118, 124
Morganella, 272
Proteus, 248–249
Providencia, 286, 287*t,* 288
Shigella, 69
Urine
E. coli screening, 37
Hafnia isolation, 328
Morganella alkalinization, 280
Morganella morganii isolation, 272
Proteus alkalinization, 249, 259
Providencia isolation, 286
Salmonella Typhi isolation, 104
Urolithiasis, 263
Urosepsis, 62, 289
Urothelial cells, 262–263
UTIs. *See* Urinary tract infections
Uveitis, 223

V
Vaccines
Salmonella Typhi, 108
Shigella, 77
Vascular prosthesis infection, 313
Ventriculitis, 186*t,* 187
Vero cells, 142, 262
Verocytotoxins, 21
Vertebral osteomyelitis, 69, 161
Vi antigen, 105
Vienna sausages, *Proteus,* 247
Virulence plasmids, 94*t, 95*
Vitek identification system. *See* Microbial identification systems
Voges-Proskauer test, 39
Citrobacter, 191
Enterobacter, 164*t*
Kluyvera, 340
Morganella, 276
Voles. *See* Rodents
Vomiting, *Edwardsiella* enteritis, 308

W
Water
Budvicia, 361
Citrobacter, 184
E. coli, 14, 25–26
Edwardsiella, 305
Hafnia, 324
Pragia, 365
Serratia, 132, 133
Shigella, 67
Watery enteritis. *See* Secretory enteritis
Weil-Felix reaction, 257

Wound infections, 9
  Citrobacter, 189
  Edwardsiella, 311–314
  Enterobacter, 160–161
  Hafnia, 325*t,* 327–328
  Morganella, 273, 274
  Photorhabdus, 356–357
  Proteus, 252–253, 290
  Providencia, 290
  Serratia, 138
  Xenorhabdus, 356–357
  Yersinia, 220–221

X
Xenocoumacins, 358
Xenorhabdins, 358
Xenorhabdus, 345*t,* 346*t,* 355–358, 365*t*
Xylose-lysine-desoxycholate agar, 70, 85, 314
Xylose-lysine-tergitol 4 agar, 86*t,* 87

Y
Yersinia, 23
  antimicrobials, 237–238
  arthritis, 219–220
  biotyping, 228–230
  cardiovascular infections, 218
  cellular invasion, 234–235
  chronic infections, 223
  enzymatic activities, 236
  epidemiology, 205–210
  erythema nodosum, 222–223
  gastroenteritis, 210–213
  hepatic disease, 222
  identification, 226–230
  immunodiagnostics, 230
  isolation, 223–226
  lipopolysaccharides, 237
  pathogenesis, 231–237, 331
  pharyngitis, 219
  respiratory infections, 221–222
  septicemia, 213–218, 222
  taxonomy, 202–205
  typing, 230–231
  wound infections, 220–221
Yersinia aldovae, 209*t,* 228
Yersiniabactin, 234
Yersinia bercovieri, 204, 205*t,* 209*t,* 228, 229
Yersinia enterocolitica
  antimicrobials, 237–238
  arthritis, 219–220
  biochemical characteristics, 228
  biotyping, 228–230
  cardiovascular infections, 218
  cellular invasion, 234
  chronic infections, 223
  enteric disease, 210–213
  epidemiology, 207–210
  identification, 227
  immunodiagnostics, 230
  iron acquisition, 234
  isolation, 224–226
  liver disease, 222
  pathogenesis, 231, 331
  pathogenic strains, 207
  pharyngitis, 219
  respiratory infections, 221–222
  septicemia, 213–217
  taxonomy, 202–204
  typing, 230–231
  urease, 236
  wound infections, 220–221
Yersinia frederiksenii, 203–204, 205*t,* 208*t,* 209*t,* 228, 230, 234, 238
Yersinia intermedia, 203, 204, 205*t,* 208*t,* 209*t,* 228, 230, 234
Yersinia kristensenii, 204, 205*t,* 208*t,* 209, 209*t,* 228, 230, 231, 234, 238
Yersinia mollaretii, 204, 205*t,* 209*t,* 228, 229
Yersinia outer membrane proteins, 235–236
Yersinia pestis
  antimicrobials, 238
  biochemical characteristics, 228
  distinguishing from *Ewingella,* 346
  epidemiology, 205–206, 209
  identification, 227
  iron acquisition, 232
  pathogenesis, 231, 232
  pneumonic plague, 222
  septicemia, 218
  taxonomy, 202, 204, 205*t*
Yersinia philomiragia, 204, 205*t*
Yersinia pseudotuberculosis
  antimicrobials, 238
  biochemical characteristics, 228
  cellular invasion, 234
  epidemiology, 206–207, 209
  gastrointestinal syndromes, 213
  identification, 227
  immunodiagnostics, 230
  iron acquisition, 234
  isolation, 226
  liver disease, 222
  pathogenesis, 231, 232, 331
  reactive arthritis, 220
  septicemia, 217–218
  taxonomy, 202, 203, 204, 205*t*
  typing, 230
Yersinia rohdei, 204, 205*t,* 209*t,* 228, 238
Yersinia ruckeri, 204, 205*t,* 209*t,* 228
Yersinosis
  duration, 212
  epidemiology, 207–210
  manifestations, 210–212
  mesenteric lymphadenitis, 212–213
Y media, 224
Yokenella, 345*t,* 353–355